ENERGY EFFICIENCY IN PROCESS TECHNOLOGY

Proceedings of the International Conference on Energy Efficiency in Process Technology, held in Athens, Greece, 19–22 October 1992; organized by the Commission of the European Communities (DG XII) with the participation of the Centre for Renewable Energy Sources, the European Federation of Chemical Engineering, the European Chemical Industry Council, the Aluminium of Greece and Eurotherm.

ENERGY EFFICIENCY IN PROCESS TECHNOLOGY

Edited by

P. A. PILAVACHI

Commission of the European Communities, Brussels, Belgium

ELSEVIER APPLIED SCIENCE
LONDON and NEW YORK

ELSEVIER SCIENCE PUBLISHERS LTD
Crown House, Linton Road, Barking, Essex IG11 8JU, England

WITH 156 TABLES AND 709 ILLUSTRATIONS

British Library Cataloguing in Publication Data

Energy Efficiency in Process Technology
I Pilavachi, P A
660

ISBN 1-85861-019-2

Library of Congress CIP data applied for

Publication No EUR 14594 EN of the Commission of the European Communities Dissemination of Scientific and Technical Knowledge Unit, Directorate-General Information Industries and Innovation, and Telecommunications. Luxembourg

PREFACE

The main objective of the Community's energy policy consists in securing a sufficient energy supply to meet the present and future demand of its Member States, and in reducing the Community's dependence on imported energy. This is the reason why, since 1975, the Commission has been stimulating universities, industries and national laboratories of the EC Member States to perform R & D work aimed at energy saving. In spite of the temporary fall in oil prices, energy-saving measures will continue to remain imperative, as resources will continue to diminish rapidly. Research, development and demonstration effort in this area must therefore be maintained.

In this context, the Commission of the European Communities organized an International Conference on Energy Efficiency in Process Technology, with the participation of the Centre for Renewable Energy Sources, the European Federation of Chemical Engineering, the European Chemical Industry Council, the Aluminium of Greece and Eurotherm.

The Conference objective was to provide an international forum for the presentation and discussion of recent R & D relevant to energy efficiency, taking into account environmental aspects, in the energy-intensive process industries. A wide range of industrial sectors was covered, including new processes and equipment. Projects carried out within the present European Community JOULE Programme were included. The following is a partial list of topics addressed:

- Chemical Reactors
- Heat Exchangers
- Separation Processes
- Furnaces, Kilns and Ovens
- Combustion
- Process Integration
- Dynamic Simulation and Batch Processes
- Exergy Analysis
- Thermodynamic Cycles
- Efficient Production and Use of Electricity
- Sensors and Instrumentation

P. A. Pilavachi

Conference Organization
Dr P. A. Pilavachi, DG XII, Commission of the European Communities

Scientific Committee
Prof. R. W. K. Allen, UKAEA—Harwell Laboratory, UK
Prof. D. Behrenst, Dechema, Germany
Prof. A. E. Bergles, Rensselaer Polytechnic Institute, USA
Dr E. N. Carabateas, General Secretariat for Research and Technology, Greece
Prof. M. da Graça Carvalho, Instituto Superior Técnico, Portugal
Prof. K. Cornwell, Heriot-Watt University, UK
Mr R. C. Dumon, CEC expert, France
Prof. G. Froment, Rijksuniversiteit Gent, Belgium
Prof. M. Groll, Universität Stuttgart, Germany
Prof. E. P. Gyftopoulos, Massachusetts Institute of Technology, USA
Prof. B. Kalitventzeff, Université de Liège, Belgium
Prof. N. Koumoutsos, National Technical University of Athens, Greece
Prof. N. N. Kulov, Russian Academy of Sciences, Russia
Prof. P. Le Goff, Institut National Polytechnique de Lorraine, France
Prof. Z. Leszczynski, Industrial Chemistry Research Institute, Poland
Prof. E. Macchi, Politecnico di Milano, Italy
Mr A. Mercer, ETSU, UK
Prof. H. L. J. Meunier, Faculté Polytechnique de Mons, Belgium
Prof. S. Pierucci, Politecnico di Milano, Italy
Dr P. A. Pilavachi, DG XII, Commission of the European Communities
Prof. K. E. Porter, University of Aston, UK
Prof. D. A. Reay, CEC expert, UK
Prof. S. S. Stecco, University of Florence, Italy
Mr B. P. ter Meulen, TNO, The Netherlands
Dr P. Trambouze, Institut Français du Pétrole, France
Dr P. Tzannetakis, Motor Oil, Greece
Prof. K. R. Westerterp, University of Twente, The Netherlands
Mr P. Zegers, DG XII, Commission of the European Communities

Guests of Honour
Mr Ioannis Paleokrassas
Minister of Industry, Energy, Technology and Commerce, Greece

Mr Georgios Contogeorgis
Former Minister for the National Economy, Greece
Former Member of the Commission of the European Communities

Administrative Issues
Mr G. Weidenbach, DG XII, Commission of the European Communities

From left to right: Mr G. Gorgias, Mr G. Contogeorgis, Dr P. A. Pilavachi, Mr I. Paleokrassas, Mr C. Nicolaóu, Mr N. Tavoularis. In the background: Dr E. Carabateas.

Conference Hall.

Plenary Session Dr P A Pilavachi

Plenary Session Mr P Zegers

Plenary Session: Mr I. Paleokrassas.

Plenary Session: Prof. N. Chrysochoides.

Plenary Session (front view)

Plenary Session (rear view)

CONTENTS

Session 3: Drying
Chairman: Prof. R. W. K. ALLEN (*UKAEA—Harwell Laboratory, UK*)

Session 4: Sensors and Instrumentation
Chairman: Prof. F. DURST (*Universität Erlangen-Nürnberg, Germany*)

Session 5: Separation Processes
Chairman: Prof. P. LE GOFF (*Institut National Polytechnique de Lorraine, France*)

Distillation

* Text in French.

Session 7: Heat Exchangers (continued)
Chairman: Prof. A. KARABELAS (*Aristotle University of Thessaloniki, Greece*)

Fouling

Session 8: Furnaces, Kilns and Ovens
Chairman: Mr R. DUMON (*CEC Expert, France*)

Session 8: Furnaces, Kilns and Ovens (continued)
Chairman: Prof. H. L. J. MEUNIER (*Faculté Polytechnique de Mons, Belgium*)

Session 9: Combustion
Chairman: Prof. H. L. J. MEUNIER (*Faculté Polytechnique de Mons, Belgium*)

Session 10: Process Integration
Chairman: Prof. S. PIERUCCI (*Politecnico di Milano, Italy*)

Session 10: Process Integration (continued)
Chairman: Mr J. KOSMADAKIS (*Motor Oil, Greece*)

Session 11: Dynamic Simulation and Batch Processes
Chairman: Prof. D. TSAHALIS (*University of Patras, Greece*)

Dynamic Simulation

Batch Processes

Session 12: Efficient Production and Use of Electricity
Chairman: Prof. E. MACCHI (*Politecnico di Milano, Italy*)

Session 13: Chemical Reactors (continued)
Chairman: Prof. K. R. WESTERTERP (*University of Twente, The Netherlands*)

Session 14: New Process Routes
Chairman: Prof. K. R. WESTERTERP (*University of Twente, The Netherlands*)

Session 15: Exergy Analysis
Chairman: Prof. E. P. GYFTOPOULOS (*Massachusetts Institute of Technology, USA*)

PLENARY SESSION

Chairman: Dr P.A. Pilavachi

INTRODUCTION

Dr P.A. Pilavachi
Directorate General for Science, Research and Development
Commission of the European Communities, Brussels, Belgium

Honourable Ministers,
Ladies and Gentlemen,

I have pleasure in welcoming you all to this Conference which is taking place in this historical city of Athens. As a Greek myself this pleasure is even greater.

We are honoured to have present Mr Ioannis Paleokrassas, the Minister of Industry, Energy, Technology and Commerce, and Mr Georgios Contogeorgis, Former Minister for the National Economy and Former Member of the Commission of the European Communities.

Mr Paleokrassas will give the Opening Address, to be followed by a Welcome address, given by Mr Zegers, on behalf of Professor Fasella, Director General for Science, Research and Development of the Commission of the European Communities, who cannot be with us. Professor Chrysochoides, who is President of CRES, will then describe some of the activities of his organisation in energy efficiency.

I take this opportunity to thank the Scientific Committee for their support in assisting with the preparations for this conference, all those who contributed to make this event possible, and you, the delegates, for making the journey to Greece to attend. I hope that you will find these few days both stimulating and enjoyable.

For your information, I am very pleased to tell you that we have here over 200 delegates from a total of 23 countries and from countries as distant as the USA, Canada, Australia, India and Japan. 156 papers were submitted and 109 papers will be presented in the Technical Sessions.

It is now, an honour and a great pleasure for me to invite Mr Paleokrassas to open the Conference.

OPENING ADDRESS

Mr Ioannis Paleokrassas
Minister of Industry, Energy, Technology and Commerce, Greece

Mr Minister,
Mr Chairman,
Ladies and Gentlemen,

Let me first of all express my deep appreciation for the decision of the Commission of the European Communities to organise this very important Conference in Athens. As the Minister of Industry, Energy and Technology, I have the honour to open this Conference and I trust that its results would lead to better and more innovative uses of energy.

It is not my intention, speaking to a highly specialised audience, to present the many available statistics on energy. Let me only emphasise a few general points which influence the efficient use of energy.

Energy, especially in the developed countries, is the backbone of the economy. Energy is by far the most important "commodity" on a world scale. Energy influences and is influenced by geopolitical factors, strategic considerations, political and social changes and technology progress.

Improved energy efficiency means planning. Planning means to assess the future. A very complicated and difficult procedure when we know how difficult it is to explain the past or, more difficult, to understand the present.

Efficient use of energy could only be achieved on the basis of a genuine energy strategy, and the future of energy should be linked to more efficient, less vulnerable and environmentally sustainable energy sources.

A national strategy, part of a larger Community and international energy strategy, should promote commercial, technological and regulatory tools in order to diversify energy sources, make "transportation" of energy more flexible and provide incentives for the most efficient use by the consumer. Specifically, this energy strategy must secure competition throughout the energy sector, expand the existing technology choices, promote R&D, support the economy and protect the environment.

State intervention in the energy market should be limited in specific instances where markets cannot or do not work efficiently. Having said that, we have to admit that energy is closely linked with national security, social considerations and environmental issues which markets are unlikely to give adequate weight. This is the reason for which governments are oriented to take into account the cost of their intervention on such matters. It is important, therefore, to justify such intervention by rigorous cost - benefit analysis and rely to the maximum extent on economic incentives, allowing at the same time the economy to achieve energy security and environmental goals at the lowest possible cost.

The importance of energy efficiency is a major factor in rational use of energy. It is common sense that the right use of energy means, automatically, environmental improvement and savings on precious energy raw materials and finances. In order to achieve this, special attention should be paid in the development of appropriate technologies in the use of residential, commercial, industrial and transportation energy. In addition, a great effort must be made in improving energy raw materials for the promotion of new energy products. Energy efficiency needs a national commitment in every sector of energy use and production. This will reduce energy costs to consumers, enhance environmental quality, maintain and enhance the standard of living, increase freedom and energy security and promote a strong economy.

Specifically, in the field of electricity, it is extremely important to produce, distribute and consume electricity as efficiently and as cleanly as possible. It is our intention to open electricity to a more competitive and market oriented production. In the oil sector, it is a fact that oil will remain a critical fuel for the industrialised countries, and for that matter energy planning should promote the technological research in a way which creates possibilities for making the economies less dependent on oil in the near future.

As far as natural gas is concerned, this is the fuel which evidently maximises energy efficiency. For this, with the help of the European Community, we proceed to the construction of the infrastructure for the use of natural gas. A major part of our strategy is the construction of a main pipe line for the transportation of natural gas from Russia together with a terminal for storage and distribution of LNG supplied by Algeria. We believe that all the above will provide our industry with new choices particularly beneficial for our society and environment.

In the field of coal, a special effort must be made to develop new "clean coal" technologies which increase energy efficiency, reduce losses and improve the environment by comparison with todays practices.

Special attention must also be given to renewable energy sources. These unlimited quantities of natural energy must be converted in various ways into usable energy based on specific planning and objectives. The renewable sources technologies have experienced significant technical progress over the past ten years. It is now the time to use this knowledge in order to support the future.

At this point, it is important to note that this conference shows that the European Community is setting the foundation and strategic planning for the development of new technology processes for energy efficiency. I must also say that, I am pleased to discover that special importance is given to nuclear fusion. I strongly believe that the competitiveness of the European Community in the international markets depends mainly on the development of technological energy processes and the rational use of energy.

It is also a fact that the European Member States, as far as energy supply is concerned, depend heavily on the Gulf States, Russia and North Africa. Events in the former Soviet Union have minimised the production of energy raw materials while exports have been reduced substantially. It will take a long time for them to achieve their previous exporting performances. North Africa is currently under a deep socio-political crisis. In the Gulf, recently, a large military operation took place in order to stop one nation to take control over the large energy supplies of the area.

Consumer countries are consequently vulnerable from changes in producer countries and countries along the transportation lines.

Planning and efficiency are the European answers to these realities

On these general observations, may I wish to your conference the very best and to the participants a pleasant stay in Athens.

R&D ACTIVITIES OF THE EUROPEAN COMMUNITY

Mr P. Zegers
on behalf of Professor P. Fasella
Director General for Science, Research and Development
Commission of the European Communities, Brussels, Belgium

On behalf of the Commission of the European Communities and of Professor Fasella, who could unfortunately not be here today, I welcome you in this conference on "Energy Efficiency in Process Technology". As you know, this conference is organized in the framework of one of the specific R&D programmes of the Commission: JOULE. In this presentation I would like to give you an overview of the R&D activities of the Commission and how the activities, which you will discuss during the next days, fit in this framework.

Let me start by giving you a short description of the three most important institutions of the European Community:

The **Council of Ministers** of the twelve Members States which takes decisions.

The **European Parliament** which, at present, has mainly an advisory function, but which will get more power if the Maastricht Treaty is approved.

Finally, the **Commission of the European Communities** (CEC) which is the executive body of the European Community and which is composed of 24 Directorate Generals in which around 20,000 officials are employed (including around 2000 persons engaged in the four Joint Research Centres of the EC). The budget of the CEC is around 55×10^9 ECU/year. The main political objective of the CEC is the single European market which is expected on 1 January 1993. Furthermore, the CEC is involved in a range of activities which include research and development.

R&D in the Joint Research Centres of the European Community

In the European Community two types of R&D are carried out:

Indirect contract research where the CEC is funding research activities in industry, universities and other organizations.

Direct R&D which is carried out by EC scientists in the Commissions' four Joint Research Centres (JRC) of Ispra, Geel, Karlsruhe and Petten; the JRC Ispra being by far the largest research centre. At present, around 2000 persons are employed in the JRC. The JRC started in the early sixties in the framework of the Euratom Treaty with the objective of carrying out R&D on topics related to nuclear energy. With time, the objectives have changed and after a restructurization which took place a few years ago, eight institutions have been created within the JRC. The Institute of Prospective Studies (Ispra) is mainly involved in the preparation of future JRC programmes. Two institutions deal with topics on nuclear energy: Nuclear Measurements (Geel) and Transuranien Elements (Karlsruhe). The Institute dealing with Materials Research has sections in both Ispra and Petten. The remaining four Institutions which deal with systems Engineering, Environment, Remote Sensing and Safety are all based in Ispra.

Contract Research

The European Community has carried out contract research since the sixties. This research was generally shared cost research where the Commission pays 50% of the full economic cost and the contractors pay the other 50%. For universities the CEC made an arrangement where the Commission pays 100% of the marginal cost of the project (e.g. additional scientists which must be hired or additional equipment which must be bought); on the other hand only a small part of the overhead costs is paid by the CEC.

Funds available for contract research, in the ongoing 3rd Framework Programm (5,700 MECU for five years) is about 4 times higher than funds available for direct research carried out in the JRC (871 MECU for three years).

I should also mention the **concerted actions** which are directed mainly at information exchange on a particular topic by a number of organizations in the EC; here only the secretarial work is funded by the Commission. Concerted actions often lead to contract research projects funded by the Commission.

EC contract research is characterized by a number of features which will be briefly described. **Precompetitiveness** is an important requirement for nearly all CEC contract R&D. The projects are generally **pluriannual** and funded on a 50-50% basis as was mentioned before. Projects are also required to have at least two independent participants from different EC Member States. This **multinational** character of EC projects is a very important feature which makes a major contribution to the integration of European research and the creation of European R&D networks. This requirement forces organizations to look across the borders to see what is going on in other EC countries. **Subsidiarity** has been discussed a lot lately in particular in connection with the Maastricht Treaty. It requires that Commission actions should have an added value as compared to if these actions would have been carried out in national programmes.

It is also foreseen that different R&D programmes interact. In the case of JOULE, which carries out R&D to demonstrate the technical feasibility of new energy techniques or technologies, there is a close collaboration with the THERMIE demonstration programme which aims at demonstrating their economic feasibility. For several years organizations from EFTA countries have been able to participate in EC projects on certain conditions. After signing the European Economic Space agreement, to be implemented by the end of this year, organizations from EFTA countries will be able to participate in EC projects under the same conditions as organizations from EC Member States.

The EC Framework Programmes

Until 1987, contract research was carried out to serve the needs of other objectives agreed upon by the Member States such as nuclear energy, environment, etc. In 1987, the Single European Act brought R&D within the formal competence of the Community. It also established basic structures of the EC R&D: adoption of multiannual framework programmes and their implementation through specific programmes. The first framework programme ran from 1984 to 1987. The second (1987-1991) and third (1990-1994) framework programmes were proposed and adopted on the basis of the Single Act. In order to assure continuity, framework programmes were made to overlap after 1990. Since the first framework programme, priorities have changed from a strong emphasis on energy in the first framework programme to information technology and quality of life in the third framework programme.

The third framework programme started in 1990 and is to last five years (1990-1994). For the first time all 15 specific programmes were launched simultaneously. These programmes are listed in Table 1. The largest programme deals with Information Technologies and Communications; Non Nuclear Energy (JOULE) is one of the smaller programmes and its funds amount to 157 MECU.

Table 1: Third Framework Programme (1990-1994)

			Programmes	MECU
54.6 %	ENABLING TECHNOLOGIES	1)	INFORMATION TECHNOLOGIES	1352
		2)	COMMUNICATIONS TECHNOLOGIES	489
		3)	DEVELOPMENT OF TELEMATIC SYSTEMS OF GENERAL INTEREST	380
		4)	INDUSTRIAL & MATERIALS TECHNOLOGIES	748
		5)	MEASUREMENT & TESTING	140
36.3 %	MANAGEMENT OF NATURAL RESOURCES	6)	ENVIRONMENT	414
		7)	MARINE SCIENCE & TECHNOLOGY	104
		8)	BIOTECHNOLOGY	164
		9)	AGRICULTURAL & AGRO-INDUSTRIAL RESEARCH	333
		10)	BIOMEDICAL & HEALTH RESEARCH	133
		11)	LIFE SCIENCES AND TECHNOLOGIES FOR DEVELOPING COUNTRIES	111
		12)	NON NUCLEAR ENERGIES	157
		13)	NUCLEAR FISSION SAFETY	199
		14)	CONTROLLED NUCLEAR FUSION	458
9.1 %	MANAGEMENT OF INTELLECTUAL RESOURCES	15)	HUMAN CAPITAL & MOBILITY	518
100 %				5.700

The total available funds which amount to 5700 MECU have been committed in the period 1990-1992. Due to the fact that the fourth framework programme is expected to start only in 1995, there is a gap for the period 1993-1994 which threatens the continuity of the specific programmes. For that reason, the Commission submitted a proposal for additional funding of 1600 MECU covering the period 1993-1994. A start has also been made on the preparation of the fourth framework programme.

The emphasis of successive framework programmes has shifted; this is illustrated in Table 2 where the allocation of funds is given (in percentages) for six major R&D areas for the second and third framework programmes.

Table 2: Spending for major specific programmes in %.

Areas	2nd FP 1987-1991	3rd FP 1990-1994
Information and Communications	42%	39%
Industry	16%	19%
Quality of life	7%	13%
Biotechnology	5%	5%
Energy	22%	14%
Improvement of S/T Cooperation	5%	9%
Others	3%	1%

A major part of the EC budget for R&D is allocated to Information Technologies and Communications but it has sligthly decreased. Funding for Industry increased a few percent. Quality of Life (which includes environment) and the Improvement of Scientific and Technological Cooperation both nearly doubled. R&D funding for Energy, including non nuclear energy and nuclear fission and fusion, decreased from 22% to 14%.

The Non Nuclear Energy specific programme

The objective of the **Non Nuclear Energy (JOULE)** specific programme is to demonstrate the **technical** feasibility of technologies which lead to renewable energy, energy savings and abatement of energy related pollution. It should be mentioned that the JOULE programme is closely related to the THERMIE programme which aims at the demonstration of the economic feasibility of energy technologies. The THERMIE programme does not form part of the Framework Programmes; the available funds are about two times larger than for JOULE. In view of the fact that JOULE and THERMIE are complementary there is a close collaboration between the two programmes.

The JOULE programme consists of four subprogrammes of which Modelling is relatively small and deals with the development of computer models of energy supply and demand and of energy related environmental aspects. The three major subprogrammes of JOULE deal with:

> Fossil Fuels
> Renewable Energy
> Rational Use of Energy.

Let me say a few words about their content:

R&D on **Fossil Fuels** is based on the fact that fossil fuels and in particular coal are available in large quantities for many years to go. Energy supply could be secured for many years, if the problem of pollution emission is solved. R&D is therefore carried out on fluidized bed combustion which has low NO_x and SO_2 emission levels, due to low combustion temperatures and to absorption of SO_2 by CaO. Research on coal gasifiers is also carried out because pollutants can be more easily extracted from the exhaust gases of coal gasifiers than from exhaust gases in powder coal combustion electricity plants. In general one may say that the pollutant emissions of NO_x, hydrocarbons, CO and CO_2 can be reduced by at least one order of magnitude at a reasonable cost and with a decrease of energy conversion efficiency of only a few percent. A major problem, however, is the storage of CO_2. Different possibilities are presently being explored. Major criteria for such storage systems are that they are suitable for large scale storage of CO_2 and absolutely safe; large scale release of CO_2 from storage systems, where CO_2 has been collected for many years, could cause a global disaster.

In the subprogramme on **Renewable Energy** R&D is carried out in the fields of photovoltaic, wind and geothermal energy and biomass. Photovoltaic and wind energy deal exclusively with electricity production. Also for biomass and geothermal energy, the most interesting applications are for electricity production.

In the field of **Rational Use of Energy,** development of fuel cells has the promise of energy savings ranging from 30% in power production to 60% in road traction. In the transportation sector, R&D is carried out along two lines: improvement of combustion engines and development of electric vehicles. In both areas the eight major European car manufacturers and a large number of other laboratories are involved.

The building and industrial sector both use 35 to 40% of the primary energy; these sectors have a large potential for energy savings. In industry process integration alone is said to have an energy saving potential of 20%. Further energy savings can be obtained with improved heat exchangers, separation processes, etc.

These are the topics which you will be discussing during the next days in these beautiful surroundings. I would like to finish my presentation by wishing you a fruitful and successful conference.

ACTIVITIES ON ENERGY EFFICIENCY IN THE GREEK INDUSTRY-CONTRIBUTION OF CRES

PROF. NICHOLAOS CHRYSOCHOIDES
President of the Governing Board
Centre for Renewable Energy Sources, Greece

INTRODUCTION

The first energy Crisis in 1973, and the consequent effects, on a worldwide scale, triggered Energy Efficiency measures and Programmes in the majority of the developed Nations. The European Commission established the Non-Nuclear Energy Research and Development Programmes, un ambitious project within the Directorate General 12th for Science, Research and Development, almost right after the Crisis, in 1975.

Since then, the Energy Efficiency and the Security of Energy Supply of the major Industrial European Nations, has appreciably improved. Energy Saving through systematic Energy Efficiency projects has resulted in a considerable energy economy of more than 20% in many EEC countries.

Nevertheless, more has to be achieved, given the growing need for environmental protection, for the minimisation of oil dependence and for the reduction of the industrial energy cost.

In Greece, the efforts towards higher Energy Efficiency, started immediately after the Crisis with the creation of the National Energy Board. One of the main problems to be faced was the lack of a comprehensive strategy and the means to implement one. In the following years, several studies for Energy Efficiency measures were carried out, but the application of an integrated saving programme to the various energy consumming sectors (industrial, residential-commercial, transportation) proved to be a difficult task.

However, during this period, the dependency of the country on imported oil was reduced, since the share of oil in the gross energy supply fell from 81% in 1973 to 59% in 1990 (fig.1). At the same period the energy diversification efforts of the Public Power Corporation, produced a positive result by reducing the share of oil in electricity production from over 50% in 1973 to 14% in 1990 (fig.2).

The indicators for the rest of the Economy though, did not show match improvement in the energy field. Greek Industry, traditionally oriented towards energy intensive processes, did not respond to the challenges and kept consuming a relatively high percentage of energy. The artificially low fuel prices and the lack of an integrated National Energy Efficiency Programme were among the factors that led industry to fall behind the achievements of our major partners. Insufficient information and the absence of incentives caused private consumption to follow across the same lines.

ENERGY SITUATION IN GREECE

Before we make an estimation, of the amount of the effort still remaining to be done in Greece, in the Energy Efficiency field, let as have a look in the present energy balance of the country. All following figures correspond to 1990 values, but they are very indicative of the todays situation.

Figure 3 shows the percentage of each primary energy source contribution to the total yearly energy supply which corresponds to almost 22 Mtoe. Oil has the highest participation and natural gas participation is almost negligible (as you propably know, significant amounts of gas consumption in Greece will start in 1994, after the completion of the big pipe line, under construction now, crossing the country from Macedonia down to Athens with branches to the main industrial areas).

The utilization of each primary energy source, to the various consumption forms is shown in the next figures.
Hydro Power, fig.4, which represents 3% of the cross energy supply is of course used entirely for electricity production covering about 10% of the total electricity consumption.

Natural gas, fig.5, goes almost by 100% to the industry, since no distribution grid for residential and commercial supply actually exists for the time.

Oil, which represents 59% of the cross energy supply, is used by almost half in transportation while the rest is used for industrial, residential, commercial and electricity production (fig.6).

Finally, solid fuels, mainly lignite, which represent 37% of the cross energy supply, are primary used for electricity production by 85% and the rest 15% for industrial processes (fig.7).
It is interesting to consider the energy breakdown for the 3 end users.
The total energy consumption for Greece in 1990 was 13,71 Mtoe.

The share for the transportation sector was 42% while the share for the other two sectors, residential-commercial on one hand and industrial on the other, was approximatelly equal, 29% for each (fig. 8).

Comparing the above shares to the average shares of the European Community we come to the following conclusions.

The transportation sector in Greece is higher by ten units, indicating the fact that there is a great potential for improving energy efficiency in our transportation system.

In the industrial sector, although the percentage is appoximately the same, one should consider the fact that Greek industry is not the same type of heavy industry of most community countries and therefore a big improvement in energy efficiency could be achieved in this area.

Finally the percentage for the residential and commercial sector is lower in Greece by approximately 9 units. Heavier climate and better living conditions is probably the main reason for this higher average figer for most community countries. Nevertheles, the potential for energy efficiency improvement, in residential homes and public building is still very promising for the Greek conditions.

In the next 3 figures the form of energies used for each of the 3 sectors is shown.
In the transportation sector which consumes approximatelly 6 Mtoe, almost 100% is covered by oil (fig.9).

In the industrial sector which consumes almost 4 Mtoe, approximately 50% of the energy used is oil while the rest is almost equally covered by solid fuels and electricity (fig.10).

In the residential and commercial sector, consuming about 4 Mtoe, the break down is 64% oil, 35% electricity and 1% solid fuels (fig.11).

There is not need for special analysis to conclude that much remains to be done in Greece in the Energy Efficiency and Energy Saving field. It is estimated that a 10-15% over all saving in energy consumption could be easilly achieved by the end of this decay, if proper energy efficiency measures are applied. The yearly saving resulting from such a programme could amound to more that half a billion ECUs.

And now a fiew words about CRES and its contribution to energy efficiency and saving.

THE CENTER FOR RENEWABLE ENERGY SOURCES

The Center for Renewable Energy Sources was established by Presidential Decree and became operational in the latter part of 1988. It must be stressed once more, that the significant contribution of the European Commission in the foundation and operation of CRES, allowed it to proceed fast and become a useful partner in the frame of European Energy Cooperation and a focal point for the Community within Greece. The Centers' aims, specified in its charter, are the Rational Use of Energy (RUE), the promotion of the exploitation of Renewable Energy Sources (RES) and the Environmental Protection (EP). The Centers' activities are supervised by the Ministry of Industry Energy Technology and Commerce (MIETC). The Governing Board includes representatives of the General Secretariat for Research and Technology, the Public Power Corporation, the Commission of the European Community and the Federation of Greek industry. Legally, it is governed by Private Law and enjoys complete financial and administrative autonomy.

To meet its objectives CRES is active in several fields of applications, as discribed in Table I. Rational Use of Energy and Renewables are the two bigest activities of the center. The renewables division has 7 sections, namely: Wind Energy, Active Solar, Passive Solar, Photovoltaics, Biomass applications, Geothermal and Hydro. Two activities, with growing importance are the Organization for the Promotion of New Technologies and the Environmental protection.

In fig. 12, the budget distribution is given. It is obvious that Rational Use of Energy and a large part of Biomass dealing with oil substitution absorb almost 50% of the total budget.

The change of the number of the personnel is shown in fig.13 while the number of projects for the same period is shown in fig.14. The 98 projects, currently in force, are almost by 100% competitive and other Community programmes as indicated below:

CRES' Projects

-	Valoren	33
-	Joule	21
-	Demonstration-Thermie	6
-	Regional Development	12
-	OPET (31 actions)	1
-	Special Studies (Sprint, Value etc.)	11
-	Training	3
-	Stride	3
-	Other Contracts	8
	Total	98

In fig.15, CRES inflow for the period 1988-1991 with deflated value referred to 1988, is shown.

THE PROMOTION OF RUE

The Centers' activities towards Industry, are channeled mainly through the RUE Division. A number of projects, categorised broadly by their content as Basic Research, Technical studies, Energy Audits, Policy Information and Applied Research, were initiated, with the aim of promoting the Energy Efficiency of Greek Industry.

Still, a program for improving the energy efficiency of Greek Industry, can not be the sole work of the Rational Use of Energy Division. The Centers' ambition is to utilise its personnel, its structure and its international contacts, in order to act as a catalyst for the mobilisation of all the interested parties towards the common goal.

Some of CRES's activities connected specificaly with energy efficiency in industries are presented below:

Basic research

Basic research activities relevant to the RUE sector, cover a range of fields and industries. The Cente is active in the areas of Heat Transfer Enhancement, Combustion and Emission Control, Heat Exchanger Fouling, etc, in collaboration with Greek Industry and European partners, in the framework of the JOULE program. Some of these projects are approaching their completion, and results from the relevant ones, will be presented in this conference. The participation of local industrial partners and other European Organaizations in research proposals for the JOULE Programme was secured, along the guidelines of DG XII.

Technical studies

The great importance of an Integrated National Energy Efficiency Programme and the necessary infrastructure to support it, was already mentioned. CRES secured a contract from the MIETC to evaluate critically the effects of past policy measures in Greece and internationally, to carry out a review of the existing relevant studies and draw on the know-how and help of European experts, to produce a final report containing a range of recommendations for a National Energy Efficiency Programme. The significance of this comprehensive study is obvious for Greece, bearing in mind that it is intented to be approved by the actual policy makers. It will be ready by the end of this month. A similar study on least cost planning, financed by the European Commission under the supervision of the National Energy Board, in collaboration with Public Power Corporation, the Public Oil Corporation and the Public Gas Corporation and managed by CRES, is also now under way.

Energy Audits

Small and Medium Size industries in Greece constitute a large part of Greek Industry. Usually they suffer both, from lack of scientific personnel capable of following the latest achievements in energy efficient technologies, as well as from financial means that would allow them to hire expert consultants. Our experience shows, that the most cost-effective way to introduce and adopt energy management principles as part of their everyday routine, is to start by offering solutions, based on actual on -the- spot measurements. The Energy Bus project of the RUE section is a major tool for the achievement of these aims. The bus was designed and manufactured in Holland, according to the NOVEM standards, modified for the local conditions to carry out energy audits. The vehicle is equipped with an infrared camera and electric energy analyzer, in addition to the normal equipement. Short, one-day energy audits are performed and depending on the preliminary analysis, longer and more complete energy flow analysis can be carried out. During the initial phase of the project, the cost is heavily subsidised by CRES and MIETC, but a fee reflecting the full economic cost is going to be charged in the future. A good sign for the success of this project is the fact that the requests from industry for the Energy Bus visits are multiplying.

Training

Information and education in energy efficient practices is central to any RUE program. To this end CRES proposed and secured two contracts from MIETC for the production of materials for two publicity campaigns, targetting the groups of primary school children and home users respectively. Also, the Centre has organised a number of seminars for the introduction of current energy efficiency practices to groups of engineers and scientists from industry, technical firms and local authorities.

Applications

In partnership with Greek industries in the Cement Sector, in the Textile Sector and in the Oil Refinery Sector,13 pilot-demonstration large projects were launched last year and they will all be completed by the end of the year.

The capital investment for these projects was of the order 7 MECU while the expected energy saving is of the order of 12 Ktoe per year.

CONCLUDING REMARKS

Having presented an outline of our Centers' current activities and strategy for the promotion of Energy Efficiency, we recognise that there is a major challenge ahead.

CRES is ready now to play its role, in collaboration with the Commission of the European Community and in close partnership with all interested parties from Government, Academic Institutions and Industry, for the realization of a long-sighted National Energy Efficiency Programme.

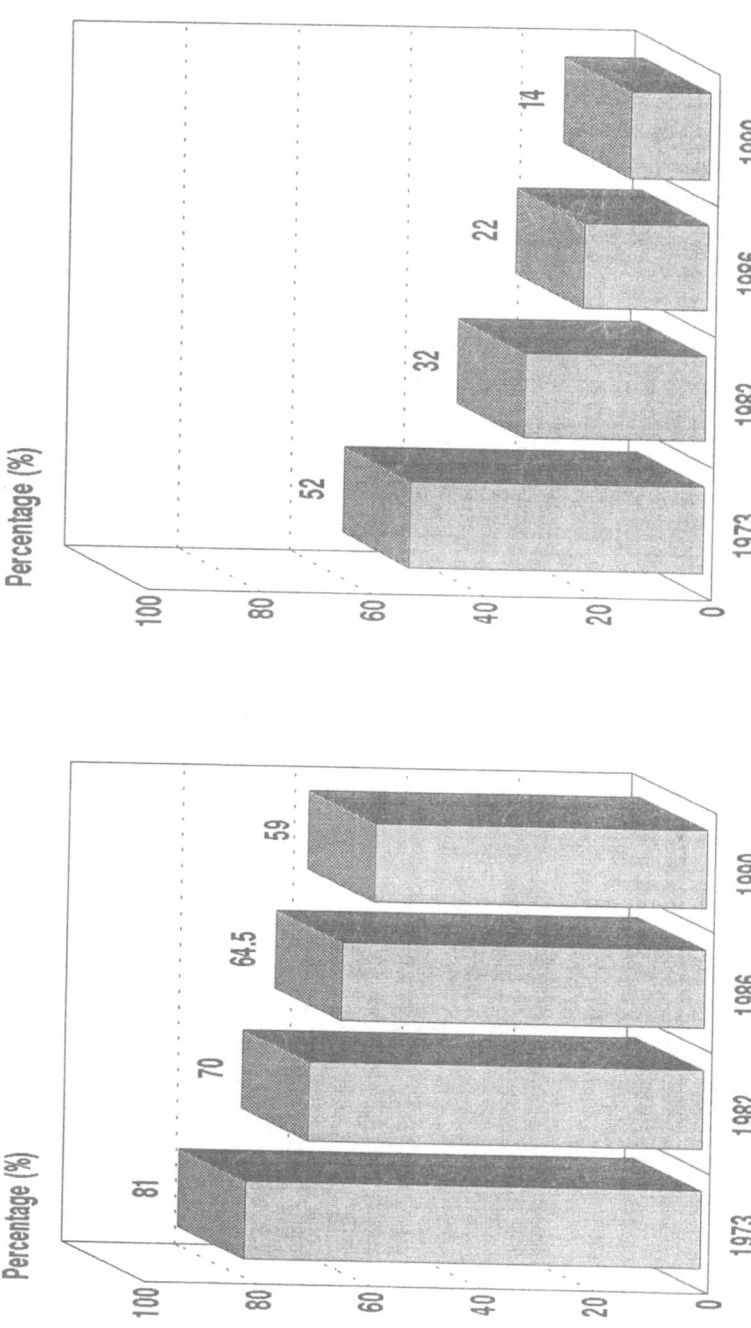

Fig. 1 : Share of Hydrocarbons in Gross Energy Supply (Greece 1973-1990)

Fig. 2 : Share of Hydrocarbons in Electricity Production (Greece 1973-1990)

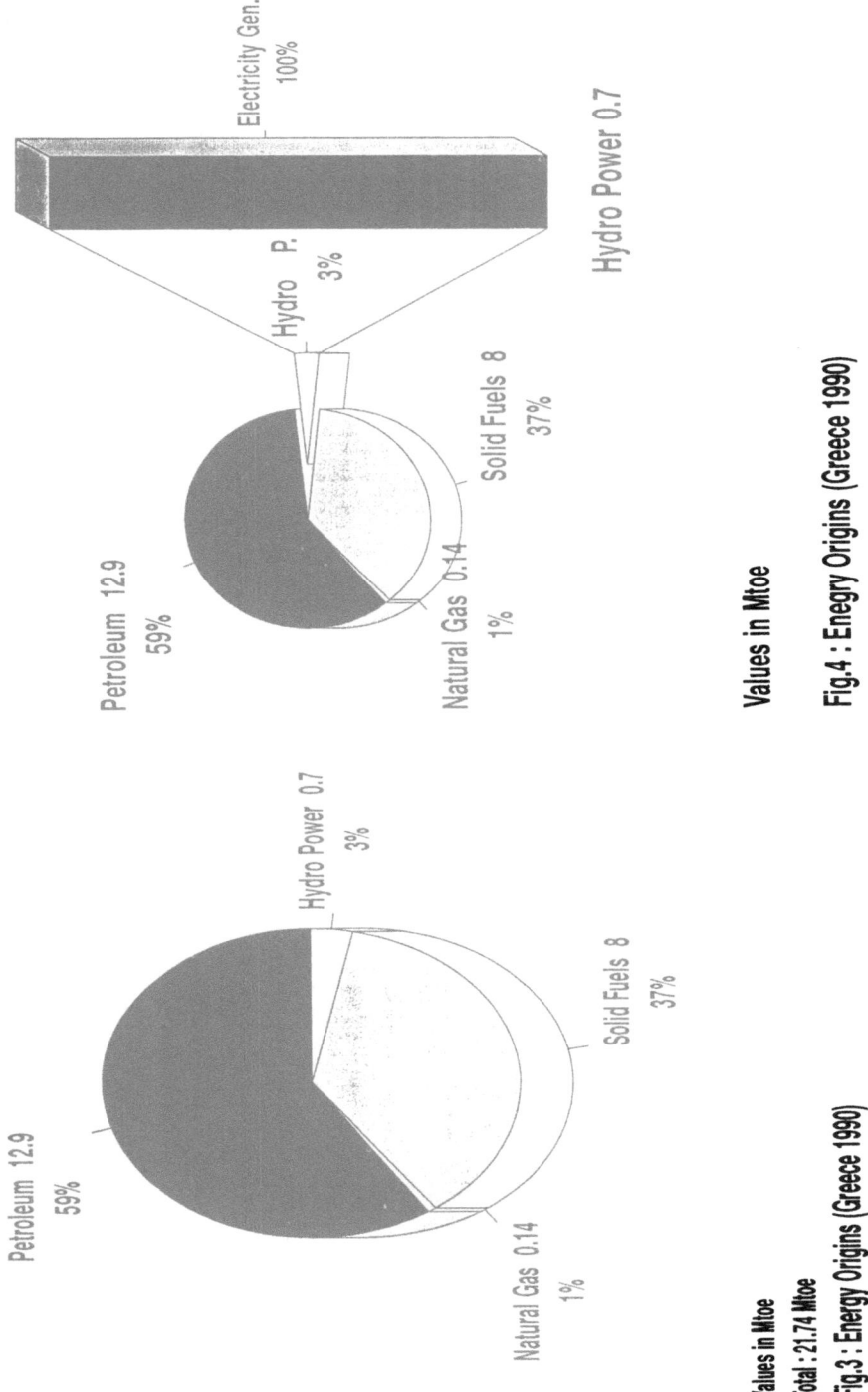

Electricity Gen.
100%

Hydro Power 0.7

Hydro P.
3%

Petroleum 12.9
59%

Solid Fuels 8
37%

Natural Gas 0.14
1%

Values in Mtoe

Fig.4 : Enegry Origins (Greece 1990)

Hydro Power 0.7
3%

Petroleum 12.9
59%

Solid Fuels 8
37%

Natural Gas 0.14
1%

Values in Mtoe
Total : 21.74 Mtoe
Fig.3 : Energy Origins (Greece 1990)

24

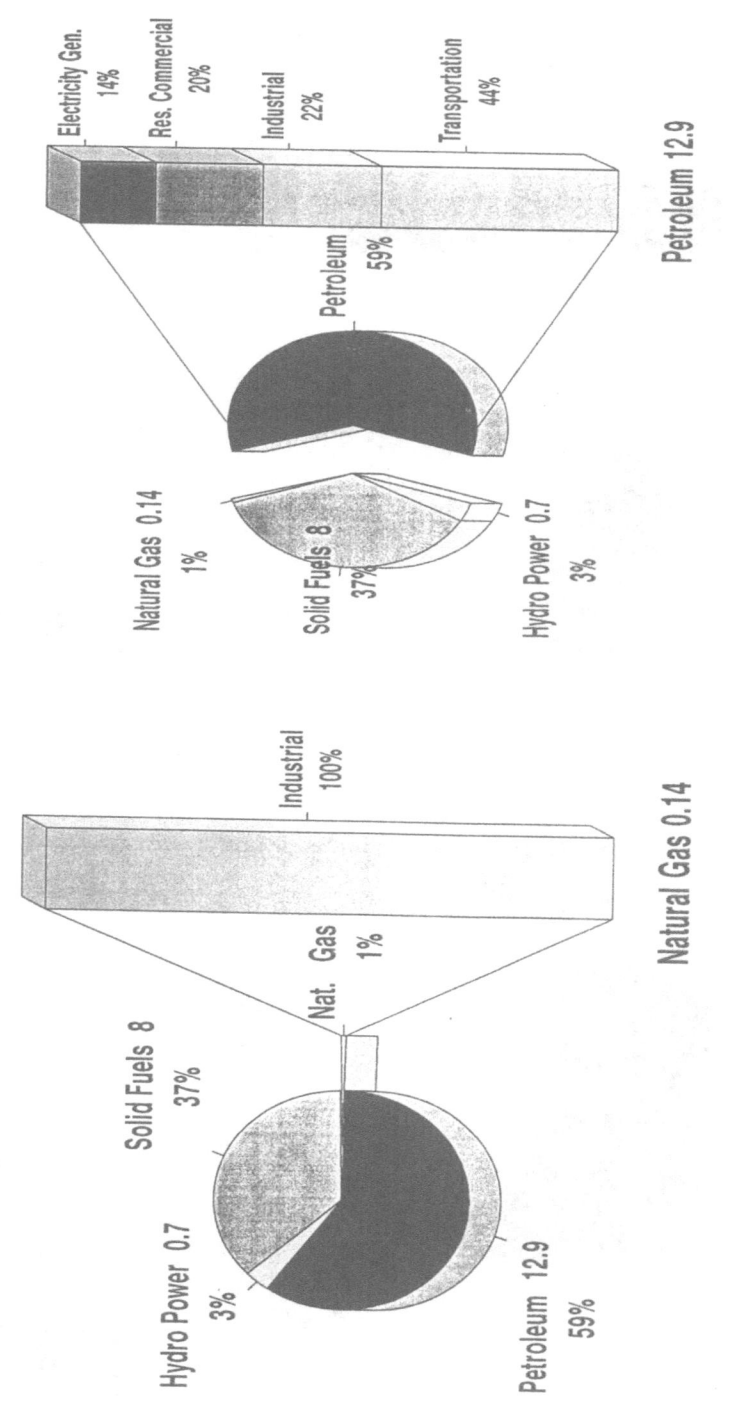

Electricity Gen. 14%

Res. Commercial 20%

Industrial 22%

Transportation 44%

Petroleum 12.9

Petroleum 59%

Natural Gas 0.14 1%

Solid Fuels 8 37%

Hydro Power 0.7 3%

Values in Mtoe
Fig. 6 : Energy Origins (Greece 1990)

Industrial 100%

Natural Gas 0.14

Nat. Gas 1%

Solid Fuels 8 37%

Hydro Power 0.7 3%

Petroleum 12.9 59%

Values in Mtoe
Fig.5 : Energy Origins (Greece 1990)

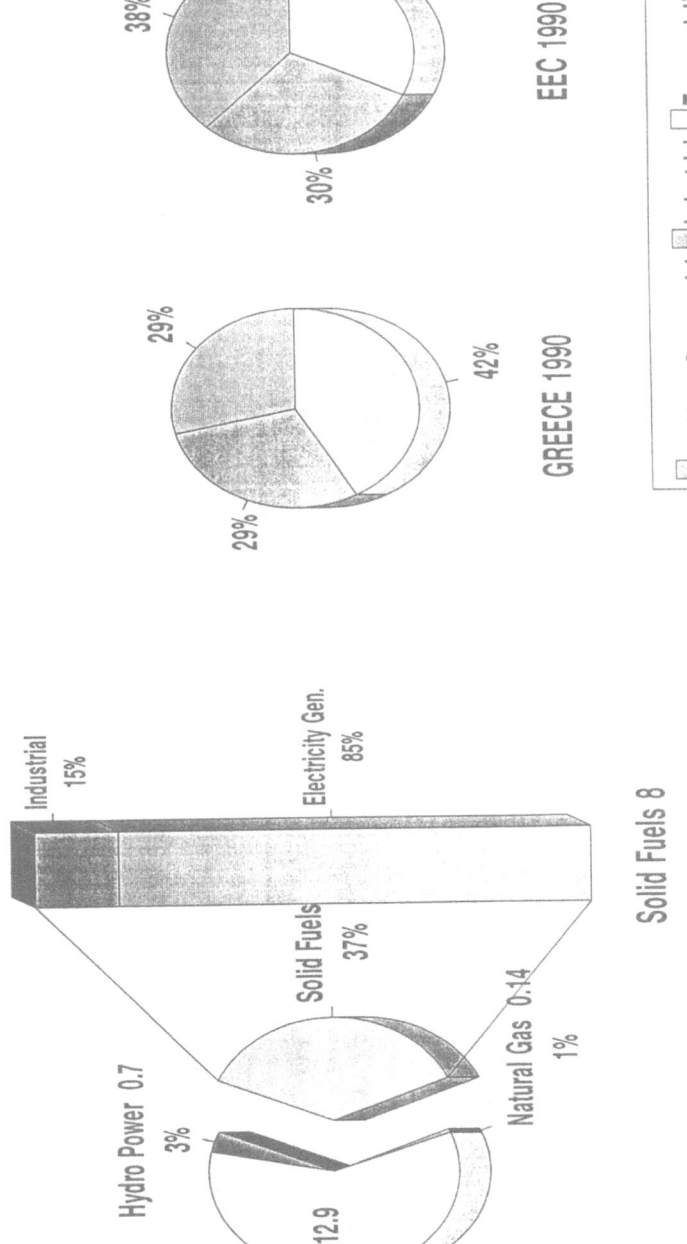

ENERGY BREAKDOWN

The 3 End Users

38%

32%

EEC 1990

30%

29%

29%

42%

GREECE 1990

Resident. Commercial ▨Industrial ☐Transportation

Fig. 8 : Energy Breakdown, The 3 End Users

Industrial 15%

Electricity Gen. 85%

Solid Fuels 37%

Solid Fuels 8

Natural Gas 0.14 1%

Hydro Power 0.7 3%

Petroleum 12.9 59%

Values in Mtoe

Fig. 7 : Energy Origins (Greece 1990)

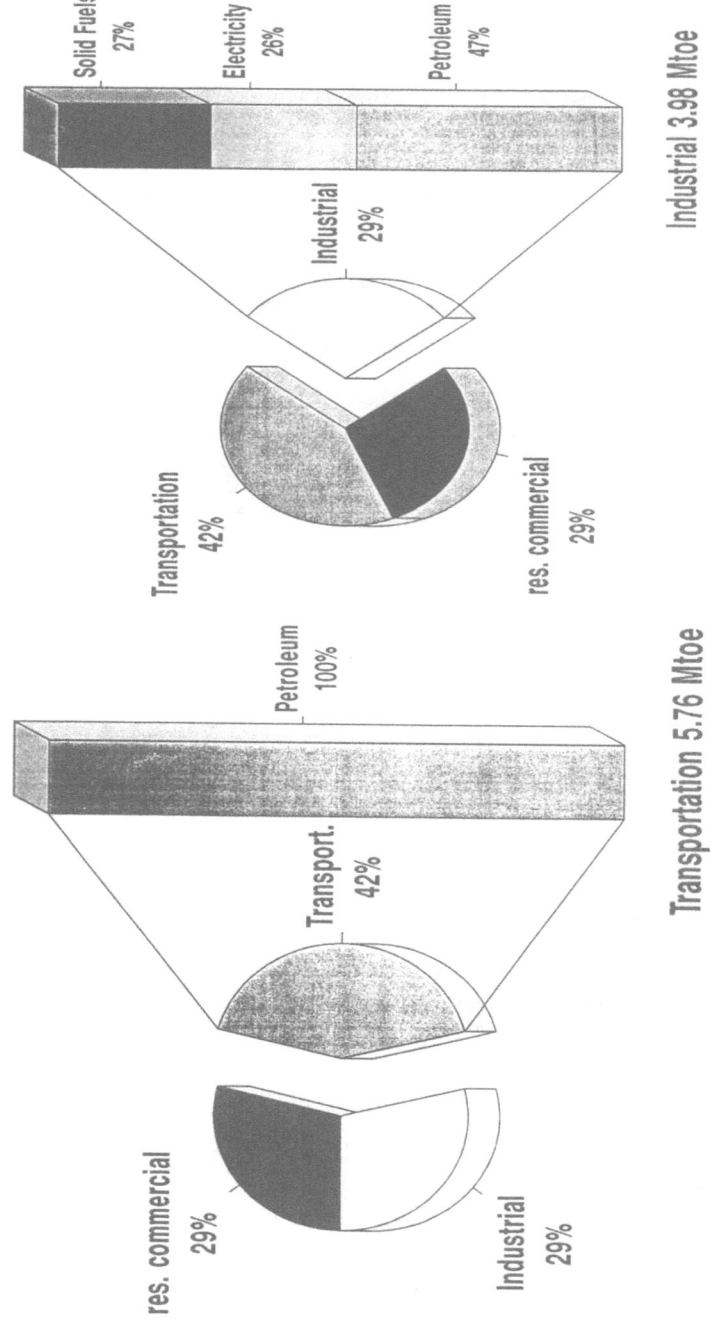

Fig. 9 : The 3 End Users (Greece 1990)

Fig. 10 : The 3 End Users (Greece 1990)

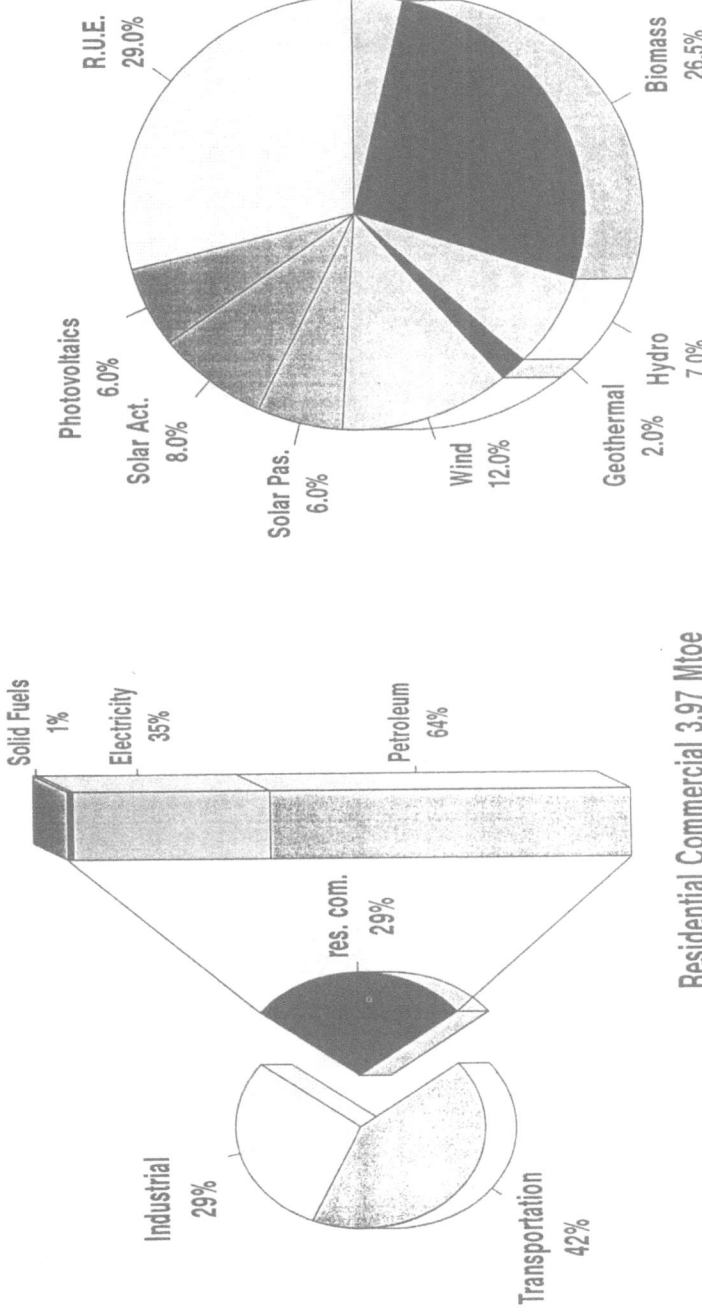

Fig. 12 : TOTAL BUDGET OF PROJECTS, Breakdown % (Sep 1992)

Fig. 11 : The 3 End Users (Greece 1990)

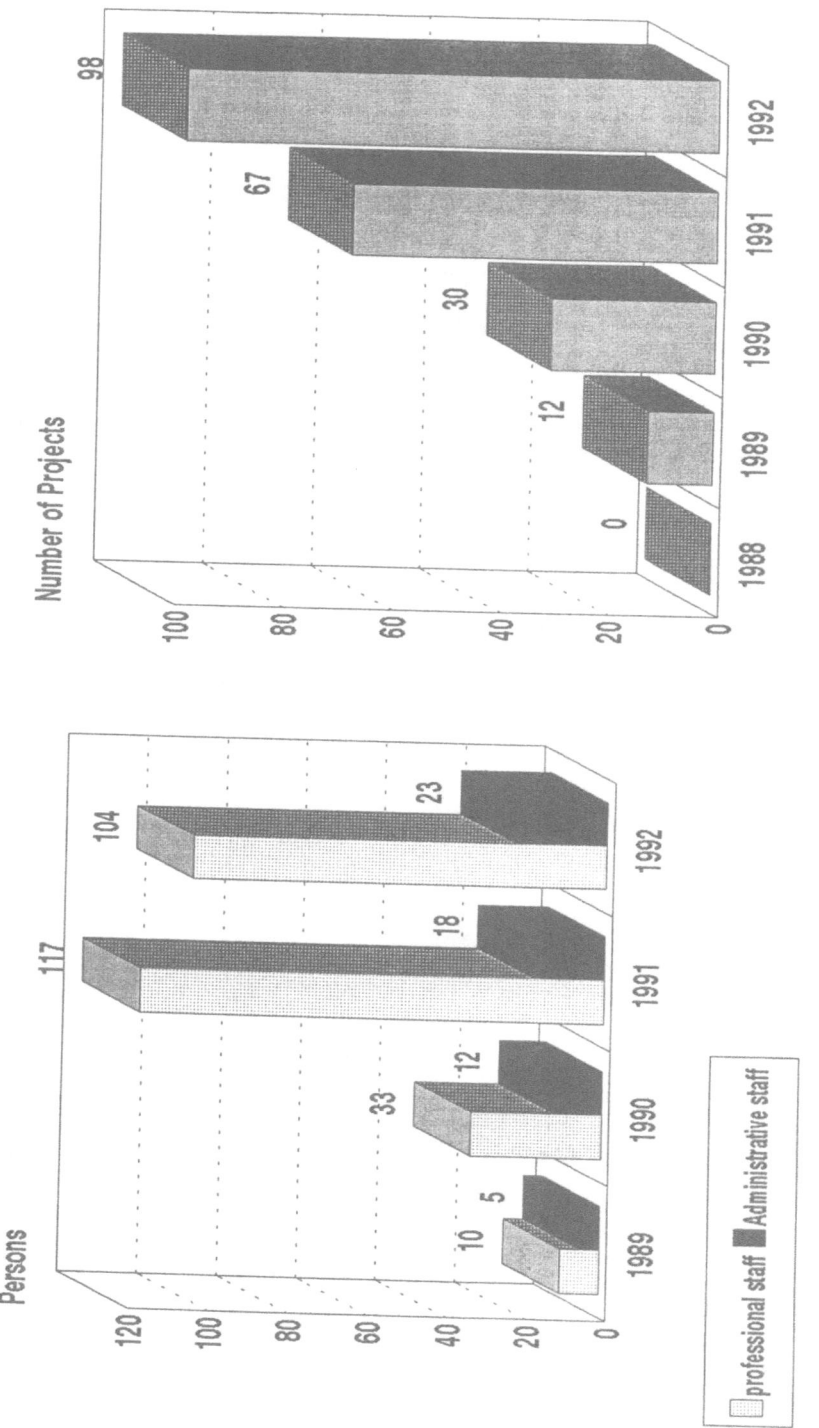

Number of Projects

Fig.14 : C.R.E.S. Projects

Persons

professional staff ■ Administrative staff

Fig. 13 : C.R.E.S. Personnel

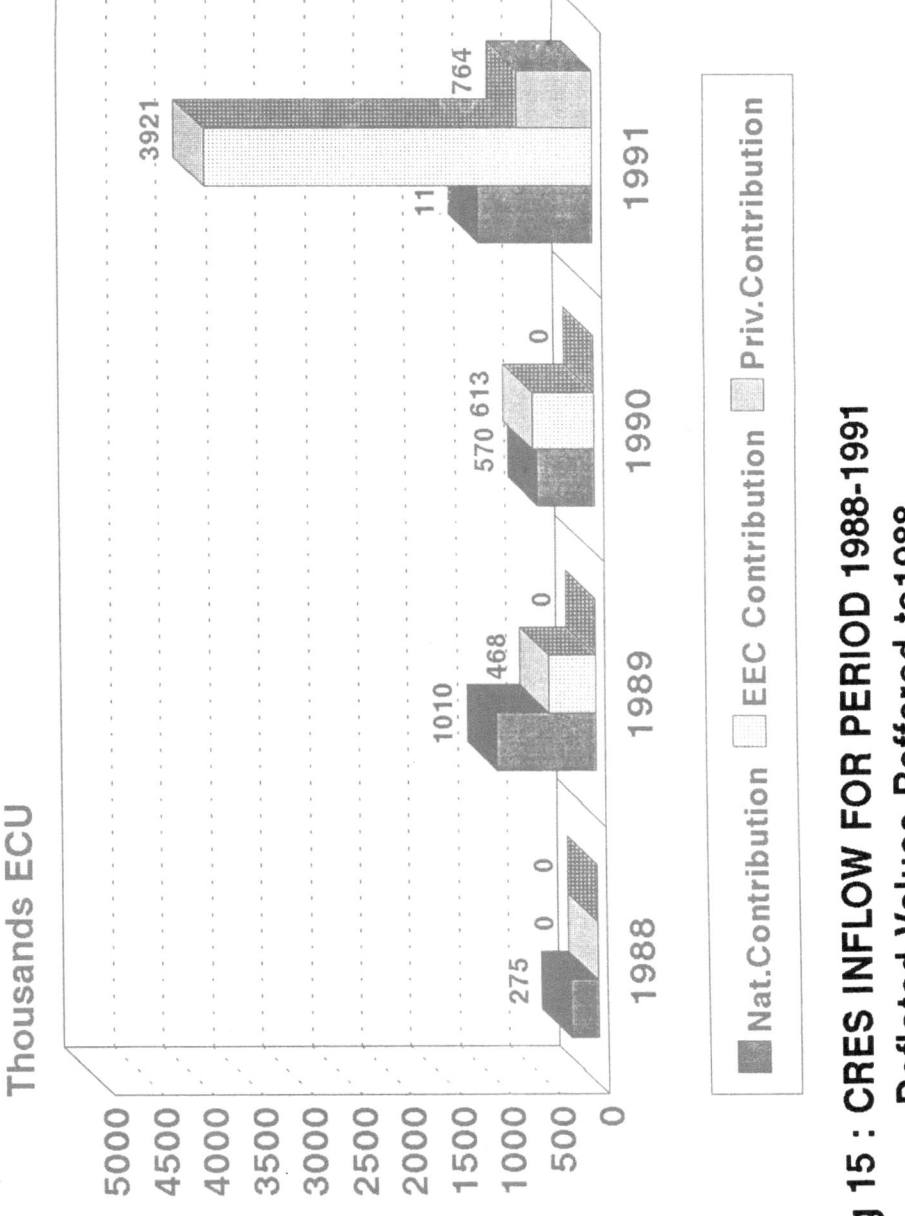

Thousands ECU

5000
4500
4000
3500
3000
2500
2000
1500
1000
500
0

1988
1989
1990
1991

275
0
0

1010
468
0

570 613
0

3921
764
11

Nat.Contribution   EEC Contribution   Priv.Contribution

Fig 15 : CRES INFLOW FOR PERIOD 1988-1991
Deflated Values Reffered to1988

SCIENTIFIC COMMITTEE DINNER ADDRESS

THE MAASTRICHT TREATY ON EUROPEAN UNION - THE FUTURE OF EUROPE

Mr George Contogeorgis
Former Minister for the National Economy, Greece
Former Member of the Commission of the European Communities

Mr Minister,
Mr Secretary General,
Ladies and Gentlemen,

The Treaty on European Union signed at Maastricht on February 7th 1992, has been an important step, I would say a leap forward, to the achievement of the European Unification.

It has been ratified, so far, by four Member States i.e. Luxembourg, Ireland, Greece and France. The remaining Member States are expected to have finalized the ratification procedures by the end of this year, except Denmark. The position of this country in the Union is still uncertain. Hopefully it will be cleared soon. Otherwise there will be a legal problem to be solved by the other eleven Member States on how to put into effect, the Treaty not ratified by a signatory country. Nevertheless, these Member States have already expressed their strong political will to go ahead, even without Denmark.

From the ratification procedure, so far, scepticism and unexpected difficulties have emerged, not uncommon in the long life of the European Community. We hope that after the results of the recent European Council in the United Kingdom the common interest of the European people will prevail over short term national interests and the procedures of unification will be accelerated, so that the targets set by the Maastricht Treaty be reached without delays.

The institutional structure of the European Union after Maastricht is based on three pillars:

First: The European Community and its institutional framework.

Second: The Common Foreign and Security Policy, taking progressively, according to the Treaty of Maastricht, a Community dimension and form, without however being communitised for the moment.

and Third: Cooperation in the fields of justice and home affairs, a new activity of an intergovernmental form.

Substantial progress towards achievement of European integration is reflected in the rules introducing for the first time new Community activities in the fields of industry, consumer protection, trans-european networks, education and high quality professional training, health, culture and environment as well as the new rules strengthening existing community activities, in the social field, on economic and social cohesion, on promoting community policy in the sphere of development and cooperation and on the promotion of research and technological development.

All these arrangements contained in the Maastricht Treaty, predetermine the course of Europe to the end of the 20th century.

For the Economic and Monetary Union the rules are binding and they lead clearly, through Community procedures to the progressive creation of a single economic and monetary area with the adoption, beginning 1999 at the latest, of a single european currency which will be the confirmation and the guarantee, of the viability of the Union.

The United Kingdom, only, shall not be obliged or committed to move to the third stage of economic and monetary Union, without a separate decision to do so by its government and Parliament. The United Kingdom shall notify the Council whether it intends to move to the third stage in proper time. We hope that the decision will be positive. The United Kingdom too, shall not take part to the development of a solid Community social policy, based on the Social Charter adopted in 1989.

In the field of Common Foreign and Security Policy including the eventual framing of a Common Defense Policy, which might in time lead to a common Defense, remarkable progress was made compared to the present intergovernmental Political Cooperation. However, many arrangements are not binding or clear and many important issues have been left to be defined later.

This will be the main task of a new intergovernmental Conference scheduled for 1996, by the Treaty of Maastricht.

As for the third pillar of the Union, i.e. Justice and Home Affairs, it is the start of pure intergovernmental cooperation within the frame of the Union. A lot has to be done in this field so that we reach, eventually, the point of a common policy enacted through the procedures and Institutions of the Union.

I would like to mention particularly the fact that the Maastricht Treaty emphasizes, for the first time the interest of the Union to promote its cultural aspect as well as the need to strengthen the Community's activities in the field of Science, Research and Technology.

Special provisions of the Treaty aim at strengthening the conscience of the European citizen, through the recognition of the European identity of the Union and the introduction of the citizenship of the Union, for the nationals of its Member States.

With these provisions it is confirmed and accepted for the first time that the economic and political union of Europe would not survive for long unless it is based on common cultural characteristics creating the European identity of the peoples who constitute and will constitute in the future the European Union.

Historically, coalitions and alliances among countries with peoples of different civilizations have appeared many times to serve concrete targets, usually for a short period of time. But the real economic and political union of Europe presupposes common cultural roots and common ideals. That is why Jean Monnet was saying: "Nous ne coalisons pas des Etats, nous unissons des hommes" (We do not form a coalition of States, we unite people).

The arrangements of the Treaty of Maastricht, by which the Community's action in the field of science, research and technology is strengthened, are also extremely important for the future of the United Europe.

Science and technology, in the forms we know them, were born in Europe. The roots of scientific thought go back to ancient Greece. Through Rome the classical Greek philosophy and science extended to western and northern Europe where for centuries they developed and flourished. From Europe, they were transferred to the U.S.A.

Europe for centuries was the undisputed leader of Science and Technology in the World.

Today although Europe together with the U.S.A. and Japan are the leaders, however in many fields of modern technology Europe has lost its leading position in favour of the U.S.A. and Japan, despite the fact that Europe does not lack research scientists of high scientific caliber and total spending in this field is quite important.

Then comes the question: Why has Europe lost its leading role in science and technology?

The answer is simple.

The inevitable dispersion of the available funds into twelve countries, the lack of collaboration and reduced mobility and exchange of research scientists and knowledge, isolated research work in Member States and the inevitable duplication of programs result in the waste of economic resources and human potential of the Community and consequently in reduced efficiency and inadequate benefits.

Close cooperation among the Member States and concerted action at Community level for projects which would be run more efficiently than at national level, combined with increased Community financial support, would offer Europe the possibility of regaining the lead which it has now lost or risks losing in some sectors of research and technology. The time has come for Europe to understand that science, research and technology is a key area for the future of European Union and for the European people.

Any delay in this field will undoubtedly result in reduced competitiveness of the European industry and in weakening the European economy. But a weak European economy would result in a weaker political strength for Europe. Such a development would mean a marginal position for Europe in the world's affairs and the loss of a leading role which the European Union deserves, must and can play in the world, in the coming 21st century. Europe has to win the battle of scientific research and technological progress, because the very future of European people is at stake.

Therefore, the creation of a European scientific area is a necessity for Europe and an essential element for its economic and political union. But if technological progress is an essential element for the economic development of Europe as a whole, it is an urgent priority for individual Member States too, especially those lagging in technology.

For these States, like Greece, the proper planning and organisation of the whole effort in this field is a priority which would enable them to use more efficiently the possibilities offered by the Community in order to accelerate their progress and fill their technological gap.

Although the Maastricht Treaty did not meet the expectations of the majority of the European people nevertheless it is a decisive leap forward and a turning point in the long process to the economic, monetary and political union of Europe, a Union with no alternative for the European people, which started fourty years ago. The Maastricht Treaty really creates the framework of a European Union founded on the principles of the respect for the individual, on democracy, on solidarity and on the principle of subsidiarity for action in areas which do not fall within its exclusive competence.

Indeed actual political and economic developments in Europe after the changes in Central and Eastern Europe and the Soviet Union, as well as all over the World, make it necessary, more than ever before, for the European people to accelerate the procedures of their full economic, monetary and political union. There is no other choice.

European people must put aside short-sighted, professional or regional or even national interests. They should face the future with vision and should realize that their long term interests lie with a united Europe. Any thought that with the Maastricht Treaty their national independence is shrunk or even eliminated is unrealistic. On the contrary the Maastricht Treaty creates the requirements and offers to them the unique opportunity to achieve progressively their union within which they secure collectively their independence. We must realize that our past and our destiny are common, the problems we face are common, the solution to our problems is common and there is only one answer: European Union. Our Country is our origin. The notion, the sentiment of our nation is human, inherent in our life. It is acceptable and should be respected. However, our future is in the European Union. There is no contradiction. Both can go together, based on the solid principles on which the European Union is built and within the framework of its institutional structures.

Undoubtedly the road to the union of Europe is difficult, full of obstacles, for both the large and the small countries. Great economic and political adjustments are necessary, which will change our way of life and sometimes long established habits. Inevitably these adjustments will meet inertia and opposition and sometimes even crises will appear. But we should not forget what Jean Monnet was saying after the Treaty of the European Defence Community was not approved by the French Assembly on August 30, 1954: "Europe will be built within crises and will be the aggregate of the solutions to these crises".

Years later Jean Monnet at his retreat home at Houjarray, South of Paris, on the occasion of discussions about the difficulties, the European Communities were then facing, was saying to his visitors "Continuez, continuez, il n'y a pas pour les peuples d'Europe d'autre avenir que dans l'Union" i.e. "keep trying, there is no other future for the European peoples than their Union". Jean Monnet was seeing in this Union a decisive element for the reconciliation of all the peoples of the World, and the European Community itself, as a stage to the forms of organisation of tomorrow's World.

With this vision and this approach we should face the recent Community crisis in the process of ratification and implementation of the Maastricht Treaty and we must overcome any possible resistance which on the contrary, I should say, must convince us that we are following the right path.

Certainly, the establishment of the economic monetary and political union requires a tremendous effort from all Community countries, big or small. It requires even sacrifices. But we must realize that this effort is the price of an investment, a price which we have to pay now in order to enjoy in the future the benefits of our economic, political and human union. If we don't respond positively to this challenge we risk losing our way of life and even our independence.

Difficulties for some smaller countries are of course much greater and the effort of adaptation becomes sometimes unbearable because the problems they face are more or less structural, and to be solved they need much more effort.

Greece is among these countries.

Currently Greece is facing many economic problems such as the large budget deficit, the high rate of inflation, the high public debt as a percentage of GDP and more difficult structural problems as the relative technological gap, the lower rate of productivity and competitivity of its production, the largeness of the public sector, the deficiencies of the infrastructure. In order to deal with all these problems Greece has to follow an austere economic policy, combined with a great effort to put the country in the way of further development.

Otherwise, Greece risks the danger of not being able to follow Europe in the way of unification, as provided by the Treaty of Maastricht.

And for Greece too as for the other European countries there is no alternative to a United Europe.

Fortunately, the Greek people are not alone in this effort of adaptation to the requirements of the Maastricht Treaty.

The European Treaties based on the principle of solidarity have provided policies to help the economic and social convergence of all areas of the Community. For this purpose major financial assistance is already offered by the Community. New funds will become available under the new Delors Plan II. Economic and social cohesion is a fundamental element of the united Europe. It has to be achieved.

The overwhelming majority of the Greek people has accepted the challenge. The Greek Parliament with a majority of 285 votes out of 300 Deputies has approved the Treaty of Maastricht.

This strong political will must be transformed into moving power for the promotion and implementation of the proper policies so that Greece meets the timetable and the requirements of the Maastricht Treaty together with the other countries. It can be done. The historic challenge must and will be answered.

Thank you.

SESSION 1:

Overview of Energy Intensive Sectors

Chairman: Prof. N. Koumoutsos

THE NATIONAL ENERGY STRATEGY OF THE UNITED STATES: INDUSTRIAL ENERGY USE

Elias P. Gyftopoulos
Massachusetts Institute of Technology
Cambridge, Massachusetts, USA

ABSTRACT

The industrial energy strategy of the United States is discussed. Past achievements and current research in progress are summarized.

GENERAL REMARKS

Improvement in energy end-use efficiency offers the largest opportunity of all alternatives to meet the energy requirements of a growing world. At the current rate of energy consumption, a universal one percentage point improvement in end-use efficiency, a perfectly do-able accomplishment, would be equivalent to discovering every five years forever a new oil supply equal to the 1990 oil proven reserves of the entire United States.

Presently, the satisfaction of the residential, commercial, transportation, and industrial energy needs of developed and developing societies is very inefficient. Thermodynamic analysis yields that the average efficiency of energy utilization is just about one tenth in industrialized nations, and even less than one tenth in developing countries. From the engineering standpoint, this is a very low efficiency, and the theoretical potential for improvement is enormous. Of course, energy end-use efficiency will never approach unity. Nevertheless, the present low value underscores the opportunity for large savings. No scientific barriers exist to prevent overwhelming improvements. For example, changing the average efficiency from about 10 to about 20 percent, not an unreasonable goal over the next few decades, would reduce energy consumption by one half without curtailing energy services.

Many cost-effective, energy-saving technologies exist for use in space lighting, heating, and air-conditioning, in new designs of large and small vehicles, and in industrial manufacturing processes. Examples are more luminous fluorescent lamps, 60- to 100-miles-per-gallon automobiles, cogenerators of heat and electricity, and radically new process technologies, such as hardening of metal surfaces by laser irradiation rather than heat-treating.

In this paper, I will not discuss the generic and specific methods that can be used for cost-effective energy savings. They have appeared in many publications of the 1970s and 1980s. Instead, I will restrict my remarks to the current programs of the Department of Energy of the United States in industrial energy conservation. These programs are described in References 1 and 2, which are the sources for this paper.

ENERGY CONSUMPTION IN THE UNITED STATES

In 1990, the United States consumed over 80 quads of primary energy (Figure 1). About one quarter of this energy was consumed in the transportation sector, and the remaining three quarters were shared about equally between the industrial and buildings (commercial and residential) sectors.

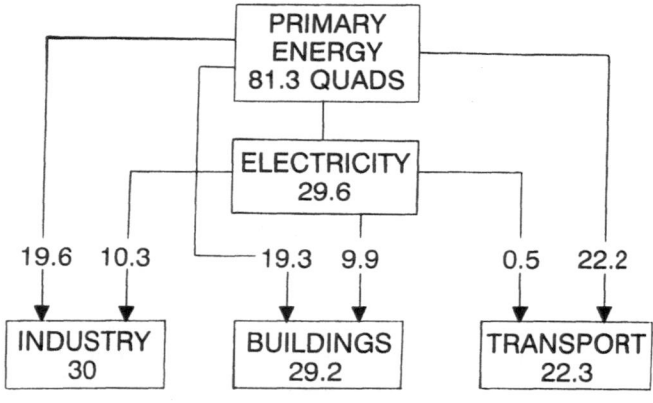

Figure 1

The industrial sector accounts for approximately one quarter of all U.S. petroleum consumption, with more than one half of this consumption being used for such materials as plastics and petrochemicals. In 1990, excluding fuels used as nonenergy feedstocks (but including the energy used to generate and deliver electricity), the industrial sector accounted for 37 percent of all primary energy consumption.

Great progress has been made in the efficient use of energy in the industrial sector. Energy consumption per $ 1982 of industrial output has been steadily decreasing. From 1973 to 1985, the decrease in this important ratio is illustrated in Figure 2. It is seen from the figure that the decrease was about one third or, equivalently, that about 35 percent more industrial output was generated by the U.S. economy from 1973 to 1985 without an increase in energy consumption.

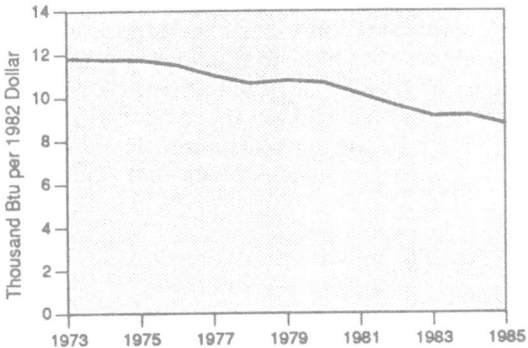

Source: Energy Information Administration, *Energy Conservation Indicators 1986 Annual Report*, February 1988.

Figure 2

It is estimated that about one half of the improvement of energy consumption per constant dollar of industrial output was due to cost-effective efficiency improvements made by industry in response to higher energy prices. The other half was due to structural shifts in U.S. demand, such as reduced consumption of energy-intensive products, and reduced production of energy-intensive goods — steel and automobiles. The reduced production was achieved by substituting imports for goods that were previously manufactured in the United States.

Past achievements by no means have exhausted the currently available opportunities for continued improvements in cost-effective energy utilization. In fact, the opportunities become even broader and larger and more challenging when considered together with the problems created by wastes. Currently, industrial practices in the United States produce more than 600 million tons of solid hazardous wastes per year, together with millions of tons of waste gases that contain chemicals worth $500 million per year. The Environmental Protection Agency estimates that approximately 11 billion tons of nonhazardous solid wastes and wastewaters are also produced per year.

These waste streams are potential feedstock sources. Presently, however, they are a serious environmental problem, and burden industry with growing disposal costs. To meet environmental regulations, industry spends about $46 billion per year on pollution controls. Transforming wastes into usable feedstocks can reduce the requirements of the United States for energy and improve the quality of the environment.

THE NATIONAL ENERGY STRATEGY

To continue the improvement in cost-effective energy use, the National Energy Strategy of the Bush administration has adopted three goals for industrial energy use: "(1) encourage increased energy efficiency and fuel flexibility in the industrial sector to reduce petroleum dependence; (2) encourage cost-effective measures to reduce energy costs; and (3) reduce industrial waste generation, increase recycling of wastes, and increase the use of plant- and consumer-generated wastes as process feedstocks."

"The Strategy estimates that industrial energy use would be about 55 quads in 2030 without the National Energy Strategy initiatives. With the initiatives, the industrial sector is expected to be 5 percent more energy-efficient by the year 2005, 10 percent more energy-efficient by the year 2010, and 15 percent more energy-efficient by the year 2030. These improvements culminate in a reduction of about 9 quads per year by 2030."

Energy Efficiency and Fuel Flexibility

The National Energy Strategy claims that "energy efficiency and fuel flexibility will improve because of increased support for research, development, and demonstration, use of industry-Government cost sharing, and increased information and technology dissemination efforts through the Energy Analysis and Diagnostic Centers and other outreach projects. Industrial energy research and development will stress reduction of waste energy in industry and advanced industrial processes. Moreover, increased energy audits of industrial plants through the Energy Analysis and Diagnostic Centers will reduce energy use and help disseminate the results of Government and industry research and development."

R&D on Waste Energy Reduction. "The Department of Energy will increase its support of research and development on equipment that will improve energy conversion efficiency, recover energy from industrial waste heat, and provide higher temperature structural materials and related technical information that support these advances. Opportunities exist in all industries to develop technologies that deliver energy services to industrial processes at higher efficiencies and with greater fuel flexibility than at present. Energy that is wasted in the manufacturing process can be captured and used, thereby reducing energy requirements to produce goods and services and increasing plant production. Therefore, savings are achieved from lower energy costs and more efficient plant utilization. The Nation will also derive substantial environmental benefits by reducing the energy that it requires to produce goods and services."

"Department of Energy research and development support will stress advanced chemical and mechanical heat pumps; process heat exchangers and ceramic recuperators; advanced materials such as continuous fiber-ceramic composites; advanced combustion systems for industry; and industrial cogeneration technology. These technologies will address a mix of near-, mid-, and long-term opportunities to save energy in the industrial sector. Research and development will be carried out through innovative cost-sharing arrangements with industry."

R&D on Industrial Processes. "The Department of Energy will increase its support of research and development on new industrial processes that offer significant opportunities for improving energy efficiency and increasing industry's flexibility in using alternative fuels — particularly, renewable fuels. The development of new industrial processes, from raw material to final product, presents a major opportunity for industry to improve its energy efficiency and fuel flexibility. Opportunities to save energy by redefining the production process exist within specific industries, such as steelmaking, as well as within process steps that cut across industries, such as separations technology."

"The Department of Energy supports the development of new industrial processes for advanced steelmaking, sensors and controls, improved membranes for separation systems, and process electrolysis. Department research and development focuses on technologies that offer energy savings and fuel flexibility in the near, mid-, and long terms. The Department of Energy will pursue cooperative cost-sharing arrangements to carry out Federal research and development programs. The Government will also be pursuing a permanent 25-percent research and development tax credit to encourage private industrial investment."

Energy Audits

Opportunities to save energy in an industrial plant can often be identified by a careful study of the flow of energy through the plant. Energy audits of industrial facilities reveal simple ways to cut energy use at practically no cost.

"Energy Analysis and Diagnostic Centers, operated by universities for the Department of Energy, have performed more than 2,800 preliminary plant energy audits for small- and medium-sized companies. The Department's expansion plan calls for adding 3 universities in fiscal year 1992 to the 18 already participating. In addition to helping smaller firms improve their energy efficiency, this program provides hands-on audit training for engineering students. The Department will encourage similar private-sector programs, such as utility-conducted industrial audits performed as part of a demand-side management program."

Waste Reduction, Waste Recycling, and Use of Wastes as Feedstocks

"The reduction of waste generation is an important strategy to control costs and improve productivity. Potentially dischargeable waste is not produced and therefore does not require treatment and disposal. Waste reduction ensures that more raw material becomes product, thereby reducing energy requirements, saving natural resources, and lessening environmental impacts. After wastes are reduced to their technical minima, industry may use or convert unavoidable wastes to feedstocks or fuels. If use or conversion is impossible, it may treat wastes and release them into the environment. More restrictive environmental regulations, rising energy costs, and the requirement for more economic waste control require developing and investing in technologies to reduce industrial wastes. Hundreds of U.S. companies have instituted waste reduction measures that have lowered production costs and raised corporate profits while reducing energy use and environmental impacts. Nevertheless, cost-effective waste minimization can be increased."

"Incomplete knowledge of the most advanced waste management practices is an important obstacle to more effective waste management. There are a wide variety of production processes that require individualized waste management strategies. In addition, implementation of new waste management techniques may require regulatory changes."

"The Government will continue to rely on private industry to make economic choices on waste management alternatives. However, to overcome the lack of advanced waste reduction and utilization technology, information barriers, and regulatory deficiencies, the following actions are required: support research and development on advanced process technology that reduces wastes, support research and development on waste use and conversion technology, determine which regulatory changes may help foster improved waste management without compromising environmental quality, and develop an outreach program."

R&D on Waste Reduction Technologies. "The Department of Energy will increase funding of cost-shared research and technology development directed specifically at industrial waste reduction. Long-term waste solutions often involve redesign of major portions of an industrial process that may require significant research and development, and many small- and medium-sized companies do not have the necessary resources. Even firms that have sufficient resources must evaluate the relative merits of developing new production technology versus product-related research and development."

"The Department of Energy, in close coordination with industry, will target the cost-shared effort to key areas with potential for substantial energy reduction. Other criteria will include opportunities to reduce environmental impacts, to increase overall industrial productivity, and to save natural resources. Initially, the Department will target chemical processes because of the large amounts of wastes that they produce and the large investments that are being made in pollution control activities ($4.2 billion in 1988). Additional industries, such as the petroleum industry and the pulp and paper industry, will follow as additional analysis is done on their waste reduction opportunities and needs."

R&D on Waste Use and Conversion. "The Federal Government will increase its support for research and development on the innovative mechanical, biochemical, and thermochemical processes that industry needs to convert industrial wastes economically into feedstocks or fuels. All industries produce wastes at every stage, from raw material input through product distribution and servicing. Many opportunities exist for profitable recovery and conversion of some of these wastes rather than payment of continuing and generally escalating costs for their environmentally sound disposal. However, the lack of cost-effective recovery techniques to use the materials and energy content of industrial wastes efficiently limits their use."

"The Department of Energy will focus on techniques for improved recovery of metals and other materials from auto scrap, recovery of useful products from waste tires, recovery of adhesives and other useful materials from wood wastes, recovery of high-value products from food wastes, and separation and collection of useful gaseous materials. Solar technologies will be developed to decontaminate wastewater and destroy hazardous industrial chemicals. The Department will pursue near-, mid-, and long-term research and development objectives through cost-sharing with industry. Though strongly market oriented and well connected to the ultimate industrial users of the technology, the Department's approach in this area emphasizes bringing capabilities of the National Laboratories to bear on the complex technical issues involved."

Industrial Waste Regulation Reform. "The Department of Energy and the Environmental Protection Agency will determine the extend to which existing regulatory programs discourage investment in innovative waste and pollutant minimization technologies. The evaluation will include input from private industry on existing regulatory barriers and potential solutions. The Department and the Agency will then suggest legislative or regulatory changes to encourage waste minimization investments."

Waste Outreach Program Development. "The lack of good data, worker information programs, and auditing procedures may create significant barriers to widespread adoption of waste reduction practices. The Department of Energy will develop a coordinated outreach program to communicate research results, provide technical information and advice, and disseminate industrial waste stream data to the industrial sector. This effort will be coordinated with the Environmental Protection Agency, as well as with leading industry groups with interest in waste reduction and use."

RESEARCH IN PROGRESS

A bibliography of all scientific and technical reports sponsored by the U.S. Department of Energy Industrial Energy Conservation Program during the years 1988-1990 is given in Reference 3.

In fiscal year 1993, the funding of research and development projects in industry efficiency by the U.S. Department of Energy will be about $100 million. It will be shared by about 220 different projects addressing different aspects of industrial energy conservation. They are managed by the Office of Industrial Technologies. They are briefly described in Reference 4, and fall in the general program categories listed in Table 1.

CONCLUDING REMARKS

The interest in energy issues, both new sources of energy and more cost-effective energy uses, is no longer in the headlines. Reduced demand for OPEC oil has kept energy prices low. The reduced demand is due partly to energy savings, partly to world stagnant economies, and partly to the desire of oil producers to sell as much oil as they possibly can. In $ 1987, oil prices are only about 50 percent higher than in the '50s and '60s. So the desire to address the energy problem is not as acute as it was in the decade that spanned the seventies and eighties.

Despite these deficiencies, it is purposeful to continue our efforts in the area of cost-effective energy utilization because the results are important from the points of view of the environment, resource availability, and possible climatic changes.

REFERENCES

1. National Energy Strategy, First Edition, 1991/1992, U.S. Department of Energy. Available from the National Technical Information Service, U.S. Department of Commerce, 5285 Port Royal Road, Springfield, VA 22161.

2. National Energy Strategy, One Year Later, February 1992, U.S. Department of Energy. Available from the same service as reference 1.

3. Industrial Technologies Technical Reports, A Bibliography, January 1992. DOE/OSTI—3409/2 (DE92000497). Available from the Office of Scientific and Technical Information, P.O. Box 62, Oak Ridge, TN 37831.

4. Office of Industrial Technologies, Research in Progress, August 1991. DOE/OSTI-11633/3 (DE91014180). Available from the same office as reference 3.

INTERNATIONAL CO-OPERATION IN R&TD: GLOBAL INDUSTRY VIEW

Dr. David Bricknell
Director of Science & Technology
European Chemical Industry Council (CEFIC)

SUMMARY

The chemical industry plays a vital role in national and international economies both through its contribution to the prosperity of Europe and through its contribution to improving the quality of life enjoyed by modern society. It is one of Western Europe's most international, competitive and successful industries embracing a wide field of processing and manufacturing activities. It accounts for about 30% of the world's production of chemicals and employs more than two million people.

The European chemical industry has built its present day success and competitiveness in world markets on a basis of intensive and self-funded research and development and on close co-operation with academic institutions reaching back for more than 100 years.

Such commitment by industry to long-term research and development is only possible in an operational climate which encourages the pursuit of new science and technology and guarantees the ability to exploit innovations throughout the European market-place.

The industry's key objectives are:-

- an improved climate for the conduct of research and technological development and for the exploitation of innovation.
- enhanced support for educational institutions in their roles both as educators and as leading-edge pioneers in scientific research.
- selective public funding of R&TD in areas which will have great importance for future communal welfare and prosperity.

The industry has proposed FIVE areas which its judges to be particularly fertile for co-operative R&TD. These are:-

- advanced fundamental knowledge of chemical reactivity and principles of catalysis

- new environmentally benign and resource (including energy) minimising chemical process technologies
- biotechnology and life sciences
- environmental protection and remediation processes
- reliable synthesis, processing and reprocessing of functional and structural materials

INTRODUCTION

The chemical industry plays a vital role in national and international economies both through its contribution to the prosperity of Europe and through its contribution to improving the quality of life enjoyed by modern society. It is one of Western Europe's most international, competitive and successful industries embracing a wide field of processing and manufacturing activities. It accounts for about 30% of the world's production of chemicals and employs more than two million people (Figure 1).

The European chemical industry has built its present day success and competitiveness in world markets on a basis of intensive and self-funded research and development and on close co-operation with academic institutions reaching back for more than 100 years (Figures 2 and 3).

Such commitment by industry to long-term research and development is only possible in an operational climate which encourages the pursuit of new science and technology and guarantees the ability to exploit innovations throughout the European market-place.

The fundamental requirement for ensuring the enduring attraction of Europe as a base for a world-competitive chemical industry is the provision by governments of a framework of regulatory controls which are internationally harmonised, scientifically-based and applied co-operatively with the industry.

The European industry is conscious of the need to preserve its position by the continued funding of effective, and broadly-based product and process development. Competitors in the world industry - notably in the USA and Japan - are deploying ever-increasing resources towards industrial Research and Technological Development (R&TD). They benefit from strong governmental interest and are able to exploit the advantages of large home markets with uniform laws, regulations and standards (Figure 4).

The creation of the single market in the European Community (EC) by the end of 1992, the developments towards the so-called European Economic Area by the EC and EFTA countries and the emergence of free market economies in Central and Eastern Europe, afford new opportunities for improving prosperity and the quality of life.

The European chemical industry through its scientific and technological capabilities is able and willing to make a major contribution to safeguarding and enhancing communal welfare and prosperity; but its ability to contribute will be dependent on the nature of its operating environment.

In its policy paper entitled "European Policy for Science and Technology: the Position of the European Chemical Industry" published in January 1992, the industry sets out ideas and actions to maintain and improve the industry's operating environment throughout Europe.

The industry's key objectives are (Figure 5) :-

-   **an improved climate for the conduct of research and technological development and for the exploitation of innovation.**

-   **enhanced support for educational institutions in their roles both as educators and as leading-edge pioneers in scientific research.**

-   **selective public funding of R&TD in areas which will have great importance for future communal welfare and prosperity.**

THE NEED FOR CO-OPERATION IN R&TD

The chemical industry has traditionally pioneered its own R&TD programmes - partly through confidence (some would say arrogance) in its ability to know what had to be done and how best to do it, partly through an indifference (perhaps ignorance) to the opportunities for co-operation - but perhaps most importantly because of the over-riding need to compete with the other companies within the industry (Figure 6).

To-day, times are changing:-

-   companies cannot afford to fund all of the R&TD themselves

-   some issues which society recognises to be important e.g. climatic change, waste-management, environmental impacts and higher energy efficiency are beyond the power of any single company to solve

-   the need for co-operation is recognised

-   the chemical industry has the know-how and the human resources to help to solve society's problems (Figure 7)

-   the Community's R&TD Framework Programme can provide an excellent vehicle for R&TD co-operation. Co-operative actions between suppliers and customers - both vertically within an industrial sector and horizontally across industrial sectors are of increasing importance. Industry-academic institution links are vital to ensure that basic research objectives can be made relevant to long-term industrial competitiveness.

THE CHEMICAL INDUSTRY PRIORITIES

The industry has proposed FIVE areas which its judges to be particularly fertile for government-funded co-operative R&TD. These are : -

-   advanced fundamental knowledge of chemical reactivity and principles of catalysis
-   new environmentally benign and resource (including energy) minimising chemical process technologies
-   biotechnology and life sciences
-   environmental protection and remediation processes
-   reliable synthesis, processing and reprocessing of functional and structural materials

The chemical industry believes that government funding should be concentrated on three categories of R&TD (Figures 8 and 9) : -

-   fundamental research in academic institutions
-   pre-legislative research
-   targeted research on major new cross-sectoral generic industrial technologies

THE JOULE PROGRAMME

Co-operative research under the JOULE component of the Third Framework Programme is relevant to several of the key areas for R&TD identified by the chemical industry. In particular, the JOULE component of the future work on energy efficient processes, taking into account environmental aspects, needs to be explored. Community sponsored research on targeted large-scale novel technologies should undoubtedly involve new energy efficient technologies.

TASKS FOR NATIONAL GOVERNMENTS

Following the current trend towards SUBSIDIARITY in European Community matters, action must start at the national level. The chemical industry will advocate the following (Figure 10) : -

-   increased funding for basic research in academic institutions
-   better education in science and technology for all young people
-   improved integration of national and EC-R&TD programmes.

TASKS FOR THE EUROPEAN COMMUNITY

The chemical industry believes the following to be the most important tasks to be addressed by the Community's institutions (Figure 11) : -

- Strengthen support for research in academic institutions
- Focus support on those areas where industry alone cannot succeed
- Promote the European climate for innovation
- Ensure R&TD programmes meet industry's needs and that they are "user-friendly"
- Adopt a pro-active "marriage-broking" function for cross-sectoral co-operation
- Ensure DG IV takes a benign view of R&TD co-operation.

TASKS FOR CEFIC

On behalf of the European chemical industry, CEFIC will work within existing European structures (Figure 12) : -

- with the EC R&TD Framework Programme
- with EUREKA
- with COST

* To promote cooperative R&D activities in the key areas identified
* To stimulate greater chemical industry participation in public R&TD programmes.

CONCLUSIONS

To sum up : the chemical industry believes that its continuing ability to conduct R&TD and exploit innovation will be crucial to its success and to the future prosperity of the peoples of Europe.

Governments must implement imaginative policies for Science and Technology. An essential part of these policies must include funding for fundamental research in universities and institutes and for research in support of legislation. Funding for co-operative research with industry will be necessary to seek technological solutions to problems identified by society for which no commercially-viable technology exists.

CEFIC, on behalf of the chemical industry, will work to encourage participation by chemical companies in existing programmes and through a beneficial re-orientation and expansion of JOULE in the Fourth Framework Programme will try to ensure a more active and continuing participation in the future.

SOURCES

1. European Policy for Science and Technology - The Position of the European Chemical Industry, CEFIC January 1992.
2. Treaty on European Union, CONF-UP-UEM 2002/92, Brussels, 1 February 1992.
3. From the Single Act to Maastricht and Beyond, The Means to Match our Ambitions, COM(92) 2000 final, Brussels, 11 February 1992
4. Research after Maastricht : An Assessment. A Strategy, SEC(92) 682 final, Brussels, 9 April 1992.
5. Evaluation of the Second Framework Programme for Research and Technological Development, COM(92) 675 final, Brussels 22 April 1992.

*   Directly employs more than two million people

    –   About 8% of the total work-force

*   Accounts for about 30% of world chemical production

    –   Total sales in excess of 400 billion US $

*   Generates for Western Europe an annual positive
    trade balance of 28 billion US $

FIGURE 1 – THE WEST EUROPEAN CHEMICAL INDUSTRY IN
PERSPECTIVE

Source : CEFIC

*   About 22 billion US $ annually

*   About 22 % of all non-military R&D expenditure

*   Present-day competitiveness derived from and
    sustained by effective exploitation of innovation
    in global markets

FIGURE 2 – CHEMICAL INDUSTRY'S RESEARCH AND DEVELOPMENT
EXPENDITURE IN WESTERN EUROPE

Source : CEFIC

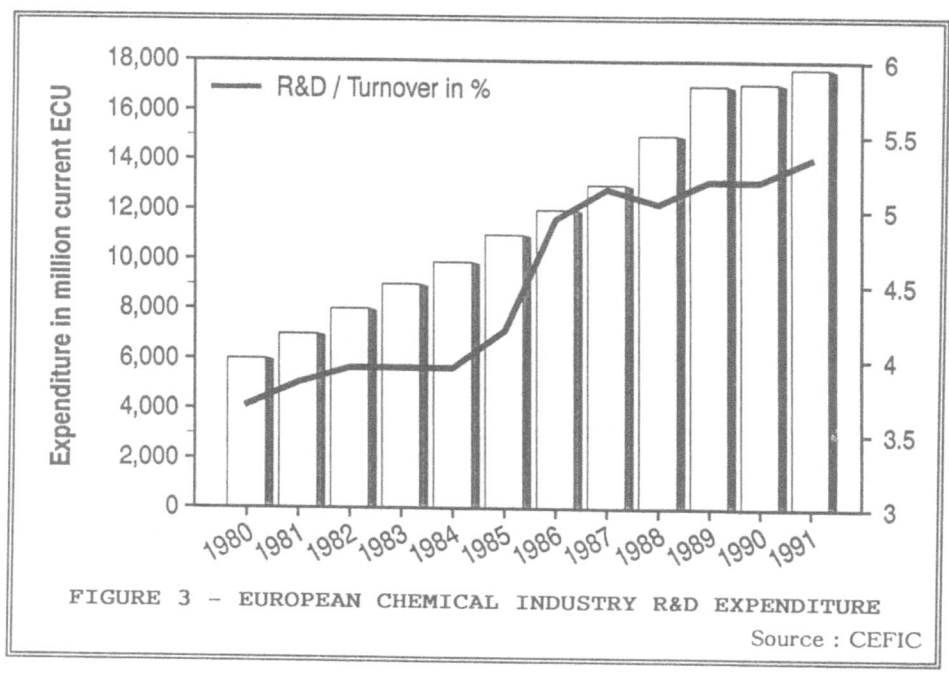

FIGURE 3 – EUROPEAN CHEMICAL INDUSTRY R&D EXPENDITURE

Source : CEFIC

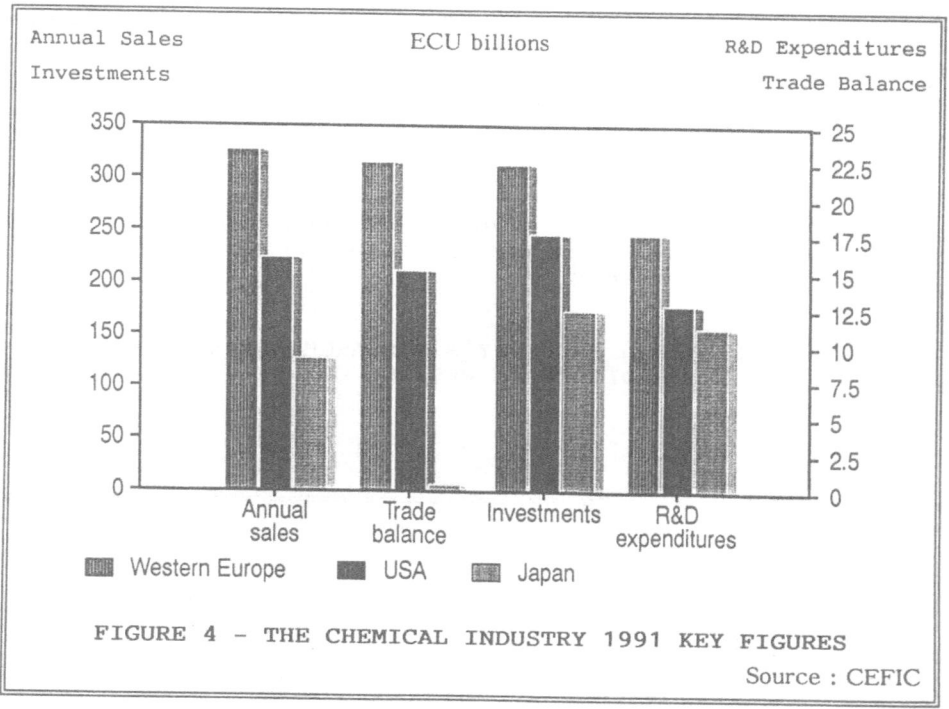

FIGURE 4 – THE CHEMICAL INDUSTRY 1991 KEY FIGURES

Source : CEFIC

*   Improve the climate in Europe for R&TD and for the exploitation of innovation

*   Encourage Stronger support for research in academic institutions

*   Improve the education of the public in science and technology and the understanding of the pivotal role of our industry in society

*   Promote governmental support for cooperative R&TD in areas critical for future communal prosperity

FIGURE 5 - SCIENCE AND TECHNOLOGY POLICY :
WHAT MUST BE DONE

*   Long history of independence and self-reliance

*   Links with academic institutions in fundamental research

*   Need to keep critical areas of R&D "in-house" and confidential

    -   Cannot sub-contract R&D in core business

*   Need to maintain competitive edge vis-à-vis competitors

FIGURE 6 - THE CHEMICAL INDUSTRY'S APPROACH TO R&D

* Concern for environmentally quality

* Threat of climatic change

* Depletion of natural resources

* Increasing public scepticism about industry and the benefits of technological innovation

The chemical industry has the knowledge, skills and motivation to help to solve these problems - but a cooperative approach with governmental support will be essential

FIGURE 7 - INDUSTRY AND GOVERNMENTS FACE CHALLENGES AND CONSTRAINTS

---

* Fundamental research in Academic Institutions

* Pre-legislative/pre-normative/quality of measurement research

* Targeted research on new cross-sectoral industrial technologies
    - New Pandolfi initiative
    - Fourth Framework Programme
    - JOULE/BRITE/Environment/others

FIGURE 8 - PUBLIC FUNDING SHOULD BE DIRECTED TOWARDS

* Fundamental research

    - Catalysis, reaction kinetics

    - Environmental impacts/protection

    - Biotechnology

* Pre-legislative research

    - Fill in knowledge gaps for existing and future legislation

* New industrial technologies and processes

    - Chemical process intensification

    - Plastics waste : re-use/recycling/ depolymerisation

    - Clean incineration/energy recovery

    - CO_2/CH_4 chemistry

FIGURE 9 - CHEMICAL INDUSTRY EXAMPLES

* Increased funding for research in academic institutions

* Better education in science and technology for all young people

* Improved integration of national and EC R&TD programmes

FIGURE 10 - TASKS FOR NATIONAL GOVERNMENTS

*   Strengthen support for research in academic institutions

*   Focus support on those areas where industry alone cannot succeed

*   Promote the European climate for innovation

*   Ensure R&TD programmes meet industry's needs and that they are "user-friendly"

*   Adopt a pro-active "marriage-broking" function for cross-sectoral co-operation

*   Ensure DG IV takes a benign view of R&TD co-operation

FIGURE 11 - TASKS FOR THE EUROPEAN COMMUNITY

CEFIC will work within existing European structures

-   With the EC R&TD Framework Programme

-   With EUREKA

-   With COST

*   To promote co-operative R&D activities in the key areas identified

*   To stimulate greater chemical industry participation in public R&TD programmes

FIGURE 12 - TASKS FOR CEFIC

ENERGY SAVING IN THE PETROCHEMICAL INDUSTRY

P. TRAMBOUZE
Institut Français du Pétrole
C.E.D.I.-BP N° 3
69390 - VERNAISON
France

ABSTRACT

A great deal of effort has already been made to reduce energy consumption in the petrochemical industry by optimizing the use of energy in petrochemical complexes. Further improvements are still possible, especially by considering feedstock material as the equivalent of energy.

Processing itself can be improved from several points of view. Integration of different steps of a process may be the source of substantial improvements. This is a way of intensifying a process. Membrane reactors or reactive distillation are examples of this type of approach. Integration of different processes must also be considered, with the dual objective of saving both energy and feedstock. Of course, improvements can also be obtained by modifying some equipment to achieve better yields and selectivities.

Chemistry may be the source of significant increases in yield and selectivity, by using new operating conditions or new catalysts. Sometimes, new chemical routes could also be considered. The substitution of one source of energy or feedstock for another must be considered for various improvements.

Likewise, the recycling of waste material will become an important way of reducing energy and feedstock consumption and should be considered as soon as any new material is put on the market.

INTRODUCTION

Since the first energy crisis in 1973, the petrochemical industry has been forced to make many changes concerning energy and feedstock consumption. In Europe, as well as worldwide, great efforts have been made in order to reduce the overall consumption of energy by this particularly important industrial sector. In 1977, this sector of the European industry consumed 110 Mtoe, which can be compared to the overall energy consumption by the EEC, namely 530 Mtoe. The second energy crisis in 1979 reinforced the need for a strong energy conservation policy. Even if the current price of crude oil is back to its price in 1976-78, there is no doubt that, in the future, the industry will have to face a great increase in energy prices, and hence efforts to save energy and feedstocks must be continued.

We shall consider the various technical means industry can take advantage of to reduce its energy consumption and therefore improve its competitiveness. The most immediate way to reduce energy consumption is to improve existing processes. Such improvements may be looked for in the various aspects of processing: chemistry, process engineering and equipment. On a long term basis, improvement might be achieved by changing the type of feedstock used, by implementing new technology, by improving the quality of the end products to reduce their consumption, and by recycling part of the end products manufactured from petrochemical basestocks.

IMPROVEMENT OF PROCESSES

In the past 20 years, a great deal of effort has been extended to reduce energy consumption by chemical processes. General methods have been proposed and developed to clearly indicate the parts of a process in which improvements could be made. Today, process integration [1], exergetic analysis [5,6], pinch technology [2,29] or calculation of the energy content [26] of a given chemical are common practice for improving the energy efficiency of a process.

It has been found that great improvements can first be obtained by optimizing the networks of heat exchange systems, namely furnaces, heat exchangers and coolers. In this line, a great deal of research has been done to integrate these networks inside a process or even between several processes.

A second important finding has been that separation processes to recover and purify products account for over 40% of the chemical process energy demand [22]. Therefore, since distillation is the most widely used separation process and the unit operation that consumes the most energy [11], the heat integration of the reboilers and condensers in distillation columns has been the focus of attention by most engineers. For this purpose the use of heat pumps or vapor recompression systems has found many useful applications [8,9,30].

Efforts have also been made to improve the efficiency of distillation, using high-efficiency packings or trays (to reduce the reflux ratio) and implementing advanced process control for operating distillation columns without any "safety margin" requiring extra reflux and energy. Energy savings in the 10-40% range have thus been obtained [8,22].

Other options have also been looked at, such as completing distillation with advanced separation processes to form energy-efficient hybrid systems or completely replacing distillation by another technique. Among these hybrid systems, the following combinations may be cited as examples:
- distillation-adsorption (pressure swing adsorption) to produce fuel-grade ethanol [22],
- distillation-membrane pervaporation for ethanol/water separation [22],
- distillation-crystallization [24].

Examples of improved performances for a variety of equipment may be found in Reference [18,39]. There are now a great number of research projects dealing with membrane separation [35,36] and supercritical extraction. In the future these separation methods could offer new additional ways of saving energy.

IMPROVEMENT IN CHEMISTRY

It is clear that feedstock materials in the petrochemical industry are generally equivalent to energy and represent a large proportion of total consumption. Therefore, for any process, energy savings can also be found in feedstock utilization as well as in the energy needed for operating. In this outlook, any increase in yield or selectivity of the chemical transformation can provide a twofold advantage. First, the gain in selectivity makes possible a reduction of feedstock consumption for a fixed end product requirement. At the same time, the production of secondary products or waste is limited. Secondly, the increase in the yield of the chemical transformation will limit the amount of raw material having to be recycled. As a consequence, the job to be performed by the separation section will be reduced, and therefore the energy requirement will be diminished.

Table 1 gives a few examples of industrial production yields for some important products [38].

TABLE 1
Examples of processes with relatively low yields

PRODUCT	FEEDSTOCK	YIELD (% of theory)
Propylene	Propane	85-87
n-Butene	Butane	75
Ethylene oxide	Ethylene	72-75
2 Ethyl hexanol	Propylene	65-70
Acrylonitrile	Propylene	72-75
Caprolactam	Cyclohexane	70
Terephthalic acid	p Xylene	85-90
Maleic anhydride	Butane	55-60
Phthalic anhydride	o-Xylene	76-80

In the same way, Table 2 gives examples of processes with

low conversion per pass [38].

TABLE 2
Examples of processes with low conversion per pass.

PRODUCT	FEEDSTOCK	CONVERSION (%)
Propylene	Propane	65
i Butene	i Butane	65
Butadiene	Butane	35-45
Styrene	Ethylbenzene	60
Ethylbenzene	Benzene+Ethylene	30
Ethylene oxide	Ethylene	25-30
Acetic acid	Methanol	70
Ethanol	Ethylene	6
Vinyl acetate	Ethylene+Acetic acid	10-25

From the above figures, it is clear that substantial improvements could possibly be made.

To compare different ways of synthesizing a given chemical, calculating the energy content of the end product issuing from different processes is an effective method [26]. For example, this type of calculation clearly shows that the best energy route for manufacturing hexamethylene-diamine is by reacting HCN with butadiene. The other two processes (dimerization of acrylonitrile and hydrogenation or from adipic acid) are less favorable from the energy point of view.

To promote the yield and selectivity of a chemical transformation, various possibilities exist. First, the operating conditions (especially temperature and pressure) can be optimized in keeping with the thermodynamics and the kinetics. For catalytic processes, improvements can often be brought about by means of new catalysts. The increase in the activity of a catalyst is not always the best objective. Enhanced activity can be useful only if the existing catalyst is not very active or if the reaction temperature can advantageously be reduced (better selectivity or stability). Likewise, the combination of a new catalyst and modified operating conditions may be the key to progress.

Improvements in yields may also be obtained by modifying the technology. A good example of this possibility is the so-called reactive distillation. The old concept of simultaneously performing the reaction and the separation of products has recently been renewed for the manufacture of MTBE (methyl *tert*-butyl ether). It is equally applicable to TAME (*tert*-amyl methyl ether) production [23]. Chemical reactions that are characterized by an unfavorable reaction equilibrium, a high heat of reaction and a significant rate of reaction at distillation temperature are good candidates for reactive distillation. The reaction may be catalytic or homogeneous. The same principle is applicable to liquid-liquid systems (extractive reaction).

Another technology, for simultaneously achieving the reaction and the separation of one of the product, is a membrane reactor [37]. Inorganic membranes are best suited for such applications, because the reaction operating conditions require relatively high temperatures. Several applications have already been investigated, such as dehydrogenation reactions [32] and the production of hydrogen [34]. In such cases hydrogen-permeable membranes are used, and the in-situ separation of hydrogen causes a shift in the thermodynamic equilibrium and therefore better conversion. Great conversion increases have been experimentally observed, e.g. from 20 to 90% for the dehydrogenation of cyclohexane [37].

Another possibility is to use the membrane as a support for a catalytic species, with the reactants being fed from each side of the membrane. In doing so, a selectivity increase could be obtained. A particularly advantageous application could be the oxidative coupling of methane leading to selectivities greater than 95% [31]. A membrane may also be used as a support for immobilized enzymes [37] or a homogeneous catalyst [33].

Since the feedstock represents a large portion of the total energy required by a petrochemical process, the origin and nature of the starting material can also have a great effect on the overall economics. In this respect, renewable materials produced locally by agriculture could provide some advantages, especially by reducing the amount of oil imported by the countries of the EEC. This is the case of products such as cellulosic or sugar containing materials which are easy to transform into ethanol. Fatty acids can also be obtained from vegetable oils. Such products are already considered as energy products and are added to gasoline or to diesel-oil. Their petrochemical applications will certainly be developed.

Similarly, it might be advantageous to develop the chemistry of natural gas, since this vector of energy as well as of carbon and hydrogen could well become more abundant in the future, at least in some area of the world. The use of electricity will also certainly be developed further, especially in countries where production by nuclear power stations will dominate. On the other hand, the implementation of cogeneration systems in petrochemical complexes affords great potential. As a consequence, the flexibility of processes concerning the type of energy used could be as beneficial as the flexibility concerning the nature of the feedstock.

The quality of the end product could also influence the need on the market, leading to lower consumption. For example, the demand for polyethylene for the manufacture of films has been reduced as a consequence of the better qualities of LLDPE.

EXAMPLES

Referring to Table 3, let us consider two of the most important processes for petrochemistry and see which important improvements have been made in the recent past. The steam-cracking process is the key link between refining and petrochemistry for the production of olefins and aromatics. On the other hand, synthesis gas manufacture is a major step for methanol, ammonia and hydrogen production [38].

TABLE 3
World demand (kt/year) for major petrochemicals
The European share is around 25%.

PRODUCT	DEMAND		
	1992	1995	2000
Ethylene	60,430	68,100	80,400
Propylene	32,530	37,500	45,400
Butadiene	6,460	7,050	8,010
Benzene	23,600	26,280	30,200
Toluene	11,680	12,640	14,200
Xylenes	14,350	16,110	18,540
Styrene	18,200	21,100	23,900
Methanol	24,000	27,800	35,500
Vinyl chloride	21,000	22,300	24,100
Ammonia	125,000	139,000	162,000

Source: IFP-Economics and Process Evaluation Department
-June 1992.

Energy consumption in ethylene plants has been reduced by approximatively 50 to 60 percent in the plants built prior 1973 and plants based on current technology [7,10]. Hereunder are some of the points where changes have been made:
• Cracking furnaces: air preheating, new coil configuration giving better yields, new high-temperature heat recovery exchangers with lower pressure drop, gas turbine cogeneration.
• Hot fractionation: lower reflux ratio, lower pressure drop in the gasoline fractionator.
• Cold fractionation: demethanizer expander, deethanizer feed optimization, ethylene fractionator side reboiler, etc.
It appears that the C_2 splitter is the most important column for energy optimization and indicates the use of minimum possible reflux versus required product specifications [4].
Of course heat integration must be at a maximum inside the steam cracking plant, but process integration must also be sought by considering the surrounding plants in the petrochemical complex and even in neighbouring refineries. In this respect, a great deal of synergy can be found in exchanging intermediate products. An important choice for the profitability of a steam-cracking plant is related to the choice of the feedstock coming from the refinery. Greater flexibility with regard to the feedstock could certainly be profitable [4]. For example, paraffinic hydrocarbons, especially straight paraffins, are the best feedstock for ethylene production. On the other hand, isoparaffinic, naphthenic and aromatic hydrocarbons are more suitable for producing high octane gasoline [3] (see Table 4).

TABLE 4
Octane Number and Ethylene Yield of Light Hydrocarbons

Hydrocarbons	Octane number (RON)	Ethylene yield wt%
n-pentane	61.7	37.0
n-hexane	24.8	40.0
n-C_7-C_8	0	43.0
iso-octane	100	24.0

With the same objective, olefins produced in catalytic cracking could be sent to the steam-cracking unit for separation. This exchange of products (mainly olefins and aromatics) between refineries and petrochemical complexes will be more and more important for an overall economical operation.

Other attempts (catalytic, thermo-catalytic or initiated pyrolysis) have been made to basically modify the steam-cracking process, so as to achieve better ethylene selectivity [17].

2- Synthesis gas manufacture.

The manufacture of a mixture of CO and H_2 is essential for the production of various very important chemicals such as hydrogen, methanol, ammonia, urea, oxo alcohols, etc. Many different feedstocks are used: natural gas, petroleum cuts (LPG, naphtha, fuel oil or residues) or coal. Depending on the specific case, a succession of treatments is performed to produce synthesis gas with the desired composition (H_2/CO ratio). There are however two basic industrial operations implemented separately or jointly:

-the partial oxidation with air or oxygen,
-steam reforming.

The following operations are then carried out to adjust the H_2/CO ratio:

-conversion of CO with steam (shift conversion),
-extraction of CO_2.

Partial oxidation can be used to treat any gaseous, liquid or solid feed. Steam reforming is a catalytic operation used for treating feeds ranging from methane to cuts with an end boiling point of 200°C.

Recently many process options have been proposed to reduce the overall energy requirement. For methanol production, combined reforming can be used. For this, a primary and secondary reformer are in series as in ammonia production. But rather than using air as the oxidant in the autothermal secondary reformer, oxygen is used [19,20]. A particularly interesting combination is proposed by Kellog with an open tube reforming exchanger process [19]. Process feed (hydrocarbon+steam) enters at the top of the exchanger and passes down axially through the catalyst-filled tubes and exits at the bottom, where it mixes with autothermal reformer effluent (see Figure 1). Similar technologies are proposed by other process licensors.

For hydrogen production, KTI proposes to add an adiabatic pre-reforming reactor and a post-reformer heat exchanger to a conventional steam-reforming furnace [21]. For some applications, such as oxo-alcohols, the H_2/CO ratio must be relatively low. Carbon formation may become a problem when the reformer is operated at higher severity. The use of a partially sulfur-poisoned nickel catalyst can be a solution, as proposed by Haldor-Topsoe in the Sparg process [25].

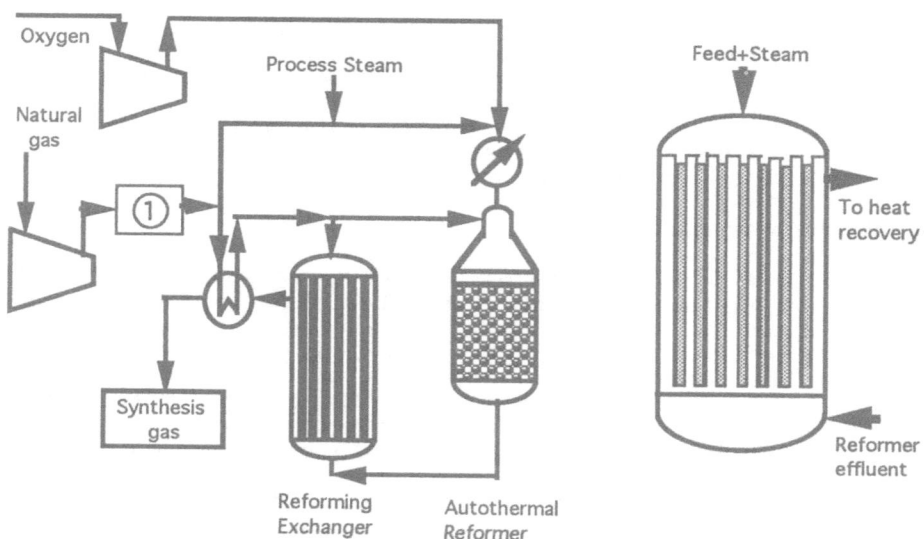

Figure 1. Combination reforming: reforming exchanger schematic diagram (on the left), open catalytic tube reforming exchanger (on the right) - (1)=sulfur removal -[from reference 19].

RECYCLING OF WASTE MATERIALS

As can be observed in Table 3, a large part of petrochemical production is aimed at furnishing plastics: polyethylene (PE), polypropylene, polystyrene, polyvinylchloride, polyethylene terephthalate, etc. The total world production in 1990 was roughly 100 million tons. Most of these materials (39% of the total) are used for packaging, leading to large quantities of wastes.

Approximately 180 million tons of municipal solid waste (MSW) is generated yearly in the US, and about 110 million tons in the European Community. In Japan, industrial and municipal waste combined totals roughly 300 million tons per year [14]. The weight percentage of plastics in MSW is approximatively 9%. Total post-consumer recycling is around 12% in the US and only 1% in Europe [15]. The current recycling rate for post-consumer plastics is slightly over 1% of virgin production. Table 5 gives the composition of plastic MSW in Europe [15].

TABLE 5
Composition of plastic municipal solid waste in Europe

MATERIAL	Weight percent
Polyolefins (HDPE, LDPE, LLDPE, PE)	65
Polystyrene (PS)	15
Polyvinylchloride (PVC)	10
Polyethylene terephthalate (PET)	5
Other	5

HDPE=high density polyethylene
LDPE=low density polyethylene
LLDPE=linear low density polyethylene

PVC receives the most attention for recycling in Europe because of its wide use in mineral-water bottles and other packaging. PET and HDPE are the two principal resins recycled in North America. In 1988 the US recycled 80,000 tons of PET from two-liter soda bottles. PET may be broken down to monomers and repolymerized to virgin resin. The US recycled 33,000 tons of HPDE in 1988. The bulk of PE currently recycled (roughly 350,000 kt/year) is LLDPE or LDPE thin film waste. In the US there are plans for recovering 115,000 tons of PS and 50,000 tons of PE film. Polypropylene from old automobile battery cases is recycled at the rate of about 50,000 tons a year in the US.

From the economic point of view, based on third-quarter 1990 prices, virgin LDPE costs just over $1,000/t to produce; recycled LDPE costs from $350/t for presorted material to nearly $700/t for unsorted plastic from MSW [12,16].

Among the formulations presented so far, cascade recycling seems to be the most rational one, i.e. recycling materials eventually in several stages into products with a reduced technological requirement and finally burning them with energy recovery [16].

National targets already set in Western Europe include 80% of plastic packaging in Germany by 1995, 40% of containers in Italy by the end of 1992, and 35% of plastics in the Netherlands by 2000 [12]. European regulations will quadruple the amount of plastics recycled by the year 2000 [13]. As a consequence of this irreversible trend, an important petrochemical mine is opening up and should contribute significantly to slowing down the growth of the energy demand.

CONCLUSION

We have mentioned various possible ways of reducing the demand for energy and feedstock materials by the petrochemical industry. It must also be pointed out that all these improvements of processes will simultaneously reduce emissions and waste materials, contributing so far largely to the protection of the environment, which is another important target for the chemical industry. The same dual result is obtained by recycling plastics.

REFERENCES

1. Smith, R. (Organizer), <u>Understanding Process Integration II</u>-IChemE Symposium Series N° 109 (1988).

2. Linnhoff, B. et al., <u>A User Guide on Process Integration for the Efficient Use of Energy</u> -The Institution of Chemical Engineers, 1982.

3. Kobayashi, N.,"Energy Conservation in Ethylene Production"-<u>Energy Developments in Japan</u>, Vol. 6, 135-155, 1983.

4. Picciotti, M. and Kaiser, V.,"Select Process Schemes for Optimum Petrochemicals" - <u>Hydrocarbon Processing</u>, June 1979, 99-105.

5. Maloney, D.P. and Burton, J.R.,"Using Second Law Analysis for Energy Conservation Studies in the Petrochemical Industry" - <u>Energy</u>, Vol. 5, 925-930, 1980.

6. Timmerhaus, K.D. and Flynn, T.M., "Energy Conservation through Use of Second Law Analysis"-<u>Erdöl und Kohle- Erdgas-Petrochemie</u>, Bd. 33, Heft 5, Mai 1980,208-214.

7. Bate, D.J.,"Fuel Saving Techniques in the Petrochemical Industry" - <u>Journal of the Institute of Energy</u>, September 1980, 124-133.

8. Kenney, W.F., "Reducing the Energy Demand of Separation Processes"- <u>Chemical Eng. Progress</u>, March 1979,68-71.

9. Tripathi, P. and Shukla, D., "Environmentally Responsible Energy Management"- <u>Hydrocarbon Processing</u>, Oct. 1991, 45-46.

10. Rhoe, A.,"Energy Saving Approach for Refining, Petrochemical and Chemical Process Plants"- <u>Chemical Engineering World</u>, Vol. XX, N° 4, April 1985, 54-60.

11. Nachod, J.E.,"Economics of Energy Conservation in the Chemical and Petrochemical Industries"- <u>A.I.Ch.E. Spring National Meeting</u>, New-Orleans, April 1986, Preprint 9B.

12. Chynoweth, E.,"Wastes Make Inroads in Plastics Markets"-<u>Chemical Week</u>, April 24, 1991, 40.

13. ECN Plastics Review,"Recycled Plastics to Grow Fourfold under Euro Rules"-<u>European Chemical News</u>, 10 June 1991, 19.

14. Newsfront, "Plastics Recycling Gains Momentum"-<u>Chemical Engineering</u>, Nov. 1990, 37-43.

15. Kirkman, A. and Kline, C.H., "Recycling Plastics Today"-<u>Chemtech</u>, October 1991, 606-614.

16. Leghissa, S., "Plastic Wastes: How to Avoid, reduce and Recycle It"- Italian Plastics Quality Trade, 1, 1991, 103-105.

17. Chernykh, S.P. et al., "Ways to Reduce the Specific Consumption of Materials in the Production of Lower Olefins"- The Soviet Chemical Industry, 16:2, 1984.

18. Pilavachi, P.A. (editor), "Improved Energy Efficiency in the Process Industries", Proceedings of a European Seminar, Brussels, 23-24 October 1990- Report EUR 13541.

19. Schneider, R.V. and LeBlanc, J.R.,"Choose Optimal Syngas Route"- Hydrocarbon Processing, March 1992, 51-57.

20. Farina, G.L. and Supp, E., "Produce Syngas for Methanol"- Hydrocarbon Processing, March 1992, 77-79.

21. Giacobbe, F.G. et al., "Increase Hydrogen Production"- Hydrocarbon Processing, March 1992, 69-72.

22. Humphrey,J.L. and Seibert, A.F., "Separation Technology: An Opportunity for Energy Savings"- Chemical Engineering Progress, March 1992, 32-41.

23. DeGarmo, J.L. et al., "Consider Reactive Distillation"- Chemical Engineering Progress, March 1992, 43-50.

24. Wynn, N.P., "Separate Organics by Melt Crystallization"- Chemical Engineering Progress, March 1992, 52-60.

25. Udengaard, N.R., et al.,"Sulfur Passivated Reforming Process Lowers Syngas H_2/CO Ratio"- Oil and Gas Journal, March 9, 1992, 62-67.

26. Le Goff, P. et al., "Energétique industrielle" , Tomes 1, 2 et 3- Editions Technique & Documentation, Paris (1979)(in French).

29. "Pinch Concept Helps to Evaluate Heat-Recovery Networks for Improved Petrochem Operation"- Oil and Gas Journal, May 28, 1984, 113-118.

30. Benchecroun, N.,"Pompes à Chaleur dans l'Industrie"- Revue de l'Institut Français du Pétrole, Vol. 40, N° 1, Janv.-Févr. 1985, 113-123 (in French).

31. Fujimoto, A. et al.,"Selective Oxidative Coupling of Methane with a Membrane Reactor"- Natural Gas Conversion, A. Holmen et al. (Editors), Elsevier Science Publishers, Amsterdam, 1991.

32. Shu, J. et al.,"Catalytic Palladium-based Membrane Reactors: a Review"- The Canadian Journal of Chemical Engineering, Vol. 69, October 1991, 1036-1060.

33. Kim, J.S. and Datta, R.,"Supported Liquid-Phase Catalytic Membrane Reactor-Separator for Homogeneous Catalysis"- A.I.Ch.E. Journal, Nov. 1991, Vol. 37, N° 11, 1657-1667.

34. Adris, A.M. et al.,"A Fluidized Bed Membrane Reactor for the Steam Reforming of Methane"-The Canadian Journal of Chemical Engineering, Vol. 69, October 1991, 1061-1070.

35. Li, N.N. et al., "Membrane Separation Processes in the Petrochemical Industry"-U.S. Department of Commerce, NTIS Report DE90-001600, Sept. 30, 1987.

36. Fouda, A.E. et al., "Membrane Separations in Chemical Engineering"- A.I.Ch.E. Symposium Series 272, Vol. 85, 1989.

37. Govind, R. and Itoh, N., "Membrane Reactor Technology"- A.I.Ch.E. Symposium Series 268, Vol. 85, 1989.

38. Chauvel, A. and Lefebvre, G., "Petrochemical Processes", Vol. 1 and 2, Editions Technip, Paris, 1989.

39. Chauvel, A. et al., "Potentialités d'innovations de la pétrochimie européenne",-Pétrole et techniques, N° 335, Sept.-Oct. 1987,51-58 (in French).

SAVING ENERGY IN ALUMINA - ALUMINIUM PLANTS

DENAXAS NICOLAOS and ZERVOS ANTONIOS
Production Department, ALUMINIUM de GRECE S.A.
Usine de Saint Nicolas. 32003 PARALIA DISTOMOU - GRECE

INTRODUCTION

High energy consumption is characteristic of the industrial production of aluminium. In the , initial, alumina production phase, THERMAL CONSUMPTION is predominant and becomes an important factor in the cost of production when the raw material used is a bauxite with a high diasporic content which, in the instance of existing bauxite reserves on our old continent, is the case. In the, second, aluminium production phase, a high consumption of ELECTRIC ENERGY is predominant. That is why aluminium production is termed an ELECTRICITY - HUNGRY INDUSTRY. Successive energy crises therefore forced this industry to reduce, in both phases of production, its energy consumption. Our plant comprises both of these procedures, that is alumina and aluminium production.
Furthermore, it should be noted that :
a) the raw material used in the production of alumina is bauxite very rich in diaspore
b) the electrical energy is supplied at a far from competitive price.
Consequently it is of major importance, in order to obtain better technical results, that continuous efforts are made to reduce energy consumption. In our highly competitive market, improvement in cost efficiency must be our immediate aim.
We can quote many examples of improvements carried out in our plant, ranging from very small ameliorations and alterations to new installations, changes in the process, automation of production units etc., which have this aim in view.
But we will limit ourselves to four examples, each of which concerns a different method of energy consumption reduction. Two of these methods were distinguished by their innovative character and have been approved as prototypes benefiting from subsidies from other Community Programs.
In the alumina production plant two of the major projects that have been carried out with the purpose, among others, of energy saving (thermal mainly but also electric) are :
a) Interstage cooling in continuous precipitation of aluminium hydroxide
b) Alumina, flash - calcining , kiln.
In the first project, reduction in the consumption of thermal and electric energy has been attained through a partial alteration of the process at the precipitation stage. This alteration was aimed at increasing the productivity of the suspension and resulted in a reduction primarily of

specific thermal consumption and, to a lesser extent, of electric consumption.

In the second project, a kiln was approved and built - using entirely new technology for that period-for the calcination of alumina, in order to partially replace the existing rotary kilns. The reduction in thermal consumption obtained was of the order of 20 %.

In the aluminium production plant two cases of energy saving could be quoted :

c) the potlines conversion and

d) the fuel consumption of the anodes bake oven

The former is a modernization investment of a heavy capital cost, while the latter is an amelioration achieved step-by-step without requiring a relatively important capital cost.

Potlines conversion project consists in revamping the three potlines, by implementation of a new alumina point feeding system, thus allowing the indroduction of alumina into the reduction cell to be effected in a perfectly controlled manner, consequently resulting in an increase in current efficiency and decrease in specific power consumption.

Anodes Bake Oven energy consumption has been significantly reduced since 1980 as a result of improving consumption control and optimizing operation parameters, as well. The profit in thermal energy saving is of the order of 30 %.

a) Interstage cooling in continuous precipitation of aluminium hydroxide

Alumina production for the needs of an aluminium smelting plant, is carried out by the Bayer process. The Bayer process of alumina production from bauxite raw material, consists in separating the alumina from the other metallic oxydes contained in the ore by selective dissolution, and afterwards in its recovery. These different operations are realised in a cyclical process.

Alumina dissolution is realised in an aluminate liquor of sodium, which contains :

> . free soda able to dissolve the alumina,
> . soda combined with the alumina,
> . soda combined with the soluble impurities of the ore.

thus, the Bayer cycle can be represented schematically in the following way :

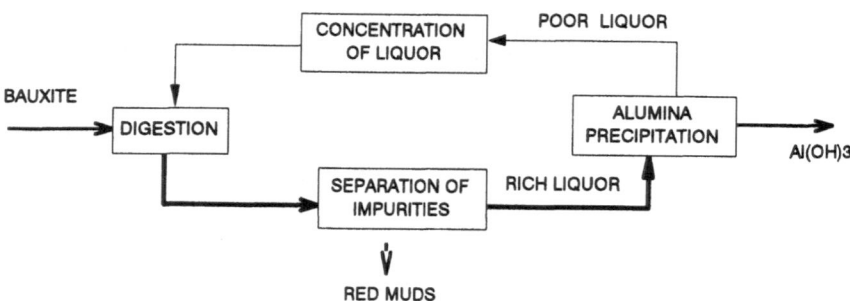

Figure 1. Schematically representation of the Bayer cycle

The fundamental operations : In the cycle two principal operations are shown :
DIGESTION : During this reaction the alumina of the bauxite is dissolved. The impurities during this reaction are not dissolved and remain solid, in suspension.

ALUMINA PRECIPITATION : This reaction is called the decomposition. The alumina is recuperated into solid form as aluminium hydroxide.

For these two operations - digestion and decomposition - we define a PRODUCTIVITY.

For the digestion, it' s the alumina quantity produced per cubic meter of poor liquor.

For the decomposition, it' s the alumina quantity produced per cubic meter of rich liquor.

In an existing factory an increase of the productivity permits an increase in production, since each m^3 of the circulating liquor can dissolve and then precipitate more alumina. Consequently, the specific consumption of energy is reduced. The proposed project helps in increasing the productivity of the decomposition, by cooling the suspension of the aluminium hydroxide during the precipitation stage.

Decomposition of the aluminate liquors : The chemical reaction of the precipitation of the aluminium hydroxide from the sodium aluminate solutions is a balanced reaction which follows conventional thermodynamic and kinetic laws. Two parameters will then play an important role, the temperature and the Rp of equilibrium, (which expresses the solubility in equilibrium of the alumina in a liquor for which the concentration of soda is given). The more the temperature reaction rises, the faster the cristallization reaction is performed. On the other hand, the precipitated alumina quality in total, will be less since the temperature at the end of the reaction will be a little higher as well. This in turn will raise the Rp of equilibrium.

The initial temperature acts also upon the granulometry of the alumina. In order to respect international specifications concerning the granulometry of alumina one must work at high temperatures (60ºC), which decreases the productivity of the liquors.

Thus the project consists in cooling the aluminate suspensions of the sodium and alumina during decomposition, while ensuring that this cooling will not have any incidence on the granulometric quality of the alumina.

Description of the decomposition process : The decomposition unit of the A.D.G. plant is represented schematically in Fig. 2. The aluminate liquor, the temperature of which has been previously adjusted in order to respect the granulometric specification, is mixed with the alumina (aluminium hydroxide) coming through the filters and used as the initial nucleus for the precipitation. This operation is realised in the initial nucleus tank. The suspension is then pumped equally into two files of agitated precipitators where the precipitation in itself is realised (tank 2 and 16). These tanks have an approximate volume of 3.000 m^3 for a diameter of 12 m.. The suspension circulates from tank to tank. The two files are joined in the tank 14 and the common suspension is pumped from the tank 15 towards the tank 1 that feeds the filtration of the feed by its major part, towards a classificator for the rest where the biggest grains are separated, in order to constitute the production of the finest grains which are driven towards tank 1. Under normal working conditions, the temperature of the suspension varies from 58 to 65ºC, and during its course we observe a natural cooling of between 2 and 5ºC depending on the exterior temperature.The aluminate flow varies from 1.000 to 1.100 cubic metres per hour, the suspension flow coming from tank 15 being 2.200 to 2.400 cubic metres per hour.

Project description : The project consists in cooling one part of the suspension during decomposition. Due to the high content of the suspension in dry materials (800 to 900 g/l) it was necessary to choose a particular technology, that of the PLATULAIRES heat exchangers supplied by the BARRIQUAND company.

Six exchangers will be installed (3 per files) between the precipitations tanks 8 and 9, 9 and 10, 10 and 11 on one hand, 21 and 22, 22 and 23, 23 and 24 on the other hand. Each of the exchangers

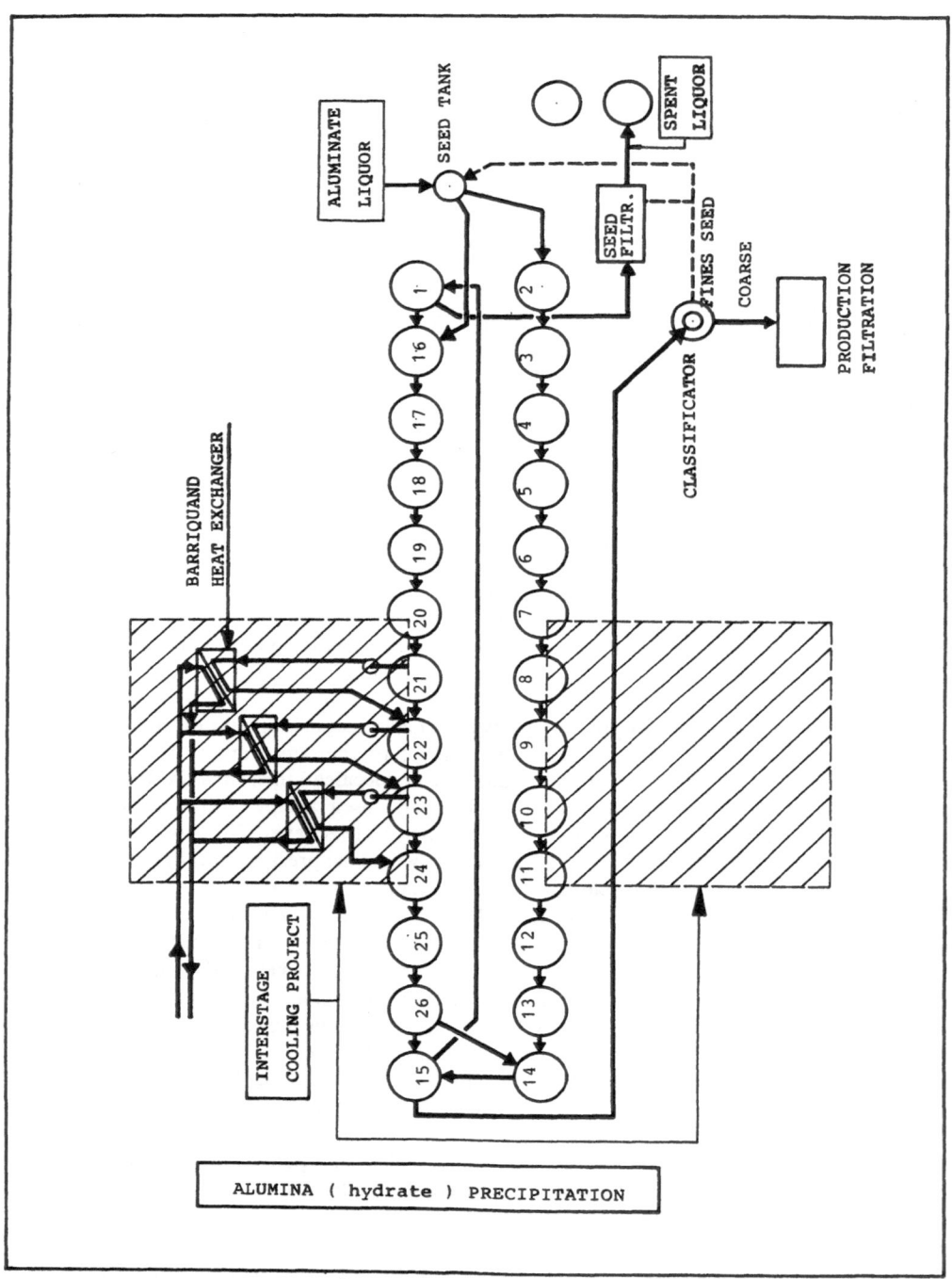

Figure 2. Scheme of the decomposition unit of A.D.G. ALUMINA plant.

has a surface of 276,5 sq.m. and will receive 25 % of the flow of the suspension (300 m³/h) which will circulate at a speed of 1,4 m/s.˝They will be fed directly from tha tank by an immersed pump (1 pump per exchanger), the cooled suspension going into the precipitator tank coming next.

The cooling is performed by closed circuit with water turning through an atmospheric refrigerant. The cooling water flow is of 240 m³/h per exchanger.

Instruments to measure flow, temperature and pressure will be installed to check the optimum operation of each exchanger together with a bank of regulators for each exchanger which will vary the flow of the coolant according to the desired exit temperature of the suspension. All this information will be processed automatically and transmitted to the screen in the control room of the decomposition unit. In order to avoid the eventual clogging of the exchanger due to accidental shut - down, an automatic disposition for washing the exchangers on the suspension side has been foreseen. At the same time, a cleaning circuit with soda has also been planned. This unit will ensure an average cooling of 5°C of the suspension, taking into account periodical halts, for the maintenance of the tanks and of the relative exchangers.

Choice of materials : The practical realisation of this operation is not an easy one at either a technological or a practical level. Actualy, the suspensions are highly charged in dry materials (800 to 900 gr/l) and are sealing, since they continue to precipitate the alumina. The conventional technologies appear to be inappropriate.

Flash cooling needs voluminous installations and permits no recuperation of heat evaporation. Tubular and multitubular exchangers are considered unfavourable, on the one hand because of their cost and space requirements, on the other because of their difficulty in distributing the suspension into the tubes and a lower coefficient of exchange.

High performance plate exchangers, at the exchange level, are very ill-adapted to the suspensions (sealing), due to their conception.

In order for spiral exchangers to bear the pressure, they need "rivets" on the suspension side, a fact which increases the sealing dangers, makes their emptying difficult and their mechanical cleaning very difficult.

The choice then presents itself of an exchanger of innovative and original technology which is much less widespread. These exchangers are called " p l a t u l a i r e s " and are manufactured by the French BARRIQUAND Company.

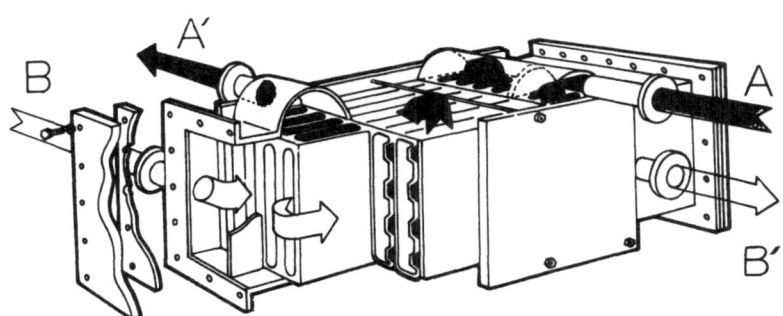

Figure 3. BARRIQUAND exchanger type "platulaire UP"

In this type of exchanger the plates are flat and smooth. It is not necessary to have rivets on the suspension side but only on the water side. Mechanical cleaning is possible. In the case of a shut -

down, the dangers of sealing by alumina settling are reduced. The exchanger is better placed above the tanks and not on the ground as is usual, as this permits a desrease in the length of the pipes carrying the suspensions and ensures that they empty by simple gravity in case of a pump shut-down.

The results : The minimum productivity gain is 0,7 kg/m³ for every degree of interstage cooling, and production increases by at least 20.000 tons per year for 5 °C on a continuous basis. The equivalent energy saving is 5.130 tons of fuel oil (thermal energy) and 3.150 Mwh of electric energy. The total annual energy saving is 5.430 tep.
The following diagram show the productivity gain and the production level increase.

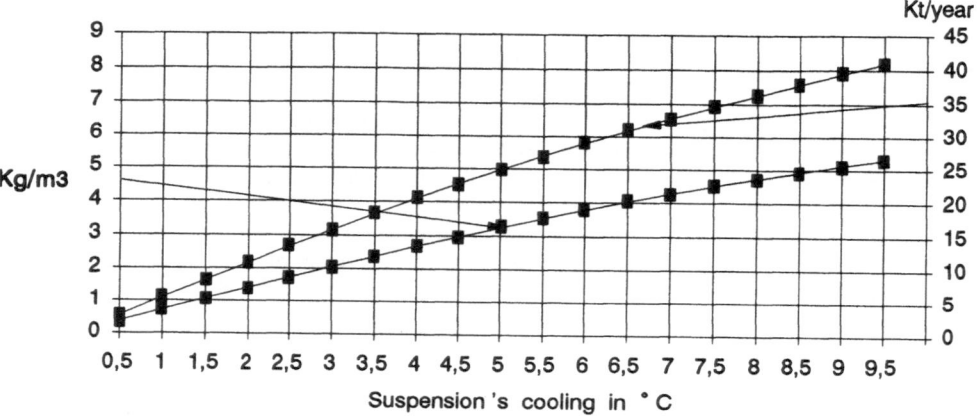

Suspension 's cooling in ° C

b) Alumina, flash - calcining, kiln

The world' s first industrial alumina suspension calciner, based on the flash calcination process, is under production since 1984 at Aluminium de Grece's St Nicolas plant in Greece.
The 950 mt/day installation, has evolved through research efforts carried out by Fives - Cail Babcock (FCB) and pilot tests conducted jointly by FCB, Aluminium Pechiney (AP) and Aluminium de Grece (ADG).

Backgound of flash calcination : Sometime earlier, three main types of kilns existed for ores shaft kilns, rotary kilns and fluid - bed kilns. The suspension flash calcining process is the outgrowth of improvements to rotary kilns in the cement - making industry to reduce heat consumption and increase productivity.
The first breakthrough was the short - kiln design with a dry-process preheater which is made up of four stages of exchange ducts and cyclones. The heat exchange in each stage takes place in parallel flow between the ground material carried by the gases and the carrying gases, whence the terms for : **suspension exchange and suspension preheater** .
The second breaktthrough was the Japanese development of precalcination.
FCB has set up many dry-process cement plants with suspension - preheater kilns and has also developed its own precalcining chamber design.

Flash calcination flow - sheet : At the same time , FCB has developed a calcining flowsheet for fine sized ores or materials (minus 1 mm), in which a suspension calcining chamber of the type used in cement manufacture is utilised. The rotary kiln is completely eliminated, and the cooling of the calcined product takes place in the same manner as its preheating (i.e. in a dry - process exchanger). The calcining zone, the heart of a suspension or flash - calcining installation, consists of a cyclone calcining chamber (CC or PC) and of an associated separation cyclone (SC). The preheated particles are dragged upwards by the hot air coming from the cooler, and this suspension is introduced tangentially into the calcining chamber. The fuel is introduced at one or several points of the chamber roof. The presence of the material absorbing the calories released dy the combustion is such, that there is no high flame temperature within the chamber, hence there is less risk of overheating. The centrifuging effect of the material on the walls increases residence time (especially for the largest particles). At the outlet a round-square transformation of the flue, homogeneizes the suspension for completing the heat exchange in parallel flow. The associated cyclone SC, separates the calcined c product from the fumes. Since the exchange is of the parallel - flow

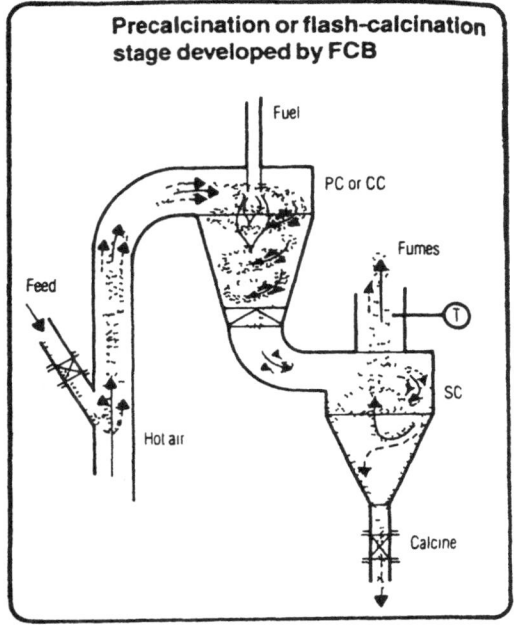

type, the temperature of the fumes is very close to the temperature of the solid particles, and consequently, it is easy to control the degree of calcination by adjusting the fuel flow to a set fume temperature. A suspension preheater makes it possible to use the hot fumes from the calcining zone to dry and precalcine the solid particles, while a suspension cooler allows the use of heat from the calcined product, to preheat the combustion air entering the calcining zone.

Process application to ALUMINA : The flow - sheet of an alumina flash - calcination installation, is only an application of the general flow - sheet (fig. 4).

The preheater consists of two stages (C1 and C2). The introduction of hydrate into the upper stage, the separating element of which is a multicyclone (C1), takes place through a drying flash venturi, which accepts the hydrate with 7% to 9% moisture content. The alumina leaves the preheater (C2) at 320 - 350 °C and with a lossin water content of 85 - 90%.

The cooler consists of three stages. The air of the cooler is preheated at about 800 °C, while the calcined alumina exits the last cyclone (R3) at about 280 - 300 °C . Hence, a final fluid - bed cooler with water circulation coils is needed to reach the required temperature (80 °C).

The air - particles flow enters the calcining chamber (CC) at 400 - 500 °C . The calcined product leaves the calcining zone (SC) at temperatures between 1 000 and 1.200 °C depending on the aimed degree of alumina calcining.

Results : The heat consumption is about 3.100 KJ/kg of calcined alumina and the annual profit is approximatively 8.700 tep.

The new type of flash - calcining kiln offers, apart from the lower heat consumption, the following advantages : **yields a better calcined alumina quality** : lower a - alumina content for

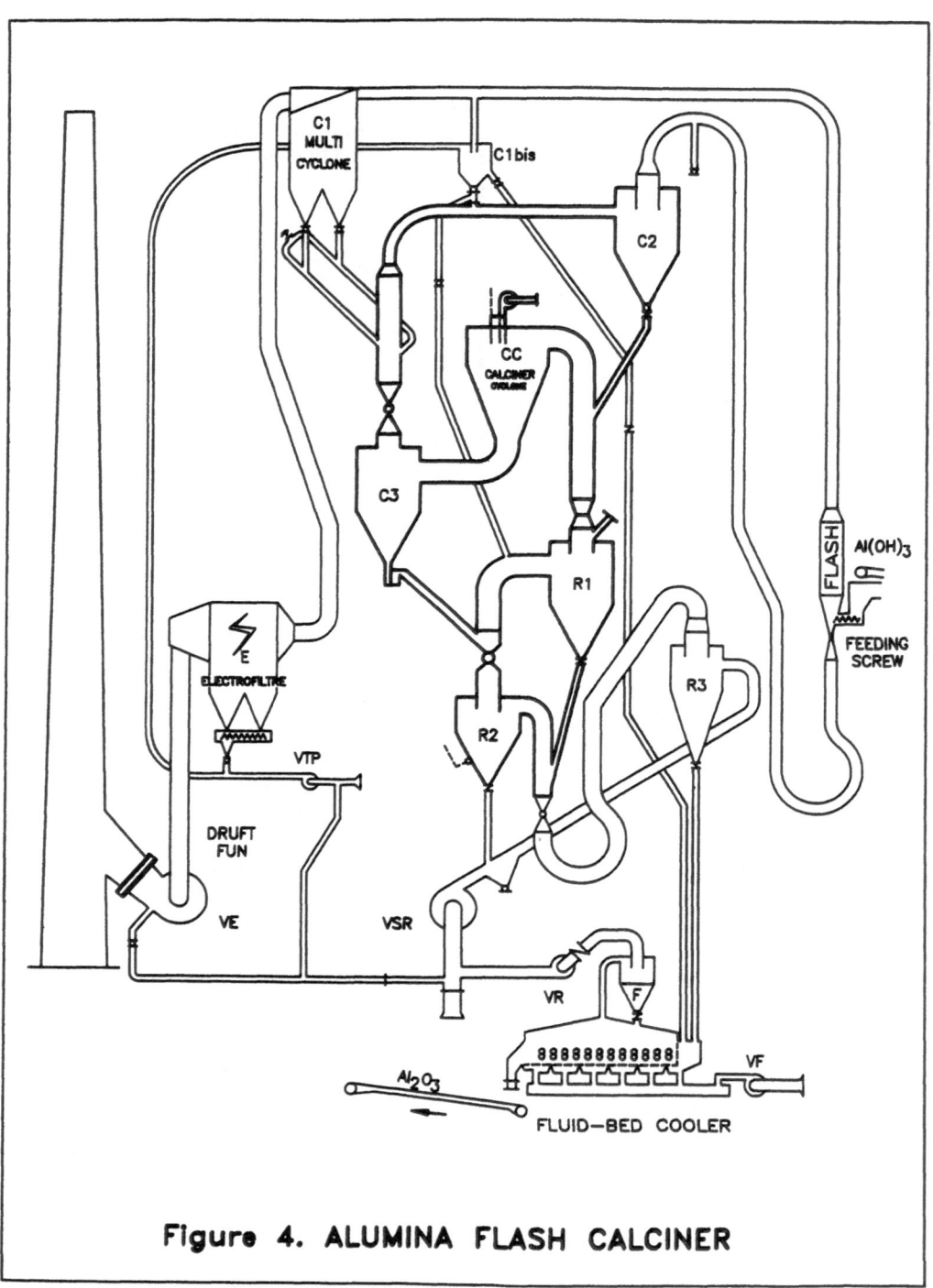

Figure 4. ALUMINA FLASH CALCINER

a given B.E.T. surface, excellent solubility and steadiness of the above calcined alunina caracteristiques, **increased utilisation flexibility.**

With the financial support of the Commission of the European Communities (JOULE II proposal PL 920135) a Hybrid Knowledge - Based System for Process Control will be developed and tested to improve the existing control system of the flash - calciner.

It is expected that this will result in a reduction of the nominal fuel-oil consumption, of the flash-calciner, up to 10 % of its present value, which means that the potential fuel-oil savings can be as high as 2.000 t/y and also a reduction in the nominal electric energy consumption at the calciner draft fan, amounting to 5 % of the current consumption.

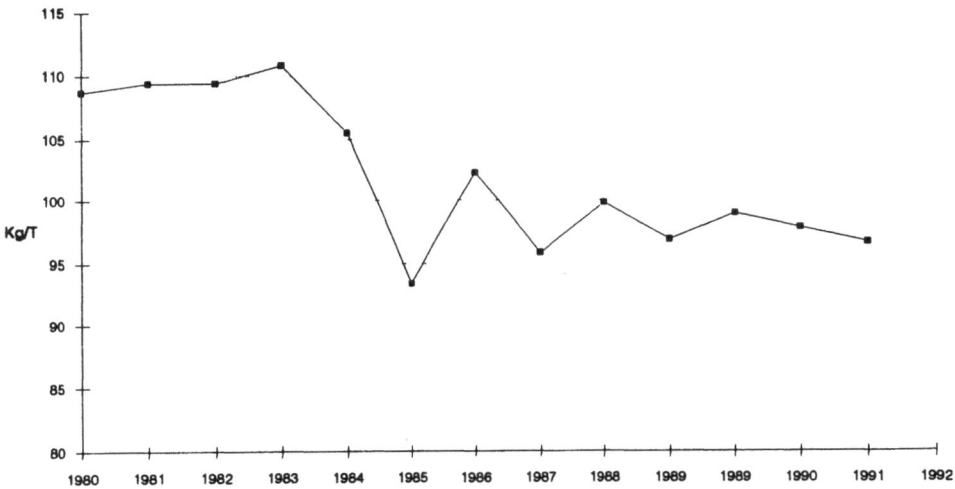

Figure 5. Calcination, specific fuel - oil comsumption with year

c) The potlines conversion

Process description: Aluminium is produced by electrolytic reduction of alumina (Al_2O_3) dissolved in a molten electrolyte (bath) based on cryolite (Na_3AlF_3). The overall process is indicated in Fig.6.

The incoming alternating current is transformed to direct current and is fed to a line of cells (pots) connected in series in a side-by-side configuration. Therefore, the operation is at constant current, but the individual voltage of each pot can be differed.

ALUMINIUM DE GRECE has three potlines of prebaked anodes design, each one comprising 260 pots installed (that is, 780 pots at total). Two potlines operate at 70 KA, while the third operates at 90 KA.

The main overall electrochemical reaction occuring at around 960 °C in the pot is represented by the equation:

$$2Al_2O_3(dissolved) + 3C_{(s)} = 4Al_{(l)} + 3CO_{2(g)}$$

The metal is deposited at the bottom of the pot, and oxygen is discharged at the carbon anode.

Oxygen and carbon react each other forming carbon dioxide.Anode gas also contain fluoride emissions from the bath.The gas collecting and treatment system is based on the use of power-operated hoods and dry scrubbing facilities.

Metal removal (vacuum tapping) is executed periodically every 48 hours using transportable ladles, by means of which is transferred to the foundry for purification and casting.

The overall annual production of our plant is 152000 t of liquid metal.

Project description: The power consumption and the current efficiency of a potline for the production of aluminium by the reduction of alumina are very closely dependent on the alumina content in the electrolytic bath.

Before this conversion was carried out, the introduction of alumina into the pot (see Fig.7) had being effected by side crust breaking by means of multi-purpose tending machines (see Fig.8). This process had taking place every 6 hours, resulting in an abrupt increase of the alumina concentration in the bath,that had leading to undermine the pot's performance.

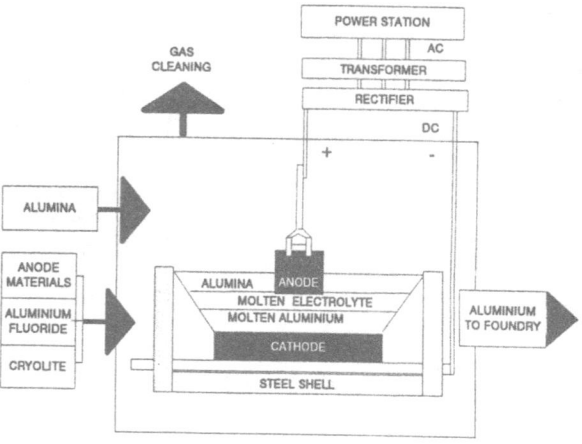

Fig.6. Flow sheet for aluminium production.

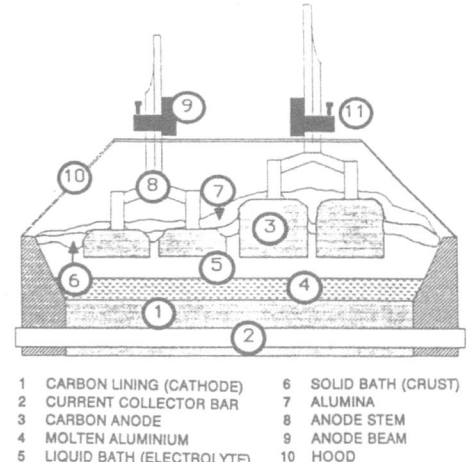

1	CARBON LINING (CATHODE)	6	SOLID BATH (CRUST)
2	CURRENT COLLECTOR BAR	7	ALUMINA
3	CARBON ANODE	8	ANODE STEM
4	MOLTEN ALUMINIUM	9	ANODE BEAM
5	LIQUID BATH (ELECTROLYTE)	10	HOOD
		11	ANODE STEM CLAMP

Fig.7. Aluminium electrolysing cell with prebaked anodes.

The plan consisted in installing on the pot's superstructure the two component units called crust breakers, situated in the pot's longitudinal axis with the purpose of breaking the solidified bath crust (see Fig.9). A feeding device linked to a conveying system despatches a dose of alumina into the hole thus created. Alumina is stored in a hopper, each pot having its own hopper. These component units are controlled by a micro-computer, which regulates the alumina content according to the pot's functioning parameters and the anode - cathode distance. This technique is called point feeding.As a result, the alumina content remains constant and is regulated more or less on the amount optimizing the current efficiency and the specific power consumption.

This technique is used on modern pots, which are conceived accordingly. The innovation consists in applying this technique to old-technology pots, while they are in function and without modifying their structure (potlining). We are the only plant in the world to have proceeded with this type of conversion. The method and technique were developed by engineers and technicians of the **ALUMINIUM DE GRECE** Company.

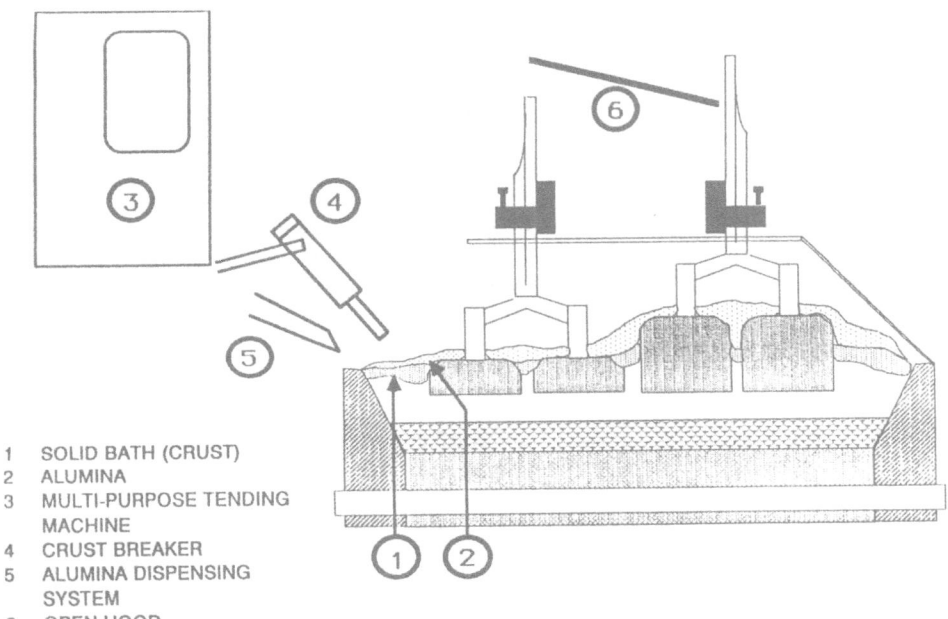

1 SOLID BATH (CRUST)
2 ALUMINA
3 MULTI-PURPOSE TENDING
   MACHINE
4 CRUST BREAKER
5 ALUMINA DISPENSING
   SYSTEM
6 OPEN HOOD

Fig. 8. Side-worked cell.

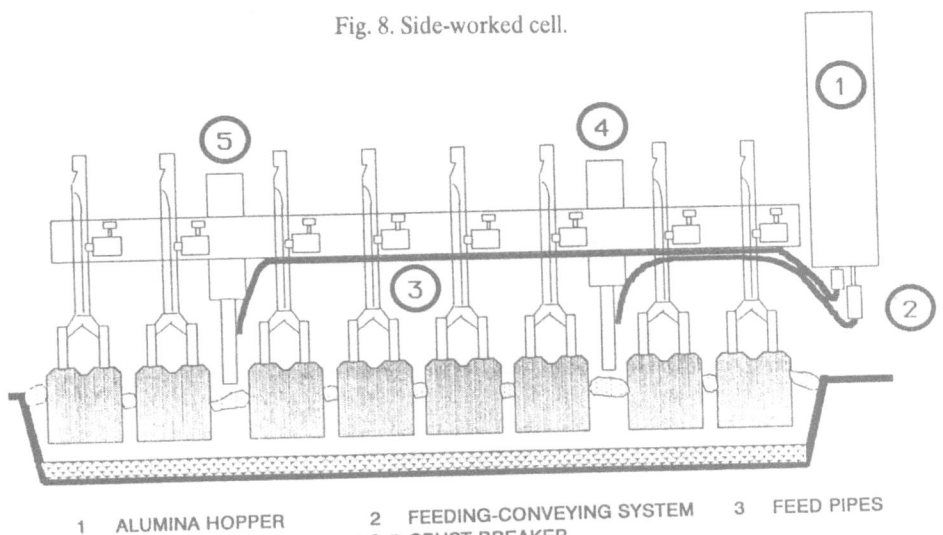

1 ALUMINA HOPPER   2 FEEDING-CONVEYING SYSTEM   3 FEED PIPES
                      4 & 5 CRUST BREAKER

Fig. 9. Point-fed cell.

Objectives and results: This revamping of our potlines has aimed at improving technical perfomances, productivity, as well as the working conditions. More precisely, the objectives to be achieved are:

1. Increase of the current efficiency, resulting in a reduction of the specific power consumption by 250 kWh per tonne of aluminium produced (i.e. a 1.18 % gain), that is, for a production of 150000 tonnes of aluminium per year, a total annual saving of 37.5 GWh.
2. Decrease of fluorine emission by 3 kg per tonne of aluminium (a 20 % gain) i.e. 450 tonnes per year.
3. Increase of the life-span of pots stopped for relining by 10 months (an 18 % gain).
4. Improvement in productivity by a staff reduction of 45 peoples (a 20 % gain).

In terms of energy consumption, the expected total energy saving is 0.0218 tep per tonne of aluminium, i.e. 3270 tep per year for an annual production of 150000 tonnes of aluminium.

Figures tabulated in Table 1. show the improvement to be expected, as far as the technical perfomances and productivity are concerned, in the case of potline C (operating at 90 KA), which was converted more recently. This conversion has been financed through E.E.C. funds

TABLE 1.Gain after conversion of potline C.

	Units	1987-89 Period of reference	Expected gain (Objectives)	July91-June92 Results recorded	Real gain
Annual Production	t Al	56000		60239	
Current Efficiency	%	89.5	1.1	91.8	2.3
Specific Power Consumption(D.C.)	kWh/t Al	13600	150	13292	308
Fluorine Emission	kg F/t Al	15.3	3.0	12.0	3.3
Number of Pots Stopped for Relining	pots/year	34	9	32	2
Average Life of Pots Stopped for Relining	months	52,9	10	60.1	7.2
Average Number of Pots Functioning		198		246	
Staff	listed		15		15
Specific energy saving	tep/tAl		0.01375		0.02698
Annual energy saving	tep/year		770		1625

The results recorded with the point feeding system are distincly far superior with regards to the results corresponding to the old feeding system.

In the case of potline C, these results summarized as following (see also Table 1.):

-Current efficiency increased by 2.3 %,resulted in a reduction of the specific power consumption by 308 kWh per tonne of aluminium produced (a gain of 2.3 %).

-Fluorine emission decreased by 3.3 kg per tonne of aluminium produced (a gain of 21 %) i.e. 186 tonnes per year.

-Life-span of pots stopped for relining increased by 7.2 months (a gain of 13.5 %).

-Indeed,the productivity increased as anticipated. Besides, a significant improvement in working conditions has been observed, with particular reference to dust-in-air concentrations in the potrooms.

-Specific energy saving amounted 0.02968 tep per t of aluminium produced.Annual energy saving totalled 1625 tep for a production of 60239 tonnes of aluminium.

d) fuel consumption of the Anodes Bake Oven

Process description: The anodes used for the electrolytic extraction of Aluminium consist of petroleum coke and pitch as a binder (see fig. 10.). After having been upgraded, these two raw materials are mixed giving a paste, which subsequently is compacted, forming the anodes (referred to as green anodes). The green anodes are then heated at 1100 °C over a period of about 8 days prior to being cooled down (referred to as baked anodes).

The target of baking the green anodes is the transition of the binder to solid coke, that leads the anode block to be good electricity conductor and strength as far as the mechanical properties are concerned (both required for using them in the electrolytic cells). The chemical processes occuring during this thermal treatment are the pitch pyrolysis and the growth of the graphite crystallites. Baking takes place in a particular type of bake oven called "continuous-fire baking furnace" or "ring furnace" or "section baking furnace".

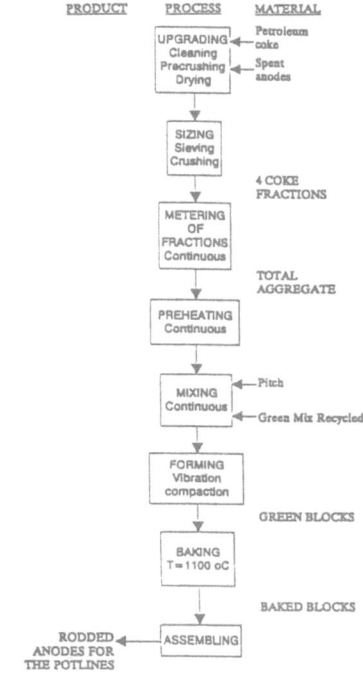

Fig. 10. Flowsheet for anodes manufacturing

Description of the anode bake oven: The baking assembly comprise two parallel rows connected at each end by cross-overs ducts, thus forming a ring (see fig 11.). Each row has a series of sections, each one separated from the next by a headwall.

The section is divided to 6 parallel rectangular pits (baking cells). The pits are separated by the heating flue walls located parallel to the bake oven centreline, through which the heat of combustion is transmitted (see fig. 13.). Each pit loads piles of anode blocks surrounded by coke in order to

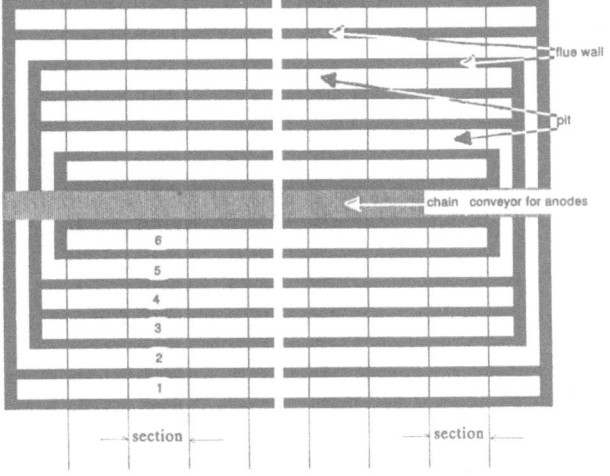

Fig.11. "Ring" Bake Oven.

maintain the anodes pre-formed shape during the pitch softening.

Flue walls are designed so as to maximize the heat transfer between the anode blocks and the flue gas (see also fig 12.), having baffles to extend the combustion gas way and distribute the heat exchange more evenly on the surface of the flue wall. They are built from aluminosilicate refractory bricks selected for their low expansion coefficient, ability to withstand the temperature and resistance to corrosion from the fluorides.

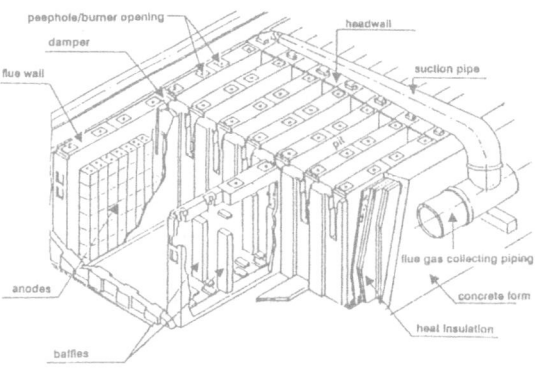

Fig.12. Perspective view of a section.

The bake oven is heated by burner ramps the length of which is equal to to the width of the sections. Each burner ramp has injectors which are placed on the openings ("peepholes") of a certain section. Upstream of the burners (with respect to the gas flow), there is a blowing pipe (ramp) to blow combustion air, which is preheated passing through the cooling sections before entering the high temperature region where burners functioning.

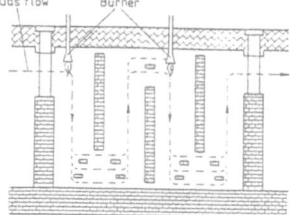

Fig.13 Cross-section of a flue wall [1]

Downstream of the burners, there is an exhaust ramp for combustion fumes. A baking zone, the so- called "fire", consists of a set of sections lying between the exhaust ramp and the blower ramp, some of them being heated (or pre-heating), while some of them being cooled. Thus, a "fire" is composed of two consecutives heat exchangers (see fig.14.).

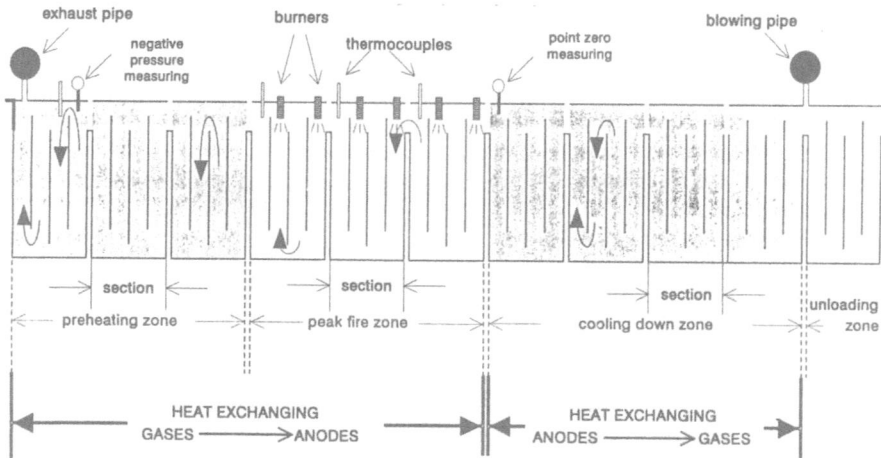

Fig.14. A group of 10 sections consisting a "fire".

At the end of the heat treatment cycle of a certain section the fuel supply is shut off, and a new section is connected for pre-heating at the other end of the fire. In this way the fires moves around the ring.

Heating is by the combustion of fuel and, for a considerable part, by the combustion of the pitch volatiles matters , released during the coking of pitch (methane, tar, hydrogen). The targeted heat-up rate of the anodes blocks is adjusted by the fuel supply to the burners and the gas flow.

Improvements : 1. Brickwork. The term "brickwork" signifies the whole refractory maintenance of the bake oven. At the early beginning of the '80s decade, an intensive campaign of improving the brickwork quality was held. This lead to the improvement of the flue wall state (deformations, slots) which consequently impeded the entry of parasitical incoming air (overall flue gas outlet flow reduced from 110000 nm^3/h to 80000 nm^3/h or specifically from 1,84 $nm^3/Mj/t$ of offered energy to 1.69 $nm^3/Mj/t$). Obviously, the thermal losses restricted, hence the heat exchange between anodes and flue walls became more efficient.

2. Burners.Replacement of the continuous fuel flow burners by high frequency impulsion burners.Continuous fuel flow burners had been using for the fuel flow regulation, the temperature taken by a thermocouple corresponding to a set of flue walls. The combustion mixture had been resulting while fuel and air being co-flowing

The impulsion burners are commanded by a micro-processor, while the temperature process value is measured by an independent sensor for each flue wall. The combustion mixture results while fuel and air being counterflowing. That is, impulsion burners provide an independent regulation and better control of the fuel flow towards the burners.

3.Optimization of the operation parameters. Continuous efforts on optimizing diverse operation parametres were made, and especially the flow adjustement as a fonction of flue gas temperature measured upstream of the exhaust ramp and the gas temperature curve (versus time) used as set-point.

Results : The above-mentioned improvements, that have been made since 1980, resulted in a reduction of the specific fuel consumption by 0.039 t fuel/t of baked anode (1980-1991) or an annual reduction of 3100 t (for an annual production of 80000 t of baked anodes). The equivalent heat energy figures are 0.0352 tep/t and 2819 tep respectively .

This gain achieved, although the heat offered by the combustion of the pitch volatiles matters decreased due to the diminution of the pitch percentage in the anodes, as a result of the better anodes characteristics (see Figure 15.). Consequently, the real (global) energy saving is to be considered the sum of the saving corresponding to the fuel consumption and of the less heat by the volatiles matters that was used. Thus, the specific energy saving obtained is 0.0470 tep/tn of baked anodes and the annual saving is approximatively 3760 tep. The cumulative saving for this period of 12 years amounts to about 45150 tep.(see table 2)

TABLE 2 Heat Energy Saving in the Bake Oven (1980-1991).

	Fuel Consumption (t)	Heat by consumed (Mj)	Energy fuel (tep)	**Global Heat Energy (Mj)	(tep)
per tonne of baked anode	0.039	1475	0.0352	1968	0.0470
*annually	3100	1.2×10^8	2819	1.6×10^8	3762

* For a production of 80000 t of baked anodes
**Including the heat energy by the combustion of pitch volatiles matters.

An additional gain resulted from the direct energy saving and the better heat transfer is the decreasing of the refractory bricks consumption, which is a crucial part of the anodes production cost (see Figure 16.).

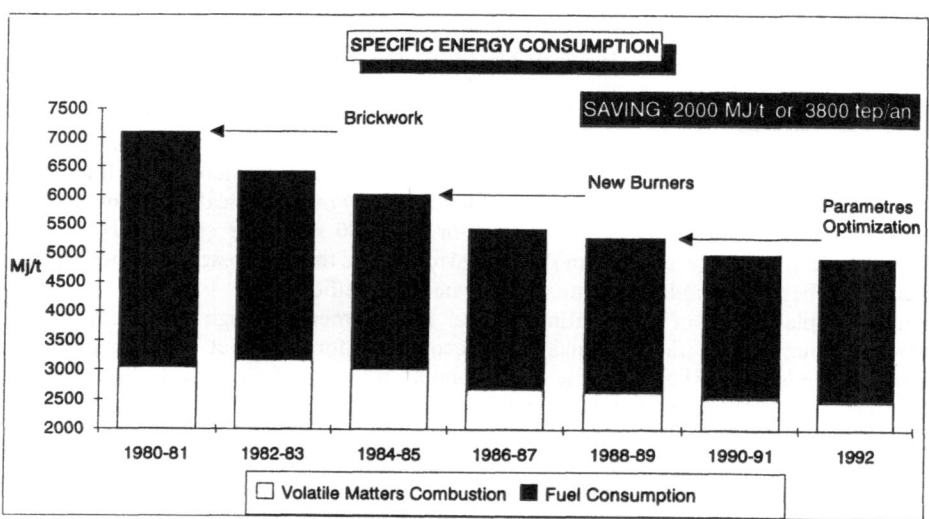

Fig. 15. Variation in specific energy consumption with year

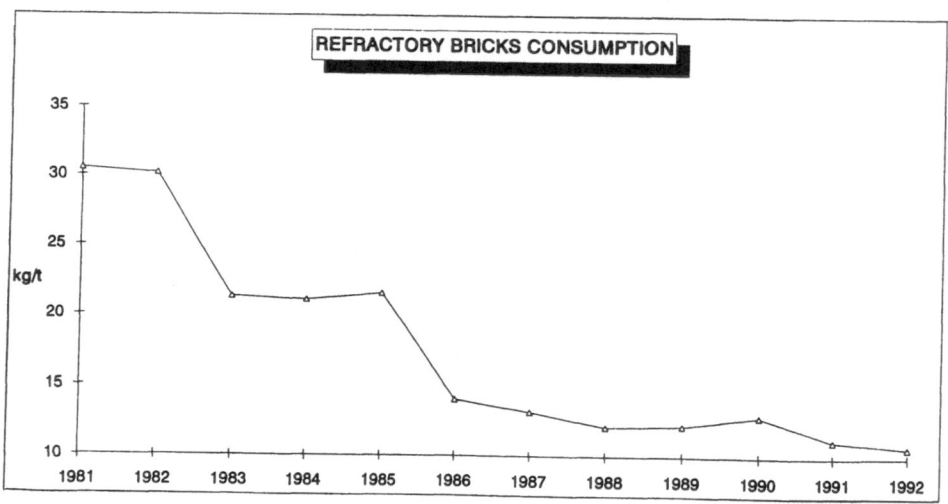

Fig.16. Refractory bricks consumption.

REFERENCES

1 & 2 K.Grjotheim and B.J.Welch & K.Grjotheim and Kvande, Aluminium Smelter Technology, & Understanding the Hall-Héroult Process for Production Of Aluminium, Aluminium-Verlag, Dusseldotf, 1988 &1986.

SESSION 1:

Overview of Energy Intensive Sectors
Continued

Chairman: Dr E.N. Carabateas

ENERGY CONSERVATION
IN THE IRON AND STEEL INDUSTRY

ERNST WORRELL, JEROEN DE BEER, KORNELIS BLOK
Department of Science, Technology & Society, Utrecht University
Padualaan 14, NL-3584 CH Utrecht, The Netherlands

ABSTRACT

The specific energy consumption of a modern integrated steel plant (using the BOF-route) and the potential and costs to improve the energy efficiency are studied in this paper. The output of the plant is defined as 25% slabs, 45% hot rolled product and 30% cold rolled product. The net energy consumption for this plant is at present 17.6 GJ of fuel/ton product and 0.85 GJ electricity/ton product (19.7 GJ primary fuel/ton product). On basis of an energy analysis 13 energy conservation measures are identified. The total *technical* conservation potential of the measures amounts to 2.9 GJ/ton product (primary energy), or 15% savings, reducing the specific energy consumption to 16.8 primary GJ/ton product. Restricting the potential to measures with a simple payback period of less than seven years 0.6 GJ/ton product (or 3%) can be saved. Implementing a carbon tax of 140 US$/ton C or an investment grant of 40% would increase the savings to 10% and 8% respectively. In the long term a larger potential for energy conservation exists, estimated at a *minimum* of 3 GJ extra compared to the technical potential savings in the year 2000, reducing the specific energy consumption to approximately 14 GJ/ton (30% savings compared to the present level).

INTRODUCTION

A considerable part of the primary energy consumption of modern society is connected with the production of basic materials. The steel and chemical industry are examples of energy intensive industries. Energy costs are a large part of the production costs of materials (like crude steel), making investments in energy conservation attractive. However, energy conservation is not only important from an economic point-of-view, but also because of the necessity to protect the environment. Energy consumption for the production of basic materials can be reduced in several ways: more energy efficient production (energy conservation) or more efficient use of materials (by improving the properties or by recycling). Among production plants differences in technology and energy consumption can be noted [1]. The aim of this article is an assessment of the potential for energy conservation in the iron and steel industry. In order to reach this, we start with determining the energy consumption of a typical modern steel plant. Subsequently energy conservation measures in the short term and their impact on the energy consumption are investigated. In the past many studies have been done to improve the energy efficiency of production processes. However, they have been limited to unit-operations. Only a few were directed to the overall improvement of the energy efficiency (e.g. 'Energy and the Steel industry'[2]). In this study we will carry out an energy analysis of modern steel production and combine it with the potential to improve the energy efficiency. We will also assess the attractiveness of energy conservation measures from an economic point of view. In this study we will use the database ICARUS [10,19] which has been constructed within our group recently.

PROCESS DESCRIPTION AND ENERGY CONSUMPTION

In this section we will describe each unit-operation of the production process, with emphasis on energy consumption, and review the possible energy conservation measures. We will use consumption data of the Hoogovens plant at IJmuiden, the Netherlands, for the year 1988 as a model. It is a modern integrated steel plant with a yearly production of 5 million ton crude steel. Electricity is produced both on site and purchased from the public grid. Efficiency of electricity production is taken as 39% (including grid losses). In table 1 the structure of the plant is summarized.

TABLE 1
Summary of the plant structure

Plant structure:

Coke plant:	100% self providing
Sinter plant:	50% of the blast furnace feed (on site)
Pellet plant:	50% of the blast furnace feed (on site)
Blast Furnace:	2 blast furnaces, with coal/coke feed
BOF-plant:	2 BOF plants, with 10% scrap addition (average)
Vacuum furnace:	Oxygen Blowing (RH-OB), treating 100% of the production
CCM:	2 twin-strand machines, treating 100% of the production
Hot Strip Mill:	75% of the production is treated
Cold Strip Mill:	30% of the production is treated

Coke production. Coke is produced on site in orthodox coke batteries. The production at Hoogovens consumes net 3.86 GJ fuel/ton coke. Electricity consumption is estimated at 41 kWh/ton coke (or 0.15 GJ$_e$). Hoogovens uses 0.42 ton coke/ton crude steel (tcs). This makes the energy consumption for coke production 1.62 GJ/tcs and 0.06 GJ$_e$/tcs.

An increase of the rate of direct coal injection in the blast furnaces at Hoogovens, will decrease the demand for coke. The savings are calculated with the blast furnace.

In the past several experiments have been done to recover heat from the coke gases, for instance in Japan and Germany. The recovered heat can be used for preheating coal. Older literature [4] mentions possible savings of 5% of the total energy consumption. However, the experiences in the experiments mentioned before were very negative, because of fouling problems of the exchanger. These problems have not been solved yet. Therefore we will not take this measure into account. It is possible to recover some waste heat (after the first cooling step) in the form of hot water, but the associated savings will be very limited.

Using dry coke quenching the sensible heat of the coke is recovered. Savings amount to 1200-1500 MJ/ton (or 31% of the fuel demand). Another advantage of this technique is the reduction of dust emissions. Investment costs are high: recent calculations give investments of 175 Dfl/GJ [5]. The figure might increase due to lay-out problems at the plants.

Sinter plant. Iron ore is mixed with breeze (or coke) and is heated with coke gas fired burners to 1300-1400 °C, to produce a homogeneous sinter with 57% iron. Heat is recovered at the sinter plant and used for preheating combustion air and raising steam in a waste heat boiler. Energy consumption is 2.36 GJ/ton sinter or, with a 50% sinter (0.73 ton/tcs) input in the blast furnace 1.23 GJ fuel/tcs and 0.21 GJ$_e$/tcs. No energy saving measures are identified.

Pellet plant. A mixture of different iron ores is ground, and mixed (wet) with a fluxing agent, and converted to pellets in the induration machine. The pellets contain 57% iron. Total energy consumption is 0.66 GJ/ton pellet (52% gas in the burners, and 48% as breeze and magnetite) [12], and 0.14 GJ$_e$/ton pellet. With 0.7 ton pellet/ton crude steel the specific energy consumption is 0.46 GJ fuel/tcs, and 0.1 GJ$_e$/tcs. Waste heat is already fed to the wind driers. No other energy conservation measures are identified.

Blast furnace. Blast furnace gas is recovered and 13 MWe top gas pressure turbines are installed [11]. In 1988 115 kg coal/ton pig iron was injected, and 0.37 ton coke/ton pig iron used. The hot blast stoves use approximately 1.92 GJ/ton pig iron [5]. Net energy consumption is 12.92 GJ/ton iron, or 11.90 GJ/tcs (0.92 ton pig iron/tcs), distributed as follows over the energy carriers: 14.70 GJ coal (or coke), -1.78 GJ blast furnace gas and -0.08 GJ$_e$ (per ton crude steel).

   Direct coal injection in the blast furnace replaces the use of coke, saving energy in the coke production. The input of 1 ton coal saves 0.8 ton of coke and hot wind from the stoves, but uses more oxygen. The net savings are calculated at 3.76 GJ/ton coal injected [5]. This can be increased to a higher input-rate in the year 2000. The maximum input rate is determined by both the type of coal and the effects of coal injection on the quality of the pig iron produced. For the year 2000 we assume a practical maximum injection rate of 180 kg/ton pig iron, saving 13.5% of the energy demand of the coke production. In 1988 the injection rate was 115 kg/ton iron. Costs are estimated at 20 Dfl/GJ [13]. The theoretical maximum of 270 kg/ton pig iron is still under investigation. It is not sure whether this rate is implementable.

   In several countries research work is being done on slag heat recovery, but no implementations are known yet. Hoogovens has evaluated this option too, but found it not feasible to implement at IJmuiden. Savings are estimated at 0.32 GJ/ton crude steel [6]. Because of the technical problems we will assume that this measure can not be implemented before the year 2000.

Basic Oxygen Furnace (BOF) steel plant The pig iron is converted to crude steel in the BOF-plant by applying pure oxygen to the hot metal in convertors. To the pig iron 10% scrap is added. In the process itself no energy is consumed, but produced in the form of BOF-gas, which can be recovered. Fuel is consumed to preheat and dry the convertors, estimated at 51 MJ/tcs (or 0.05 GJ/tcs) [3]. Electricity consumption is estimated at 23 kWh/tcs (or 0.08 GJ$_e$/tcs) [5]. This figure includes the production of oxygen which is consumed at a rate of 48 Nm^3/ton steel.

   At the BOF-plant gas is produced with a heating value of 8.8 MJ/Nm^3 and a temperature of 1200°C. The gas is cleaned and rejected. Only at one 330 ton convertor is gas recovered using repressed combustion (26% of the produced gas). It is planned to rebuild two other 330 ton convertors to the same system, saving 0.6 GJ/tcs. After rebuilding, nearly 80% of the gas will be recovered. This measure can therefore be implemented at 54% of the production. Investment costs are 50 Dfl/GJ [5].

   At Nippon Steel Corporation in Japan a closed OG-system [15] has been developed, resulting in savings of 0.98 GJ/tcs (0.38 GJ/tcs extra, relative to the standard system), and an increased molten steel production of 0.4%. The system consists of a sealed hood and control systems, and can be applied at BOF-plant 2 (80% of the production). The payback period is one year [15], and the investment costs are calculated at 10 Dfl/GJ. In this study we assume that the measure will be applied in two steps. For one convertor the closed OG-system can be applied directly. For the two other convertors we assume it will be applied simultaneously with repressed combustion.

Currently 80 kg of <u>scrap</u> is added to the BOF-plant per ton of crude steel. This can be increased to 130 kg/tcs [5], (for BOF-plant 2, accounting for 80% of the production) or even 250 kg/tcs (for BOF-plant 1, accounting for 20% of the production). The increased scrap content will decrease the temperature of the molten steel. Therefore it has to be reheated, using approximately 0.5 GJ fuel/tcs [5]. At BOF-plant 1 net savings will amount to 1.75 GJ/tcs, because of the reduced production of pig iron. At BOF-plant 2 this will amount only 0.16 GJ/tcs. Overall the net savings amount to 3.2% of the fuel demand for pig iron production.

The BOF-process is a semi-continuous process, but <u>oxygen</u> is produced continuously. The development of capacity to store oxygen (called "wechsel-speicher") and a management system has been completed in recent years. Savings amount to 10% of the electricity consumption needed for the oxygen-production [5]. No cost figures are present for this study.

The same technical problems as with the heat recovery of the blast furnace slag make the implementation of <u>BOF-slag heat recovery</u> not feasible before the year 2000. Savings could (theoretically) amount to 0.14 GJ fuel/tcs [6].

Vacuum reheating furnace (Oxygen Blowing). The quality of the steel can be improved by reheating under vacuum and improves the casting performance. At Hoogovens a RH-OB has been installed [11] with an estimated energy consumption of 0.5 GJ/tcs. No energy conservation measures are identified in this unit operation.

Continuous Casting Machine (CCM). Two continuous casting machines (twin-strand casters) have been installed at Hoogovens, responsible for 100% of the production, making ingot casting abundant. In 1989 casting has totally been shifted towards continuous casting [14]. Ladles of the caster are preheated prior to accepting a liquid steel charge so that temperature stratifications in the steel and ladle skulls are minimised. Fuel consumption is typically estimated at 0.02 GJ/tcs [7]. Electricity consumption is estimated at 12.2 kWh/tcs (or 0.04 GJ_e/tcs) [5].

Hot strip mill. The slabs of the caster are reheated in the slabbing furnace to 1200 °C, and rolled to strips (thickness 1.5 to 12.5 mm) [11]. The hot strip mills are equiped with automatic quality controls. Fuel is consumed in the slabbing furnace (1.82 GJ/ton) [5] and electricity for the mill (103 kWh/ton or 0.37 GJ_e/ton) [3]. Only 75% of the production is treated in the hot strip mill [5]. Per ton product the energy consumption is therefore calculated at 1.37 fuel GJ/ton and 0.28 GJ_e/ton.

'Hot connection' implies the transport of the cast metal with a temperature of nearly 700 °C to the hot strip mill, saving energy in the slabbing furnace. This measure requires insulation at the hot strip mill, as well as computer control and management of the transporting system (logistics). Of the cast steel a maximum of 60% can be used as hot input to the strip-mill [5]. The savings amount to 0.3 GJ/tcs [5]. This is a reduction of 18% relative to the fuel consumption, needed for reheating. No costs are charged because the investments have already been made.

At the slabbing furnaces and walking beam furnaces <u>recuperative burners</u> are used, saving 0.3 GJ/tcs. A further improvement of the recuperative burners seems feasible, saving an additional 0.1 GJ/tcs [5]. The investment costs are estimated at 30 Dfl/GJ [5].

When rejected the rolled steel is cooled by spraying water, with a temperature of 80 °C. An <u>absorption heat pump</u> (or better: heat transformer) is installed to generate steam (3.5 bar, 130 °C), which is delivered to the steam network at Hoogovens. Savings are unclear, but estimated at 5% of the energy consumption of the rolling mill. Investment costs are 42 Dfl/GJ [5]

Cold rolling mill. Hot rolled steel is rolled thinner in the cold rolling mill, which consists of a pickling line, cleaning line, tandem mills and annealing lines. Of the total production 30% is treated in the cold strip mill. The energy consumption spread over different energy carriers is typically estimated at 0.9 GJ fuel/tcs, 0.2 GJ steam/tcs and 146 kWh/tcs electricity (or 0.53

GJ_e/tcs) [3]. The net energy consumption is calculated at 0.33 GJ/ton and 0.16 GJ_e/ton (30% of the production).

Steam could be saved by process improvement of the pickling operation, and generation of hot water in a waste heat boiler. Savings are estimated at 0.01 GJ/tcs [2]. This measure is taken into account with other additional heat recovery (see further), and not separately in this specific case.

Overall. Besides the unit operations described above, energy is used for utilities. The energy savings measures which can not be ascribed to a specific unit operation are described below.

It is estimated that (overall) a 2% saving of the energy demand [5] is feasible by good housekeeping. No costs are ascribed to this measure.

Options for additional waste heat recovery exist, especially for the recovery of low temperature waste heat. However, there is no possibillity yet to use this low quality heat. The heat can be recovered at nearly all process stages: sintering, casting, rolling and finishing [8]. These streams are taken into account, when integrating processes. We will apply (a limited) 3% savings, with high investment costs (100 Dfl/GJ).

Research is done on the possibility of using an expansion turbine for the natural gas (40 bar pressure), delivered to Hoogovens, for the production of electricity. The savings are dependent of the gas flow. The gas flow is calculated at 40.000 m^3/hr, generating 12 GWh. Overall this is a saving of 0.04 PJe (or 0.7% of the total electricity demand). Investments are calculated at 98 Dfl/GJ. O&M-costs are estimated at 3% of the investments [9].

At Hoogovens speed controlled drives have been installed at BOF-plant 2. However, experiences with this technique have not yet led to the further introduction because of operational problems. Because of the problems, we assume a limited potential for speed controlled drives, with savings of 5% of the total electricity demand. Very high investment costs will make this measure less attractive (estimated at 300 Dfl/GJ_e).

The specific energy consumption for the steel production is presented in table 2.

ECONOMIC ASSESSMENT

In this section we will evaluate the energy conservation measures described above, for different investment criteria. In table 3 the energy conservation measures are summarized, together with economic data. The cost-effectiveness of each measure is calculated, using formula 1. The results are presented in table 3.

$$C_{spec} = \frac{\alpha \cdot I + OM - SEPC}{\text{annual amount of primary energy saved}} \qquad (1)$$

in which:

α = an annuity factor depending on the interest rate r and the depreciation period n: $\alpha = r/(1-(1+r)^{-n})$;

I = the initial investment, expressed in Dfl per annually saved unit of final energy consumption (fuel or electricity);

OM = operation and maintenance cost expressed in Dfl (1988) per saved unit of final energy consumption;

SEPC = saved energy purchase costs, expressed in Dfl (1988) per unit of final energy consumption saved.

Costs are expressed in Dutch guilders (Dfl) (approximately 0.5 US$ or 0.43 ECU). This quantity is negative in cases that the benefits (adverted energy costs) are larger than the costs. The annuity factor is calculated using a real interest rate of 10% and the life-time of the technique as the depreciation time. Energy prices of 1990 are used (natural gas 6.8 Dfl/GJ, coal 4.7 Dfl/GJ and electricity 14.7 Dfl/GJ$_e$). The measures can be ordered according to cost effectiveness, as shown in table 3, or in a supply curve. Figure 1 shows the supply curve of energy conservation techniques for the production of crude steel.

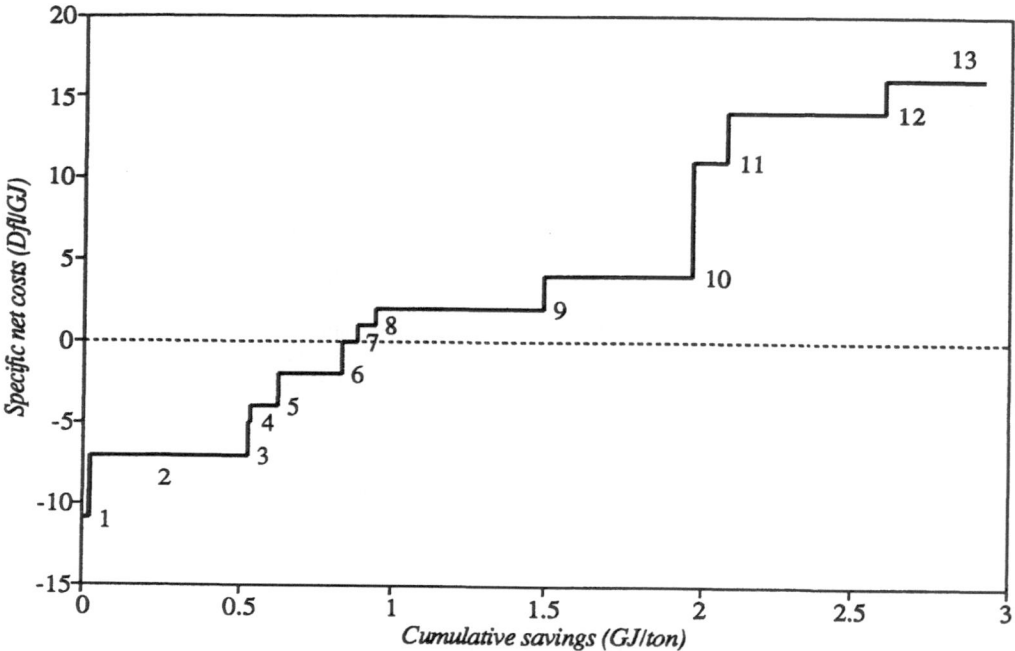

Figure 1. Supply curve of energy conservation techniques for the production of crude steel. On the horizontal axis the cumulative potential of primary energy (in GJ/tcs) saved is given. Vertically the marginal cost of the last measure is depicted for a discount rate of 10%. The numbers in the figure refer to the measures mentioned in table 3.

Decision makers on investments on energy conservation apply criteria for the profitability of an investment like the simple payback period. The simple payback period is commonly defined as the investment costs divided by the average yearly revenues. Hoogovens today use a maximum investment criteria of 50 Dfl/GJ (1988) [5], which can be compared to a simple payback period of 7 years. For each conservation technique the payback period is calculated. The results on the specific energy consumption are shown in table 2.

Policy measures of the government could stimulate investments in energy conservation. Examples of policy measures are an investment grant (for energy conservation projects) or an energy (or carbon) tax, as discussed in different countries and international organisations. These measures will decrease the payback period of the measures. Using our financial data it is possible to estimate the effect of the mentioned policy measures. In table 2 also the results are shown for the potential effects of two types of policy measures, e.g. an investment grant of 40% or a carbon-tax of 140 US$/ton C.

TABLE 3

Energy conservation measures in the iron and steel industry. Summary of the data of the individual measures and the cost-effectiveness

MEASURES	TYPE	SHARE CONSUMP. (%)	SAVING (%)	INVEST. (Dfl/GJ)	O&M (Dfl/GJ)	LIFE-TIME	SAVINGS FUEL (GJ/ton)	SAVINGS ELECTR (GJ/ton)	SAVINGS PRIMARY	COSTS-EFF. (Dfl/GJ)	SIMPLE PAYBACK PERIOD (Years)	CUMUL. SAVINGS (GJ/ton)	SEC (GJ/ton)
1 BOF-Oxygen'speicher'	e	9%	10%	0	0	20	0.00	0.01	0.02	-11	0.0	0.02	19.72
2 Hot connection	f	5%	18%	0	0	20	0.14	0.00	0.14	-7	0.0	0.17	19.57
3 Good housekeeping	f	100%	2%	0	0	20	0.35	0.00	0.35	-7	0.0	0.52	19.22
4 Expansion turbine	e	100%	1%	98	3	20	0.00	0.01	0.02	-5	3.8	0.53	19.21
5 BOF-Closed OG-system	f	25%	2%	10	1	10	0.09	0.00	0.09	-4	1.7	0.62	19.12
6 Direct coal injection	f	9%	14%	24	2	20	0.21	0.00	0.21	-2	5.0	0.83	18.91
7 Slabbing furnace	f	5%	9%	30	2	10	0.05	0.00	0.05	-0	5.7	0.88	18.86
8 Hot strip mill-heatpump	f	6%	5%	42	1	10	0.05	0.00	0.05	1	7.6	0.94	18.80
9 BOF-gas recovery + closed OG	f	52%	6%	50	3	20	0.55	0.00	0.55	2	11.6	1.49	18.25
10 BOF-Extra scrap	f	86%	3%	75	2	20	0.48	0.00	0.48	4	14.2	1.97	17.77
11 El: speed control drives	e	91%	5%	300	24	10	0.00	0.04	0.11	11	> 15	2.08	17.66
12 Waste heat recovery	f	98%	3%	100	5	10	0.52	0.00	0.52	14	> 15	2.60	17.14
13 Dry coke quenching	f	8%	24%	175	2	20	0.33	0.00	0.33	16	> 15	2.92	16.82

Notes with table 3:
- f = fuel, e = electricity
- Specific Energy Consumption (SEC) without savings is 19.74 GJ/ton product (primary energy)
- Electricity generation efficiency is 39%.
- The calculated savings have been corrected for mutual influences. It has been assumed that the most cost-effective measure is implemented first. Therefore the measures are additive, taking the effects of previous measures into account.
- All financial data is in Dutch guilders (Dfl), approximately 0.5 US$ or 0.43 ECU).
- Cost-effectiveness has been calculated using an interest rate of 10%.

TABLE 2

Specific energy consumption for the production of crude steel, divided in unit operations, applying a criterium of 7 years for the simple payback period. The potential influence of two policy measures, e.g. an investment grant of 40% and a carbon tax of 140 US$/ton C is given.

Case	Reference plant		Technical potential		7 year pay-back period		7 year pbp 40% grant		7 year pbp C-tax 140 $/tC	
Energy carrier	Fuel	Elec	Fuel	Elec	Fuel	Elec	Fuel	Elec	Fuel	Elec
Coke plant	1.62	0.06	1.00	0.06	1.59	0.06	1.38	0.06	1.38	0.06
Sinter plant	1.23	0.21	1.17	0.20	1.21	0.21	1.21	0.21	1.21	0.21
Pellet plant	0.46	0.10	0.44	0.09	0.45	0.10	0.45	0.10	0.45	0.10
Blast furnace	11.98	-0.08	11.39	-0.08	11.74	-0.08	11.74	-0.08	11.74	-0.08
BOF-plant	0.05	0.08	-1.07	0.07	-0.04	0.07	-0.59	0.07	-1.07	0.07
RH-OB	0.50	0.00	0.48	0.00	0.49	0.00	0.49	0.00	0.49	0.00
CCM	0.02	0.04	0.02	0.04	0.02	0.04	0.02	0.04	0.02	0.04
Hot strip mill	1.37	0.28	1.06	0.26	1.15	0.28	1.10	0.26	1.10	0.26
Cold strip mill	0.33	0.16	0.31	0.15	0.32	0.16	0.32	0.16	0.32	0.16
TOTAL	17.56	0.85	14.79	0.79	17.03	0.84	16.12	0.83	15.64	0.83
TOTAL - Primary	19.74		16.82		19.18		18.25		17.77	

The energy conservation potential can also be calculated for the different products of the plant: slabs, hot rolled product and cold rolled product. The specific primary energy consumption for the reference case and the the technical potential of energy conservation are presented in table 4.

TABLE 4

The specific primary energy consumption for three distinguished products and the technical energy conservation potential. Between parenthesis the energy conservation relative to the reference case.

Product	Reference case			Technical minimum 2000		
	Fuel (GJ/tcs)	Electr. (GJ$_e$/tcs)	Primary (GJ/tcs)	Fuel (GJ/tcs)	Electr. (GJ$_e$/tcs)	Primary (GJ/tcs)
Slab	15.86	0.41	16.91	13.43 (15%)	0.38 (7%)	14.40 (15%)
Hot Rolled Product	17.68	0.78	19.68	14.72 (17%)	0.72 (8%)	16.57 (16%)
Cold Rolled Product	18.78	1.31	22.14	15.77 (16%)	1.21 (8%)	18.87 (15%)

ENERGY CONSERVATION IN THE LONG TERM

In this section we will take a brief look on the potential for energy conservation on the long term. An assessment as shown above is not possible, because data concerning potential savings and financial data are missing or insufficient at the moment. Three major developments should be mentioned: Converted Cyclone (or Blast) Furnace (CCF or CBF), strip casting and direct rolling. The Converted Blast Furnace is the combination of coal gasification with the direct reduction of iron oxides [16]. In this way production of coke is redundant, saving energy and investment costs. A pilot-plant will be built on the short term, but commercial application is not expected till 2010-2020. Savings including coke production are estimated at nearly 20% of the 1988 energy consumption. Strip casting implies the direct casting of strips (coils), replacing hot rolling [17]. This technique was already proposed by Bessemer in 1891 ! [18]. The savings can be estimated from the energy needed for the slabbing furnace and the driving energy of the hot strip mill (11% of the 1988 base case energy consumption. A variation is the direct rolling of the cast steel, replacing the slabbing furnace (or walking beam furnaces). This option is technically feasible, but large reconstructions are needed in the lay-out of the steel plant. This makes direct rolling financially not feasible and only possible with temporary close-down of the complete plant.
The future integrated steel plant will -without doubt- be a compact plant, with fully integrated unit-operations. Energy consumption will be 3 till 5 GJ/tcs less, compared to the expected minimum figure in the year 2000.

CONCLUSIONS AND DISCUSSION

As a reference plant we investigated a modern integrated steel plant with a (relatively) low specific energy consumption of 19.74 GJ/ton product of primary energy (17.56 GJ fuel/ton and 0.85 GJ_e/ton) for a production package of 25% slabs, 45% hot rolled product, and 30% cold rolled product and an input of 10% scrap. In spite of the high efficiency it is shown that a potential for energy conservation still exists, of 2.8 GJ fuel/ton (16%) and 0.06 GJ_e/ton (7%). The specific primary energy consumption can be decreased to 16.82 GJ/ton (15%). The incorporated energy conservation technologies are only end-use technologies. Cogeneration (CHP) has not been taken into account. Also only proven technologies which can be implemented before the year 2000 have been investigated. The above figures apply to the technical energy conservation potential, not to the economically profitable potential. The economic potential has been analyzed, applying as an investment criterium a simple payback period of seven years, showing a saving of 0.6 GJ (primary)/ton (or 3%).
The theoretical effect of two policy measures on investments with a payback period of seven years has been assessed. A grant of 40% on the initial investments or a carbon tax of 140 US$/ton C results in a specific energy consumption of 18.2 GJ (primary)/ton (a saving of 8%) respectively 17.8 (primary) GJ/ton (a saving of 10%). In the future the specific energy consumption can decrease even more, by the application of new technologies, to approximately 14 GJ/ton (30% lower compared to the current specific energy consumption).
We have shown that the largest savings can be obtained in the BOF-plant and the hot strip mill. In this context it should be noted that we had to simplify the options for the BOF-gas recovery system, in order not to disturb the economic analysis. The installation of a sealed type gas recovery, is only possible after the installation of repressed combustion and BOF-gas recovery at the convertors. This however does not influence the results strongly.
Overall it is concluded that the approach gives good insight in the opportunities and potential of energy conservation measures on the specific energy consumption of unit operations. Also it is concluded that the approach offers the possibility to make preliminary assessments of some policy measures to stimulate the efficient use of energy.

Acknowledgements
The authors wish to acknowledge Hoogovens Groep BV, IJmuiden for their cooperation and Prof.
Dr. W.C. Turkenburg for his review of the paper.

REFERENCES

1. F.G.H. van Wees, J.A. Over, J.E. van Buuren, P.M.B. Ronde: Energy consumption for steel production. An example of energy accounting, ESC, Report 38, Petten, November 1986

2. International Iron and Steel Institute, Committee on technology: Energy and the steel industry, Brussels, 1982

3. A.R. Braun, J. Isings: Energy consumption in the basic metals industry, Hengelo, June 1990 (in Dutch)

4. M.H. Chiogioji: Industrial energy conservation (Energy, power and environment, part 4), New York, 1979

5. Personal communication with mr. R.M. van Ginkel, Hoogovens BV, 21 November 1990 and February 1992

6. ETSU: Energy use and energy efficiency in UK manufacturing industry up to the year 2000, volume 2, Harwell, October 1984

7. Caddet: A horizontal ladle preheating station fired with a self-recuperative burner improves steel production operations (project 2B.F06.141.87.UK), July 1987

8. H.M. Nolzen: Möglichkeiten der Energie-rückgewinnung in der Hüttenindustrie, insbesondere bei der Walzstahlerzeugung, Stahl und Eisen, 1984 14, pp. 671-678

9. Information about the gas expansion project, Energycompany Amsterdam, 23 April 1990

10. E. Worrell, R.F.A. Cuelenaere, J.G. de Beer, K. Blok: ICARUS: The potential for energy conservation in the Netherlands up to the year 2000, Department of Science, Technology & Society, University of Utrecht, Utrecht, May 1992

11. S. Wolthuizen: Iron, Steel and Rolling technology, Hoogovens, IJmuiden, June 1988 (in Dutch)

12. W. Buters, E. Keddeman, J. Rengersen: Pellet bed temperature control at Hoogovens IJmuiden and the effects on pellet quality, ICHEME - 5th International Symposium on agglomeration, Brighton, September 25-27, 1989

13. R.C. Brouwer, H.L. Toxopeus: Injection massive de charbon dans les hauts fourneaux d'Hoogovens IJmuiden, La Revue de Metallurgie, 1988, 4 (April 1988), pp. 323-334

14. N.H.H. Beyer, R. Boom: Hoogovens' view on steelmaking for high-productivity slab casting, Proceedings of The Sixth International Iron and Steel Congress - volume 1, Nagoya, October 21-26, 1990

15. Caddet: Closed system reduces losses from an oxygen-steel furnace, Project No. JP90.018/2B F06 (Result 80), October 1991

16. J.M. van langen, R.B. Smith: The Converted Blast Furnace, 4. Kohle-Stahl-Kolloquium, Berlin, February 21-22, 1989 (in German)

17. R.K. Pitler: Steel industry process technology trends, Materials and Society, 2[9] (1985), pp. 147-159

18. H. Bessemer: On the manufacture of continuous sheets of malleable iron and steel direct from fluid metal, Autumn Conference, Iron and Steel Institute, October 6, 1891

19. K. Blok, E. Worrell, R.F.A. Cuelenaere, W.C. Turkenburg: The cost-effectiveness of carbon dioxide emission reduction achieved by energy conservation, Energy Policy (forthcoming)

PAPER AND ENERGY: A FINNISH VIEW

ANTERO KOMPPA
The Finnish Pulp and Paper Research Institute
P.O. Box 70, 02151 Espoo, Finland

ABSTRACT

According to the present trend the industrialised countries try to decrease the waste problem through intensified recycling of fibrous material. Recycling, however, does not save energy as much as is commonly believed, at least if compared with the modern Scandinavian pulping technology, in which the energy consumption figures are fairly low and the process energy recovery rates are high. Combustion of waste paper - at least some fractions of it - should also be considered as a good alternative for recycling.

In this presentation the energy consumption in the production of various paper grades is compared. The utilization of energy in different sub-processes and the possibilities of reducing energy consumption, as a result of research into the papermaking process, are discussed. Finally, some predictions are made in regard to possible effects imposed by the future availability of energy.

BACKGROUND

A typical modern paper mill is a huge investment : a magazine paper mill with a production of 200-300 000 tons/a would cost USD 300-400 million now. The economically reasonable lifetime of a paper machine is 15-20 years. Therefore, sudden changes either in the raw material supply or in the product range are usually not possible or at least very difficult to carry out, due to technological and economical reasons.

The paper industry is very energy-intensive; for example in Finland the share of pulp and paper industry in the national electric power consumption is 31%(equal to 19 TWh/a). As a rule, big and modern mills are superior to old and smaller mills in terms of efficiency and specific consumption (i.e. calculated per a ton of paper) of energy, water and wood raw material as well as in the use of labour. Therefore a modern mill usually yields better cost efficiency, too. However, as a result of increased production capacity and improved paper quality, the total consumption of energy, both electric power and heat, is continuously increasing.

Due to customer's requirements, paper quality has to be improved all the time. Papers are to be produced with better optical properties, higher strength and smoother surface for printing. To obtain these properties, the pulp raw material has to be well processed and multi-staged, high quality papermaking and coating processes must be applied.

PAPER PRODUCTION AND PAPER GRADES

The world paper production may be roughly divided into different paper grades according to Figure 1. More than 40% of total paper production is aimed for printing and writing grades and almost 40% is made for packaging industry, various speciality paper grades covering the rest. The product range varies from country to country: e.g. in Finland more than 68% of the production is printing and writing paper grades.

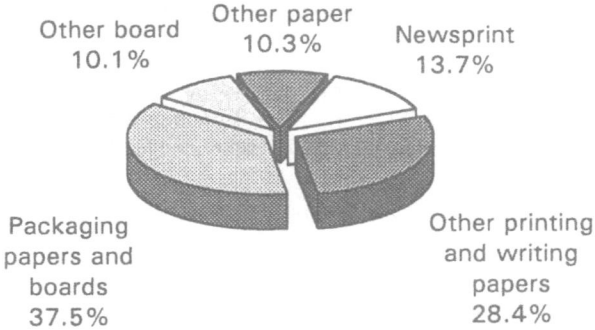

Figure 1. World production of paper grades in 1991, **totally about 241 million tons** [1].

Also the average mill size greatly varies from country to country (TABLE 1). Modern, large mills usually result in low specific energy consumption and small specific effluent and emission levels. In this development Finland is on the top of the list: during the last twenty years the Finnish paper production has been more than doubled but simultaneously the effluent load has decreased to a sixth of the previous level (Figure 2).

TABLE 1
Top producers of paper and board in the world 1990

Country	Total Capacity (1000 t)	Number of Paper Machines	Average Capacity (1000 t)
United States	**75793**	**1234**	61
Japan	25129	824	30
Canada	19800	270	73
China	13000	1200(3600)	3.6
F.R. Germany	12787	373	34
Soviet Union	12643	632	20
Finland	10067	107	**94**
Sweden	9384	127	74
France	7453	250	30
Italy	7369	456	12

Source: Valmet Paper Machinery Inc. [2]

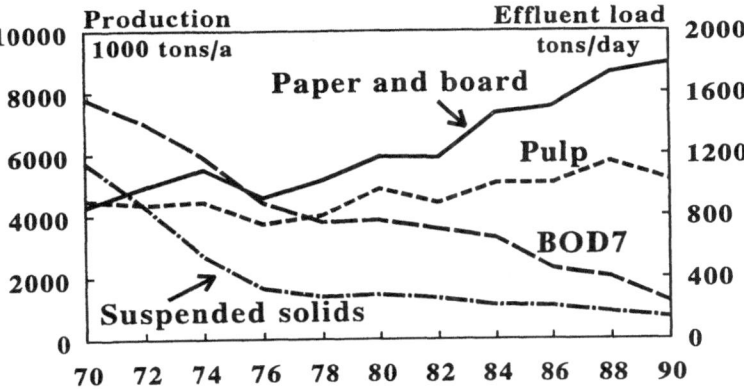

Figure 2. Production rate and effluent load in the Finnish paper industry 1970-1990 [4].

The consumption of paper is concentrated into the industrialised countries (TABLE 2)[5]. In the USA the consumption per capita exceeds 300 kg/year and in many of the European countries it exceeds 200 kg/year. It is rather obvious, that the average consumption in the world (44.8 kg per capita in 1990) cannot reach this level. Very likely, the consumption of paper per capita will not decrease, either. In the short run, the biggest growth is predicted to take place in the Eastern European countries.

TABLE 2
Apparent consumption of paper in some countries 1990

	kg per capita		kg per capita	Population (million)
USA	311	Western Europe	162	361
Sweden	245	Eastern Europe	35	427
W. Germany	232	Total Europe	93	788
Japan	228			
Finland	225	Northern America	302	276
Canada	215	Australasia	133	26
Switzerland	215	Latin America	26	439
Belgium	210	Asia	20	3111
Denmark	204	Africa	6	647

Source: PPI Annual Review [5]. In 1991 most of the above figures were slightly lower due to the worldwide economic recession.

FIBROUS RAW MATERIALS

Traditionally, paper is made of cellulosic fibres. In addition to the fibrous raw material, many paper grades contain mineral pigments and coatings, too. Most of the papermaking uses **wood based pulp**, either virgin fibre or recycled (in some countries, annual plants are used as the primary fibre source). In the Scandinavian countries the forests grow faster than felling increase. Due to this, e.g. in Finland more than a fourth of the annual wood growth remains in the forests and the increasing use of recycled fibre rapidly enhances this trend. However, it is essential to maintain the forests in a good shape through forestry and proper thinning. The <u>forest industry should be seen as an element in the carbon and oxygen cycles</u>. A growing tree takes carbon dioxide from the air and releases oxygen and hydrogen, which are available in air and water. These elements form 99.5% of the weight of wood. If trees are not utilised, they finally die and decay releasing the stored carbon dioxide back into the atmosphere without providing products of a sustainable economy for mankind.

The use of wood fibres in paper is based on the ability of fibres to bond with each other when dewatered and dried. This 'internal glue' gives very strong bonding between fibres without any additives. The pulp can be made either in a **mechanical** or a **chemical process**. Both the chemical and mechanical pulps do have the bonding ability but due to the different particle shape and size the chemical pulp yields stronger bonds. The hydrogen bonding is reversible: whenever paper is rewetted the bonds tend to open and will be created again when drying. This phenomenon allows **recycling** of paper; prior to paper making, recycled pulp is usually de-inked, too. The above three principal pulping methods exhibit quite considerable differences in regard to pulp quality, yield and energy consumption as is shown below.

As a rule, chemical and long mechanical pulp fibres give paper <u>strength</u> and <u>handling properties</u> whereas mineral pigments and short mechanical pulp particles give <u>optical</u> properties (appearance and opacity) and <u>surface</u> properties, essential in printing. Mineral pigments like china clay and calcium carbonate usually are less expensive than fibres and also the preparation of them consumes less energy than that of wood pulp.

Figure 3 schematically presents the pulping and papermaking processes. The paper making process itself is very much the same for most paper grades. The percentages in Figure 3 reveal the dry solids content of the process stream at different stages: In the forest the dry solids of a wood is about 50%. After various process stages the dry solids content, again, reaches 50% in the middle of the paper machine, i.e. at the dryer section. In between these points the pulp slush has to be diluted at multiple process stages even down to 0.5-1% solids content in order to obtain adequate paper quality. After the paper machine some paper grades will be coated with mineral pigments, dried again and finally the paper will be calendered to obtain adequate gloss of surface.

In **pulping processes** energy is required mainly for <u>heating of the process</u> (chemical pulping), for <u>refining</u> of the wood logs or chips (mechanical pulping) and for <u>pumping</u> of pulp slush and water. In the **paper making process** energy is required mainly for <u>pumping</u> of pulp slush, water and air, for <u>processing energy</u> and for <u>drying</u> of paper. Energy can be <u>recovered</u> mainly from the pulping processes and from the paper machine dryer section exhausts.

Chemical pulping
The chemical pulping process is very complicated and capital intensive (a typical new kraft pulp mill would cost more than USD 800 million, typical capacity 450 000 tons/a). In the chemical process about half of the wood raw material (i.e. the lignin from between the fibres) is dissolved; hence the <u>pulp yield is close to 50%</u>: for a ton of pulp about 2 ton of wood (5...6 cubic metres) is required.

In a modern pulp mill with cooking liquor recovery process, the dissolved wood is combusted, and therefore the <u>chemical kraft pulping process is a net producer of energy</u>, both heat (steam) and electric power. In addition to the electric power used for the pulping process itself (0.6 - 0.7 MWh/ton), a modern Finnish kraft pulp mill produces net (saleable) electric power about 0.2 - 0.6 MWh/ton of pulp (90-270 GWh/a), depending on the amount of heat required for the eventual integrated paper mill. The pulping process consumes heat about 10 - 15 GJ/t. Dissolved wood only (black cooking liquor) is required as a fuel. Today the Finnish forest industry already obtains about 40% of the total energy from dissolved wood and woodhandling residues.

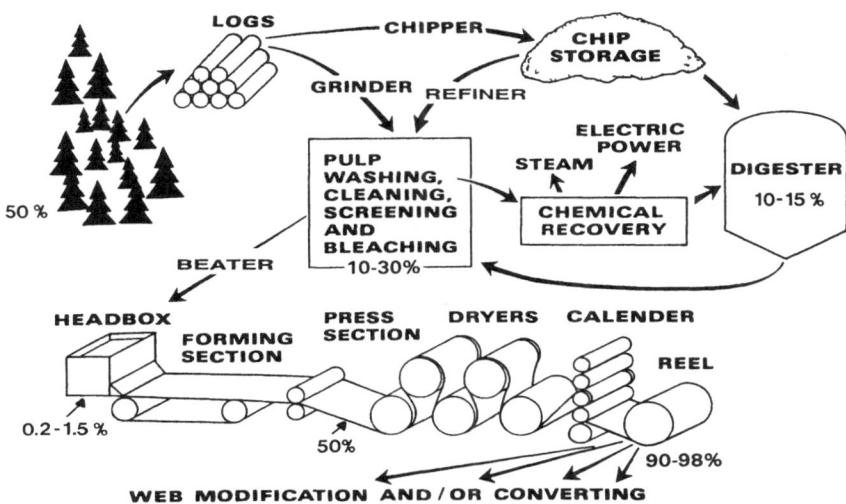

Figure 3: Schematic representation of the papermaking process. Mechanical pulping processes (grinding, refining) are simple but energy-intensive. Chemical pulping requires complicated recovery process, which produces steam and electric power from cooking liquor combustion.

Mechanical pulping

In a mechanical pulping process the wood is defiberized, either through grinding of logs or through refining of wood chips. If compared to the chemical pulping, the mechanical pulping requires a relatively simple and low-cost process (typically 25-35 million USD). The pulp yield can amount to 98%, therefore for a ton of paper only slightly above a ton of wood is required (2.5-3 cubic metres). Mechanical pulping processes conserve wood but consume a lot of electric power. However, in an up-to-date mechanical pulping process quite a high percentage of the power consumption can be recovered in the form of process steam, which greatly reduces the fuel consumption of the paper mill. (Mechanical pulping usually is integrated to the paper mill and the residues from woodhandling are combusted in the boiler house for steam generation; this, together with the heat recovery, produces steam enough for the paper mill.)

According to statistics a modern TMP (thermo-mechanical pulping) process requires electric power as an average about 2.7 MWh/t but it produces process steam for 2.7 GJ/t from the heat recovery (about 60% of the refining energy can be recovered in the form of process steam). With proper heat recovery the differences in energy consumption between different mechanical pulping methods are not very big; the TMP process allows the best recovery (with highest process pressure). The newest PGW (pressure groundwood) processes also quite well recover heat into hot water or even into process steam whereas the traditional SGW (stone groundwood) plants usually cannot be equipped with an efficient heat recovery system.

With the current prices of energy and wood raw material, mechanical pulp is typically much cheaper than chemical pulp (depending on the quality, 50...70% of the price of chemical pulp). It is therefore possible to produce high-quality wood-containing magazine papers at much lower cost than woodfree grades (made of chemical pulp only). Due to it's importance as a power consumer the development of the mechanical pulping is one of the key topics in the forest industry research. Results from a large national research project [6] show that it is possible to obtain refining energy savings of 15-20%(equal to 60-80 GWh/a for a typical magazine paper mill) through use of a novel refiner type and enzymatic additives.

Fibre recycling
The main reason for paper recycling is the increasing solid waste problem in heavily populated countries. As about one third of the landfill material in urban areas is paper and board products, the paper industry, the community and individuals do have a common interest in solving this problem. Paper is relatively easily recyclable. Recycling reduces the energy consumption of pulping, too. However, the fibres cannot be recycled eternally, as they crush and lose important papermaking properties after 4-6 processing cycles. Consequently, **virgin fibres will always be required to a varying extent**. Due to new legislation in Europe and in the USA, the amount of recycled paper available is increasing rapidly. Accordingly, the demands of the market are heavily pushing up the utilization of recycled fibres in many paper grades.

The use of recycled pulp does save energy, depending on the pulping method compared with, even up to 1 MWh/ton of pulp. In some paperboard grades, however, the use of re-cycled pulp results in higher energy consumption than possible with virgin fibres as the latter one does possess better bulk, i.e. specific volume; hence less virgin fibre pulp is required for same board properties than obtainable with recycled pulp. Also, the deinking process produces effluent sludge, about 20% of weight, consisting of deinking and mixed waste (printing inks with heavy metals, deinking soap, filler pigments and impurities coming with the recycled paper).

TABLE 3
Specific energy consumption in various pulping processes

Pulp type	Average specific energy consumption		
	Heat GJ/t		Electric power MWh/t
Unbleached kraft pulp	10 - 11	(own fuel)	Own power generation: 0.4 - 0.6
Bleached kraft pulp	12 - 14	(own fuel)	0.6 - 0.7
GW and PGW	0.2 - **-1.0** *)		Purchased power: 1.4 - 1.6
TMP and CTMP	**-1.9** - **-3.3** **)		1.8 - 3.3 (depending on product quality)
Recycled, deinked pulp	0.4		0.4

Bold negative figures = heat recovery.
*) As a rule, modern PGW processes incorporate heat recovery systems (recovery into hot water or low pressure process steam); about 30% of the grinding energy can be recovered.
**) About 60% of the refining energy can be recovered in the form of process steam.

Mineral pigments
The use of mineral pigments results in lower consumption of energy and wood raw materials in papermaking; however, the processing of pigments also requires energy. Quite often the energy required for pigment processing is neglected as in most countries the pigments are imported, i.e. the process energy is used abroad.

Calcium carbonate (chalk) and china clay are the most common natural pigments used, due to their advantageous properties and price (30...50% of the price of mechanical pulp). The use of precipitated calcium carbonate (PCC) is increasing due to it's good optical properties. When very pure fuels are used in combustion, it is possible to produce PCC from the flue gases through the reaction between carbon dioxide and calcium. So far, however, this kind of a process is very expensive and is being used for some speciality grades only.

Bleaching of the pulp also slightly increases energy consumption; the brightness of unbleached kraft pulp, however, is too low for printing papers. Recycled paper is very seldom suitable for printing papers without deinking but it is suitable for some packaging boards. Mechanical pulps usually have rather high brightness and therefore they are suitable for many products without bleaching - even if bleaching is required, it is carried out using hydrogen peroxide or dithionite, which are environmentally harmless.

RECYCLING OR COMBUSTION?

The heating value of newsprint is close to that of dry wood (14-20 GJ/t) and it's sulphur content is very low, too. Also, the recycling of some converted and laminated papers and boards - especially those made for packaging - results in low quality pulp. Therefore these could be quite well used as a fuel. According to a recent Finnish research project, waste paper combustion is both technically and economically feasible and safe in modern industrial and communal boiler houses; investment costs necessary are low and sulphur, dioxin and heavy metal emissions to air are very low. The best fractions of waste paper could be utilised for recycled pulp and the rest could replace fossil fuels that usually are environmentally harmful. At the moment the Scandinavian forest resources would allow even higher production of paper. The growing forests maintain the circulation of carbon dioxide, thus reducing the greenhouse effect. Without proper thinning the forests will deteriorate thus finally reducing the annual growth and the binding ability of carbon dioxide.

PAPER AND ENERGY

Suitably processed, strongly diluted pulp slush will be made paper web through spreading it onto the wire cloth at the wet end of the paper machine (Figure 3). Then, it will be dewatered and thereby formed to a planar fibre network. Water will be further removed through wet pressing and finally through evaporation, usually with contact drying cylinders. Thus, both electric power and heat are required in paper making.

According to the Finnish statistics the average power consumption in paper making is about 0.7 MWh/ton of paper and the process heat consumption is about 6 GJ/t. The specific energy consumptions for some paper grades are presented in TABLE 4. As can be seen the energy consumption quite strongly increases with the degree of paper quality.

Heat
In the paper making process the major part (90%) of heat is consumed for paper drying. The need of drying energy has been continuously decreased e.g. through more efficient mechanical dewatering (wet pressing).

Modern paper mills incorporate efficient process heat recovery, both from the pulping and paper making processes: heat is recovered into process steam and hot water as well as hot air for the use in paper dryer ventilation system. With improving efficiency of heat recovery and more economical new processes, all recovered heat cannot be utilised at the paper mill, any longer. Some paper mills, close to urban areas, can sell the heat for heating of surrounding houses. In most cases, however, suitable heat users are not available. Another problem of the heat recovery is that most of the heat is available during the warm season when the needs for heating are low. The excess heat could be utilised e.g. in the agriculture, if possible due to the heat transfer distances.

The closure of the white water system has increased the heat economy of the mill, too: as the amount of effluent is low also the amount of heat escaping from the process decreases.

TABLE 4
Specific energy consumption for some paper grades

Paper grade	Heat consumption GJ/t	Power consumption MWh/t
Brown kraftliner	6.0	0.5
Newsprint, SC magazine paper	5.5	0.6
Office paper	7.0	0.7
LWC magazine paper	5.5	0.8
Folding boxboard	7.0	0.8
Coated fine paper	8.0	0.9
Tissue paper	7.0	1.0

NOTE: these figures do not contain the pulping energy, nor the energy required for product transporting from the mill to customers.

Electric power

Roughly, the electric power at paper mill is consumed at various process stages as follows:

Raw material and water pumping	25%
Vacuum pumps, compressors, ventilation	25%
Process energy (beating, electric drives)	38%
Infrared dryers	9%
Others	3%

Hence, most of the electric power is used for various pumping purposes (raw material, water and air) in the process. A big consumer of pumping energy is the so-called short circulation (12%) in which pulp slush is strongly diluted prior to cleaning, screening and spreading onto the paper machine wire. In principle, it would be possible to decrease the utilisation of pumping energy through an increase of the pulp consistency (i.e. with less of dilution). However, even the slightest increase in forming consistency results in significant loss of paper quality (evenness), because the wood fibres tend to strongly flocculate in a water suspension; this would be a very potential topic for energy research.

The vacuum system of the paper machine is the biggest single pumping power consumer (15%). Vacuum is required at various stages of the process for efficient water removal, for cleaning and for supporting purposes. The need of vacuum is strongly increasing due to increasing operational speed (increase of speed is the most efficient way to improve production at existing and also at new mills). At the present, a large research project [7] aims at better process runability with vacuum system energy savings of 20-30%.

The ventilation of paper machine, heat recovery, air compressors and blowers consume about 9 % of paper mill power. If the value of heat would increase, the heat recovery could quite easily be made more effective through the use of larger heat exchange capacity: at the present price ratio, however, the increase in heat recovery is not profitable.

About 9% of the process energy is used for processing (beating) of pulp prior to forming into paper. The modern beaters already are very efficient and the power losses are rather low. Various electric drives of paper mill machinery consume about 29% of power and the increasing vacuum system power also increases the driving power consumption due to increasing friction losses.

Lately, one of the fastest growing single power consumers at the paper mill is the use of infrared (IR) dryers for paper drying and moisture profiling. In practice, these devices are still quite poor in energy efficiency (20-30% only). A recent research project shows that the efficiency can be raised quite easily up to 50%[7].

The improval of production efficiency is a very straightforward way to reduce the use of energy. Modern paper mills already do have extensive process and quality control systems through which it is possible to eliminate production and quality disturbances that fastly decrease the energy efficiency, too (the production of broke consumes the same amount of energy as the production of saleable product). Old fashioned paper mills usually do use lots of energy due to inefficient processes and losses; these also have big effluent and air emissions, not only due to old fashioned design but also due to improper manual operation.

FUTURE VIEW OF ENERGY CONSUMPTION

Thanks to the research and development work, the specific energy consumption in paper industry will continuously decrease. However, the increase in the degree of quality seems to slightly increase at least the mechanical pulping energy consumption. Due to increasing amount of paper production the total energy consumption of pulp and paper industry, very likely, will increase rather fastly at least during the next ten years: According to predictions the paper and board consumption in the world would annually grow at least 2.5% and would rise from the present 241 million tons up to 310-330 million tons, i.e. by 70...90 million tons until year 2000. More than a half of this would be covered by recycling (which will be strongly intensified at least in the Western European countries and in USA). 21% of the growth is predicted to be chemical pulp, 7% of mechanical pulp and 17% of mineral filler and coating pigments [3]. Using the above energy consumption figures and paper grade production shares the expected growth in production would require some 60-80 TWh/a extra electric power and 600-700 000 TJ/a heat.

The availability of energy versus paper production
If the price or the availability of energy do change, the effects on the paper industry will be significant. In various countries, the effects on the paper industry would be different, depending on the fibrous raw material and energy resources available. The following prediction holds true for Finland but could be applied to many other countries, as well:

Shortage of purchased electric power : If the availability of purchased electric power would decrease, the pulp and paper industry could, to some extent, compensate the lack of electricity by own power generation, through increasing the production of back pressure power and finally condensing power, too. As a result of the power shortage and the increasing power price, the production of energy intensive mechanical pulps would decrease. If the price of electricity would get very high, the utilisation of wood as a fuel finally would be more profitable than paper production. Then, the production of lower quality (=less profitable) paper grades would decrease and the freed wood supply would be utilised for power production.

If, for some reason, the **consumption of electric power in paper industry would be restricted**, the production of mechanical pulps would be decreased. Then, the groundwood (SGW and PGW) processes could overcome the TMP process as they do consume less energy. This kind of a change cannot be very fast, however, due to expensive and laborious investments necessary. Furthermore: the groundwood processes require the raw material in the form of logs whereas the TMP process can utilise wood chips; the availability of raw material obviously would also affect the process choice. A major change from mechanical pulping to the chemical one is unlikely, due to the very high investment cost necessary and due to the smaller pulp yield.

Shortage of market fuels : Due to the increasing use of recycled pulp, more and more of the annual wood biomass growth remains in the forests. This would allow an increase in the use of wood as a fuel, if economically possible. Finnish pulp and paper industry already obtains 40% of energy from wood-based residues. However, excessive use of wood for energy would decrease all pulp and paper production.

At the beginning, the industry could relatively well adjust with the limitations by changing to other fuel types and sources of energy and - to a smaller extent - to other processes. If the utilisation of electricity would be free and advantageous enough, TMP pulping process would overcome the other mechanical pulping processes, which would increase the amount of process steam, indirectly generated by electric power. If, simultaneously, the use of purchased electric power would be limited, the production of mechanical papers would be replaced by production of woodfree and recycled paper grades.

The decrease in mechanical pulping would also cut down a broad range of paper grades necessary for the present European life style: the production of wood-containing (mechanical) papers would decrease, in the order of value added in the production. The first loser would be virgin fibre newsprint, followed by the uncoated wood-containing printing and writing papers. Finally also the production of coated paper and paperboard grades would decrease.

REFERENCES

1. PPI Annual Review, World trends and trade. PPI, 1992, **7**, 32-115.

2. Pesonen, J., Opening remarks, Valmet Paper Machine Days 1990, Jyväskylä, Finland, 14-15 June 1990.

3. Hakulin, L., Nightmare ahead for pulp producers. PPI, 1991, **10**, 44-54

4. The Finnish Forest Industries, Facts and Figures 1991, Central Association of Finnish Forest Industries, Helsinki, 1991.

5. PPI Annual Review, World trends and trade. PPI, 1991, **7**, 21-109.

6. Sundholm, J., KUITU, Energy-efficient mechanical pulping, Interim report 1988-1990, Ministry of trade and industry, Energy Department, Reviews B:97, 1991, 45 p.

7. Paulapuro, H., Komppa, A., RAINA, Energy-efficient paper production, Interim report 1988-1990, Ministry of trade and industry, Energy Department, Reviews B:95, 1991, 26 p.

LATEST ALUMINIUM RECYCLING AND ENVIRONMENTAL TECHNOLOGY IN EUROPE

O.H.PERRY AND W. BATEMAN
Stein Atkinson Stordy Limited,
Ounsdale Road, Wombourne, Wolverhampton, WV5 8BY, U.K.

ABSTRACT

Recycling of aluminium requires 5% of the energy needed to produce new aluminium from ore. Currently, there is increased market pressure to produce a cleaner and more cost effective product from scrap aluminium whilst minimising emissions to the environment and maximising fuel efficiency. The case for decoating is, therefore, overwhelming.

As the growth in the use of aluminium products, in general and coated products in particular, has proceeded so has the technology of decoating.

This paper presents and discusses the technology involved in successful decoating and the types of plant currently in use.

MARKETS FOR RECYCLING

To illustrate the size of recycling market in U.S.A. the amount of metal produced by secondary smelters (recycle melters) equalled 1,414,000 metric tonnes in 1990, which was larger than the entire primary smelting production in 1958.

In Europe can production and recycling is below that of U.S.A. but is growing fast.

DECOATING.

The objective of decoating is a straightforward one: to remove the coating, which may be paint, lacquer, plastic paper or oil with minimal interference to the aluminium surface. Whilst chemical solvent treatments are possible the most cost effective method is by thermal processing. There are three types of thermal processing plant in use for decoating, but the process technology is very similar and, therefore, the information given in this paper is generally applicable to all three types.

COATINGS.

It is not possible to give a single definition for the coloured coatings. A variety of chemical compounds are used which are predominantly organic and very volatile. Benzene and toluene are often present in small quantities. The inorganic component is usually restricted to the colouring and for added mass. Paint and lacquer suppliers have their own formulations and these are proprietary information.

Packaging material coatings also vary widely. Paper laminates may be 50% coating and plastic laminates significantly higher.

Increasingly, due to a large extent to tighter emission controls, oily scrap is being processed in a controlled environment. Oil contents be as high as 10% by weight and very often contain water from machining operations. Typically, the oils are of light grade used in forming operations.

TYPES OF SCRAP.

It is now possible to recycle the following types of aluminium scrap:-

Used beverage cans	UBC
New can stock	NCS
Used food cans	UFS
Mixed low copper	MLC
Painted siding	
Extrusion and Swarf.	

Of these materials MLC, Painted Siding, NCS and Extrusions have traditionally proved difficult to process effectively.

MLC contains high levels of tramp materials yet it is the cheapest source of scrap material. Environmental regulations now preclude feeding it directly into the smelter.

Painted Siding is the product of coil coating lines. The coating contains higher than average levels of volatiles.

NCS is the spoil from beverage can production. NCS is degrated by lubricating oil used in the forming process.

Extrusions are typified by aluminium window frames. There is an increasing practice of bonding in polyurethane insulation foam.

Other difficulties can arise from the use of vinyl coatings in the latest generation of soft drink containers.

PACKAGING MATERIALS.

This group of materials ranges from household foil to food containers. Food containers have been processed albeit in small quantities. Work is currently in hand to extend the overall range of materials to include foil products of all types.

THERMAL TREATMENT.

Thermal treatment to remove the paint and lacquer from an aluminium surface requires careful control of both temperature and oxygen The upper temperature limit is the melting point of aluminium but below this temperature aluminium is oxidised at a rate dependent on the oxygen content of the atmosphere.

Figure 1 shows the weight loss from samples of beverage cans heated in an atmosphere containing 11% O_2 in nitrogen.

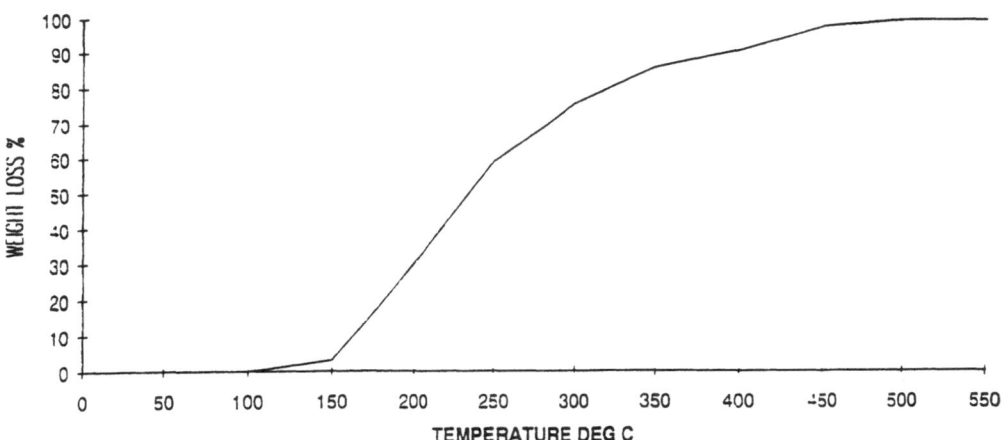

Figure 1.   Effect of temperature on coating loss.

TEMPERATURE

It will be noted from Figure 1 that at a temperature of 500°C all the coating has been removed. Heated above this temperature oxidation is apparent from the gain in sample weight.

EFFECT OF COATING (VOC)

The coating is normally referred to as volatile organic compounds (VOC). The percentage by weight of VOC is very variable. Used beverage cans are typically 1-3% by weight, new can stock up to 5%, coloured foil 20% and laminates 50%.

The decoating process removes the coating and the products of the reactions are carried away to an afterburner or thermal oxidiser, where they are combusted. A proportion of the heat released in the afterburner is returned to the decoating system, thus increasing thermal efficiency.

For all coatings there is a critical level at which the whole system becomes autothermic, i.e. no external heat is required. This is approximately 5% by weight of coating as illustrated in Figure 2.

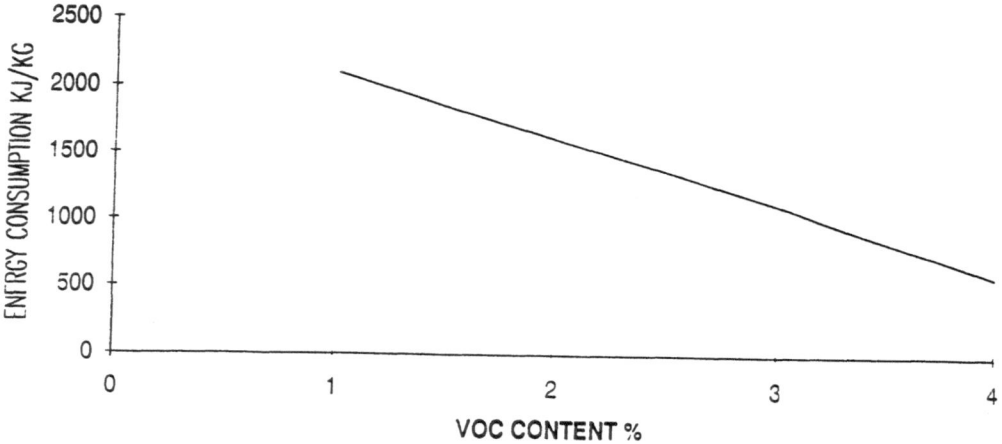

Figure 2   Effect of coating mass on energy consumption.

Since the coatings are volatile the oxygen available has to be restricted otherwise uncontrolled combustion would take place which would increase the temperature of aluminium scrap to the point where melting would occur. This situation arises increasingly at oxygen levels above 12%. Below 4% there is insufficient oxygen available for effective decoating. Thus, the optimum level is 6-8%.

EFFECT OF MOISTURE.

Moisture content of the scrap is normally in the range of 0-5%. This can vary from batch to batch and from one season to another.

Any moisture in the scrap reduces the overall efficiency of the process resulting in increased energy consumption. High moisture content also has a dilution effect producing a consequent reduction in the oxygen level.

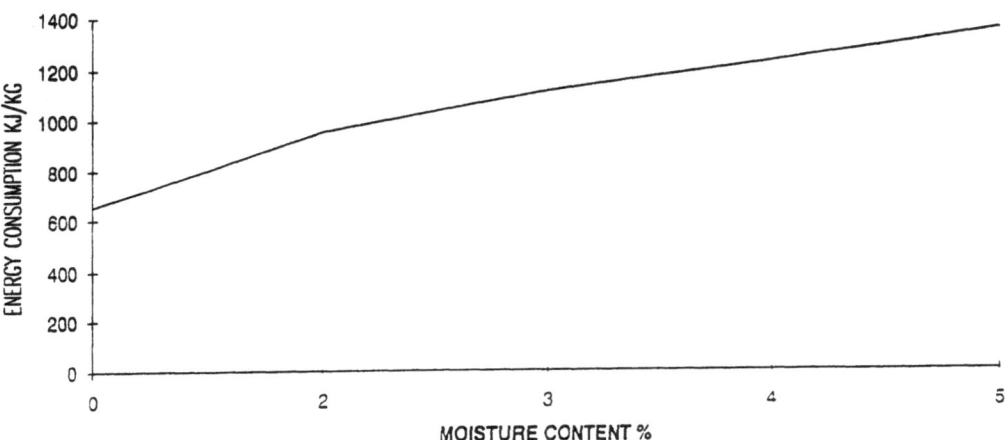

Figure 3.   Effect of moisture content on energy consumption.

HEAT TRANSFER.

Heat transfer in the decoating process is predominantly by convection. Prior to decoating, the material is shredded into small pieces producing a high surface area to mass ratio. In the rotary kiln there is a critical velocity at which, if exceeded, the aluminium shreds are carried away by the waste gas. The gas velocity through the kiln is therefore, of necessity, low. Most shredded aluminium products are processed at velocities below 3 m/s. Consequently, heat transfer coefficients rarely exceed 25 W/m²K.

Rotary kilns have helical flights on the inner surface which pick up the scrap and subsequently allow it to fall freely in the hot gas flow. The kilns operate on an inlet to exit gas temperature difference of 300-450°C. Therefore, to maintain acceptable throughputs of material a high mass flow of gas is necessary. Since the critical velocity cannot be exceeded, kiln diameter is varied according to throughput.

EMISSIONS:

The gas leaving the decoating process contains the following regulated pollutants:-

> Particulate matter
> Carbon monoxide and hydrocarbons
> Nitrogen oxides
> Hydrogen chloride
> Hydrogen fluoride
> Sulphur dioxide.

PARTICULATE MATTER.

Particulate matter in the kiln exit gas is a mixture of organic and inorganic compounds, plus aluminium fines produced in the kiln by abrasion. Approximately 50% is organic and, hence, combustible.

CARBON MONOXIDE AND HYDROCARBONS:

These combustibles are produced during the low temperature degradation of the organic coating. The concentration of these compounds varies widely during the decoating process. UBC is not segregated and the amount of paint and lacquer on the can varies according to the print design. The maximum carbon monoxide concentration in the kiln exit gas is about 4000 ppmv (0.4%) and hydrocarbons 500 ppmv (0.05%).

ACID NEUTRALISATION:

The gas entering the cleaning system contains mainly inorganic particulate matter, hydrogen chloride and hydrogen fluoride and very small quantities of organics. The acids are neutralised by treatment with lime. Depending on the levels required the lime is either injected directly into the waste gas stream or neutralised in a reactor tower which is more

The waste gas finally passes through a bag filter system before exhausting to atmosphere. The levels of pollutants in the stack gas are given below:-

Particulate:	<	25 mg/m^3
Carbon Monoxide	<	100 mg/m^3
Organics	<	20 mg/m^3
Hydrogen Chloride	<	10 mg/m^3
Hydrogen Fluoride	<	2 mg/m^3
NOx	<	100 ppm v.

PROCESS PLANT:

A complete decoating system includes three main areas of operation:

* Metal Preparation
* Waste Gas Recirculation
* Air Pollution Control System.

METAL PREPARATION:

Careful metal preparation is the key to producing a quality product accompanied by low energy consumption. The purpose of metal preparation is to take infeed scrap, containing all kinds of unwanted material and present it to the rotary kiln with the highest possible amount of aluminium and the lowest amount of tramp elements.

Infeed material is accepted in the form of bales, briquettes or loose product with a maximum moisture content of 10%, iron content of 30% and 10% coating by weight.

Prior to entering the kiln, the infeed material is subjected to a number of mechanical and electrical operations:-

SHREDDER MILL.

The shredder will break the bales open, where appropriate, and typically shreds the scrap into pieces 50 mm square, ensuring the material is not crushed and the maximum surface area exposed. A bottom grate segregates the oversized material which is recycled back to the mill via an associated recirculation conveyor. Alternatively, a grate with larger apertures can be fitted for processing automotive, light iron and MLC scrap.

A negative pressure is maintained within the shredder to prevent dust emissions escaping. Dust emissions arising within the shredder are collected by the bag filter system.

STORAGE HOPPER.

The shredded material is carried to a storage hopper where product discharge is controlled by a vibratory conveyor. This forms a "live" bottom base to the hopper and regulates the rate of feed to the upstream system.

ELECTROMAGNETIC SEPARATOR:

The material is carried by a toughened rubber belt conveyor from the storage hopper and passed under an inline overband electro-magnetic separator. This removes free iron contaminants including the remnants of steel strip used to band the bales. The maximum free iron content of the material leaving the electro-magnetic separator is 0.5%.

GYRATORY SCREEN:

To remove fine material and dirt from the scrap, it is passed over a gyratory screen separator, fitted with a deck of 3mm perforations. All material of less than 3 mm is removed with a separation efficiency of 90% +.

MAGNETIC DRUM SEPARATOR:

A magnetic drum separator is used at the same time to remove irony-aluminium. Any irony-aluminium is held on the drum whilst the remaining ferrous material falls on to the product conveyor below. At this stage, the ferrous content is below 0.1%.

SCRAP CHARGE.

Upon exiting the kiln at 500°C, the decoated material is delivered to the melting furnaces. Mineral residual carbon is present on the surface of the shreds, thus preventing smoking and flaming.

The melt rate is optimised by the use of preheated scrap. Scrap heated to 482°C melts some 70% faster than at room temperature reducing to 45% faster at 370°C.

WASTE GAS RECIRCULATION SYSTEM:

The waste gas recirculation system comprises the rotary kiln, recirculation fan and afterburner unit.

Waste gas, containing scrap generated contaminants, passed from the kiln into a charge plenum where it mixed with ambient air ingress introduced with the charge from the surge hopper.

RECIRCULATION FAN:

The waste gas exits the plenum at around 350°C. The recirculation fan performs the twin tasks of circulating the gas within the recirculation loop and providing the negative pressure for the kiln itself.

Incorporated in the ductwork upstream of the fan is an automatic water dilution system which is activated in the event of an upset condition causing an overtemperature condition, thus, the fan, the driving force of the system is protected.

AFTERBURNER UNIT:

The recirculation fan discharges the waste gas into the refractory line afterbuner unit. Waste gas enters the unit tangentially initially causing it to circulate round the burner. Baffles are incorporated into the design to encourage thorough mixing of volatiles and oxygen. The afterburners primary purpose is to combust the volatiles and carbon particles released in the kiln. Destruction efficiencies of 95-97% are commonly obtained.

The afterburner is also the heat source for the decoating operation and is partly regenerative. A large proportion of the heat required by the process is obtained from the combustion of the V.O.C's, thus reducing operating costs.

The amount of volatiles in the waste gas continually varies. To ensure that the oxygen content of the gas leaving the afterburner is kept constant at 6-8% the afterburner burner is equipped with a dual air/fuel ratio control. For start up the air/fuel ratio is preset to 10:1 for natural gas. During processing, the burner can be switched to 7:1 or additional secondary air can be supplied thereby covering the cyclic ratio of the volatiles.

The afterburner is normally controlled at 820°C. In the event of a high level of volatiles entering the afterburner and providing a substantial source of additional heat, the process control system compensates. Once an overtemperature alarm is registered, the burner is turned down, the recirculation fan speed is slowed, to reduce the volume of volatile laden waste gas entering the burner, and the feed rate is backed off to reduce throughput rate. Thus, whilst aluminium scrap with a 3% VOC coating might run at 10 tonnes/hour, material with a 12% coating would flow at say 7 tonnes/hour for the period of the alarm.

AIR POLLUTION CONTROL SYSTEM (A.P.C):

At the spill off junction with the recirculation loop, some 40% of the waste gas enters the air pollution control system. The major components of the system are:-

* Recuperator
* Acid neutralisation
* Dust collector.

RECUPERATOR:

A recuperator is installed in the APC line for two reasons. Firstly, to lower the waste gas temperature closer to the acid dew point and secondly to make preheated air available to the afterburner.

The recuperator is of a multi-tube parallel flow design in which the hot gas flows inside, the tubes and the cooling air flows outside. The gas entering the recuperator is at 820°C and leaves at 250°C.

ACID NEUTRALISATION:

Acid gas control is accomplished by injecting a neutralising agent into the gas stream. The waste gas enters the reactor tower of a dry acid gas/lime scrubbing system and is injected with a mixture of fresh hydrated lime and recycled lime. The gas and suspended particles rise vertically up the tower at a velocity designed to give the necessary residence time for maximum acid neutralisation.

The lime neutralising agent also acts as a filter bag precoat material in the dust collector, thereby further increasing neutralisation efficiency.

DUST COLLECTOR (BAGHOUSE FILTER SYSTEM).

The gas enters the pulse jet-type dust collector unit, via a full width inlet/distribution plenum; this configuration allows a uniform distribution of the dust-laden gas, across the multiple filter sleeves. The gas moves through the filter media, the dust being trapped on the outer surface of the sleeves and the cleaned gas being collected in the outlet plenum.

The uniform lime dust cake forming on the outer surface of the filter sleeves enhances the filter effect; electronically-timed cleaning, by means of short bursts of compressed air, maintains the pressure drop without pre-set limits.

FUTURE TRENDS.

Future trends moving down the product line. We have carried out tests on recycling toothpaste tubes, aluminium pet food cans, lithoplate (for printing), painted foil, yogurt tops etc.

So far, all these materials can be recycled economically and complying to latest environmental requirements.

SESSION 2:

Thermodynamic Cycles

Chairman: Mr R. Dumon

NONCONVENTIONAL THERMODYNAMIC CYCLES FOR THE NINETIES: COMPARISONS AND TRENDS

Sergio S. Stecco
ASME Fellow
Department of Energy Engineering
Università di Firenze
Via di Santa Marta, 3
50139 Firenze, Italy

ABSTRACT

Ways to improve the performance of power cycles by using different working fluids will be a major objective in coming years. One of these is to modify the Brayton cycle by using water and steam injection. Such applications are comparable to gas-steam combined cycles (GSCC) in which the air and steam flows are separated. The introduction of water-steam in the typical "dry" process of the Brayton cycle affords notable improvement to system performance in terms of power, efficiency, and/or NOx emissions. Some of these solutions, among them STIG and Cheng, have been patented, with plants already onstream worldwide.

A new cycle, the humid air turbine cycle (HAT), has recently been proposed and patented. Its main innovation is that steam is produced along the airflow, thus eliminating the heat recovery boiler. A special component of this cycle is a multistage saturator, where the air and hot water mix to produce vaporization at a variable temperature, with some beneficial thermodynamic effects.

This paper presents the state-of-art of the possible modifications to the gas turbine cycle by means of water-steam adduction. The trends and developments of the Eighties will be discussed and the advantages and disadvantages of the various solutions will be pointed out. Special attention is devoted to a) increase in power output, b) variations in efficiency, and c) effects on NOx release. Both energy and exergy approaches are employed. In addition, the problem of water consumption, or recovery, is examined, since water shortages and rising costs are aspects that will become increasingly important in the near future.

In addition to Brayton cycles, a second group, Kalina cycles, will be discussed. In these, heat recovery is enhanced by means of a suitable mixture of water and ammonia.

The main difference between these cycles and the conventional Rankine cycle as a bottomer is related to different composition in the water-ammonia mixture in the various plant components, which optimize heat transfer and reduce exergy losses. An analysis of the practical possibilities of these cycles is presented and discussed.

NOMENCLATURE

H = Enthalpy
M = Mass flow
p = Pressure
T = Temperature
TIT = Turbine inlet temperature
W_{el} = Electric output
W_h = Thermal output
β = Compression ratio
n = Efficiency
Φ = Water content

WATER STEAM INJECTION IN SIMPLE CYCLE GAS TURBINES

The injection of steam or water in gas turbine combustors has become normal practice in many applications. Its main advantages are: a) power increases and b) reduction in the NOx emissions.

Historically, water and steam injection for terrestrial applications was first proposed by Nicolin in 1951 at Stal-Laval. Commercial application, however, came about thirty years later with Allison's 501-KH (related to Cheng cycle) and General Electric's LM-5000 (related to the STIG cycle), both aeroderivative engines.

A schematic of the plant is shown in fig.1.

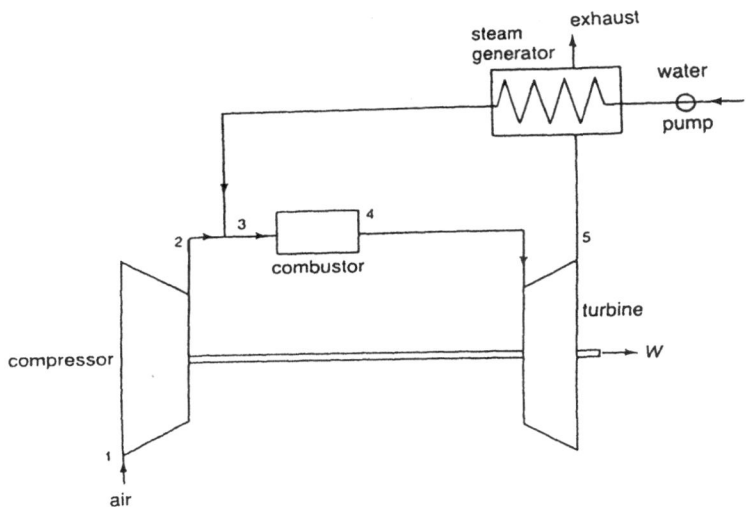

Figure 1: Schematic of simple-cycle water-injected gas turbine

Effects on Power Output

The first effect of steam or water injection is, as mentioned, the increase in power output with a greater mass flow across the turbine stages. The major difficulty (and consequently limitation in the added steam fraction) is encountered in blade design characteristics, in order to enable a mass flow increase without sensible losses.

At nominal conditions, most turbines are designed to have one or two stages close to sonic. As a consequence, the flow section per unit mass flow is reduced, and thus weight and dimensions, thereby accruing beneficial effects in terms of centrifugal stress and costs. To ensure greater mass flow, it is necessary to accept substantial increases in pressure at turbine inlet (an consequently at turbine discharge). Rotational speed must be increased, with the result that the surge line is approached. Under this aspect, aeroderivative engines are preferable, since they are designed with ample margins with respect to the surge line.

Generally speaking, we may assume that the increase in mass flow, DM, is related to the new compression ratio, β, by a linear relationship, namely

$$\beta' = \beta(1+DM/M) \qquad (1)$$

Effects on NOx Production

NOx production in gas turbine emissions can run very high, especially as a result of the high combustion temperatures, the presence of nitrogen in the fuel (typically in the case of methane), and extensive oxidation. Particularly dangerous are the temperature peaks in the combustion gases. Steam injection has an "averaging" effect on the three-dimensional temperature pattern in combustor, reducing thus the presence of such peaks.

Effects on Efficiency

The effects of injection on efficiency have been amply debated. The injection process does not present outright positive effects considering that mixing water and steam at temperatures lower than compressed air contributes negatively to efficiency but can be very promising in regenerative cycles.

Some details are presented in fig.2.

THE STEAM WATER INJECTED REGENERATIVE CYCLE

In the regenerative configuration, an additional gain results from the enhancing effect of regeneration and the greater effectiveness of exhaust-gas heat recovery. A typical plot is presented in Figure 3. In the scheme, pressurized water is injected by special devices into the airflow coming from the compressor. This atomized water has a cooling effect on the flow as it vaporizes and removes heat. Inlet conditions at the regenerative RHE are cooler, and the heat from the exhaust gases can be recovered at much lower temperatures than in conventional schemes. Moreover, the temperature difference between hot and cold flows (exhaust gases and air-steam) is increased, thereby reducing the heat exchanger surfaces.

As the refrigerating effect of water can be considerable, saturation of the steam may be reached. This condition can represent the upper limit in the injectable water mass fraction. By contrast, saturation in the case of steam injection is never encountered in actual practice.

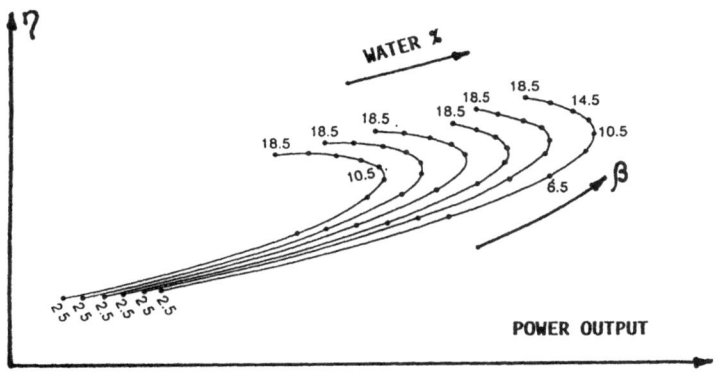

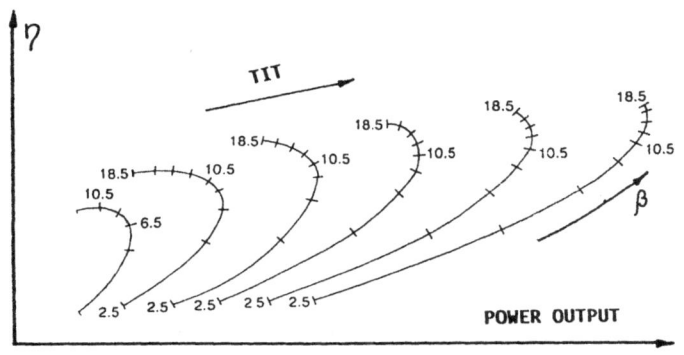

Figure 2: Performances of water/steam injected gas turbines in simple cycle

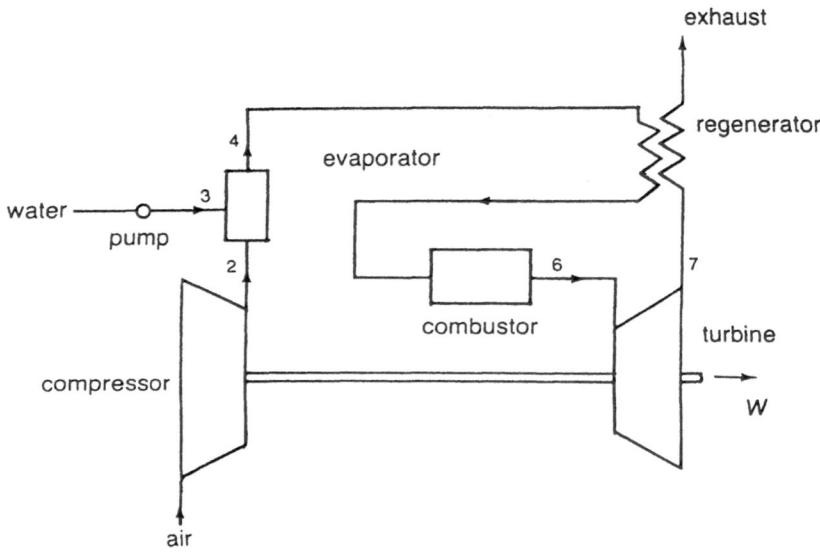

Figure 3 Schematic of water/steam injected gas turbines in regenerative cycle

With respect to a water-steam injection simple cycle, the regenerative cycle with water injection achieves higher efficiency at a lower compression ratio. The biggest problems result from corrosion and water deposition in the RHE, which clearly explains the necessity of limiting the water mass fraction. NOx emissions are naturally the same as in nonregenerative cycles. Some additional data are presented in Figure 4.

THE CHENG CYCLE

The Cheng cycle, a natural development of steam-water injection in regenerative cycles, was proposed in 1982 (Lloyd Jones et al., 1982) and saw its first application in 1985 (Lloyd Jones et al., 1985). The Cheng cycle series 7 engine consists of an Allison 501-KH gas turbine combined with a heat recovery steam generator (HRSG) which are closely matched to assure high cycle efficiency. Steam from

the HRSG is injected into the gas turbine combustion region
to increase the power output of the basic engine. The KH
version of the 501 engine incorporates modifications made
by Allison specifically for operation of the Cheng cycle.

A schematic of the Cheng cycle is presented in fig. 5

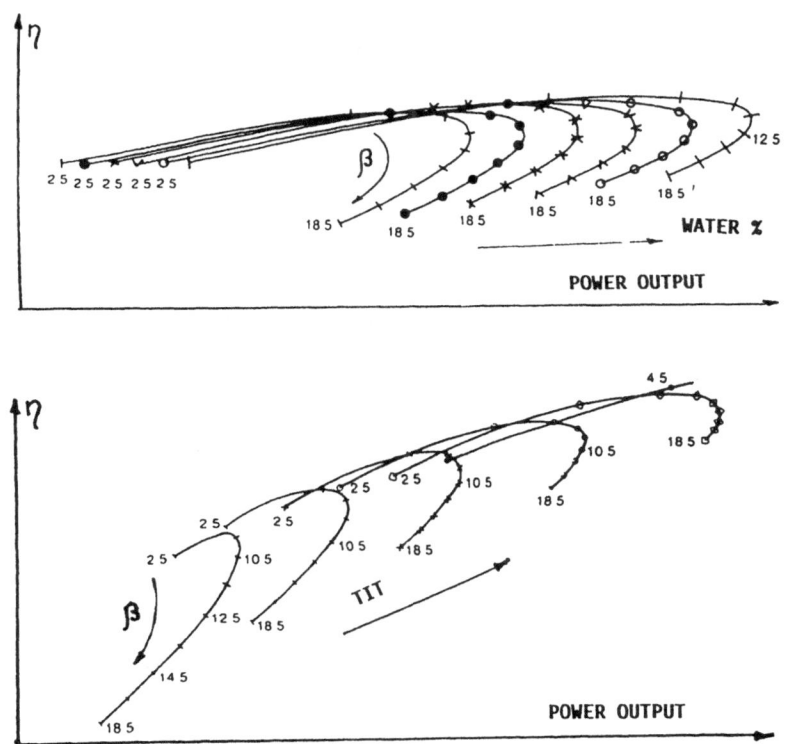

Figure 4: Performances of water/steam injected gas turbines
in regenerative cycle

   The HRSG consists of:

 - Superheater. The superheater, located immediately
downstream of the turbine exhaust diffuser section, raises

the temperature of the injection steam to the highest possible value prior to injection. The increase in temperature reduces the amount of fuel to be added to the gas turbine to raise the steam temperature to turbine inlet temperature, thereby enhancing the engine heat rate.

- Secondary combustor. The secondary combustor, located immediately downstream of the superheater, is followed by a combustion duct section. The secondary combustion provides the additional steam to be used either for increased power generation or as process steam, depending upon the desired operating condition.

- Evaporator. The evaporator, located immediately downstream of the combustion duct, serves for steam generation.

- Feedwater heater.

The operating regime of the Cheng cycle series 7 engine cogeneration is illustrated in Figure 6. Point 2 indicates the power generated by the basic gas turbine driving the generator

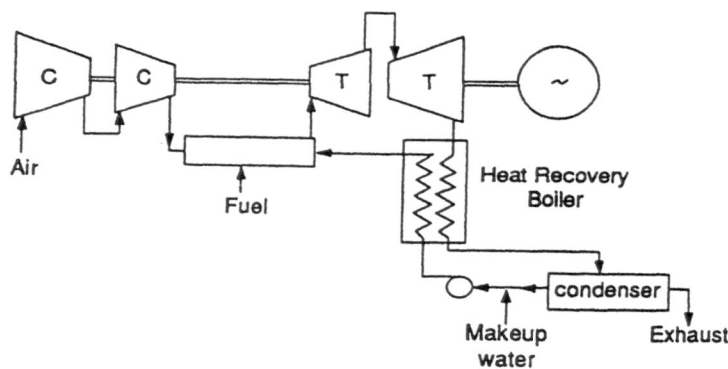

Figure 5: Schematic of the Cheng Cycle.

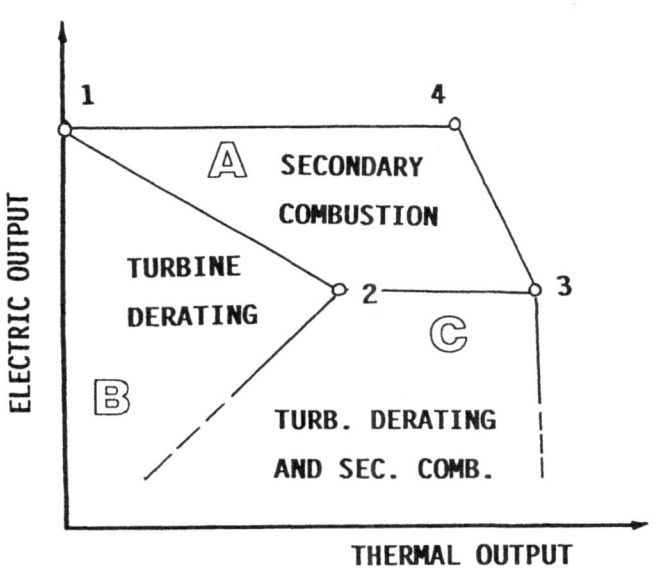

Figure 6: Operational regimes for the Cheng cycle.

with no steam injection, and the amount of process steam generated by the HRSG with no heat added to the exhaust gas by the secondary combustor. At such conditions, the super-heater operates dry. As steam is directed from process to the gas turbine, the power output of the turbine and gener-ator increases along line 2-1, which represents the con-stant turbine inlet temperature of 1895°F (1308 K), the highest allowable firing temperature for continuous duty. At point 1, all the steam generated by the gas turbine exhaust heat energy is injected into the turbine, with none available for process. Use of the secondary combustor ex-pands the operating regime of the cogeneration system from line 1-2 to encompass all of area A. Line 3-4 represents the firing of the secondary combustor to obtain a tempera-ture of 1600°F (1144 K) at evaporator entry.

In most cogeneration installations, the process steam demand is given highest priority. If, for example, the process steam demand were 20 million BTU/hour (5.8 MW), the electrical power output would be limited to about 4 MW. If a higher power output were required, firing of the second-ary combustor would provide additional steam for injection to allow raising the electrical power to any level up to a

maximum of 6 MW, while maintaining the process steam flow
of 20 million BTU/hour. Alternately, if the process steam
demand increased, firing of the secondary combustor would
permit generation of additional process steam, while maintaining the electrical power output of 4 MW. Obviously,
combinations exist that allow operation anywhere within
area A. Operation in areas B and C is possible by derating
the gas turbine operation to lower turbine inlet temperatures. In the first commercial installation at California
State University at San José, operation is initially being
limited to a turbine inlet temperature of 1800°F (1255 K)
and a secondary combustion firing rate to provide a temperature of 1400°F (1033 K) at evaporator entry.

THE HUMID AIR CYCLE (HAT)

The humid air cycle (Rao, 1990) is an intercooled gas
turbine cycle having an air-water mixing evaporator before
the combustion chamber and an exhaust gas recovery system.
The solution has several advantages, i.e., increase in
efficiency and in power output, reduction in NOx, which are
similar to those encountered in STIG and CHENG plants.
However, efficiency can be further enhanced by special
nonisothermal vaporization, as illustrated in Figure 7.

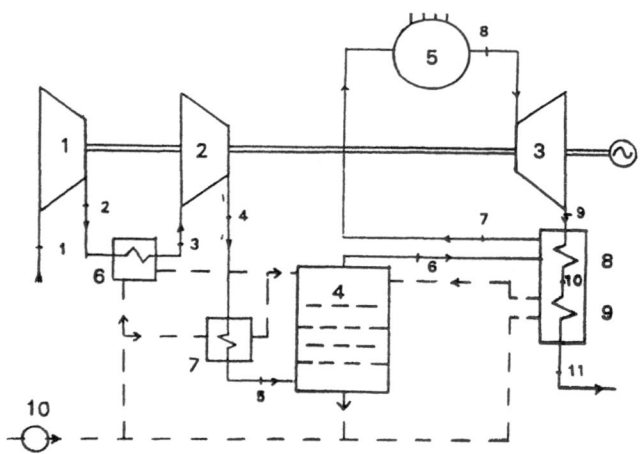

Figure 7: Schematic of the HAT cycle.

The components that distinguish the humid air cycle
from a traditional turbogas cycle are the two cooling

stages after the two compression stages, the mixture evaporator, the surface recuperator between the mixture and exhaust gases, and the economizer before the gas discharge. The main components are:

- First refrigerator (6). The first refrigerator is a water-gas exchanger that cools the airflow leaving the first compression stage, at the same time providing heat for the subsequent evaporation. Obviously, the lower input temperature at the second compression stage leads to decreased compression in the second stage.

- Second refrigerator (7). The second refrigerator, an exchanger resembling the first, supplies heat for water evaporation, while decreasing the temperature of the air entering the evaporator to enhance the regenerative effect.

- Economizer (9). The economizer is a water-gas exchanger that recovers the low-temperature exhaust heat, returning it to the cycle to allow evaporation of the water in the evaporator.

- Evaporator (4). The evaporator is a water-gas mixture exchanger that operates in multistage countercurrents causing water evaporation at variable pressures (given the partial steam pressure in the mixture) and therefore at variable temperatures. The heat needed for evaporation is supplied by the hot water coming from the refrigerators, the economizer, and the incoming air which, in addition to being humidified, is also heated via contact with the hot water.

- Recuperator (8). The recuperator is a gas-gas exchanger which recovers heat from the high-temperature exhaust gases and heats the air-steam mixture before it enters the combustion chamber.

The results in terms of efficiency look very promising and significant research is currently under way (Lindgren et al., 1992; Rao and Joiner, 1990; Stecco et al., 1993a).

THE KALINA CYCLE

The Kalina cycle is a closed cycle whose working fluid is a mixture of water and ammonia with a different composition in the boiler and the condenser. The use of a nonazeotropic mixture of fluids with different boiling points decreases the exergy losses in the heat recovery boiler when the hot flow is a source of sensible heat, i.e., a fluid whose temperature changes during heat transfer. If this kind of heat source is used to evaporate and superheat a pure substance, there is notable limitation to the high-efficiency heat recovery on account of the isothermal heat exchange occurring during the change of state. In a mixture of two fluids with different boiling points, the temperature of a saturated liquid is not the same as that of a saturated vapor.

As a consequence, the exergy losses in a boiler with the mixture flows are much lower than those in a boiler for a pure substance (Figure 8).

The analysis carried out at the Department of Energy Engineering of the Università d Firenze (DEF) has given promising results (Figure 9). Accordingly, the possibilities of practical application have been examined (Olsson et al., 1991; Desideri et al., 1991), with the first operational powerplant going onstream in 1991.

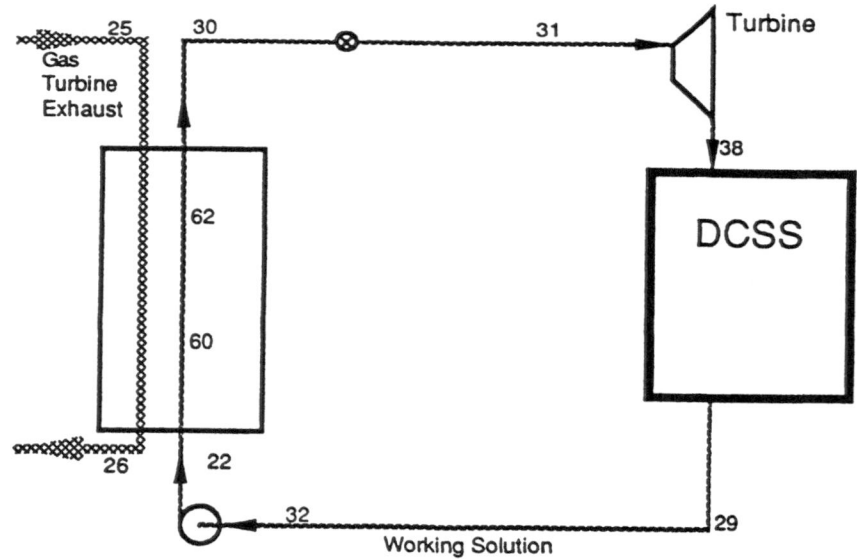

Figure 8: Schematic of the Kalina Cycle

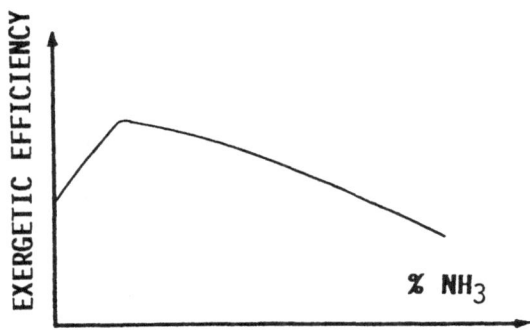

Figure 9: Exergetic Efficiency vs. Basic Composition

Improvements to the Kalina cycle deserve some additional comment. Kalina cycle systems may be considered as a combination of two highly interactive subsystems: one containing the heat acquisition process plus the power producing device (denoted CS1, KCS6, or KCS8) and one for distillation-condensation (denoted by D2 or D3, depending on the design). A combination is described as KCSxDy, where x and y denote the subsystems. Our description is limited to one solution, the KCS1D2.

When water is heated by a hot gas vapor, there is a tremendous destruction of exergy that cannot be reduced either by technological advances in components or by spending more on the heat exchangers. What is required is a fluid with different thermodynamic properties. An obvious solution first attempted over a century ago is to replace the water by a mixture since, at constant pressure, mixtures boil at variable temperatures.

If the sole change is to use a mixture as the working fluid, what has been gained in the boiler is lost in the condenser. At the condenser's lower pressure, the change in temperature during condensation is greatly increased. If the back pressure on the turbine is high enough to ensure complete condensation, efficiency is diminished. To circumvent this limitation, the composition of the fluid entering the condenser must somehow be made leaner than that in the boiler and turbine. The leaner mixture will condense at a lower pressure at the given coolant temperature.

136

One of the distinguishing features of the Kalina cycle
is the use of distillation-condensation mixing processes to
create fluids of different thermodynamic properties in
different parts of the cycle. While the flow chart for the
heat acquisition portion of the system much resembles a
Rankine cycle, its thermal behavior during heat acquisition
differs considerably because of the 75% ammonia-water
mixture of the Kalina cycle system. The temperature profile
in the oncethrough boiler is shown in Figure 10a, with same
information in the exergetic temperature coordinates in
Figure 10b. The exergetic temperature is defined by Kalina
as the product of temperature by the Carnot factor.

As can be seen, there is a substantial reduction in
exergy losses during heat acquisition for this Kalina cycle
in comparison to a Rankine cycle.

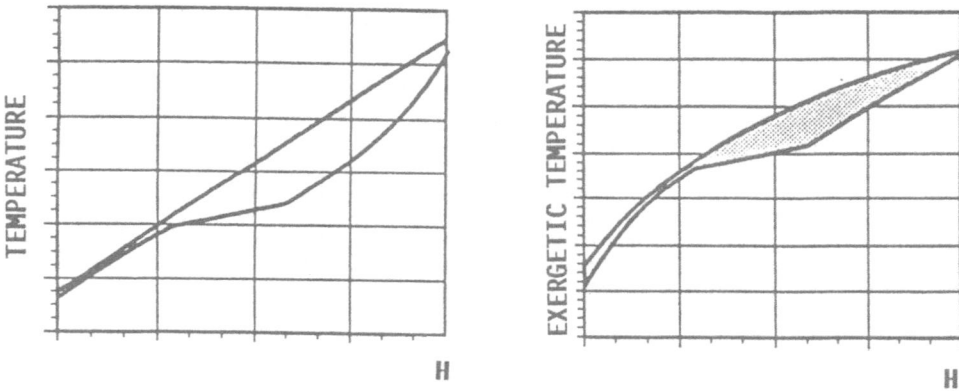

Figure 10: Temperature and Exergetic Temperature Distribution in the Waste heat Boiler of Kalina System KCS1D2.

The major difference in structure between system
KCS1D2 and a single-pressure Rankine cycle occurs in the
distillation-condensation subsystem, illustrated in Figure
11. After expansion in the turbine (point 38), the vapor is
at too low a pressure and too high a concentration (75%
ammonia) to be entirely condensed at the temperature of the
available coolant. Therefore, the working fluid can only be
partially condensed in the recuperative heat exchanger

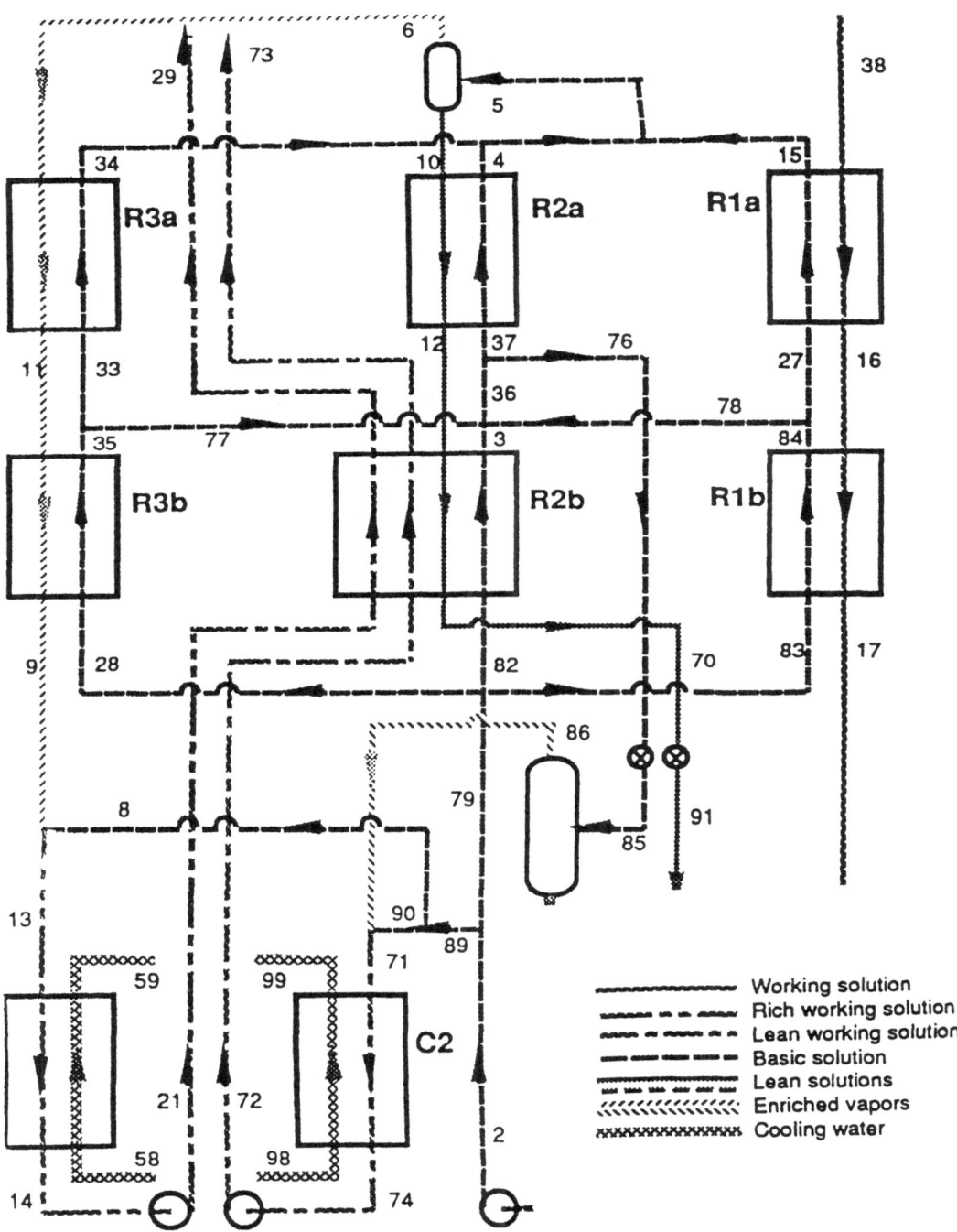

Figure 11: Distillation-condensation subsystem.

(R1). At the heat exchanger exit, a lean solution (42% ammonia, from R2) is mixed with the two-phase flow from the recuperator (R1) to provide a 50% ammonia solution which can be condensed at the coolant temperature.

The condensed liquid from condenser C1 is pumped to a higher pressure and divided into four parts: Three are heated recuperatively in heat exchangers R1, R2, and R3 and one is used to adjust the composition of the very rich fluid coming from R3 (station 9), thereby re-creating the working fluid (75% ammonia) to be condensed in the high-pressure condenser (C2). The three parts flowing to R1, R2, and R3 are heated to form a two-phase mixture at a predetermined temperature (station 5) which is separated (not flashed) into a rich vapor phase (98% ammonia) and a lean liquid ((40% ammonia). Enough of the concentrated vapor is returned to the lean liquid emerging from the separator to raise its concentration to 43% ammonia, as required by recuperator R2.

System KCS1D2 has a second-law efficiency of approximately 70%. The gains are due to both the reduced exergy losses in the heat acquisition portion of the cycle and the increased pressure drop across the turbine made possible by the change in composition in the DCSS. This performance represents a significant improvement over the second-law efficiency of a double-pressure Rankine cycle with the same thermal boundary conditions (64.0 to 65.3%), but it is still short of the goal of 80% (as stated by the authors, 1992).

REFERENCES

Bidini, G. and Stecco, S.S., 1991, "A Computer Code Using Exergy for Optimizing Thermal Plants," ASME Transactions, Journal of Engineering for Gas Turbines and Power, Vol. 113, pp. 145-150.

Boyce, M.P., 1982, Gas Turbine Engineering Handbook, Gulf Publishing Company, 1989

Burnham, J.B., Giuliani, M.H., and Moeller, D.J., 1987, "Development, Installation, and Operating Results of a Steam Injection System STIG in a General Electric LM 5000 Gas Generator," ASME Journal of Engineering for Gas Turbines and Power, Vol. 109, pp. 257-262.

Desideri, U., Olsson, E., Stecco, S.S., and Svedberg,

G., 1991, "The Ammonia-Water Mixture Rankine Cycle: Considerations on Its Applicability as a Bottoming Cycle," Proceedings, IECEC, Boston, MA.

El Sayed, M. and Tribus, M., 1985a, "Thermodynamic Properties of Water-Ammonia Mixtures: Theoretical Implementation for Use in Power Cycle Analysis," Proceedings, ASME Winter Meeting, Miami Beach, FL.

El Sayed, M. and Tribus, M., 1985b, "A Theoretical Comparison of the Rankine and Kalina Cycles," Proceedings, ASME Winter Meeting, Miami Beach, FL.

Kalina, A., 1984, "Combined Cycle System with Novel Bottoming Cycle," ASME Journal of Engineering for Gas Turbines and Power, Vol. 106, pp. 737-742.

Kalina, A. and Tribus M., 1992, " Advances in Kalina cycle technology (1980-1991)" FLOWERS 92 Proceedings, Florence, Italy, pp.97-124.

Larson, E.D. and Williams, R.H., 1986, "Steam Injected Gas Turbines," ASME Paper 86-GT-147.

Lindgren, G., Eriksson, J., Bredhe, K., and Annerwall, K., 1992, "The HAT Cycle: A Possible Future for Power and Cogeneration," Proceedings, FLOWERS '92, Florence, Italy.

Lloyd Jones, J., et al. 1985, "Design and Construction of the First Commercial Cheng Cycle Series 7 Cogeneration Plant," Proceedings, IGTI, ASME Paper 85-GT-122.

Lloyd Jones, J., Flynn, B.R., and Strother, R., 1982, "Operating Flexibility and Economic Benefits of a Dual-Fluid Cycle 501-KB Gas Turbine Engine in Cogeneration Applications," ASME Paper 82-GT-298.

Olsson, E., Desideri, U., Steccco, S.S., and Svedberg, G., 1991, "An Integrated Gas Turbine-Kalina Cycle for Cogeneration," Proceedings, IGTI, Orlando, FL.

Rao, A.D. and Joiner, J.R., 1990, "A Technical and Economic Evaluation of the Humid Air Turbine Cycle," Proceedings, Electric Power Research Institute Contractors Meeting, Palo Alto, CA.

Stecco, S.S. and Desideri, U., 1989, "A Thermodynamic Analysis of the Kalina Cycles: Comparisons, Problems and Perspectives," Proceedings, IGTI, Toronto, Canada.

Stecco, S.S. and Desideri, U., 1991, "Considerations on the Design Principles for a Binary Mixture Heat Recovery Boiler," Proceedings, ASME COGENTURBO, Budapest, Hungary.

Stecco, S.S., Desideri, U., and Bettagli, N., 1993a, "A Modified Humid Air Cycle: Parametric Analysis," Submitted to 6th ASME COGENTURBO, Bournemouth, U.K.

Stecco, S.S., Desideri, U., and Bettagli, N., 1993b, "The Humid Air Cycle: Some Thermodynamic Considerations,", Submitted to IGTI, Cincinnati, OH.

OPTIMIZATION OF A COMBINED-CYCLE PLANT WITH THERMODYNAMIC, ECONOMIC AND ENVIRONMENTAL CONSIDERATIONS

Christos A. Frangopoulos[*] and Vasilios A. Bulmetis [#]

National Technical University of Athens
Department of Naval Architecture and Marine Engineering
P.O. Box 640 70, 157 10 Zografou, Greece

ABSTRACT

Environmental considerations are combined with thermodynamics and economics for the design optimization of a double-pressure combined-cycle plant, which produces electricity. Two pollutants are examined in this particular example, SO_2 and NO_x, and a penalty policy is considered.

Three objectives are studied in separate and the optimization results are compared with each other: (i) maximization of the plant thermal efficiency, (ii) minimization of the capital and fuel expenses, (iii) minimization of the total cost including capital and fuel expenses as well as penalties for the unabated emissions of SO_2 and NO_x. For convenient reference, the three objectives are called *thermodynamic*, *thermoeconomic* and *environomic* respectively.

In addition to the effect of certain parameters on the optimal design of the system, a sensitivity analysis reveals a critical value for the penalty of each emission of the plant. A penalty higher than the critical one forces the system design towards higher values of the degree of abatement. Thus, the procedure is useful not only to the designer in setting the design specifications of the system, but also to the society in setting environmental standards and implementing a policy to achieve those.

NOMENCLATURE

c_c capital cost factor
c_f price of fuel
c_{pg} specific heat capacity of exhaust gases
c_N unit penalty for NO_x emissions
c_S unit penalty for SO_2 emissions
$\dot{H}_f = \dot{m}_f H_u$ energy flow rate with the fuel
H_u low heating value of fuel
L Lagrangian
$\dot{m}$ mass flow rate
$\dot{m}_{Ni}$ NO_x mass flow rate before abatement
$\dot{m}_{Si}$ SO_2 mass flow rate before abatement
P_{s2} low pressure of steam
p pollution measure
$r_C = P_2/P_1$ compressor pressure ratio
s mass content of sulphur in the fuel
T_1 temperature at the inlet of the compressor
T_3 temperature at the exit of the combustor

T_{s12} temperature of low-pressure steam
t time
$\dot{W}$ electric power of the system
w net power density
x set of independent decision variables
y_r the function of unit r
Z annualized capital cost

Greek letters

Γ cost of goods or services used, or pollution penalty
δ degree of abatement
η cycle thermal efficiency
λ Lagrange multiplier
$\tau_3 = T_3/T_1$

---

[*] Asst. Professor,   [#] Student

Subscripts

ch	chemical	rm	the m th abatement element of unit r
env	environomic	S	sulphur dioxide, SO_2
f	fuel	s1	high temperature steam
g	flue gas at the turbine exit	te	thermoeconomic
N	nitrogen oxides, NO_x	th	thermodynamic
r	the r th unit (r = 0 : environment)		

Superscripts

*	optimum value	$\cdot$	per unit time

1.   INTRODUCTION

One of the primary concerns during the '70s and '80s was the depletion of energy (exergy) resources. The combined thermodynamic and economic analysis of thermal and chemical plants, known as *thermoeconomics* [1, 2], helps to increase the plant efficiency and at the same time achieve an economic optimal design and operation.

In recent years, the decline of the environment due primarily to our energy-related activities has become severe and raises serious concern too. For this reason, the analysis and optimization of energy-intensive systems has to deal not only with energy (exergy) consumption and economics, but also with the pollution and degradation of the environment.

A method has been developed for this purpose recently [3, 4], which is briefly described in Section 2 for the reader's convenience. In Section 3, the method is applied for the optimal design of a combined cycle plant. The sensitivity of the optimal solution to the values of selected parameters is studied also.

2.   A BRIEF DESCRIPTION OF ENVIRONOMIC ANALYSIS AND OPTIMIZATION

Environomic analysis is an extension of thermo-economics [1, 2], which takes into consideration not only energy, exergy and costs, but also pollution and degradation of the environment. One particular approach of analysis is described in the following, which is a further development of earlier work [5, 6, 7].

2.1 Environomic Functional Analysis

A complex system is considered as a set of interrelated units. Each unit has one particular function (purpose, product). In order to perform its function, the unit is supplied with goods and services from the environment as well as with functions from other units. The functional diagram is used also, which establishes the relationships among units as well as between the system and the environment, although it is not shown here in order to save space. The following additional features are introduced for the environomic analysis.

Each pollutant p_{rm} emitted by unit r is shown by a dashed line with the arrow pointing towards the unit (Fig. 1), which indicates that (i) p_{rm} is a function of the product (output) and the technical characteristics of the unit, and (ii) p_{rm} is, in general, something the system has to pay for. Except of emitted pollutants, other environmental or social adverse effects may be represented by p_{rm}'s, if properly quantified. If equipment for abatement of the m th pollutant of unit r is installed, an element rm is shown beside the unit (Fig. 1). In that case, the initial quantity p_{rmi} (before abatement) of the pollutant is reduced to p_{rm}. The effectiveness of the equipment is measured by the *degree of abatement*:

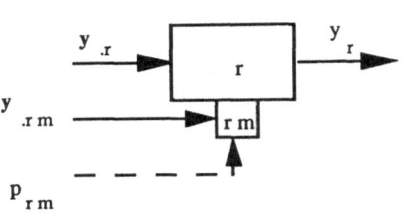

Fig. 1. A unit with a pollution abatement element.

$$\delta_{mm} \equiv \frac{p_{mmi} - p_{mm}}{p_{mmi}} \qquad (1)$$

Alternatively, the pollution abatement equipment may be considered as a separate unit (abatement unit), which is treated like any other unit of the system. The operation of the pollution abatement equipment requires goods and services coming from the environment and/or from other units (vector $y_{.rm}$ in Fig. 1). Each input to a unit (Fig. 1) is expressed as a mathematical function of the output (product) and the design and operation characteristics of the unit. The function of a unit is distributed to other units, abatement elements and the environment.

2.2 Measures of Pollution

Pollution causes a disturbance to the environment, i.e., a departure from its initial (or desirable) state of equilibrium. Exergy (essergy), as a measure of departure from equilibrium, is a very appropriate measure of pollution too. Thus, thermal, chemical and other forms of exergy can be used as measures of thermal, chemical and other forms of pollution respectively.

Exergy (essergy) offers a theoretically elegant measure of pollution. Since resources consumed can be measured in terms of exergy, a uniform measure is established for both the inputs and the outputs of the system (including the pollutants). However, such a measure may not be always convenient in practice because (i) it usually requires additional computations, (ii) pollution regulation standards and limits are usually given in g/kJ, g/kWh, g/Nm^3.

Another measure of pollution, which is computed more easily than exergy and makes the comparison of the effected pollution with established standards or limits more direct and convenient, is the following:

$$p = \frac{\alpha - \alpha_o}{\bar{\alpha} - \alpha_o} \cdot A \qquad (2)$$

where

α an intensive property characterising the pollutant
α_o the intensive property of the same pollutant in the environment
$\bar{\alpha}$ *harmfulness limit* of the intensive property: if it is exceeded, the pollution becomes particularly harmful to the environment
A quantity of the pollutant.

Equation (2) has been obtained by elaborating on some ideas presented by Gaivao and Jaumotte [8]. Thermal and chemical pollution are given as two examples of applying Eq. (2):

$$p_Q = \frac{T - T_o}{\bar{T} - T_o} Q \qquad (3)$$

$$p_{ch} = \frac{\chi - \chi_o}{\bar{\chi} - \chi_o} \chi m = \frac{c - c_o}{\bar{c} - c_o} cV \qquad (4)$$

where

T temperature at which heat is rejected
Q heat quantity
χ mass content of the pollutant in the rejected stream
m total mass of the stream
c concentration of the pollutant in the rejected stream (e.g. in g/Nm^3)
V total volume of the stream (e.g. in Nm^3)
$\bar{T}, \bar{\chi}, \bar{c}$ harmfulness limit for T, χ and c respectively
T_o, χ_o, c_o the values of T, χ and c in the environment.

Of course, Eqs. (2)-(4) are valid with flow rates too: $\dot{Q}, \dot{m}, \dot{V}, \dot{p}$, etc.

The term $(\alpha - \alpha_o)/(\bar{\alpha} - \alpha_o)$ may be called the *harmfulness factor* of the particular pollutant. Environmental considerations may impose an upper limit on each pollutant of the system:

$$p \leq \hat{p} \qquad (5)$$

Careful environmental studies have to be conducted in order to determine the values of $\bar{\alpha}$ and $\hat{p}$ for each pollutant in a particular region. Pollutant movement (one particular aspect being the transboundary pollution) and decay or assimilation speed have to be taken into consideration in these studies.

2.3 Statement of the Environomic Optimization Problem

If there were no possibility of abating pollutants at any stage, the objective of system optimization would be to maximize its efficiency (subject to certain economic constraints), since increasing the efficiency and reducing the pollution are parallel objectives. However, reducing the pollutants not only by increasing the efficiency but also by changing other design and operation characteristics of the system (process abatement) or by removing those (pre-process and post-process abatement), requires expenditure in both capital (equipment) and resources (fuel, electricity, etc.). Thus, the objective becomes more involved.

An example of objective function is the total installation and operation cost of the system, which contains the following terms: capital cost of the system (with explicit representation of the pollution abatement equipment), cost of resources consumed, environmental and social cost due to pollution, benefit to the system (e.g., revenue) from products and services it provides to the environment. All costs (including capital) and benefits may be measured either in monetary or physical units (e.g., exergy).

In addition to the objective function, equality and inequality constraints are taken into consideration, which may be imposed by the operability of the system, safety considerations, state regulations, etc. Among the inequality constraints, an upper limit imposed on p_{rm} may be included, Eq. (5). If the operating conditions of the system change with time, the objective function is evaluated by an integration over the time period of the analysis.

2.4 Solution of the Optimization Problem

In simple cases, the environomic optimization problem can be solved by direct use of a nonlinear programming algorithm, in the same way as it has been done for the thermoeconomic optimization of a simple system [9]. It is relatively easy to use such an approach, however it does not offer any information regarding the internal economy of the system. Furthermore, if the system is complex, it may fail to determine the optimum point.

If a better understanding of the system and its internal economy is required and/or if the direct approach fail to solve the problem, then the Intelligent Functional Approach can be used, the general formulation of which has been presented in earlier work [7]. It uses the method of Lagrange multipliers. Each multiplier is an economic indicator (marginal cost).

Lagrange multipliers associated with pollution penalties are of particular importance for the environomic optimization, e.g.

$$\lambda_{rm} = \frac{\partial \Gamma_{rm}}{\partial p_{rm}} \tag{6}$$

where Γ_{rm} is the penalty imposed because of the pollutant p_{rm}.

The optimum value of the degree of abatement, δ^*_{rm}, is determined by the equation

$$\frac{\partial L}{\partial \delta_{rm}} = 0 \tag{7}$$

where L is the Lagrangian. Increasing values of λ_{rm} move the optimum point toward a higher abatement efficiency. Γ_{rm} is a function of p_{rm}, which can be formulated by a governmental agency or the society so, that the resulting values of λ_{rm} lead to favorable values of δ^*_{rm}. In formulating Γ_{rm}, expenses for neutralizing the pollution and restoring the environment in its previous state may be taken into consideration (depollution costs). In such a case, λ_{rm} may be called the *marginal cost of depollution*. Also, if an environment of appropriate standards is the social objective, Γ_{rm} may be given such a form that the operation of the system complies with the standards.

3. OPTIMIZATION OF A COMBINED-CYCLE PLANT

3.1 Description of the Plant and Main Assumptions

A double-pressure Joule-Rankine combined-cycle plant with no supplementary firing has been studied in an earlier work [10], where the thermoeconomic optimum has been compared with two thermodynamic optima: maximum of the efficiency and maximum of the net power density of the cycle. The same plant will be studied here and the environomic optimum will be compared with the thermodynamic and thermoeconomic optima.

In this particular example, two pollutants are considered (SO_2 and NO_x), although the method does not place any restriction. A flue gas desulphurization (FGD) unit for SO_2 abatement and steam injection in the gas-turbine combustion chamber to reduce NO_x formation are envisaged. These two facilities are not included in the plant, when thermodynamic or thermoeconomic optimization is performed.

The size (and consequently the cost) of the FGD unit depends largly on the flue gas flow. For this reason, it is less expensive to desulphurize a partial flow at the maximum technically possible degree, $\delta_{S,max}$, than the total flow at a lower degree [11, 12]. Therefore, a by-pass of the FGD unit is installed.

Undue complication is avoided by the following assumptions: (i) the system operates at steady state with total power output $\dot{W}$, which is known; (ii) thermal and mechanical losses are considered negligible; (iii) power consumed by the auxiliary equipment is not taken into consideration, although it may be significant; (iv) air and exhaust gases are considered ideal gases of constant specific heat capacity; when steam is injected, appropriate average values are determined for the properties of the mixture.

3.2 Objective Functions

Under pure thermodynamic considerations the following objective functions are examined: (a) maximization of the cycle thermal efficiency;   (b) minimization of the fuel consumption; (c) maximization of the net power density; (d) minimization of the total irreversibility rate of the system; (e) maximization of the second-law efficiency of the system.

It can be proved that, under the assumptions stated in Section 3.1, the five objectives (a)...(e) correspond to only two different optimization problems: (a) and (c) [10]. The corresponding objective functions are written :

$$\max \eta = \frac{\dot{W}}{\dot{H}_f} = \frac{\dot{W}}{\dot{m}_f H_u} \tag{8}$$

and

$$\max w = \frac{\dot{W}}{\dot{m}_g c_{pg} T_1} \tag{9}$$

Analytic expressions for η and w in terms of design and operation characteristics of the system are given in Ref. 10. Three of those characteristics are selected as the independent variables of each thermodynamic optimization problem

$$x_{th} = (r_C, P_{s2}, T_{s12}) \tag{10}$$

for the reasons explained in Ref. 10.

Among several thermoeconomic objectives, the annual cost of owning and operating the system is selected as the objective function:

$$\min Z_{te} = c_c \sum_r Z_r + c_f \dot{m}_f H_u t \tag{11}$$

Analytic expressions for Z_r are given in Ref. 10. The capital cost factor c_c is introduced in order to facilitate the sensitivity analysis: its nominal value is equal to one. For convenient comparison between the results of the thermodynamic and thermoeconomic optimization, we keep the same independent variables:

$$x_{te} = (r_C, P_{s2}, T_{s12}) \tag{12}$$

The minimization of the annual cost of owning and operating the system is selected as the environomic objective also, which now includes additional terms:

$$\min Z_{env} = Z_{te} + Z_{FGD} + \Gamma_{el} + \Gamma_{w,S} + \Gamma_{ls} + \Gamma_{w,N} + \Gamma_S + \Gamma_N \tag{13}$$

where Z_{FGD} is the annualized capital cost of the FGD equipment including operation and maintenance, Γ_{el}, $\Gamma_{w,S}$ and Γ_{ls} are the annual costs for electricity, water and limestone used for desulphurization, $\Gamma_{w,N}$ is the annual cost of water for steam injection, Γ_S and Γ_N are the penalties imposed on the system for emitted SO_2 and NO_x (environmental cost). The degrees of SO_2 and NO_x abatement are included in the independent variables:

$$x_{env} = (r_C, P_{s2}, T_{s12}, \delta_S, \delta_N) \tag{14}$$

Analytic expressions for Z_{FGD}, Γ_{el}, $\Gamma_{w,S}$, Γ_{ls} and $\Gamma_{w,N}$ are presented in Ref. 4, but are not included here because of space limitations. A linear relationship is assumed between a measure of pollution and the annual penalty:

$$\Gamma_S = c_S \dot{p}_S t \quad , \qquad \Gamma_N = c_N \dot{p}_N t \qquad (15)$$

The mass flow rates of SO_2 and NO_x emissions (after abatement) are used as measures of pollution. Before abatement, the mass flow rate of SO_2 and NO_x are given by the equations

$$\dot{m}_{Si} = 2s\dot{m}_f \qquad\qquad \dot{m}_{Ni} = \kappa_N \dot{m}_f H_u \qquad (16)$$

where κ_N is a constant. A model developed by Toughton [13] has been used in a simplified form to correlate the degree of NO_x abatement and the steam injected

$$\delta_N = \delta_N (r_{sf}) \qquad (17)$$

where r_{sf} is the injected steam to fuel ratio: $r_{sf} = \dot{m}_{st,N}/\dot{m}_f$.

3.3 Optimization Results and Comments

A nominal set of parameter values has been selected (Table 1) and the four optimization problems have been solved. The optimum points are given in Table 2. Figures 2-7 present the results of sensitivity analysis with respect to some important parameters.

Table 1. Nominal set of parameter values.

W	= 60 MW	P_1	= 1 bar	P_{s1}	= 100 bar
H_u	= 42500 kJ/kg	T_1	= 293 K	T_{s1}	= 813 K
τ_3	= 5.0				
δ_{Smax}	= 0.95	s	= 0.025 kg S/kg	κ_N	= $0.3868 \cdot 10^{-6}$ kg NO_x/kJ
c_c	= 1.0	c_S	= 1.7 \$/kg SO_2	c_N	= 2.0 \$/kg NO_2
c_f	= $4 \cdot 10^{-6}$ \$/kJ				

Table 2. Optimum values of variables for the parameter values of Table 1.

Variable	Objective			
	max η	max w	min Z_{te}	min Z_{env}
$r*_C$	11.30	4.60	8.92	9.40
$P*_{s2}$	10.45	16.64	20.50	19.14
$T*_{s12}$	584.11	813.00	540.26	556.64
$\delta*_S$	-	-	-	0.95
$\delta*_N$	-	-	-	0.7164
η	0.4942	0.4611	0.4912	0.4646

The effect of the temperature ratio τ_3 on the optimum values of the objective functions is depicted in Fig. 2. The optimum efficiency and net power density increase continuously with τ_3. The optimum total annual cost with or without environmental considerations have their minimum value at $\tau_3 \approx 5$. Consequently, the thermoeconomic and environomic optimum values of τ_3 are revealed by Fig. 2.

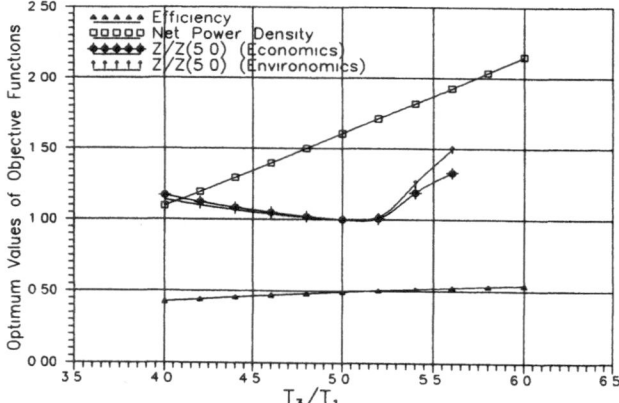

Fig. 2. Variation of the optimum values of the objective functions with the temperature ratio T_3/T_1.

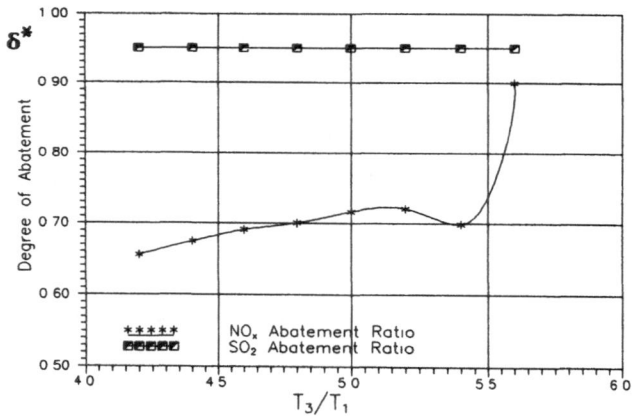

Fig. 3. Optimum degrees of SO_2 and NO_X abatement as functions of the temperature ratio T_3/T_1.

The temperature ratio τ_3 has no effect on the SO_2 degree of abatement, but a strong one on the NO_X degree of abatement (Fig. 3), which is explained as follows. Higher values of the combustor exit temperature T_3 result in higher NO_X formation. In order to keep the penalty Γ_N on NO_X emissions low, the degree of abatement has to increase. For $\tau_3 > 5.2$, the capital cost of the gas turbine increases abruptly. As a consequence, more and more power is produced by the steam turbine, the thermal efficiency of the whole system decreases and the fuel consumption increases. The steep rise of the NO_X degree of abatement is due to the combined effect of high NO_X formation (because of high temperature T_3) and high fuel consumption.

In this example, the Lagrangian is linear with respect to δ_S. Therefore, δ^*_S can not be determined by Eq. (7), but by the following procedure [3, 4]: A critical value c^o_S is obtained if Eq. (7) is solved for the unit penalty c_S. The optimum value δ^*_S is obtained by the following rule:

If $c_S > c^o_S$, then $\delta^*_S = \delta_{S,max}$.

If $c_S < c^o_S$, then $\delta^*_S = 0$.

The critical values c^o_S are given in Table 3. Since the selected value for the SO_2 penalty is higher than the critical one (Table 1), the optimum degreee of SO_2 abatement, δ^*_S, is always equal to the maximum possible (Fig. 3).

Table 3. Critical values of c_S for several values of τ_3.

τ_3	4.2	4.4	4.6	4.8	5.0	5.2	5.4	5.6
c^o_S ($/kg SO_2)	0.3308	0.3095	0.2951	0.2898	0.3044	0.3530	0.4769	0.9496

The optimum degree of NO_x abatement, δ^*_N, takes a steep rise for $\tau_3 > 5.4$, because the total efficiency of the plant decreases and consequently the fuel consumption and NO_x emissions increase. The decrease in efficiency is due to the fact that for high values of τ_3 the gas-turbine unit becomes prohibitively expensive and the system degenerates to a steam-cycle one.

The effect of the SO_2 and NO_x unit penalties on the optimum values of the decision variables r_C, and P_{s2} is shown in Figures 4-5. For $1 \leq c_N \leq 5$, a value of c_S higher than c^o_S, results in a drop of the optimum compressor pressure ratio and a rise of the optimum low pressure of steam, for the following reason. With no SO_2 abatement, the optimum values of r_C and P_{s2} are the result of a balance between capital expenses, (which increase with increasing r_C), and fuel expenses, (which decrease with increasing r_C). If $c_S > c^o_S$, the FGD unit is required, the capital and operation cost of which increases with the flow rate of exhaust gases, including the injected steam. Consequently, the optimal solution is driven towards a system with lower flow rate of exhaust gases, i.e. higher net power density, which (as it is shown in Table 2) has a lower compressor pressure ratio and a higher pressure for the low-pressure steam.

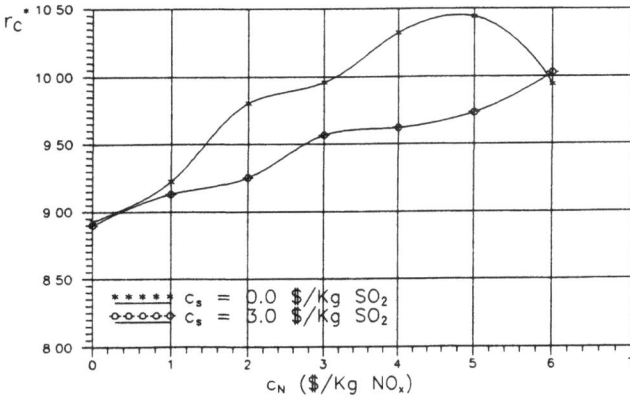

Fig. 4. Optimum compressor pressure ratio as a
function of the unit penalties for SO_2 and NO_x emissions.

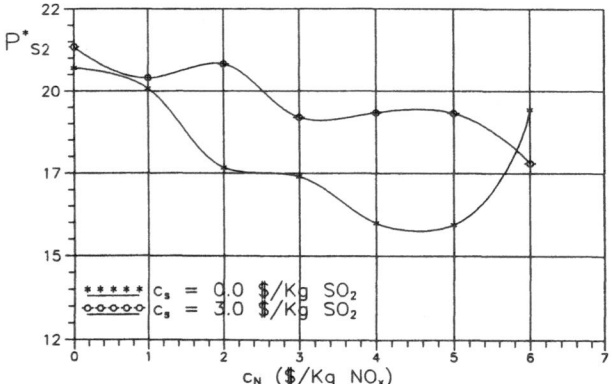

*Fig. 5. Optimum low pressure of steam as a
function of the unit penalties for SO_2 and NO_x emissions.*

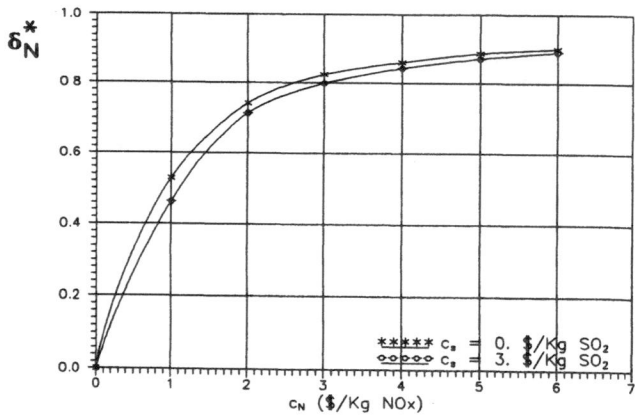

*Fig. 6. Optimum degree of NO_x abatement as a
function of the unit penalties for SO_2 and NO_x emissions.*

The optimum NO_x degree of abatement increases continuously with increasing values of the unit penalty (Fig. 6). It reaches the value of 80% for $c_N = 3$ \$/kg NO_x, while for a further increase by 10% it is necessary to double the value of c_N.

The fuel price has no effect on the optimum SO_2 degree of abatement, as it is shown in Fig. 7, because in this example c_S is higher than the critical value and δ^*_S is always equal to the maximum possible value. However, increasing fuel prices result in decreasing values for the optimum NO_x degree of abatement. At a first glance, the result may seem strange, but it is explained as follows.

Increasing fuel prices drive the optimum point towards higher values of efficiency. Since steam injection, which is used here for NO_x abatement, has an adverse effect on efficiency, the quantity of injected steam is reduced, when the fuel price increases. Consequently, the optimum NO_x degree of

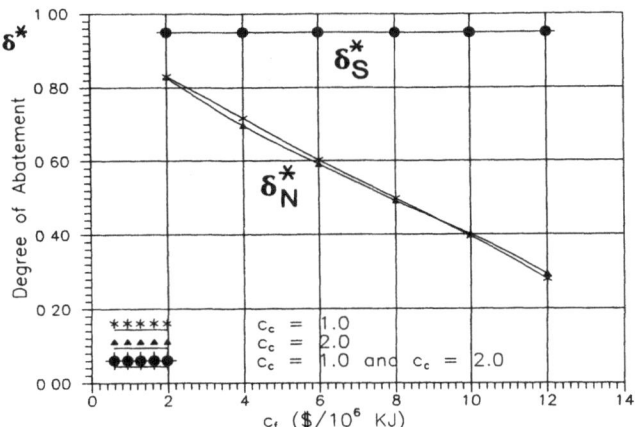

*Fig. 7. Optimum degrees of SO₂ and NOₓ abatement as a
function of fuel price and capital cost.*

abatement is reduced. Instead of steam injection, other techniques (e.g. dry low NO_x combustion) might be more favorable from this point of view.

CLOSURE

The design optimization problem of a combined-cycle plant has been solved by combining thermodynamic, economic and environmental considerations. Although the numerical results may not be generally applicable, because they are strongly dependent on the cost functions and coefficients, the procedure is of general applicability.

In addition to the effect of certain parameters on the optimal design of the system, the analysis reveals a critical value for the penalty of each emission of the plant. A penalty higher than the critical one forces the system design towards higher values of the degree of abatement. Thus, the procedure is of significant help not only to the designer in setting the design specifications of the system, but also to the society in setting environmental standards and implementing a policy to achieve those.

In order for the environmental aspects to be taken into consideration quantitatively, the environmental and social cost of pollution is needed, which can be obtained by a study at a regional, national or inter-national level.

REFERENCES

1. Tribus, M. and Evans, R.B., "A Contribution to the Theory of Thermoeconomics," UCLA Report No. 62-36, University of California at Los Angeles, Los Angeles, Ca., 1962.

2. El-Sayed, Y.M. and Evans, R.B., "Thermoeconomics and the Design of Heat Systems," *J. of Engineering for Power,* Vol. 92, No. 1, 1970, pp. 27-35.

3. Frangopoulos, C.A., "Introduction to Environomics," in *Second Law Analysis - Industrial and Environmental Applications,* G.M. Reistad et al., eds., AES-Vol. 25/HTD-Vol. 191, ASME, New York, 1991, pp. 49-54.

4. Frangopoulos, C.A., "An Introduction to Environomic Analysis and Optimization of Energy-Intensive Systems," International Symposium on Efficiency, Costs, Optimization and Simulation of Energy Systems, ECOS '92, Zaragoza, Spain, June 15-18, 1992.

5. Frangopoulos, C.A., "Thermoeconomic Functional Analysis: A Method for Optimal Design or Improvement of Complex Thermal Systems," Ph.D. Thesis, Georgia Institute of Technology, Atlanta, Ga., 1983.

6.   Frangopoulos, C.A., "Thermoeconomic Functional Analysis and Optimization," *Energy*, Vol. 12, No. 7, 1987, pp. 563-571.

7.   Frangopoulos, C. A., "Intelligent Functional Approach: A Method for Analysis and Optimal Synthesis-Design-Operation of Complex Systems," in: *A Future for Energy*, S.S. Stecco and M.J. Moran, eds., Pergamon Press, Oxford, 1990, pp. 805-815.

8.   Gaivao, A. and Jaumotte, A.L., "Evaluation Économique de la Pollution de l'Environment par une Activité Industrielle. Application aux Centrales Électriques," *Entropie*, No. 121, 1985, pp. 5-11.

9.   Frangopoulos, C. A., "Optimal Design of a Gas Turbine Plant by a Thermoeconomic Approach," *2nd International Symposium on Turbomachinery, Combined-Cycle Technologies and Cogeneration*, G.K. Serovy and T.H. Fransson, eds., Montreux, Switzerland, 1988, pp. 369-375.

10. Frangopoulos, C.A., "Comparison of Thermoeconomic and Thermodynamic Optimal Designs of a Combined-Cycle Plant", *International Conference on the Analysis of Thermal and Energy Systems*, D. A. Kouremenos, G. Tsatsaronis and C.D. Rakopoulos, eds., Athens, Greece, 1991, pp. 305-318.

11. Rentz, O., *"Techno-Ökonomie Betrieblicher Emissionsminderungsmaßnahmen"*, Berlin, 1979.

12. Welsh, H., "A Cost Comparison of Alternative Policies for Sulphur Dioxide Control", *Energy Economics*, October 1988, pp. 287-297.

13. Touchton, G.L., "An Experimentally Verified NO_x Prediction Algorithm Incorporating the Effects of Steam Injection," J. of Engineering for Gas Turbines and Power, Vol. 106, October 1984, pp. 833-840.

SESSION 3:

Drying

Chairman: Prof. R.W.K. Allen

MODELLING RADIATIVE HEAT TRANSFER
IN INDUSTRIAL ENCLOSURES

J. ADNOT, W. CAI
Ecole des Mines de Paris
60, Bd. Saint Michel
75272 - Paris Cedex 06

P. LIEVOUX, P. HENNIG
Gaz de France/CERUG
361, av. du Pdt Wilson
BP 33
93211 - La Plaine St Denis

B. TUCKER
British Gas
Wharf Lane, Solihall
UK - B91 2JW
West Midlands

J. HAMEURY J.R.FILTZ
LNE
5, rue E. Fermi
78190 - Trappes

J.J. Ph. ELICH
TU Delft/Applied Physics
Lorentzwed 1
NL - 2628 CJ Delft

I. GUGLYURTLU
LNETI-DEC
Azinhaga dos lameiros a
estrada do Paco do Lumiar
P-1699 Lisboa Cedex

ABSTRACT

In the frame of the JOULE program, a group of six teams work on the RADHEAT project for the modelling and experimentation of the thermal radiative industrial enclosures used in the drying and heating processes. The teams (Ecole des Mines de Paris, LNE, Caz de France/CERUG, TU Delft, British Gas, LNETI-DEC) have gathered the whole modelling and experimental capacities and can really characterize the radiative behavior of the processes. Real cases representing common interest for the teams are also studied in common according to the projects of each institution.

INTRODUCTION

In the paper and the textile industries, where drying and surface treatment are important processes, it is becoming more and more advantageous to consider a change from a non radiative process with low efficiency (vapor, direct contact ...) to a radiative one (infrared, micro wave). However, the implementation requires careful examination of the processes based on numerical analysis and experimental data.

Radiative heat transfer in an enclosure depends on several parameters : geometry, directional and spectral radiative properties of the surfaces and media, temperature, etc ... There are several ways of modelling radiative heat transfer. A model should be chosen according to the radiative properties of materials and media and other conditions such as the temperature level. Thus, it seems very difficult, at the beginning, to have a general calculation method which can treat all possible cases.

Within the frame of the JOULE a project named RADHEAT gathered a group of six research laboratories work together at the European level to derive modelling guidelines for industrial enclosures, especially the radiant systems, and to propose high quality experimental services in order to back up eventually industry. They tried to associate their modelling capacities as well as experimentation ones, in order to model and characterize the real industrial enclosures in the drying or heating processes. At the first stage of the project, each team exploits its possibilities for the modelling as well as for the measurements. Some common cases are defined and should be tested by the teams. These common cases will serve as bench marks for the entire processes when the other transfer modes should also be accounted for.

Besides these 'academic cases', the teams regroup between them to study some real cases. One infrared gas emitter used in a paper drying process and radiant tube systems in heating and baking process are being studied.

An experimental cube enclosure simulating the gas burner configuration has been defined and is in construction. The radiative properties of the enclosure materials are to be measured. The irradiance distribution as well as the temperature field obtained by the experimentation will be compared with the modelling results carried out by the teams.

MODELLING METHODS AND EXPERIMENTAL APPARATUS

Thermal radiation contributes substantially to heat transfer in numerous industrial applications: :furnaces, combustion chamber and the radiant systems. Thermal radiation, different from the conduction and convection, is an action at distance by its electromagnetic nature. The spectral and spatial variations of the radiative properties involve the resolution of the integral-differential equations for the modelling. Only few analytical solutions can be obtained limited to some academic configurations. Most of radiative problems are treated by numerical methods which are often specific and adapted according to the relative importance of phenomena.

No method can be at the same time simple, accurate and general. Having in mind the industrial applications, the main types of radiative problems can be defined, in order of increasing difficulty : surface exchanges in non participating media, surfaces in absorbing media, surfaces in absorbing and scattering media. So it is desirable to have general models for each of these classes in order to answer the industrial needs.

For the real industrial applications, the radiative property variations should be taken into account in order to characterize the radiative behavior and treatment quality (irradiance distribution, efficiency, etc.) which can be obtained also by measurements.

THE MODELLING METHODS

The methods used by the teams are principally the ray-tracing methods and zone methods. The Monte Carlo method and the discrete transfer method are also used. For non-grey gas treatment, the models of weighted-sum-of-grey-gases and wide band models are used.

General Numerical Structure by Ray-tracing Method

The development of a general numerical radiative modelling framework, taken in charge principally by the team EMP, is based on the following philosophy : (a) The modelling procedure is sufficiently general to cover a wide range of industrial applications; (b) The radiation coexists almost always with other thermal phenomena like conduction and convection, so it should be easy for the procedure to be incorporated in the combined system ; (c) The radiative modelling should not be mathematically so sophisticated as to obscure the physical background of the whole system.

With this modelling philosophy, one ray-tracing method is developed for the thermal radiative modelling. It consists of simulating the thermal radiation by following the discrete rays in a deterministic way (figure 1). The whole trajectory information obtained by ray-tracing constitutes a numerically interpretable data base which characterizes the geometrical and directional behavior of thermal radiation in the enclosure. Different kinds of useful modelling information can then be deduced (exchange factor, transfer factor, net exchange fluxes, etc.).

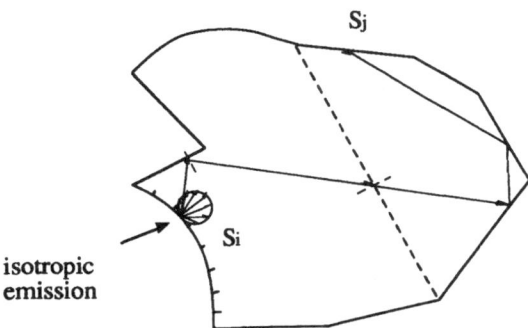

FIGURE 1 - Ray tracing procedure in an enclosure

The numerical frame developed is suited for a wide range of configurations. The surfaces can be diffuse, specular or semi-transparent with spectral variation, the media can be transparent, absorbing or isotropically scattering. The geometries can be bi-dimensional as well as tri-dimensional.

The numerical frame is organized in five independent modules (figure 2). The first module is the enclosure description which describes the radiative system to be studied. Both the physical conditions and geometric parameters are described. Geometric modelers are used for the geometric meshing if the geometry is complex. The second module is the most important part in the software. It simulates the thermal radiation in the enclosure by a ray-tracing method. By a complete ray-tracing procedure, we can obtain an interpretable data base of the radiative exchange behavior in the enclosure. The theoretical details of radiative model can be found in [5]. The third module can translate the ray-tracing data base in different ways. The first useful results are the exchange factors between the surface and volume elements of the enclosure which can take into account the specular reflection and the transparency of surfaces.

The other derived useful results are equivalent radiant plans construction which are useful for the study of angular irradiance distribution. The irradiance distribution can be also obtained directly from the data base. The fourth module is for radiative heat transfer calculation. The transfer factors can be calculated by taking into account the diffuse aspect of surface radiative properties and the isotropic scattering of the media. For the non grey media, the models of weighted-sum-of-grey-gases is used.

The radiative fluxes can then be easily calculated if the temperature field of the enclosure is known. In the case of the entire system study which need to consider other heat transfer modes like conduction and convection, we establish the interface with other engineering softwares and study the whole system in a iterative way if necessary.

> **module I. enclosure description**
> - physical decription
> - geometric description
>
> **module II. ray-tracing procedure**
>
> **module III. information treatment**
> - exchange factor calculation
> - equivalent radiant plan construction
> - irradiance distribution calculation
>
> **module IV. radiative heat transfer calculation**
> - transfer factor calculation
> - radiative flux calculation
>
> **module V. thermal coupling study by solver**

FIGURE 2 - Organizational structure of the general numerical frame

The ray-tracing method converges to accurate solution if the ray number is sufficiently great since it simulates the physical nature of the phenomenon.

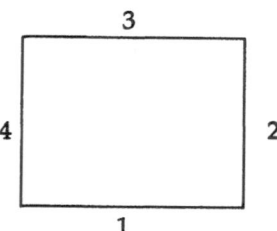

FIGURE 3 - a square with S1=25, S4=20.

If we take an example of a square (figure 3). The exchange factors are calculated between the surfaces. We can see from figure 4 that by increasing the ray number, the result approaches to analytical one ($F_{14,ana.}$ = 0.2597). That means that the ray-tracing method converges to analytical solution.

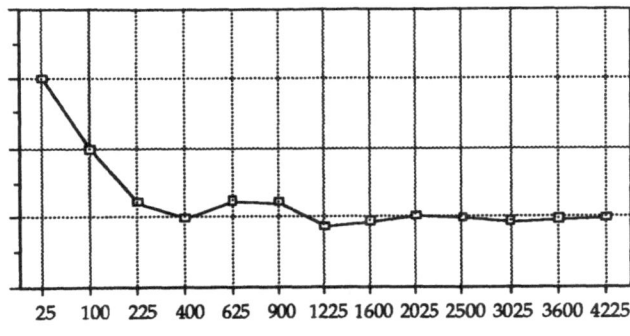

FIGURE 4 - The ray-tracing result of exchange factor F_{14}

Zone Method and Non Grey Gas Treatments

The zone method, which is used and adapted by the teams T.U. Delft and British Gas, is based on a discretization of the enclosure space and the boundary surface into a number of elements of uniform properties. Radiative exchange can then be described in the form of a linear matrix equation. The method, proposed by Hottel [13 [14] is relatively old and well-known. However, new developments still take place [17] [16] [15]. In general, the zone method is an accurate method provided a sufficient number of zones is used, but especially the calculation of exchange factors can be quite lengthy. The zone method is known for its accuracy, but it is extremely cumbersome to handle with it a location-dependent absorption coefficient. If this is needed, another method has to be chosen, for instance the Monte Carlo method or the Discrete Transfer Method.

At TU Delft, attention has been paid to the application of the zone method in situations where a spectral approach is needed. A sketch of a numerical radiative transfer model is given in [23], closure relations have been also discussed. The coupling study of zone method with 3D flow computation is also exposed in [18]. When the zone method is combined with complete 3D simulation of turbulent flow, one volume of zone method can consist of a number of grid cells of flow calculation. these may be located entirely or partly inside a zone. As a consequence, emissive powers of the cells have to be averaged to obtain the emissive power of the volume. Likewise, the radiative source term of the zone method calculation has to be distributed over a number of cells.

British Gas has many years experience in the use of the Hottel zone method for radiation transfer calculation and has combined this with a Monte Carlo technique to calculate the exchange areas required to simulate realistic geometries[19] [20]. British Gas is involved in the practical design and monitoring of radiant drying, space heating and heat treatment processes.

For the non-grey gas, a wide band model is used [3] at TU Delft. British Gas has been using a simple model of weighted-sum-of-grey-gases[22].

THE MEASUREMENT CAPACITIES

For the radiative measurements, the workers at LNE calibrate their infrared radiometers and pyrometers as well as infrared cameras. The radiative properties of the materials like emissivity, reflectance, transmittance can be measured. For the characterization of infrared emitters, two apparatus are developed. One is the spectral hemispheric irradiance measurement apparatus (figure 5) , the other is directional total irradiance measurement apparatus (figure 6). The technical details of these measurement apparatus are described in [12].

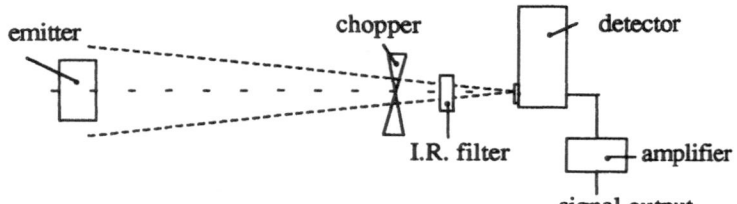

FIGURE 5 - Spectral irradiance measurement apparatus

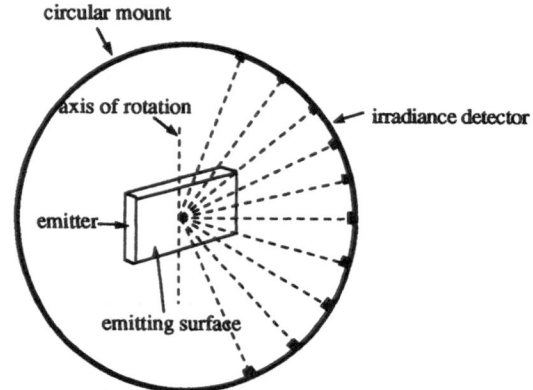

FIGURE 6 - Spatial irradiance measurement apparatus

At TU Delft, measurement facilities of emissivity and reflectivity are also available for certain materials. The schematic diagram of a specular reflectance unit is shown in figure 7. The apparatus and its performance are well described in [11]. The temperature measurement can be carried out by LNE, British Gas, Gaz de France as well as LNETI[1] [2].

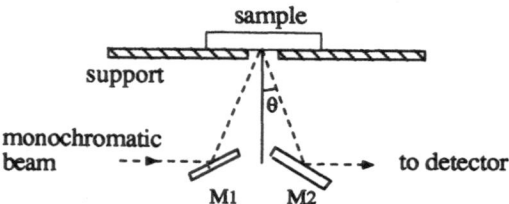

FIGURE 7 - Schematic diagram of a specular reflectance unit

CASE STUDIES

With the assembled modelling and measurement capacities, some industrial cases in the drying or heating processes are studied. Experimental simplified devices representing the characteristics of burners or dryers are being constructed and will be modelled and studied by all the teams.

CASE 1 - AN INFRARED CERAMIC GAS EMITTER USED IN PAPER DRYING PROCESS

One ceramic burner used in a paper drying process is studied (figure 8) by EMP, GdF. In this system, the premixed gaz enters in the canals and heated by the hot ceramic plaquette. The combustion takes place in the alveolus. The heated surfaces of the plaquette and the bars constitute a radiant screen and heat the paper placed in front.

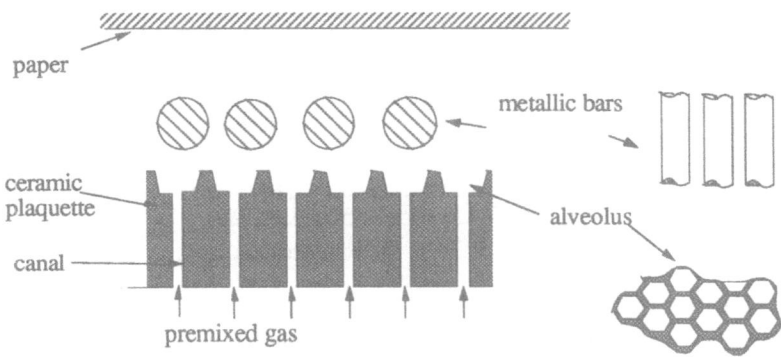

FIGURE 8 - Description of the burner

The thermal radiation plays an important role in this process. The combustion gas participates in the radiative transfer. The system is modeled in a three-dimensional enclosure with a participating non grey gas. The ray-tracing method is used to calculate the exchange factors between the elements in the enclosure. The radiative flux is obtained by a flux-temperature model. The gas is treated by using a model of weighted-sum-of-grey-gases [6].

CASE 2 - A RADIANT TUBE SYSTEM USED IN FURNACES

A typical case of radiant tube system is defined by British Gas. The modelling work that will be carried out by British Gas(figure 9). TU Delft and EMP are going to simulate radiant flux distribution under radiant tube. The influence of furnace design will be studied. And the results will be compared with common rules of thumb.

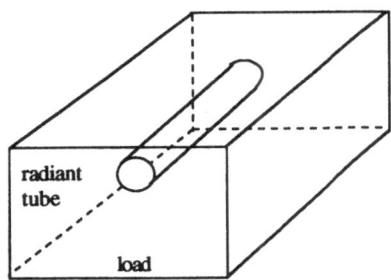

FIGURE 9 - A radiant tube case

CASE 3 - A RADIANT TUBE FURNACE FOR THE BAKING OF ALIMENTARY PRODUCTS

The radiant tube furnaces studied principally by LNETI are of the long furnace type in which hot gases travel through pipes that are long compare to their diameter (figure 10). The gas temperature is not uniform along the direction of the flow. the load is heated almost entirely by radiation from radiant tubes above or below the load. The system will be modelled and compared with the experimentation results in order to define : (a) the influence of load emissivity; (b) the optimum configuration for the tube; (c) the limitation of the heat input to assure the treatment quality.

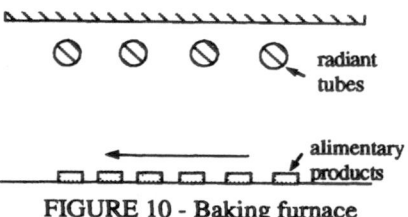

FIGURE 10 - Baking furnace

CASE 4 - EXPERIMENTAL RADIANT CUBES

The experimental cubes simulating metallic fiber burner in drying applications are being constructed at Gaz de France (figure 11). The top wall of the cube simulates the radiant plate using a metallic fiber material, the bottom wall simulates the charge using a painted coat material. The other walls are isolated and they can be either diffuse or specular. The enclosure is filled with a non homogeneous combustion gas.

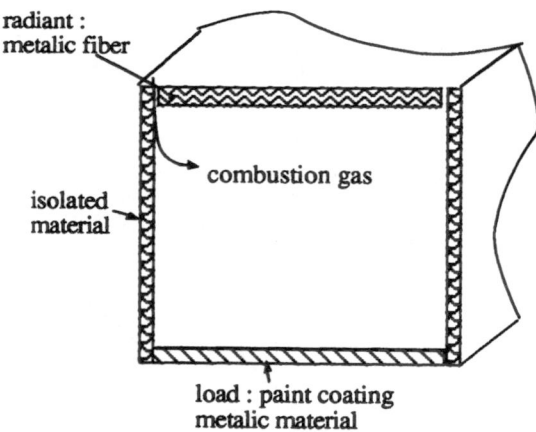

FIGURE 11 - A experimental cube

The radiative properties (emissivity, reflectivity) of the wall materials are to be measured. the temperatures distributions will be also measured. The enclosure will be modelled by EMP, TU Delft on taking into account the spectral radiation property variation of the wall material and gas absorption. The results obtained by simulation (irradiance, heat flux) will be compared with those of experimentation.

The experimental radiant cubes can characterize the radiation behavior of the industrial enclosures. By performing the sensitivity studies, we can validate the modelling method and well understand the functionality of the radiant systems in order to foresee the amelioration in the actual processes.

CONCLUSIONS

The working frame for the modelling of the industrial radiative enclosures has been described. With the modelling and experimental capacities developed for the drying or heating processes, a certain number of applications are described and studied. The case studies allow us to develop the relevant knowledge and models necessary for the radiative process routes in the field of drying and heat treatment. The acquired knowledge an then help us to prepare the characterization and eventual normalization of performance of new types of radiant systems.

ACKNOWLEDGMENTS

Support from CCE under program JOULE 1 is acknowledged as well as support from ADEME, GdF, Solaronics Co., CTP, Novem and the presenting institutions.

REFERENCES

[1]   J. Adnot, Overall experimental capacity of the RADHEAT group for radiant heating and drying, *RADHEAT report CWD4*, 1992.

[2]   J. Adnot, Overall modelling capacity of the RADHEAT group for radiant heating and drying, *RADHEAT report CWD3*, 1992.

[3]   L. Alexander, e.a. Surface emissivities of furnace linings and their effect on heat transfer in an enclosure, *Proc. First Eur. Conf. on Industrial Furnaces and Boilers (Lisbon)* 1991.

[4]   W. Cai, Developpement et applications de modèles d'échanges radiatifs par suivi de rayons, *PhD thesis*, Ecole des Mines de Paris, 1992.

[5]   W. Cai, Radiative exchange model by ray-tracing. *RADHEAT technical report EMP01*, 1992.

[6]   W. Cai, Radiative exchange modelling of a gas burner element. *RADHEAT technical report EMP04*, 1992.

[7]   W. Cai, INFRAYS - a radiative transfer modelling software, *RADHEAT technical report EMP02*, 1992.

[8]   W. Cai, Geometric interface INFRAYS/PATRAN, *RADHEAT technical report EMP05*, 1992.

[9]   W. Cai, Numerical limitation of ray-tracing method, *RADHEAT technical report EMP03*, 1992.

[10]   J. Elich, Survey of the possibilities in Delft to measure radiative properties of materials, *RADHEAT technical report TUD02*, 1992.

[11]   H. Haitjema, Spectrally selective tinoxide and indiumoxide coatings. *PhD thesis*, Delft University of Technology , Netherlands, 1989.

[12]   J Hameury and J.R. Filtz, Measurements apparatus in radiative industrial enclosures. *RADHEAT technical report LNE01*, 1992.

[13]   H.C. Hottel and E.S. Cohen, radiant heat exchange in a gas filled enclosure : allowance for nonuniformity of gas temperature, *AIChE Journal*, Vol. 4, no. 1, pp. 3-14, 1958.

[14]   H.C. Hottel and A.F. Sarofim, Gaseous radiation with temperature gradients, allowance for isotropic scatter, *Fundamental Research in heat Transfer* (J.A. Clark, ed.), pp. 139 - 159, Pergamon Press, 1963.

[15]   M.E. Larsen and J.R. Howell, The exchange factor method : an alternative basis for zonal analysis of radiating enclosures, *J. of heat Transfer*, Vol. 110, pp. 456-462, 1985.

[16]   M.H.N. Naraghi, B.T.F. Chung and B. Litkouhi, A continuous exchange factor method for radiative exchange in enclosures with participating media, *J. of Heat Transfer*, Vol. 110, pp. 456-462, 1988.

[17]   J.J. Noble, The zone method : explicit matrix relations for total exchange areas, *Int. heat Mass Transfer*, Vol. 18, pp. 261-269, 1975.

[18]   L. Post, Modelling of flow and combustion in a glass melting furnace. *PhD thesis*, Delft University of Technology, Delft, Netherlands, 1988.

[19]   R.J. Tucker, Direct exchange areas for calculating radiation heat transfer in rectangular furnaces, *ASME J. of Heat Transfer, 1986.*

[20]   ]R.J. Tucker and J. Ward, Use of a Monte Carlo technique for the determination of radiation exchange areas in long furnace models, *Proc. 8th. In. Heat Transfer Conf.*, 1986.

[21]   J.M. Rhine, R.J. Tucker, Modelling of gas fired furnaces and boilers, *British Gas Technical Monograph, McGraw-Hill,* 1990.

[22]   Tucker, Amelioration of weighted-sum-of-grey-gases model in the zone method, *RADHEAT technical report BG01*, 1992.

[23]   J.A. Wieringa, J.J. Ph. Elich, The application of the zone method in non-grey situations with participating gas, *RADHEAT technical report TUD01*, 1992.

DEVELOPMENT AND EVALUATION OF AN ADVANCED MODEL FOR SPRAY DRYERS

P Hutchinson*,J Pereira,[†]M Sommerfeld [‡]

* School of Mechanical Engineering, Cranfield Institute of Technology, Cranfield, Beds MK43 0AL England.

[†] Instituto Superior Technico, Av, Rovisco Pais, 1096 Lisboa Codex, Portugal.

[‡] Lehrstuhl fur Stromungsmechanik, Technische Fakultat, Freidrich-Alexander-Universitat, Erlangen-Nurnberg, Cauerstasse 4, D8520 Erlangen, Germany.

ABSTRACT

The object of the work described here is the development and evaluation of a computational fluid dynamics model which will assist the design of spray dryers.

The approach provides a new basis for the design of spray dryers which takes into account the detailed interaction between the drying gas and the product. As a result, the design can be optimised so as to reduce the energy consumption for a given through-put, while maintaining the product quality. In addition, the insights into the drying process provided by the model should allow retrofit modifications to existing dryers to improve their energy performance by adjusting the interaction between spray and gas.

INTRODUCTION

The research described here is supported by the Joule project of the CEC . The work comprises a sub-project within the advanced flow modelling group I, co-ordinated by the National Technical University of Athens and is a collaborative effort between Cranfield Institute of Technology, the Universitdt of Erlangen Nurnberg, Imperial College of Science and Technology, and the Instituto Superior Technico. Technical Co-ordination of this group is provided by Cranfield Institute of Technology.

Computational Fluid Dynamics has great potential for application in the process industries, but so far this has only been realised to a relatively limited extent. Many key issues in design for the process industries are related to the behaviour of fluids in turbulent flow often involving a second phase and reaction or heat transfer. The use of computational fluid dynamics methods to analyze these processes can be of great help to the designer by reducing the need to resort to "cut and try" approaches for the design of complex equipment. The object of the work described here is the development and evaluation of a computational fluid dynamics model which will assist the design of energy efficient spray dryers. There is a substantial scope for improvement by ensuring that the input heat of the drying gas is more efficiently transferred to the liquid droplets, but lack of detailed knowledge of the drying process presently obstructs the realisation of this greater energy efficiency.

The approach taken is well established in the CFD field and is to construct appropriate models for each of the important physical processes of a spray drier. In order to identify limits of accuracy and to evaluate, refine and validate the models, each is then tested against appropriate detailed data.

The physical processes involved in spray dryers may be classified into four stages, namely, atomization of the feed into a spray, spray air contacting (mixing and flow), drying of the spray (evaporation of moisture and volatiles) and the separation of the dried product from the air.

The work reported here is concerned with the first two stages identified above and is the result of the first six months of the Joule contract.

The physical processes which are modelled are turbulent flow of the gas and the processes of spray mixing. The theoretical approach is described below and which discusses the selection of turbulence model and modelling of the spray behaviour. This last area is particularly challenging and two different, but related, approaches are described. The first comprises the established Lagrangian simulation method which is based on tracking particle motion through the gas flow. The second approach is more novel and relies on an encapsulation of Lagrangian aspects of the particles behaviour into parameters which form a basis for an Eulerian representation of the particle dynamics. Next we give a brief account of numerical methods used in the CFD calculation, followed by a description of the experimental measurements and, finally, conclusions.

MATHEMATICAL MODEL

Single Phase Turbulent Flow

The continuity and momentum equations for steady, two-dimensional incompressible flow at high Reynolds number are:

$$\frac{\partial U_j}{\partial x_j} = 0,$$

(1)

$$\frac{\partial U_j U_i}{\partial x_j} = \frac{1}{\rho} \frac{\partial P}{\partial x_i} - \frac{\partial \langle u_i u_j \rangle}{\partial x_j}$$

(2)

Upper and lower cases relate to mean and turbulence quantities, respectively; $\langle \rangle$ denotes temporal average of the variable, P represents pressure.

To close the above equations both a second-moment model and the standard k - ε eddy viscosity (EVM) model were considered. For the Reynolds Stress Model (RSM), the transport equations for the components of Reynolds stress tensor can be written as:

$$\frac{\partial U_k \langle u_i u_j \rangle}{\partial x_k} = D_{ij} + P_{ij} + \phi_{ij} - \varepsilon_{ij}$$

(3)

where the terms on the RHS of Eq. (3) represent the diffusion, production, pressure redistribution and dissipation rate of $\langle u_i u_j \rangle$. Apart from the P_{ij} term all the remaining ones in the RHS of (3) need to be modelled. The diffusion term is represented by a generalised gradient diffusion model [1] (Daly and Harlow 1970): The pressure-redistribution term is modelled following [2] Rotta (1951) and [3] Naot et al (1970) "slow" and "rapid" models:

For the basic model, the dissipation rate of $\langle u_i u_j \rangle$ is represented by an isotropic tensor as $\varepsilon_{ij} = < 2/3\varepsilon\delta_{ij}$. The scalar dissipation rate is computed by solving a transport equation of the form

$$\frac{\partial U_j \varepsilon}{\partial x_j} = \frac{\partial}{\partial x_i} \left(C_\varepsilon \frac{k}{\varepsilon} \langle u_i u_k \rangle \frac{\partial \varepsilon}{\partial x_k} \right) + C_{\varepsilon 1} \frac{\varepsilon}{k} P_k - C_{\varepsilon_2} \frac{\varepsilon^2}{k}$$

(4)

where $P_k = P_{kk}/2$. The value of model constants are taken as for previous studies set out below in see table 1.

Together with the above model the standard form of the k - ε eddy-viscosity model was considered [4] (Launder and Spalding, 1974).

Table 1: Model Constants

C_1	C_2	C_3	C_ε	$C_{\varepsilon 1}$	$C_{\varepsilon 2}$
1.8	0.6	0.24	0.15	1.44	1.92

Spray Model

The earliest models for droplet motion in turbulent fluids fall into two extreme categories. The mean trajectory model omits the dispersive effects of gas turbulence and hence tends to over-predict spray penetration. At the other limit the no slip model assumes that the droplets follow the flow and turbulence and neglects slip between the droplet and gas.

Subsequently a stochastic approach was developed [5] (Gosman and Ioannides 1983) which simulates the interaction between droplets and turbulence by tracking particles through the flow field and simulating the effect of turbulence by creating random flow field fluctuations, consistent with the statistics of the flow.

While the stochastic model offers a clear way forward and can be elaborated to include additional physical effects in dense sprays such as collision and coalescence, it is, by its nature, time consuming in implementation and difficult to integrate with the flow calculations. In this paper we argue that the stochastic effects of the flow field fluctuations can be embodied in a diffusive model for particle motion which is an elaboration of earlier work by one of us [6] (Hutchinson, Hewitt and Dukler 1970). In this earlier model the motion of droplets suspended in an annular two-phase flow was treated as a diffusion process and gave a reasonable account of the variation of droplet mass transfer coefficient with the physical properties of the media and those of the flow. The essential assumption, which is valid in many stochastic systems, is that the random motion of the droplets may be well described as a diffusion process. Indeed after two or three random collisions, many processes are well described by a Gaussian distribution [7] (Papoulis 1965).

The earlier diffusion model for particle motion was restricted to describe only the concentration field, since the effects of inertia at injection were unimportant. However, this is clearly unacceptable for a description of spray formation in dryers. Thus, in this paper we extend the diffusion model to treat the distributions of both concentration and momenta of particles.

Once the model is established it is simply necessary to define a diffusion coefficient and appropriate boundary conditions in order to fully define the problem. The diffusion coefficient is obtained from a one dimensional simulation of the particle motion as described below. The boundary conditions are determined by the wall conditions appropriate to a diffusion process, eg concentration equals zero at the wall for an absorbing boundary and grad c = 0 for a fully reflecting boundary. A definition of the inlet mass flux completes the specification.

Mass Conservation

Consider the normal mass conservation equation:

$$\frac{\partial C(t)}{\partial t} + \frac{\partial}{\partial x_j} \left[(U_j(t) C(t)) \right] =$$

(5)

Reynold's averaging gives:

$$\frac{\partial C}{\partial t} + \frac{\partial}{\partial x_j} (U_j C + \langle U'_j C' \rangle) = 0$$

(6)

The diffusion approximation implies:

$$\langle U_j C' \rangle = - \Gamma \frac{\partial C}{\partial x_j}$$

(7)

which applied to equation 2 yields:

$$\frac{\partial C}{\partial t} + \frac{\partial U_j C}{\partial x_j} = \Gamma \frac{\partial}{\partial x_j} \frac{\partial C}{\partial x_j}$$

(8)

where C is the mean number density of particles, U_j is a mean velocity component, U_j is a fluctuating velocity component, and C a fluctuating component of concentration. t and x_j are the time and space co-ordinates respectively and Γ is the diffusion coefficient. Equation (13) is the well known description for the development of the concentration distribution in a flow in with both mean motion and diffusion.

Momentum Conservation

In order to develop the analagous equation for the distribution of mean velocity we consider the momentum conservation equation:

$$\frac{\partial C(t) U_i(t)}{\partial t} + \frac{\partial}{\partial x_j} [C(t) U_i(t) U_j(t)]$$

(9)

where F is the force per unit volume on the particles arising from drag and external sources.

Again, Reynold's averaging, repeated application of equation 7 above and the diffusion approximation gives:

$$\frac{C \partial U_i}{\partial t} + C \, U_j \, \frac{\partial U_i}{\partial x_j} - \Gamma \frac{\partial C}{\partial x_j} \, \frac{\partial U_i}{\partial x_j} = F$$

(10)

The mean force is set from the mean drag interaction between the particles and the fluid, but can also take account of other external force fields.

The momentum exchange term also provides a means for incorporating momentum transfer from the particle field into the gas flow. It is also noteworthy that setting the diffusion coefficient to zero recovers the familiar trajectory model.

Equations 8 and 10 provide a set of convection diffusion equations capable of describing the particle flow field in a form similar to that commonly used in computational fluid dynamics calculations for single-phase fluids. Hence it is reasonably straightforward to integrate this approach for description of the droplet distribution into such calculations.

The Diffusion Coefficient

The diffusion coefficient is computed by simulation of the random motion of a particle in one dimension by repeated solution of:

$$\frac{\partial V_{p_i}}{\partial t} = \frac{C_D \Gamma_p^2}{2 M_p} \rho_g \, |V_G - V_{p_i}| \, (V_G - V_{p_i})$$

(11)

where V_{pi} is the particle velocity at interaction i, V_{Gi} is the gas velocity, C_D is the drag coefficient r_p is the droplet mass, P_G is the gas density, and M_p is the droplet mass.

V_G is chosen as the product of the velocity scale for the turbulence and a random sign.

The simulation proceeds by repeated solution of equation (11) in which the displacement in each interaction is obtained by integration of the equation of motion for an interaction time, Ti, determined from three competing physical effects. If the particle is small, it may be captured by the eddy and remain with it until the eddy itself decays after a time, T_{EDDY}. If the particle is heavy it may pass through the eddy in a time T_{SLIP} which is determined by the relative displacement of the eddy and particle and is limited by the eddy length scale.

Finally, if the mean velocity of the droplets and the fluid is very different, the interaction may be limited by a mean penetration time T_{cd} where $_{cd}$ implies crossing trajectories. This last term is neglected in the present work and hence the interaction time T_i is determined as the minimum of T_{EDDY} and T_{SLIP}.

To complete the prescription for the simulation it is necessary to chose a length scale, a velocity scale and a time scale and these are taken from the turbulence model for the gas flow.

Finally the diffusion coefficient is calculated using the ratio of mean square displacement to mean time of displacement.

$$\Gamma_\rho = \frac{1}{2} \frac{<1_{p_i}^2>}{<T_{p_i}>},$$

In practice 10^4 steps are sufficient to give a converged calculation of the diffusion coefficient to a repeatability better than .1%.

Earlier calculations [8], P Hutchinson, et al have shown that the model works well for isothermal flows when compared with established data such as that of [9] Schneider and Lumley.

The model is also capable of extension to deal with dense sprays by introducing the processes of collision and coalescence in the form of transfer cross sections for scattering between different droplet size groups.

In order to be useful for modelling flow with heat transfer it is also necessary to extend the model to take account of the effects of vaporisation on the droplet. This is readily achieved by extending the drag equation to account for the effects of mass flux to or from the droplet surface. Predictions of residence time moisture content and temperature histories are also important for spray dryers. Normally these values are difficult to obtain from Eulerian models. However, the diffusion equation represents a Gaussian Markov process and, hence, the full probability distribution of all particle trajectories follows from a knowledge of the concentration and distribution [7] Papoulis 1965, and this in turn yields the desired quantities.

Numerical Methods

A finite-volume method using a staggered variable grid arrangement was used for solving the system of equations.The pressure and mean velocity fields were coupled using the SIMPLE algorithm [9] (Patankar and Spalding, 1972). The 13-point quadratic upstream weighted scheme [10] (QUDS; Leonard, 1979) was used for convection discretization in all transport equations. At each control volume face, the dependent variable was interpolated by a quadratic surface taking into account grid non-uniformity. The strongly implicit method, [11] (Stone, 1968) was used to solve the system of algebraic equations including only five diagonals in the coefficients matrix. The other coefficients were appropriately incorporated in the source term.

Finally, the stabilising approach suggested by [12] Huang and Leschziner (1985) was used for a numerical solution of the second moment closure.

EXPERIMENTAL MEASUREMENTS

Two sets of experimental studies have been conducted. The first from ISTP is concerned with evaluating the choice of turbulence model and the second from Erlangen with providing data on the development of a spray in a turbulent co-flowing gas stream.

Measurements of Mean Velocity and Turbulence Characteristics in an Unconfined Wake Behind a Disk

The overall objective of these measurements is to assist the development of numerical models for spraying processes. In this first phase, current one-point turbulence closures of first and second order are examined by comparison with experimental data obtained from the flow in the isothermal turbulent near wake of a disk with a central jet.

The measurements have been carried out with a one-component dual beam LDV in forward scattering mode.

The air in both streams was seeded with nominal $1 \mu m$ diameter TiO_2 particles. The seed particles were introduced into the piping system by cyclone seeders. Prior to commencing the main programme of measurements, tests were performed to identify long-term repeatability of the results, influence of seeding conditions, counter operating conditions and data averaging procedures. Simple ensemble averaging or transit time weighted averaging did not result in accurate mean values due to the non-uniform particle density in the separated region. Therefore in order to reduce velocity bias the so called "controlled processor" algorithm was used. This algorithm divides time into equally spaced intervals and only the first particle measured during each time interval is validated for the data processing. Between 10^4 to 5.10^4 individual samples were used for the calculation of the statistics depending on the distribution of the velocity histogram. The data rate ranged between 0.5 to 1 kHz inside the separated flow region and up to 5 kHz in the flow main streams.

The transmitting and receiver optics were both held stationary while the test section was mounted on a traversing table. This allowed access to any measuring position in the flow centre plane. Mean axial and radial velocity components and their variances were

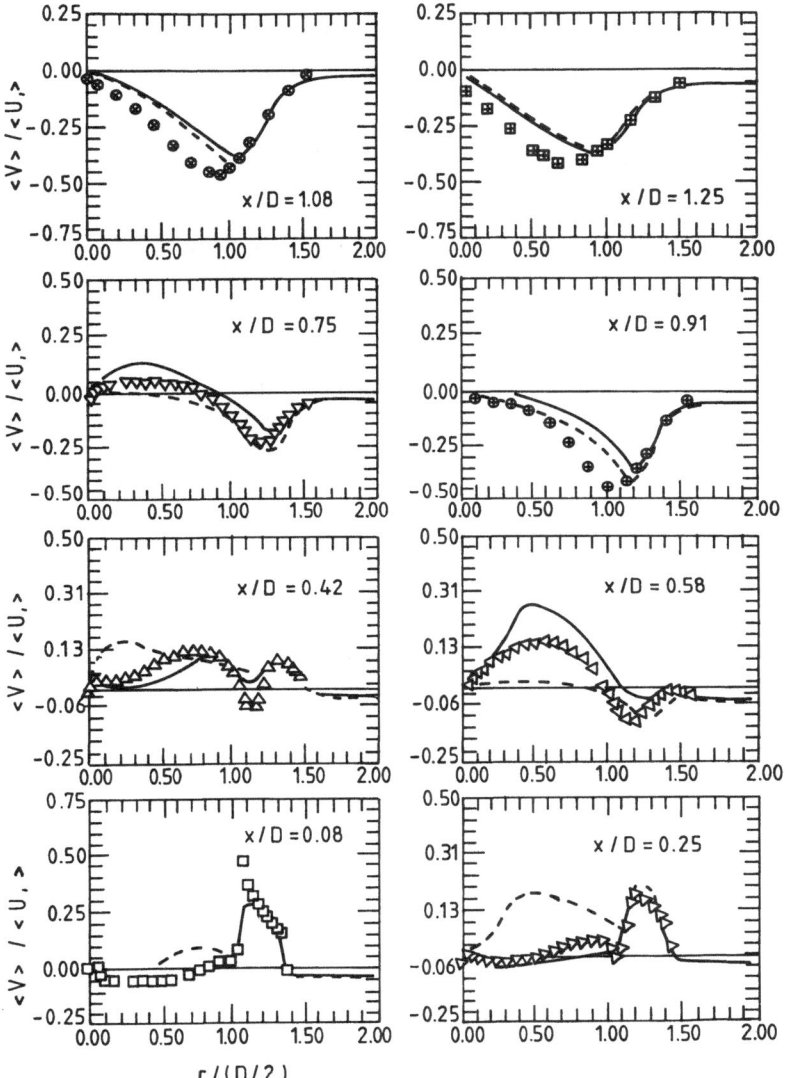

Figure 1   V-velocity Radial Development.   Predictions: RSM:, --- EVM:
Measurements: Symbols.

calculated from 10 to 50 thousand samples, as well as velocity histograms. The turbulent Reynolds stress, u v was evaluated by a commonly used approach for one component LDV system, from $u\,v = -n^2 + 1/2(u^2 + v^2)$, where n refers to the component at 45 degrees between the two normal co-ordinates directions.

The velocity measurements may be affected by non-turbulent Doppler broadening due to gradients of mean velocity across the measuring volume and transit broadening.

The number of individual velocity values used to form the averages was always above 10^4. As a result, the mean and variance values are subject to random errors of 2% respectively, for a 95% confidence interval.

A sample set of comparisons between measurements and the predictions of a Reynolds stress model and eddy viscosity model are shown in Fig 1. These results indicate that although the flow is basically pressure-driven at the initial stages, a Reynolds stress mean field prediction is globally in closer agreement with experiments than the eddy viscosity model. However, neither model is able to capture an increase in turbulence energy level near stagnation regions.

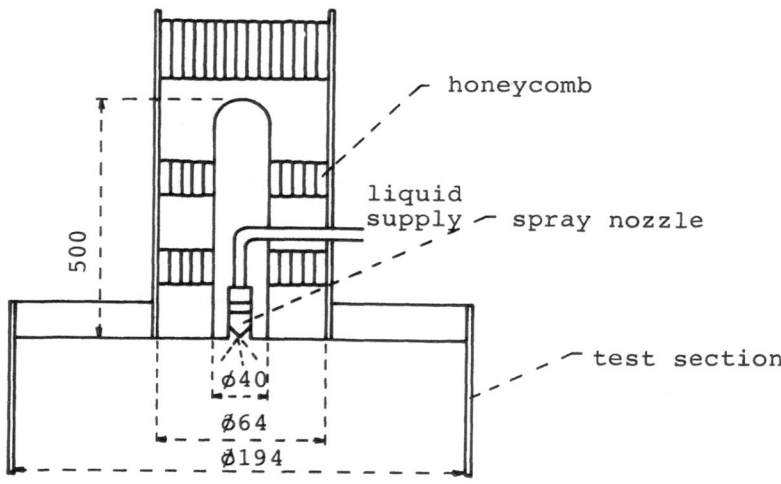

Figure 2   Test section for experimental studies (all dimensions in mm).

Spray evaporation in a Co-flowing Annular jet

The experimental studies at LSTM Erlangen aim to provide data to validate numerical models for two-phase flow systems with heat and mass transfer between the phases. Recent numerical studies have shown [13] (Berlemont et al. 1991) that turbulent effects on droplet evaporation are very important for prediction of the proper development of droplet size. Experimental results on spray evaporation which are necessary to evaluate such models for turbulent droplet evaporation are, however, very rare.

The evaporation of a spray issuing into a co-flowing turbulent, heated air stream was measured using Phase-Doppler anemometry. The experimental set-up with dimensions is shown in Fig 2 where an annular air jet at 80°C is entering a test section of 194 mm inner diameter.

In the centre of the annulus a hollow cone pressure atomizer was mounted to deliver a spray of isopropyl alcohol.

Phase-Doppler anemometry allowed measurement of the change in the droplet size distribution throughout the flow field and the local correlation between droplet size and velocity. Furthermore, accurate droplet mass flux measurements were possible using a recently developed method [14] (Qui and Sommerfeld, 1992).

At present is has only been possible to measure the properties of the droplets. Conventional approaches to obtain the gas flow velocities by seeding the flow with small tracer particles are not applicable in evaporating sprays. The addition of a solid tracer would result in a contamination of the windows in the test section and, additionally, would considerably influence the evaporation characteristics. The addition of small liquid droplets in the annular air stream on the other hand would influence the mass balance of liquid mass flux and may cause interactions of tracer an droplets (ie coalescence). Furthermore, small droplets added to the air flow will evaporate very fast, mainly in the inlet pipe. Hence, addition of larger droplets would be required, which would considerably influence the vapour fraction in the flow field. This conflicts with the need to specify detailed inlet conditions for the numerical predictions. All these problems can be avoided if the existing small droplets in the spray are used as a tracer. This involves the solution of two major problems. The number density of small droplets (ie $D_p < 3$ um) is very small in some regions. Doppler-bursts from very small droplets fall within the noise level of the electronic signal which additionally results in a decrease of the data rate for tracer droplets. This is a result of the commonly applied amplitude trigger method and will be solved by the development of a new trigger method which identifies Doppler-bursts which fall within the noise level of the electronic signal from the photo detectors.

Axial velocities and the associated velocity fluctuations, averaged over all droplet sizes have been measured for seven cross-sections. The profiles of the axial droplet velocities shown in Fig 3 exhibit a rather symmetric shape which demonstrates the proper alignment of the flow. In regions where no velocity data are plotted the number density of the droplets was too low to acquire data within a reasonable measuring time. The maximum averaged axial velocity just downstream of the spray nozzle (x = 0) is about 16 m/s and then decreases to about 11 m/s at z = 25 mm. Going further downstream

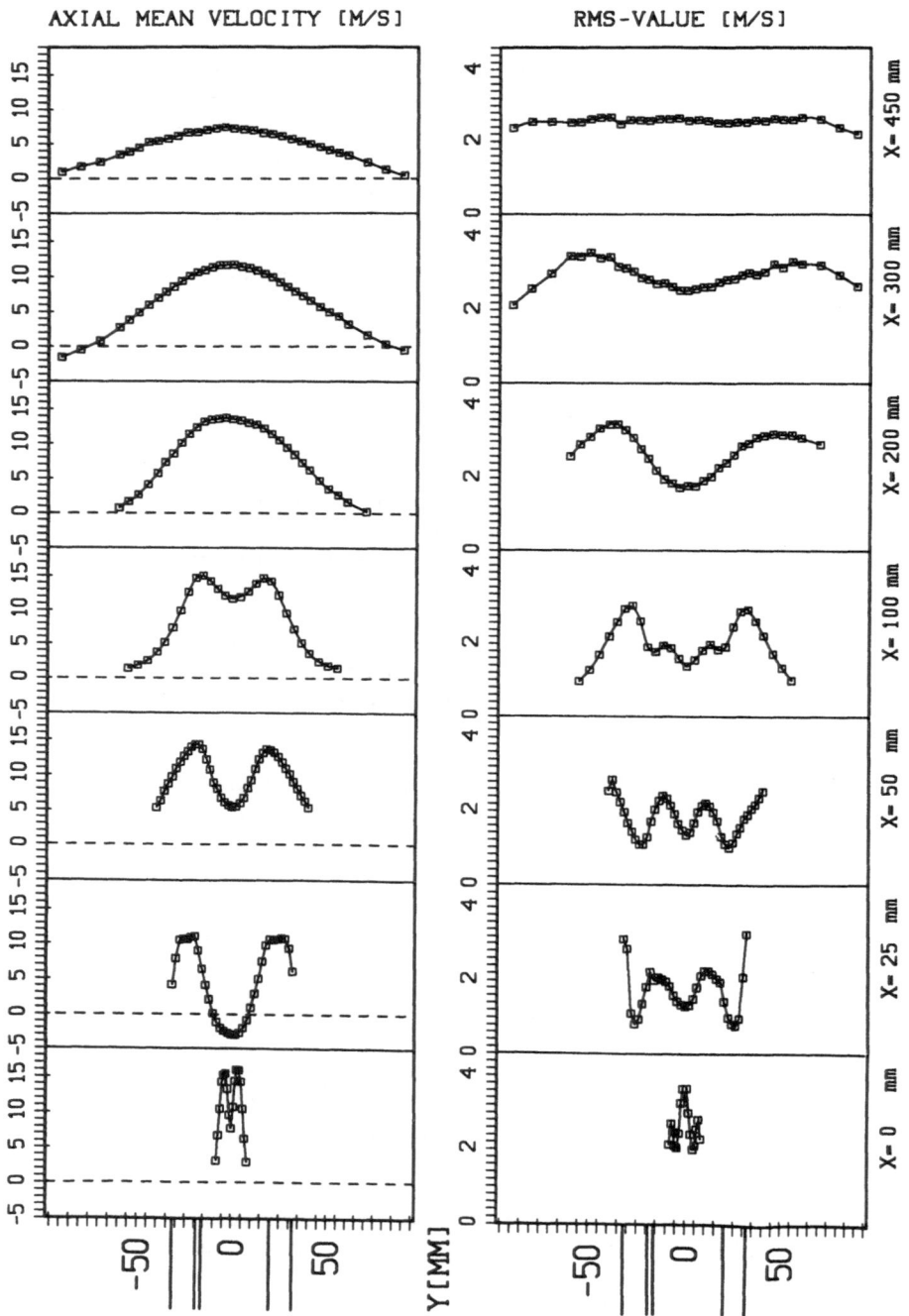

Fig 3   Profiles of axial mean velocities and velocity fluctuations of the droplet phase.

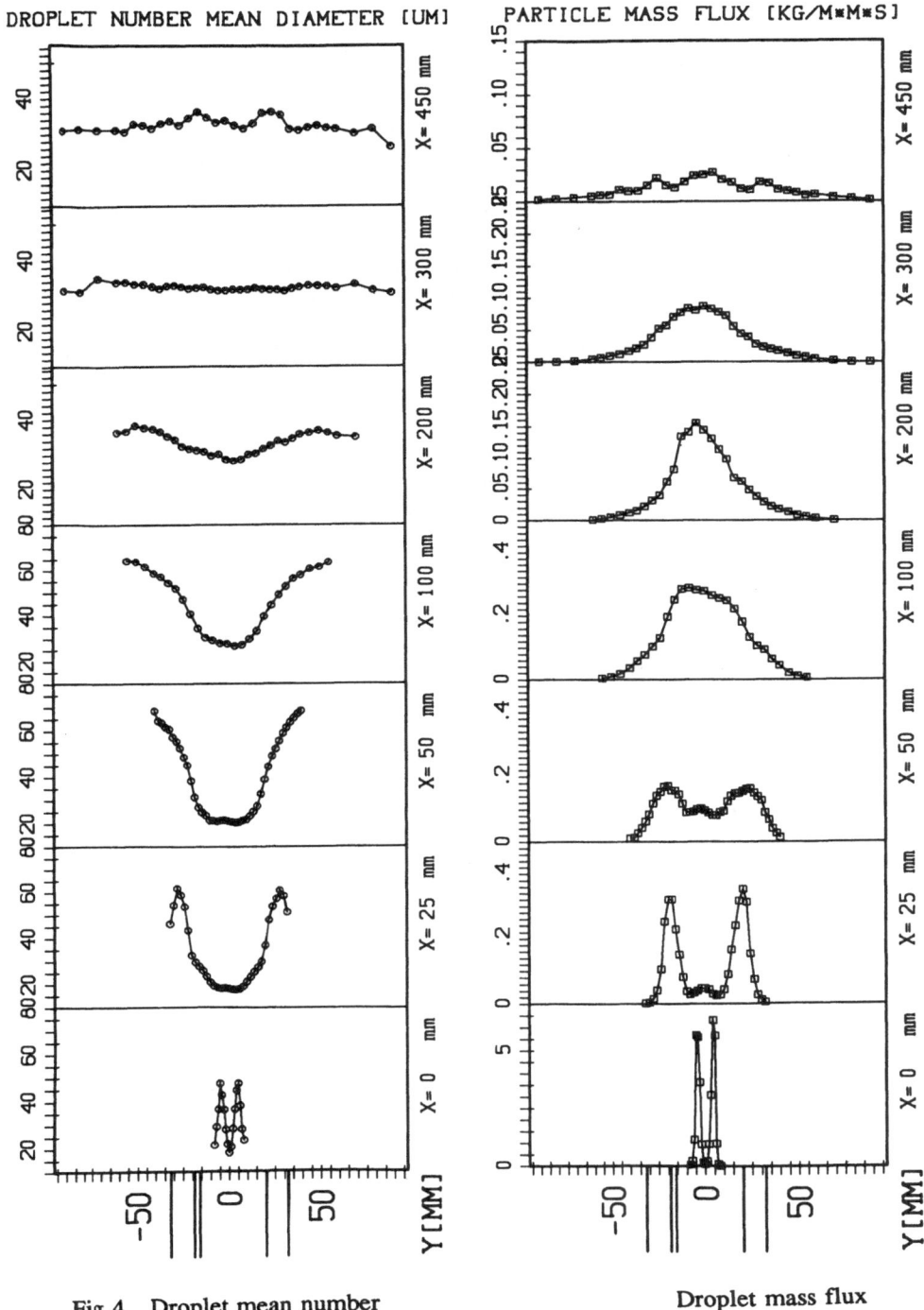

Fig 4   Droplet mean number
Diameter

Droplet mass flux

(x = 50 and 100 mm) again an increase of the maximum droplet velocity is observed which may be associated with an acceleration of the flow as a result of evaporation and the consequent increased vapour mass fraction. From x = 200 the maximum axial droplet velocity continuously decays due to the heat losses required for the droplet evaporation. The mean droplet diameter and droplet mass flux are shown in Fig 4. Some interesting effects are illustrated which demonstrate the complexity of the flow. It is quite obvious that the droplet mass flux increases considerably from z = 50 to 100 mm downstream the inlet. This is also demonstrated by integrating the mass flux profiles to yield the droplet mass flow rate in each cross-section. Initially a decrease is observed in the droplet mass flow rate and thereafter a continuous increase up to z= 200. Further downstream the droplet mass flow rate decreases as expected. This phenomenon may be associated with several effects. It may be that the evaporation causes a cooling of the flow and re-condensation on droplets in the core of the spray. However, this effect is not seen in the preliminary numerical calculations, and further work is required to establish its source.

Isopropyl vapour is convected upstream in the recirculation region of the pipe expansion and diffuses into the spray core where it condenses on the cooled droplets. Also a number of small droplets are moving in the recirculation region and may coalesce with droplets within the spray cone.

The above described effects need further exploration. In a situation where not all the flow parameters can be measured it is clear that a combined approach based on experimental and numerical studies is required to develop an understanding of the physical phenomena occurring in such a complex flow.

CONCLUSIONS

A new formulation for the description of spray development based on a diffusion model has been represented and shown to have some advantages over the established Lagrangian droplet tracking approximation.

Data taken in an unconfined wake behind a disk suggest that a Reynolds stress model is a more appropriate choice than an eddy viscosity approach for representation of the turbulent flow in geometries typical of spray dryers.

Detailed measurements of spray vaporisation in a co-flowing jet have indicated that complex vaporisation and condensation phenomena can occur in heated sprays and that a combination of detailed measurements and modelling is essential if a fundamental understanding of spray dryers is to be obtained.

REFERENCES

1   Daly, B J & Harlow, F H, (1970): "Transport Equations in Turbulence," Phys,

Fluids, 13, No 11, pp 2634-2649.

2   Rotta, J C (1951): "Statistsche Theorie Nichthomogener Turbulenz," Z Phys., 129, pp 547-572.

3   Naot, D A, Shavit & Wolfshtein M (1970): "Interaction Between Components of the Turbulent Velocity Correlation Tensor," Israel Journal of Technology, 88, pp 259.

4   Launder, B E & Spalding D B (1974): The Numerical Computation of Turbulent Flows, Comp. Meth. Appl. Mech. Eng., 3, pp. 269-289.

5   A D Gosman & E Ioannides (1983) "Aspects of Computer Simulation of Liquid-fuelled Combustors". J Energy, volume 7, number 6, pp 482-490.

6   P Hutchinson, G F Hewitt & A E Dukler (1970) "Deposition of Liquid or Solid ispersions from Turbulent Gas Streams: a Stochastic Model", Chem Eng Sci Volume 26, pp 419-439.

7   A Papoulis, "Probability, Random Variables and Stochastic Processes", McGraw Hill (1965).

8   P Hutchinson, J S C Tan, & M E Gill, (1990) " Droplet Dispersion in an Isotropic, Homogeneous Turbulent Flow Using a Diffusion Stochastic Method" Fifth Workshop on Two Phase Flow Predictions proceedings, Erlangen, pp 186-194.

9   Patankar, S V & Spalding D B (1973): " A Calculation Procedure for Heat, Mass and Momentum Transfer in Three-dimensional Parabolic Flow," Int. J. Heat Mass Trans., 15, pp, 1787-1806.

10   Leonard, B P (1979): "A Stable and Accurate Convective Modelling Procedure Based on Quadratic Upstream Interpolation," Comp. Meths. Apl. Mech. Engng., 19 p 59-98.

11   Stone, J L (1968): "Iterative Solution of Implicit Approximations of Multi-dimensional Partial Differential Equations," SIAM J. Numer. Anal. 5, No 3 pp 530-558.

12   Huang, P G & Leschziner, (1985): "Stabliszation of Recirculating Flow Computations Performed with second-moment Closures and Third-order Discritzation", Proc. 5th Symp. Turbulent Shear Flows, Cornell University, 20.7-20.12.

13   Berlemont A, Grancher, M-S & Gousbet, G 1991: "On the Lagrangian Simulation of Turbulence Influence on Droplet Evaporation." Int. J. Heat and Mass Transfer, 34, 2805-2812.

14   Qui, H-H, & Sommerfled, M. 1992: A reliable method for determining the Measurement Volume Size and Particle Mass Fluxes using Phase-Doppler Anemometry, in press, Experiments in Fluids.

NEW DRYING METHODS FOR AQUEOUS SOLUTIONS AT LOW TEMPERATURES

GUNNAR MINDS/JESPER NYVAD
Danish Technological Institute (DTI)
Dept. of Energy Technology/Refrigeration and Heat Pumps
Teknologiparken, DK-8000 Aarhus C

ABSTRACT

A new method of drying aqueous solutions at low temperatures is presented. The new method is based on water vapour compression and is operating in vacuum. The energy savings compared to traditional spray drying systems are up to 90%.

A test plant with a capacity of 100 kg solution per hour has been operating since the autumn of 1991 at DTI. Test results are presented and a cycloid water vapour compressor is described.

The test plant shows satisfactory results. The drying tests have shown that with the use of mechanical water vapour compression it is possible to dry substances that until now has not been possible to dry.

INTRODUCTION

The demand for research in and development of more environmentally friendly and energy saving methods for solving several tasks within the field of:

- cooling, heat pumps and waste heat recovery systems
- drying and evaporation of aqueous solutions of:
  - articles of food
  - industrial sewage
  - sludge and liquid manure
  - etc.

has caused DTI in co-operation with several large Danish firms to start a research and development programme.

The programme will in the near future result in process equipment, which in a quite new way exploits a very energy efficient and environmentally desirable technique.

For cooling of water, production of ice, heat pumps and waste heat recovery systems, it can be an advantage by means of mechanical water vapour compression to use water as refrigerant. Both condensation and evaporation takes place in direct contact heat exchange without any heat transfer surfaces. In this way energy consumption is considerably decreased compared to traditional refrigerating plants with e.g. ammonia or other refrigerants.

Drying and Evaporation

The method, which makes it possible to separate these aqueous solutions into a solid/dry part and a distillate, is in principle to reuse the energy used for evaporation of the water by means of a mechanical water vapour compressor.

This material will only in details describe the processes and components for drying of aqueous solutions of substances like e.g. proteins, yeast, enzymes, eggs, and egg whites. Characteristic for all these substances is that they are destroyed at temperatures above 60°C - some even at lower temperatures - which is the reason why the drying has to take place in vacuum.

Drying of Temperature Sensitive Aqueous Solutions

The principle of the drying process can be explained by a couple of simple sketches.

At direct heating, approx. 2500 KJ/kg is used for the evaporation of water. See fig. 1.

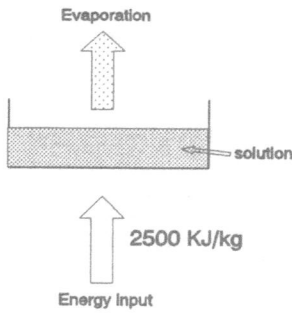

Fig. 1 Direct evaporation

If a water condenser is mounted instead of the heat source, and a water vapour compressor "sucks" the vapour away from the solution, compresses it, and leads the vapour to the condenser, the power to the compressor and the energy from the evaporation is lead back to the process. See fig. 2.

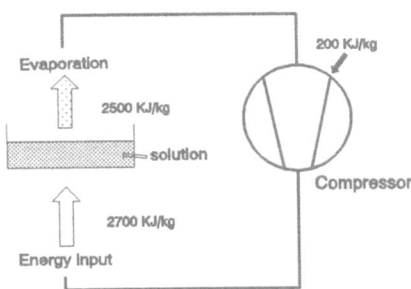

Fig. 2 Evaporating by water vapour compressor

The consumed energy (approx. 200 - 600 KJ/kg evaporated water) is in this way only 8 - 24% of the energy without water vapour compression. At spray drying, the energy consumption is approx. 7500 KJ/kg evaporated water. Seen in relation to spray drying, the energy consumption is only between 3 and 8%.

DESCRIPTION OF TEST PLANT

A plant with a capacity of approx. 100 kg solution per hour has been tested at DTI since autumn 1991. The plant is in principle constructed as depicted in Fig. 3

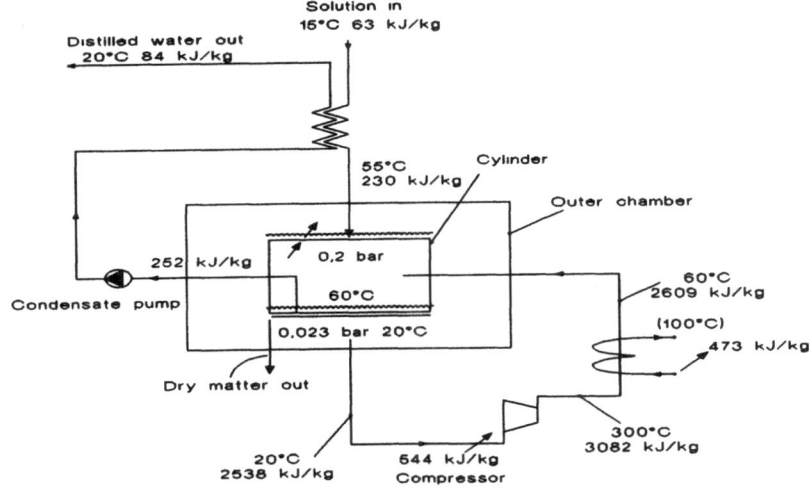

Fig. 3   Process diagram for vacuum dryer

The power to the compressor is in principle a surplus in the process. In practice the surplus power is removed by:

1. Injection of water in the compressor during the compression in order to avoid too high internal temperatures.

2. Heat transmission to the surroundings.

The test plants consist of an external vacuum vessel with a diameter of 2100 mm and a length of approx. 2000 mm. The vessel is designed for full vacuum and 2 bars absolute pressure for the sake of the sterilization taking place with steam at 120°C.

The cylinder diameter is 1500 and the length approx. 1000 mm.

Fig. 4   Photos of the test plant

The solid matter tap takes place by means of a screw conveyor that is leading the matter out through a vacuum sluice.

The mechanical water vapour compressor in the test plant is a cycloid compressor.

A cycloid compressor is a positive displacement compressor characterized by having a solid integrated volume ratio, and in this way also a solid compression ratio (pressure ratio).

A cycloid compressor can however work at alteration pressure ratios. If the actual pressure ratio is lower than the one built-in, an overcompression takes place before it is opened to the exhaust side of the compressor. At higher pressure ratios,

a back stream flow will take place when it is opened to the
exhaust side. In both cases a loss will arise, but when evaluating these, the following must be taken into consideration:

-   The cycloid compressor contains no valves, which means that
    suction and exhaustion pressure drops are avoided.

-   As the cycloid compressor operates in vacuum (< 1 bar), the
    leakage losses internally in the compressor will be very
    small, as the difference pressure over the compressor will
    always be smaller than 1 bar, while the pressure ratios can
    be up to 10 - 12.

-   The efficiency of the cycloid compressor is fairly constant
    near the integrated pressure ratio, see fig. 5. The effici-
    ency drops at higher pressure ratios than the one
    integrated.

These conditions mean that the cycloid compressor can be used
for drying of different substances each with their set of temperature, without changing the energy consumption considerably.

THEORETICAL COMPRESSION WORK FOR CYCLOID COMPRESSORS

The compression process in a cycloid compressor can, until the
point where it is opened to the exhaust side, be expected to
follow an irreversible polytrop which is characterised by:

$$p \cdot v^n = p_1 \cdot v_1{}^n = p_2 \cdot v_2{}^n \tag{1}$$

where
p is the pressure of water vapour in Pa.
v is the specific volume of water vapour in m^3/kg.
n is the polytropic index for water vapour.

When opened to the exhaust side, the pressure rises (drops)
momentarily to the back pressure, and the water vapour is
exhausted at constant pressure. This displacement work is
expected to take place without any losses.

With the assumption of an ideal gas the compression work from
condition 1 to condition 2 can be calculated as follows:

$$W_{12} = \frac{p_1 v_1}{\eta_i}\left(\frac{1}{n-1}\left(v_i^{n-1} - 1\right) + \frac{p_2}{p_1}v_i^{-1} - 1\right) \tag{2}$$

where
W_{12} is the compression work in J/kg.

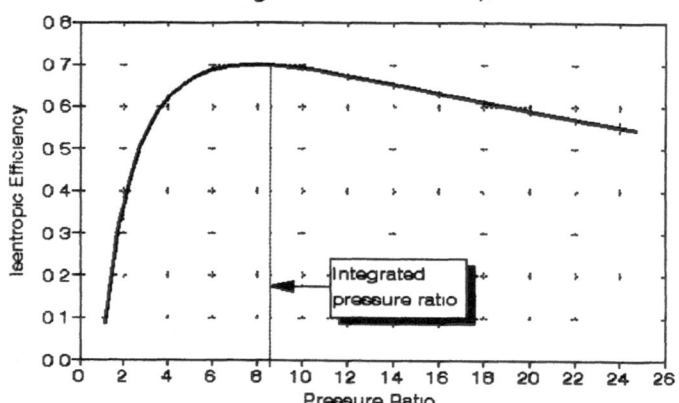

Fig. 5 Isentropic efficiency for the cycloid compressor

η_1 is the polytropic efficiency.
v_1 is the integrated volume ratio in the compressor.

The polytropic efficiency depends on the conditions of the vapour at the inlet of the compressor. With the assumption of an ideal gas it can be determined as:

$$\eta_1 = \frac{n}{n-1} \frac{\gamma-1}{\gamma} \tag{3}$$

where
γ is the specific heat capacity ratio of water vapour.

The capacity of cycloid compressors for the drying plant can according to (1), (2) and (3) be determined theoretically with the following specifications:

Swept volume: 5000 m³/h
Integrated volume ratio (v_1):   4.3
Superheat of water vapour at inlet   0°C
Subcooling of water vapour after condenser: 5°C

The total capacity diagram for the compressor can be seen in figure 6.

THEORETICAL CALCULATIONS OF THE ENERGY CONSUMPTION AT DRYING PROCESSES

As there at the present level are not enough measured values from the drying plant, the capacity calculations are based on measured test values, which are described in reference [1].

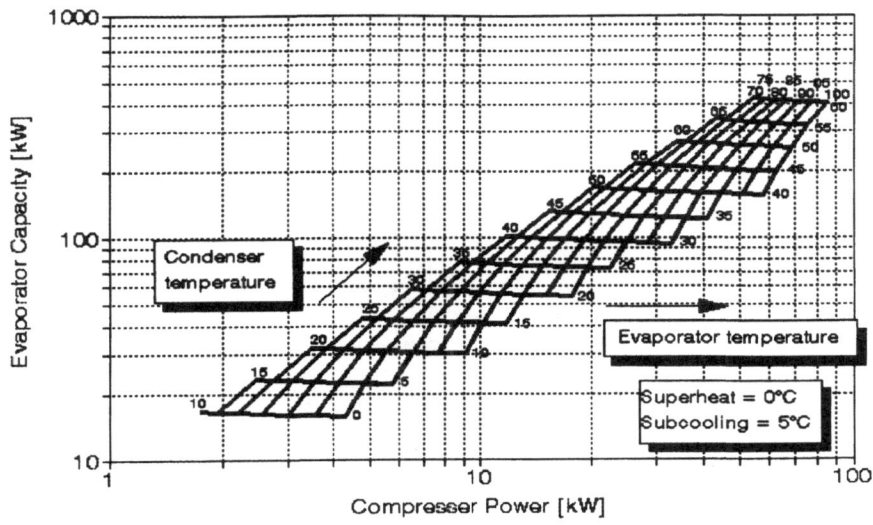

Fig. 6   Cycloid compressor diagram

It is to be noticed that the measured values stated in reference [1] are for high condensation temperatures (102-130°C), and the temperature difference across the cylinder is also higher (64-80°C) than in the present work.

It is assumed that the measured values can be used at the low condensation temperatures and small temperature differences for which the present plant is constructed.

The preliminary test results however indicate that this is no bad assumption.

In figure 7 the capacity of the plant is shown for two different products - egg whites and milk.

Where the characteristic of the compressor and the product cross, the point of operation is given. On the dotted curve the power consumption of the compressor can be read.

PRELIMINARY TEST RESULTS

As tests are still going on and are not expected to end until at the beginning of 1993, more detailed measurements will be available. The results will be presented to the extent that the participating firms accept.

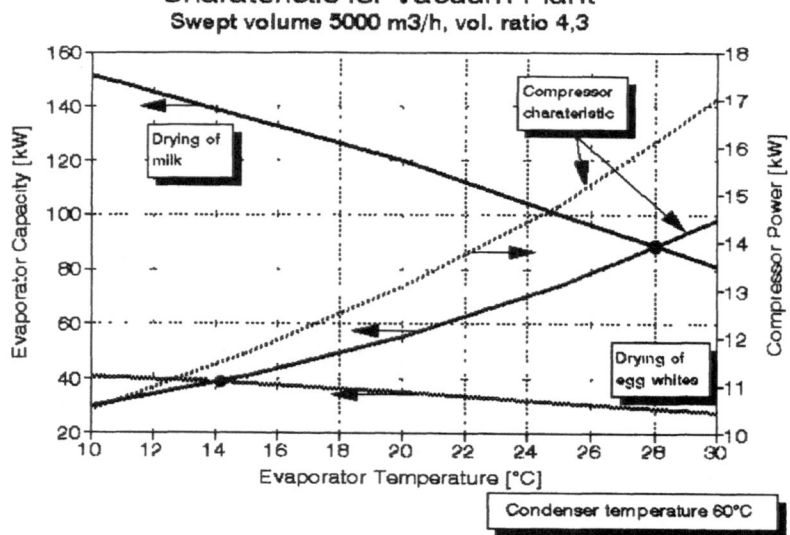

Fig. 7   Characteristic for vacuum plant for two different
products

Product Structure and Quality

The dry matter has another appearance and structure than the
one of drying in a spray drying plant.

Where the powder formed in a spray drying plant has a characteristic uniform ball-shape, the structure of the powder from
the vacuum drying system is very different - both with regard
to the porosity and the solubility in water.

Fig. 8 shows desiccated whole eggs (egg whites and yolks in a
natural ratio of mixture).

The structure is broken flakes with smooth clean surfaces and
little porosity. The product is easily soluble in water. The
test baking result is like fresh eggs.

Fig. 9 shows vacuum dried egg whites without yolks.

Here the structure is quite different from the mixture of
whites and yolks. The surface is rugged and uneven with typical
viscous fractured surfaces. The grains are porous. It is easily
soluble in water. The test baking shows the same result as for
fresh egg whites.

Fig. 8   Desiccated whole eggs

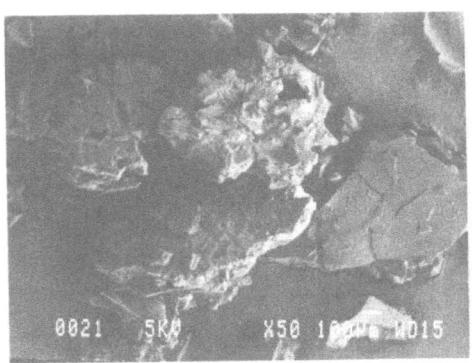
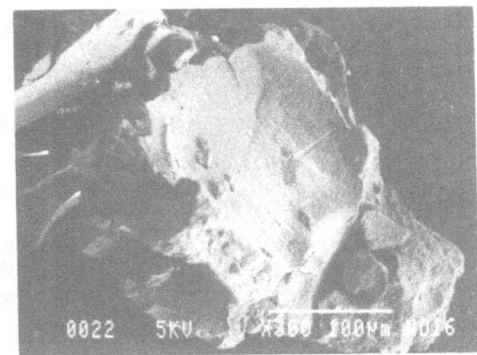

Fig. 9   Egg whites without yolks

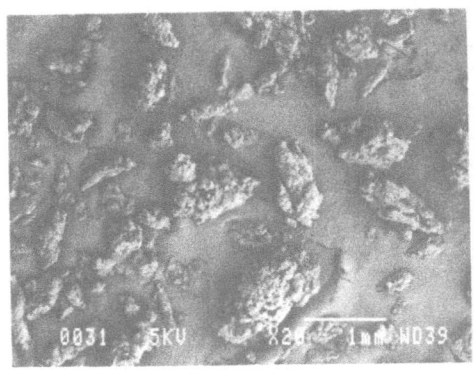
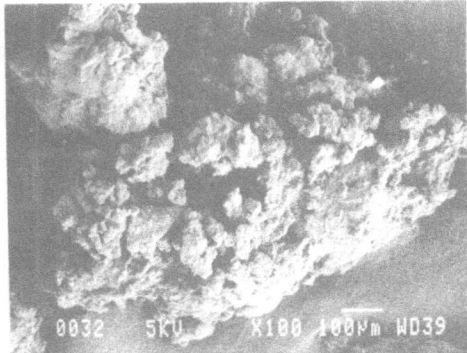

Fig. 10   Vacuum dried protein

Whipping to a froth: The preliminary results indicate that the whipping demands a little higher effect, but the difference to fresh egg whites is small.

Fig. 10 shows a protein formed by cultivation of bacteria on the basis of methane.

Here the structure is very porous, and there is no fractured surfaces like on the eggs. It was not either expected as this is a suspension of the solid matter in water. The product is insoluble in water.

ECONOMIC ASPECT

Energy savings compared to spray drying.

Solution: 22% dry matter (78% Water)
Evaporated water: 3.546 kg/kg dry matter
Power plant efficiency: 0.45 (Denmark)
Heat of combustion for oil: 42,300 kJ/kg
Boiler efficiency (spray drying): 0.89

Spray drying:

Energy consumption: 7,500 kJ/kg evaporated water
Oil consumption: 0.20 kg oil/kg evap. water
Oil consumption: 0.71 kg oil/kg dry matter

Water vapour compression:

Energy consumption: 544 kJ/kg evaporated water
Oil cons. (power plant): 1,209 kJ/kg evaporated water
Oil cons. (power plant): 0.029 kg oil/kg evap. water
Oil cons. (power plant): 0.101 kg oil/kg dry matter

Energy savings in equivalence to oil consumption:

   = 0.71 - 0.101 = 0.609 kg oil/kg dry matter

Equivalence to 86%

Investment calculations seems to be at the same level for both systems.

CONCLUSION

The test plant, which has the size of a small production plant, has until now shown that it can fulfil the expected demands.

Drying in vacuum in connection with the use of mechanical water vapour compression has created new possibilities of treating substances that until now have not been possible to dry. - Either the expenses have been too high, or the temperatures which have been necessary during the work have been unacceptable for the substance.

A full-scale production plant is expected to be operating in 1993.

REFERENCES

1. Baumann R., Leistung von Walzentrocknern, Chemie-Ing.- Techn. 10/1952, page 607-617 (in German).

2. Frank Elefsen and Karsten Pedersen, Energiøkonomisk Metode til Inddampning af Koncentrerede Vandige Opløsninger af Temperaturfølsomme produkter. EFP87 in Danish with English summary.

ABSORPTION - DEHUMIDIFICATION
A UNIT OPERATION FOR HEAT RECOVERY FROM DRYERS

COLIN PRITCHARD and JOHN CURRIE
Department of Chemical Engineering,
University of Edinburgh,
King's Buildings, Edinburgh EH9 3JL, Scotland

ABSTRACT

Drying operations account for 10% of primary energy consumption of the main industrial sectors in Europe. Energy recovery is uncommon, and most drying processes discharge hot, humid air to atmosphere. The purpose of this work was to develop an absorption heat transformer (A.H.T.) to simultaneously dehumidify and reheat dryer exhaust streams for recycle to the drying process.

Experiments have been carried out on a single-stage, direct-contact A.H.T. utilising aqueous Lithium Bromide as absorbent. It was concluded that, in order to maximise the drying capacity of the recycled airstream, the dehumidification and reheating processes must be carried out as separate operations. A two-stage variant has been constructed and operated with a wide range of inlet temperatures and humidities. Using a 'typical' spray dryer exhaust stream at 100°C and humidity 0.2 kg/kg, outlet air temperatures of 170°C and humidities of 0.04 kg/kg may be obtained: the recycling of such a stream to the spray dryer could give rise to energy savings of 20%.

Specification of A.H Ts. for a variety of dryer types indicates that absorption dehumidification may be accorded the status of a unit operation applicable to any hot, humid exhaust gas stream. The design of an A.H.T. for use with a large industrial spray dryer is developed as a case study.

INTRODUCTION

Drying of solid products is one of the most widespread of the unit operations of the chemical and process industries and accounts for around 10% of the primary energy consumption of the main industrial sectors in Europe (1). Around 90% of dryers employ hot air as the heat transfer medium, discharging hot, humid (and sometimes dusty) exhaust streams directly to atmosphere. These exhausts retain a high proportion of the energy supplied to the dryer, mostly in the form of latent heat whose recovery by conventional means is infeasible. However, three features of the industrial drying context add impetus to the search for viable means of energy recovery: the cost of the wasted energy itself, the environmental impact of plumes and dusty exhausts, and the drive to achieve higher throughputs from existing dryer plant.

Absorption- dehumidification offers the possibility of returning a proportion of the enthalpy of the exhaust into the dryer inlet gases by upgrading this waste heat to a useful temperature. Our laboratory work and plant studies have demonstrated the feasibility of this concept for spray dryer exhausts, and its applicability to the exhaust gas streams from a wide variety of dryer types. This leads us to the conclusion that such an energy recovery system should always be considered at the design stage, as a unit operation associated with dryers.

This unit operation utilises an open-cycle absorption heat transformer to upgrade the energy of the dryer exhaust. The principle of a heat transformer is illustrated in Figure 1, which shows a conventional, closed-cycle heat transformer in which part of a moderate-temperature waste heat stream is upgraded to a higher temperature at the expense of downgrading the remainder, usually to ambient temperature. The principles and application of closed-cycle heat transformers have been extensively documented (2).

In the open-cycle variant, Figure 2, the refrigerant is water, absorbed from the humid airstream by a solvent (e.g. concentrated LiBr solution) having a very low vapour pressure. As water is absorbed it releases its latent heat of vaporisation. Both the absorbing liquid and the air stream in direct contact with it are thereby heated, to a temperature at which the partial pressure of residual moisture in the airstream equals the vapour pressure over the hot solution. The LiBr solution must be regenerated for reuse. This is effected in a low-pressure evaporator supplied with low-grade waste heat, which may come from a process stream or possibly from a portion of the exhaust gas stream itself.

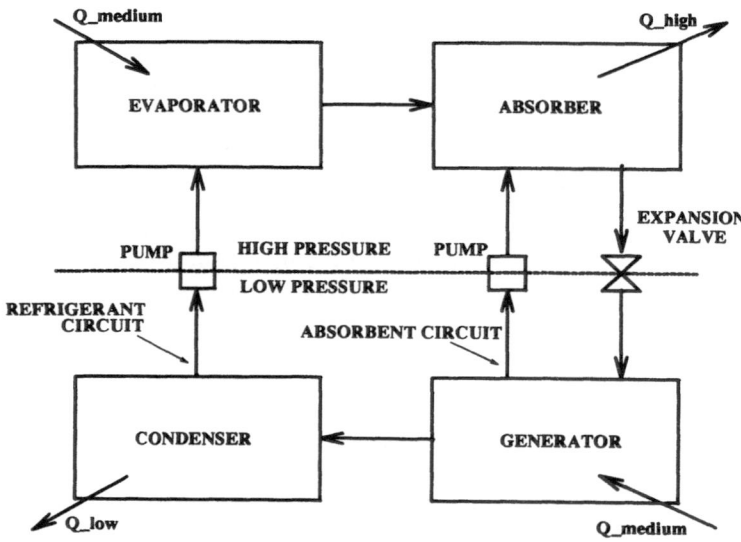

Figure 1: Absorption heat transformer- closed cycle.

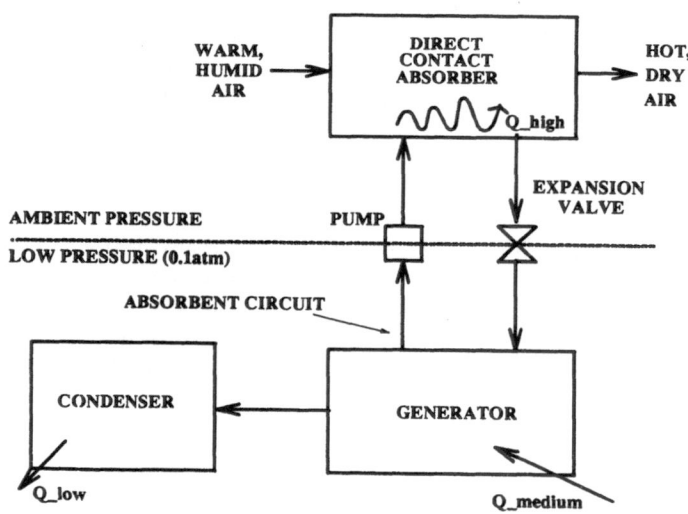

Figure 2: Absorption heat transformer- open cycle.

EXPERIMENTS

Experiments using a single-stage A.H.T. and a laboratory-generated hot, humid airstream have been reported elsewhere (3). In a sequence of experiments using LiBr as absorbent, a maximum temperature of 150°C was obtained, using an extremely humid exhaust flowing cocurrent to the absorbing liquid. For realistic dryer conditions, the limiting temperature attainable is 120°C. This limit represents the temperature of equilibrium between the concentrated salt solution and the dehumidified airstream. Countercurrent flow of the airstream yields higher mass transfer rates than cocurrent flow, but at the expense of a lower outlet temperature and of a larger column diameter necessitated by flooding conditions. (Table 1.)

A substantially greater level of heat recovery is attainable if the dehumidification and reheating operations are separated. Dehumidification by direct contact with LiBr at low temperatures yields a very dry airstream ideal for recycling. This stream may then be reheated by **indirect** contact with a steam/LiBr mixture: using 1 atm. steam at 100°C, temperatures in excess of 170°C may be obtained, giving a worthwhile temperature lift and a significant recovery of exhaust heat.

Flow	LiBr conc %w/w	Gas humidity g/g	$T_{LiBr.in}$ °C	$T_{G.in}$ °C	$T_{G.out}$ °C	Heat trans rate U_o kW/m^3	Mass trans rate K_o g/s.m^3
co-	65	0.1	102	101	108	11	25
co-	56	1	102	103	122	22	53
co-	63	1	101	100	130	35	88
co-	70	1.1	101	103	142	45	87
counter-	65	0.1	100	61	99	33	41
counter-	65	0.3	101	104	110	12	70
counter-	58	0.5	100	107	105	14	52
counter-	58	0.8	130	100	116	17	55
counter-	58	1	100	100	110	12	54

Table 1: Comparison of mass and heat transfer rates for direct contact absorber.

A laboratory-scale 2-stage absorption heat recovery unit has been built and is shown in Figure 3. This utilises an aqueous solution of LiBr (62-68 wt% LiBr) at a set inlet temperature) to absorb water from an airstream of up to 8 m^{3}hr^{-1} having humidity in the range 0.1-0.5 kg/kg dry air and temperature in the range 60- 100°C, characteristic of many spray dryer exhausts (4). The dehumidified, reheated airstream (0.04 kg/kg, 170°C) is suitable for recycling to a dryer inlet; the diluted LiBr solution is regenerated at 90-100°C in an evaporator operating at 0.1 bara. Simultaneous measurements of temperature and pressure in this evaporator are logged by PC and used to compute the composition of the resulting solution on-line, and thus to control the (electric) heater so as to provide a composition suitable for return to the absorber.

Using type K thermocouples (±0.5degC) and a Farnell pressure transducer (±0.02 bara) to determine the conditions in the evaporator, the concentration of the regenerated LiBr may be controlled to ±1%. A Lee-Integer high temperature humidity probe is used to determine inlet and outlet gas humidities to permit a check on the mass and enthalpy balances of the system.

RESULTS and MODELLING

In parallel with this experimental work, a computer program has been developed to simulate the absorption- dehumidification process. This is based on a method for interfacial heat- and mass- transfer calculations proposed by Treybal (5) and utilises physical properties of LiBr solutions computed from the equations and data of Brunk (6) and Liu (7). It involves the simultaneous solution of mass and enthalpy balance equations and heat and mass transfer rate equations for each successive slice of the absorption column. For countercurrent flow the entire profile must be computed iteratively to match stream inlet and outlet conditions. Figure 4 compares an experimentally determined absorber temperature profile with a computer simulation of the simultaneous heat and mass transfer process, for cocurrent operation. It may be seen that, for cocurrent flow of the absorbent solution and humid airstream, a thermal equilibrium between the two is established very rapidly.

Medium	Operation	Air flow g/s	$T_{G.in}$ °C	$T_{G.out}$ °C	$T_{i.in}$ °C	$T_{i.out}$ °C	Re_{air}	h_o W/m²K
LiBr	Heating	2	21	68	70	58	4200	180
LiBr	Heating	2	60	81	75	85	4200	180
LiBr	Cooling	2	68	31	31	37	4200	210
LiBr	Heating	3	21	71	73	59	6400	240
LiBr	Heating	3	26	96	100	77	6400	260
LiBr	Cooling	3	58	40	40	46	6400	270
Steam	Heating	3	21	96	98	97	6400	240
Steam	Heating	4	25	101	104	103	8500	295
Steam	Heating	4	32	97	101	100	8500	270
Steam	Heating	4	23	97	100	100	8500	295
Steam	Heating	4	15	99	102	100	8500	290
Steam	Heating	4	11	93	104	100	8500	185
Steam	Heating	2.17	64	97	101	101	4600	125
Steam	Heating	2.67	69	98	101	100	5700	160

Table 2: Heat transfer coefficients for reheat column.

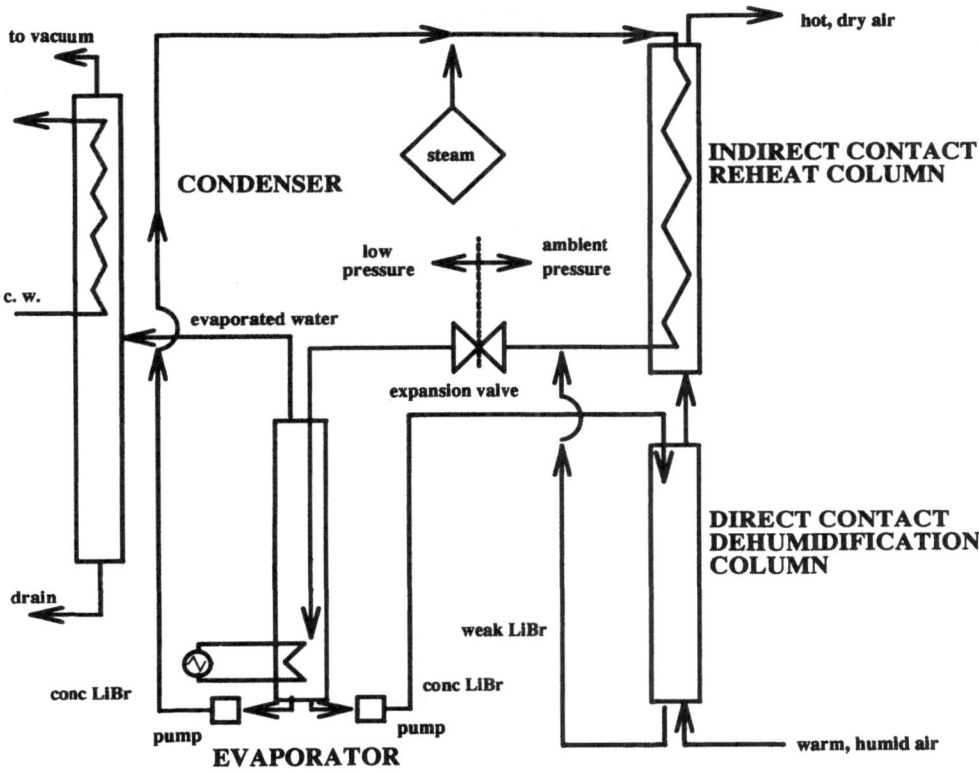

Figure 3: Two stage heat transformer pilot plant.

Experimentally determined volumetric heat- and mass- transfer rates in the direct-contact absorber are given in Table 1. The packing used in this part of the experiment had a specific surface area of 300 m^{-1}.

The indirect-contact reheater utilises a novel shell-and-tube exchanger with extremely extended surface on the shell (air) side and atmospheric pressure steam/LiBr solution as heating medium on the tubeside. Overall heat transfer coefficients based on the plain tube area are given in Table 2. Since the shell-side resistance predominates, it is possible to derive an experimental relationship for the gas-side film coefficient, based on the plain tube o.d. and annular flow of gas:

$$Nu = 0.23 \ Re^{0.80} \ Pr^{0.33}$$

for Re in the range 4200-8500.

Comparison with the predictions of the Sieder-Tate equation indicates an approximately tenfold augmentation of the plain tube area. Characteristic temperature and humidity profiles in the absorption and reheat columns are shown in Figure 5.

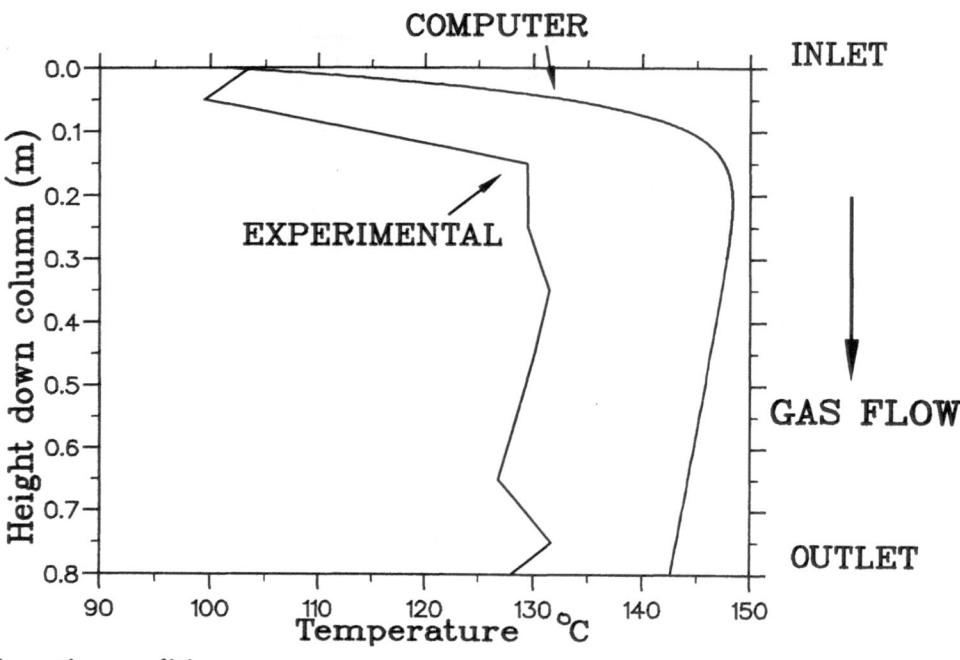

Operating conditions:
Air 0.6g/s Water 0.55g/s @101.2°C   Mode : Cocurrent
(humidity: 0.917g/g)   Temperature lift: 33.7degC
LiBr 66%w/w, 2ml/s @101°C

Figure 4: Temperature profiles in the direct- contact absorber.

INDUSTRIAL APPLICATION

A detailed study has been made of the application of the 2-stage heat transformer to an industrial spray dryer exhaust. Currently this dries 0.50 kg/s of a 20% wt. solution of organic product, consuming fuel at 1.87 MW. The combustion gases are mixed with dilution air to give a gas stream of 7.7 m^3s^{-1} at 550°C at the dryer inlet. 80% of the energy consumed leaves as latent and sensible heat in the exhaust airstream; losses from the uninsulated dryer and ductwork account for the remaining 20%. Current practice is to cool the exhaust stream by direct contact with water in a venturi scrubber prior to its discharge to atmosphere. This has the effect of eliminating dust emissions, but produces a warm and near-saturated exhaust stream which creates a major plume at discharge. An investigation was made of the potential for recovering a portion of this lost energy, utilising a waste heat stream available on site.

Figure 6 shows the plant layout and design for the incorporation of a 2-stage absorption heat transformer. Part of the cool, humid exhaust stream from the induced draught (exhaust) fan is diverted to the A.H.T. where it is dehumidified, reheated and returned to the dryer in place of the dilution air in current use. In addition to fuel savings, a number of advantages would accrue through this process modification. By eliminating dilution air, the entire dryer circuit operates essentially under nitrogen - an advantage when handling oxidisable products. Reduction of stack gas flow also reduces the visible plume. Further, recycling of reheated air allows higher dryer inlet temperatures which, if acceptable for the product concerned, lead to a higher drying capacity.

The spray dryer study indicated that the last two advantages named above - plume reduction and increased capacity - would enable the company to meet environmental and production targets and would themselves justify the heat transformer installation as a retrofit. Fuel savings equivalent to 0.37 MW (20%), worth c.£19k. p.a., are in themselves insufficient justification for the investment. No significant energy saving could be achieved by the use of the waste heat stream alone.

Tunnel dryers, drum dryers and fluidised bed dryers, which utilise lower temperature airsteams (150-200 °C), could benefit correspondingly more from an A.H.T. energy recovery system. In some cases the entire heating load could be met from the low- grade heat used in the A.H.T. The capital required for direct-fired heaters or H.P steam plant could also then be saved. The greatest advantages accrue when the A.H.T. heat recovery is specified at the dryer design stage, where the heat transformer becomes a Unit Operation in its own right. In any particular installation the justification will depend on local factors - fuel costs, environmental considerations and the availability and temperature of waste heat streams - but the principles of design remain common across the whole range of applications.

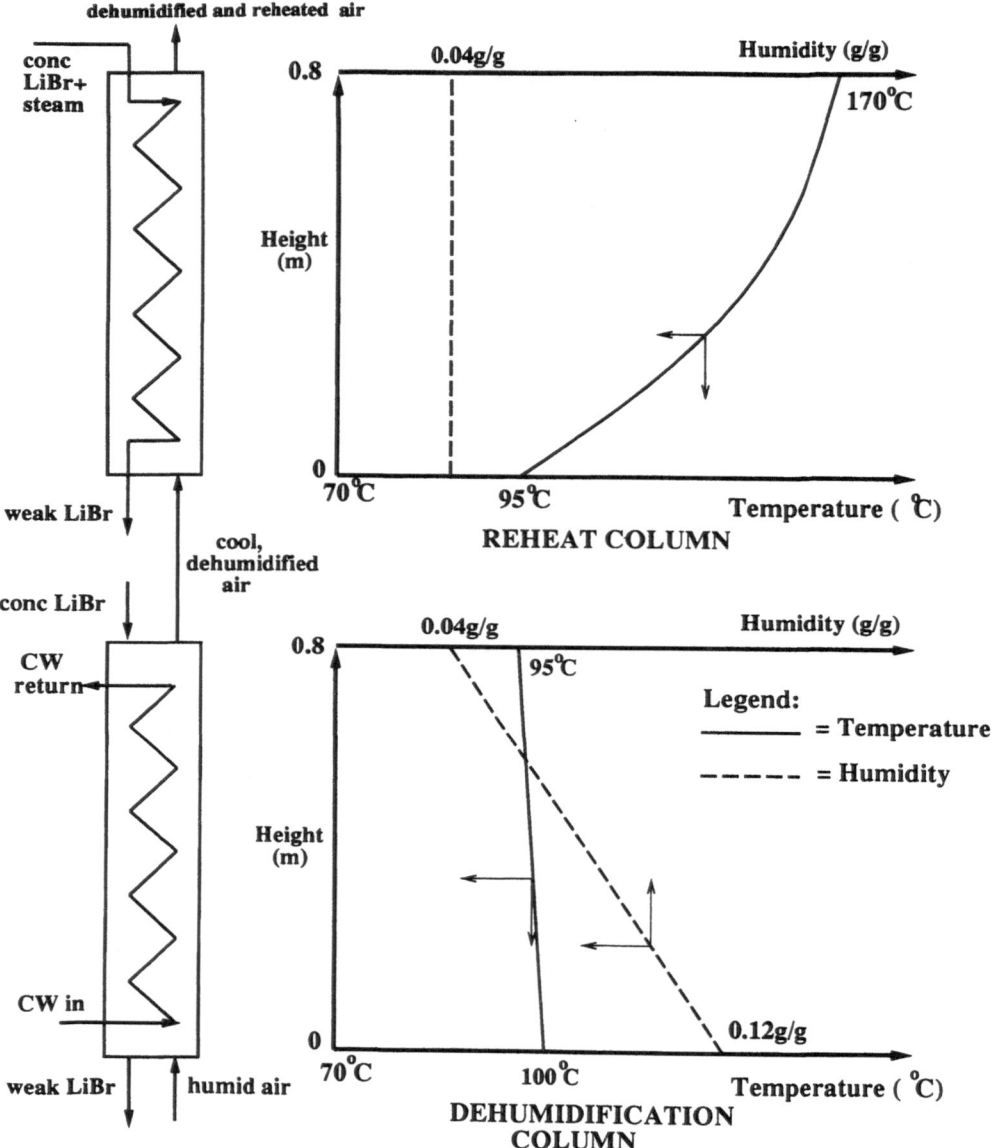

Figure 5: Typical temperature and humidity profiles for air in a 2-stage A.H.T.

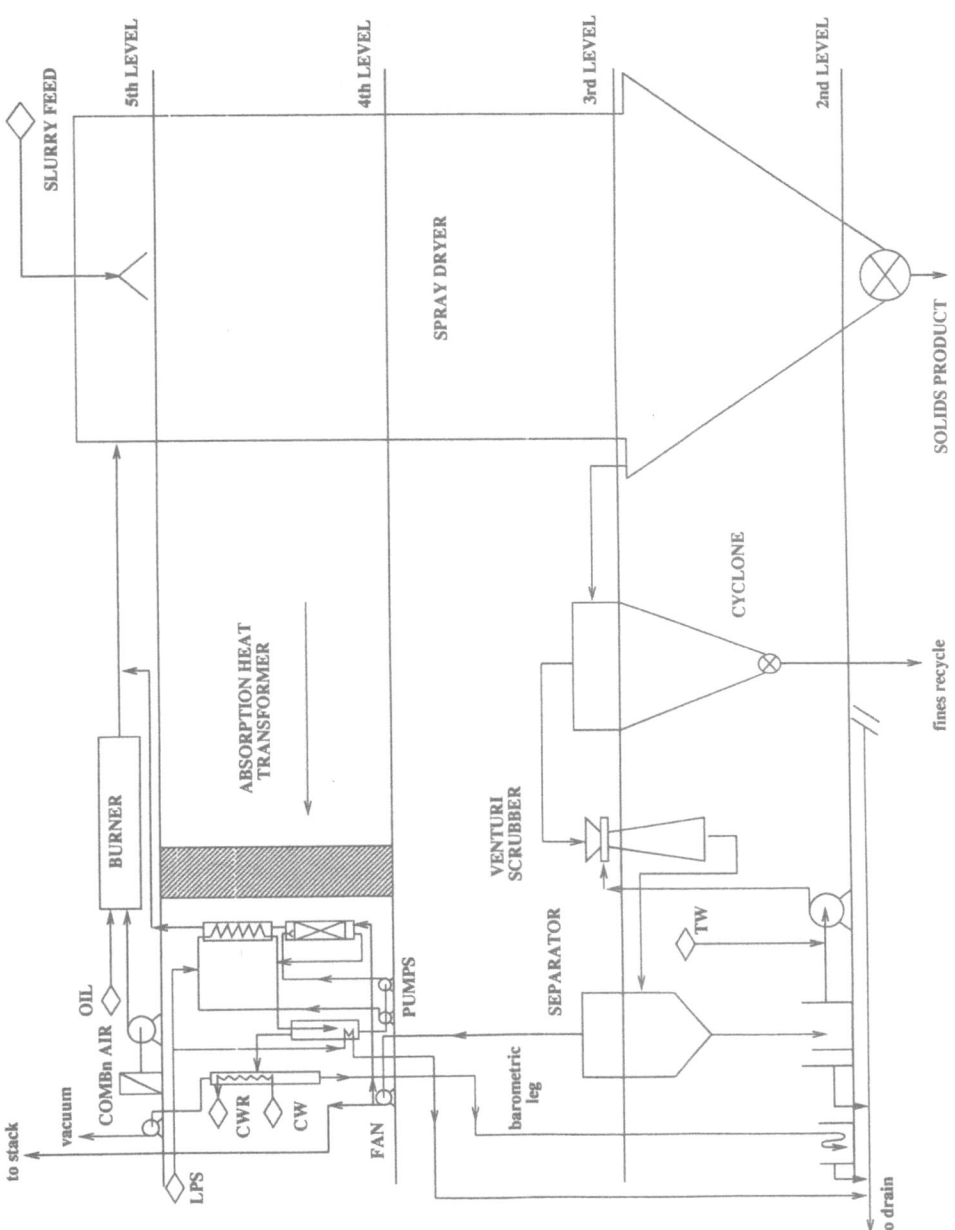

Figure 6: Layout for incorporation of A.H.T in an industrial spray dryer plant.

REFERENCES

(1) Bahu, R., "Dryers" in "Energy Efficiency in Industrial Processes: Future R & D Requirements", 1988, CEC report EUR 12046 EN.

(2) Berntsson, T. M. *et al*, "Heat Transformers in Industrial Processes", 1989, CADDET.

(3) Pritchard, C. L., "An Open Cycle Heat Transformer for Waste Heat Recovery from Dryers", 1990, Proc 3rd Intl Workshop on Research Activities on Advanced Heat Pumps, Graz.

(4) Masters, K., "Spray Drying Handbook", 1991, Longman.

(5) Treybal, R., "Adiabatic Gas absorption and Stripping in Packed Towers", Ind. Eng. Chem. 1969 **61**, (7), 36-41.

(6) Brunk, M.F., "Thermodynamische und Physikalische Eigenschaften der Losung LiBr/H_2O als Grundlage fur die Prozessimulation von Absorptions Kaltenlagen", 1982, Ki Klime Kalte Heizung, 365-376.

(7) Liu, B-q., 1990, PhD Thesis, I.N.P. Lorraine.

NOMENCLATURE

Symbol	Meaning	Units
U_o	Volumetric heat transfer rate in absorber	kWm^{-3}
K_o	Volumetric mass transfer rate in absorber	$gs^{-1}m^{-3}$
h_o	Gas htc based on outside area of pipe	$Wm^{-2}K^{-1}$
$T_{G.in}$	Inlet gas temperature	$^{\circ}C$
$T_{G.out}$	Outlet gas temperature	$^{\circ}C$
$T_{i.in}$	Temperature at inlet to inner tube in reheat column	$^{\circ}C$
$T_{i.out}$	Temperature at outlet from inner tube in reheat column	$^{\circ}C$
$T_{LiBr.in}$	Inlet temperature of LiBr	$^{\circ}C$
Nu	Gas Nusselt Number	-
Re	Gas Reynolds Number	-
Pr	Prandtl Number (humid air)	-

ENERGY-EFFICIENT BATCH DRYING
OF STRUCTURAL CERAMICS AND CONSTRUCTION TIMBER

ALBERT DALHUIJSEN
TNO Institute of Applied Physics
P.O. Box 155, 2600 AD Delft, The Netherlands

BRIAN NORTON
University of Ulster
Newton Abbey, BT37 0QB, Northern Ireland, United Kingdom

ABSTRACT

In the production of structural ceramics and construction timber the drying process not only is an important factor in the overall energy consumption, but also greatly influences product quality and production costs. Optimizing the current drying process is necessary and possible, based on a thorough understanding of the interaction between process parameters. The modern approach is to use computer simulation models to gain this insight and to aid in optimizing the process. In this paper a JOULE project is described, aimed at optimizing batch drying processes for structural ceramics and construction timber through a modular computer-based simulation environment. Some modelling assumptions are discussed and some preliminary results are presented.

INTRODUCTION

Drying of solid products consumes about 10 % of the energy used by the main industrial sectors of Europe [1]. Convection type dryers, employing hot air as the energy source, account for about 90 % of the dryers used in industry at present. Thus it is clear that improved operation and design of industrial convection type dryers will contribute to significant energy savings in Europe and enhance industrial competitiveness.

Structural ceramics are very common in every day life. Products as bricks and roof tiles are employed virtually everywhere. The same goes for construction timber. In the manufacturing of these products, however, the drying process is extremely energy consuming. Moreover, the product quality and production costs are also greatly decided by the drying process. Therefore, optimizing the current drying process is of great concern. For this a thorough understanding of the interaction between process parameters is necessary. Computer simulation models are very useful in obtaining this insight, revealing various interaction mechanisms of importance to the process. Moreover, such models offer the opportunity of testing various equipment design modifications or new process operation strategies without the necessity of experimental tryouts or interruption of manufacturing processes.

Thus, a JOULE project was proposed and accepted, aimed at optimizing batch drying processes for structural ceramics and construction timber, by developing and using a modular computer-based simulation environment. Since convective batch dryers are the most common dryer types in manufacturing structural ceramics and construction timber in many European countries, the project is limited to convection, batch-type dryers. In close cooperation various research groups in Europe are currently contributing to an advanced simulation model of the drying process. The partners involved are TNO (NL), University of Ulster (UK), Staffordshire University (UK), University of Perpignan (F) and Cenertec (P).

STRUCTURAL CERAMICS

The manufacturing process of structural ceramic products, as bricks and rooftiles, can be schematically described as follows: pretreatment of the raw material, forming of products, drying of products, firing of products.

The products are normally dried by placing them on shelves in specially designed chambers and then circulating a current of hot air through the chamber for a given amount of time (typically 30-40 hours), following a specified temperature and humidity schedule. Waste air of the kiln very often serves as a source for hot air. Depending on need, waste air is mixed with fresh air and heated with burners. The water content of the products typically ranges from 30% (on dry basis) before drying, to less then 5% after drying. The main parameters that determine the optimum drying time for this type of process are:
- the maximum permissible rate at wich the fastest drying products within the chamber can be dried, in order to prevent the formation of drying cracks;
- the time taken for the slowest drying products to reach a predetermined moisture level;
- the dimension of the products and their characteristic drying behaviour.

Current practice in industry shows that maintaining uniform drying conditions in the drying chambers is rarely achieved. Non-uniform drying conditions lead to a longer than necessary drying time, higher energy consumption and unequal product quality.

Current practice also indicates that waste heat from the kiln can often be used in a more efficient manner. Moreover, modern kilns often have greatly reduced waste heat due to optimizations, thus requiring more auxiliary heating of the drying air with burners. Lastly, environmental considerations currently often require the usage of different, uncommon clays, thus requiring process modifications and other drying strategies.

Optimization of the drying process, with respect to drying uniformity, optimal conditions for specific raw materials and energy consumption, therefore is desirable.

CONSTRUCTION TIMBER

For nearly all applications, timber has to be dried after cutting and sawing. One method still practised in a few countries is to dry it by air-seasoning outside or in unconditioned rooms. With this method serious problems arise with unequal moisture contents and uneven quality. The drying time, depending on the species, thickness and initial moisture content, takes several months to more than a year.

Therefore, nowadays timber-drying kilns dry the timber at elevated temperatures by circulating a current of hot air (60-80°C) through the kiln following a specified temperature and humidity schedule. Drying times are, however, still in the order of 2 to 8 weeks depending on timber species, thickness and initial moisture content. The consequence is a high energy consumption for heating and ventilation.

One possible solution to overcome both, long drying time and high energy consumption, is thought to be the method of "high temperature drying", developed in Australia, New Zealand and USA for permeable species. The basis for this method is to use temperatures around 110-130 °C, as in this range the timber becomes much more flexible than at low temperatures. Besides the advantage of short drying times (expected to be around 12-24 hours for permeable species) this process requires less energy than conventional methods. However, problems with some species show that high temperature drying is not recommendable for all timber species due to checks and warp under these extreme conditions. Sharp cross-sectional moisture gradients are imposed in the wood, depending on the external drying conditions and the characteristics of the species, thus in some cases causing large, unpermissible stress build-ups within the timber.

For the application of high temperature drying in the timber industry to be widespread in Europe, systematic research that will lead to the confident optimal practical application of this process must be undertaken.

PROJECT DESCRIPTION

In a joint project supported by JOULE, which recently started, a computational environment for batch drying processes is developed. The industrial proces is simulated on various levels of detail and a distinction is made between simulation of the process and the product.

In the product model the drying behaviour of the material itself is simulated. This is shown schematically in figure 1.

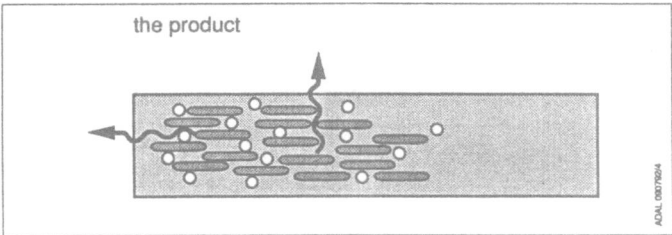

Figure 1. Schematic representation of the drying behaviour
within the product, as modelled in the product model.

In the simulation of the process various levels of detail are distinguished.

In an airflow model the airflow patterns in the drying chamber and the local temperatures and humidities are calculated, schematically shown in figure 2. This model describes heat transfer and flow phenomena in great detail and consists of a customized, advanced air flow model (CFD), based on Patankar's [2] method, for the batch drying process. These modelling efforts give insight into the drying uniformity or alternatively, to optimization measures to existing drying chambers or for the design of new equipment.

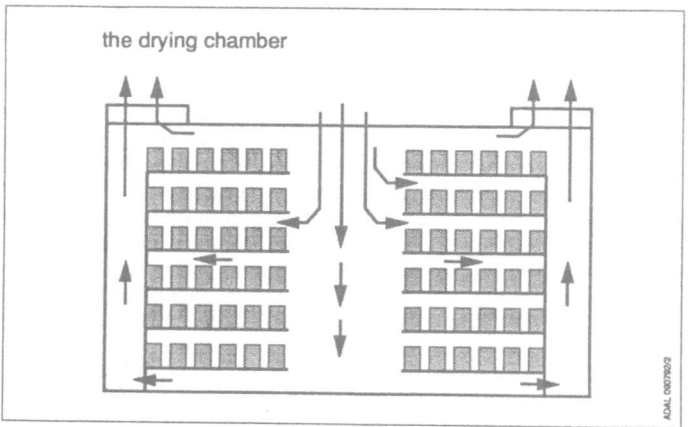

Figure 2. Schematic representation of the airflow in the drying chamber, as modelled in the airflow model.

In the <u>dryer model</u> a more 'macroscopic' approach towards the simulation of drying chamber is applied. The drying rate of the products in the drying chamber, given the material characterization and the supply-air conditions, is calculated. This model aims at providing a quite accurate simulation of the different processes taking place in dryers. Drying times and energy requirements can be derived from a knowledge of the temperature and humidity of the drying air, heat and mass transfer coefficients and from the moisture content of the products to be dried. In order to optimize process conditions, the dryer model is used in conjunction with the installation model. The <u>installation model</u> simulates an entire drying installation, including drying chambers, air blowers, burners, heat exchangers and the related pipework. The model is specifically designed for optimally integrating energy flows in industrial drying systems. Installation and dryer model are schematically shown in figure 3.

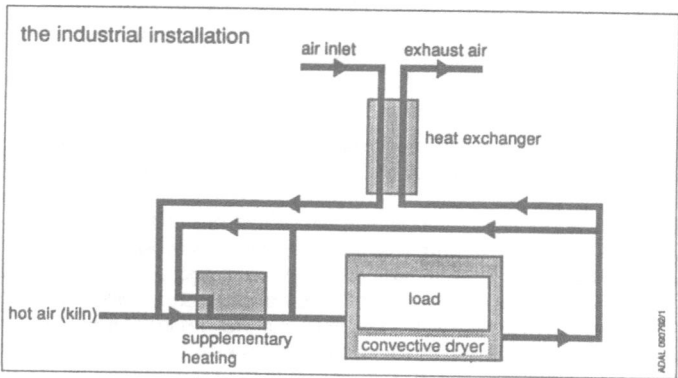

Figure 3. Schematic representation of the drying chamber as part of the industrial installation.

Of course the models do not stand alone but interact, as is schematically shown in figure 4.

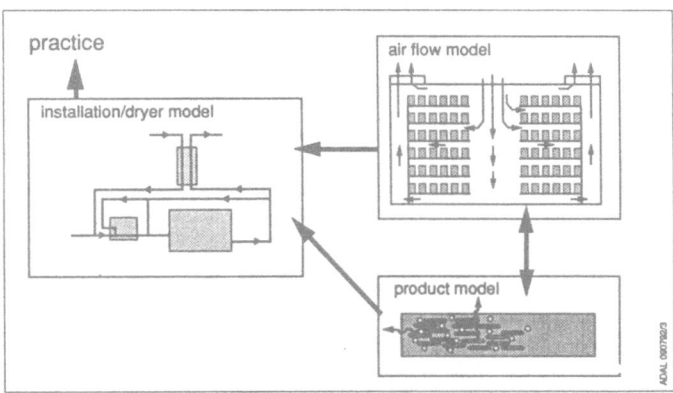

Figure 4. Interaction between the developed models.

With the various levels of approach as described above it is envisaged that the necessary insight into the drying process and the important physical mechanisms can be obtained. Although attention sofar has mainly been focussed on the modelling efforts, experimental input will be necessary for all of the models and experimental validation is of great importance. This is treated in the JOULE project.

PRELIMINARY RESULTS

The JOULE project as described is still in a starting phase, making it difficult to extensively describe the models and obtained results. However, some preliminary information can be given, in part based on previous work of the authors.

For timber drying the moisture transport in the wood is modelled in great detail to study the sharp moisture gradients due to the extreme external conditions imposed in high temperature drying. Water exists in wood either as:
- free water (water that fills the wood cell cavities above Fibre Saturation Point (FSP));
- bound water (water associated with the cell walls by hydrogen bonding);
- water vapour (water present in the air that fills the void spaces not occupied by free water).

While free water is still present within the cell lumen, water vapour is in equilibrium with liquid water at the local temperature. When moisture content drops below FSP, the wood sorption isotherms are used to relate moisture content, local temperature and water vapour pressure.

Several transport mechanisms of moisture and best transfer have been identified [3]. Heat is transported either by conduction through the wood solid matrix or by convection through fluid transport mechanisms. Liquid and gas movement occur either by diffusive or convective transport. A gradient in the mole fraction of each component of the gaseous phase (water vapour and dry air) induces a diffusion process within the gas phase. A total pressure gradient gives risk to bulk flow of the gaseous mixture. Free water flows in bulk form driven by a pressure gradient within the liquid phase. Bound water diffusion through the solid wood matrix arises from the existence of a chemical potential gradient within the adsorbed water molecules.

The modelling process flows closely the one described by Plumb and co-workers [4]. Conservation laws are formulated for total mass, energy and chemical species mass conservation;

$$\frac{\partial \rho}{\partial t} + \overline{\nabla}(\rho \overline{v}) = 0 \tag{1}$$

$$\rho \frac{Dh}{Dt} = -\overline{\nabla} \ \overline{q} + \frac{Dp}{Dt} + \nabla \ \overline{v} : \tau + \phi \tag{2}$$

$$\frac{\partial \rho_1}{\partial t} + \overline{\nabla}(\rho_1 \overline{v}_1) = 0 \tag{3}$$

where ρ=density, v=velocity, h=enthalpy, q=heat flux, p=pressure, τ=shear, Φ=source term. The resulting equations are volume averaged according to the technique described by Whitaker [5]. The use of constitutive equations derived from the use of Darcy's laws of liquid and gaseous phase velocities (equations 4-5), definition of the appropriate conditions at the liquid-solid, liquid-gas and solid-gas interfaces and the adoption of the several assumptions listed in table 1 [6] allow for the conservation equations to be reduced to two coupled differential equations which describe mass and energy transfer within the wood (equations 6-7)

$$v_1 = -\frac{KK_{1r}}{\mu_1}\left(\frac{\partial P_1}{\partial z} - \rho_1 g\right) \tag{4}$$

$$v_g = -\frac{KK_{gr}}{\mu_g}\left(\frac{\partial P_g}{\partial z} - \rho_g g\right) \tag{5}$$

where K=intrinsic permeability, K_{lr}=relative liquid phase permeability, K_{gr}=relative gaseous phase permeability, μ=dynamic viscosity, p=pressure, g=inertia.

TABLE 1
Assumptions used in modelling the high-temperature, pressurised wood drying process.

---

1 The wood cell wall is rigid with constant density.
2 Within the averaging volume the specific heat is constant (hence, enthalpy is a linear function of temperature).
3 Thermal conductive fluxes are expressed by Fourier's law of heat conduction.
4 Compressional work and viscous dissipation are negligible.
5 Thermal conductivity is constant within the averaging volume.
6 The product of deviations is negligible when compared with the product of the averages.
7 Thermal equilibrium is assumed, implying the temperatures in the different phases within a given averaging volume are equal.
8 Diffusive transport of thermal energy is the gas phase is negligible.

$$\frac{\phi}{\Delta MC}\frac{\partial MC}{\partial t}=-\frac{1}{\Delta MC}\frac{\partial}{\partial z}\left[\frac{K_l}{\mu_l}(C_s'-C_s)\frac{\partial MC}{\partial z}\right]+\Delta T\frac{\partial}{\partial z}\left[\frac{K_l}{\mu_l}(C_T'-C_T)\frac{\partial \theta}{\partial z}\right]+\frac{\partial}{\partial z}\left(D_{MC}\frac{\partial MC}{\partial z}\right) \tag{6}$$

$$\rho\, c_p\frac{\partial \theta}{\partial T}-\frac{\rho_l C_{pl}K_l}{\mu_l}\left[\frac{C_s'-C_s}{\Delta MC}\frac{\partial MC}{\partial z}+\Delta T(C_T'-C_T)\left(\frac{\partial \theta}{\partial z}\right)\right]\left(\frac{\partial \theta}{\partial z}\right)=\frac{\partial}{\partial z}\left(k_{eff}\frac{\partial \theta}{\partial z}\right) \tag{7}$$

where MC=dimensionless moisture content (dry basis), D_{MC}= bound water diffusion coefficient, k_{eff}=effective thermal conductivity and C_x's are transport coefficients.

The transport coefficients, C_s, C_s', C_T, C_T', are defined in equations 8-11, $\Delta T=T_1-T_0$ and $\Delta MC=MC_{max}-MC_{FSP}$.

$$C_s=\frac{\partial P_g}{\partial g} \tag{8}$$

$$C_s'=\frac{\partial P_g}{\partial g} \tag{9}$$

$$C_T=\frac{\partial P_g}{\partial T} \tag{10}$$

$$C_T'=\frac{\partial P_g}{\partial T} \tag{11}$$

Additional simplifications include a one-dimensional approach, elimination of the air-species continuity equation and gravitational effects, the formulation of an identity between water vapour and bound water diffusion and the use of the heat and mass transfer analogy to identify heat and mass transfer coefficients at the surface. Additionally, radiation heat transfer was accounted for in the description of the energy exchange at the surface. The transport coefficients are related to the dependent variables (moisture content and temperature) by a mechanistic model first developed by Comstock [7].
Appropriate sets of boundary and initial conditions are required to undertake a simulation. The differential equations are approximated by a finite difference technique. A solution was obtained using an implicit numerical upwind scheme according to the formulation described by Patankar [2]. The numerical solution describes temperature, internal pressure and moisture distribution as functions of a spatial coordinate and time. Furthermore, the model is used to predict drying rates and heat and mass transfer coefficients according to drying conditions.
Results of calculations in comparison to experiments are presented in a companion paper at this conference: "High temperature presssurized wood drying; experiment and simulation" by B. Norton and F. Neto da Silva. Further work will be performed in the framework of the JOULE project, extending the model to include drying induced stresses.

For clays, a product model is under development giving the drying rate in the first and second drying phase, depending on air temperature and humidity around the products, taking into account the shrinkage of the product. This model is incorporated into the dryer model. Necessary input to the product model, which will be different for different clays, is obtained from relatively simple laboratory drying experiments with one or two products. With this information, along with information concerning the industrial drying chamber (air flow through the chamber, air supply temperatures and humidities in time, geometry of the chamber, etcetera), the drying rate of a complete charge of bricks, for instance, can be predicted. A preliminary result, relating to an existing industrial dryer is given in figure 5. The figure shows the maximum and minimum drying rate in the drying chamber which is operated at "plug flow" conditions (in contrast to an ideally stirred chamber). Condensation occurs at the beginning of the drying process. The figure also shows the maximum allowable drying rate in the first phase of the drying process as determined in a laboratory dryer.

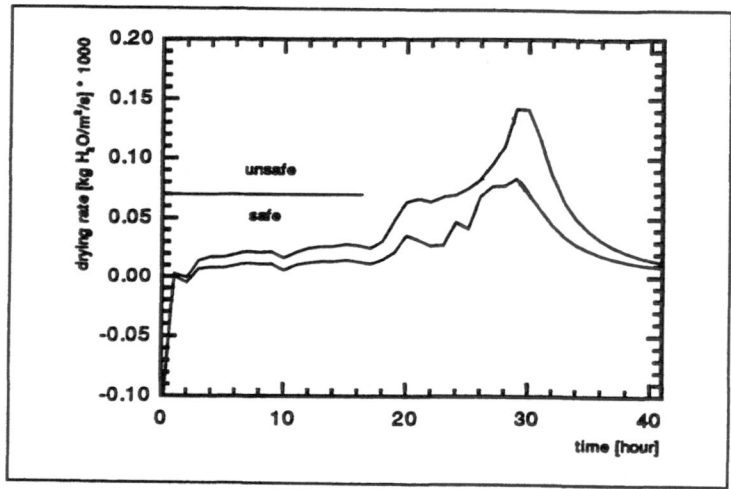

Figure 5. Maximum and minimum drying rate in the chamber when operated at plug flow conditions (as calculated).

The results indicate that the performance in the first phase of the drying process can be improved. Although plug flow conditions give rise to non-uniform drying in chamber, the overall drying time is less than in case of an ideally stirred chamber, see figure 6. This indicates that by combining the advances of plug flow and measures to improve drying uniformity, for example alternating the current of air, a more economical drying process can be designed.

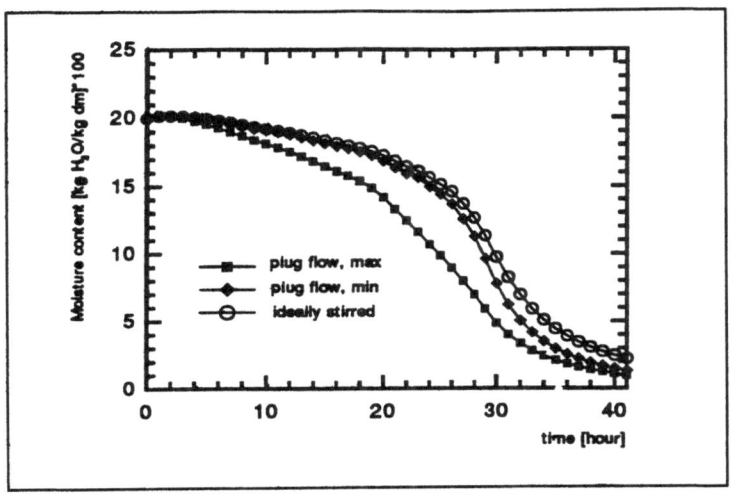

Figure 6. Drying curve with the chamber operated at plug flow conditions and ideally mixed conditions.

DISCUSSION

Numerical modeling of the drying of structural ceramics and construction timber in batch, convective dryers is the main subject of the recently started JOULE project. The process is simulated on various levels: a macroscopic approach (dryer model and installation model) and a more microscopic approach (airflow and detailed product model) is employed throughout the project. Preliminary results in simulating the drying behaviour within the naterial (timber drying) and the drying rates on an industrial scale (bricks) are promising and will be extended in the very near future.

REFERENCES

1. Bahu, R., Dryers. Energy Efficiency in Industrial Processes, Future R & D Requirements. Proc. CEC Seminar, 30 June-1 July 1988, Ed. P.A. Pilavachi, Commission of the European Communities, p.183-199
2. Patankar S.V., Numerical Heat Transfer and Fluid Flow, Hemisphere Publishing Corporation, 1980.
3. Stanish M.A., Schajer G.S. et Kayihan F., Mathematical Model of Wood Drying form Heat and Mass Transfer Fundamentals, Drying 85, pp 360-367, 1985.
4. Plumb, O.A. Spolek, G.A. Olmstead, B.A. Heat and mass transfer in wood during drying, Int.J.Heat Mass Transfer, 28, 9, 1669-1678, 1985
5. Whitaker S.A., Simultaneous Heat, Mass and Momentum Transfer in Porous Media: a Theory of Drying, Advances in Heat Transfer, 13, 119-203, 1977.
6. Neto da Silva, F.J., High-temperature, pressurized wood drying: experimental and simulation tests, Ph.D Thesis, Cranfield Institute of Technology, U.K., 1992.
7. Comstock J.L., Directional Permeability of Softwoods, Wood and Fiber, 1, 4, 183-189, 1970.

HIGH TEMPERATURE, PRESSURIZED WOOD DRYING: EXPERIMENTAL AND SIMULATION RESULTS

Brian Norton,
PROBE: centre for Performance Research On the Built Environment,
Department of Building and Environmental Engineering,
University of Ulster,
Newtownabbey BT37 0QB Northern Ireland

Fernando Neto da Silva,
Escola Superior de Tecnologia e Gestào
Instituto Politécnico da Guarda
6300 Guarda, Portugal

ABSTRACT

High Temperature, high pressure drying of wood is more rapid, makes more efficient use of energy and ensures good final product quality. It is shown that a newly-developed simulation model predicts the drying process at temperatures above 100°C and pressures greater than one atmosphere.

BACKGROUND

Recently felled logs can have moisture contents above 170% (dry basis) or more. The required final moisture content varies with the environment where wood will be used; however, even for garden furniture, the recommended level of moisture content should not exceed 16% (dry basis). The world annual energy consumption during 1980 for wood drying operations was 2×10^{17} J [1] and drying operations consume about 70% of the total energy used in the manufacture of most wood products [2]. At a given pressure equilibrium moisture content decreases with an increase in temperature. However, higher operating pressures lead to an overall **Increase** in the equilibrium moisture content; it is then possible to accelerate the drying process using temperatures above 100°C.

First attempts to reduce wood drying time using high temperatures date back to the middle of the 19th century [3]. Though subsequent practical studies, some at pressures above atmospheric, have been undertaken [4-11], the lack of an adequate simulation model has inhibited process optimisation.

EXPERIMENTAL DESIGN

A 202mm diameter cylindrical drying chamber held a single sample of sawn timber of maximum dimensions 1000 x 150 x 25mm. The drying fluid, superheated steam, was recirculated in a closed loop by a centrifugal fan. Given the high operating temperatures, the fan bearings were water cooled. A three-phase, 12kW electrical heater, installed in a connecting pipe, was connected to a proportional controller which kept the temperature of the drying fluid at a constant value. The maximum operating temperature was 220°C.

The wood sample was suspended from two 1.6mm thick stainless steel cantilevers each of which incorporated strain gauges. On-line recording and analysis of the strain-gauge output voltages facilitated continuous monitoring of the wood mass and of the drying rate. Pitot pipes linked to a differential micro-manometer were installed to monitor local velocity at the wood surface. Humidity was measured using a dry and wet bulb hygrometer. Wet bulb temperatures were measured by inserting a thermocouple within a water spray emerging from a nozzle located at an inner wall. Pressure variation drying the drying process was recorded using a pressure transducer. Metal-sheathed type-T thermocouples with a diameter of 1mm were inserted into the wood sample at the surface and at depths from the surface extending from 3mm to 12mm. The energy consumed by the heater was measured.

EXPERIMENTAL RESULTS

At the start of the drying process, the drying fluid was essentially air. As moisture evaporated from the wood, the pressure within the rig increased until the required pre-set value was achieved The set of experimental conditions employed are shown in Table 1. Drying conditions were attained after an initial heating and pressurisation period of one hour. Equilibrium moisture contents varied between 1.3% for 150°C at atmospheric pressure to 8% for 130°C at 202650 Pa.

TABLE 1

Drying conditions and initial moisture content for all samples

Experiment	Dry bulb (°C)	Total pressure (Pa)	Free Stream Velocity (m/s)	Initial Moisture Content of Wood (%)
1	130	101325	7.2	209.7
2	130	202650	5.1	205.8
3	135	101325	7.3	219.3
4	135	151988	6.1	208.2
5	135	202650	5.2	243.9
6	135	253313	4.6	203.3
7	150	101325	7.5	235.8
8	150	151988	6.3	260.5
9	150	202650	5.4	267.1
10	150	303975	4.2	265.0

The required drying times, as shown in Table 2, varied between 5.7 h (for atmospheric pressure and a dry bulb temperature of 150°C) and 17 h (for a total pressure of 253313 Pa and a dry bulb temperature of 135°C). Drying time increased with an increase in the total pressure. The required drying time was also reduced if it is conducted at a higher dry bulb temperature.

TABLE 2

Duration of drying in hours depending on drying conditions

Pressure (Pa)	Dry bulb temperature (°C)		
	130	135	150
101325	7.5 hr	9.9 hr	5.7 hr
151988		11.5 hr	7.5 hr
202650	16.3 hr	11.8 hr	7.3 hr
253313		17.0 hr	
303975			8.7 hr

The most common defects in the dried wood were cracked knots and darkening with surface staining being associated with higher applied pressures; a summary of the defects observed is given in Table 3.

TABLE 3
Drying defects dependence upon drying conditions.

Pressure (Pa)	Dry bulb temperature (°C)		
	130	135	150
101325	Cracked knots. Surface darkening.	Cracked knots. Surface darkening and throughout darkening.	Cracked knots. Internal splits. Mild bowing
151988		Surface and throughout darkening. Severe bowing.	Cracked knots. Surface darkening and throughout darkening. Mild cupping.
202650	Cracked knots. Severe bowing. Blue staining	Cracked knots. Surface and throughout darkening. Blue staining. Mild cupping.	Cracked knots. Surface darkening. Mild cupping.
253313		Cracked knots. Surface darkening and throughout darkening. Internal splits. Warping. Blue staining. Collapse.	

TABLE 4
Heating energy consumption (MJ) dependence upon drying conditions

Pressure(Pa) [bar]	Device	Dry bulb temperature (°C)		
		130	135	150
101325 [1]	Heating	120 2 MJ	141.0 MJ	
151988 [1.5]	Heating		261.2 MJ	128.0 MJ
202650 [2]	Heating	221 3 MJ	144.3 MJ	115.4 MJ
253313 [2.5]	Heating		213.5 MJ	
303975 [3]	Heating			124.4 MJ

Higher drying temperatures and pressures in these particular experiments gave similar heating energy usage as shown in Table 4. Energy consumed in recirculating the drying fluid increased as pressure increased; fan energy was 74.7 MJ at 130°C and 101325 Pa but 179.8 MJ at 150°C and 303975 Pa. However, as optimum drying conditions were not necessarily applied, these values cannot be compared with data obtained from different experimental or commercial installations.

COMPARISON WITH SIMULATION RESULTS

A description of the mathematical basis and underlying assumptions in the simulation model is provided in a companion paper in these proceedings [12].

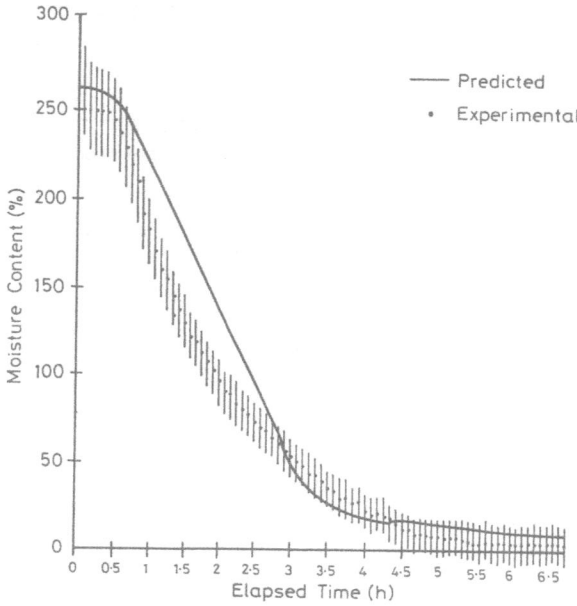

Figure 1:   Predicted and measured average moisture content for the drying conditions of experiment 8 as described in Table 1.

The predicted and measured average moisture content evolution for a dry bulb temperature of 135°C and a total pressure of 151988 Pa is shown in figure 1. The vertical bars at each measured value represent experimental error. The model underestimates water evaporation during the initial drying period; the straight line suggests that the duration of the constant drying rate period has been over-predicted by the simulation model. The almost continuously decreasing slope of the experimental curve reveals that a constant drying rate period persists for only a short period at the beginning of the drying process. During the final stages the average moisture content decreased rapidly towards zero. The measured decline of the drying rate is faster than predicted which leads to a value of the average final moisture content being lower than predicted.

Figure 2 represents the predicted and the measured drying rate for identical values of the dry bulb temperature and of the total pressure. The experimentally measured drying rate during the initial period is slightly higher than predicted by the simulation model, surface moisture content thus decreased more rapidly than predicted. Whilst difficult to delineate from the measured data, the simulation model predicts the existence of a constant drying rate period. A decreasing drying rate period is described both experimentally and numerically. However due to the different duration of the constant drying rate period, the magnitude of the experimental and predicted drying rate during this period is different. The agreement between measured and predicted results is shown in figure 2.

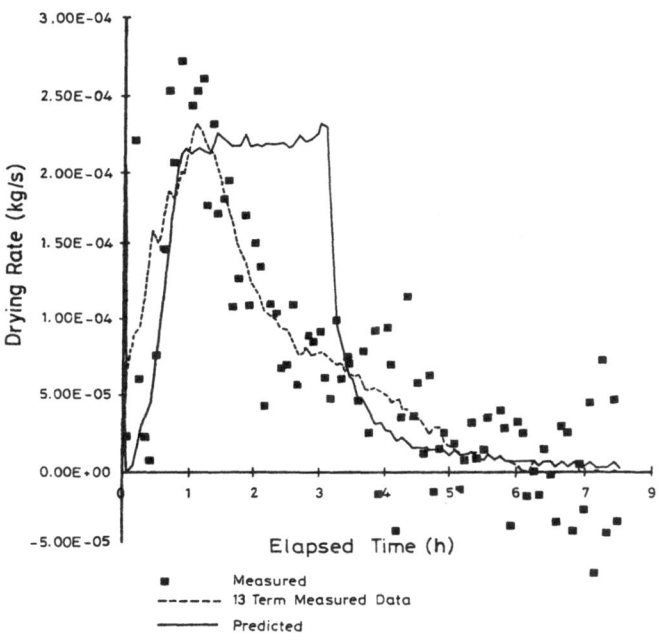

Figure 2:   Predicted and measured drying rate for the drying conditions of experiment 8 as described in Table 1..

The moisture content evolution predicted for several locations within the sample has been plotted in figure 3. During an initial phase, moisture content drops steadily at all locations and the moisture content gradient between the surface and the centre is not pronounced since relative liquid permeability does not differ significantly. At this stage, the lack of a significant moisture content gradient indicates that the rate of evaporation is controlled solely by the amount of moisture that can be evaporated at the surface.

As surface moisture content decreases, the value of the relative liquid permeability drops rapidly and the rate at which liquid water reaches the surface decreases. The moisture content gradient between the surface and the centre increased progressively. The surface zones enter the hygroscopic domain and the flux of liquid water to the surface is interrupted: water is transported to the surface solely by diffusion mechanisms.

As drying continues, water is no longer present as liquid and diffusion mechanisms, developed in response to a moisture content gradient, then control the drying process. Surface moisture content attains an equilibrium value with respect to the drying conditions within the kiln and moisture content within the inner regions decreases continuously leading to a progressive decrease of the moisture content gradient.

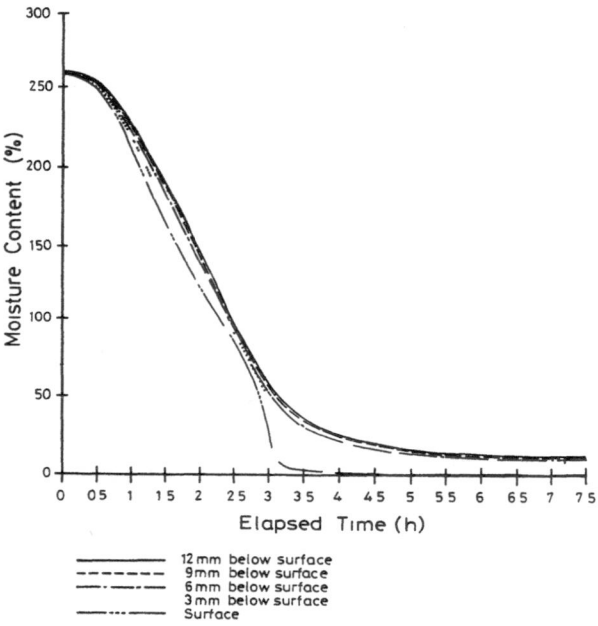

Figure 3:   Predicted moisture content evolution for the drying conditions of experiment 8 as described in Table 1..

As may be seen from figure 4 the model describes correctly the temperature increase during the warm up period and part of the residence period in which the temperatures remained at a constant value equal to the wet bulb temperature. However, the model does not predict the existence of a thermal gradient during the last drying stages which clearly exists in the measured temperature profiles. This is because the model does not account for energy transport within the diffusion movement of bound water.

Temperature profiles describing more faithfully the occurrence of thermal gradients during the last phase of the wood drying process can only be provided by more complex models (e.g. [13]) which are unsuitable for process optimisation.

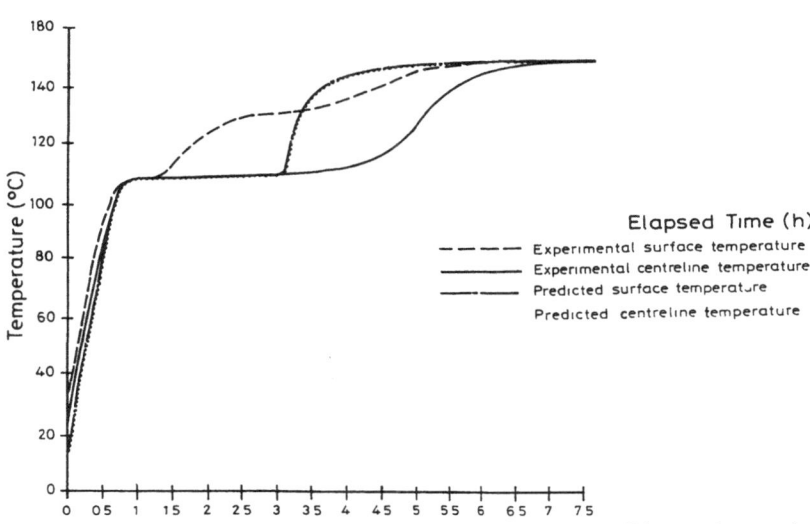

Figure 4: Predicted and measured temperatures for the drying conditions of experiment 8 as described in Table 1..

Figure 5 illustrates the development of pressure gradients within the wood sample. The boundary conditions at the surface require the surface pressure to be identical to the total pressure within the drying chamber. Hence, as free water leaves the cells increasing the effective void fraction and surface pressure increases above the atmospheric value, a small partial vacuum develops within the wood during the initial stages.

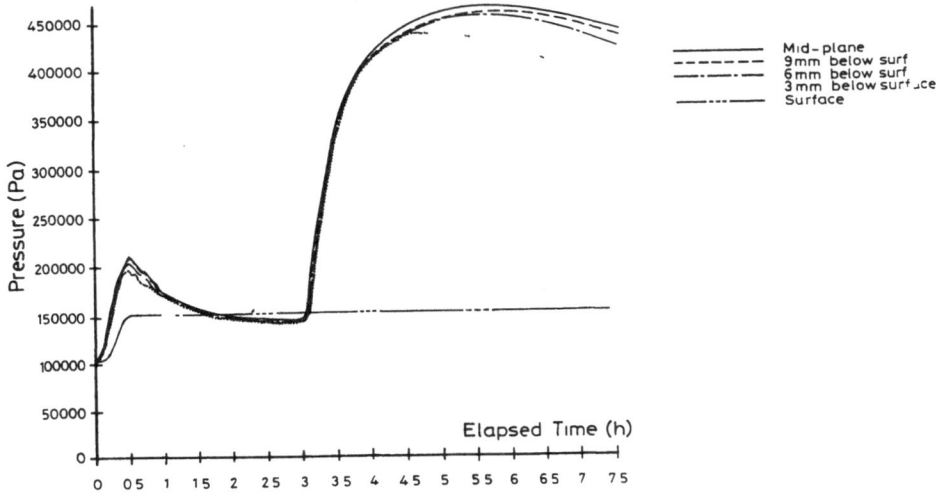

Figure 5: Predicted total pressure evolution for they drying conditions of experiment 8 as described in Table 1.

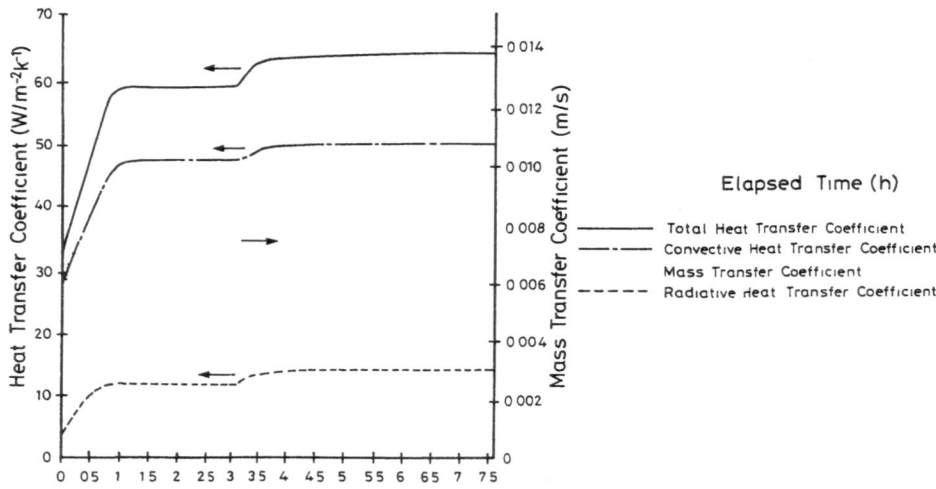

Figure 6:   Predicted surface transfer coefficients for the drying conditions of experiment 8 as described in Table 1.

Figure 6 reveals that both the convection and the radiation heat transfer coefficients increase with dry bulb temperature. Furthermore, the predicted results reveal that surface transfer coefficients do not depend upon surface moisture content which is a direct consequence of the assumption that the heat and mass transfer analogy could be extended to hygroscopic regions. The figure provides also a comparison of the convection and radiation contribution towards the energy transport into the wood, showing that radiation heat transfer contributes to 20% of the total heat transferred to the wood surface. This figure illustrates that in high temperature wood drying simulation models radiation heat transfer should not be neglected.

CONCLUSION

A newly-developed numerical model has been shown to simulate the wood drying process at temperatures above 100°C and pressures above 1 atm. Though the rate of energy consumption is similar for drying at high temperatures and pressures, the duration of the process is shorter thus allowing greater utilization of the capital, and initial energy, investment in the plant. The model can be used to design optimal drying schedules aiming at increasing the final product quality, minimising energy consumption and required drying time.

ACKNOWLEDGEMENTS

The financial support of the Commission of the European Communities through their Mobility of Researchers programme is acknowledged

REFERENCES

1. Simpson W.T. "Drying Wood   A Review - Part 1" Drying Technology, 2, 2, 235-264, 1984.

2. Rosen H.N. "High Temperature Initial Drying of Wood·   Potential for Energy Recovery", Forest Products Journal, 34, 3, 10-18, 1984.

3. Kimball K.E. and Lowery D P   "High-Temperature and Conventional Methods for Drying Lodgepole Pine and Western Larch Studs", Forest Products Journal, 17, 4, 32-40, 1987.

4. Villiere, A. Novvelles techniques de sechage artificiel   sachage a haute temperature. Sechage par contrifugation   Revue Internationale du Bois,195-197, 214, 1953.

5. Kauman W.G. "Equilibrium Moisture Content Relations and Drying Control in Superheated Steam Drying", Forest Products Journal, 6, 9, 328-332, 1956.

6. Kollmann F.F.P. "High Temperature Drying   Research, Application and Experience in Germany, Forest Products Journal, 11, 11, 508-515, 1961.

7. Wengert E M "Review of High-Temperature Kiln, Drying of Hardwoods", Southern Lumberman, v.225, n 2794, pp 17-21, 1972

8. Christensen F.J. and Barker L S 'High Speed Drying Research and the Development of an Experimental Continuous Feed Mechanical Kiln for Sawn Timber", Australian Forest Industries Journal, v 39, n 6, pp 29-37, 1973

9. Fung P Y "High Temperature Drying of Australian Hardwoods", Australian Forest Industries Journal, pp 46-50, 1976

10. Boone R.S. "High-Temperature Kiln Drying of 4/4 Lumber from 12 Hardwood Species", Forest Products Journal, 34, 3, 10-18, 1984

11. Rosen H.N., Bodkin R E et K D Gaddis "Pressure Steam Drying of Lumber" Forest Products Journal, 33, 1, 17-24, 1983

12. Dalhuijsen A., and Norton B , "Batch Drying of Construction Timber and Structural Ceramics" Proc. of Energy Efficiency in Process Technology, Athens, Greece, October 1992.

13. Stanish M.A , Schajer G S et Kayihan F "Mathematical Model of Wood Drying form Heat and Mass Transfer Fundamentals", Drying '85, pp 360-367 1985.

TECHNO-ECONOMIC EVALUATION OF TEXTILE BOBBINS DRYING PROCESSES

J.RIBEIRO (*), I.CABRITA(*), J.VENTURA(**), J.M.FIADEIRO(***)

(*) Departamento de Energias Convencionais
Laboratório Nacional de Engenharia e Tecnologia Industrial
Azinhaga dos Lameiros, 1699 Lisboa Codex - PORTUGAL

(**) Departamento de Engenharia Mecânica
Instituto Superior Técnico
Avenida Rovisco Pais 1096 Lisboa Codex - PORTUGAL

(***) Centro Tecnológico das Ind. Têxtil e do Vestuário de Portugal
Delegação da Covilhã
Rua Conde da Ericeira 6200 Covilhã- PORTUGAL

ABSTRACT

Drying of textile bobbins was carried out on two kinds of laboratory-scale dryers; in a cross circulation drying rig, atmospheric pressure heating air is used, while in the through circulation drying apparatus, the heating air is pressurized before the drying process takes place.

In both processes, air was electrically heated; different air temperatures and velocities were used. Inside the bobbin, seven thermocouples were placed, in order to provide temperature data during the drying process.

In through circulation drying, plots of temperature against time indicated the presence of an evaporation front, moving from inside the bobbin to the outside. In cross circulation drying, two moving fronts were observed, one starting from inside the bobbin, another from the outside.

Comparison of the two drying processes shows 'through circulation' drying to be much faster (about eight times) than 'cross circulation', although the first process needs extra energy for drying air pressurisation.

Experimental results so far indicate temperature to be the most important factor; although the cost to produce higher temperature is higher, drying times are shorter, and these two opposed effects result in an overall economy. However, temperature values are limited by the nature of the material, the limit commonly used in industry being 90°C.

INTRODUCTION

Drying process of textile bobbins is normally carried out in two steps, centrifugal extraction of water followed by heat input that gives rise to evaporation of moisture. Mechanical drying processes, are more economical than thermal ones, however, they are not sufficient to reach desirable drying levels. Hence, mechanical drying must be usually followed by thermal drying

Thermal processes used in the textile industry, particularly in the wool industry, consist of passing a hot air stream over the surface of the material to be dried. Air flow transfers heat to the material by forced convection which, at the same time, carries away evaporated water. The process continues until equilibrium is attained, depending upon drying air temperature and humidity.

Drying rate depends upon several factors, that primarily include the material structure, air temperature, air humidity, air flow turbulence, material thickness and exposed surface area.

Data on textile bobbins drying are insufficient. Nissan *et al.* (1959) did some work with woollen flannel of closed structure. The material was wound as cylindrical bobbins and was dried in a wind tunnel in which hot air flowed along the axis of the cylinder. Heat transfer was observed to take place solely from the outer surface. A more recent study using cans of scoured wool, was conducted by Walker (1969). This work aimed at determining the drying rate curve and the temperature distribution.

In the present work, experiments were carried out using Cheviot wool bobbins (500 g net weight) with an approximate conical form. Internal diameter was varied from 0.035 to 0.06m. External diameter was kept between 0.13 and 0.16m. The bobbins were 0.15 m long and were supported in a polyethylene structure with holes.

In industrial drying processes, textile wool bobbins are mounted one on top of the other, forming a long cylinder. In the work reported in this paper, only one bobbin was used with bottom and top covered by stainless steel plates. The textile bobbin was arranged with 7 copper/constantan thermocouples located in radial direction and equidistant from both the lower and upper edges, as shown in Fig. 1. Preliminary experiments showed that temperature differences along the bobbin were not important.

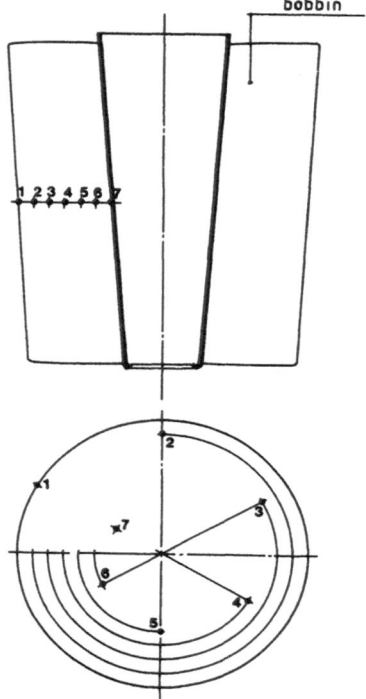

Fig. 1 - Sketch of bobbin with thermocouples in position.

THE EXPERIMENTAL SET-UP

As stated before, two types of dryer were used: A cross circulation drying rig, where heating air was at atmospheric pressure, and a through circulation apparatus, with pressurized drying air.

Cross circulation drying.

Hot air, at the required temperature and velocity, was directed to the bobbin located axially inside a 20 cm i.d. duct. An electronic scale allowed continuous weighing of the bobbin. Temperature and weight data were monitored and acquired by a datalogger, which sent them to a file on a microcomputer.

Air velocity was monitored by an orifice plate located on the air duct. Air was electrically heated and its temperature controlled by a PID regulator. Values of air temperature, air humidity and temperatures inside the bobbin, were fed at previously defined time intervals to the datalogger, to be recorded in the data file. Figure 2 shows a sketch of the experimental set up.

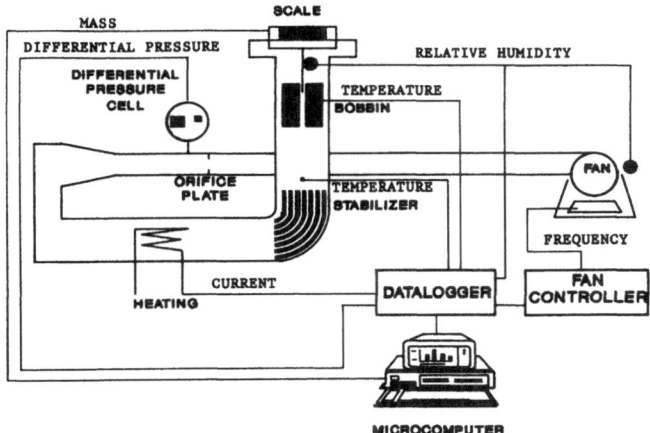

Fig. 2 - Schematic diagram of the wind tunnel.

Words in small type denote measured properties.

Through circulation drying.

Further experiments were carried out on a pressurized rig, in which air was forced to move through the bobbin from inside to outside. Five bobbins were put top to top and the second bobbin was fitted with thermocouples. Air was electrically heated and its temperature set by a PID controller. Air velocity was controlled by two valves and monitored by a rotameter. Fig. 3 shows a sketch of the experimental rig.

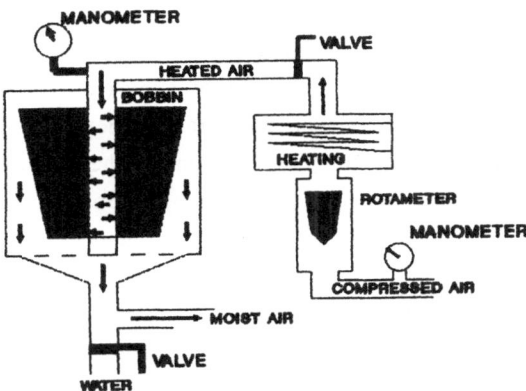

Fig. 3 - Schematic diagram of the pressurized rig.

EXPERIMENTAL RESULTS.

Cross Circulation Drying.

In these experiments, temperatures of 70, 80 and 90 °C and air velocities ranging from 2 to 4 m/s were used. From the raw data, profiles of temperature and moisture content against time were obtained, a. typical result being shown in Fig. 4.

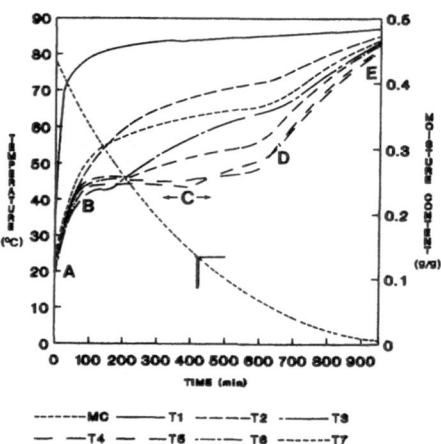

Fig. 4 - Temperature and moisture content along the time. Duct drying.

The curve labelled MC refers to the bulk moisture content of the bobbin.
Curves labelled Ti refer to temperatures inside the bobbin (T2 to T6) and on
exterior surface (T1) and interior surface (T7).

It appears that, after the first stage corresponding to the heating of the bobbin (Region AB), a levelling of temperature is observed for all the interior points (Region BC). The value at which the levelling in temperature occurs, and its duration, depend on the position of the thermocouple and correspond to a state of equilibrium reached between the amount of heat transferred to the material and the heat lost by evaporation. The constant value of the temperature at this stage is referred to as "pseudo-wet-bulb temperature" (Nissan *et al.*, 1959).

At the end of this constant period for each location on the interior, the bobbin becomes dried, hence there is no more water to evaporate and as a result the equilibrium can no longer be maintained and the temperature rises again. However, there is still water vapour diffusing outward from inner and cooler layers, which slows down the rate at which the temperature rises (Region CD). When the bobbin is dried in all points, the rising of temperature becomes stronger (Region DE) and the experimental test run is completed.

<u>Evaporation front</u>. The points C (end of the constant temperature zone) for each thermocouple location can be associated to an evaporation front which is the boundary between two zones, namely the wet, liquid moisture diffusion zone, on the interior, and the dry, vapour diffusion zone, on the exterior. As air convection drying takes place on both sides of the bobbin, two evaporation fronts develop progressing towards each other until they meet (Point D). This point is used to define the end of drying.

Table 1 shows drying times for each thermocouple location, for different air temperatures and velocities.

TABLE 1
Drying time in duct experiments (min.)

		Air Temperature (°C)									
		90			80			70			60
		Air velocity (m/s)			Air velocity (m/s)			Air velocity (m/s)			Air V. (m/s)
Point Position	Radius (cm)	4	3	2	4	3	2	4	3	2	3
1	6.65	0	0	0	0	0	0	0	0	0	0
2	5.97	44	45	57	52	54	64	69	58	82	63
3	5.29	194	224	237	244	258	297	271	272	308	321
4	4.61	387	473	480	488	520	591	597	626	641	797
5	3.93	445	479	477	508	552	562	567	662	653	817
6	3.25	115	147	158	164	178	166	217	224	223	289
7	2.57	0	0	0	0	0	0	0	0	0	0

From the values in Table 1, Figure 5 was drawn, which shows the position of the evaporation fronts with time.

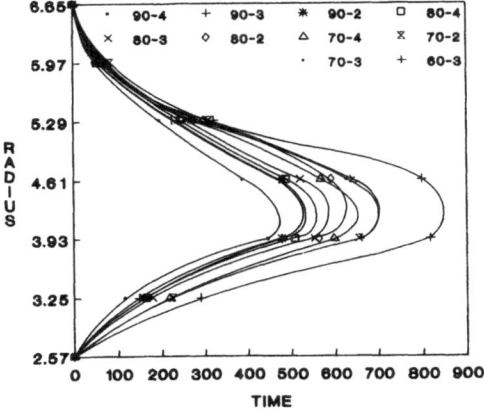

Fig. 5 - Evaporation Front Disappearance. Duct Experiments.

The slight non-symmetry observed in the curves in Fig. 5 can be explained by the geometry of the bobbin, with different heat transfer areas on the outside and inside. From Fig. 5 drying times can be obtained, and these are shown in Table 2.

This Table also shows the final moisture content of the bobbin, when the drying criterion is the disappearance of the evaporation fronts. However, wool is considered to have an equilibrium moisture content of 18.25% on dry basis, for standard temperature and humidity conditions (*i. e.* 20 °C and 65% RH). This means that beyond 18.25% it is not necessary to continue the drying process. This implies that the above criterion will in general cause overdrying of the bobbins. Energy used to evaporate 1 kg of water, neglecting blower consumption, is shown in the last column of Table 2.

TABLE 2
Evaporation front disappearance and energy consumption. Duct experiments.

Air Temperature	Air Velocity	Drying Time	Initial Moisture Content	Final Moisture Content	Energy consumption
oC	(m/s)	(min.)	(kg/kg)	(kg/kg)	(MJ/kg)
90	4	475	0.449	0.097	1620
	3	521	0.497	0.087	1307
	2	539	0.510	0.115	935
80	4	575	0.481	0.096	1755
	3	607	0.459	0.087	1438
	2	632	0.490	0.096	942
70	4	661	0.476	0.109	1764
	3	718	0.470	0.110	1465
	2	707	0.483	0.123	962
60	3	846	0.449	0.110	1466

Figure 6 plots this energy data against air velocity for the temperatures used.

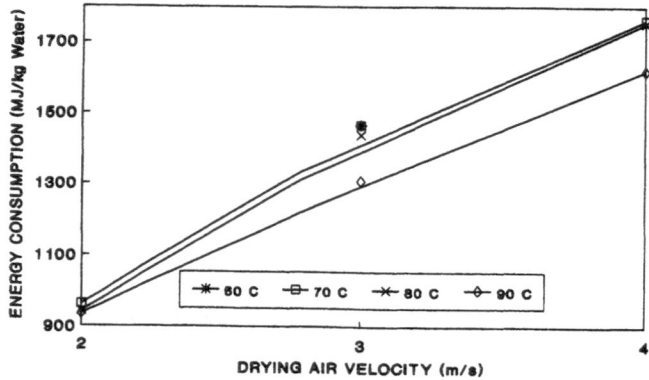

Fig. 6 - Energy consumption against air velocity and temperature.

It may be seen that higher temperatures correspond to lower energy consumption for the same air velocities and higher velocities correspond to higher consumption for the same temperatures.

Through Circulation Drying.

In this experiments there is only one drying front which progresses from inside to outside. Fig. 7 presents a typical result. The end of drying was considered to be when the temperature of the most exterior point reached half of the total temperature rise.

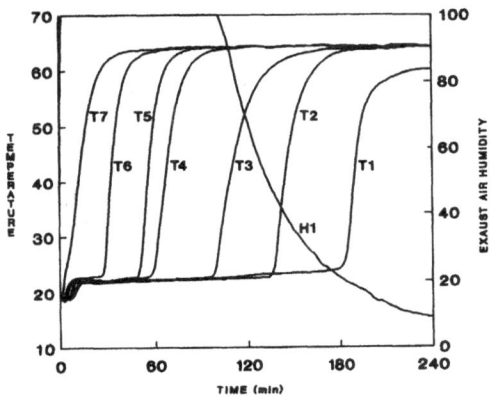

Fig. 7 - Temperature and air exhaust humidity along the time. Pressurized drying.

Each curve in Figure 8, which represents the evolution of the evaporation front, was obtained from a set of date like the one in Figure 7.

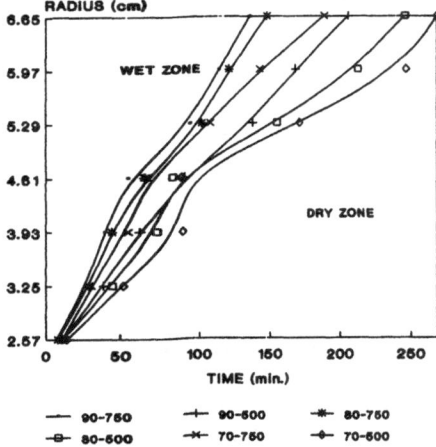

Fig. 8- Evaporation Front Disappearance. Pressurized Experiments.

Energy consumption in pressurized drying, can be calculated based on drying time and energy need to heat and move the air . Considering power consumption in heating air as thermal energy and power consumption in moving air as electrical energy, total energy consumption to evaporate 1 kg of water, on drying five bobbins was calculated. Table 3 shows the results of the calculations.

TABLE 3

Energy consumption on pressurized drying

Temperat. (°C)	Flow Rate (m³/min)	Drying Time (min.)	Water Evaporated (kg)	Power Used (kW)	Energy Consumption (MJ/kg)		
					Thermal	Electric	Total
90	0.750	137	1.2677	7.50	7.4	48.6	56.0
	0.500	211	1.2682	4.95	7.6	49.4	57.0
80	0.750	149	1.2645	7.50	6.9	53.0	59.9
	0.500	239	1.2505	4.95	7.5	56.8	64.3
70	0.750	189	1.2435	7.50	7.4	68.4	75.8
	0.500	289	1.2456	4.95	7.5	68.9	76.4

Figure 9 shows energy consumption versus temperature and flow for the cases studied.

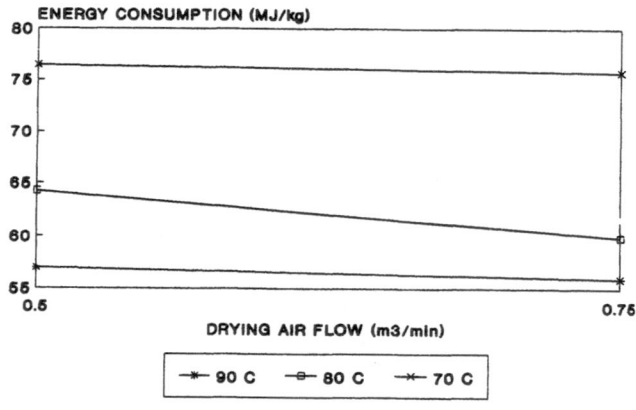

Fig. 9- Energy consumption against air velocity and temperature.

DISCUSSION AND CONCLUSIONS

Energy consumption values obtained in the present experiments are too high, due to the following reasons:

i) Cross circulation experiments were carried out for a single bobbin, the air at the exit being still able to perform further drying.

ii) In through circulation drying, five bobbins were dried each time; in this case, the high energy values resulted from the compression consumption. Although air pressure upstream of the bobbin only needed to be about 10 cm H_2O, the air had to be pressurized to a much higher value due to limitations of the flow rate measurement device.

iii) For the above reasons, detailed comparisons between the two drying processes cannot be carried out, although it is possible to obtain some conclusions with respect to the values of the experimental parameters.

iv) Some experiments carried out in an industrial plant led to the conclusion that, in energy consumption terms, both processes have similar results, with observed values of 9 800, and 7 600 kJ per kg of water evaporated, for "through circulation" and "cross circulation", respectively. But, as the first method is much faster than the cross circulation (about 5 times faster), it becomes much better in an overall evaluation. This difference in drying time was also observed in our laboratory experiments, although not so strongly, due to the reasons given in i) and ii).

v) As general conclusions from the present experiments
    a) higher temperature gradients lead to shorter drying time.
    b) higher air velocities lead to shorter drying time.
    c) in cross circulation experiments higher drying air temperatures can lead to lower energy consumption whilst higher velocities result in higher energy consumption.

So we conclude that, as far as the experimental work is concerned, through circulation drying (pressurized drying) is more economic than cross circulation drying (duct drying). However, due to the size of the experimental rig only qualitative conclusions are allowed.

REFERENCES

Nissan, A.H.; W. A. Kaye; T. V. Bell (1959) Mechanism of Drying Thick Porous Bodies During the Falling-Rate Period. I The Pseudo-Wet-Bulb Temperature. *AICHE J.*, 5, 103-110.
Walker, B:V: (1969) The Drying Characteristics of Scoured Wool.N.Z.J.Sci., 12, 139-164.

STATE OF THE ART IN DEWATERING PROCESSES DURING PAPER MANUFACTURE

VINCENT Jean-Paul
Centre Technique de l'Industrie des Papiers, Cartons et Celluloses
B. P. 7110
38020 GRENOBLE Cedex (FRANCE)

ABSTRACT

The European paper and board industry has currently to pursue an important investment effort in order to improve its competitiveness and to meet the challenge of *an increasing product demand both* in quantity and quality.

Papermaking is essentially based on water removal from a wet fibers and fillers network formed on a continuous wire, and then mechanically dewatered in the press section and thermally dried in the drying section.

The different dewatering devices of the machines have to be improved in order to work *with higher energy efficiency, at higher speed and at constant or improved paper quality.*

The ratio of the energy costs for removing 1 T of water at the **press** *section or at the* **drying** *section is presently :* **1 for 6.**

Due to the fact that *most paper machines are dryer limited, an increase of the sheet dryness after the press section of 1 %*, brings in practice *marginal gains of 2.1 % in production rate,* which means savings in the range of **0.64 billion ecus/year** in the E.E.C. This clearly stresses the interest of *developing mechanical dewatering methods in paper industry.*

Recent developments have led to the design of *long nip presses* of various type with an achievable exit dryness of the paper sheet changing from about 45 % to about 55 % in a 10 year period.

Further progress includes the use of a certain amount of *heat transfer* at the press level:

Steam boxes, used for heating the incoming sheet, are in current use in industry now,

Hot presses (in the range of 100°C) give good results on slow machines manufacturing heavy weight papers.

The present developments in course in the world concern two main techniques :

- *Impulse-drying* , using a roller heated at temperatures *in the range of 150 to 500°C.*

- *Press-drying,* with temperatures ranging from *150 to 250°C.*

The main draw-backs of these techniques are sheet delamination, sheet sticking to the roll and paper color changes.

Together with the **continuing technological adaptation** of these techniques, further progress in the field of sheet dewatering in paper manufacture, require now the **refinement to the knowledge of the basic phenomena** which take place in the pressing nip, in the porous media (paper and felts) and in the press itself.

ECONOMICS

The European paper and board industry has currently to pursue an important investment effort in order to improve its competitiveness and to meet the challenge of an increasing product demand both in quantity and quality.

Table 1 shows with a few figures an over view of this industrial sector.

In the E.C. Countries, the per capita consumption of paper has increased during the last 5 years at **3,8 %** per year at a level of **145 kg.** In *Scandinavia,* it has increased at **0,3 %** per year at a level of **225 kg.**

There is *a large potential market* in Central Europe.

In the E.C. countries, the average mill capacity is still in the range of **40 000 T per year** whereas in Scandinavia it reaches **180 000 T per year.**

There is *a high level of rationalizing to be achieved* in the E.C.

If we take the example of the *French paper and board industry* (Table 2), we see that *the number of running machines is decreasing by 1,7 % each year with a production increasing by 6,4 % and constant staff.* The average machine capacity is increased by 8,1 % each year, with an average staff per machine increasing by 1,7 %.
There is still *a large potential of machine replacement* especially in Southern Europe, but the fast increasing *demand for higher capacity machines operated with a limited crew,* put the challenge on the development of faster and wider machines.
The 10 m width already achieved seems to be a limit from the mechanical point of view.

The developments are mainly sought, for the next few years, towards **faster machines.** Water is the indispensable vector of cellulose fibers for sheet forming on the wire of the paper machine. The sheet has then to be progressively dewatered using mechanical pressing and thermal evaporation.

In this water elimination process, *speed increase means* in particular *a lengthening of the dryer section* which increases proportionally the investment level.

TABLE 1 : EUROPEAN PAPER AND BOARD INDUSTRY

	Populat. Million	Number of mills	Average mill product. 1000 T	Apparent consumpt. Million T	Per capita consumpt. kg
E.C. COUNTRIES					
1985	272.2	877	30.0	36.1	133
1990	328	1064	35.9	51.9	158
Average annual increase(%)			3.9		3.8
NORDIC COUNTRIES					
1985	17.4	125	128.8	3.9	224
1990	17.6	115	167.0	4.0	227
Average annual increase(%)	0.2	-1.6	5.9	0.5	0.3

TABLE 2 : FRENCH PAPER AND BOARD INDUSTRY

	1985	1990	Average annual variation
Number of companies	119	115	- 0.7
Number of mills	154	149	- 0.6
Number of running machines	275	252	- 1.7
Staff	24 542	24 560	0.0
Production (million T)	5.34	7.05	6.4
Apparent consumption (million T)	6.56	8.76	6.7
Per capita consumption (kg)	118.7	154.8	6.1

In fact all along the production line *process intensification has to be achieved,* in order to meet these requirements, but the improvement of the **pressing and drying sections of the paper machine and mainly the mechanical dewatering part, is essential.**

The basic reason might look very trivial, but *it remains the fact that water removal by mechanical means leads to operating costs which are lower in the ratio 5 to 10, compared to water removal by evaporation.*

On an average paper machine with a production of about 10 T/H, the energy costs are in the range of :

- for pressing :
. 4.4 ecus/T of paper or
. 2.5 ecus/T of removed water in the press section.

- for drying :
. 17.0 ecus/T of paper or
. 14.8 ecus/T of evaporated water in the dryers.

The ratio of the energy costs for removing 1 T of water at the press section and at the drying section is presently :

1 to 6

Should we improve by **2 %** the sheet dryness after the press section of all the paper machines in the E.E.C., savings could be achieved, in the range of :

1.28 billion ecus/year

This is in no way unrealistic as several percents of dryness have already been gained with modern press designs in the past few years.

In terms of constant production, **the oil consumption** could be reduced by :

0.505 MT/year

or :

43.3 Mecus/year

EXISTING TECHNOLOGIES

Paper and board are manufactured from cellulose fibres which are suspended in water at approximatively 1 % consistency. This suspension is formed into a web on a endless fabric. Water is then removed via a sequence of drainage (gravity or vacuum), mechanical pressure (pressing) and evaporation, by passage of the web over steam heated cylinders. As we have seen, pressing is more energy efficient than drying.

In pressing, a thin web of paper is passed through a nip, which essentially is formed by two rolls one on top of the other. The web is supported by a felt or felts which collect the water eliminated by the mechanical press.

The press section fulfils a dual role. As well as the removal of water, it also has an important bearing on the development of product characteristics. These can be bulk or surface characteristics.

Conventionally, pressing has been carried out at ambient temperature but recent developments have seen the introduction of steam boxes to raise the web temperature prior to pressing, or of heated press rolls. The latter developments have the potential to improve the energy efficiency of paper manufacture but this potential is not fully realised because of process related problems. In particular there is a tendency for the web to adhere to the press roll or felt. This causes web breaks and machine downtime or a deterioration in product properties.

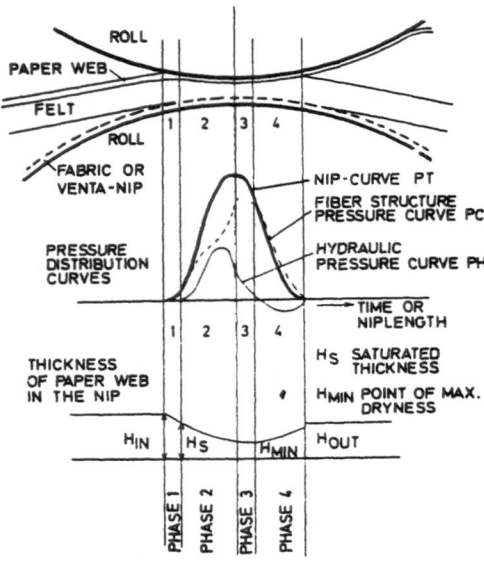

FIGURE 1 : THE PRESSING NIP AFTER P.B. WAHSTROM
P.P.M.C. Oct. 1969

The basic understanding of the pressing process distinguishes four distinct phases (Figure 1), in which sheet structure, felt structure, surface energies and pressure all play differing roles.

In the *first phase* as the nip closes, the sheet starts to saturate.

The *second phase* is from the point of saturation up to the centre of the nip where the loading is at a maximum.

The *third phase* begins at the point of maximum paper web compression and ends when the hydraulic pressure reaches zero.

The *fourth phase* occurs at the exit from the press as the paper web starts to expand. Water begins to flow backwards from the felt into the paper web.

These phenomena give *sheet rewetting* which reduces the efficiency of pressing and the global energy efficiency of paper manufacture.

Understanding pressing requires *knowledge of fundamental physical principles* :

- *mechanics of paper sheet and felts,*
- *filtration through the porous media,*
- *rewetting in the expansion phase of the nip,*
- *behaviour of felts and rolls rubber covers,*
- *heat transfer and concomitent deformation,*
- *heat transfer and evaporation inside the sheet,*
- *sheet delamination and crushing,*
- *color changes,*

and their application to a *dynamic situation.*

The loading pressure is opposed by a structural pressure and an hydraulic pressure in the different points within the paper web (Figure 2).

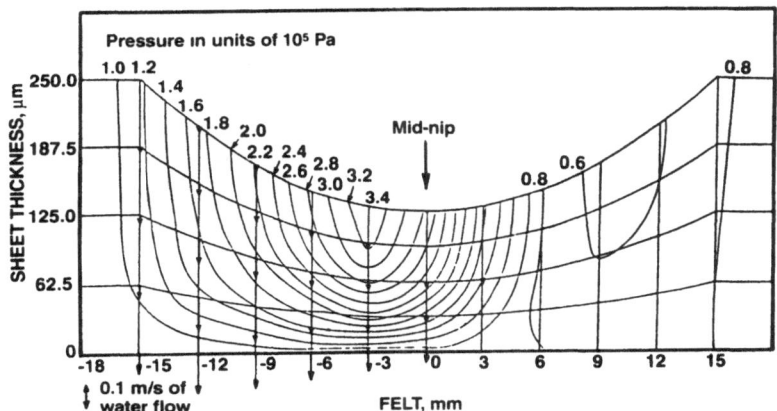

FIGURE 2 : HYDAULIC PRESSURE DISTRIBUTION IN THE NIP
(VINCENT, LEBEAU, ROUX, P.T.I., August 1988)

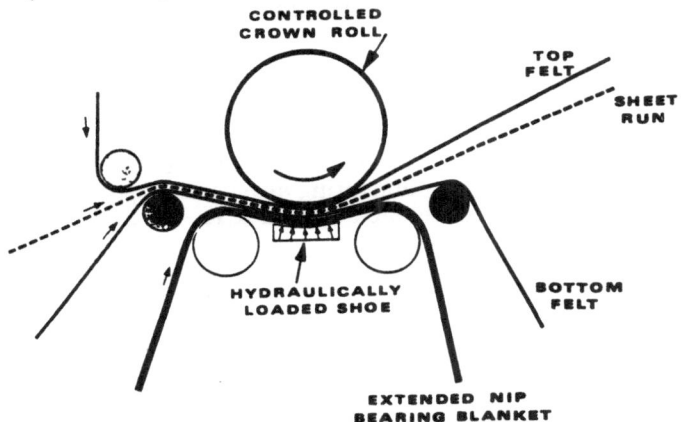

FIGURE 3 : PRINCIPLE OF SHOE PRESSES
BELOIT Extended nip press

RECENT INDUSTRIAL DEVELOPMENTS

Paper machine press sections have developed rapidly during the last years due to a better understanding of occuring phenomena in the press nip, based on concepts firstly acquired by experience and then confirmed by mathematical analysis.

SHOE PRESSES

This has led to the development of **shoe presses** (Figure 3) of various design with an achievable exit dryness of the paper sheet changing **from about 45 % to about 55 % in a 10 years period.**

About *200 such presses have been installed,* in all positions, for three main grades : linerboard, test liner, corrugated medium.

The machines manufacturers are able to modulate the machine direction pressure profile, in order to adapt to the type of paper to be pressed. In some cases, cross profile control is also possible.

The main reasons which limit the use of shoe presses for printing and writing papers, are *the loss of stiffness and the risk of sheet marking.* In the case of newsprint, the high percentage of mechanical pulp is limiting this risk.

For chemical pulp containing grades, sheet densification may be reduced :

- by reduction of number of nips,
- by reduction of rewetting (special felts),
- by nip length increase,
- by sheet temperature increase.

The result is still not satisfactory for all cases.

HEAT TRANSFER

Further progress includes the use of a certain amount of **heat transfer** at the press level.

Steam boxes, used for heating the incoming sheet, and reducing water viscosity, are in current use in industry now. The gain in dryness is in the range of 1 % for 10 °C.

TEM-SEC **hot presses,** (50 delivered) *using a steam heated (100°C) central roll* give good results on slow machines, manufacturing *heavy weight papers.*
The same principles are now developed by BLACK - CLAWSON for other grades, under the name Direct Action Hot Presses.

In general, press section engineering, goes towards more and more *compact* designs, in *order to limit heat losses.*

Remaining problems include :

- *design of roll covers,* bearing ever increasing temperatures,
- *design of non marking felts,*
- *selection of the more favorable location of heating devices.*

FELTS

Taking into account the requirements of the shoe presses, *felt characteristics* are more and more difficult to achieve :

- necessity of *higher water transport capacity*, due to the lengthening of nips, leading to thicker and heavier structures,
- *good resistance to compaction,*
- *low resistance to internal water flow,*
- *fine and non marking surface,* limiting rewetting, and leading to multilayer designs.

R. & D. IN THE SECTOR

Energy savings in the papermaking processes may be achieved through intensification of the mechanical dewatering part.

New pressing technologies including heat transfer are to be designed and refined and *quality and operational problems* are to be overcome.

The following general scheme may be proposed :

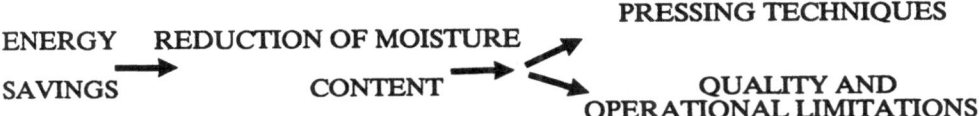

ENERGY   REDUCTION OF MOISTURE   PRESSING TECHNIQUES

SAVINGS   CONTENT   QUALITY AND OPERATIONAL LIMITATIONS

TECHNOLOGICAL RESEARCH

The present developments in course in the world are concerning two main techniques :

- **Impulse-drying presses,** using a roll heated at temperatures *in the range of 150 to 500°C.*

Water at the paper surface in contact with the heated roll *vaporize,* and drive the remaining liquid water through the sheet to the felt, leading to very high efficiency.

The main advantages of this thechnique are:

- *very high achieved dryness,*
- *low global energy consumption,*
- *lower investment costs,* due to reduction of the dryer length,
- *modification and often improvement of the sheet physical properties :* higher mechanical resistance due to sheet densification, smoothness of the side in contact with the heated roll, which may *avoid subsequent calendering.*

The main draw-backs are :

- *delamination,*

Several types of delamination are observed, from invisible weakness, only traceable through mechanical testing, to apparent crushing.

This defect appears at 150 °C and is a function of operating conditions. The solutions may be found in:

. a longer residence time in the press,
. in a perforated heated roll in order to avoid steam pressure build-up,
. in a better control of the temperature at the end of the nip, by using a special ceramic roll.

-energy consumption,
Comparative studies have shown that Impulse drying energy consumption was in the ratio 1 to 2, compared to conventional drying. Still, to be meaningful, these comparisons have to take into account the final costs of the types of energy used for heating the presses (induction, infra-red).

- felts,
Felt behaviour has to be carefully followed in the Impulse drying process. In case of paper break for example, elevated temperatures may very rapidly destroy the felts, or at least, change largely their surface characteristics. New types of felts, with higher temperature resistance, which are of course more expensive, become necessary.

- sheet sticking to the rolls,
Sticking to the heated rolls starts at about 80 °C, and increases rapidly with temperature. Pilot trials have shown that at 200 °C, sticking becomes completly out of control. New roll surfaces have to be developed.

- design,
High temperatures require high quality materials and design for a good cross-direction repartition of the tranfers.The grammage, moisture, and temperature profiles of the in-going sheet itself have also to be very good.

- Press drying, the principle of which is near conventional multi-cylinder drying but where the sheet is kept under restraint during the whole drying period. The temperatures used in this case are in the *range of 150 to 250°C.*

The main advantages are :

- *increase of the evaporation rate in the ratio 1 to 20* compared to conventional drying,
- *improvement of surface smoothness,*
- *machine- and cross- direction* sheet parameters *equivalent* due to drying under restraint,
- *mechanical resistance increase,* due to softening of lignin in wood containing grades,
- *sheet densification,*
- *better sheet dimensional stability,*
- *decrease of refining energy requirements,* at equivalent physical properties.

Refining is a mechanical treatment applied to the fibers to enhance their bonding abilities. This might be a very important energy saving source even if it is difficult to evaluate it at this point. We may remind that refining energy is in the order of 150 kWh per T of paper. *A 10 % saving means in Europe* :

22.9 M ecus/year

A sheet made from raw pulp and dried by Press-drying, has the same properties as a sheet conventionally dried and refined at 100 KWh/T.

The main problem remains sheet marking, especially for fine papers.

As a conclusion we may say that **the shoe press technique has a tendency to widen to all grades,** and that **a general increase of temperature, before and in the press, is also observed.**
Recent studies **combining the two techniques** seems very promising.
Among **the techniques in development, Impulse drying,** due to its **relative simplicity and its remarkable performance,** seems, despite important problems remaining to be solved, to be the way which has the greatest chance, in the future, of achieve industrial application.

MAIN RESEARCH POLES

Experimental work is in progress in several laboratories on these techniques, trying to cope with the draw-backs of the methods concerning sheet quality and ease of operations.

The developments in course with **the machine builders** are made in :

BELOIT (USA)
Interest : Impulse drying,
Pilot : Extended nip press, 400° by induction, 2000 m/min.

BLACK-CLAWSON (USA)
Interest : Hot pressing,
Pilot : 2 inverted D.A.H.P., 1800 m/min.

TAMPELLA (Finland)
Interest : Press drying,
Pilots : CONDABELT system,
      .145 °C, 100 m/min,
      .not heated, 1200 m/mn.

VOITH (Germany)
Interest : Impulse drying and Press drying.

VALMET (Finland)
Interest : Impulse drying and Press drying,
Pilots :   . heated shoe-press 100 °C, 1500 m/min,
        . heated press 200 °C.

ALBANY International (U.S.A.) (10)
Felt manufacturer.

The developments with **the research centres** are made in :

University of MAINE (U.S.A.) (7)

C.T.P. (France) (8)

I.P.S.T. (U.S.A.) (9)
Interest : Pioneers of Impulse drying,
Pilot : 240 m/mn, infra-red and induction heating.

PAPRICAN (Canada) (11)
Interest : Impulse drying,
Pilot : . press, 250 °C, 300 m/min,
    . machine, 350°C, 1200 m/min, 2 inverted and induction heated E.N.P.

K.C.L. (Finland) (12)

S.T.F.I. (Sweden) (13)
Interest: Impulse drying and Press drying,
Pilot : FEX machine, 2000 m/mn,
Project of installing an Impulse drying section.

P.I.R.A. International (U.K.) (15)
Interest : Press drying and Impuse drying,
Pilot : Press, 350 °C, 400 m/min.

BASIC RESEARCH REQUIREMENTS

From a fundamental point of view, **the description of the basic phenomena which characterize press behaviour,** has to be largely completed in order to implement a **mathematical model** of the dewatering processes, leading *to a better knowledge of phenomena occuring inside the sheet and felts, and the presses themselves.*

Such a **model** could be used as a **software engineering tool** by the **different industrialists of the sector concerned with press construction and operation** *(paper makers, press section builders, felt manufacturers, roll cover manufacturers).*
This would give them the possibility of making all kind of *design and extrapolation work.* It should in particular help engineering the **future high speed machine.**

CONCLUSIONS

The different dewatering devices of the paper machines have to be redesigned, in order to be able to work *with higher energy efficiency, at higher speed and at constant or improved paper quality.*

From the point of view of energy saving, water removal using mechanical methods leads to **5 to 10** times lower costs than heat transfer and evaporation.

Based on present practice, should new developments lead to an improvement of the sheet dryness after the press section of about 2 % in 5 years, potential savings would be in the range of **0.5 millions T of oil per year** in the E.E.C. **or 43.3 millions ecus.** In terms of increased production the savings would reach **1.3 billions ecus per year.**

Wet pressing on paper machines has then become a topic of large interest in the recent years, and several research teams in the World are interested in this development.

From a technological point of view, we may say that **the tendency is to the widespread use of the shoe press technique, and to a general increase of temperature in the presses.**

Among **the techniques in development, Impulse drying,** due to **its relative simplicity and its remarkable performances,** seems, despite important problems remaining to be solved, the way of the future.

Fundamental progresses are based on **the understanding of basic phenomenons** taking place in the press nip, in the porous media (paper and felts) and the press itself.**The development of a press section engineering tool**, based on a *flexible mathematical model* and *a valid parameters data base*, would be of high interest to all industrialists of the sector.

The combination of **existing technologies intensification** and **development of new concepts** will mean in the future **large energy savings in the European paper industry.**

BIBLIOGRAPHY

1. WAHSTROM B. "A long term study of water removal and moisture distribution on a newsprint machine press section" Pulp and Paper Magazine of Canada, volume 60, n°8 and 9, pp. T379 - T401 and pp. T418-451 / 1960

2. NILSSON P. LARSSON K.O. "Paper web performance in a press nip" Pulp and Paper Magazine of Canada, volume 69, n° 24, pp. T438 / 1968

3. WESTRA H.A. "A new contribution to press nip analysis" Paper Technology and Industry, volume 16, n°3, pp 165-171 / 1975

4. CARLSSON G., LINDSTROM T., FLORENT T. "Permeability to water of compressed pulp fiber mats" Report/Department of Paper Technology, the Royal Institute of Technology, Stockholm / 1983

5. CAULFIELD D.F., YOUNG T.L., WEGNER T.H. "The role of web properties in water removal by wet pressing. Characterization of dewatering time constant" TAPPI Journal / February 1982.

6. MUKHOPADHYAY A.L., KINGSURY H.B. "A study of two-dimensional flow in the press nip" TAPPI Journal / July 1980

7. JEWETT K.B. "The application of a model for two phases flow through a compressible porous medium to the wet pressing of paper" Ph. D. Thesis, University of Maine (Orono) 1984.

8. VINCENT J.P., LEBEAU B., ROUX J.C. "A simulation tool for the press section of the P.M." Paper Technology and Industry, August 1988.

9. LINDSAY J.D. "The Physics of impulse drying : new insights from numerical modelling" Fundamentals of paper making - Cambridge - vol. 2 /Sept. 1989.

10. EL-HOSSEINY F. "Mathematical modelling of wet pressing of paper" Nordic Pulp and Paper Research Journal n°1 / 1991.

11. KEREKES R., Mc DONALD D. "A decreasing permeability model of wet pressing : theory" TAPPI Journal / Dec. 1991.

12. SZIKLA Z. PAULAPURO H. "Compression behaviour of fiber mats in wet pressing" Fundamentals of Papermaking, Cambridge, vol. 2 / Sept. 1989.

13. BACK E. "Why is press drying/impulse drying delayed ?" TAPPI Journal /Mars 1991.

14. Mac GREGOR M. "Wet pressing research in 1989" Fundamentals of papermaking, Cambridge, vol. 2 / Sept. 1989.

15. DANILEWICZ A. "Increased water removal rates - an alternative approach" Proceedings from P.I.T.A. effective water removal conference, York, U.K., December 1991.

SESSION 4:

Sensors and Instrumentation

Chairman: Prof. F. Durst

DEVELOPMENT OF ADVANCED SENSORS

Franz Durst
Lehrstuhl für Strömungsmechanik
Universität Erlangen-Nürnberg
Cauer Str. 4, D - 8520 Erlangen
Germany

ABSTRACT

There are not many key technologies left in which Europe is leading but the field of sensor and microsystem technologies is still a European domain. This is stressed in the introduction of the present paper and it is pointed out that joint efforts are needed to maintain the European lead in this field.

The work reported summarizes research and development efforts carried out within the JOULE 1-Project No. CT 900056 on "Advanced Sensors". Work is reported on flow rate metering using modern optical and thermal techniques, as well as an ultrasonic measuring method. Furthermore, the development and completion of a Coriolis mass flow meter for granular material is described together with measuring results that demonstrate its correct functioning. Work on the miniaturisation of phase-Doppler anemometers is summarized, together with optical measuring techniques applicable for concentration measurements in combustion systems. The report also includes a summary of development work on a humidity sensor, based on zinc oxide doped with lithium in order to get an extremely good response to humidity. Unique test equipment is presented which was employed in the present study. Suggestions for further developments are made.

INTRODUCTION

The present paper summarizes research and development work in the field of advanced sensors. Sensors are widely used in all fields of engineering and science in order to sense physical and chemical properties of materials and/or to measure quantities such as flow rate, viscosity, pressure, temperature, etc. Hence, sensors are essential elements of measuring systems and allow us to obtain quantitative, physical, chemical, medical, etc. information on systems and system performances. The JOULE I-PROJECT "Advanced Sensors" was initiated to further advanced sensors in such fields as:

o   low flow rate measurements with laser-Doppler anemometers and thermosensors.

o   mass flow rate measurements of granular material.

o   humidity sensors on lithium doped and sintered ZnO-material.

o   miniaturized phase-Doppler systems for industrial applications.

o   optical techniques for measurements of scalar properties in combustion flows.

The present paper summarizes the outcome of the research and development work carried out within the JOULE Programme on "Advanced Sensors" and provides the major results. The participating institutes of the project were: ATZ-EVUS, Germany; University of Aveiro, Portugal; Ecole Centrale Paris, France; Imperial College London, Great Britain.

Looking at these results it is clear that major progress has been achieved in areas relating to low volume rate measurements of liquids and mass flow rate measurements of granular material. Through the work carried out in the present project, new instruments resulted that can be employed in areas where no other instruments are presently applicable. This will be of help to the process industry yielding more accurate information on the volume and mass flows, e.g. into chemical reactors.

The work on the humidity sensors has led to new material on which highly sensitive sensors can be build. This work has led to first humidity sensors and their applications in laboratory tests. These have been encouraging indeed and it is urgently suggested to continue this work even beyond the present JOULE-project. A new generation of humidity sensors will result applicable in the process industry. Being of ceramic material, applications in harsh environments can be foreseen.

The application of semi-conductor lasers allowed the minituarization of phase-Doppler anemometers. Small instruments resulted out of the work that are readily applicable in various fields of chemical process engineering. There have been extensive applications to liquid sprays but also to liquid-solid two-phase flows and also liquid-gas flows (bubble flows). In all these applications, the minituarized systems have worked as successfully as PDA-systems as they are employed these days. The minituarization will help the PDA-measuring technique to enter into new areas of industrial applications.

The work on optical techniques for measurements of scalar properties in combustion flows has mainly concentrated so far on the measuring itself rather than the development of sensors. The applications in combusting systems are difficult and require time to yield the knowledge for handling optics in the form of sensors.

DEVELOPMENT OF NEW SENSORS FOR LOWEST FLOWRATE MEASUREMENTS

The need for precise flow rate measurements in industry is well recognized, especially for the reduction of energy consumption. There are two major problems with common flow rate meters:
a) There are no instruments available for high accuracy measurements of low and very low flow rates.

b) There are no instruments available which start functioning from low flowrates and have a large dynamic range. Two new instruments, which solve one or both of these problems were built at the University of Erlangen-Nürnberg.

LDA flow rate meter

The working principle of the LDA flow rate meter is based on the determination of the fluid velocity on the axis of a circular pipe at an axial position where a fully developed velocity profile is known to exist (Fig. 1). The velocity is measured with a simple Laser Doppler anemometer (LDA), and the only range limitation for such a system arises due to the electronic Doppler- frequency determination. The overall accuracy is about 1% of the measured value. The fluid must be transparent and contains tracer particles. The measuring principle is independent of pressure, temperature and the orientation of the sensor.

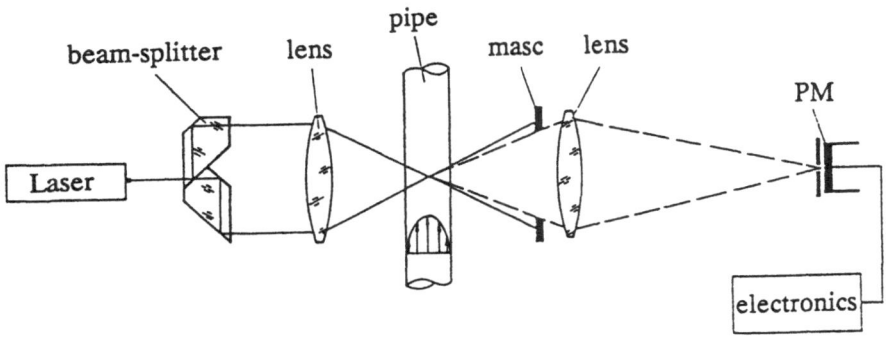

Figure 1. Working principle of LDA-flowrate meter

By using laser-diodes, semi-conductor detectors and holographic beam-splitters a compact and reasonably priced unit was manufactured, see Fig. 2. Results were reported in ref (1) and (2) that demonstrate the high precision of the instrument and its wide dynamic rage.

Measurements become possible from 0.5 cm³/min to 1.000 cm³/min. Over the entire measuring range, an accuracy of $\pm$ 1% of the actual measured values is achieved. At the lower end, no other instrument is available to allow flow rate measurements with such an accuracy.

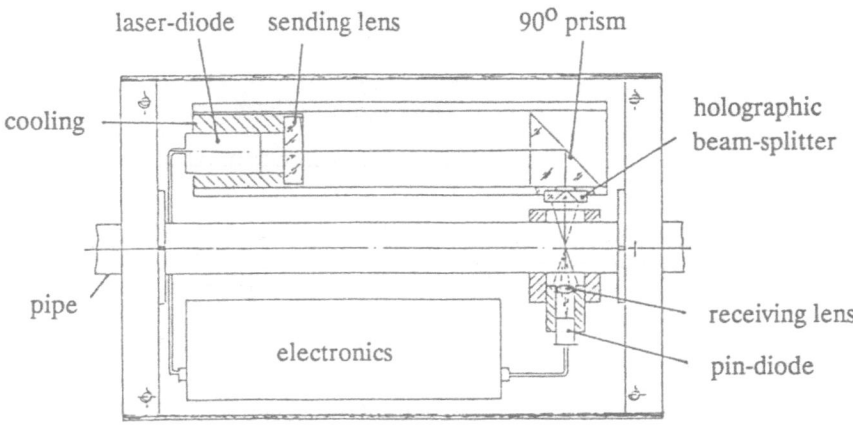

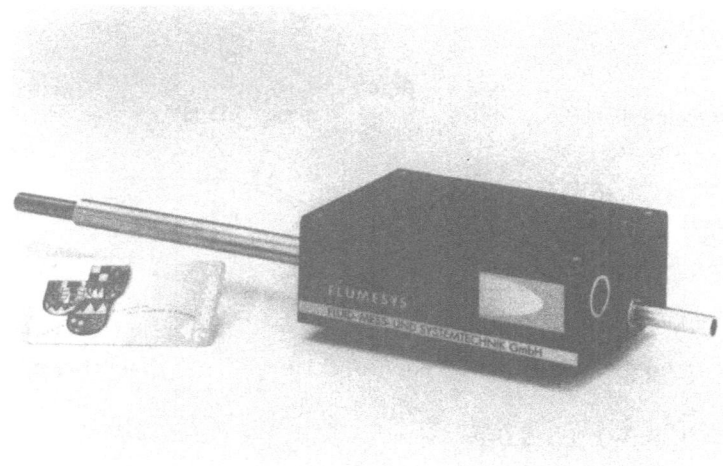

Figure 2. Prototype of LDA-flowrate meter with semi-conductor components

New Vortex Meter for Low Flow Rate Measurements

The vortex-meter principle is well-known: a bluff body produces a periodic separation of vortices with a separation frequency, which is proportional to the upstream velocity. The flow rate can then be determined by measuring the separation frequency. The linear relationship between velocity (and also flow rate) and separation frequency is only valid for the high Reynolds number region (>10,000), whereby the Reynolds number is defined with the characteristic length of the bluff body. Vortex meters have the advantage that they do not have moving parts and are also insensitive to dirt or dust (the vortex separation still takes place). In contrast, thermal flow rate meters, that are normally used for measurements of low flow rates of gases, are very sensitive to dirt because the thermal conductivity coefficient changes; recalibration is, therefore, necessary. At LSTM-Erlangen the extension of the vortex-meter principle to lower Reynolds numbers was investigated. The main result was that the relationship between shedding frequency and flow rate is unique but not linear in this region. Thus, if the relationship is known, a vortex-meter can work even at low Reynolds numbers.

A prototype was then built (Fig. 3) with the aim of developing a sensor, which works for low flow rates, has a large dynamic range, and which is insensitive to disturbances (because of the measuring principle). In the prototype a split-fiber probe was used as the bluff body, which generates the vortices and monitors them as well. The split fiber was connected to a standard constant temperature anemometer.

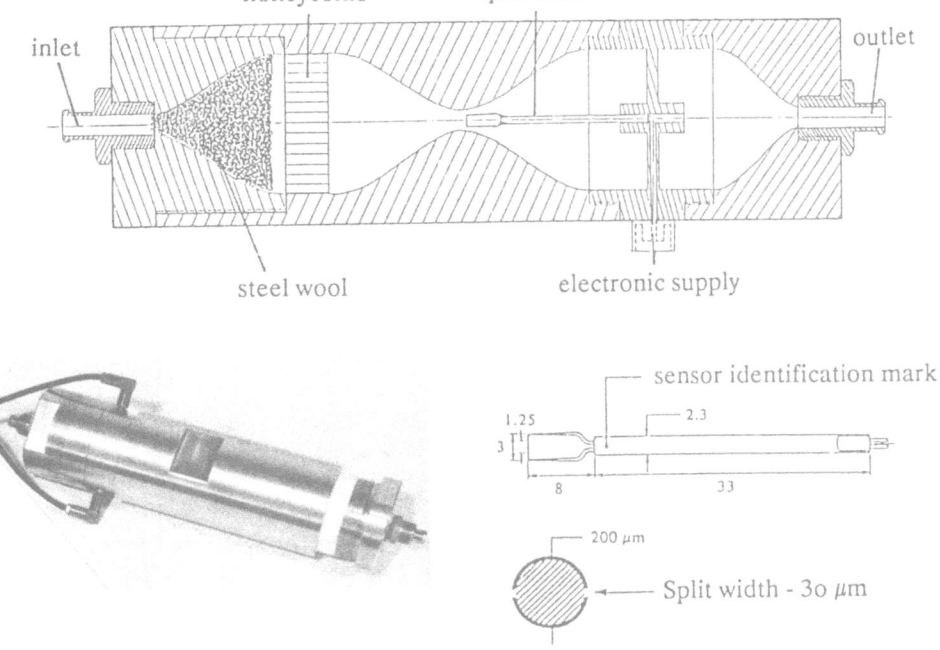

Figure 3. New vortex meter prototype

The first measurements with this small sensor show encouraging results (Fig. 4). A future task will be to use the hot film sensor as a calibrated anemometer in the region where no separation of vortices occurs. As data for a self- calibration, the frequency-flow rate information in the higher Reynolds-number region could be used.

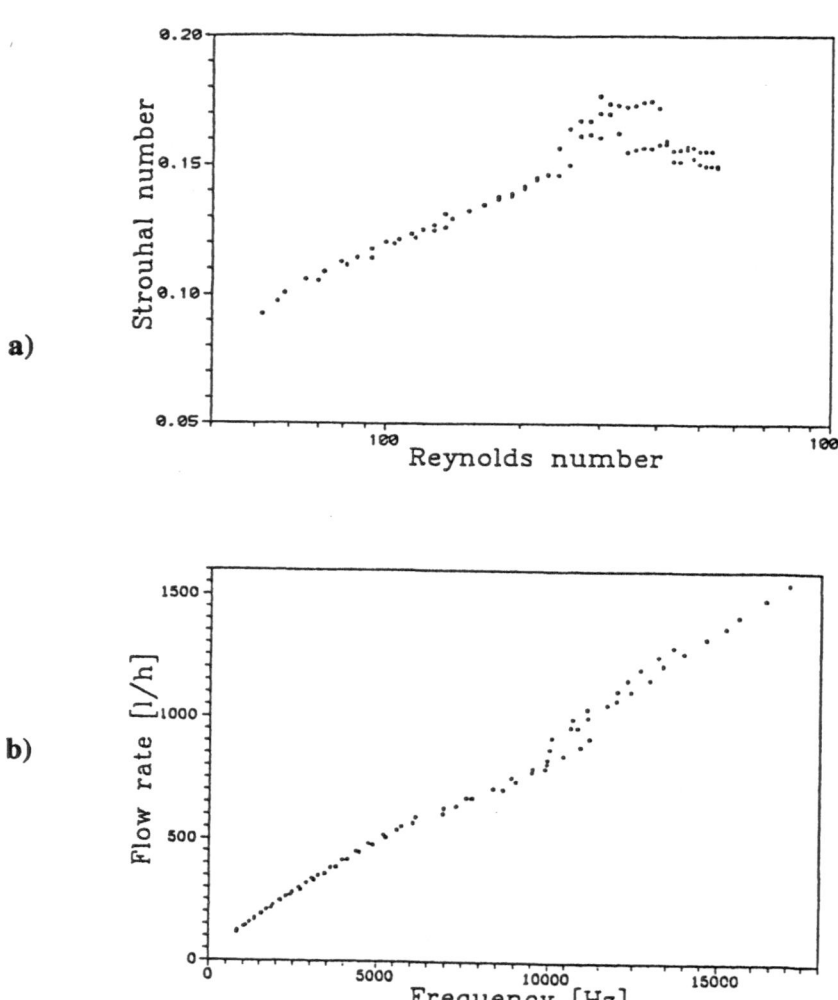

Figure 4.   a) Strouhal number over Reynolds number
b) Flowrate over frequency

CORIOLIS MASS FLOW METER FOR GRANULAR MATERIALS

In many fields of process engineering granular materials are pneumatically transported. To controll the processes, the instantaneous mass flow rate needs to be measured. Development work carried out at ATZ-EVUS in Sulzbach-Rosenberg, Germany demonstrated that the Coriolis force caused by the motion of the granular material rotated in a turbine can be used to measure the mass flow rate of these materials. This had already been demonstrated for large flow rates and it was the aim of the present study to build an instrument that could also measure at flow rates as low as a few hundred grams per minute. In addition to reducing the flowrate, it was intended to decrease the time response of the instrument so that instantaneous mass flow rate measurements became visible in situations with rapidly varying mass flow rate.

The newly developed mass flow meter uses the Coriolis effect on the basis of a radial turbine. This turbine is shown in the photograph presented in figure 5 in which the central opening can be seen, through which the granular material enters the rotating turbine. Six radially arranged channels in the turbine wheel serve as an outlet for the mass flow and take on the force that results in a moment on the turbine drive given by:

$$M = \dot{m} \cdot \omega \cdot R^2$$

M = torque [Nm], $\dot{m}$ = mass flow [kg/s], ω = angular velocity [2π/s]
R = radius of turbine [m]

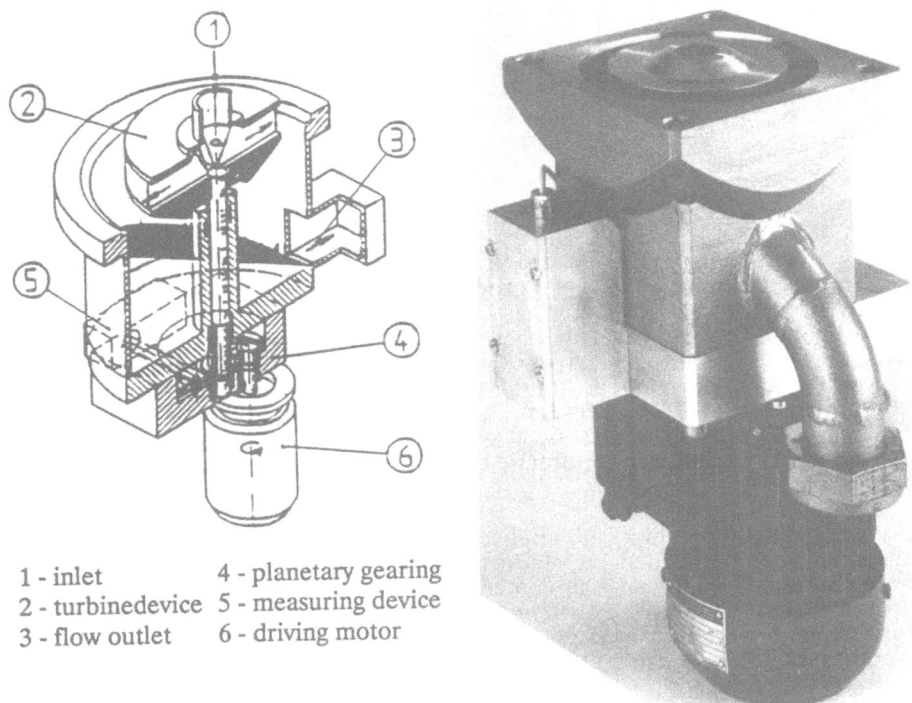

1 - inlet   4 - planetary gearing
2 - turbinedevice   5 - measuring device
3 - flow outlet   6 - driving motor

Figure 5. Coriolis meter

Control and signal processing devices for the Coriolis meter have been developed and successfully tested. The flow sheet is shown in Fig. 6. These devices are now miniaturized. For the main control purpose a micro-controller chip is used.

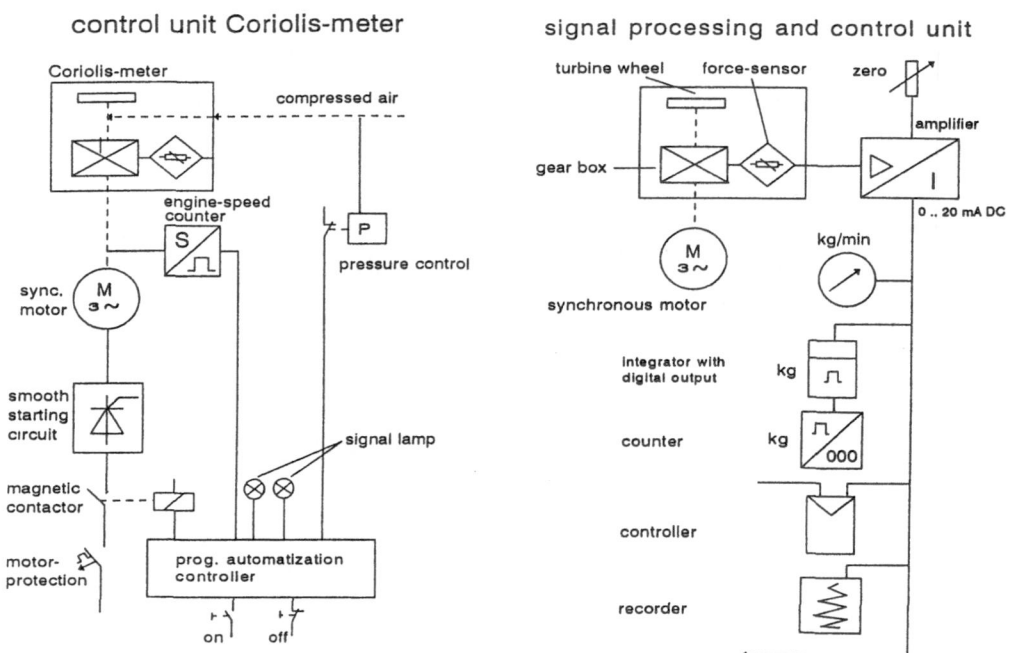

Figure 6. Control and signal processing devices for Coriolis meter

The experimental verification of the measurement accuracy is carried out in a large pilot plant erected at the technical laboratories at ATZ-EVUS. In order to calibrate the measuring device developed (Fig. 5), (type A 0 - 10 kg/min), extensive test-runs have been carried out. Measurements of the mass flux were performed in a pilot plant (see Fig. 7) consisting of a storage bin, a screw feeder, pneumatic conveyer lines, a weighing bin and fully automatic control systems. The result for quartz sand is shown in Fig. 8.

At present, a wide range of applications are under test at ATZ-EVUS. Measurements are performed for products of the food industry, such as sugar, salt, corn, etc., powder products such as cement, powder sugar, detergent, etc. and so far the instrument has shown excellent performance for all the products. Test measurements for carbon powder have been less successful because of the agglomeration of the material at the turbine walls.

Figure 7. Pilot plant of the Coriolis meter

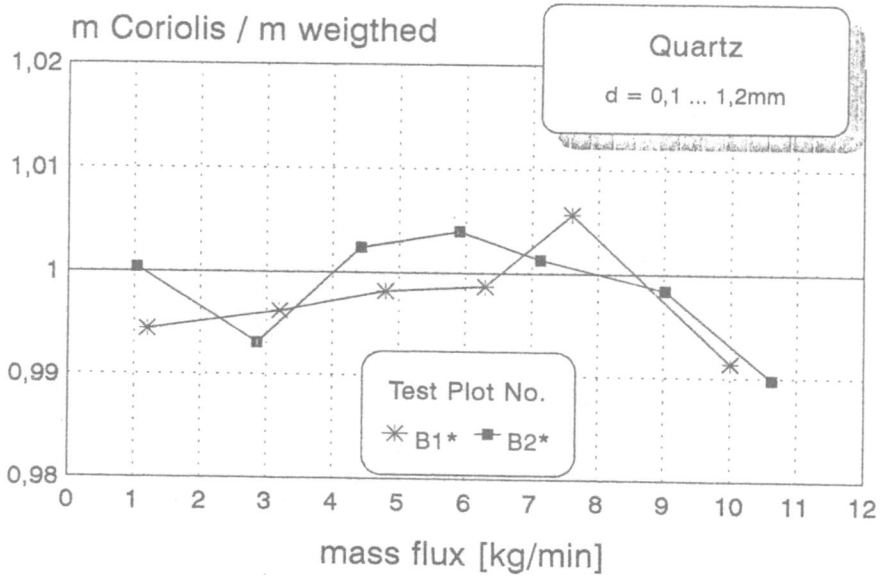

Figure 8. Test measurements

This figure demonstrates that the Coriolis meter can measure the mass flux of granular material with an accuracy of + 1 % based on the actual measured value. Extensive experimental investigations in the pilot plant also have shown a response time of 0,5 second for the Coriolis meter. In power plant experiments the measurement device has shown no erosion problems, leading to reliable operation. The input of thousands of tons of material have proved the reliable operation in harsh industrial environments. Preliminary tests applying gas-fluid two-phase flows have shown that it will be possible to measure the fluid mass fraction by means of a Coriolis meter.

The application of Coriolis meters in process engineering industry for on-line measurements of mass flux rates are foreseen in the following application fields of

o   feeding from non-pressurized storage bins
o   feeding from pressurized storage bins
o   mass flux in pneumatic conveyer lines

The combination of Coriolis meters with highly accurate feeding or dosing devices will result in new feeding systems, allowing an accurate pulsation-free feeding of granular materials in the process engineering industry (Fig.9).

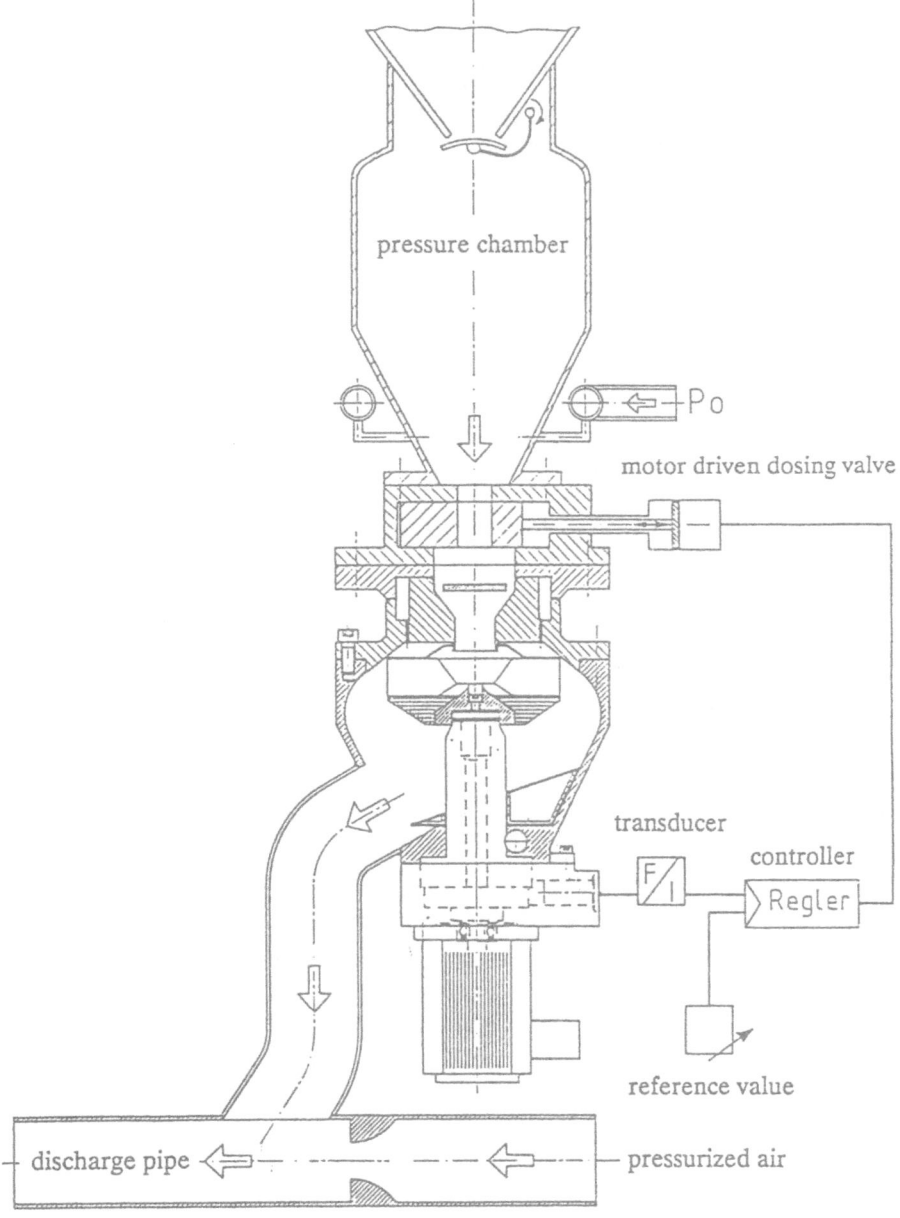

Fig. 9: Coriolis meter included in a feeding system

ADVANCED HUMIDITY SENSORS

Part of the work within the JOULE I-programme "Advanced Sensors" has focused on the development of a new humidity sensor. This sensor is a ceramic type with a resistive readout. The main objectives of the work have been fast response, wide dynamic range of the sensor signal and stable behaviour of the sensor, even after lengthy exposure to harsh environments.

This new sensor is made of ZnO and doped with Li. The raw materials are milled, pressed together and sintered. Different methods of production and different combinations of the proportions of materials, sintering times and temperatures had been tested during the last months. Now there is an established and reliable way to produce the humidity sensor with good performance.

The outstanding features of this humidity sensor are:

o   The dynamic range of the resistive signal covers about 4 orders of magnitude, see figure 10.

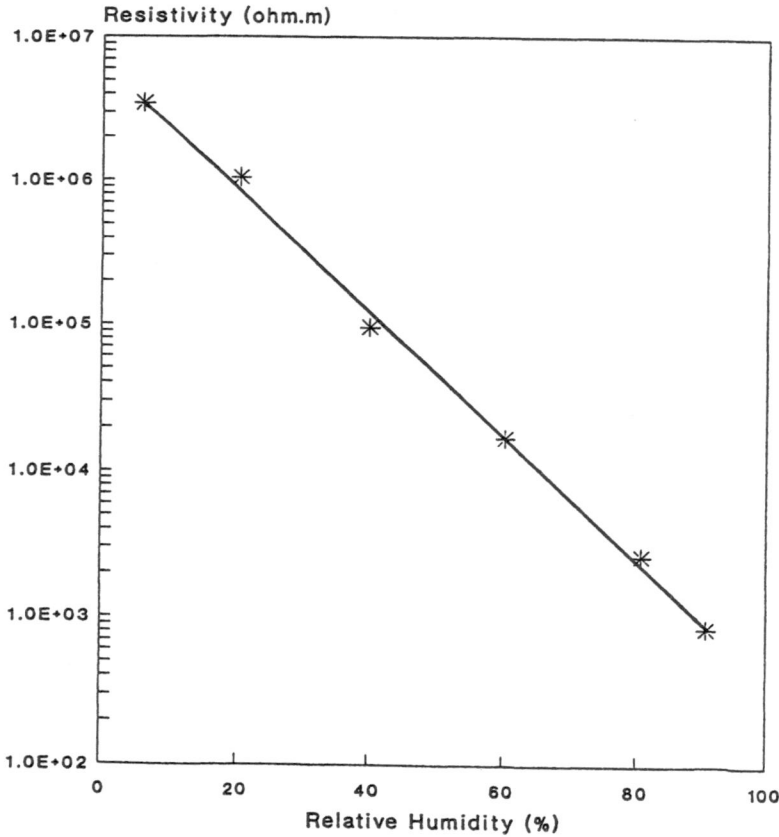

Figure 10. Sensor resistivity as a function of relative humidity

o   Performing a change of humidity from 5 % to 95 % RH and back at an ambient temperature of 25 degree C, the signal shows no hysteresis, see figure 11.

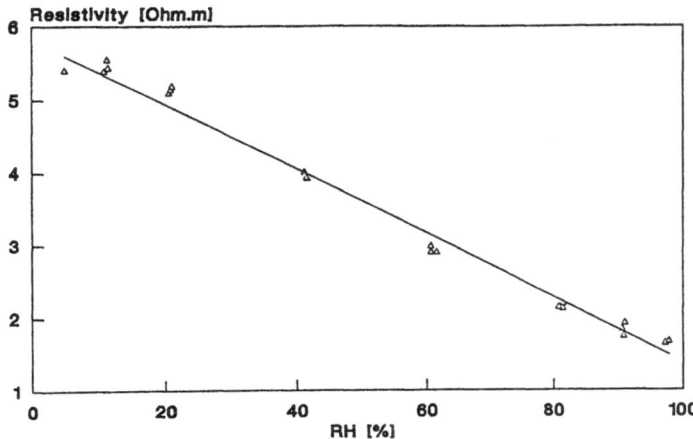

Figure 11. Change of resistivity shows no hysteresis with relative humidity changes

o   The first step towards a sensor with a capability of regeneration after a long time exposure to high humidity is shown in figure 12.

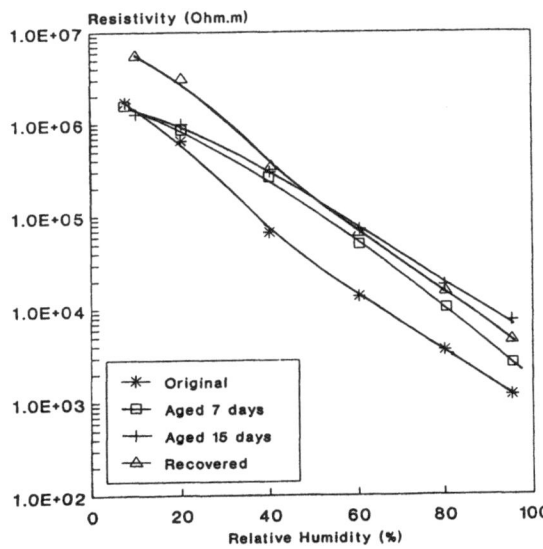

Figure 12. Change of resistivity with relative humidity after aging and regeneration

Looking at these data, the sensor appears to be very interesting as a low cost sensor for applications in harsh environments or as the heart of a standard protection circuit to protect electronic circuitry against an excess of humidity.

During the last months, there have, however, been some problems. The University of Aveiro has no facilities to test the new ZnO humidity sensor at different temperatures. Also the effort to manufacture the sensor as a thick film with a thickness of about 25 μm failed.

To overcome these problems, a cooperation between the "Lehrstuhl für Strömungsmechanik" (LSTM), University of Erlangen and the University of Aveiro, concerning the humidity sensors, has been started. A recently established group at Erlangen, working on the field of humidity measurement, will start the tests, which need advanced humidity generation at different temperatures.

Exact measurement and control of relative humidity (RH) are a pre-condition for the development of a new humidity sensor. The main problem is the temperature dependence of RH. Therefore generation of specified RH is only possible, if one can fix the temperature of the test-chamber at a constant level. Using simple saturated salt solutions for humidity generation, which usually serve as a RH-standard at room temperature, is not sufficient for testing the prototype sensor at different temperature conditions, as planned by the group at Aveiro.

The best way for exact generation of different RH levels would be the use of a NBS traceable 2-pressure-generator, but this RH generator costs about $ 100,000 to $ 150,000 and is therefore not acceptable. At LSTM a temperature controlled humidity reference generator has been installed, providing facilities to produce a relative humidity of 5 % RH to 95 % RH with an overall accuracy of +/- 1 % RH. For improved accuracy and for measurement at temperatures far away from room temperature a self-designed mixing unit of cryogenic dried and wet air is available. This setup provides also an extended dewpoint range and is controlled by a cryogenic dewpoint-mirror hygrometer. This combines an overall fast response, even at very low levels, and an exact dewpoint measurement.

Figure 13. Vaportron (RH-generator for climate conditions)

Figure 14. CR-1 (cryogenic dewpoint hygrometer for very low dewpoints)

The first aim is to create a fast responding environment for settling an accurate humidity for the range of climatic RH-values as well as conditions of dry air. Besides calibration purposes for the Portuguese reference sensors, the apparatus will allow improved measurement of the new ZnO humidity sensor at different conditions. Especially, tests with an extended range of temperature will take place in Erlangen.

The second aim is directed to a better understanding of the physical and chemical processes that occur during the adsorption and desorption of water from the surface and probably from the bulk of the sensor. Also a model of an equivalent electronic circuit of the sensor has to be developed. This would allow a re-design of the sensor and by this a more simple electronic circuitry to read out the sensor signal. For a simple integration in any kind of application, manufacturing in a thick film technique would be desirable.

In the scope of the new installed Centre for Microsystem Technologies at Erlangen, the new humidity sensor will be tested in a practical device, using an integrated electrical circuit and the thick film humidity sensor on the same board or even on the same chip. To provide an application for demonstration, the final goal will be a practical demonstration unit of the sensor in form of a low-cost device.

CONCLUSIONS AND FINAL REMARKS

There are not many key technologies left, in which Europe is leading in comparison to the United States and Japan. Sensor technology is still a high-tech field where Europe has a lead. Considering that sensors are the basis for many measurements and also the basis of microsystem technology, it is apparent that special efforts are needed in the field of sensor technology in order to ensure that Europe keeps its leading position. This will have far reaching implications in all fields of engineering and science.

In the present project, only four sensor systems were described. For these systems far reaching advancements could be achieved through collaboration between several European research groups. The work and the outcome of it demonstrated that University researchers have the knowledge and expertise to further advance existing flow rate measuring instrumentation, to develop mass flow meters for granular materials, to develop new humidity sensors and to miniaturize phase-Doppler anemometers. Further work is needed in this field and other fields of sensor technology in order to ensure that European industry and science has the sensors and microsystems needed to advance our knowledge in engineering and science.

REFERENCES

1. S. Bopp, F. Durst, M. Teufel, H. Weber, "Volumetric flow rate measurements in oscillating pipe flows with a laser-Doppler sensor"; Meas. Sci. Technol. 1 (1990) 917 - 923

2. M. Teufel, "Entwicklung von Durchflußmeßgeräten zur Erfassung kleinster Volumenströme"; Dissertation, Techn. Fak. Univ. Erlangen-Nürnberg, Erlangen 1992

3. E. Joanni, J. L. Baptista, "Lithium doped Zinc Oxide Humidity Sensors" Paper presented at the SENSOR 91 Congress, Nuremberg, May 1991

4. A. Naqwi, X, Liu, F. Durst, "Extended phase-Doppler system for characterization of multiphase flows", Part. Part. Syst. charact. 8 (1991) 16-22.

5. A. Naqwi, F. Durst, R. Müller, P. Zeller, "LDA and PDA systems based on diode-pumped Nd:YAG laser" May 1991, Sensor Conf., Nuremberg

SELECTION OF INSTRUMENTATION TO MONITOR ENERGY EFFICIENCY OF A PROCESS HEATING SYSTEM

K. Urbaniec and P. Zalewski
Płock Branch of the Warsaw University of Technology
Jachowicza 2/4, PL-09-400 Płock, Poland

ABSTRACT

This paper is concerned with developing a method of rational selection of instrumentation that is needed to supply input data to a computer-aided energy monitoring system, and applying this method in the planning of a monitoring system in a sugar factory. The monitoring of energy usage in the sugar factory is based on the estimation of heat balance of the multiple-effect evaporator. The heat balance is calculated using input data from on-line measurements of a number of pressure, temperature and flow values. A prerequisite for reliable energy monitoring is that the errors of estimation of heat flows and other important variables are kept under control. When designing an energy monitoring procedure, the influence of measuring errors of the individual instruments on the resulting uncertainty of calculated variables is studied using "Monte-Carlo" simulation. Drawing conclusions from the simulation results, a configuration of measuring instruments is selected to facilitate high-quality estimation of the heat balance.

INTRODUCTION

The energy is supplied to a beet sugar factory via a power plant featuring a back-pressure steam turbine, and a process heating system. The steam from the turbine exhaust is supplied to a multiple-effect evaporator in which sugar solution is concentrated to a definite dry substance content. The vapors separated from the solution are in turn supplied to various heat receivers in the process, thus making multiple use of heat possible. A high energy efficiency can be attained, but the process heating system is rather complicated and to avoid unnecessary energy expenditure, it must be monitored to provide energy-related data for the process operator.

The monitoring of energy usage in a sugar factory involves repetitive estimation of the heat balance of a multiple-effect evaporator. The heat balance is typically calculated using input data from pressure, temperature and flow measurements, and laboratory-determined concentrations of juice between evaporator effects. As this is a tedious and time-consuming procedure, computer-aided energy monitoring becomes increasingly popular. When applying computerized estimation of the evaporator heat balance, one should preferably avoid using laboratory-determined data, acquiring all the necessary input data from on-line measurements.

The estimated heat balance, expressed by the values of quantities like vapour flows, heat flows and heat-transfer coefficients, is incomplete unless the uncertainty intervals of these values are evaluated. Whether applying a conventional energy monitoring procedure, or operating a computer-aided monitoring system, the uncertainty intervals of the estimated values of heat-balance quantities depend on the errors in input data. These, in turn, depend on the accuracy of measuring instruments.

In order to keep the uncertainty intervals of the estimated values under control, one should carefully plan the selection of measuring instruments. It should be emphasized that this is usually not the question of minimizing the uncertainty of the estimated heat balance, but rather of selecting least accurate instruments which still ensure a sufficiently high quality of estimation. The cost of instrumentation can thus be minimized.

MATERIALS AND METHODS

Mathematical model

Using the relationships reflecting the laws of mass and energy conservation [1], one can arrive at a system of equations describing the state of the multiple-effect evaporator

$$f_1(\underline{x}, \underline{y}) = 0$$
$$f_2(\underline{x}, \underline{y}) = 0 \qquad\qquad (1)$$
$$\dots$$
$$f_n(\underline{x}, \underline{y}) = 0$$

where $\underline{x} = x_1, x_2, \dots, x_m$ is a vector of measured parameters, $\underline{y} = y_1, y_2, \dots, y_n$ is a vector of estimated heat-balance quantities, and $f_1, f_2, \dots, f_n$ are real-valued functions (m and n are numbers of measured parameters and estimated quantities, respectively).

Assuming that the above equations have an unambiguous solution, they can be symbolically rewritten as

$$
\begin{aligned}
y_1 &= F_1(\underline{x}) \\
y_2 &= F_2(\underline{x}) \\
&\cdots \\
y_n &= F_n(\underline{x})
\end{aligned}
\tag{2}
$$

where $F_1, F_2, .., F_n$ are real-valued functions. It should however be noted that some of these functions may be given only implicitly in their algorithmic representation as they cannot be expressed in an algebraic form.

When relying on inaccurate measurements, these equations yield the values of estimated quantities which differ from the accurate ones

$$
\begin{aligned}
y_1 + \delta y_1 &= F_1(\underline{x} + \delta\underline{x}) \\
y_2 + \delta y_2 &= F_2(\underline{x} + \delta\underline{x}) \\
&\cdots \\
y_n + \delta y_n &= F_n(\underline{x} + \delta\underline{x})
\end{aligned}
\tag{3}
$$

where $\delta\underline{x} = \delta x_1, \delta x_2, .., \delta x_m$ is the error vector of measured parameters and $\delta\underline{y} = \delta y_1, \delta y_2, .., \delta y_n$ is the resulting error vector of estimated heat-balance quantities.

Using the above equation systems, the planning of measuring instruments can be aided by the evaluation of uncertainty intervals in which the resulting errors are contained. At first, it is necessary to adopt a reference vector $\underline{x}$ of the measured parameters and, using the equation system (2), to find the corresponding reference vector $\underline{y}$ of the estimated quantities. If the functions $F_1, F_2, .., F_n$ were given explicitly, then the equation system (3) would make it easy to study the resulting errors. As this is usually not the case, the evaluation of uncertainty intervals must rely on extensive computations.

Assuming that the error probability distributions are known for all the measuring instruments, one can generate multiple vectors $\underline{x} + \delta\underline{x}$ representing inaccurate measurements. For each vector of "inaccurate" parameters, a vector of resulting estimation errors $\delta\underline{y}$ can be determined from the equation system (3). Multiple vectors $\delta\underline{y}$ can be finally analyzed to evaluate the uncertainty intervals for all the heat-balance quantities.

As measures of the uncertainty interval, mean error and maximum error can be applied. After generating sufficiently many vectors of "inaccurate" parameters and substituting them into equation system (3), the calculated vectors of resulting estimation errors constitute a data base for the final evaluation procedure. For each heat balance quantity, the corresponding data are reviewed to recover maximum error, and to calculate the mean error.

Evaluation of uncertainty intervals

The nature of parameters and heat-balance quantities may vary depending on the requirements of energy monitoring and the details of instrumentation. Two practical cases are reviewed in Tables 1 and 2.

When the monitoring problem under consideration is related to a new thermal system, the reference vector of measured parameters is defined using the design data on the heat balance. When preparing the monitoring of an existing system, results of measurements can be used; however, this may require arranging special measurement sessions during which high-accuracy instrumentation is used, as routine industrial measurements are usually insufficient to attain a high accuracy of the reference data.

The reference vector of estimated heat-balance quantities is calculated from the equation system (2).

From the characteristics of the measuring instruments, the uncertainty intervals of measured parameters can be identified. It is assumed in this study that the actual errors may vary within these intervals and that the error probability distributions are uniform. This is a pessimistic assumption leading to the dispersions of parameter values which are greater than those resulting from Gaussian distributions which can be expected to hold in reality.

The essential part of the evaluation of heat-balance uncertainty is performed using the "Monte Carlo" method. Using a random number generator, the vectors of "inaccurate" parameters are repetitively calculated and used as input data to the equation system (3). Each time, a vector of resulting errors in heat-balance quantities is determined by comparing the actual solution of the equation system with the reference vector of estimated

TABLE 1

Summary of parameters measured to determine the heat balance of a thermal
system with a quadruple-effect evaporator; after [2].

Quantity	No. of measurements
Evaporator parameters	
Steam or vapour pressure	8
Thick juice concentration	1
Steam, condensate or juice flow	7
Juice or water temperature	4
Heat receiver parameters	
Vapour, condensate or juice flow	12
Condensate or juice temperature	6
Manual inputs	
Thin juice concentration	1
Valve settings (for estimating flows of noncondensible gases)	8

TABLE 2

Summary of parameters measured to determine the heat balance of a thermal
system with a quintuple-effect evaporator.

Quantity	No. of measurements
Evaporator parameters	
Steam or vapour pressure	10
Juice concentrations	3
Steam, condensate or juice flow	12
Juice or water temperature	3
Heat receiver parameters	
Vapour, condensate or juice flow	13
Condensate or juice temperature	16
Manual inputs	
Valve settings (for estimating flows of noncondensible gases)	10

heat-balance quantities. After repeating this procedure sufficiently many times, the data base for the final error evaluation is generated.

Taking into account that the dimension of vector $\underline{y}$, i.e. the number of calculated heat-balance quantities is of the order 30-40, it would be impractical to consider all the resulting errors. Instead, one can concentrate on the most important quantities, combining their mean errors into some synthetic index which could be regarded as an over-all measure of the resulting heat-balance uncertainty. The investigation of maximum errors can also be limited to the quantities selected as most important, but these errors should be considered separately for each quantity.

There are two most important groups of heat-balance quantities, namely the mass flows of water evaporated G_i, and the effective temperature differences Δt_i between heating medium (steam or vapour) and juice, in the individual evaporators numbered $i = 1, 2, .., N$. As the values of the quantities named are used to calculate heat transfer coefficients k_i, the mean errors of heat transfer coefficients in the evaporators ζk_i seem to be particularly well suited to serve as indicators of the heat-balance uncertainty. Two synthetic measures of the over-all uncertainty can be considered, namely the arithmetic average of mean errors

$$A = (\Sigma \; \zeta k_i)/N \qquad\qquad (4)$$

and the weighted average of mean errors, with mass flows of water evaporated in the individual evaporators assumed as weights

$$W = (\Sigma \; G_i \; \zeta k_i)/(\Sigma \; G_i) \qquad\qquad (5)$$

In both expressions, the summation is performed for $i = 1, 2, .., N$.

Computer program

The work reported here can be seen as an extension of studies on industrial energy monitoring systems [1,2]. It is therefore possible to use a collection of mathematical models, algorithms and program modules originally developed for other purposes. Most notably, a complete library of thermodynamic functions and evaporator heat-balance routines programmed in TURBOPASCAL (version 5.5) is available. This makes it possible to generate new TURBOPASCAL programs for the evaluation of heat-balance uncertainty mainly by assembling a number of ready-to-use modules.

A complete program consists of three main parts:
- data input and manipulation of data files,
- calculations and creation of the data base,
- data base analysis and result output.

The first part is so designed that once a certain data file corresponding
to a set of measuring instruments has been written to the computer memory,
one can easily generate new files by modifying the data on selected
instruments.

A typical evaluation of heat-balance uncertainty intervals for a given
set of measuring instruments involves generating thousands of random error
vectors, calculating as many solutions of the equation system (3) and
writing them into the data base. After completing the final analysis, the
values of synthetic measures of the over-all uncertainty, calculated
according to (4) and (5) are output along with the values of mean errors
and maximum errors of heat-balance quantities.

Looking at the evaluation results, one can conclude whether or not the
proposed measuring instruments are sufficiently accurate. If the accuracy
parameters of certain instruments need to be corrected, then one can edit a
copy of the data file, thus generating a new file, and repeat the
evaluation run.

When analyzing the evaluation results, the instrumentation designer
may also conclude that the heat-balance uncertainty can be reduced by
modifying the set of measuring instruments (i.e., redefining the vector $\underline{x}$)
and changing the equation system (1) accordingly. This would of course
require modifications in the data input and calculation parts of the
computer program. Using the modified program, the designer can evaluate the
heat-balance uncertainty characteristic of his new instrumentation
proposal.

RESULTS

Two solutions were studied for energy monitoring in a sugar factory
equipped with a quadruple-effect evaporator: a conventional approach based
on refractometric determination (in the laboratory) of juice concentrations
between the evaporator effects, and a computer-aided monitoring procedure
based on low-cost radiometric concentration meters. In both cases,
following error values were assumed for other parameters measured (in per

cent of the measuring range): pressure 1.5, flow 1 and temperature 1.5.

Assuming the error of refractometric determination of juice concentration at 0.1% DS, the resulting uncertainty was evaluated in terms of heat transfer coefficients in the evaporator effects, using the formulae (4) and (5). The arithmetic average of mean errors was 2.5% and the weighted average was 2%.

The calculations were repeated assuming the error of concentration meters at 1% DS. The arithmetic average of mean errors increased to 16.6% and the weighted average to 14.4%. This must be seen as an indication that this type of concentration meter is unsuitable for heat-balance monitoring. It can be added that an alternative monitoring method was finally designed for this sugar factory, using the measurements summarized in Table 1.

The second example is related to a sugar factory equipped with a quintuple-effect evaporator. A computer-aided energy monitoring system was designed using on-line measurements reviewed in Table 2, and the selection of measuring instruments had to be studied with the aim to minimize their cost. For each instrument except concentration meters (assumed error 0.5% of the measuring range), the designer proposed two alternatives with different accuracy parameters and different prices. Typical error values were identical in both cases: pressure 0.2%, flow 0.5% or 0.75%, and temperature 0.2%. The low-accuracy alternative was based on standard (specified by respective manufacturers) measuring ranges of all the instruments. For the high-accuracy alternative, specially adjusted measuring ranges (specified by the designer) were assumed.

The values of error indices characterizing the high-accuracy and low-accuracy instrumentation alternatives turned out to be very close. In the former case, the arithmetic and weighted averages of mean errors were about 1.2% and 1.1%, respectively; in the latter case, the corresponding figures were 1.5% and 1.4%. The maximum uncertainty of heat transfer coefficients (worst case recovered from a data base containing 5000 vectors of resulting errors) was found in the fourth evaporator effect: 5% for high-accuracy instrumentation and 7% for low-accuracy instrumentation. Both alternatives are therefore acceptable and there is no need to apply costly instruments.

On the basis of a few additional simulation runs, it was concluded that even if the accuracy of about 1/3 of the instruments (mainly flow meters) becomes worse than specified in the low-accuracy design proposal, the resulting errors still remain at an acceptable level. This was interpreted as an indication that some of the instruments proposed can be replaced by less costly ones. The instrumentation alternative which was finally accepted turned out to be several per cent cheaper than the original low-price proposal. For the sugar factory in question (processing capability 6000 tons of beet per day), the saving in relation to the high-cost proposal was about 80 000 ECU.

DISCUSSION

The nature of the method of selection of instrumentation presented above is speculative. Arbitrary assumptions were made on the distributions of measuring errors, uncertainty measures to be applied as evaluation criteria, volume of data to be acquired from simulation computations, and allowable values of average and maximum errors of estimated variables. Although these details can be discussed, the method proved effective in the systematic evaluation of qualitative differences between instrumentation alternatives considered.

It seems that the proposed principle of studying and evaluating various instrumentation alternatives is applicable to process heating systems in general. When designing an energy monitoring system which is required to provide high-accuracy information, control engineers usually tend to select very accurate and very costly instruments even if certain measurements have little effect on the resulting accuracy of crucial monitoring variables. Using the simulation approach, the designer is able to evaluate and compare instrumentation alternatives on a cost effectiveness basis.

ACKNOWLEDGEMENT

This work was financed by the State Committee for Scientific Research, Warsaw, Poland, under research grant No. 9 0552 9101.

REFERENCES

1. Urbaniec, K., Modern Energy Economy in Beet Sugar Factories, Elsevier, Amsterdam-Oxford-New York-Tokyo, 1989.
2. Urbaniec, K., Developments in computer-aided monitoring of energy balances of sugar factories. Zuckerindustrie, 1991, 116, 509-512.

A DEVICE FOR THE MEASUREMENT OF QUALITY IN AN EVAPORATOR TUBE

McNeil D.A., Grant I. and Cornwell K.
Department of Mechanical Engineering
Heriot-Watt University, Edinburgh EH7 4AS, Scotland

ABSTRACT

A technique for estimating the mass flow rate and inlet and outlet mass dryness fractions of flow within a tube of a direct expansion evaporator is described. The technique involves using two sensors, consisting of insulated coils of wire, within a tube; one at the inlet and the other at the outlet. These sensors are used to measure the coil to fluid heat-transfer coefficient, the fluid temperature and the fluid-wall temperature difference. When the heat load between the two sensors has been estimated, Chen-type correlations of the coil's heat-transfer performance allow the mass dryness fractions at the sensor locations and the mass flow rate through the tube to be estimated.

NOTATION

A : Area (m^2)
b : Temperature coefficient of resistance $(°C^{-1})$
C : Specific Heat (J/kgK)
h : Enthalpy (J/kg)
I : Current (A)
L : length of wire used in sensor construction (m)
P : Electrical power (w)
Q : heat-transfer rate (w)
R : Electrical resistance (Ω)
T : Temperature (K)
x : Mass dryness fraction or quality $(-)$
α : Heat-transfer coefficient (w/m^2K)
ϕ : Heat-transfer coefficient relationship
ρ : Material resistivity (Ω/m^3)

Subscripts

c : coil
f : fluid
fg : from liquid to vapour
p : at constant pressure

r : reference
s : surface
$1,2$: locations

INTRODUCTION

The actual heat transfer and pressure drop experienced by a fluid in a multi-pass, direct-expansion evaporator may differ from the design values if the proportion of the mass flow rate carried by each tube and the mass dryness fraction entering each tube is not uniform across a given pass. To establish if the performance is significantly affected, the extent of this mal-distribution would have to be measured. This is a difficult task which requires a new measurement technique to be developed. This measurement technique must allow a sensor located within an evaporator tube to pass a signal to a display located outside the heat exchanger. This sensor should not interfere excessively with either the mass dryness fractions being estimated nor the mass flow rate passing through the tube. Thus, such an instrument would require; (a) an electrical output dependent on the mass dryness fraction, and (b) a body shape with a flow resistance which does not significantly alter the quantities being measured. A sensor which was reported to measure mass dryness fraction in the range 0.6 to 0.9, was invented as part of work undertaken for York International (1) and patented by one of the present authors (2). This paper describes the development of a sensor which operates in the range 0.0 to 0.7.

SENSOR CONCEPT

The sensor has three operational modes; one for measuring fluid-wall temperature difference, another for measuring fluid temperature and a third for measuring the local heat-transfer coefficient.

The sensor, as shown on Figure 1, consists of a polypropylene, cylindrical body; shaped so as to minimize flow resistance. Along the body is placed an alumel wire coil. This wire contacts the tube wall at one end and a copper lead at the other. The copper lead passes along the inside of the tube, through the end plate and onto an energizing-come-measurement electrical circuit. A copper-Alumel thermocouple is formed with the cold junction in the fluid and the hot junction on the tube wall; facilitating the first mode of operation.

The second mode of operation is established when a small electrical current is passed through the wire. The resistance of the Alumel wire is dependent on its temperature, which, if the current is small, will be the fluid temperature.

The third mode of operation is established when a larger electrical current is passed through the wire. The resistance of the Alumel wire is dependent on the wire temperature, which, in turn, is dependent on the heat-transfer coefficient between the wire and the two-phase flow.

SENSOR DEVELOPMENT

Development of the sensor required two avenues of investigation; one to establish the maximum allowable fluid flow resistance to give an acceptable tolerance to the quantities being measured and another to ascertain an electrical technique which would permit the quantities to be measured.

Limits to Flow Resistance

A limit to the fluid flow resistance induced by the sensor was established by a computer simulation of the tubes within a pass. Two types of simulation were performed; one with and one without the sensors in position. The error induced by the sensor was taken as the change in reading associated with the physical presence of the sensor.

The actual mass dryness fraction occurring at the proposed sensor location was established by solving the one-dimensional momentum and energy equations assuming that each tube contained equal quantities of liquid and vapour and passed equal mass flow rates of mixture.

These equations required considerable empirical input before a solution could be sought. The momentum equation requires the acceleration pressure gradient, for which the Morris method (3) was used; the frictional pressure gradient, for which the Chisholm method (4) was used; and the slip ratio, for which the CISE correlation (5) was used. Additionally, the energy equation requires the tube-side, heat-transfer coefficient, for which the Chen correlation (6) was used, and a shell-side coefficient. The HTFS program TASC3 was used to estimate the shell-side coefficient for the shell in which the sensor is to be inserted. The method used to solve these equations has been described previously by McNeil (7).

A proportion of the tubes were assumed to be instrumented as shown on
Figure 2. This was achieved by injecting a form loss at a specific
location within a tube. This form loss was modelled by giving the sensor
a specified single-phase loss coefficient which was corrected for
two-phase effects using the Morris multiplier (3). Typical results of
the change in quality at the sensor location and the mass flow rate
through the tube are shown as a function of sensor resistance and
proportion of tubes instrumented in Figures 3 and 4. These calculations
indicated that the sensor should have a single-phase loss coefficient
not exceeding 3.

Coil Design

When the coil, which has a reference resistance R_r at a temperature T_r,
is subjected to a current I, it rises to a temperature T_c. P, the power
dissipated by this coil, is given by

$$P = I^2 R_r [1 + b(T_c - T_r)] \qquad (1)$$

where b is the temperature coefficient of resistance.

The sensitivity of this sensor is dP/dT_c, which, for a coil with a wire
length L, a wire cross-sectional area A and a resistivity ρ, can be
written as

$$\frac{dP}{dT_c} = I^2 R_r b = I^2 \frac{\rho L b}{A} \qquad (2)$$

This can be optimized by paying attention to material selection (ρb),
wire geometry (L/A) and the supplied current.

Several materials were tested; the preferred one being Alumel, a Ni-Al
alloy which is the negative leg of a K-type thermocouple. This has the
advantages of a large Seebeck effect (20 $\mu V/^{0}C$ with respect to the tube
material of copper) and a large power dissipation effect. The coil has
to be short enough for the evaporator tube heat-transfer surface not to
be significantly reduced and should also be as thin as possible. The
wire diameter used was 0.193 mm and the coil length sufficient to occupy
30 mm of heat-exchanger tube.

SENSOR CALIBRATION

The sensor was calibrated by placing it in the loop shown in Figure 5.
R113 was circulated from the reservoir, through the flow meter and
electrical heaters and delivered to the horizontal leg where the sensor
was located. The fluid flowed from this horizontal leg into the
separator from where the liquid was returned to the reservoir and the
vapour passed to the condenser where it was condensed and returned to
the reservoir. The fluid temperature and flow rate were measured and the
quality determined from the electrical load. The heat-transfer
coefficient α between the coil and the two-phase flow was then
determined from

$$\alpha = \frac{Pb}{\left(\dfrac{P}{P_r} - 1 + b(T_r - T_f) \right) A_s} \tag{3}$$

where P_r is the power dissipated by the coil at its reference
temperature and A_s is the area of the coil surface exposed to the fluid.

A typical curve of heat-transfer coefficient against quality is shown on
Figure 6; on which is included the correlation of Liu and Winterton (8).

APPLICATION

To apply the method to a heat-exchanger, sensors will be placed at the
inlet and exit of an evaporator tube. The fluid temperature, the
fluid-wall temperature difference and the heat-transfer coefficient can
then be determined at both sensors. For a fluid with a enthalpy of
evaporation h_{fg} and a liquid specific heat capacity C_p in a tube of
known diameter, the unknown entrance mass dryness fraction x_1 and the
unknown mass flow rate $\dot{M}$ are related to the measured heat-transfer
coefficient α_1 by

$$\alpha_1 = \phi \left(\dot{M} , x_1 \right) \tag{4}$$

where ϕ is the function specified by correlating the sensor response
typified by Figure 6. Similarly, at the tube outlet,

$$\alpha_2 = \phi \left(\dot{M} , x_2 \right) \tag{5}$$

If the heat transferred to the fluid between these sensors is Q, then a heat balance gives

$$Q = \dot{M} \left(C_p (T_{f2} - T_{f1}) + h_{fg} (x_2 - x_1) \right) \tag{6}$$

where T_{f1} and T_{f2} are the fluid saturation temperatures at the tube inlet and exit respectively. This, providing that the heat transferred to the fluid as it passes between the two sensors can be estimated, generates 3 equations involving 3 unknowns.

The heat transferred between the sensors can be estimated by integrating the heat-transfer equations along the tube. This requires the shell- and tube-side heat-transfer coefficients to be known. The local shell-side values can be made by passing a single-phase fluid through the tube and measuring the distribution of wall temperature. The tube-side values can be determined from a correlation, such as that due to Liu and Winterton (8). Comparison of the tube-side values and the values obtained from the sensors can be made with some alterations being made to the values used in the analysis until the best agreement is obtained.

DISCUSSION AND CONCLUSION

The problems associated with mal-distribution of both phase and mass flow rate in evaporators are not well understood. One of the major obstacles to improved understanding is a quantitative measure of these effects. Measuring mal-distribution is not an easy task, however, a method has been devised which permits some estimate to be made. The method is not simple and involves several sources of error. Some errors will arise because the method makes implicit measurements of inter-related quantities which then have to be deduced. The largest uncertainty, however, is probably the heat transferred to the fluid between the sensors. With careful preparitory work, using both correlations, measurements and cross checks, a reasonable estimate may still be obtained. This data should be good enough to at least make possible a first order, quantitative study of mal-distribution in a direct expansion evaporators. Further work may be necessary before more precise data can be obtained.

REFERENCES

(1) McCafferty, J.B. (1990), 'Refrigerant Distribution in Shell and Tube Evaporators', Ph.D. thesis, Heriot-Watt University, Edinburgh.
(2) Cornwell, K. (1989), "Sensor for Conduit Wall Temperature and Vapour Fraction Measurement", British Patent Application Number 8912906.8.
(3) Morris, S.D. (1984), A Simple Model for Estimating Two-Phase Momentum Flux, 1st National Conference on Heat Transfer, University of Leeds.
(4) Chisholm, D. (1983), Two-Phase Flow in Pipelines and Heat Exchangers, London: Goodwin, ISBN 0-7114-5748-4
(5) Premoli, A., Francesco D. and Prima A. (1970), An Empirical Correlation for Evaluating Two-Phase Mixture Density Under Adiabatic Conditions.
(6) Chen, J.C. (1966), A Correlation for Boiling Heat Transfer to Saturated Fluids in Convective Flow, Ind. Engng. Chem. Proc. Des. Dev. 5, pp 322-329.
(7) McNeil, D.A. (1990), Pipeline Design for Two-Phase Flow, Paper 4 of Multi-Phase Flow in Pipeline Systems, HMSO: London.
(8) Liu, Z. and Winterton, R.H.S. (1991), A General Correlation for Saturated and Subcooled Flow Boiling in Tubes and Annuli, Based on a Nucleate Pool Boiling Equation, Int. J. Heat and Mass Transfer, Vol. 34, No. 11, pp 2759-2766.

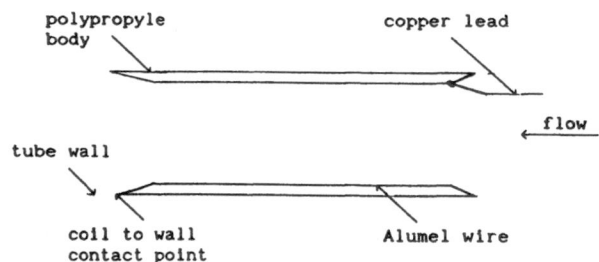

Figure 1: Sensor Design

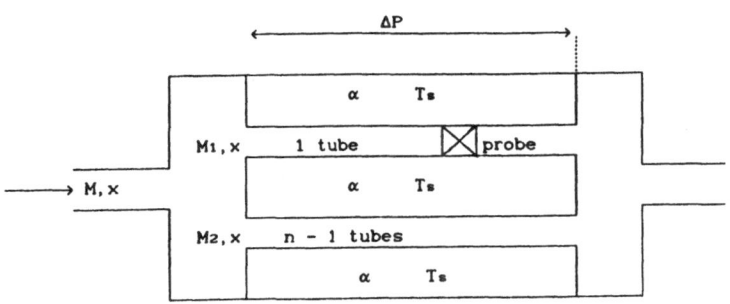

Figure 2: Flow model for sensor effect on quality and mass flow rate

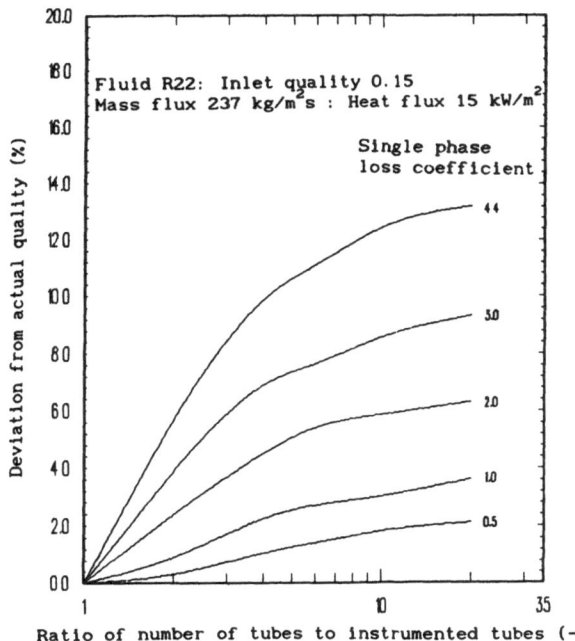

Figure 3: Effect of sensor on the quality measured

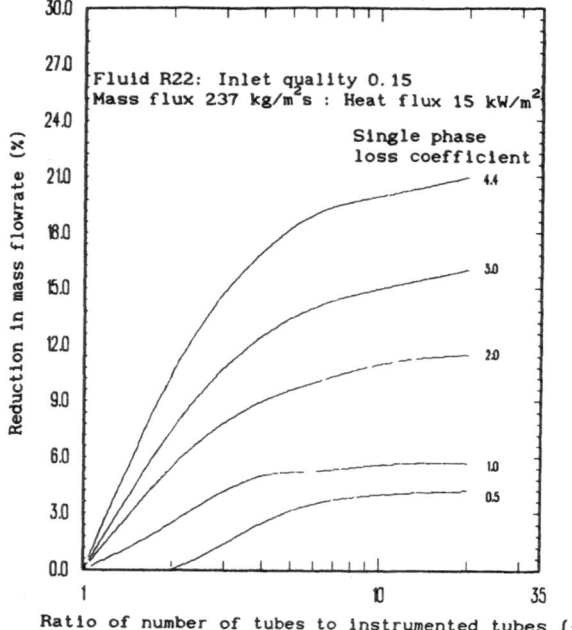

Figure 4: Effect of sensor on the tube mass flowrate

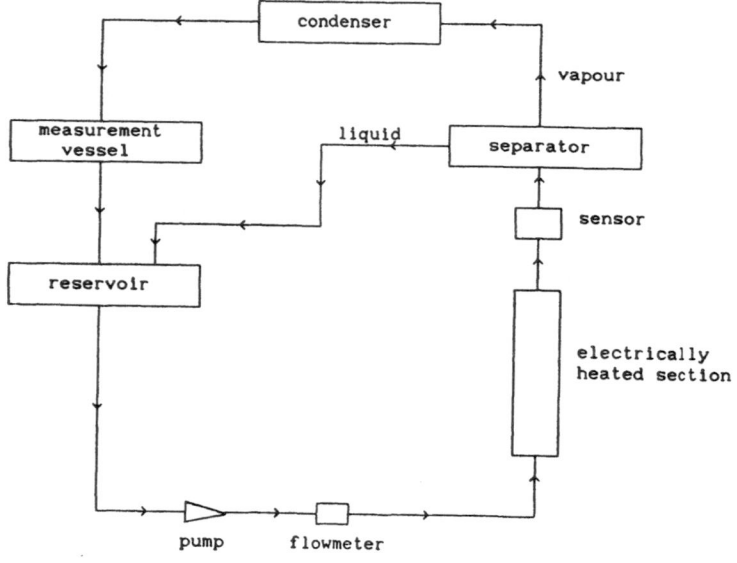

Figure 5: Probe calibration loop

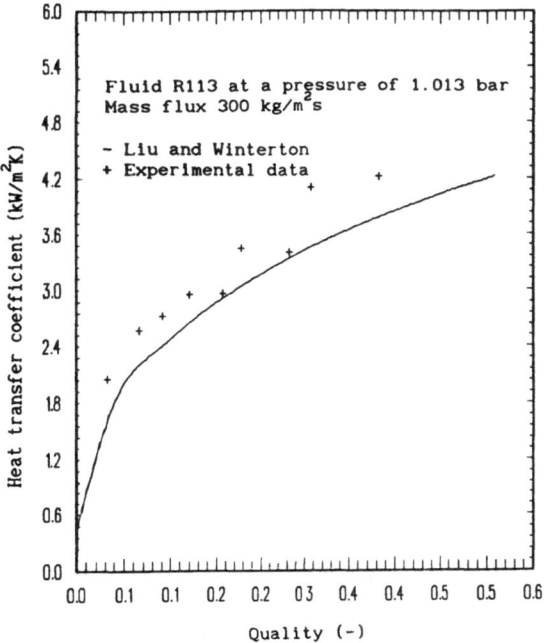

Figure 6: Sensor response

SENSORS FOR MONITORING AND CONTROL OF FURNACES, KILNS AND OVENS

GEROLD DIMACZEK

HANS-GEORG FASSBINDER, DRAGAN STEVANOVIC, CUNG VAN TRAN

Applikations- und Technikzentrum für
Energieverfahrens-, Umwelt- und Strömungstechnik
Kropfersrichter Straße 6-8 8458 Sulzbach-Rosenberg, Germany

SUMMARY

Sensors for videomatic observation, temperature sensors, mass flux sensors and humidity sensors for the application in harsh environmental conditions are described. Literature studies, laboratory and industrial experiments for excisting sensors and new developments in the field of mass flux metering were carried out. The results of these studies are summarized and special emphasis is given to the applicability and the reliable operation of these sensors in furnaces, kilns and ovens, in order to clarify whether these sensors can serve as fast and accurate sensing devices for effective control systems minimizing energy consumption and pollution emission.

INTRODUCTION and OBJECTIVES

Advanced control systems are required to reduce energy consumption and decrease pollutant levels in industrial furnaces, such as those used in glass-ceramic-, cement- and baking ovens. For these purposes accurate sensors are needed, withstanding the conditions in furnaces, kilns and ovens. Although there is a lot of experience in application of temperature-, mass flux-, humidity- and gas sensors, a lack of information remains in the field of application under high temperature or harsh environmental conditions. Applied research work in this important area is performed in the CEC JOULE I-project No. 0051 C "Energy saving and pollution abatement in glass-making furnaces, cement kilns and baking ovens" in the subtask "Sensors for harsh environmental conditions". Instituts in the Netherlands (Institute for Cereals, Flour and Bread - TNO-Cereals, Division of Technology and Society - TNO-Tech.) as well as in Germany (University Erlangen-Nuremberg - LSTM, Applikations- und Technikzentrum - ATZ-EVUS) are working in this area. In order to investigate sensors applicable in harsh environmental conditions, literature studies, experimental investigations and sensor developments were carried out in the present project covering the fields of

o   high temperature video furnace probes   *ATZ-EVUS/LSTM*
o   temperature sensors, mass flux sensors   *ATZ-EVUS/LSTM*
o   humidity and gas sensors   *TNO-Techn.*
o   sensor systems for baking ovens   *TNO-Cereals*

The aim of this research work is to establish guidelines for application of sensors, to select sensors suitable for control systems and to develop new sensors for e.g. mass flux metering and over all furnace monitoring. The application of these sensors in industry shall prove the reliable operation and deliver verification data for numerical simulations of furnace operation. In four selected examples the progress in research is presented.

RESULTS OF THE R+D ACTIVITIES

Videomatic fire probe

Energy consumption and pollution emission of furnaces and kilns will depend on the flame structure which is affected by convection, diffusion and combustion processes. In order to observe the turbulence structure of technical and laboratory flames a video fire probe has been developed in close cooperation of ATZ-EVUS and GRUNDIG [1] which allows videomatic and cinematographic studies of the flame front in the vicinity of the flame and not only through windows in the wall of a furnace. The video fire probe consists of a special wide-angle lens-system up to 94°, a miniaturized CCD-colour camera (Charge Coupled Device) and an amplifier for the video-signal (CCIR-PAL standard). The electronic equipment is housed in a CrNi-steel tube (Fig. 1), which is water-cooled and has also a connector for pressurized air in order to clean the lens in front of the probe.

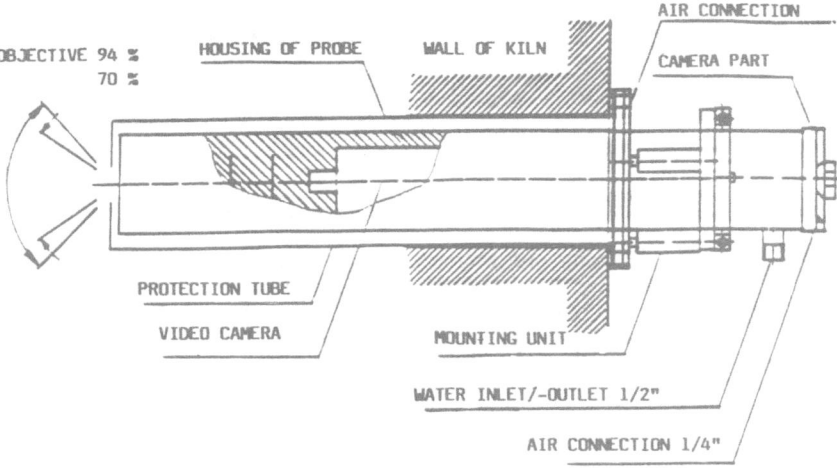

Fig. 1: *Sketch of assembly of video fire probe GRUNDIG FRS*

In order to have the possibility of adjustment of several camera features a special remote control unit was developed. It is necessary not only to get global visual information of an industrial flame but also to resolve brightness, colour and turbulence structures in the combusting zone. With this information a flame could be adjusted to low pollution conditions. The adjustable features of the control unit are:

- *colour compensation (internal, automatic, manual)*
- *white compensation (permanent, single shot)*
- *colour temperature control*
- *shutter   on/off   (1/60 - 1/1000 S)*
- *amplification   on/off   (+ 6 dB)*
- *aperture   on/off   (- 6 dB)*

The furnace probe system was successfully tested in a laboratory and in a real glass-melting furnace (end port fired). For industrial application the probe was installed in the crown of the furnace in ambient temperatures of 1800°C. Cooling rates of 30 m³/h potable water (housing of the probe) and 10 m³/h air (protection tube), 1 m³/h air (air scavening of interior) had been sufficient to ensure a reliable operation of the video camera. Long term stability tests over 3 months have proved that the video probe system is able to withstand the high temperature and the aggressive atmosphere in the furnace. The adjustable features of the remote control unit lead to suitable pictures of well defined colour and brightness (Fig. 2).

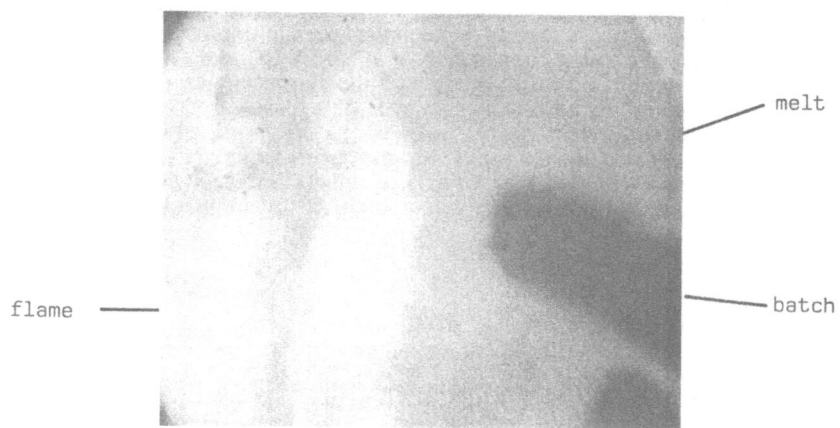

Fig. 2:   *Video print: Glass-melting furnace in operation*

By observing the flame stability and colour and by monitoring the furnace operational parameters, it has been possible to control the optimal batch feeding rate and to adjust the air/fuel ratio for combustion and pressure inside the furnace by visual control to an optimum. Deviations in quality of the molten glass occuring non-periodically, could be assigned to an overload in batch material and an instable flame structure due to incorrect combustion control. For service and cleaning purposes the axial and radial single adjustment permits the removal of the probe without problems.

CORIOLIS-mass-flow meter

ATZ-EVUS developed a new mass-flow meter which uses the Coriolis effect on the basis of a radial turbine. In Fig. 3 the details of the instrument are shown. The bulk-material is entering the turbine wheel in axial direction. Six radially arranged channels in the turbine wheel serve as an outlet for the mass flow. The turbine is driven at constant speed by a synchronous motor. The radial flow of the bulk material and the rotation of the turbine result in a tangentially directed Coriolis force acting as a negative torque on the turbine shaft. The physical relationship can be described by the following formula:

$$M = \dot{m} \cdot \omega \cdot R^2$$

M = torque (Nm), $\dot{m}$ = mass flow (kg/s), ω = angular velocity (2π/s)
R = radius of turbine (m)

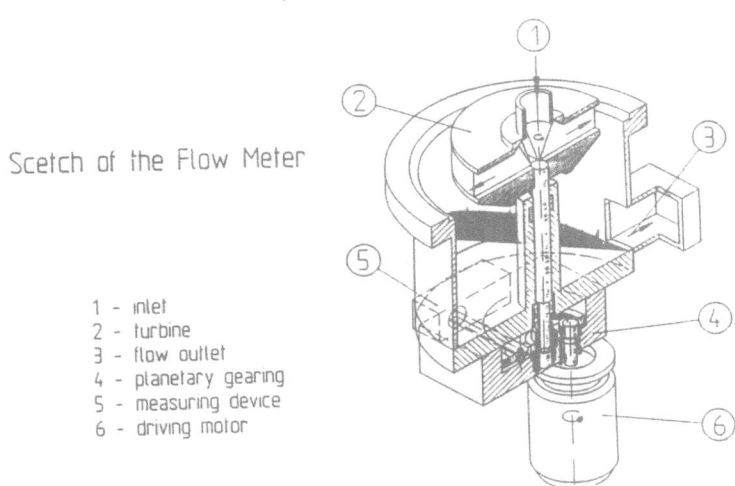

Scetch of the Flow Meter

1 - inlet
2 - turbine
3 - flow outlet
4 - planetary gearing
5 - measuring device
6 - driving motor

Fig. 3: *Main parts of the Coriolis mass-flow meter*

The reaction force on the turbine shaft is transferred to a moveable gear wheel and sensed by a strain gauge transducer. This gear arrangement compensates influence of the mechanical friction on the metering signal, so that any interference of the measuring signal by disturbances depending from time and temperature is avoided. One benefit of this special construction is, that there is no influence of physical parameters of the material measured. The application in pressurized systems is also possible (up to 16 bar). In order to overcome the problems of erosion the 6 measurement channels are protected by SiC platelets. The experimental verification of the measurement accuracy was carried out in two large pilot plants erected at the technical laboratories at ATZ-EVUS. In order to calibrate the two measuring devices developed (type A 0-10 kg/min; type B 0-100 kg/min), extensive test-runs had been carried out. Measurements of the coal-dust mass-flux were performed in a pilot plant (Fig. 4) consisting of a storage bin, full-scale screw feeder and bucket wheel of a power plant, pneumatic conveyer lines, a weighting bin and full automatic control systems. In this pilot plant two of the most important applications, i.e. charging from storage bins and direct measurement in pneumatic conveyer lines, by separating the mass-flux by means of a low pressure-loss cyclone had been tested. The results of the measurements are shown in Fig. 5.

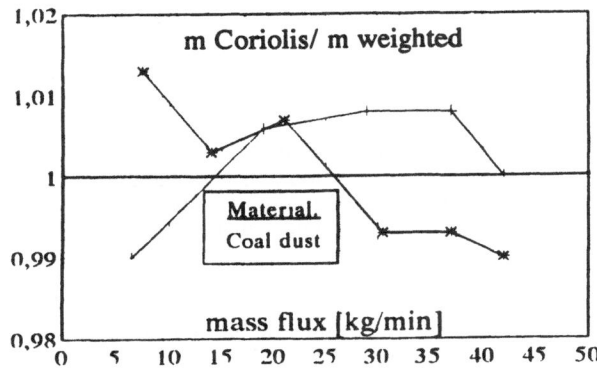

Fig. 5: *Calibration diagram*

This figure demonstrates that the Coriolis meter can measure the mass-flux of granular material with an accuracy of $\pm$ 1 % based on the actual measured value. Extensive experimental investigations in the pilot plant have also shown, that for the Coriolis meter a response time of 0,5 seconds for load reversals has been realized. In real plant experiments (power plants) the measurement device has been tested successfully in its long term stability, withstanding coal-dust temperatures of 120°C, emergency cut downs and rapid load reversals. It was shown that no erosion problems occur and reliable operation is ensured. In Fig. 6 a photo of a type B Coriolis meter installed at the power plant Lausward, Düsseldorf, Germany, is given. The throughput of thousands of tons brown coal- and coke-dust has proved the reliable operation in harsh industrial environment.

Fig. 4: *Pilot plant at ATZ-EVUS*   **Fig. 6.:** *Coriolis-meter at power plant*

Thermocouples

Response time: The dynamic behaviour of a thermocouple is usually characterized by time constant or by response time. By definition, the time constant is the time required to reach 1/e (or 36,79%) of __ T (difference between the instantenous and the initial temperature). The response time is much more vague term and there are several different definitions. The IEC (International Electrotechnical Commission) recommends that the time required to cover half the temperature step (t=To/2) to be taken as the response time *(Crovini [3])*. The most important parameters which influnce time constant are:

- presence of protective tube,
- intensity of heat transfer between the ambient and thermocouple,
- position of thermocouple junction inside protective tube,
- mass of thermocouple junction, i.e. the diameter of the wires,
- range of the step-wise temperature change and other temperature conditions.

In the literature it is usually stated that thermocouples have shorter time constants than any other type of conventional thermometers. On the other hand, different data of thermocouple time constant may be found: from 60ns for a thin-film Pt/Ir thermocouple *(Ricolfi & Scholz [4])*, up to few hundreds of seconds, for shielded thermocouples made of very thick wire. In aim to have more possibilities for analysing dynamic behaviour of thermocouples, the numerical code TCTRT (ThermoCouple Transient - Response Time) has been developed. It enables the evaluation of thermocouple junction temperature during a given transient of the ambient temperature. It inlcudes the calculation of the boundary conditions; applying the following models:

o   convective *(Gnielinski [5] and Rheinländer [6])* and radiative (Vortmeyer [7]) heat transfer between the ambient and protective tube,
o   heat conduction through the protective tube *(Stevanovic & Studovic [8])*,
o   heat transfer through the gap between the inside surface of the tube and the thermocouple junction *(Schneider [9])*,
o   heat capacity of a thermocouple junction.

The developed code enables to perform the parametric study of influence on the thermocouple response time, which is more effective than to perform the experimental measurements for a variety of conditions. On the other hand it enables the evaluation of a thermocouple dynamic behaviour in real conditions expected in the practical applications, such as gradual change in ambient temperature, velocity change of the ambient gases, change of their composition, etc.. For verification of the code TCTRT two S-type thermocouples (PtRh10Pt, 0,35 mm) were applied. One thermocouple was shielded by a protection tube (Alsint 99,7; d_i = 3 mm; d_a = 8 mm). A high accurate calibration furnace (ISOTECH) was heated up to 990°C and then both thermocouples (T = 20°C) were plugged in. The temperature signal was recorded by means of a data aquisition system. To cool down, the thermocouples were abruptly pulled out of the furnace and cooled down in ambient temperature (T = 20°C), without any significant forced convection. The good agreement between experimental and numerical data is shown in Fig. 7.. In the same Fig. the case of heating up the thermocouples from the ambient to 990°C is shown. Several other tests with different temperatures have been performed, also in good agreement with the code predictions.

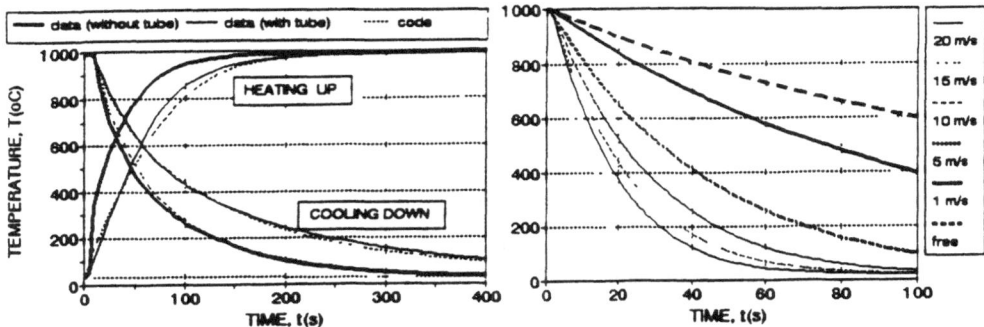

Fig. 7: *Verification*   **Fig. 8:** *Influence of gas velocity*

The code has been used for investigating the influence of some parameters on the response time. Fig. 8. shows the very important influence of the gases verlocity. On the other hand, if there is a direct contact between the thermocouple junction and protective tube, the intensity of interface conductivity is not of great influence (Fig. 9). The heat capacity of a thermocouple junction is also not very important if it is protected by a tube (Fig. 10).

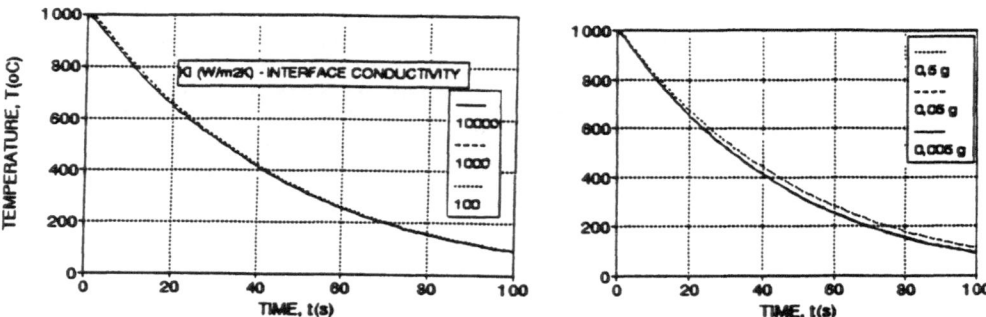

Fig. 9: *Influence of interface conductivity*   **Fig. 10:** *Influence of junction mass*

Long-term thermocouple stability: The aging of thermocouples is a general term for all kinds of changes in the thermal electromotive force which occurs while they are in operation. Those processes may be caused by different factors like oxidation, diffusion of impurities, recrystalisation, etc. and are much more intensive at higher temperatures. Aging is strongly temperature dependent as well as on the chemical and physical characteristics of the wire materials and the atmosphere. Table 1 gives the temperature limitations for three most suitable types of thermocouples for high temperature measurements.

Table 1:   *Temperature limitations for application of selected high temperature thermocouples (upon NBS Monograph 125, 1975)*

Index ANSI	Composition	Long term limit	Short term limit
S	PtRh10% - Pt	1400°C	1772°C
B	PtRh30% - PtRh6%	1500°C	1810°C
K	NiCr - NiAl	1260°C	1350°C

The most aggressive poisons for platinium thermocouples are silicon and phosphorus. Silicon is usually present in the atmospheres of different furnaces (the presence of quarz may be a source of silicon). So, if a thermocouple should be used at higher temperatures, the protective tube should be made of Alsint - almost pure Al2O3. As the electromotive force of PtRh alloys is much less sensitive to the absorption of Si, the type B thermocouple should be used in aggressive atmospheres at elevated temperatures. At least it should be mentioned that thermocouples made of thinner wires are more sensitive to the conditions which influence the process of aging. Very comprehensive data concerning long-term stability of thermocouples are given by *Körtvélyassy [2]*. Meanwhile, all data relating the aging of thermocouples may be applied just to the precisely the same operational conditions. Even slight change in gas composition can change considerably the intensity of aging. In spite of that, some data from *Körtvélyassy* are given in Table 2, having in mind that they have just illustrative character. Those data have been obtained by computer simulation, plotting regression curves of exponential type through some experimental data.

Table 2:   *Aging of a thermocouple B-type (PtRh30%-PtRh6%)*

Measured differences ΔT, °C	Real temperature, °C						
	1000	1100	1200	1300	1400	1500	1600
	Elapsed time, y-years d-days h-hours						
-1	> 10 y	1691 d	330 d	84 d	24 d	8 d	55 h
-2			1419 d	270 d	64 d	17 d	108 h
-3			3331 d	536 d	111 d	26 d	7 d
-4				872 d	166 d	37 d	9 d
-6				1731 d	290 d	58 d	13 d
-8					few y	81 d	17 d
-10						105 d	21 d
-12						129 d	25 d
-14						154 d	29 d

Thickness of wires: 0,50 mm (in protective tube), [2]

It is obvious that for high temperature measurements a certain compromise between the aging intensity and response time has to be selected: thicker wire and thicker protective tube mean higher durability but worse dynamic behaviour of a thermocouple. In practise the selection of thermocouples depends mostly on the temperature range, type of the atmosphere (oxidising or reducing) and on the place of application - as it is shown in *Fig. 11* and *Table 3*.

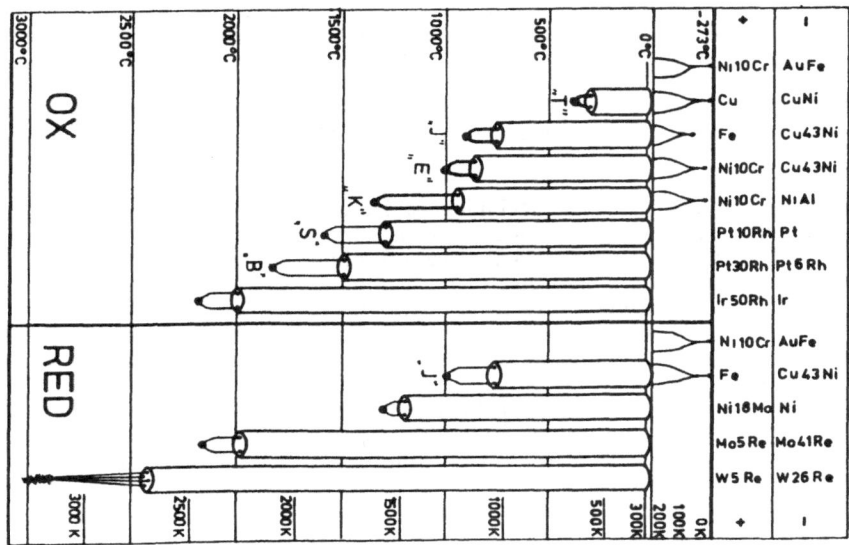

Fig. 11:   *Selection of thermocouples for oxidising (OX) and reducing (RED) atmosphere [10].*

Table 3:   *Recommendation for shielded thermocouples in industry [2]*

Industry	Location	Max. temp.	ANSI index	Protection tube / diameter
Ceramics	annular kiln	1100°C	S	ALSINT 99,7% / 5mm
	annular kiln	1200°C	B	ALSINT 9,7% / 5mm
	tunnel kiln	1300°C	B	ALSINT 99,7% / 5mm
Glass	regenerator top	1300°C	B	ALSINT 99,7% / 7mm
	regenerator bottom	600°C	K	ALSINT 99,7% / 7mm
	walls, crown,...	1550°C	B	ALSINT 99,7% / 7mm
Cement	rotary kiln	900°C	S	ALSINT 99,7% / 5mm
	hot chamber	900°C	S	ALSINT 99,7% / 5mm
	drying chamber	400°C	K	shell type

Humidity Sensors

TNO-Institute of Environmental and Energy Technology (Department of Heat and Refrigeration Technology) has concentrated on the selection and testing of moisture sensors applicable under the harsh environments of the Ceramic Industry. The following, commercially available sensors have been selected and tested:

- *1 Wet and dry bulb sensor;*
- *2 Dew point sensors;*
- *6 Polymer Humidity sensors;*
- *1 Zirconia oxygen analyzer.*

In order to test these ten sensors under industrial circumstances, a representative clay dryer was selected. All moisture sensors have been assembled around a conus where the drying air is blown into the drying chamber so that a uniform temperature and moisture distribution is guaranteed.
The moisture sensors are compared on relative humidity and as a reference a calibrated accurate mirror dew point sensor is used. The reference sensor has been cleaned regularly every week.

The experiments have started at the beginning of December 1991 and were finished in June 1992. By means of a data acquisition system all measured values were registrated and saved every minute to a computer. After every 6 minutes the average value of the last 6 measurements is saved on the hard disc of the computer. The drying cycle takes place every 43 hours. The temperature varies from 20 to 80 degrees and the relative humidity varies from 7 to 100 %. The moisture sensors are working in an atmosphere of fine clay dust and sometimes in a condensating atmosphere.

The results of the long term experiments have shown, that generally none of the moisture sensors tested worked stable during the 7 months experiment under harsh environmental conditions. Representative for the sensors tested the results of a polymer humidity sensor and a dew point sensor are presented.

Mirror dew point sensor (sensor 7): The operation principle of the optical dew point sensor is shown in Fig. 12.

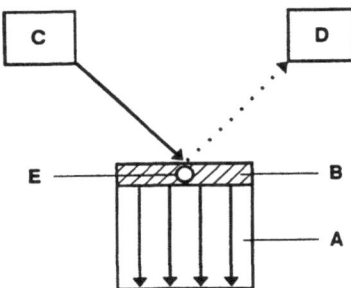

Fig. 12: *Operation principle miror dew point sensor*

Referring to Fig. 12 this comprises a cooling module (A) based on the Peltier principle and which is entirely solid state, requiring no external cooling medium. A mirror (B) is fixed to the cooling module and a platinum resistance temperature element is embedded in the mirror (E). A light source (C) is directed at the mirror and a detector (D) collects reflected light. The light detector and cooling module operate in a closed loop arrangement. The variation of light intensity dedected due to the dew/frost formation on the mirror signals the cooling module to continuously maintain the mirror, temperature in the *"just dew condition"* which is by definition the dewpoint temperature of the gas under test. The platinum resistance element measures the mirror temperature continuously and displays it digitally. An automatic balance compensation (ABS) compensates the gradual loss of sensitivity due to pollution of the sensor optics during the operation period.

Comparing the results of the long term experiments (see Fig. 13) it is shown, that the mirror dew point sensor tested has operated very accurate at the beginning of the experiment (1st drying cycle). Under the influence of increasing pollution of the mirror in time, the accuracy decreased and at the end of the experiment (last drying cycle) serious deviations occur for measurement of high and low humidity. The ABS was no more able to compensate this effect, which will be mainly due to premature condensation caused by mirror pollution.

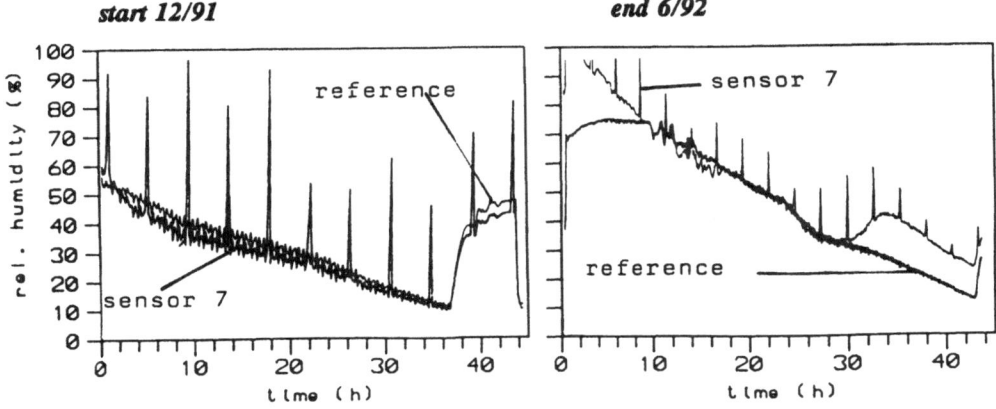

Fig. 13: *Results for mirror dew-point sensor*

Polymer humidity sensor (sensor 5): The polymer humidity sensor is a capacitance-type humidity sensor and consists of an upper and lower electrode and a humidity-sensitive polymere (Fig. 14) fixed at a glass substrate. The capacity of the condensator changes in relation to the relative humidity. The signal is converted to a linear output current signal 0 (4) ... 20 mA corresponding to a relative humidity 0 ... 100 %. This sensor has the advantage of reasonable accuracy for low price. The complete measuring head is shown in Fig. 15. A porous metal sinter filter (pore diameter 20 µm) protects the sensor against dust etc..

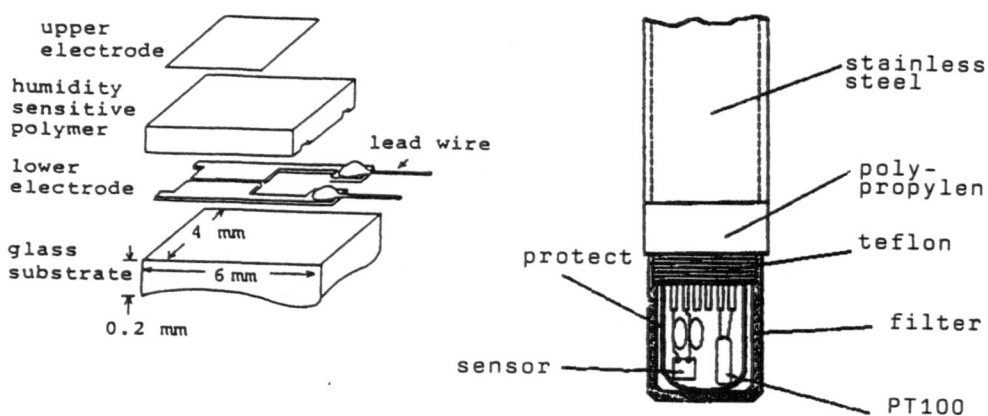

Fig. 14: *Polymer humidity sensor*   **Fig. 15:** *Measuring head*

The results of the investigations show clearly, that these polymer sensors are more stable in time (see Fig. 16). At the beginning of the experiments a slight under prediction of the relative humidity occures while at the end the humidity is slightly over predicted. In both cases the measurements for relative humidities below 40 % are very accurate.

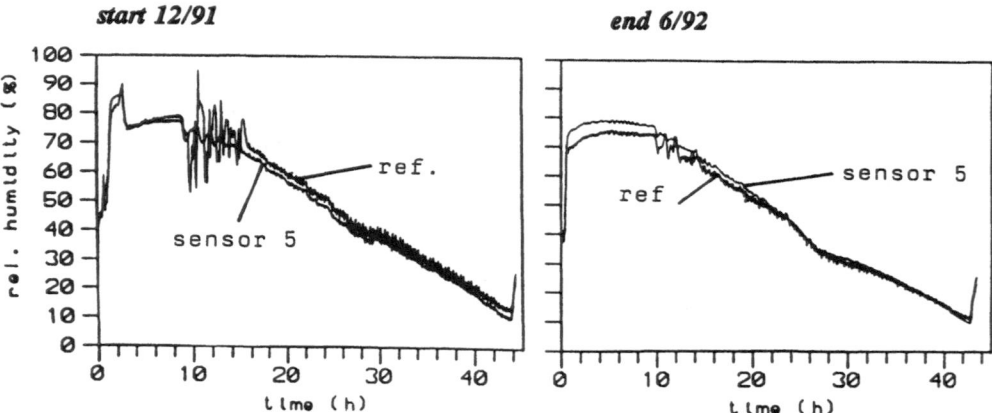

Fig. 16:   *Results for polymer humidity sensor*

CONCLUSIONS AND FURTHER OBJECTIVES

The results of the sensor research work have shown, that suitable achievements have been realized. The newly developed Coriolis-meter and the video probe are ready for industrial application. Even under harsh conditions the sensors work reliably. The combination of Coriolis-meters with high accurate feeding or dosing devices will result in new feeding systems allowing an accurate pulsation free feeding of granular materials in process engineering industry. The on-line determination of the correct mass flux in pneumatic conveyer lines, etc., is i.e. necessary for adjusting the air-/fuel-ratio in order to obtain an efficient low pollutant combustion. Choosing the optimum thermocouples and determine the response time in protection tubes by numerical simulation, the possibility of an improper operation of a regulation system provoking some malfunctions such as: increased energy consumption, decreased product quality, shortened life-time of the whole furnace or of its parts, etc., is minimized. Although the evaluation of data from the humidity sensors experiments is not completely finished, it has been shown, that none of the sensors applied operates reliable in long term. The relatively low-cost polymer humidity sensor appears to be the most suitable one for the conditions investigated. The research activities in IR-sensors and multi-sensor systems for baking ovens are not completed, but will be presented in the future.
In order to establish sets of sensors as effective tools for furnace monitoring the following activities in sensor research are foreseen in the future:

o   extensive measurements in industrial furnaces to obtain experimental date
o   application of Coriolis-meters to feeding systems
o   investigations on gas sensors (NO_x, etc.)
o   post-processing of video flame pictures for control of furnaces
o   detailed investigations on the reliability of IR-temperature sensors

REFERENCES

1.   GRUNDIG, *"Technical Description Furnace Probe System"*, SG 82, Fürth, Germany
2.   Körtvélyassy, L.v., *"Thermoelement Praxis"*, 2. Ausgabe, Vulkan-Verlag, Essen, 1987
3.   Crovini, L., *"Resistance thermometers"*, Sensors, Volume 4: Thermal Sensors, VCH, Weinheim, 1990
4.   Ricolfi, T., Scholz, J. (1990), Sensors, Volume 4: Thermal Sensors ,VCH, Weinheim 1990
5.   Gnielinski, V. (1984), *"Wärmeübertragung bei erzwungener einphasiger Strömung"*, VDI-Wärmeatlas, 4th Edition, VDI-Verlag, Düsseldorf, 1984
6.   Rheinländer, J. (1984), *"Konvektive Wärmeübertragung bei Auftriebsströmung an senkrechten, geneigten und horizontalen Platten, horizontalen Zylindern und Kugeln"*, VDI-Wärmeatlas, 4th Edition, VDI-Verlag, Düsseldorf, 1984
7.   Vortmeyer, D., *"Strahlung technischer Oberflächen"*, VDI-Wärmeatlas, 4th Edition, VDI-Verlag GmbH, Düsseldorf, 1984
8.   Stevanovic, D.; Studovic, M., *"Method for transient fuel element response evaluation"*, 5th Nat. Conf. on Heat and Nuclear Eng. Problems on PRB, Varnas, Bulgaria, 1981
9.   Schneider, P.J., *"Conduction"*, Section 3 in: Handbook of Heat Transfer, (Edts. Rohsenow, Hartnett), McGraw-Hill, 1973
10. Sauer K., Jumo-Information Nr. 90.170/017/84, pp. 1-4, Jumo, Fulda, Germany, 1984

LASER DEFLECTION, ABSORPTION AND DIFFUSION IN A LAMINAR FLAME : APPLICATION TO LOCAL DENSITY, TEMPERATURE AND CONCENTRATION MEASUREMENTS OF HYDROCARBONS

J.-P. MARTIN, M.-Y. PERRIN, J.-C. ROLON

Laboratoire d'Énergétique Moléculaire et Macroscopique, Combustion

C.N.R.S. et École Centrale Paris

92295 Châtenay-Malabry, France

ABSTRACT

We report on three complementary optical (non intrusive) techniques used to measure local density, concentration and temperature in laminar reacting or non-equilibrium flows and in flames. Density and density gradients (temperature can also be deduced from these measurements) are measured by laser beam deflection, concentration of hydrocarbon by laser absorption in the infrared and temperature by Rayleigh scattering. These three complementary informations give us a map of the local quantities in reacting or non-equilibrium flows and in flames. Laser beam deflection has also been used to study the extinction limits of propane/air and hydrogen/air counterdiffusion flames.

INTRODUCTION

Basic research in chemical reactions, radiation heat transfer, and fluid flow for combustion systems has been performed for many decades. Mathematical, physical and numerical models to predict velocity, temperature, and species concentrations in laminar or turbulent (premixed or diffusion) flames have been developed. However, the need to increase operating pressure and temperature in gasoline engines, diesel engines, gas turbines for industry or aeroplane, rockets or thrusters, while simultaneously improving safety and pollution standards require good confidence in the results of the models. Validation of the prediction of the models can be performed by comparing it to reliable experimental results. Another strategic aspect is the development of "intelligent" combustors able to automatically adjust the burning conditions (heat release, temperature level, stability or pollutants production,...) provided they receive an adequate signal from a sensor. Thus, it is important to develop reliable, sensistive sensors to measure density, temperature and chemicals concentrations in reacting flows.

LASER BEAM DEFLECTION TOMOGRAPHY

We describe here the development and use of a laser beam deflection tomography method to measure the local density in axysimetric and non-axysimetric flows and deduce the associate temperature. The inversion technique and the experimental set-up have been described in detail in references [1 - 4]. Thus, we will describe here only the main key points and discuss the experimental results in order to illustrate the capabilities, accuracy and limitations of the method by presenting experimental results in the interacting region between two supersonic free jets and in laminar diffusion propane/air or hydrogen/air flames.

Principle of the method and experimental procedure :

The method is based on the deflection of a laser beam crossing a medium of variable density $\rho(x, y, z)$, thus of variable index of refraction $n(x, y, z)$ [5]. Let us consider the geometry displayed on figure 1.

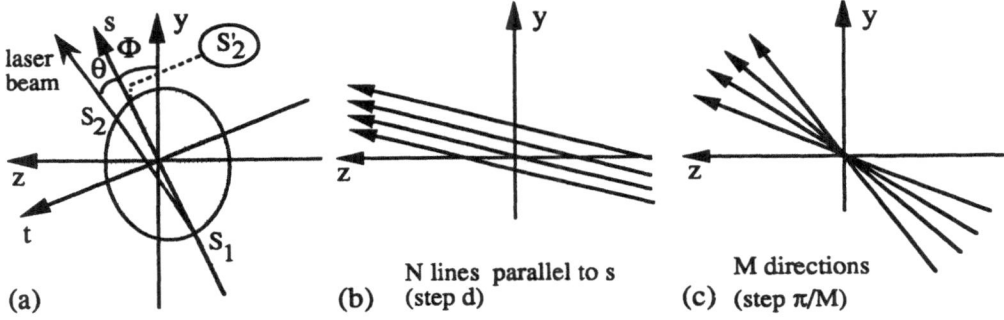

Figure 1. (a) Laser beam deviation by a density gradient; S_1 and S_2 are the laser beam input and exit points in and out of the flow, S_2' is located on the s axis; (b) and (c) principle of beam deflection tomography : $M \times N$ deflection measurements are taken.

The (O, x, y, z) referential is linked to the geometry of the flow. Ox axis is parallel to the direction of the flow. An incident laser beam enters the flow, in the yz plane defined by $x = x_0$, along the Os axis making an angle Φ with Oy. The axis Ot is orthogonal to Os. The entering and exit points of the beam into and out of the probed medium are respectively S_1 and S_2 . The total deflection is given by :

$$\theta(x, y, z) = \int_{S_1}^{S_2} \boldsymbol{\nabla}_\perp ln[n(x, y, z)] \, d\sigma \tag{1}$$

where $d\sigma$ is the infinitesimal path length along the laser beam. As the deflections are small, the paraxial approximation gives ($S_2 \cong S_2'$); the laser beam follows the s axis so that :

$$\theta(x, t, \Phi) \approx \int_{S_1}^{S_2'} \boldsymbol{\nabla}_\perp ln[n(x, y, z)] \, ds \tag{2}$$

where ds is along the s axis. The total deflection can be decomposed into two components $\theta_x(x, t, \Phi)$ (along Ox) and $\theta_z(x, t, \Phi)$ (radial deflection, in the yz plane). The later will be the only signal taken into consideration in the reconstruction procedure.

The problem to be solved is to inverse the measured deflections $\theta_z(x, t, \Phi)$ in order to reconstruct the local values of n(x,y,z). The tomography algorithm used here is known as the Back Projection of Filtered Projection algorithm [6]. It requires to know the evolution of the deflexion $\theta_z(x, t, \Phi)$ against t and Φ.

The experimental procedure consists then in recording the radial deviations along N s-directions separated by a distance equal to d ($t_i = i \cdot d$; $-\frac{N}{2} \leq i \leq +\frac{N}{2}$), and in M different s-directions ($\Phi_j = j\frac{\pi}{M}$; $j = 0, M$). This gives $M \times N$ integrated deviation values on which the inversion algorithm is applied.

Precision of the method :

For a given flow, the accuracy of the reconstruction is determined by the choice of the number of measuring points $M \times N$, the type of interpolation procedure, the width of the probe beam inside the flow, the filter used in the reconstruction algorithm and the signal to noise ratio in the measurements.

The choice of N is related to the highest frequency ω_{max} in the Fourier transform of the sampled object $n(x, y, z)$. From the sampling theorem, a function can be uniquely recovered from its samples if the sampling rate is larger than $2\omega_{max}$. As a consequence, d has to verify $d < \dfrac{\pi}{\omega_{max}}$. If this criterion is not met, the high frequency components are undersampled and their contributions is added to the lower frequency components and systematic errors are introduced. The band-limit ω_{max} can be determined by calculating the energy spectrum $S_{xx}(\omega)$ of the discrete Fourier transform of $\theta_z(x, t, \Phi)$ and be defined as :

$$\int_0^{\omega_{max}} S_{xx}(\omega) \, d\omega = 0.99 \int_0^{+\infty} S_{xx}(\omega) \, d\omega \tag{3}$$

M has to be chosen as a function of N. For a given spacing d, the object is reconstructed without loss of resolution if $M > pN/2$. A more accurate analysis shows that a reconstruction without artefact requires $N/2$ projections; thus :

$$M > N/2 \tag{4}$$

In order to test the validity of M, numerical tests can be performed; a given density field $n_0(x, y, z)$ is assumed, deviations $\theta_z(x, t, \Phi)$ are numerically simulated and a density field $n_r(x, y, z)$ is reconstructed, by using the described procedure, from a set of $M \times N$ data where M can be varied; if M is too small, artefacts appear on the reconstruction. In general, we have measured the radial deviations at 60 angles ($M = 60$; $\Delta\Phi = \pi/60$).

As the measured deviation is an average on all deflections of the beam of light composing the laser beam, the accuracy depends also on the width of the probe beam, and the spatial resolution is either limited by the spacing d or the size of the laser beam. In the described experiment, the probe beam is relatively thick (80 μm), so that it is not usefull to reduce the spacing d below 50μm. As a result, the spatial resolution is equal to this latter value.

Different filters can be used in the reconstruction algorithm. An additional filtering is likely to reduce the noise at the cost of the spatial resolution, we chose a rectangular window of width $2\omega_{max}$.

This method can be used as long as the radial deflections are not too large (paraxial approximation) : the paraxial approximation is valid only if the beam of light trajectory can be considered as a straight line. Numerical tests, similar to that used to choose M, have been performed to determine the maximal deviation acceptable : up to 250 mrad, the relative error of reconstruction on $(n - 1)$ does not exceed 0.6 %.

Experimental set-up and results :

In order to validate and test the technique, it has been first applied to the investigation of interacting free jets in laboratory E.M2.C, Châtenay-Malabry and in the D.L.R., Göttingen. The two experimental set-up have already been described [1-4,7]. When validated the method has been used to study a laminar diffusion flame.

The He-Ne laser probe beam is focused by a 100 mm focal length lens on the flow axis, which leads to a beam width of 80 μm. The deviations are measured by of a lateral effect position sensing detector, which delivers voltages proportional to the deviations (3.0V per mm). The laser beam is fixed ; the nozzles generating the flows are mounted on a xz translation stage (1μm accuracy at E.M.2.C., 10μm at D.L.R.) and can be rotated step by step (0.01°) around the Ox axis.The acquisition of the radial deviation and the motor displacement are controlled by a PC.

<u>Density in the interaction region between two supersonic free jets</u> : Such a configuration, in which strong and sharp density gradients associated to the shock structures existing around and near the axis of the flow field take place, is a good test case for the described technique. Let $\rho^*(x, y, z)$ be the reduced density, defined as :

$$\rho^*(x, y, z) = \frac{\rho(x, y, z) - \rho_1}{\rho_0} \qquad (5)$$

where ρ_0 and ρ_1 are the densities in the stagnation chamber and in the background gas respectively. A 2D representation of the reduced density is reproduced on Figure 2, the structure of the two jets and of the interaction region is clearly understood.On curve (c), the shock structure,which thickness can be evaluated to 80μm, characterizing the interaction is well reproduced.

We have also demonstrated that the experimental data reproduce that of existing models within a few per cent.

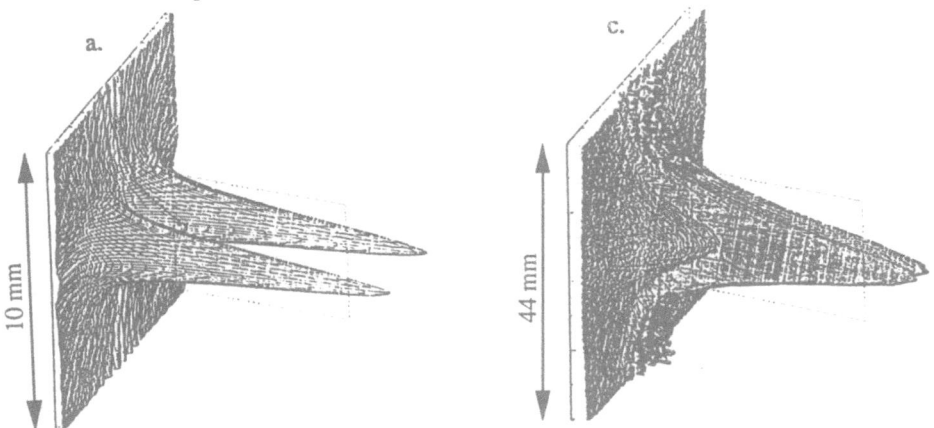

Figure 2. 2D reconstruction of the local density measured in the interaction region between two parallel free supersonic jets. (a) near the nozzles exits the two jets begin to interact but are still separated ; (c) further in the flow, the density in the interaction zone is larger than in the jets.

Density in a counter flow diffusion flat flame : We have used the technique to measure the density in a counterflow flat diffusion flame, the burner has been described in references [8]and [9]. The burner is of type I diffusion flame configuration, described by Tsuji [10]. The velocity field has been measured by Laser Doppler Anemometry (the velocity of the two opposed flows at the outlet sections of the nozzles is equal to 0.6 m s^{-1}) using the blue line of an Ar$^+$ laser.

The flat diffusion flame takes place near the stagnation point of this particular flow, in a region where the velocity of the gases is radial; perpendicular to the axis of the burner; the gases flow from the center to the outside; thus the flame is stressed along its plane. At the side of the flow, the gases go up due to buoyancy and mix with the surrounding nitrogen.

The green line of the laser has been used for the density measurements. In this configuration, due to the diameter of the flame, the laser beam is focused by a 300 mm focal length lens.

Figure 3 shows the reduced density field measured in a methane/air flame near the flame location, as the flame is warmer than the surrounding atmosphere, the density is reduced in the flame. Near the flow axis (-10mm $\leq r \leq +10$ mm), the flow is parallel to its axis [8, 9] so that the density is constant. Due to the particular geometry algeady described, the temperature is higher on the side of the flow, the density is smaller for larger r values. Then it increases to reach that of the surrounding gas.

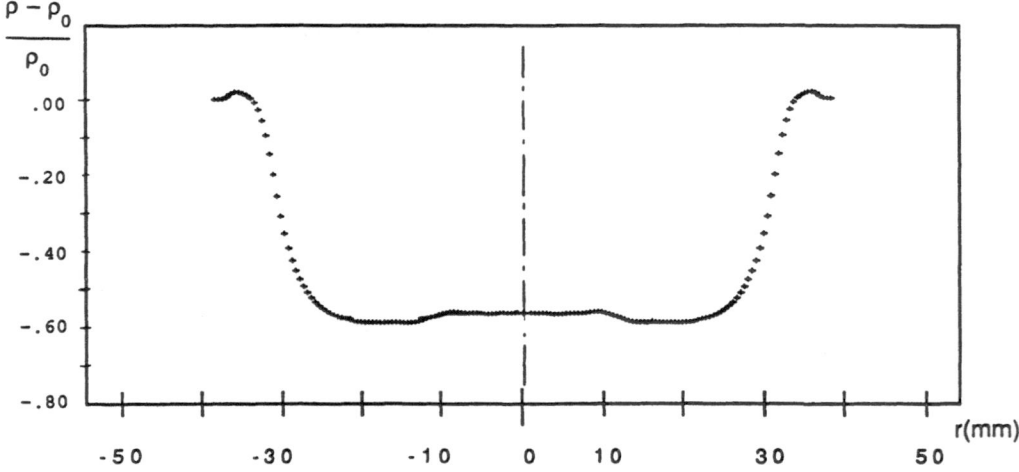

Figure 3. Reduced density field in a methane/air counter flow diffusion flame.

In such a flame, one can assume that the pressure is equal to atmospheric pressure all over and around the flame, thus it is possible to extract the local temperature which is presented on figure 4 in the flow field over the flame. Close to the flame, the temperature reaches its maximum value (2400 K) for small distances ($r \approx 10$ mm). Further out the flame plane, increasing z; the hot gases expand towards the outside of the flow, are diluted by the surrounding cold nitrogen flow, the temperature reaches smaller values at larger distances from the axis of the burner.

In the case of hydrogen/air flames, the difference of density between H_2 and the air or the reaction products is large, we have not been able to do measurements, large oscillations occur when recording the laser beam deviation so that the inversion technique cannot be performed.

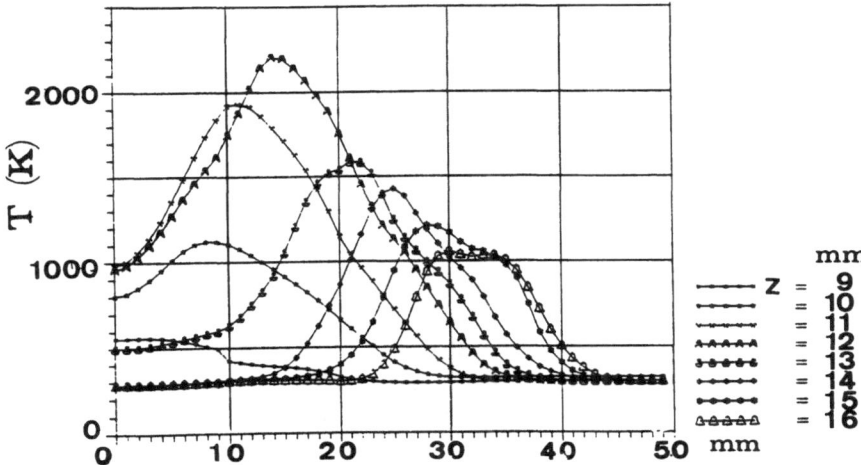

Figure 4. Temperature field in a methane/air counter flow diffusion flame. z is the distance from the flame to the measuring point and r the radial distance from the flow axis (only one half of the flow is shown).

Determination of the extinction limits of a flame : The laser detection technique has also been used to study the extinction limits in the flat diffusion flame as a function of the strain rate, of the molar fraction of nitrogen and of the mass flow rate of the fuel. To do so, the laser beam is adjusted .1 mm over the flame where the density gradient is maximum; the laser beam is strongly deviated towards the flame. The position of the laser spot on the position diode is recorded as a function of time. Some preliminary tests are made to determine the approximative flow rate of the fuel leading to extinction. The flow rate is then reduced to a value for which the flame is stable. In such a way, the flame position remains the same when the flow rate is modified (reduced) to obtain extinction. The fuel flow rate is slowly increased up to the value where the flame becomes unstable, and vanishes. At this time, the laser beam is suddenly no more deviated by the density gradient and the extinction time is noted together with the value of the fuel flow rate. The mass flow rate, at extinction, of the fuel is plotted as a function of the strain rate on figure 5. We have obtain similar curves for the evolution of the mass flow rate of the fuel, the molar fraction of the nitrogen diluting the fuel anf of the flame strength as function of the strain rate in propane/air and hydrogen/air diffusion flames. This allows us to study the extinction properties of strained flames and to compare these experimental results to that of modelization of such flames when taking more or less complexe flame chemistry into account.

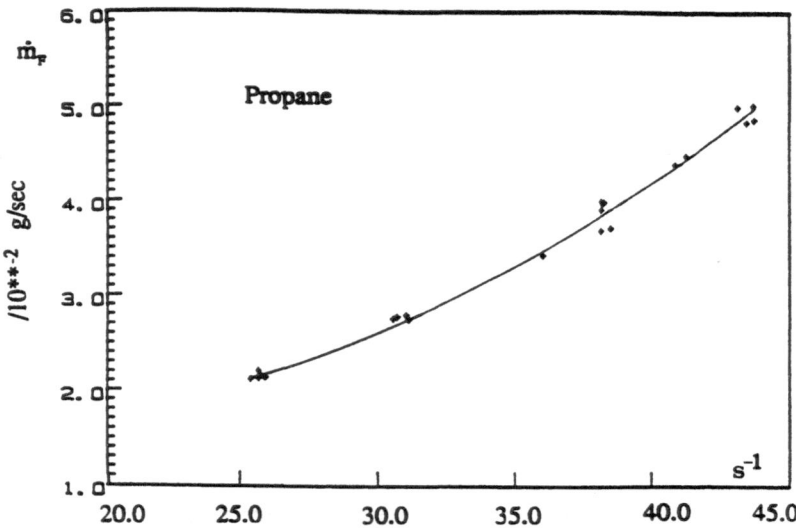

Figure 5. Mass flow rate of the fuel at extinction of a flat counterflow diffusion flame as a function of the strain rate.

CH_4 CONCENTRATION MEASUREMENTS BY LASER ABSORPTION

It has been demonstrated that the 3.39μm laser line of an He-Ne laser coïncides with an absorption line of CH_4 molecules [11]. The measurement consists in measuring, with low noice infrared (InSb or GeAu) detectors, the intensities I_0 and I_1 of the laser beam respectively before and after the flow field around the flame. The integrated absorption coefficient $\overline{\alpha}$ is calculated by using the Beer's law. The local absorption coefficient $\alpha(x, y, z)$ is proportional to the local concentration of the absorbing molecules; but it also depends on the local temperature. In order to extract, the local value of the CH_4 concentration one has either to be able to measure the absorption coefficient for different absorption lines [3] or to know the local temperature (by the former laser beam deviation technique for example). Tomography technique has also been used with a tunable diode laser for CO detection [3]. As a first step of the technique, we present the evolution of the integrated absorption coefficient of CH_4 in a flat counterflow diffusion CH_4 - air flame on figure 6.

The measurement of the local concentration by laser absorption looks to be very complicated, nevertheless as far as only integrated values are concerned this absorption technique can be easily used to detect the presence of CH_4 molecules at very low level (ppb detectivity have been achieved) and its evolution versus time in a reacting flow. As the response time of the infrared detectors is short (10^{-6} s); high frequency measurements are possible.

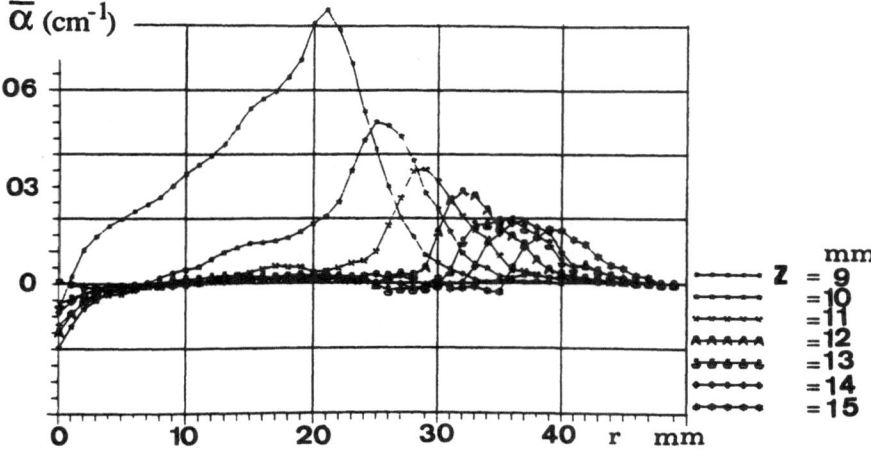

Figure 6. Evolution of the integrated absorption coefficient of CH_4 in a flat counterflow diffusion flame.

CH_4 TEMPERATURE MEASUREMENTS BY RAYLEIGH SCATTERING

The former techniques are based on integrated measurements, the local density or temperature are extracted, through a computer inversion algorithm, from a set of measurements recorded along different lines of sight crossing the flow. On-line and high frequency response cannot be obtained. Different optical techniques, such as (linear or non linear) Raman scattering, Laser Induced Fluorescence, Mie or Rayleigh scattering, ...) directly give access to local quantities. Some of it can also give time resolved response. We are presently testing an experimental set-up to perform temperature measurements by Rayleigh scattering. Results will be presented at the conference.

CONCLUSION

Different optical methods have been developed and tested to measure local density and temperature in nonequilibrium flows. Local time averaged values can be measured. Time resolved laser beam tomography and laser absorption can give access to the frequency content of the time evolution of the total density or of the concentration of hydrocarbons in a reacting flow. The tomography code and optical set-up will be made available at the end of the Joule contract.

ACKNOWLEDGEMENTS

The research has been supported by C.N.R.S. grants under the "Actions Incitatives Europe "; the French-German program "PROCOPE"; and the E.E.C. programs "Stimulation "and "JOULE ". The presented results are part of the PhD works of our students: D. Boscher, L. Philippe, J.C. Mombo-Caristan and M. Carlier-Cohat and of the Diplom-Arbeit works of E. Dornberger and G. Jacob from L.S.T.M. Erlangen. We want also to thank Pr. F. Durst (L.S.T.M.-Erlangen), Dr. G. Dettleff and S. Döring (D.L.R.-Göttingen) for their collaborations.

REFERENCES

1 - Mombo-Caristan, J.C., Philippe,L.C., Perrin, M.-Y., Martin,J.-P., Laser beam deviation as a local density probe. Application to supersonic free jets, Experiments in Fluids 7, 303 (1989)

2 - Mombo-Caristan, J.C., Philippe,L.C., Chidiac,C., Perrin,M.-Y., Martin, J.-P., Measurements of Free Jet Densities by Laser Beam Deviation, Vol 117; Progress in Aeronautics and Astronautics , AIAA, ED. E.P. Muntz, D.P. Weaver, D.H. Campbell, Washington (1989).

3 - Philippe, L.C., Perrin, M.-Y., Evolution of local rotational populations in CO-He and C0-Ar free-jets, Rarefied Gas Dynamics. Ed. A.E. Beylich.VCH (1991)

4 - Cohat, M., Perrin, M-.Y., Martin, J.-P., Dettleff, G., Döring, S., Beam deflection tomography. Application to density measurements in two interacting parallel freejets. Sixth International Symposium on Applications of Laser Techniques to Fluid Mechanics, July 1992, Lisbon, Portugal.

5 - Born, M., Wolf, E., Principles of optics. Pergamon. Press. Oxford.(1975)

6 - Kak, A. C., Tomographic imaging with diffracting and non-diffracting sources. Array signal processing. S.Haykin, Engelwood Cliffs. 351-427 (1985)

7 - Dankert, C. Flow in the interaction region of two parallel free jets. Proceedings of the 16th Rarefied Gas Dynamics (ed. by V. Boffi and C. Cercignani), pp. 486-494. Grado (1988).

8 -Rolon, J.-C., Veynante, D., Martin, J.-P., Dornberger, E., Durst, F., Jakob G.; Laser velocity and density measurements of a flat flow diffusion flame, IUTAM Symp. on Aerodynamics in Combustors, III-50; III-52, Tapei (1991)

9- Durst, F., Whitelaw, J.H., Martin, J.-P., Fuhrer, U., Baptista, J.L., Dimaczek, G., Development of Advanced Sensors, JOULE Meeting, Bruxelles, (October 1990)

10 - Tsuji, H, Counterflow Diffusion Flames, Prog. Energy Combustion Science, 8, 93 (1982)

11 - Perrin, M.-Y., Hartmann, J.M., High temperature absorption in the $3.39\mu m$ He-Ne laser line by methane, J.Q.S.R.T. 42, 459 (1989)

PROBE AND OPTICAL SENSORS FOR THE ANALYSIS OF TURBULENT HEAT TRANSFER IN RECIRCULATING FLAMES

PAULO FERRÃO AND MANUEL HEITOR

Instituto Superior Técnico, Technical University of Lisbon, Av. Rovisco Pais,
1096 Lisboa Codex, Portugal

ABSTRACT

Optical and probe techniques for measurements of velocity and scalar characteristics have been combined and improved to allow the quantification of the turbulent heat flux generated in recirculating flames. The technique consists on the extension of a conventional laser-Doppler velocimeter through its combination with either laser Rayleigh scattering or digitally-compensated fine-wire thermocouples and measurements are presented in a strongly sheared disc-stabilized propane-air flame. The results are used to assess the extent to which turbulence mixing in flames of practical relevance is altered by the accompanying heat release and reveal the existence of large zones in these flames characterized by non-gradient scalar fluxes.

INTRODUCTION

Turbulent combusting environments are widely used as energy release processes and, in particular, bluff-body flameholders are commonly used to maintain a steady flame in a high-speed turbulent combustible mixture. Through variations in design and operating conditions, these burners can produce flames of varying stability, efficiency, pollutant formation and heat transfer. Their effective application, for example in the development of low-emission combustors, requires however a better understanding of the phenomena involved and of the extent to which turbulent mixing is altered by the accompanying heat release. It should be noted that bluff-body stabilized flames, such as that schematically represented in Figure 1, are curved along their length [1], and this curvature may impose mean velocity effects on the turbulence field through the interaction between the mean adverse pressure gradient typical of these flows and the large density fluctuations associated with the turbulent flames [2, 3]. Similar effects have been observed in other recirculating flames [e.g., 4] and, for example, can be used to retard mixing and combustion in practical combustors [5].

An understanding of mixing in reacting flows is, therefore, technically important since it would provide guidelines for the design of combusting equipments. Also the strain associated with turbulent mixing may alter the rate of chemical reaction and knowledge of this interaction is important to allow the control and the physical simulation of the practically relevant phenomena of flame extinction and lift-off. [6, 7].

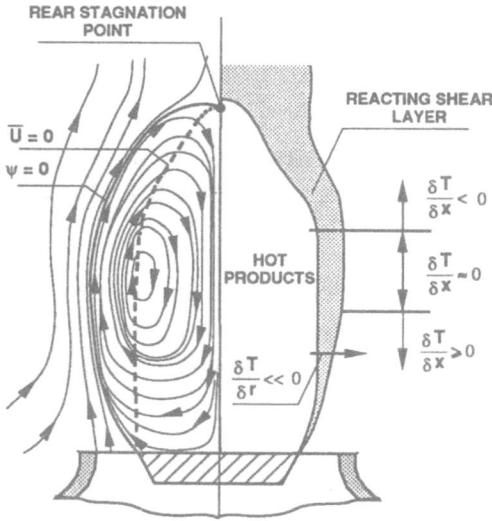

Figure 1.   Schematic diagram of baffle-stabilized premixed flame.

To achieve these objectives in premixed flames, the major interest is on the analysis of the reaction progress variable, $C=(T-T_0)/(T_{ad}-T_0)$ and its correlation with velocity components, [8]. The obvious choice for the velocity measurements is a laser-Doppler velocimeter, LDV [9]. It is non-intrusive, has good spatial resolution, is fast, accurate and fairly simple to use. The choice of a technique for the time-resolved analysis of the reaction progress variable, or of any other scalar, is however more varied and the interest here is on those methods which are most suited for simultaneous use with a LDV, [10]. Probe and optical techniques can be used and we selected thin digitally-compensated thermocouples and laser-Rayleigh scattering to achieve these objectives. The simultaneous measurements are however difficult to implement and the solution for this and the demonstration of their utilization to improve understanding of bluff-body stabilized flames is the main motivation of the present paper.

The work is an extension of that reported by Ferrão and Heitor [11] and includes the complete mapping of the turbulent heat fluxes generated in the flame studied, as well as the comparison between results with the probe and optical techniques selected for the work. The optical lay-out and the data reduction system were based on those initially proposed by References [12, 13], which were optimized to allow the characterization of flames with practical interest.

The remainder of this paper includes three sections, which present the flow configuration considered throughout the work, the two experimental techniques used and the corresponding results obtained in the flame, and a summary of the most important conclusions.

THE RECIRCULATING FLAME STUDIED

The analysis considered in this paper considers a premixed propane-air flame stabilized on a disc of D = 0.056 m diameter located at the exit section of a contraction with a diameter of 0.080 m. The flame is open to the atmosphere and is only partially premixed, but offer the advantage of easy access. The annular bulk velocities at the trailing edge of the disc were equal to 40 m/s, resulting in Reynolds numbers based on the disc diameter of 1.43 x 10^5: for the experiments reported here the equivalence ratio was 0.55 to allow analysis of lean flame conditions typical of those used in advanced low-emissions combustor concepts. The corresponding adiabatic flame temperature T_{ad} is 1644 K and the heat release parameter, $t_H = T_{ad}/T_0-1$, is 4.4 for T_0, the temperature of the approaching reactant stream, of 300 K.

The air flow was filtered (sub-micron filtering) and then seeded with powdered aluminum oxide (nominal diameter below 1.0mm before agglomeration) making use of purpose built cyclone generators.

EXPERIMENTAL METHOD AND RESULTS

Measurements of turbulent heat flux were obtained by combining a conventional laser velocimeter with bare-wire thermocouples or, alternatively, with a laser Rayleigh scattering system. The laser velocimeter was operated on the dual-beam, forward-scatter mode from the green light (514.5 nm) of a 5W multiline argon-air laser, and included sensitivity to the flow direction provided by a rotating diffraction grating (TNO, R/H). The resulting frequency shift was set up to 11 MHz. The transmitting optics comprised a 300mm focal length lenses and the half-angle between the beams was 4.8°: the calculated dimensions of the measuring volume at the e^{-2} intensity locations were 1.498 mm and 109 μm. The scattered light was collected in the forward-scatter direction are focused into the pinhole aperture of a photomultiplier (EMI 9817-B), and the resulting signal processed in a commercial frequency counter (TSI 1980-B) operated in the "single-measurement-per-burst" mode.

The simultaneous scalar measurements can be affected by the particles used as light scatterers for the LDV and subsequent joint statistical analysis further limited by the random characteristics of the Doppler signals. It is obvious that the thermocouples were located outside the measuring volume of the velocimeter and this may limit the spatial resolution of the combined LDV/thermocouple measurements. Other limitations include the need to compensate the frequency response of the thermocouples. The laser Rayleigh system has the potential to overcome these problems, although it requires the separation of the Mie scattered light associated with the Doppler signals. The details of these procedures, together with results achieved with the two measuring systems used are discussed in the following paragraphs.

The Combined Thermocouple/LDV Measuring System

The Instrumentation: The experimental apparatus developed for the combined LDV/Thermocouple measurements is represented in Figure 2, together with the block diagram of the data acquisition for simultaneous measurements. The thermocouples were made from Pt/Pt: 13%Rh wires of 40μm diameter, and their output was amplified (x 100) and digitized by a 12-bit analog-to-digital converter at sampling rates of up to 100 KHz, but normally set at 30 KHz. The noise level could be kept below 0.1% of full-scale deflection, corresponding to a maximum temperature error of ± 2 K.

The thermal inertia of the fine-wire thermocouples gives rise to first-order damping of their frequency response, for which numeric compensation was performed. The algorithm

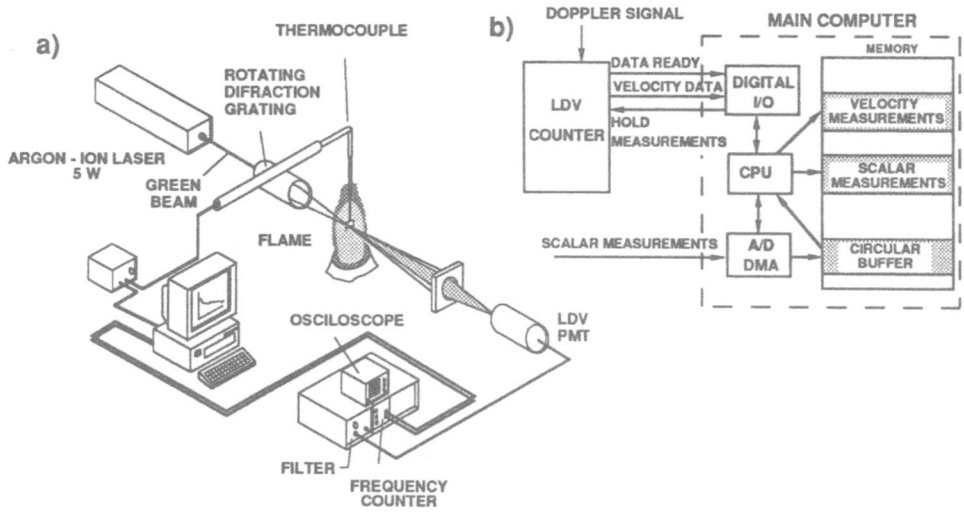

Figure 2.   Schematic diagram of the combined LDV/Thermocouple system
a) diagram of instrumentation;   b) block diagram of data acquisition system

for compensation required a temperature record of at least 0.4 ms duration, sampled at around 30 KHz before and after the occurrence of a velocity measurement, so that the temperature derivative does not reflect the effects of single bit changes of the least significant bit of the analog-to-digital converter. This imposition requires the recording of more than just two temperature samples (one just before, and one just after the occurrence of a velocity measurement) and, in practice, each velocity measurement was bracketed with six temperature samples before and after the occurrence of velocity, which were linearly interpolated previous to the calculation of the temperature derivative. The process was repeated until a statistically significant population has been achieved and this was usually set at 10000 data points and, therefore, considerably larger than the values used in previous works [10]. The procedure was implemented in a 16-bit, 80386, personal computer making use of digital and analog input/output boards (respectively, an Amplicon PC 14-AT and an Amplicon PC 30), as shown in Figure 2b). The thermocouple signals were continuously sampled through a DMA channel and stored in a circular buffer memory. The occurrence of a velocity measurement was signaled by the data ready output of the LDV counter, which was hold until the transfer of a group of 13 temperature samples (see above) from the circular buffer to a separate storage area of the computer memory was performed.

Once the measured temperature derivatives were evaluated for each pair of simultaneous velocity-temperature data, the determination of the time-resolved true gas temperature requires the knowledge of the thermocouple time constant. In the present work the variable time constant procedure of [14] was used for measurements of the axial heat flux, while a mean time constant procedure, [15], was used for the radial heat flux. The dependence of the time constant on velocity and temperature was determined within the combustion gases based on time-averaged forced convection decays of the temperature of the junction after the removal of a direct-current overheating pulse. The instrumentation built for measurements throughout the flame is represented in Figure 3 and consists of a thermocouple, a mercury relay, a square wave generator with a controlled frequency, a power supply for heating the thermocouple and a computer equipped with an A/D acquisition board using a D.M.A. channel. The process was digitally controlled, although the rate of heating pulses was set in the wave generator which controlled the mercury relay to determine the passage of current from the power supply to the thermocouple or, alternatively, the thermocouple signal to the data acquisition system.

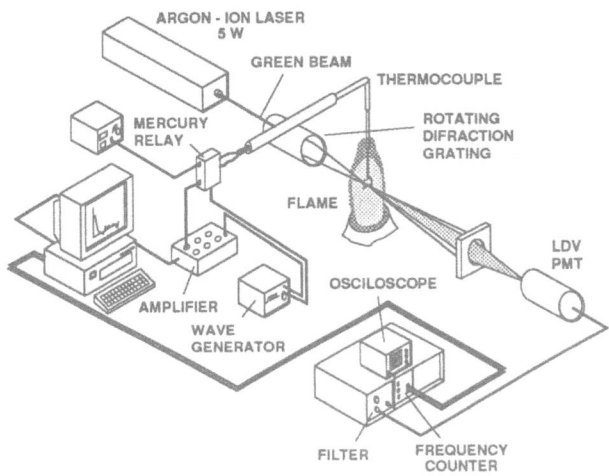

Figure 3.   Schematic diagram of the instrumentation used for the determination of the time constant of the thermocouple as a function of the flame properties.

In general more than 100 decays were used in each point and the averaged and non-dimensional curves, such as that of Figure 4, were used to evaluate the value of time constant of the thermocouple as a function of the local time-averaged velocity and temperature. The procedure involved the use of two methods, [14], namely: i) the time taken from the temperature to decay to 1/e of its initial value; and ii) the "plateau" of the curve obtained by plotting the "instantaneous" time constant along the decay. This second approach has been shown to give rise to more reliable results because it avoids the influence of secondary effects on the thermocouple response [16], and for the present flame conditions values could be successfully repeatable within 10%. The results lye in the range 9 to 20ms for gas velocities between 40 and 5m/s respectively, following very closely the empirical law of Collis and Williams in a way that are practically independent of temperature.

The accuracy of the compensation of the output of the thermocouple making use of the procedure defined above was assessed in terms of the highest frequency components in the flame, and shown to provide reliable values [11]. The corresponding uncertainty associated with the measurement of turbulent heat flux $\widetilde{u_i'' c''}$, where the tilde denotes a density averaged quantity, is likely to be an underestimation of about 10%. An important contribution to this error is due to the deterioration of the thermocouple bead due to seeding particle accretion and, for the present conditions, maximum particle fluxes of about 2000 Hz could be used, with the thermocouples replaced at the most every 10 minutes.

The largest random errors incurred in the values of velocity-temperature correlations are due to the spatial separation of the measurement locations of temperature and velocity because the thermocouple junction must lie outside the measuring volume of the anemometer and it is difficult to place, reliably, the two measurement locations closer than about 1 mm.

However, the dependence of the velocity/temperature correlation on the spatial displacement between the two measuring zones is weak along directions characterized by shallow temperature gradients, such as along the streamwise coordinates. In contrast, the influence of errors in the radial positioning of the thermocouple may be large and reach more than 20% of the maximum value of the local time averaged heat flux per each additional millimeter of separation between the two measuring points. The absolute magnitudes of the errors are proportional to the thickness of the reaction zone.

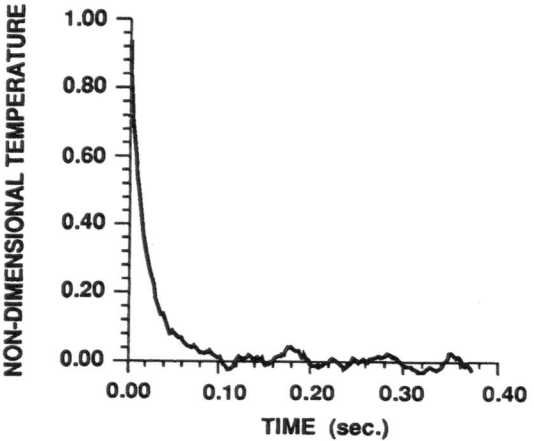

Figure 4. Typical time-averaged decay of thermocouple response after the removal of a direct-
current overheating probe.

Sample Results and Discussion: Figures 5 a) and b) show the mean velocity and temperature characteristics of the flame studied, with the latter quantified in terms of the reaction progress variable. The results are similar to those found in other baffle stabilized recirculating flames [e.g., 1, 17] and exhibit regions of reverse, accelerating and near uniform velocity. The rear stagnation point, which is associated with a pressure maximum as a result of the converging of streamlines towards the centreline, occurs at $x/D = 2.2$ and, therefore, the present recirculation zone is longer than those previously analysed in detail, as a consequence of the high Reynolds number of the present flow. The isotherms are highly curved and reveal a non-planar, and strongly sheared flame oblique to the oncoming reactants with the turbulent heat fluxes restricted to a thin zone along the shear layer, Figures 5 c) and d). These quantities represent the exchange rate of reactants responsible for the phenomenon of flame stabilization around the recirculation zone and of flame propagation downstream of this zone. The axial heat fluxes, Figure 5 d), are considerably higher than the radial values and, therefore, turbulent heat transfer is essentially directed along the isotherms rather than normal to them, as would be expected from gradient-transport models of the kind used in non-reacting flows. It should be noted that the radial fluxes are always positive, as expected in a recirculating flame. The flame is then established by the heat transfer between the hot products and the cold reactants with the sign of the radial heat flux in qualitative agreement with gradient-transport models ($\widetilde{v'' c''} = -v_c . \partial \widetilde{C}/\partial r$, where v_c is a scalar turbulent thermal diffusivity). Similar behaviour has been observed in other turbulent premixed flames [1] and has also been predicted analytically [3], and is expected to be due to the interaction between the gradients of mean pressure typical of the present flow and the large density fluctuations that occur in the flame.

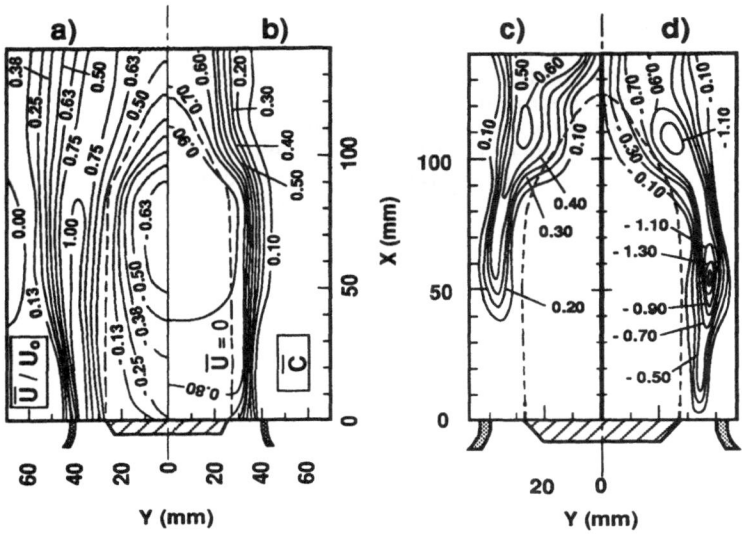

Figure 5. Time-averaged characteristics of disc-stabilized premixed flame.

a) Mean axial velocity, $\bar{U}/U_O$ b) Mean reaction progress variable, $\bar{C}$

c) Radial heat flux, $\widetilde{v''c''}/U_o \times 100$ d) Axial heat flux, $\widetilde{u''c''}/U_O \times 100$

The Combined Rayleigh/LDV Measuring System

The Instrumentation: The blue line of the laser system reported in Figure 6a) was used for Rayleigh scattering measurements with the beam focused to overlap the control volume of the LDV system. The optical measuring volume is defined by a rectangular slit at the collecting optics with 0.100 mm in width and 1.0 mm length. The Rayleigh scattered light was collected at 90° from the laser beam direction with a magnification of 1 and using an 100 mm f#2 lens. The collected light is combined and filtered by a 1 nm bandpass filter centered at 488.0 nm (Reynard Inc.) and passed through a polarizing beam splitter (Melles Griot 03 PBS 013) before being refocused on an EMI 9817-B photomultiplier. The signal was amplified and low-pass filtered at 10KHz (Khron-Hite, 3200) before digitalization using the A/D converter described above for the thermocouple measurements (i.e., Amplicon PC-30).

The present propane/air flames may affect the accuracy of the present system [e.g., 18] due to the variation of the Rayleigh cross-section from the reactants to the combustion products, but the consequent implication is that the velocity-temperature correlations presented below are overestimated up to a maximum uncertainty of 10%. A major limitation of the technique derives from its sensitivity to the presence of small particles such as soot, [10], but relatively low concentrations of larger seeding particles may be allowed if discrimination between the Mie and the Rayleigh signals is considered, as specified in Figure 6b).

For the work described here the scheme of Figure 2b) was also used for the combined LDV-Rayleigh measurements, although it was not necessary to bracket each valid Doppler realization with scalar measurements. The Rayleigh signal was continuously sampled and stored in a circular buffer, which was triggered by each valid Doppler signal and, then, analysed to select the last Rayleigh signal free of Mie contamination. In general, the time interval between the valid velocity and Rayleigh signals was between 33 and 99 µsec., although the maximum combined data rate obtained was around 130 Hz. This value represents a considerable improvement relative to other systems described in the literature and should be considered as the optimum compromise between the data rate of valid Doppler signals and the fraction of validated LDV/Rayleigh pairs of data. It should be noted that the present low seeding rates could be conveniently controlled making use of reverse cyclone seeders and the forward scatter LDV optics described before.

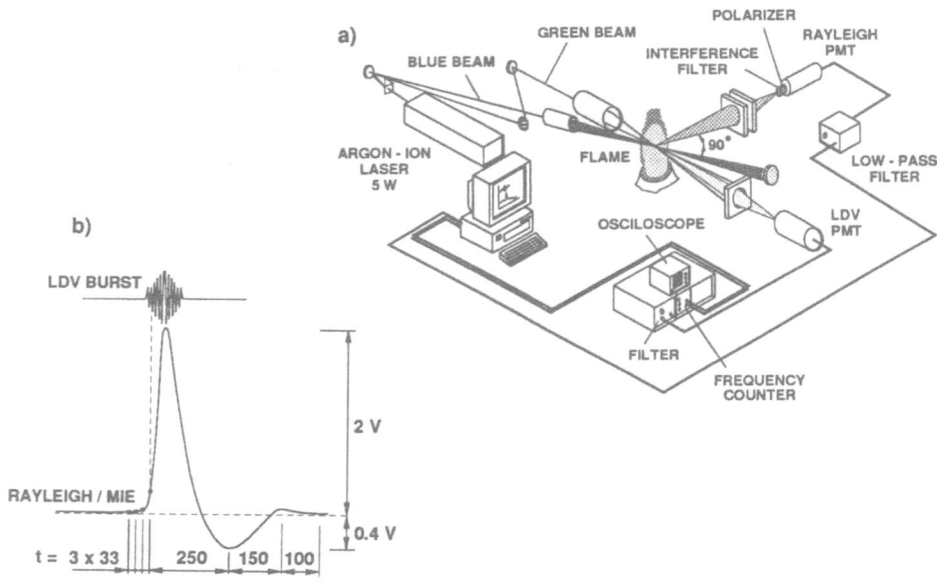

Figure 6.   Schematic diagram of the combined LDV/Rayleigh scattering system

a) diagram of instrumentation;

b) time evolution of simultaneous LDV and Rayleigh signals (time in µsec.)

Sample Results: Figure 7 compares vectors of turbulent heat transfer rate across the reacting shear layer measured with the two systems described above, which agree qualitatively and show the expected quantitative trends. The use of thermocouples is expected to underestimate the velocity-temperature correlations due to the uncertainties mentioned above, while the LDV/Rayleigh system may overestimate the results in the present air/propane flame. The results confirm that the turbulent heat fluxes are restricted to the reacting shear layer and that the turbulent heat flux vectors do not intercept the isotherms normally, but indicate, instead, a strong diffusion flux along the flame with the radial heat flux measured by any of the systems always positive.

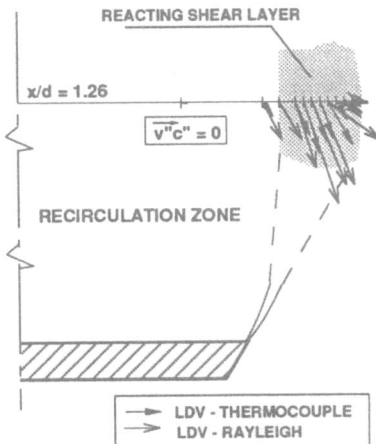

Figure 7. Representation of vectors of turbulent heat transfer rate across the recirculating flame, as measured by the combined LDV/Thermocouple and LDV/Rayleigh systems.

CONCLUSIONS

The combination of a laser velocimeter with digitally-compensated fine-wire thermocouples or, alternatively, with laser-Rayleigh scattering is shown to allow the successful characterization of the turbulent heat fluxes generated in flames with practical relevance, although the use of the thermocouples may underestimate the measured fluxes due to errors associated with the need to compensate their thermal inertia. On the other hand, the combined LDV/Rayleigh scattering system is limited to non-luminous flames and requires a careful choice of the reactants in order to minimize the variation of the Rayleigh cross-section between reactants and products.

A recirculating premixed flame with a heat release parameter of 4.6 is used to demonstrate the techniques and the results show that the turbulent heat flux is, as expected, large in the thin shear layer surrounding the hot recirculation zone where temperature fluctuations are large, but its direction is not aligned with the direction of the gradient of the mean temperature and exhibits a counter-flowing nature. The evidence is that current calculation methods for uniform density flows should not be directly used to simulate reacting flows, at least those involving zones of recirculation. This observation suggests that the interaction between density fluctuations and the mean pressure field associated with recirculating flames must be accurately simulated if the correct fluxes are to be determined.

ACKNOWLEDGMENTS

The authors are indebted to Mr. Paulo Simões for his assistance in the design and development of instrumentation and to Messrs. Carlos Carvalho and Mr. Jorge Coelho in the preparation of the manuscript. Financial support was provided by the Science Programme of the Commission of the European Communities under the Contract SC1-0459 entitled "Turbulent Mixing and Flame Extinction in Flames stabilized by Recirculation Zones".

REFERENCES

1   Heitor, M.V, Taylor, A.M.K.P and Whitelaw, J.H. (1987). The Interaction of Turbulence and Pressure Gradients in a Baffle-Stabilized Premixed Flames. J. Fluid Mechanics, 181, pp. 387-413.

2   Libby, P. and Bray, K.N.C. (1981). Counter Gradient Diffusion in Premixed Turbulent Flames. AIAA J., 19, pp. 205-213.

3   Bray, K. N. C, Libby, P. A and Moss, J. H.. (1985). Unified Modelling Approach for Premixed Turbulent Combustion - Part I: General Formulation. Comb. Flame, 61, pp. 87-102.

4   Heitor, M.V. (1992). On the Analysis of Turbulent Heat Transfer in Recirculating Flames. Intl. J. Heat and Fluid Flow, to appear.

5   Takagi, T., Okamoto, T., Taji, M. and Nakasuji, Y. (1985).Retardation of Mixing and Counter-Gradient Diffusion in a Swirling Flame, Twentieth Symposium (International) on Combustion, pp. 251-258, The Combustion Institute.

6   Milosavjevic, V., Taylor, A.M.K.P. and Whitelaw, J.H. (1989). The Influence of Burner Geometry and Flow Rates on the Stability and Symetry of Swirl-Stabilized Non-Premixed Flames. Comb. and Flame, 80, pp. 196-208.

7   Mansour, N.S., Bilger, R.W. and Dibble, R.W. (1991). Turbulent Partially Premixed of Nitrogen Diluted Methane Near Extinction. Comb. and Flame, 85, pp. 215-231.

8   Bilger, R.W. (1991). Experimental Methods in Combustion Flows, - Basic Considerations. In: "Experimental Methods for Flows with Combustion". Ed. A.M.K.P. Taylor. Academic Press, London.

9   Heitor, M.V., Stärner, S., Taylor, A.M.K.P. and Whitelaw, J.H. (1991). Velocity, Size and Heat Flux Measurements by Laser-Doppler Velocimetry. In: "Experimental Methods for Flows with Combustion", Ed. A.M.K.P. Taylor, Academic Press.

10 Ferrão, P. and Heitor, M.V. (1992). Probe and Optical Techniques for Simultaneous Scalar-Velocity Measurements. Combusting Flow Diagnostics (Eds. D.F.G. Durão et al.), Kluwer Academic Publ. NATO/Asi series, pp. 169-232.

11 Ferrão, P. and Heitor, M.V. (1992). Simultaneous Measurements of Velocity and Scalar Characteristics for the Analysis of Turbulent Heat Transfer in Recirculating Flames. Proc. Sixth International Symposium on Applications of Laser Techniques to Fluid Mechanics, Lisbon 20-22 July.

12 Gulati, A. and Driscoll, J. (1986). Velocity-Density Correlations and Favre Averages Measured in a Premixed Turbulent Flame. Comb. Sci. and Tech., 48, pp. 285-307.

13 Gladnick, P.G., LaRue, J.C. and Samuelsen, G.S. (1990). Simultaneous Optical Measurements of Velocity and Temperature Using Laser Anemometry and Rayleigh Scattering. Proc. 5th Intl. Symp. Appl. Laser Tech. to Fluid Mechanics, Lisbon, July 9-12.

14 Heitor, M.V, Taylor, A.M.K.P and Whitelaw, J.H. (1985). Simultaneous Velocity and Temperature Measurements in a Premixed Flame. Exp. in Fluids, 3, pp. 323-339.

15 Lockwood, F.C. and Moneib, H. (1981). A New On-Line Pulsing Technique for Response Measurements of Thermocouple Wires. Comb. Sci. Tech., 26, pp. 177-181.

16 Heitor, M.V. and Moreira, A.L.N. (1992). Probe Measurements of Scalar Properties in Reacting Flows. In: "Combusting Flow Diagnostics", eds D.F.G. Durão et al., Kluwer Acad. Publ., pp. 79-136.

17 Scheffer, R.W., Namazian, N. and Kelly, J. (1987). Combust. Sci. and Tech., 56, pp.101-138.

18 Namer, I. and Schefer, R.W. (1985). Error Estimates for Rayleigh Scattering Density and Temperature Measurements in Premixed Flames. Experiments in Fluids, 3, pp. 1-9 .

DEVELOPMENT OF GAS SENSORS FOR COAL GASIFICATION PROCESSES

A. Jacobs, F. De Schutter, J. Vangrunderbeek, J. Luyten
VITO, Boeretang 200, B-2400 Mol, Belgium

R. Van Landschoot, J. Schoonman,
Laboratory for Inorganic Chemistry, Delft University of Technology,
Julianalaan 136, 2628 BL Delft, The Netherlands

ABSTRACT

In the framework of the European program on non-nuclear energy (JOULE), on-line Nernst-type hydrogen and oxygen sensors for coal gasification systems are under development. The hydrogen sensor is based on Yb-doped $SrCeO_3$ as proton conducting solid electrolyte. Electrical characterisation and material compatibility have been studied and emf measurements are presented. A prototype hydrogen sensor was designed and constructed. The prototype oxygen sensor, based on ZrO_2 and developed at British Coal was selected for further testing in a synthetic test rig.

INTRODUCTION

It is expected that coal will remain an important fossil fuel for power production in the future [1]. Clean coal technology will have to minimise the strain on the environment by coal fired power stations. The Integrated Coal Gasification Combined Cycle (IGCC) is presently the most attractive option for the conversion of coal into electricity as it shows a minimal environmental hazard and optimal economical conditions combined with a high energy efficiency [2]. Consequently, many IGCC demonstration units are in construction or in the design phase. As an example, in 1989 the Dutch Electricity Generating Company N.V. Sep decided to build a 250 MW demonstration unit in Buggenum, that is planned to come on stream in 1993.

In order to be able to optimise the IGCC electricity production, an adequate process control based on an on-line, real time monitoring of the gas composition after the coal gasifier is essential [3]. Proper on-line measurements should allow to quickly detect variations in the gas composition and thus suitable adjustments of the fuel supply and gasification agents (coal, oxygen or air, steam) become feasible using appropriate control equipment. An on-line analysis of the gas composition under actual process conditions of a gasifier is, however, not possible because of the severe and aggressive environment usually existing inside the gasifier.

Considering these issues, a joint project between KEMA (NL), TU Delft (NL), VITO (B) and ECN (NL) was started in 1990 in the framework of the European R&D program on non-nuclear energy (JOULE I). This project aims at the development of direct tools to measure the gas composition downstream of a gasifier, at a location as close as possible to the gasifier.

The objective of the project is to develop an on-line gas analysis method that allows to measure the major components of product gas (i.e. H_2, CO, CO_2 and H_2O and detection of O_2 for safety reasons) with a pursued response time of about 10 seconds.

In the project, two different approaches are selected. One concerns the in-situ gas analysis using solid electrolyte based electrochemical sensors, for measuring oxygen and hydrogen respectively, while the other deals with the development of a gas sampling and gas analysis system (mass spectrometry) for measuring the main gas components.

In this paper the current status of the project is presented. Special emphasis will be devoted to the development of the on-line electrochemical hydrogen sensors.

MONITORING AND PROCESS CONTROL IN A GASIFIER

In order to illustrate the working of an IGCC installation, a process flow diagram is presented on figure 1. In essence the gasification of coal consists of the reaction between carbon in the content of the coal and water, present as steam, resulting in the formation of CO and H_2 as reaction products. In order to make the process autothermal a partial combustion is needed. Therefore oxygen or air is introduced in the reaction mixture. An inflammable gas mixture will be generated, consisting mainly of CO, H_2 and CO_2, together with an excess of the initially added H_2O and inert gas components. To prevent severe corrosion in the turbine, this raw gas has to be cleaned with respect to different contaminations (fly-ash particles, sulphur compounds like H_2S, COS, halogen compounds like HCl , HF and nitrogen compounds like NH_3, HCN). After gas cleaning, the chemical energy in the fuel gas is converted into electrical energy in the gas turbine unit. Finally in a combined cycle the gas turbine is followed by a steam turbine to convert the remaining heat in the flue gas into electricity along a steam cycle. A promising type of gasifier is the entrained bed gasifier. In the Shell gasifier dry coal is fed, while in the Texaco process a coal slurry is fed to the gasifier resulting in an excess of steam at the outlet. Typical process conditions for a gasifier block of 250 MW_e net are given in table 1 [3].

Gas sampling as near as possible to gasifier reactor zones is the ultimate goal of the project. However, due to the slagging conditions, high temperatures and safety reasons this goal is difficult to reach. Moreover, very specific information will be needed about construction details of the gasifier. In practice, the scope of the project was limited to conditions existing after the first gas cooling step, this means at temperature levels of 900°C and 700°C for the Shell and the Texaco process, respectively.

TABLE 1
Process conditions for 250 MW_e entrained-bed gasifiers

process	Shell	Texaco
coal feed (kg/s)	25	25
reactor exit gas temperature (°C)	1600	1600
reactor pressure (bar)	30	40
gas flow (kg/s)	50	50
gas cooling system		
quench with cool raw gas to(°C)	900	
radiant cooler to (°C)		700
convection cooler to (°C)	250	250
typical raw gas composition		
CO (mol %)	65	45
H_2 (mol %)	30	30
CO_2 (mol %)	2	8
H_2O (mol %)	2	18
O_2 (mol %)	$< 10^{-21}$	$< 10^{-21}$

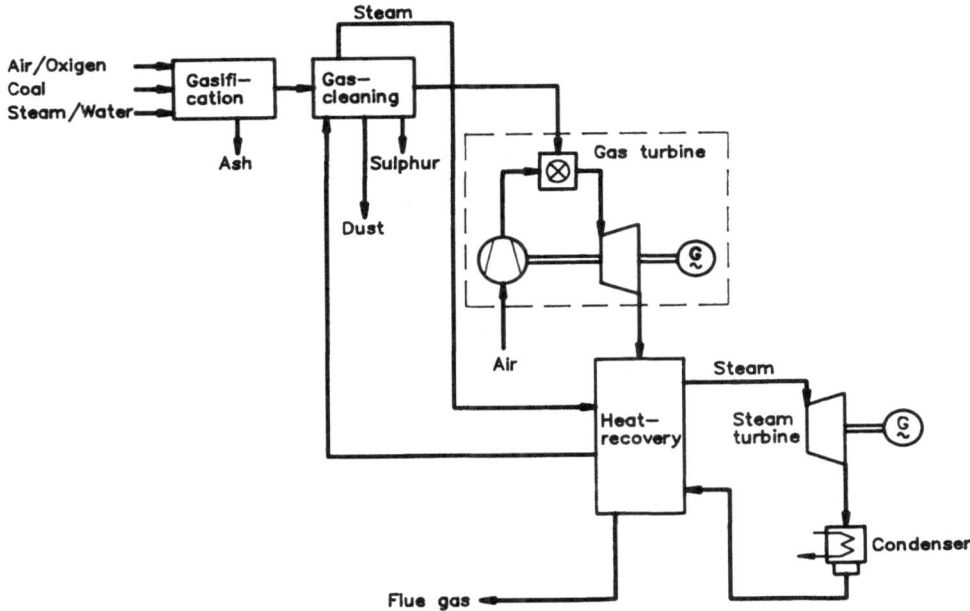

Figure 1. Scheme of an IGCC installation.

SOLID STATE NERNST TYPE GAS SENSORS

Even in the less severe conditions mentioned earlier, only a few gas sensor principles are applicable. A typical sensor widely accepted in industry is based on a potentiometric sensor for the electrochemical determination of the oxygen activity in the flue gases from combustion processes [4]. Therefore, this sensing principle was selected for constructing the oxygen as well as the hydrogen sensor. Basically, the Nernst sensor is a solid state electrochemical cell, consisting of a porous measuring electrode, a solid electrolyte and a gas or solid state reference electrode (figure 2).

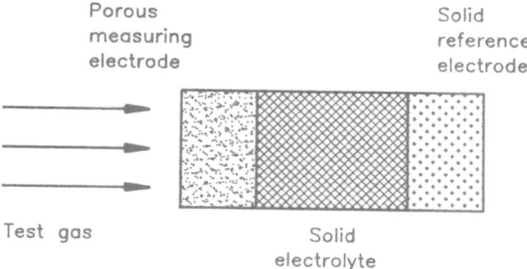

Figure 2. Schematic configuration of a solid state electrochemical sensor.

The two electrode compartments should be separated perfectly from each other by the solid electrolyte. In sensor applications the cell acts as a concentration cell, i.e. at both electrode interfaces the same equilibrium is involved. The gas activity at the sensing electrode can be obtained from the open circuit potential E (emf), if a well defined gas activity at the reference electrode is established. This cell potential (or open circuit potential) is obtained using the Nernst formula derived from the Nernst equations describing the concentration (activity) dependence of the electrode potentials. The emf of the sensor is given by

$$E = \frac{RT}{nF} \ln \frac{P(x)}{P(r)}$$

with R the gas constant, F the Faraday constant, T the temperature (K), P(r) the reference partial gas pressure, and P(x) the partial gas pressure in the sample gas mixture and n the number of electrons involved in the basic electrochemical reaction of the sensor (n=2 for hydrogen and n=4 for oxygen).

Specific advantages of this type of sensor are the potential for miniaturization, the rugged construction, its accuracy and sensitivity and the wide concentration range (logarithmic activity dependence of the signal). A high operating temperature is often required in order to obtain sufficient low resistivity of the solid electrolyte. Low resistivities (high conductivity) are a prerequisite for thermodynamically and kinetically reversible interface reactions at the electrodes of the sensor. Other problems may arise due to ageing, while selectivity may be influenced by interfering reactions.

HIGH TEMPERATURE OXYGEN SENSOR

With respect to the high temperature oxygen sensor an evaluation of commercially available industrial sensors was performed in order to select the most appropriate one and to adapt it to the needs of the coal gasification process in view of functioning in a process control system. The commercial monitoring systems of oxygen in combustion gases at high temperatures is usually performed using an electrochemical sensor built up of a solid electrolyte acting as a separator between the combustion gases and a reference gas. Of the two types of electrochemical sensors, potentiometric or amperometric, the potentiometric is mostly used in commercially available oxygen sensors.

At higher temperatures (i.e. 700°C) zirconia-calcia or zirconia-yttria are generally used as the electrolyte. As to the reference electrode, ambient air is mostly used as well as a metal-metaloxide (e.g. Pd-PdO) mixture.

Within the project, it was decided to test the prototype oxygen sensor developed at British Coal, in the synthetic material test rig. This approach has several advantages in comparison with an adaptation of commercial oxygen sensors. It allows to incorporate the experience obtained at British Coal in this project. After testing two prototypes in a test rig the key problems appeared to be the gas tightness and the attachment of the signal wires to the electrodes.

HIGH TEMPERATURE HYDROGEN SENSOR

With respect to the development of a high temperature hydrogen sensor the following material problems were envisaged :
- selection and synthesis of a suitable high temperature proton conductor (HTPC)
- electrical characterisation of the HTPC
- compatibility tests of the HTPC
- investigation of the ability to join the HTPC to a metal

In a later stage the results of these investigations, combined with the results of a search to find a suitable reference system, were used to design and to construct a first prototype hydrogen sensor.

Material aspects

With respect to the **synthesis of the HTPC**, an Yb-doped strontium cerate was selected to be used as the ion conducting element. This recently developed material has become attractive for many electrochemical systems because of its chemical and thermal stability, high proton conductivity and low conductivity activation energy.

In order to synthesise these cerates, the procedure as described by Iwahara et al. [5], was followed. The ceramic was then characterised in depth using several standard ceramic analysis techniques including an investigation of the mechanical properties. The Sr-cerate pellets had at least a density of 96% theoretical density and no open porosity. Moreover, a homogenous perovskite structure and composition was observed throughout

the pellets. The Sr-cerate pellets were relatively chemical stable for desorption at temperatures up to 900°C in ultra high vacuum.
To show the influence of dopant on the mechanical properties, different Yb containing compositions were manufactured. The results of Hg-porosimetry indicated the specimens show an increasing density at higher Yb content. With respect to the shaping of the $SrCeO_3$ solid electrolyte, pellets as well as closed-end tubes with collar (BOSCH-shape) have been fabricated. Finally, it should be remarked that the mechanical tests yielded rather low values for the mechanical strength, a result in a later phase confirmed by other investigations [7].

Concerning the **electrical characterisation** of the Yb-doped Sr-cerates, an in-depth impedance spectroscopy analysis was executed in the temperature region 200°C to 950°C in reducing as well as in oxidising ambient air. More details are presented and discussed in [8].
At moderate temperatures three relaxation processes are observed in the impedance spectra. At low frequencies electrode polarisation phenomena predominate, while at intermediate frequencies grain boundary polarisation phenomena are manifest, and at high frequencies the bulk response occurs. At increasing temperature the grain boundary polarisation effects can no longer be distinguished as a separate phenomenon.
From these curves, the bulk conductivity is deduced and, from its temperature behaviour, an activation energy of 0.64eV is calculated. This value agrees very well with the reported values in literature. The impedance spectra, measured with platinum electrodes show important differences in the electrode effects in relation to the gas composition of the ambient. In air substantial electrode blocking effects are observed, while in hydrogen containing atmospheres, a curvature to the real axis is present. This difference can be attributed to the hydrogen diffusion through the platinum layer, thereby reducing the electrode blocking effects.
The electrical characterisation is finalised with emf measurements on an air (reference) versus hydrogen concentration cell. For a fixed temperature a linear behaviour of the measured emf with the logarithm of the hydrogen partial pressure is observed, indicating a Nernstian behaviour.
With respect to the electrical properties, it can be concluded that, although all transport mechanisms are not known in detail, the measured electrical characteristics of the produced ceramics are comparable with data reported in the literature and, consequently, the material should be applicable in a hydrogen sensor.

In a first approach to test the **compatibility of Sr-cerates**, a 100h test experiment was performed in a synthetic coal gasification atmosphere. In this approach, a gas mixture consisting of Ar, H_2 (5 vol%), CO (0.45 vol%), H_2O (0.08 vol%), and H_2S (0.0033 vol%) at 1073 K and atmospheric pressure was used. Impedance measurements performed after the exposure, yield almost the same spectra at all temperatures. An Arrhenius plot of the bulk ionic conductivity yields an activation energy of 0.63 eV which is equal to the activation energy of an uncorroded specimen within the experimental error.
Although the specimens remained intact, a change in colour from green to black was detected. EDAX and XRD-analysis of the exposed specimen revealed the presence of a SrS and CeO_2 corrosion layer on the surface of the pellets, while the interior of the specimen remained unchanged. The thickness of the layer was estimated to be about 7 μm.

An ultimate compatibility test was performed by placing some Sr-cerate samples in an actual coal gasifier of British Coal. As a result, the samples degraded dramatically into powder. This effect may be attributed to a combined action of flying-slag, ash and chemical attack of the aggressive gas. The erosive action of fly ashes can be highly reduced by placing the sensor behind a filter element. A way to eliminate the chemical attack is to protect the vulnerable electrolyte by a non-porous, but hydrogen permeable and compatible material to be used as a protective cap or if electron conductive and catalytic, directly as outer electrode. As a result of a search for hydrogen permeable materials, different candidates were found. Pure Nb metal and pure Pd metal were used in a first compatibility test. The Nb sample demonstrated a good compatibility towards the CO_2 and H_2S gases, but the hydrogen permeability in the presence of H_2O was not satisfying. In spite of the slightly poorer permeability characteristics of Pd, this material showed better compatibility and was finally selected as protective material.

A final material aspect consisted of the **realisation of a ceramic/ metal joining**. The basic requirements of this joining can be summarised as follows :
- gas tight, realized by joining $SrCeO_3$ and a metal (e.g. stainless steel)
- mechanical strength better than strength of $SrCeO_3$
- operating temperature maximum 900°C
- compatible with the coal gasification atmosphere

In order to solve this problem, several attempts were made using different joining materials. However, a direct joining between the metal and the Sr-cerate is not realised yet. Joinings were realised but cracks at the inner part of the Sr-cerate pellet were observed. This phenomenon was attributed to the poor mechanical strength of the solid electrolyte, which needs to be improved. In absence of a good joining, a gold ring between the metal and the cerate, was preliminary used, a method - although not completely satisfying - mentioned in literature.

Prototype sensor

Using the results of the material tests, a prototype hydrogen sensor, shown in figure 3, was designed and constructed. The sensor is capable to accept sensing elements of the BOSCH-type shape as well as pellet-shaped solid electrolytes, which may offer important advantages concerning the choice of reference material. The probe also provides the possibility of encapsulating eventually not only a hydrogen sensor, but also the British Coal oxygen sensor based on the ZrO_2 BOSCH-tube.

The lack of an accurate metal/ceramic joining for closing the sensing tube is bypassed using a gold ring pressed between the metal and the solid electrolyte. A spring in the upper part of the probe produces the necessary strain to ensure a good fitting of the gold ring at high temperature. Furthermore a protection cap is provided in order to avoid corrosion of the solid electrolyte caused by the coal gas.
This first construction of the prototype was designed to be mounted in a tube furnace, containing the coal gas simulation gasses. An appropriate experimental test rig, for testing the prototype hydrogen and oxygen sensor as well as performing cell measurements in a synthetic gasification atmosphere, is currently under construction. These experiments are conducted in an exhaust hood, where also the preconditioning of the gas,

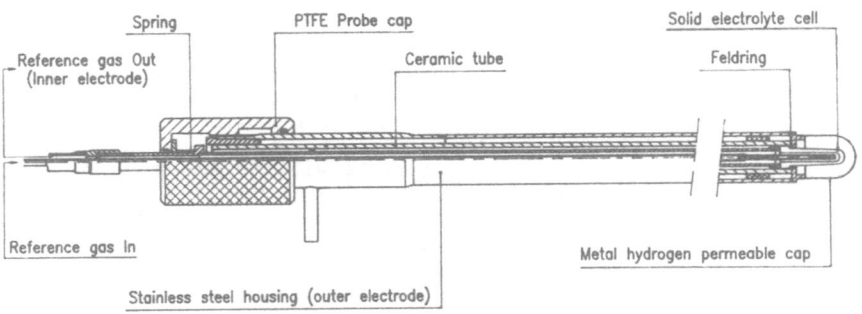

Figure 3. Design of the prototype hydrogen sensor.

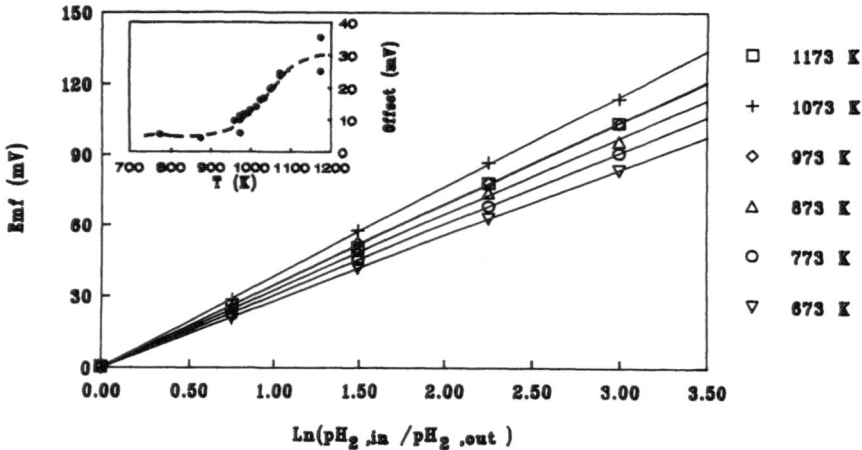

Figure 4. EMF response of the prototype hydrogen sensor in a flowing Ar/H$_2$ mixture. The outside hydrogen partial pressure is fixed at 2.5 mbar. The small offset (see insert) has been subtracted from the corresponding curves.

i.e. mixing and humidifying of the gases, will take place. In addition a gas analysis system consisting of a battery of monitors for coal gases, a gas chromatograph and a mass spectrometer can be connected to the outlet of the furnace, in order to evaluate on-line the actual gas composition. In a first test campaign, the prototype hydrogen sensor is currently tested in an H$_2$/Ar atmosphere. In the second test run the selectivity of the probe will be examined in the synthetic coal gasification atmosphere, and in the near future it will be submitted to an actual coal gasification environment.

A first test of this probe was conducted in an Ar-0.25 % H_2 mixture, while Ar with a variable hydrogen concentration was offered at the reference side. The results of this prelimary test is shown on figure 4. The main features of these curves are a Nernstian behaviour of the emf versus hydrogen concentration, and a small offset, varying from 5 mV at 700 K to 30 mV at 1200 K (see insert) when the same amount of hydrogen is offered at both sides of the electrolyte. Most probably due to electronic conduction, the slope of the lines are smaller than the theoretical ones (RT/2F) and this deviation increases with temperature.

Concerning the reference materials to be used, two possible sensor configurations can be put forward. One option is a gaseous reference system, e.g. an Ar/H_2 internal reference. The main drawback of this choice is that the reference gas has to be brought to the front of the sensor in a reference compartment without inducing a gas leak. Nevertheless most commercial systems use this type of reference system. The alternative consists of a solid reference material providing a constant hydrogen partial pressure (concentration cell mode) or a constant oxygen partial pressure (fuel cell mode). As to the constant hydrogen partial pressure systems, potential reference materials are Th-ThH_2, NbH-NbH_2, Nb-Ti-H, CaH_2-LiH, etc. The major disadvantage of the hydride systems is the strong reducing nature of the metal/metal hydride binary mixtures necessary for a stable hydrogen pressure, eventually causing compatibility problems with the cerate electrolyte.
From compatibility considerations, CeO_2 in slightly reduced condition was initially selected as a candidate for a constant oxygen pressure reference material. It was experimentally verified that already reduction by annealing in argon results in a good electrically conducting material. Alternatively the system In_2O_3-$PrO_{1.83}$-ZrO_2 contains a number of binary phase mixtures, of which some show a high conductivity. The 80:10:10 mixture has been prepared and the conductivity tested. This material has been selected as the first candidate to prepare sandwich samples, in order to study the electrolyte-reference interface. But because of a lack of confident experiments with these solid reference systems, in a first approach a gas reference will be used.
Relating to the reference material, the design of the probe allows from different reference configurations to be built in as shown in figure 5, in a first approach (5a), the probe uses a gaseous reference and a electrolyte pellet coated with a protective Pd film. In the second case (5b) a solid reference film is deposited on the electrolyte pellet. When a BOSCH-type solid electrolyte is used the design of the sensitive element can be scheduled according to possibility 5c or 5d, using a solid reference or a gaseous reference system, respectively. In both cases a protective Pd cap can be added.

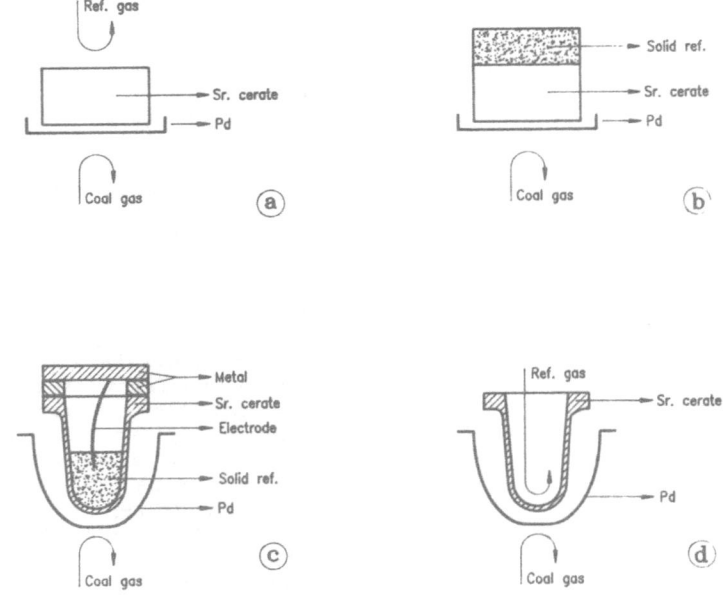

Figure 5. Four different configurations of the combination of the solid electrolyte and a reference material.

CONCLUSION

In this paper, the development of on-line electrochemical sensors for monitoring hydrogen and oxygen in coal gasification systems has been described. In an intensive material development program, the high-temperature proton conducting $SrCeO_3$ was synthesised. The electrical characterisation showed its usability in a hydrogen sensor. However, due to chemical attack of the coal gas, the material has to be protected using a hydrogen permeable metal. Also, the currently poor mechanical strength of the material did not allow a successful joining of the ceramic to a metal. In a preliminary test of the first prototype hydrogen sensor in H_2/Ar atmospheres a theoretically expected (Nernstian) response was demonstrated.

REFERENCES

[1] Sens P.F., Strategies for Future R&D in Combustion Processes. First International Conference on Combustion Technologies for a Clean Environment, Vilamoura (Algarve), Portugal, 3-6 September 1991.

[2] Novem Nederlandse Maatschappij voor Energie en Milieu B.V. Systeemstudie hoge temperatuur gasreiniging bij KV-STEG systemen. Systeemstudie in het kader van het Nationaal onderzoekprogramma Kolen, uitgevoerd door ECN, KEMA, Stork Ketels en TNO. Novem. Ref. nr. 90-310, Dossiernr. 8725-21421/500, November 1990.

[3] "Sampling and Analysis of Product Gas Produced by Coal Gasifiers Under Actual Process Conditions (High Temperature and Pressure)". Proposal for participation in the research and development programme "Joules".

[4] De Schutter F., Jacobs A., Vangrunderbeek J., "Proc. Int. School on Solid State Ionics for Sensors and Electrochromics", Erice (It), July 1992, to be published.

[5] Iwahara H., Uchida H. and Tanaka S., Solid State Ionics,9-10 (1983) 1021.

[6] Luyten J., De Schutter F., Schram J., Schoonman J., Solid State Ionics 46 (1991) 117.

[7] Bonanos N., private communication (1992).

[8] Kosacki I., J.G.M. Becht, R. Van Landschoot, J. Schoonman, "Electrical Properties of $SrCe_{0.95}Yb_{0.05}O_3$ in Hydrogen Containing Atmospheres", Submitted to Solid State Ionics.

SESSION 5:

Separation Processes

Chairman: Prof. P. Le Goff

TOWARDS THE OPTIMUM ENERGY EFFICIENT SEPARATION PROCESS

PROFESSOR K.E. PORTER

Department of Chemical Engineering and Applied Chemistry,
Aston University,
Birmingham,
U.K.

ABSTRACT

A method is required for identifying the optimum sequence of different separation methods for separating a multicomponent mixture of fluids. This will take into account that there may be a different optimum concentration change for each alternative method, and that the cost of using methods such as membrane gas separation, adsorption, and gas adsorption depends on the properties of the separating material. It is unlikely that much more progress will be made until we obtain a greater understanding of the nature of the optimum sequence of distillation columns. A brief review of the present state of the art provides the starting point for developing the ideas noted above.

INTRODUCTION

In recent years there has been much work on devising methods of reducing the energy used in separation processes. The most frequently used method of separating fluid mixtures is by distillation which uses significant amounts of energy. It has been estimated that 11% of the energy used in chemical processes is used for distillation, thus there has been work on reducing the energy used in distillation and on replacing distillation by another method of separation, by extractive distillation, absorption liquid-liquid extraction or freeze crystallisation. In general, these developments have offered solutions to particular problems. We do not yet have a comprehensive methodology for identifying the optimum energy efficient separation process for any fluid mixture.

This methodology would be able to optimise the separation of a multicomponent mixture both by the appropriate choice of separation method for each separation and provide the optimum sequence of separations.

In what follows, after a brief review of the present state of the art, some specific problems are identified which must be solved before such a methodology may be developed.

A REVIEW

"The best way of saving energy in chemical plants is to design distillation columns "leaner and harder." This quotation from a senior manager of the Du Pont company was used by Rush (1) in 1979 as the main conclusion of a paper in which he presented cost evaluations from several studies of replacing distillation by an alternative method of separation. His results, summarised in Table 1., show no case in which the replacement of distillation can be justified in economic terms. This seems less surprising when we note that the "Separations Business" is a mature business. At least 20 different methods of separation are available but they have all been used commercially for more than thirty years in competition one with another. It is most probable that distillation (the market leader) will remain the most frequently used method (2).

The alternative methods are used in situations where the cost of distillation is unusually high. That is for separating small molecules or relatively large molecules. Here small molecules are defined as mixtures in which any component has a critical temperature less than about 50°C. such that it may not be condensed by cooling water even when the pressure of the distillation column is raised. For these mixtures, the more complicated and expensive cryogenic distillation must be used. Thus substances with a molecular weight less than about 40 are sometimes separated by absorption into a solvent; adsorption on to a solid, or by gas membrane separators. These pressure driven methods have been used for many applications in which the gas mixture is already pressurised. For example, for separations in natural gas processing, ammonia plants or gasification processes.

TABLE 1 COST OF REPLACING DISTILLATION BY OTHER METHODS FOR THE SEPARATION OF DIMETHYLFORMAMIDE (1)

	liquid extraction	freeze crystalisation	vapour recompression	multiple effect distillation
Net savings,M$y	120	75	180	88
New money required, M$	3200	1600	2000	1400
Net return on new money,%	3.8	4.7	9.0	6.3
Payback time, yrs	27	21	11	16
DCF rate of return, %	-5.8	-3.2	5.9	0.6

Large molecules are defined as those mixtures which boil at say above 150^0 C even when the absolute pressure is reduced to about 20 m bar . Vacuum distillation is costly both because of the increased size of the equipment and because of the relatively high temperature of the energy required.   Mixtures with components of a molecular weight greater than about 130 come into this category. Liquid-liquid extraction or possibly freeze crystallisation may be used. Distillation is also costly when used to separate components of a low relative volatility, and impossible to use to separate azeotropes. Here solvents may be added to change the liquid phase activities (extractive distillation and azeotropic distillation).

For many separation problems, distillation remains the preferred solution and progress has been made in designing distillation columns "leaner and harder", and in providing procedures to identify the optimum sequence of columns to separate a multicomponent mixture. The obvious way to reduce the energy used in a distillation column is to increase the number of theoretical plates in the column and so reduce the required reflux ratio. Reflux ratios are now usually set at the safe minimum within the expected accuracy of the vapour-liquid equilibria data, say ten percent greater than the theoretical minimum reflux ratio. The need to revamp existing columns provided the incentive for the successful research and development on distributors which has resulted in the widespread use of packed distillation columns of a large diameter (eg 4 m) which were unknown ten years ago.

A large number of papers have been presented which describe methods of identifying the optimum sequence by means of advanced mathematical techniques incorporated into computer programmes. Methods are available which will identify either the lowest energy sequence or the"best" sequence for energy integration. Distillation is well suited for energy integration. it often requires heat at a relatively low temperature and the heat used in the reboiler is available from the condenser for reuse, albeit at a somewhat lower temperature. (Distillation columns work on temperature drop rather than on energy). A review of much recent work is given by Porter and Momoh (3) in a paper which presents a simple equation for finding the optimum sequence.

THE COST OF ENERGY

In practice the decision on whether to use a proposed process is made by an economic evaluation which takes into account the operating cost and the annualised capital cost to determine the Total Annualised Cost (TAC). Energy costs are part of the operating costs, thus the choice of a "minimum energy" separation process depends on the cost of energy relative to that of the cost of money (e.g Bank Interest rates). Thus energy saving schemes are justified more readily if the cost of energy is high and interest rates are low. At this time, in the U.K (low energy costs, high interest rates) many energy saving processes will not be justified on economic grounds. If, as a matter of longer term policy it is desired to put more emphasis on energy saving, then processes may still be chosen by the established methods of economic evaluation, but with an artificial, increased ratio in the cost of energy to the cost of money. Thus in what follows it is assumed that choices between processes will be made by the established methods of economic evaluation.

TOWARDS THE ENERGY EFFICIENT SEPARATION PROCESS - THE PROBLEM DEFINED

The brief review above, mentioned separations for which distillation has been replaced by membrane separators, adsorption or absorption and etc., and noted the work that has been done to identify the optimum sequence of distillation columns to separate a multicomponent mixture. We now seek the basis for developing a methodology which will identify the optimum energy efficient separation process. This will tell us a) Which separation method to use for a particular separation b) When and how to combine methods to make a hybrid process to use for a particular separation c) How to design the optimum sequence of separation methods (any method) to separate a multicomponent mixture.

SOME UNSOLVED PROBLEMS

All separation methods (except distillation, freeze crystallisation and the centrifugal separation of gases) are based on the properties of a separating material. This may be a membrane, an adsorbent or a solvent. Unlike distillation, (which achieves almost complete separation at little additional cost), several separation methods are relatively cheap for a partial separation but very expensive to use for a complete separation. Thus in comparing different separation methods as alternative solutions to a particular separation problem, it must be noted that:

1.   The optimum amount (or degree) of separation may be different for different methods.

2   The cost of separation depends on the properties of the separating material.

The implications of considering these two points are shown below in a discussion of choosing between a membrane gas separation and cryogenic distillation for the separation of hydrogen and carbon monoxide for the manufacture of acetic acid.

Very little work has been done on developing procedures for selecting the optimum sequence of separations for separating a multicomponent mixture, where these include several different methods of separation. A good starting point is likely to be the extensive work on selecting the optimum sequence of separations by distillation as the only method. However, previous work on the distillation sequence is all on how to find the optimum sequence and not on why separations done in a particular order result in the optimum. Thus a third unsolved problem is

3.   To develop a conceptual understanding of the optimisation of the   distillation sequence.

THE SEPARATION OF HYDROGEN AND CARBON MONOXIDE FOR THE MANUFACTURE OF ACETIC ACID

The process flowsheet is shown in Fig 1 . The separation is between an "upstream" reactor where methane and oxygen are reacted at 30 bar to produce a mixture

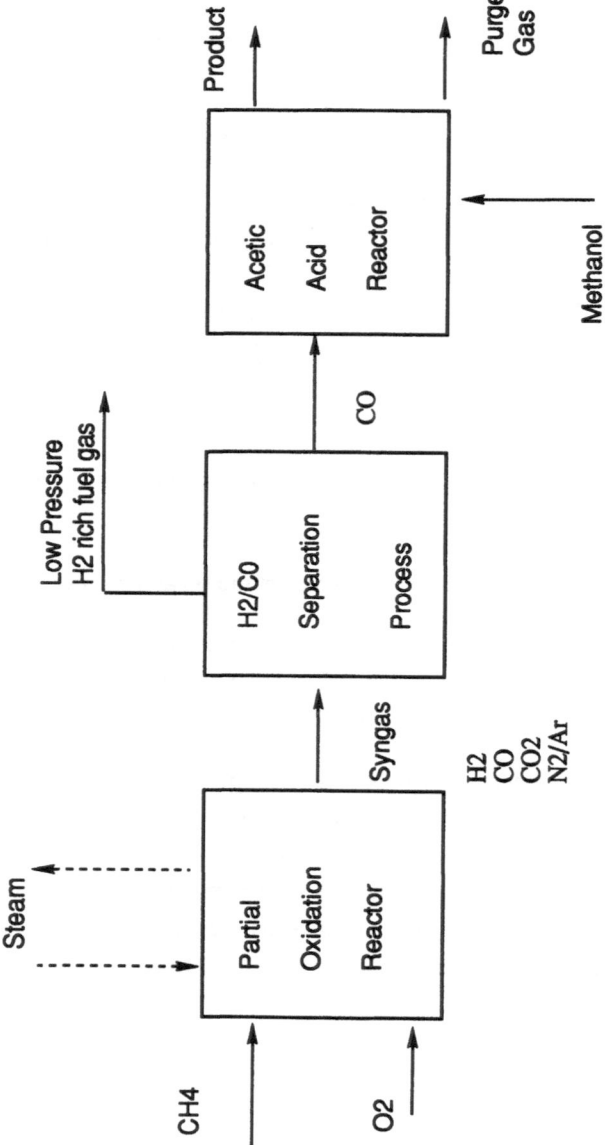

FIG 1. PROCESS FOR THE MANUFACTURE OF ACETIC ACID
AND SEPARATION COSTS FOR A PARTIAL SEPARATION

of carbon monoxide and hydrogen, and a "downstream" reactor where the separated carbon monoxide is reacted (30 bar) with methanol to produce acetic acid. It is required to calculate whether a membrane gas separator will be less costly than cryogenic distillation. The membrane is to work between the operating pressure of 30 bar and 1.5 bar. The hydrogen diffuses through the membrane much faster than the carbon monoxide, so that the high pressure gas leaving is enriched in carbon monoxide and the low pressure gas is enriched in hydrogen, The low pressure gas may be recompressed and recycled to reduce the loss of carbon monoxide with the hydrogen.

The relevant separating properties of the membrane are:

a)   The Permeability (of the fast gas)

b)   The Selectivity i.e. the ratio of the fast gas permeability to the slow gas permeability.

The capital cost of the membrane separator depends largely on the permeability which determines the membrane area required. The operating cost depends largely on the selectivity which determines the gas recycle and the loss of carbon monoxide.

Both the capital cost and the operating cost also depend on the concentration of carbon monoxide leaving the separator. As carbon monoxide concentration increases, the required membrane area increases and the recovery of carbon monoxide decreases. This is illustrated in Fig 2 for a single stage membrane (no recycle). Thus the cost of the membrane separator increases rapidly as the concentration of the carbon monoxide product is increased to high purities.

If the process design is based on a high purity carbon monoxide (99.5 %w/w) entering the acetic acid reactor, it is found that cryogenic distillation provides a lower cost of separation than the membrane separator. If however, the required purity is reduced to 95% w/w, the membrane separator provides a lower cost of separation than cryogenic distillation.

As illustrated in Fig 3 , where the cost of manufacturing acetic acid is plotted against the concentration of carbon monoxide leaving a membrane separator, there is an optimum, minimum cost of separation for the membrane (in this case 95.5% w/w). The costs plotted in Fig 3 include not only the capital cost of the membrane and the cost of the gas recycle, but the additional process costs, upstream and downstream of the separator, resulting from the incomplete separation. The cost of the membrane separator increases

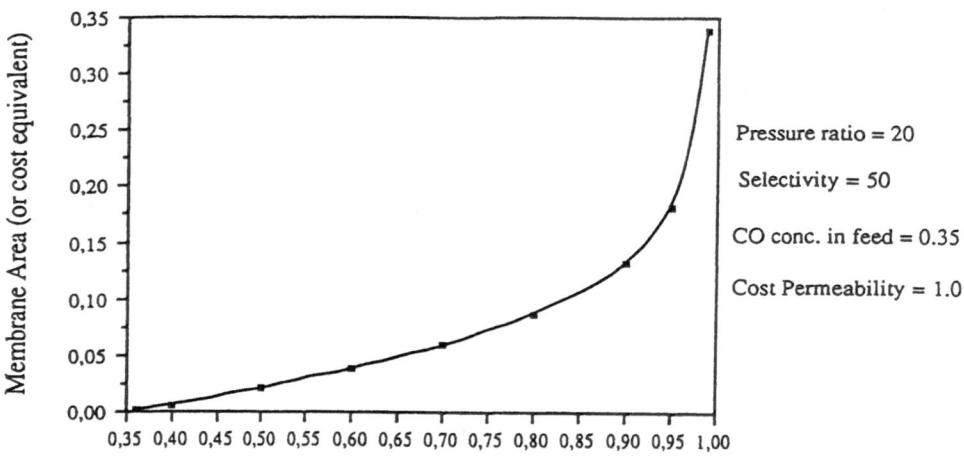

Membrane area vs product purity.

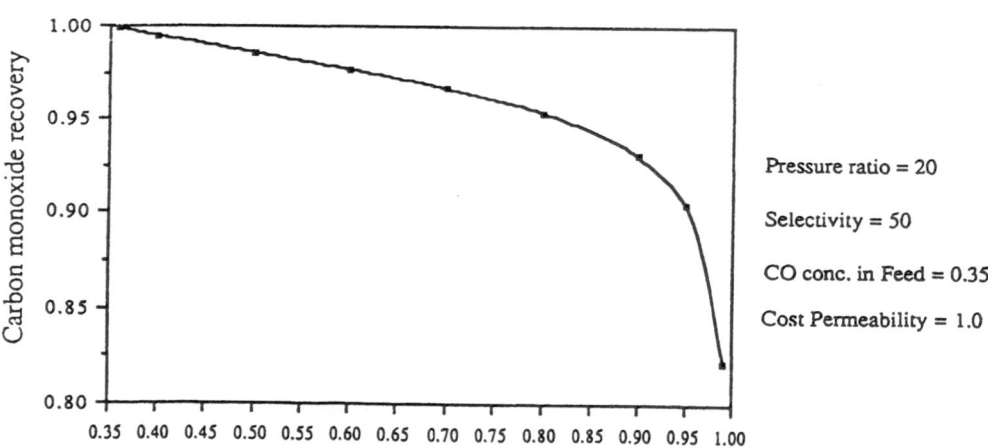

Carbon monoxide recovery vs products purity

FIG 2.   VARIATION IN MEMBRANE AREA AND CARBON
MONOXIDE PURITY FOR THE SEPARATION OF CARBON MONOXIDE
AND HYDROGEN

with carbon monoxide concentration, but the additional costs go down, to reach zero for a complete separation. Thus the optimum separation is found by cost evaluation of the whole process, not just of the membrane separator.

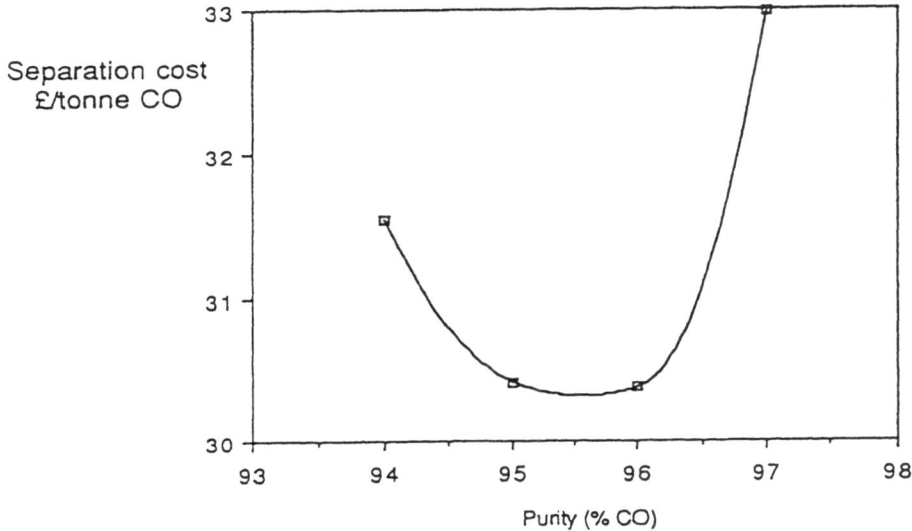

FIG 3.VARIATION IN THE TOTAL COST OF SEPARATING CARBON MONOXIDE AND HYDROGEN IN THE ACETIC ACID PROCESS WITH PURITY OF THE CARBON MONOXIDE PRODUCT

A comparison of different methods of separation should include the amount of separation as a variable and be done at the optimum concentration change for each method.

This chemical process may also be used to demonstrate how the cost of separation depends on the properties of the separating material. It was noted above that the capital cost of the membrane separator depends strongly on the fast gas permeability of the membrane, while the operating cost of the separation depends on the selectivity of the membrane. It then follows that there will be many combinations of permeability and selectivity which may produce the same cost of separation. This is shown in Fig 4 where iso cost lines are shown as plots of permeability against selectivity. This figure was produced to provide targets for molecular scientists involved in developing new membranes (4).

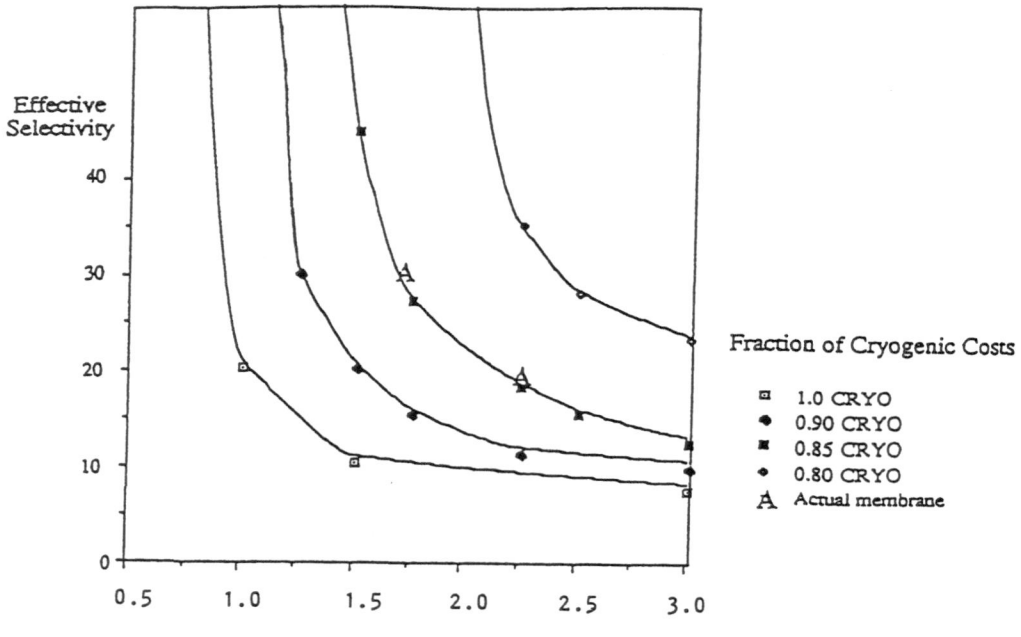

FIG 4. SELECTIVITY PERMEABILITY ISO-COST LINES
FOR THE TOTAL COST OF CARBON MONOXIDE PRODUCTION

Similar analyses may be made for other processes which involve separation and for other separating materials such as adsorbents or gas absorption solvents. An adsorbent may be defined in terms of its capacity and selectivity.

The analysis of a gas absorption solvent is more complicated in that the cost of separation depends not only on the solubility in the solvent of the gas to be removed, and the selectivity of the solvent with respect to other gases, but also on the vapour pressure and viscosity of the solvent.

THE HYBRID PROCESS

This term is used to describe a separation process in which two separation methods are used to achieve one separation. The examples shown in Fig 5 are all concerned with separating oxygen and nitrogen (air separation) by various combinations of membrane separators, adsorbers and cryogenic distillation columns.

The optimisation of such a process will require establishing how much of the separation is to be done by each method. It seems likely that a similar approach to that described above for the acetic acid process might be used for this purpose. It may be that a targeting procedure could be established to show how the properties of the separating material determine the cost of separation. A more difficult problem is that of optimising a sequence of separations using several different separation methods.

THE SEPARATION OF MULTICOMPONENT MIXTURES

Where each of the components must be separated from a multicomponent mixture, a sequence of separators is required. The number of possible sequences, (each capable of achieving the required separation), goes up rapidly with the number of components in the mixture. When all the separations are by distillation, there are 2 possible sequences for 3 components, 42 for 6 components; and 58,786 for 12 components. If alternative separation methods are to be a part of the sequence, then the number of possible sequences will be even greater. The problem is to choose the optimum sequence, i.e a) the lowest cost sequence or b) the sequence with the lowest energy consumption or c) the sequence most suitable for energy integration.

Almost all the previous work has been about finding the optimum sequence when all the separations are by distillation. For this case the lowest cost sequence, a) is likely to be that with the lowest energy consumption, b). In general, for distillation sequences, the results of design and optimisation studies by numerical methods tend to support some long established 'rules of thumb' listed below. These rules were proposed for finding the lowest cost/lowest energy sequences.

1. Favour 50/50 splits
2. Remove the most plentiful component first
3. Do the easiest separation first and the hardest last.

DIRECT HYBRID FOR N2, RECYCLE FOR ARGON

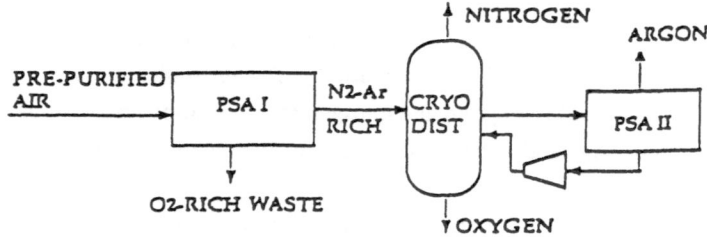

DIRECT HYBRID PROCESS

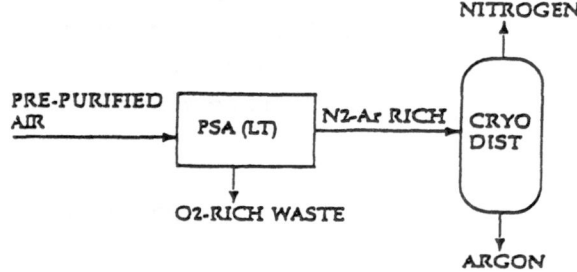

RECYCLE HYBRID PROCESS

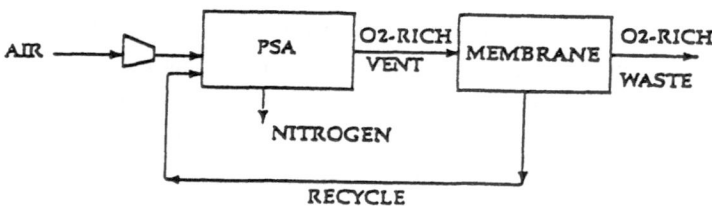

RECYCLE HYBRID PROCESS

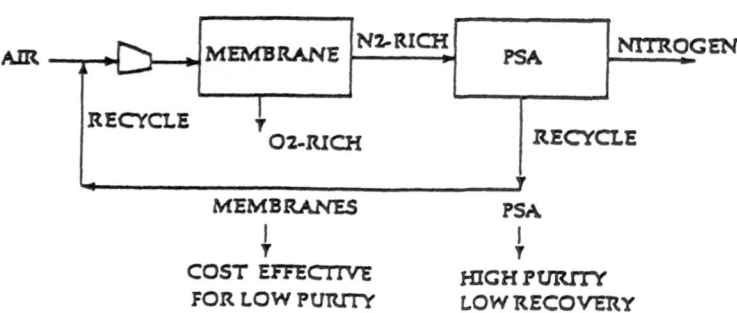

FIGURE 5
Examples of Hybrid Processes for the Separation of Air

4.   Favour the direct sequence (i.e remove components in order of their vapour pressure or volatility)

Another "rule of thumb" has been suggested on the basis of studies on energy integration. This is

5.   The lowest cost, lowest energy sequence is likely to be the sequence most suitable for energy integration.

Application of the first two rules has the effect of reducing the rate of flow to subsequent separations downstream. It seems likely that these rules will apply to any combination of separation methods.

Rules 3) and 4) (which sometimes contradict each other) may only apply to distillation. Recent theoretical work has shown that rule 3) "do the easiest separation first and the hardest last" is very often wrong.The lowest vapour load sequence is often that with the hardest separation first.

The new rule 5) that " the lowest cost,, lowest energy sequence is that most suitable for energy integration" is unlikely to be true. The sequence most suitable for energy integration must take account of energy degradation, (i.e the difference in temperature between the top and bottom of each distillation column), whereas the lowest energy sequence need not. The support for rules of thumb 3) and 5) from the results of computerised optimisation studies may be due to the use of equations for the cost of energy which make it linear in temperature. For many relatively low temperature distillations (below 100^0 C) the cost of energy will be constant.

Thus it may be concluded that despite all the recent work on finding the optimum distillation sequence, we still do not understand why the chosen sequence is the optimum.

This lack of understanding is one of the main reasons we do not yet have methods for identifying the optimum sequence of separations where alternative separation methods are to be included as well as distillation.

CONCLUSIONS

In choosing between different methods of separation, the choice should be made at the optimum degree of separation for each method, and allow for the properties of the separating material. That is both the concentration change to be achieved and the separating material may be considered to be variables in the optimisation procedure. Implementation of these ideas is a matter of technique only.

Progress in developing procedures for identifying the optimum sequence of different types of separator is held up by the present lack of understanding of the nature of the optimum sequence of distillation columns. This provides an intellectual challenge to the academic engineer to identify those ideas and simplifying concepts to take us further towards designing the optimum energy efficient separation process.

REFERENCES

1.   Rush F. E. Chem.Eng.Prog. 72, 44-49, July 1980.

2.   K.E. Porter and J.D. Jenkins Separation Processes in The Chemical Industry Chap.41, 544, Ed D. Sharp and T.F West Soc of Chemical Industry Ellis Horwood 1982

3   K.E. Porter and S. Momoh, The Chemical Engineering Journal, 46 97 (1991).

4.   K.E. Porter, A.B. Hinchcliffe and J. Pardoe Gas Separation and Purification 4, 185, December 1990

INTENSIVE HYDRODYNAMIC REGIMES IN ABSORPTION TOWERS AND FRACTIONATING COLUMNS

N.N.KULOV

Laboratory of Theoretical Fundamentals of Chemical Technology/Kurnakov Institute of General and Inorganic Chemistry of the Russian Academy of Sciences, Leninsky Prospekt, 31, Moscow 117907, Russia

Yu.N.LEBEDEV

Department of Mass Transfer Column Equipment/ VNIINEFTEMASH, 4 Roshinsky Proezd, 19/21, Moscow 113191, Russia

ABSTRACT

The following three main ways of high gas and liquid flowrates realization to reduce the size of the equipment for absorption or distillation are discussed: performance under condition of phase inversion, cocurrent flow of phases on the stage and use of a centrifugal field. The special technique for comparative estimation and choice of the optimum construction of the apparatus to arrive at a proper economic balance of investment and operating costs is developed. Industrial examples of the effective contacting devices working in the intensive regimes are given

INTRODUCTION

A promising way to reduce the size and cost of equipment for absorption and distillation is the application of intensive regimes of hydrodynamic phase interaction. The designer's goal is to arrive at a proper ecomonic balance of investment and operating costs, since the former usually increases as the latter decreases.

CHOICE OF THE OPTIMUM CONSTRUCTION OF THE CONTACTING MASS TRANSFER DEVICE

For comparative estimation and choice of the optimum construction of the apparatus applied in various technological processes, we have to determine the minimum cost of separation

$$C_s = \sum_j C_{oj} + I \qquad (1)$$

where $\sum_j C_{oj}$ are operating costs connected with the transport of material flows and heat supply to the column and I - the investment.

As we need a certain amount of energy to produce the metal and the equipment, the minimum cost would, speaking broadly, correspond to a minimum energy consumption.

The required expenditure on mixture separation in the mass transfer apparatus depends on the type of the contacting device and on the hydrodynamic regime of operation, i.e. on the gas flowrate in the column, which determine the efficiency $\eta = f(F_s)$ and pressure drop $\Delta p = f (F_s)$. In the correlations, F-factor $F_s = w_G \sqrt{\rho_G}$ is used instead of the average gas velocity w_G, which allows us to take into consideration the influence of gas density ρ_G.

As industrial columns operate under varying liquid and gas loads, the choice of the optimum construction of the contacting device should be carried out, taking into account the necessary range of effective column operation $[n]$ (turn down ratio). The $[n]$ value is usually between 1.3 and 3.0.

The column should operate effectively in the whole range of loads, therefore η_{min} and Δp_{max} should be taken into account, being determined as shown on Fig. 1a and 1b. These efficiency and pressure drop values serve to calculate C_s (η_{min}, Δp_{max}) as a function of F-factor. Figure 1c illustrates the procedure of determining the minimum cost of separation.

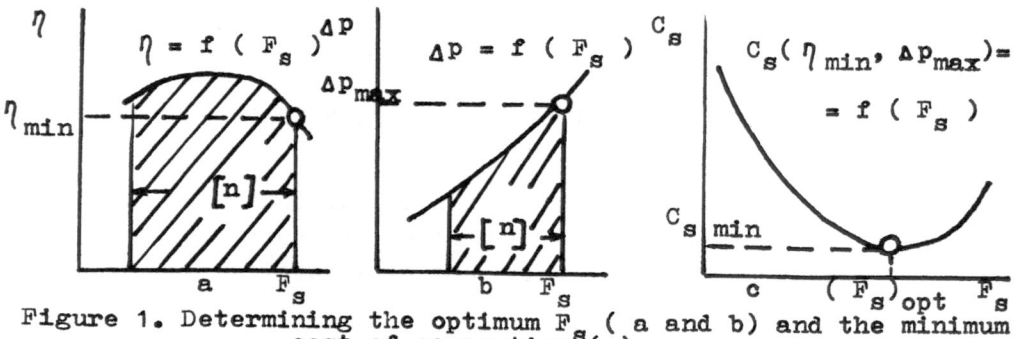

Figure 1. Determining the optimum F_s (a and b) and the minimum cost of separation (c).

To choose the optimum construction of the contacting device, the values of minimum cost for each apparatus $C_{s,min}$ are compared to the standard construction in accordance with the relationship

$$\Psi = (C_{smin})_{STD} / C_{smin} \qquad (2)$$

An analysis of numerous technological separation processes in oil processing, petrochemical and chemical industries has shown that the variety of processes may be reduced to the three main types: 1) absorption, 2) distillation and 3) distillation of thermosensitive mixtures including steam distillation.

For example, high phase flowrates can be achieved in columns with regular packing under the conditions of countercurrent flow. Investigation of these regular packings have become very popular lately.

Figure 2. The regular rhombic packing of VNIINEFTEMASH.

The regular rhombic packing of VNIINEFTEMASH, presented on the Fig. 2, has a hydraulic resistance of only 0.3 mm Hg for a layer 1 m in height, at an F-factor of 1 $Pa^{0.5}$, and of 1.35 at an F-factor of 2.6. It permits reduction of the steam supply to the bottom of the column under vacuum, i.e. to reduce the energy consumption. This results from the opportunity to provide more theoretical plates at the same pressure drop and use a lower reflux ratio.

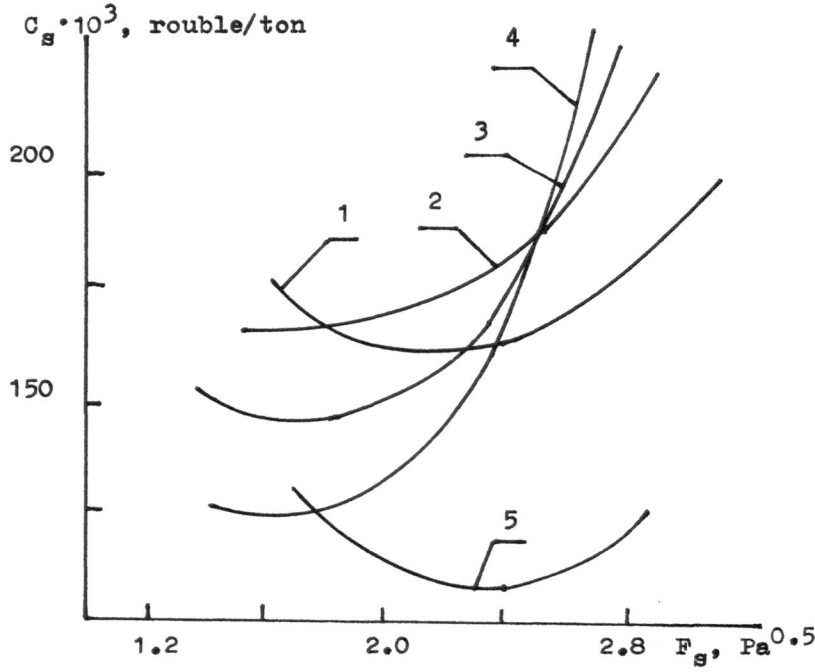

Figure 3. Vacuum distillation of mazut: 1 - sieve tray of VNII-
NEFTEMASH; 2 - S-valve tray; 3 - Glitsch valve-cap
tray; 4 - valve -cap ballast tray; 5 - Rhombic regular
packing.

On Fig. 3, if we compare the rhombic packing (see line 5) with other contacting devices,we can see that the former is the best one for operation under vacuum. The liquid flows through the bed is realized in the film-droplet regime. The packing can also operate in the emulsification regime. A column 4.2 m in diameter with the rhombic packing is operating in Ufa (Russia); a similar one, 9.0 m in diameter, is operating in Chimkent (Kazakhstan).

The present paper deals with the three main ways of attaining intensive regimes: 1) emulsification or phase inversion, 2) cocurrent flow of phases on the stage, 3) use of a centrifugal field.

PHASE INVERSION REGIME

Inversion of phases occurs with the formation of a movable emulsion system which is permeated with gas, liquid and gas-liquid eddies of numerous sizes rotating and mixing with each other. It is the hydrodynamic regime at which countercurrent flow is preserved but the nature of phase interaction changes sharply and the velocity w_G in the tower equals the flooding rate w_{GINV}.

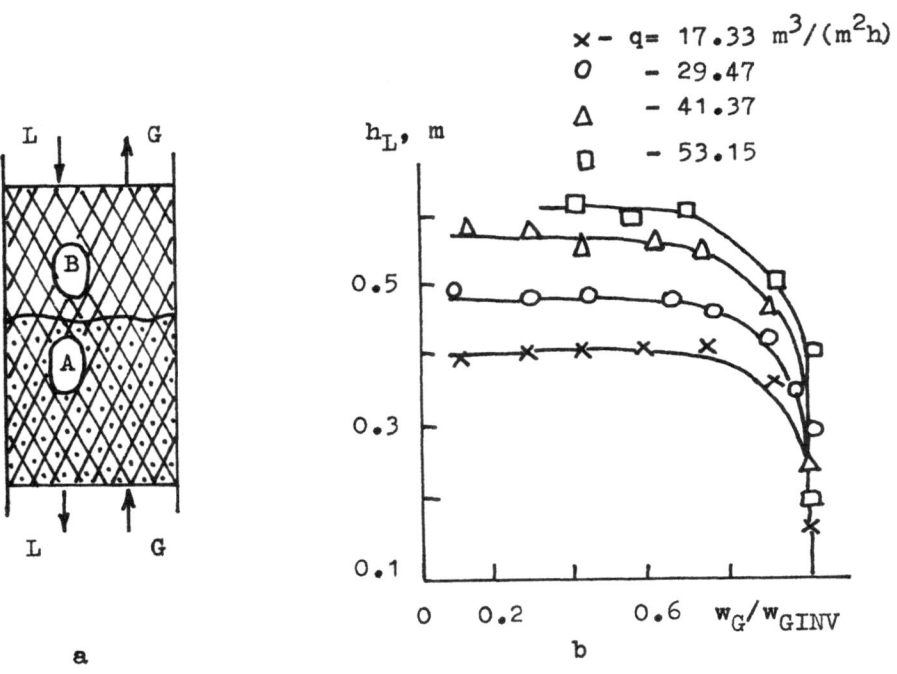

Figure 4. Phase inversion regime: (a) a shematic sketch of an irrigated packed tower and (b) desorption of CO_2 from water to air in packed tower (Raschig Rings 15 x 15 x 3 mm).

Figure 4,a shows a packed bed in which two zones exist simultaneously. The lower zone A is occupied by a layer of gas and liquid emulsion, while in the upper zone B a separate film flow of the phases is achieved. The curves on the Fig.4, b clearly reveal the known hydrodynamic models of gas and liquid interaction in countercurrent flow. The horizontal parts of the curves correspond to a film regime and the transition to the emulsification regime rate is accompanied by a decrease of the height of transfer unit h_L (depending on the liquid flow rate) of 2 to 2.6 times. Recent works [1,2] have produced developments that permit us to maintain the regime of emulsified flow. At a chemical plant producing iodine and bromine out of sea brine, a number of apparatus based on the emulsion regime have been successfully launched in particular, a desorber with a cross section of 1 m^2 and a brine capacity of 200 m^3/h. Cost of equipment would be reduced in having a larger flow.

An increase of heat and mass transfer efficiency under conditions of emulsification can be achieved also on a sieve tray in the weeping-overflowing regime (Fig. 5).

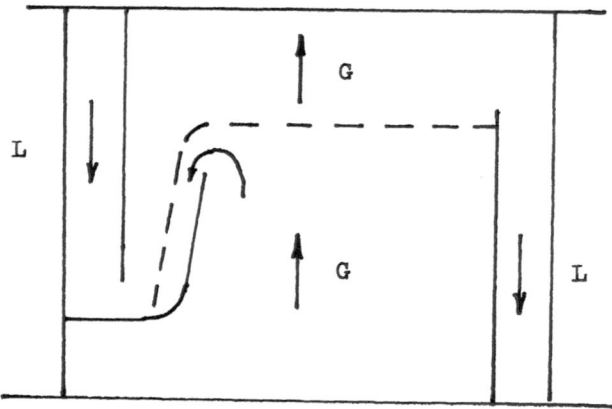

Figure 5. Sieve tray: liquid preemulsification.

Emulsification of liquid in the vicinity of the lower edge of the downcomer baffle by small part (5-25%) of the gas flow and organization of weeping-overflowing liquid transport allow us to expand limiting throughputs, to lower the "liquid gradient" on tray, and to creat an additional contact zone.

COCURRENT FLOW IN THE CONTACT STAGE

Cocurrent operation of trayed, as well as packed column has a limit to the maximum gas flow due to flooding phenomena.

Flooding can be largely avoided by cocurrent operation of the contacting equipment. For an intense cocurrent flow in the form of liquid film-mist, it is essential that liquid drops are present in the gas flow. These drops originate due to spraying of the crests of disturbance waves which cover a negligibly small fraction of the mass transfer surface area. But when being precipitated, the drops uniformly cover all the surface, the spots of surface renewal produced by the droplets are significantly larger than the diameter of the droplets. Let us consider the surface renewal model in the form

$$\beta_L = \sqrt{D_{AL} \, s} \tag{3}$$

Here β_L is the liquid phase mass transfer coefficient, D_{AL} is the liquid phase diffusion coefficient, s is the fractional surface renewed per unit time.

It is assumed that the renewal is produced by the impact of droplets falling down from the gas flow onto the surface

$$s = N_s / \sigma \tag{4}$$

where σ is the surface tension.

The energy N_s transferred to the surface by the droplets encountering the unit area of the interphase per unit of time and expended on the surface renewal can be predicted theoretically [3]. The calculated values of Sherwood number for liquid phase (see the dotted line) and the experimental data on the rate of desorption of oxygen from water in wetted-wall column are compared on the Fig. 6. The Reynolds number for gas flow Re_G^* and Re_G^{**} in Fig. 6 are referred to as the regime boundaries.

Application of surface with a certain regular roughness allows us to intensify heat and mass transfer in two-phase flows up to 2.5 - 3.0 times, as compared with the smooth surfaces. The experimental data on absorption of carbon dioxide by water in film columns with smooth (see lines 1 and 2) and rough (lines 3 and 4) walls are compared for downward cocurrent flow on Fig. 7. The lines 2 and 4 describe intensive flow regimes with entrainment. It is obvious that a combination of high flow rates and a rough tube wall allows a relative increase of the Sherwood number up to 3 times and even more.

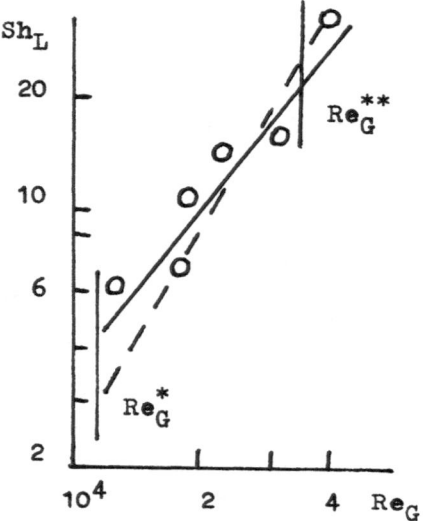

Figure 6. Downward cocurrent gas-liquid flow in the wetted-wall column: desorption of oxygen from water to air flow.

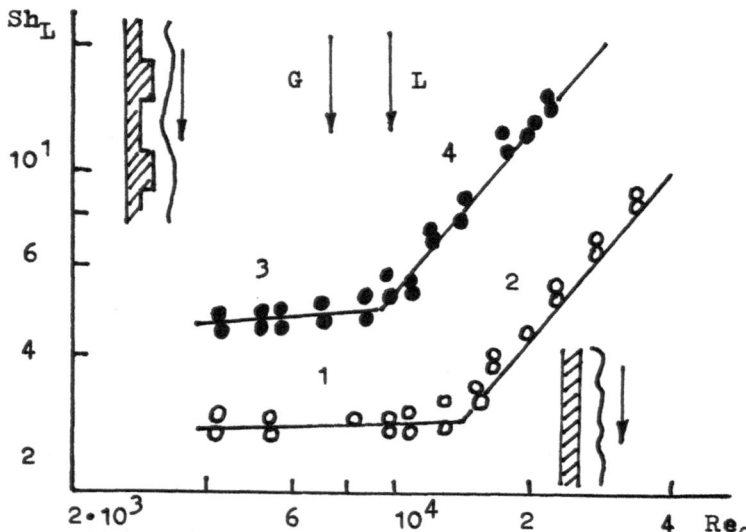

Figure 7. Absorption of CO_2 by water in film column with smooth (light dots) and rough (dark dots) walls.

TUBES WITH SWIRL

High gas velocities in cocurrent gas(vapour)- liquid flow may
result in excessive entrainment. The tendency to form entrainment can be reduced in a centrifugal field achieved by vortex
flow in contacting devices. Rotational motion in the gas phase
not only increase the mass transfer coefficient but also makes
easier the separation of liquid from gas, at the tube exit. The
development of tubes with swirling flow has resulted in several
different types of "swirl-type plates" suitable for industrial
application. Note that although cocurrent flow occures in the
tubes, but conventional countercurrent flow is obtained in the
distillation column.

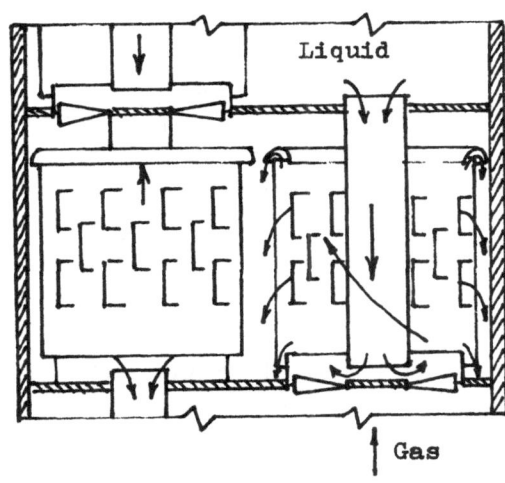

Figure 8. Centrifugal tray.

The centrifugal tray shown on Fig. 8 has been developed
for high pressure distillation and absorption. For example,
for the systems butane-isobutane, pentane-isopentane where more
than 100 theoretical plates one required. The substitution of
valve-cap trays by centrifugal ones allowed an increase in the
output up to 2.6 times. The specific consumption of metal was
reduced by 30%.

The operating limits shown in Fig. 9 of the centrifugal
tray are wide enough. In the case of rectification (low liquid

flow rates), the F-factor values are approximately 7 to 9. Such tray can be used also for absorption with the liquid flowrate up to 200 $m^3/(m^2h)$. These columns with centrifugal trays have been successfully tested in the industrial processes of pressure distillation of liquified gases, and it has been found that the Murphree vapour efficiency E_{MV} is about 0.7 in the range of the F-factor values from 2.8 to 5.6.

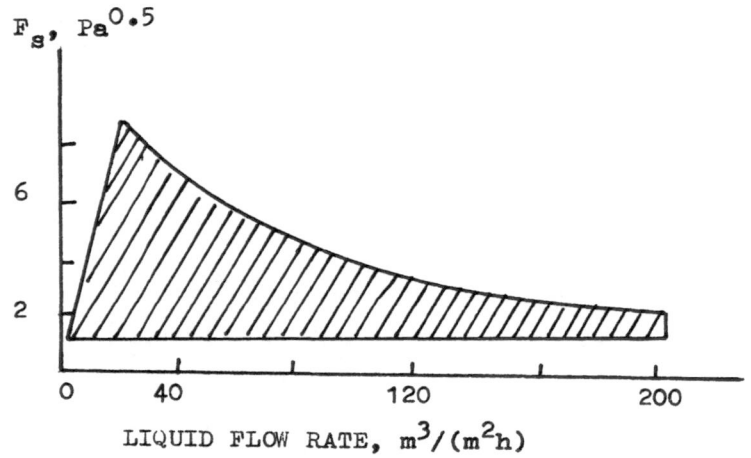

Figure 9. Operating limits of centrifugal tray.

It has also been found, that when the trays are used in pressure distillation and absorption, the relatively high pressure drop produces no increase in operating cost. Reduced expenditure values on liquified gases separation under pressure in columns with valve-cap trays and centrifugal trays are presented on Fig. 10.

The analysis of the given cost of separation values has shown that the centrifugal tray is the optimum solution for the pressure distillation discussed above (see curve 4).

CENTRIFUGAL CONTACTORS

In the equipment described previously, mass transfer is enhanced by the energy introduced in the contactor via the gas and liquid flows. In another class of phase contacting equipment the energy is supplied from an external source.

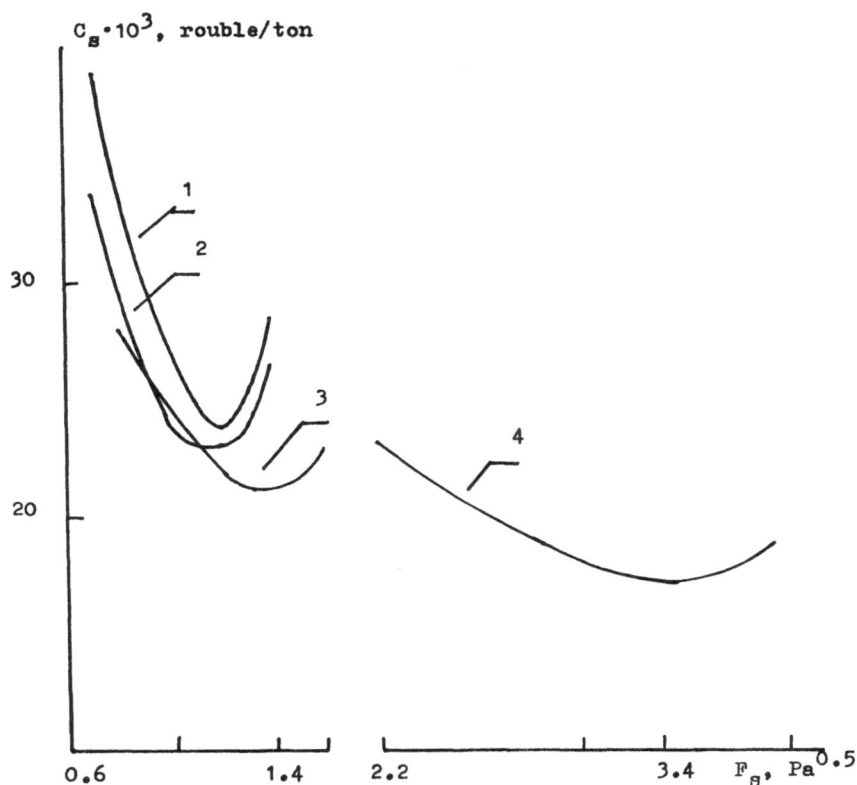

Figure 10. High pressure distillation: Deethaniser: 1 – Glitsch
valve-cap tray, 2 – valve-cap ballast tray, 3 –
S-valve-cap tray, 4 – centrifugal tray.

Devices of this type were described for the fist time by
Colin Ramshaw (1983). A centrifugal field is obtained by rotating a specially designed bed in the shape of disc with a hole
in its centre. The liquid is sprayed into the centre of the
disc, and it flows radially outwards because of the centrifugal
force. The gas is introduced at the outer surface of the disc
and is forced radially inwards countercurrent to the liquid by
the pressure driving forces.

Not much information is available in the literature, concerning the mass transfer coefficients in the liquid and gas phases.
As an example, for a rotating bed in the form of a steel net
with a surface area of 400 m^2/m^3 and a voidage 95.5%, the height

of the transfer unit based on gas phase h_G and liquid phase h_L resistance are given by the correlations

$$h_G = 1.522 \, d_1 \, Re_G^{0.39} \, Re_L^{-0.3} \, Fr^{-0.11} Sc_G^{0.66} \quad (5)$$

$$h_L = 0.000118 \, d_1 \, Re_G^{0.59} \, Fr^{-0.42} \, Sc_L^{0.5} \quad (6)$$

which can be recommended for practical application [4]. Here d_1 is the equivalent diameter, Re_L is liquid phase Reynolds number, Sc is Schmidt number and $Fr = a_\omega/g$ is modified Froude number where a_ω is the centrifugal acceleration and g is the acceleration due to gravity.

CONCLUSION

It should be underline that the main point of intensifying the separation processes which was discussed, is reducing power consumption or the cost of equipment and increasing productivity. The data considered cover a wide range of different contacting devices working under conditions of intensive regimes. They allow for the prospect of intensification to be estimated while choosing the optimum apparatus for a specific process, taking into consideration the optimum energy consumption.

REFERENCES

1. Il'inykh, A.A., Memedlyaev, Z.N., Kulov,N.N. and Maljusov, V.A., Hydrodynamics of stable regime of emulsification in packing column. Theor. Found. of Chem. Engng , 1987, 21(2), 111-117.

2. Il'inykh, A.A., Memedlyaev, Z.N. and Kulov,N.N. Mass transfer in wetted packing in regimes of hanging and emulsification. Theor. Found. of Chem. Engng , 1989, 23(5),355 - 360.

3. Kulov, N.N., Mass transfer in thin film type apparatus. In World Congress III of Chemical Engineering, 1986, Tokyo, Japan, volume 2, 710 - 713.

4. Syrenko, V.I., Kulov, N.N. and Tyutyunnikov, A.B., Hydrodynamics and mass transfer in the apparatus with rotating bed. Theor. Found. of Chem. Engng , 1992, 26(2), 173-186.

THE DESIGN AND OPTIMIZATION OF DIVIDING WALL DISTILLATION COLUMNS

C TRIANTAFYLLOU* and R SMITH

Centre for Process Integration
Department of Chemical Engineering
UMIST, Manchester, U K

ABSTRACT

For most separations fully thermally coupled distillation columns require significantly less energy than conventional arrangements. This paper describes a design model which provides a basis for investigating the degrees of freedom to minimise the energy consumption. The optimisation of fully thermally coupled columns is also discussed.

NOMENCLATURE

F   : total flowrate of the feed
q   : liquid fraction of the feed
r_i : recovery of component i relative to the feed
R   : reflux ratio
V   : vapour flow

Subscripts

ABOVE = above the sidestream
BELOW = below the sidestream
HK   = heavy key
LK   = light key
MAIN   = main column
min   = minimum
1   = Column 1
2   = Column 2
3   = Column 3

INTRODUCTION

Process integration has proven to be very successful in reducing the energy costs for conventional distillation arrangements [1,2]. However, the scope

* Present address : Exxon Chemical France, Nôtre Dames de Gravenchon, France.

for integration of conventional distillation columns into an overall process is often limited. Also, practical constraints often prevent integration of distillation columns with the rest of the process.

If the column cannot be integrated with the rest of the process, or, if the potential for integration is limited by the heat flows in the background process, then we must turn our attention back to the distillation operation itself and look at non-conventional arrangements.

Figure 1 shows two conventional arrangements for the separation of a three-component mixture (usually known as the direct and indirect sequence). One of the most

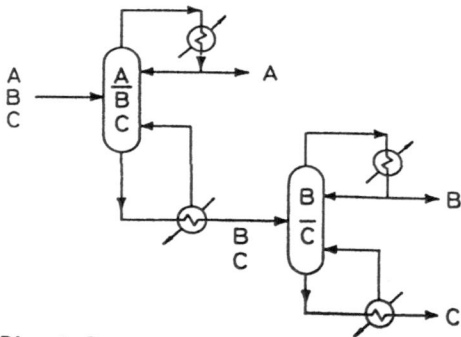

(a) Direct Sequence

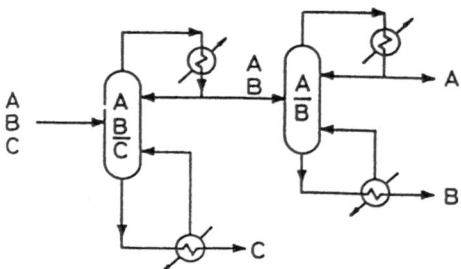

(b) Indirect Sequence

Figure 1. Conventional arrangements.

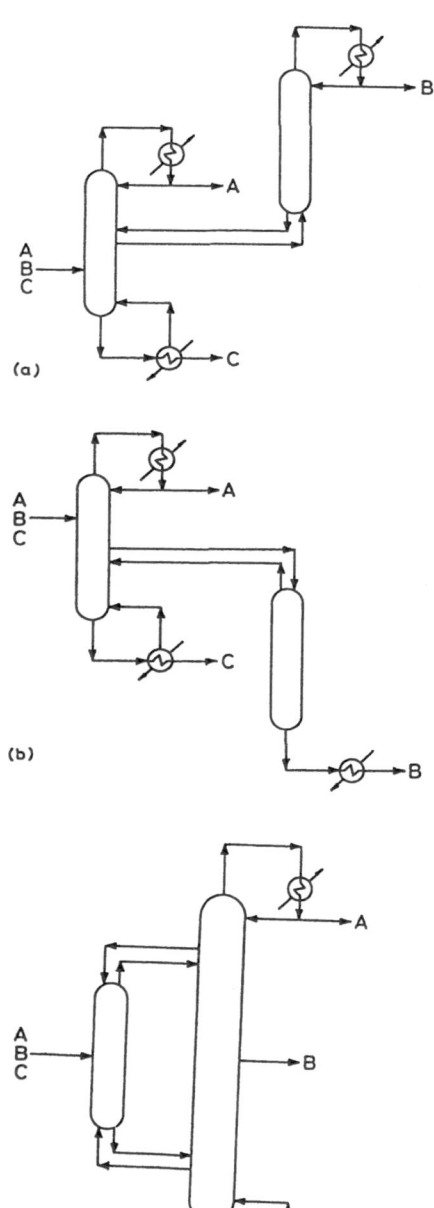

Figure 2. Non-conventional arrangements which use thermal coupling. (a) Sidestream rectifier. (b) Sidestream stripper. (c) Fully thermally coupled column.

significant non-conventional arrangements involves "thermal coupling".
Figure 2 shows a number of non-conventional arrangements which use thermal
coupling. In thermal coupling part of the heat transfer necessary for the
separation is provided by direct contact via the material flows.

The fully thermally coupled arrangement shown in Figure 2c (sometimes known
as the Petlyuk column) has been known for over 50 years [3]. Theoretical
studies [4,5] have shown that it can save, on average, around 30% of energy
costs compared with a conventional arrangement. The fully thermally coupled
arrangement shown in Figure 2c can be constructed in a single shell with an
internal dividing wall [6,7], Figure 3. If there is no heat transfer across
the dividing wall, then the arrangements shown in Figures 2c and 3 are
thermodynamically equivalent. The special advantage of the dividing wall
column is the achievement of the energy savings of full thermal coupling,
together with the capital savings from the use of a single shell for the
distillation. Capital savings also result from the use of a single reboiler
and condenser compared with a conventional arrangement. Despite their
advantages, designers have been reluctant to use fully thermally coupled
columns.

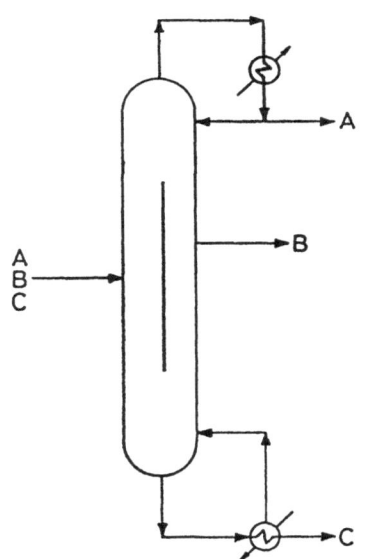

Figure 3. The dividing wall column.

At first sight it might seem a simple matter
to design an arrangement as shown in Figures
2c and 3 using a computer simulation
package. Figure 2c would provide the basis
of the simulation model for both
arrangements. On closer examination the
simulation is not straightforward. There
are more degrees of freedom than a
conventional arrangement and more variables
to be fixed before a simulation can be
performed.

One approach to designing such columns could
therefore be to develop a short-cut design
procedure which provides:

(a) a tool to rapidly assess alternative
designs and perform preliminary
optimisation

(b) an initialisation for rigorous
simulation.

A number of such models have been suggested
previously but all have limitations in their
ability to provide an initialisation for a
rigorous simulation and for preliminary
optimisation.

A CASE STUDY

One of the more difficult separations encountered in refineries requires the
separation of a mixture of close boiling C_4's. The feed specification for
such a separation is given in Table 1. The primary objective is to separate
1-Butene with a purity of 99% and a recovery of 95%. 1-Butene is one of
the middle boiling components of the feed mixture. Utility and cost data

are given in Table 2. The separation duty is considerable and the process tends to operate in isolation.

One way to separate 1-Butene from the feed mixture is to use the conventional arrangements of Figure 1. Table 3 compares the energy and capital cost for each of the conventional arrangements of Figure 1. These were obtained by using the Fenske-Underwood-Gilliland shortcut model with sizing and costing methods presented previously [8]. In each column a trade-off was performed between the reflux ratio and the number of plates, in order to design for minimum total cost (energy plus capital cost).

Table 1. The feed specification for our case study.

Component	Molar flow rate (Kmol/hr)	Mole fraction	Relative volatility
i-butane	24.50	0.0490	1.4997
1-butene	253.55	0.5071	1.3234
n-butane	34.75	0.0695	1.1086
*trans*2-butene	47.30	0.0946	1.0642
*cis*2-butene	139.90	0.2798	1.0000
Total	500.00	1.0000	—

Table 2. Utility and cost data for our case study.

Hot Utility
Steam at 150°C
Hot utility cost = 90 [£/KW·yr]

Cold Utility
Cooling water = 25°C–27°C
Cold utility cost = 8 [£/KW·yr]
Plant life time = 5 years
Interest rate = 10%

Table 3   Energy and capital cost of the conventional arrangements applied to our case study

	DIRECT SEQUENCE	INDIRECT SEQUENCE
Total Energy Cost (£/yr)	2,261,126	2,025,975
Total Column Capital (£/yr)	520,428	503,463
Total HEN Capital (£/yr)	174,356	163,828
Total Overall Cost (£/yr)	2,955,910	2,693,266

The energy requirement of both the direct and the indirect sequence can be reduced if we apply heat integration between the reboiler of one column and the condenser of the other. One of the two columns will have to operate at a higher pressure in order to provide the temperature driving force required for heat integration. Column 2 would have to operate at higher pressure for the direct sequence and Column 1 for the indirect sequence. Unfortunately, for this separation problem, both increases in pressure cannot be achieved without excessive fouling in the reboiler. Consequently, heat integration cannot be applied because of this practical difficulty.

THE DESIGN MODEL

The dividing wall column can be modelled in the same way as the two column arrangement shown in Figure 2c. A simplified model can be created if the main column is itself represented as two simple columns, Figure 4. In order to model the prefractionator we must assume a partial condenser and a partial reboiler. In this way the short-cut method calculates the vapour flow and the number of theoretical plates for every section of the fully thermally coupled column. Therefore, the short-cut solution can then be used as initialisation for the rigorous simulation.

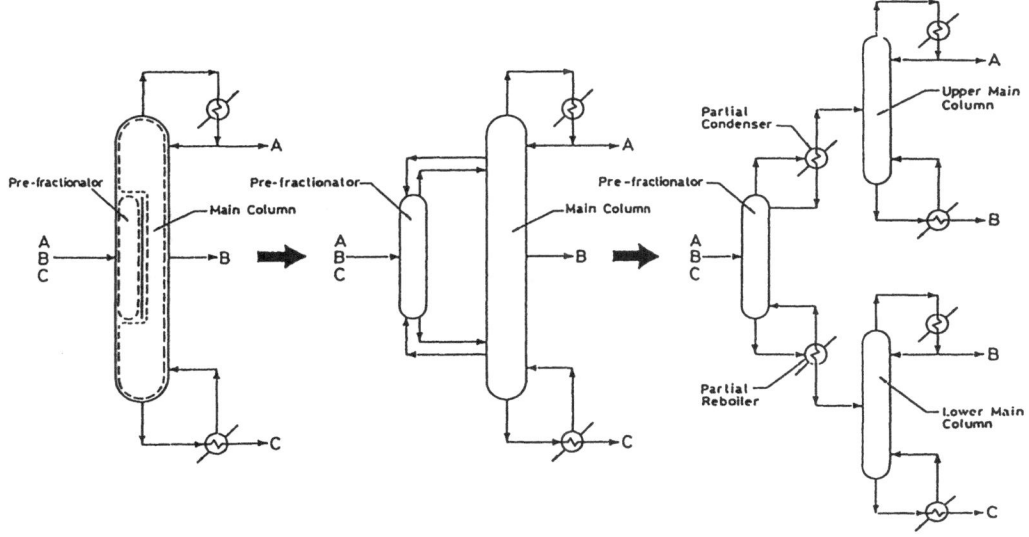

Figure 4. Simplified models for the dividing wall column.

In the short-cut model presented here we use the three-column model of Figure 4, together with the standard Fenske-Underwood-Gilliland short-cut techniques for multi-component mixtures with non-sharp separations. The basis of the model is shown in Figure 5 and the details are presented elswhere [8]. In the prefractionator (Column 1 of Figure 5), a partial condenser and a reboiler are assumed. In a ternary system the light key component (LK_1) in the prefractionator is A and the heavy key (HK_1) is C. The values for the recovery of A at the top and the recovery of C at the bottom must be specified. As we will discuss later, the specification of these two recoveries is of critical importance. For now let us note that values must be supplied for the calculation to proceed.

After calculating the mass balance, the vapour flow and the number of theoretical plates for Column 1, the flow rate and the composition of streams 2 and 4 in Figure 5 are calculated assuming vapour-liquid equilibrium in the condenser. Streams 2 and 4 will be the feed and the sidestream of Column 2 respectively. Similarly streams 3 and 5 are calculated assuming vapour-liquid equilibrium in the reboiler. Streams 3 and 5 are the feed and the sidestream of Column 3. Column 2 is designed as a sidestream column separating A (LK_2) from B (HK_2). All the C in the feed to Column 2 is assumed to go to the column bottom. Finally, Column 3 is designed as a sidestream column separating B (LK_3) from C (HK_3).

Here the A in the feed to Column 3 is assumed to go to the column top.

Linking Columns 2 and 3

After calculating the reflux ratio and the number of plates for each of the three columns, we link columns 2 and 3 by equalising the vapour flows at the bottom of Column 2 and the top of Column 3. The column with the higher vapour flow will have the same number of plates as before. However, when we increase the vapour flow of the column with the lower initial vapour flow, we must recalculate the number of plates.

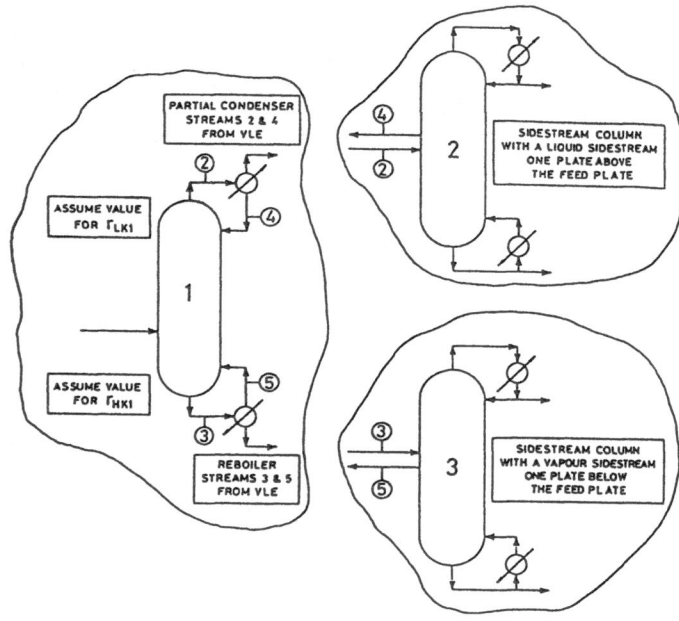

Figure 5. The shortcut model.

SELECTION OF THE PREFRACTIONATOR RECOVERIES

We now have a short-cut design model for the dividing wall column. However, we have not investigated all the degrees of freedom yet. For the prefractionator calculation (Column 1 in Figure 5), we needed to set the recovery of the light key at the top (r_{LK1}) and the recovery of the heavy key at the bottom (r_{HK1}). Specifying r_{LK1} and r_{HK1} in the prefractionator is the same as setting the vapour and liquid draw-off rates from the main column to the prefractionator. These are important design variables and should not be set arbitrarily.

If we wish to minimise energy requirements, r_{LK1} and r_{HK1} should be set so as to minimise the overall vapour flow.

Minimising the Vapour Flow for a Given (Non-Minimum) Reflux Ratio

Consider the vapour flow in Column 2 above the sidestream (V_{2ABOVE}) and the vapour flow in Column 3 below the sidestream (V_{3BELOW}). The overall vapour flow ($V_{OVERALL}$) for the dividing wall column is given by:

$$V_{OVERALL} = max [V_{2NET}, V_{3BELOW}]$$

where

$$V_{2NET} = V_{2ABOVE} - (1-q) F_1$$

In this equation we need to distinguish between V_{2NET} and V_{2ABOVE} to allow for partial vaporisation of the feed to Column 1. Any vapour in the feed to Column 1 will find its way to the top of Column 2. This vapour does not need to be generated in the reboiler and should not be included in the assessment of $V_{OVERALL}$. Note that V_{3BELOW} must be large enough to satisfy the vapour flow requirements of Column 3 as well as Column 1 [8].

The $V_{OVERALL}$ will be dominated either from the vapour flow in Column 2 (V_{2NET}) or from the vapour flow in Column 3 (V_{3BELOW}). When Column 2 dominates we can manipulate the recoveries in the prefractionator so as to make the separation in Column 2 easier [8]. At the same time the separation in Column 3 will become harder, hence V_{2NET} will decrease but V_{3BELOW} will increase. $V_{OVERALL}$ will decrease up to the point where $V_{2NET} = V_{3BELOW}$.

Alternatively, when Column 3 dominates the overall vapour flow we can manipulate the recoveries in the prefractionator in order to facilitate the separation in Column 3 and hence decrease V_{3BELOW}. At the same time the separation in Column 3 will become harder and V_{2NET} will increase.

In fact, V_{2NET} and V_{3BELOW} can be equalised for different pairs of r_{LK1} and r_{HK1}. One of these pairs will also give the minimum $V_{OVERALL}$. The recoveries can be manipulated between their maximum value which is 100% and a minimum value which is determined by the purity of the middle product. For a ternary mixture LK1=A, HK1=C, LK2=A, HK2=B, LK3=B, HK3=C, hence:

$$r_{Amin} < r_A < 1.0$$
$$r_{Cmin} < r_C < 1.0$$

where r_{Amin} and r_{Cmin} are determined by the purity of B.

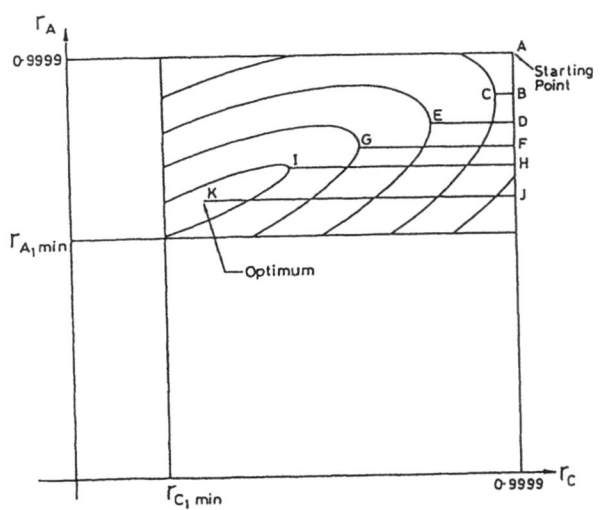

Figure 6. Contour lines of constant $V_{OVERALL}$. Column 2 dominates at the starting point.

Figure 6 shows r_A and r_C as coordinates in which contour lines of constant $V_{OVERALL}$ can be identified. We start at the point where $r_A = r_C = 1.0$. In practice we can never achieve a recovery of 1.0, hence we use a practical limit of 0.9999. This is point 'A' on Figure 6. From there onwards the recoveries can only decrease. We next identify whether Column 2 dominates Column 3 or vice versa (i.e. identify the column with the larger vapour flow). If Column 2 dominates, as is the case in Figure 6, we reduce r_A by an increment (point 'B' in Figure 6) and manipulate r_C until $V_{2NET} = V_{3BELOW}$ (point 'C' in Figure 6).

After equalising V_{2NET} and V_{3BELOW} we reduce r_A by another increment (point 'D' in Figure 6.) and manipulate r_C until $V_{2NET} = V_{3BELOW}$ (point 'E' in Figure 6). This strategy is repeated, changing r_A by increments and at each step equalising V_{2NET} and V_{3BELOW} until the changes in $V_{OVERALL}$ are no longer significant. Figure 6 shows that V_{2NET} and V_{3BELOW} are equalised at an extreme point of each contour.

For most cases the optimum will correspond with a pair of r_A and r_C which equalise V_{2NET} and V_{3BELOW}. However, sometimes the lower limit for r_A or r_C is reached before V_{2NET} and V_{3BELOW} are equal. In this case the optimum corresponds with one of the limits for r_A or r_C. Beyond this limit there is too much A (or too much C) in the middle product hence we can no more achieve the purity of B.

OPTIMISING THE REFLUX RATIO

In conventional distillation the reflux ratio can be optimised by a trade-off between energy and column capital. In the dividing wall column there are *effectively* two reflux ratios, one for the prefractionator and one for the main column.

The reflux ratio of the prefractionator (R_1), together with r_{LK1} and r_{HK1}, determine the vapour and liquid draw-off rates from the main column. As R_1 increases, more vapour and liquid must be drawn from the main column (hence the minimum reflux in the main column increases), but fewer plates are required in the prefractionator. Also, the intermediate components at the top of the prefractionator will increase and this will change the number of plates required in the main column.

The model can be used to investigate the energy/capital trade-off for the dividing wall column and to determine the best reflux ratio for a given cost scenario. We do this by varying reflux ratio in the main column, $(R/R_{min})_{MAIN}$. For each $(R/R_{min})_{MAIN}$ we vary the prefractionator reflux ratio $(R/R_{min})_{PREFR}$ and for each pair of $(R/R_{min})_{MAIN}$ and $(R/R_{min})_{PREFR}$ we select the r_{LK1} and r_{HK1} which minimises $V_{OVERALL}$. In this way we can determine those values of $(R/R_{min})_{MAIN}$ and $(R/R_{min})_{PREFR}$ which minimise the total cost (energy plus capital cost).

THE CASE STUDY REVISITED

Rather than use either of the conventional arrangements detailed in Table 3 for the separation of the mixture in Table 1, let us consider the possibility of using a dividing wall column.

Table 4 contains results of three different solutions for our case study for a dividing wall column. Solution 1 was obtained by an initial setting of both $(R/R_{min})_{MAIN}$ and $(R/R_{min})_{PREFR}$ to 1.10 and the r_{LK1} and r_{HK1} to 0.9999. In Solution 2 $(R/R_{min})_{MAIN}$ and $(R/R_{min})_{PREFR}$ were again set to 1.10 but the recoveries selected for minimum vapour flow.

In Solution 3 of Table 4 the total cost was minimised. This was done by varying $(R/R_{min})_{MAIN}$ between 1.01 and 1.10. For each $(R/R_{min})_{MAIN}$, $(R/R_{min})_{PREFR}$ was varied between 1.01 and 1.12. Finally, for every pair of values of $(R/R_{min})_{MAIN}$ and $(R/R_{min})_{PREFR}$ the recoveries were selected for minimum energy.

Comparison of Solutions 2 and 1 of Table 4 indicates the value of optimising the recoveries. Solution 2 requires 8.2% less energy than Solution 1.

Table 4 Alternative solutions for the dividing wall column
separating a mixture of close boiling C_4's

	Solution 1	Solution 2	Solution 3
$(R/R_{min})_{MAIN}$	1.10	1.10	1.05
$(R/R_{min})_{PREFR}$	1.10	1.10	1.02
r_{LK1}	0.9999	0.9640	0.9937
r_{HK1}	0.9999	0.9999	0.9937
Energy Cost (£/yr)	1,441,888	1,323,686	297,841
HEN Capital Cost (£/yr)	96,577	95,674	95,524
Column Capital Cost (£/yr)	575,262	519,300	507,571
Total Cost (£/yr)	2,113,727	1,938,660	1900,936

Minimum Energy   Minimum Total Cost

In Table 5 we compare the cost of the conventional arrangements with the
dividing wall column for our case study. Relative to the indirect sequence,
the dividing wall column requires 35.9% less energy.

Table 5 Cost comparison of the conventional arrangements and the dividing
wall column for our case study

	Direct Sequence	Indirect Sequence	Dividing Wall Column (Solution 4)	% Savings of the Dividing Wall Column Relative to Indirect Sequence
Energy Cost (£/yr)	2,261,126	2,025,975	1,297,841	35.9
Total Cost (£/yr)	2,955,910	2,693,266	1,900,936	29.4

In terms of total cost the dividing wall column is 29.4 % cheaper than the
indirect sequence.

CONCLUSIONS

For most separations the fully thermally coupled distillation column is
thermodynamically more efficient than the conventional arrangements and, as
a consequence, has lower energy requirements. Furthermore, the dividing

wall column achieves the energy savings of full thermal coupling together
with capital savings from the use of a single shell, single reboiler and
condenser, except in extreme cases.

ACKNOWLEDGEMENTS

The authors would like to express their appreciation to the UK Department of
Energy, Energy Efficiency Office for financial support of this project. Our
appreciation is also expressed to Exxon for providing the example used in
this paper and to ICI Engineering Department for advice and assistance in
developing the capital cost estimation methods. The authors would also like
to express their gratitude to Bruce Pretty for his contribution to the
project.

REFERENCES

1. Linnhoff B., Dunford H., and Smith R., 1983, Chem. Eng. Sci.,38
    (8): 1175.
2. Smith R. and Linnhoff B.,1988, IChemE, ChERD 66: 195.
3. Brugma, 1942, US Patent 2,295,256.
4. Tedder D.W. and Rudd D.F., 1978, AIChE J., 24: 203.
5. Glinos K. and Malone M.F., 1988, ChERD, 66: 229.
6. Wright R.O., 1949, US Patent 2,471,134.
7. Kaibel G., 1987, Chem. Eng. Technol. 10: 92.
8. Triantafyllou, C. and Smith, R., 1992, Trans IChemE. ChERD.

THE METHODOLOGY OF GAS ADSORPTION PROCESS DESIGN

N JORGENSEN AND E K MACDONALD
AEA Industrial Technology
Harwell Laboratory (UK)

ABSTRACT

This project addresses the strategic R&D needs of the European process industries in the field of low energy separation processes, specifically the topic of gas adsorption. The energy requirement of separation processes in the chemical and allied industries can account for as much as 40% of total manufacturing costs, therefore any reduction in the energy consumption of separation stages has a major impact on the overall cost of final products. This programme of work on gas phase adsorption consists of a co-ordinated series of research projects aimed at improving the energy efficiency of methods used to separate and recover gaseous products, or to eliminate pollutants before gases are discharged to the atmosphere. The projects are essentially pre-competitive, fundamental research, the results of which could have applications in a wide range of process industries.

The work falls into two categories; the development of improved methods for obtaining reliable basic data, and the development of improved methods to aid process design. The projects concerning basic data investigate the equilibria and kinetics of single-phase and multi-component systems, to develop new techniques for measuring these properties, and to provide better information for process design models. The work on process design is aimed at a better understanding of discrete steps in the adsorption cycle, and includes studies of pressure swing systems and non-isothermal systems. The data generated by the first set of projects provide valuable input for these models. The process design studies also include experiments to validate models, both under controlled laboratory conditions and at pilot plant scale.

INTRODUCTION

European industry uses more energy in separating and purifying products than in almost any other process operation, with the possible exception of chemical reactions. This is particularly so when separating products to high purity from very dilute mixtures, or where two products are present with similar physical properties but widely differing market values. It is important that these separations are performed as efficiently as possible to maximise yield and selectivity at minimum energy cost. Many processes involve the use of energy intensive separations:

-       gas separation, purification and enrichment using cryogenic distillation;

- recovery of volatile organic compounds (VOCs), or gas drying, using cryogenic condensation;
- separation of isomeric, or otherwise closely related chemicals using fractional distillation;
- azeotropic distillation of pure compounds from mixtures with water.

There are significant benefits to be realised in developing and adopting lower energy separation routes such as adsorption. This was identified, in a recent CEC study [1], by industrialists and research workers as a priority topic for pre-competitive research. Already, applications of adsorption processes to effect air separations and solvent recovery have been proven to be energy efficient alternatives to conventional technology and in some cases an economic benefit from reduced capital cost is also clear.

The adsorption technique can be used to remove trace impurities from a gas stream thus purifying it, or to recover valuable components present in a mixture at low concentrations. In the latter case, the regeneration step is crucial to the economics of the process, while in the former it may be economic to discard the spent adsorbent and replace it with fresh supplies. The desorption, during adsorbent regeneration, can be accomplished physically, by changing the temperature or pressure (or both) or chemically by applying another compound which replaces the previously adsorbed species. Clearly, the performance of an adsorption process depends on a ready supply of adsorbing material, the selectivity and capacity of the adsorbent and the ease of regeneration.

In addition to the benefits described above, adsorption also has an environmental protection role to play in the control of emissions. VOC emissions, odour emissions from food processing plants, petrol vapour emissions from refineries and vehicle filling stations and oxides of sulphur and nitrogen from combustion processes are all major sources of environmental pollution for which control measures such as incineration, scrubbing or refrigeration are moderately effective, but also energy intensive. Adsorption processes to remove pollutants to a very low level have the potential to reduce atmospheric pollution significantly while at the same time reducing the energy cost of implementing control measures.

Although adsorption processes are not new and have been used with some success to achieve difficult and previously energy intensive separations, the underlying science is not well understood. As a result, the design of adsorption processes has been a very empirical task, which has been a barrier to more widespread use. A better understanding of the surface chemistry is needed to reduce the time and effort required to select an adsorbent to achieve a given separation, or to design a tailor-made material for a specific task. More experimental data is needed on the thermodynamics and kinetics of adsorption processes, so that better models can be developed for process design and optimisation of plant operating conditions. Also, the theories and models developed through fundamental research must be validated within the context of real industrial scale processes and refined as required to account for scale-up factors, so they may be used with confidence by industry.

It was recognised that there was a need for a co-ordinated research programme linking the fundamental physics and chemistry of adsorption with process and equipment design. The CEC are funding a project, under the JOULE programme, on "The methodology of gas adsorption process design". Additional support has come from L'Air Liquide, Shell-KLSA, BP Research and Dommick Hunter Filters. The research partners are the Universities of Bath, CNRS-ENSIC Nancy, CNRS-LIMSI Orsay, Munchen, Madrid and Oporto and the Research Laboratories AEA Industrial Technology, Institut Francais du Petrole and Rhone-Poulenc Recherches.

OBJECTIVES

The overall objective of the JOULE research project is to devise a comprehensive methodology for the development and design of adsorption processes. This includes two distinct goals as follows:

1. To reduce the dependence on empirical methods for the process of adsorbent design, column design and process design, by enhancing the scientific understanding of adsorption processes, developing improved models and validating them on a range of scales.

2. To bring together existing and future knowledge and data on adsorption thermodynamics, kinetics and adsorbent properties in a structured, standardised and accessible form.

In order to achieve these goals, the research has encompassed the whole range of scales involved, from molecular studies of surface phenomena to pilot plant trials to verify design conditions. It is hoped that the benefits to industry will be seen through effective use of adsorption processes in appropriate applications and greatly improved energy efficiency of separation compared with traditional gas separations.

RESEARCH PROGRAMME

In order to achieve the objectives outlined above, the research programme has been divided into two principal categories, namely 'Basic data' and 'Process design'.

Basic Data

This facet of the programme has made up the bulk of the work. The emphasis of the work packages is on the development and validation of alternative methodologies for obtaining multi-component equilibrium and kinetic data and reconciling this data with different models for predicting the results.

CNRS-ENSIC, Nancy is developing a method for obtaining multi-component equilibrium and kinetic data using a chromatographic method. This task aims to develop a reliable, standardised method than can be implemented rapidly. Data from other research groups is being used for cross comparison and for validation of the new methodology.

University of Porto is developing techniques for the measurement of intra-particle diffusivity, permeability and thermal conductivity of zeolites and large pore adsorbents. This task aims to develop better mathematical relationships relating kinetics of adsorption and regeneration to the structure of adsorbent materials. Thermal conductivity data is also valuable in the modelling of adsorbers by computational fluid dynamics.

AEA Technology is investigating the kinetics and mechanisms of desorption processes. These studies are providing basic data on adsorption and regeneration phenomena which is being used to refine theoretical models. This will reduce the requirements for experimental measurement at the early stages of process development.

University of Madrid is studying the equilibria of gases on zeolites molecular sieves. Different zeolite structures can lead to ideal gas mixtures becoming non-ideal under certain conditions, and vice-versa, which affects the choice of theoretical models

Pressure Swing Adsorption

The pressurisation and blowdown of an adsorption bed (Figure 1) with a mixture of one inert and one adsorbable species was studied using three models: equilibrium, intraparticle diffusion and intraparticle diffusion/convection.

The presence of intraparticle mass transfer resistances increased both the pressurisation and blowdown times relative to the equilibrium situation. The profiles of the adsorbable species became more dispersive during pressurisation, while at the end of

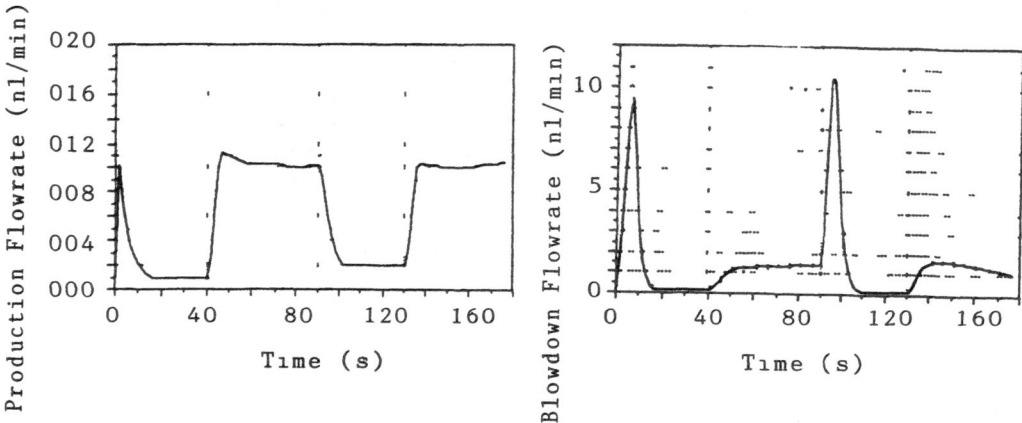

Figure 1. Flowrate during a cycle for a) production and b) blowdown.

blowdown the bed was less regenerated. This resulted in earlier breakthrough in the feed step, following pressurisation or the need for a longer purge step for regeneration after blowdown. The overall effects were loss of productivity, purity or recovery.

Using smaller particles results in lower intraparticle resistances, but at the expense of longer pressurisation and blowdown times and hence is seldom a solution. There are also likely to be associated mechanical problems.

If large-pored materials are used, a convective flow develops inside the adsorbent particles, enhancing the intraparticle mass transfer rate and improving the response of the bed in pressurisation and blowdown, thus improving the efficiency.

A good perception of the pressure dynamics during bed pressurisation and depressurisation in the absence of adsorptive effects is essential for understanding the dynamics of rapid cycling PSA systems. Mathematical models of the pressurisation and depressurisation steps without adsorption have been developed from the material balance and Darcy's law or Ergun's equation.

Experiments have been carried out to determine the effect of packing characteristics, bed length, pressure ratio and gas physical properties on the dynamics of pressurisation and depressurisation. Values of bed permeability and coefficients in the Ergun equation, required as inputs to the computer simulation, were experimentally determined for three packings, namely 2.5mm acrylic diamonds, 3mm polyethylene Rigidex and 0.2mm sand.

From experimentally measured pressure profiles (Figure 2) for different bed permeabilities, bed lengths, fluid viscosities and high to low pressure ratios it can be seen that the pressurisation and blowdown times vary inversely with bed permeability, but proportionally with bed length and fluid viscosity and are only slightly affected by the pressure ratio. The model currently predicts the pressurisation time quite closely (Figure 3) but not the depressurisation time. The model is being further refined.

Equilibrium and Kinetics

The basis of the chromatographic method for obtaining multicomponent adsorption equilibria is to add a perturbation to the chromatographic feed and study the resulting exit times and peak areas. Pure gases were introduced through gas driers and mass flow meters and blended to produce the desired composition. This was then split into two parts, one passing through a reference column and the other through the adsorption column, both in a gas chromatograph. Both flows then passed through a thermal conductivity detector (TCD). Pulses of gases of interest were injected into the adsorption column gas flow and the response of the column registered using the TCD.

Careful choice of the adsorbent particle size, column packing and gas flow characteristics was necessary for successful implementation of the technique. Of all the problems, the most difficult to overcome was the detection of peaks for highly retained components. For the adsorption of carbon dioxide on zeolite at ambient temperature, the retention is high and the mass transfer is slow, therefore the resulting peaks were hardly measurable. The measurement of Henry's constant was thus very inaccurate and it was difficult to construct an isotherm.

For less strongly retained components, such as O_2, N_2, H_2, CH_4, ethane and propane, maximum deviation of retention time and peak area measurements was of the order of 3%. Reliable predictions were possible for single or binary systems except where there were strongly adsorbed components, such as CO_2 or H_2S on zeolites.

Experiments have been carried out using a thermo gravimetric analyser to measure the adsorption and desorption of carbon dioxide from 5A zeolite in powder and pellet form. It was necessary to remove the last traces of water from the system to prevent preferential adsorption. Adsorption and desorption cycles (Figure 4) were carried out under near-isothermal conditions by switching the purge gas from nitrogen to CO_2/N_2 for the adsorption step and monitoring the change in weight with time of the adsorbent as sorption equilibrium was reached with the purge gas atmosphere. Desorption was a slower process than adsorption and for pellets the desorption rate constant was highly temperature dependent, typical of a mass transfer rate controlled by an intrapellet diffusive process. Lower desorption rate constants were observed for powder samples and there was a less marked temperature dependence.

Ternary adsorption equilibria for the mixture $CO_2/C_2H_4/C_2H_6$ in dry air over 5A zeolite were determined. The experimental results were compared with predictions from the most common theories, with only one based on the Generalised Statistical Thermodynamic Approach (GSTM) able to predict the pressure dependencies of the binary mixtures from single component equilibria alone (Figure 5).

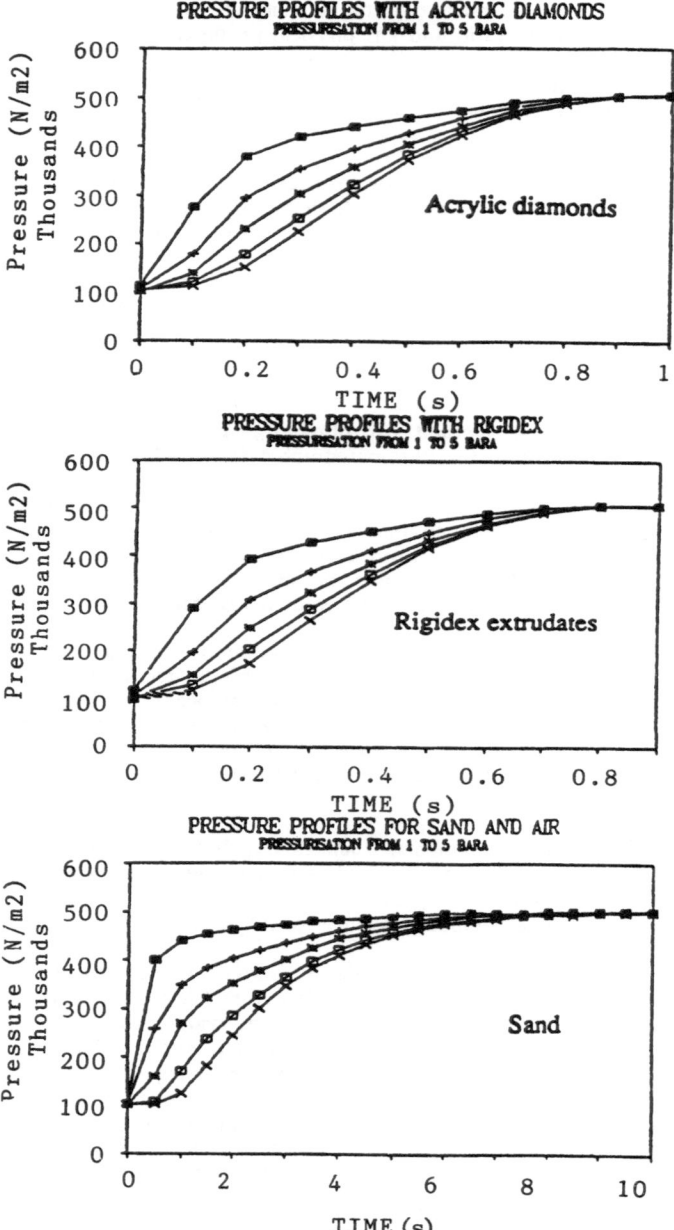

Figure 2. Pressure Profiles During Pressurisation with Air from 1 to 5 Bara for Acrylic Diamonds, Rigidex Extrudates and Sand.

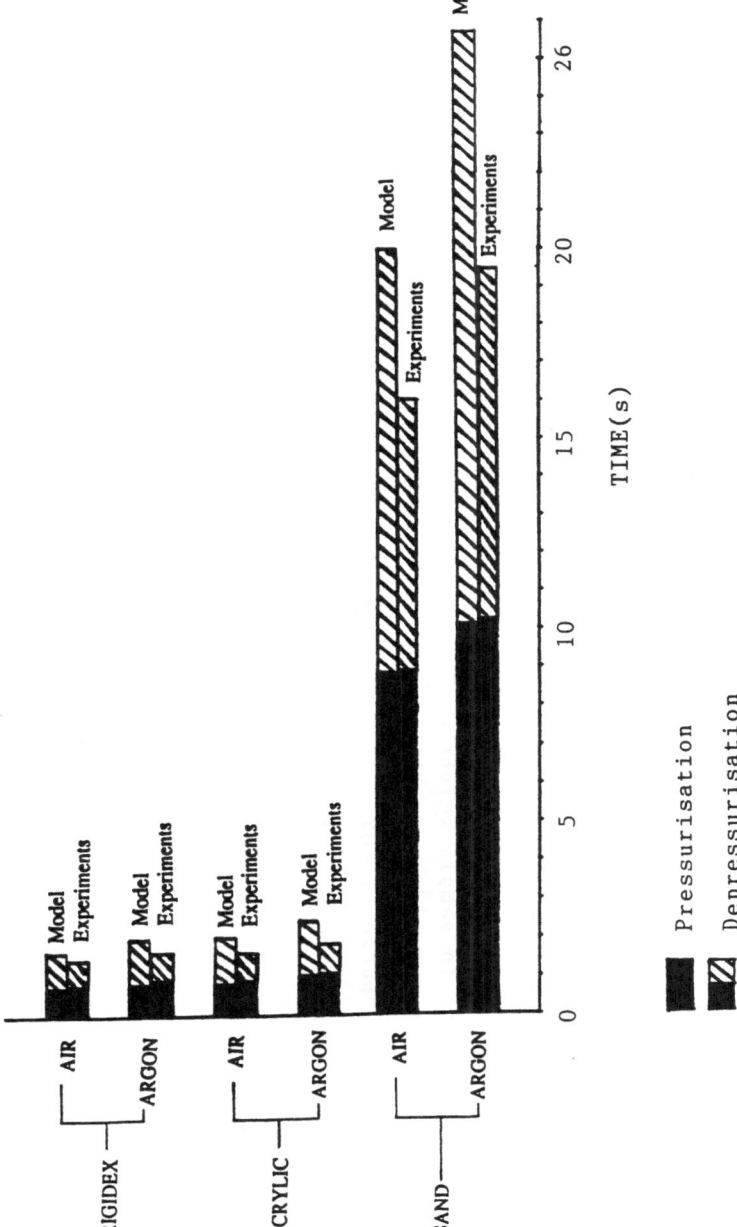

Figure 3. Bed Pressurisation and Depressurisation Times for Different Systems of Fluid and Packing.

106
105
104
103
Weight (%)
101
100
99
Temperature (C°)
980
1000
1020
1040
1060
1080

Figure 4. The change in sample weight and temperature during the adsorption and desorption by N_2 purge of 10% CO_2 in N_2 on RP £A zeolite powder at 100°C.

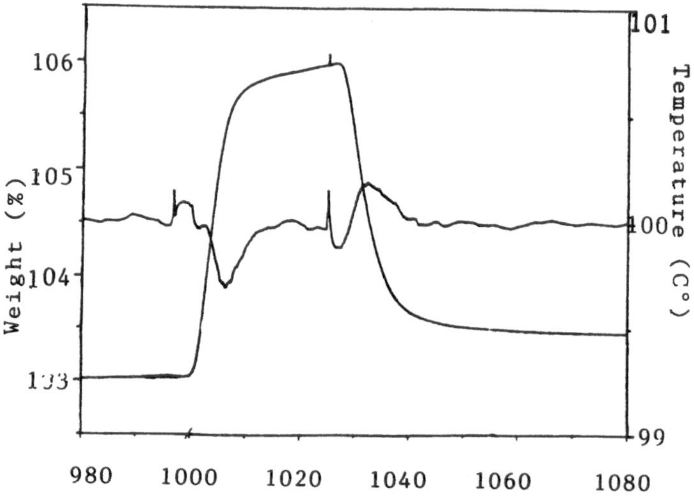

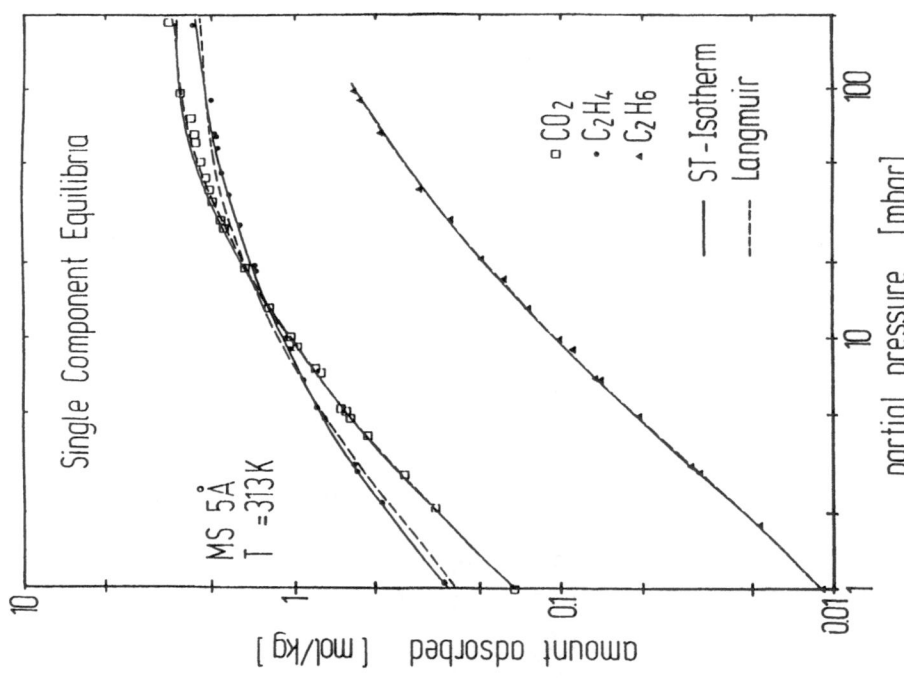

Single Component Equilibria
MS 5Å
T =313K
amount adsorbed [mol/kg]
10
1
0.1
0.01
□ CO2
• C2H4
▴ C2H6
ST-Isotherm
Langmuir
1
10
100
partial pressure [mbar]

Figure 5. Fit of the Langmuir Isotherm and the Statistical Thermodynamic Isotherm to the experimental data at T = 313K.

A multicomponent adiabatic macropore diffusion model has been developed to predict the breakthrough of multi-component mixtures in an isothermal or adiabatic adsorbent fixed bed. This model has no restrictions on validity, except for mixtures where the kinetics of one or more components might be governed by the resistance to mass transport the micropores. For mixtures which exhibit a high release of heat during adsorption the heat conduction within the bed might become the main limitation of the adsorption process.

Adsorption isotherms for CO_2, ethylene and propane on ZSM-5 zeolite with two different ratio of SiO_2/Al_2O_3 (29 and 56) have been obtained at temperatures of 281, 293 and 308K and at pressures up to 100kPa. The experimental data was fitted to the theoretical equations of Langmuir, BET, Prausnitz, Toth and Unilan. Deviations were as low as 4% with some models, but as high as 15% for those of Langmuir and BET. A decrease in SiO_2/Al_2O_3 ratio improved the adsorption of polarisable molecules, like CO_2 and ethylene, but did not have much effect on the adsorption of ethane and propane.

From binary equilibrium data, it was deduced that there was a small decrease in selectivity with increasing pressure, as well as a strong displacement of ethane molecules by CO_2 and ethylene (Figure 6) in the respective binary mixtures. At elevated pressure, the Real Adsorbed Solution model gave a better fit with experimental results than the Ideal Adsorbed Solution model. The fit was not so good as at lower pressure.

Heat and Mass Transfer

In gravimetric experiments, the diffusion coefficients are traditionally identified using simple isothermal models that consider only mass diffusion. Large discrepancies between predicted values and those from other techniques (particularly nmr) have revealed possible parasitic effects. To avoid erroneous determination of diffusion coefficients a complete model has been developed for studying the adsorption kinetics in bidispersed adsorbent pellets. This model includes mass diffusions and interface transfers at the pellet and crystallite scale, thermal conduction through pellets, heat exchange at the pellet surface and the effect of crystallite size distribution.

The model describing non-isothermal adsorption of a multi-component system, based on the Darcy or Ergun law, has been solved by an adaptive finite difference method. This numerical method presents much less numerical diffusion and oscillations than currently used centred and first order upstream schemes.

This is a complex mathematical method, but predictions give good agreement with pilot plant results under constant total pressure conditions. With a knowledge of crystal size distribution, macropore size distribution and heat of adsorption, a diffusion coefficient for propane in 5A zeolite will be determined.

Pilot Plant Evaluations

Existing adsorption models are generally based on adsorption isotherms and kinetic data for single components at mild conditions. It is necessary to generate data in conditions more representative of industrial operations and use this to validate the models developed in this project. Such conditions include high temperature and high pressure in the presence of hydrocarbon mixtures containing several adsorbable species, for example in octane number improvement of light gasoline.

A pilot-plant bed of adsorbent is operated alternatively in adsorption and desorption. The concentration profiles of the different constituents of a feed at the bed outlet, during adsorption and desorption, were determined. From the very sharp profiles

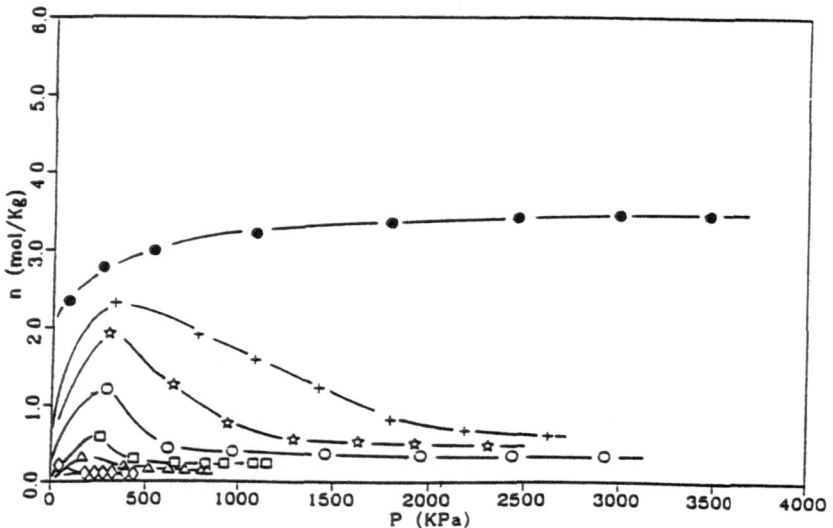

Figure 6. Individual adsorption isotherms of ethane from the binary mixture ethane(1) – CO_2(2) on 5A zeolite at 293K. Initial volumetric ratio are: V_1/V_2 = + 1/8, ✿ 1/4, ○ 1/2, □ 1/1, △ 2/1, ◊ 4/1.

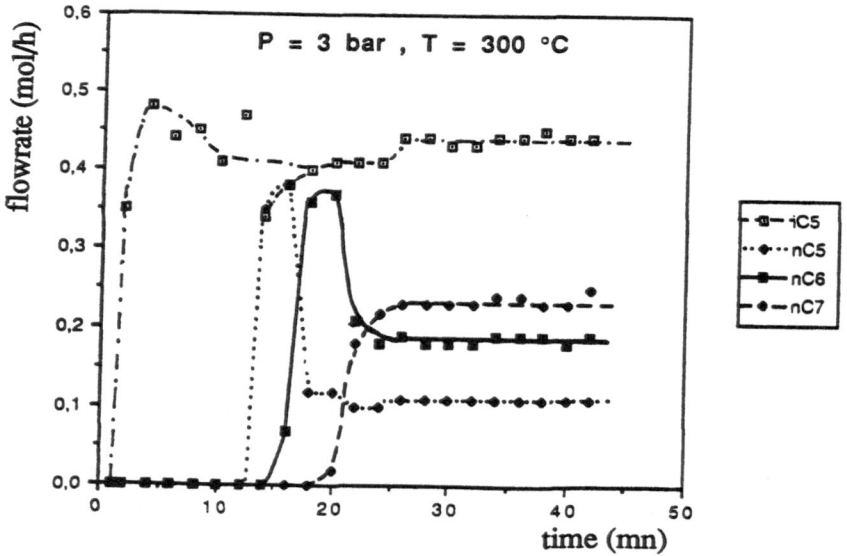

Figure 7. Adsorption of iC5-nC5-nC6-nC7

appropriate for modelling of multi-component equilibria. Understanding how this behaviour can be influenced is providing knowledge from which guidelines on adsorbent selection can be based.

Technical University of Munich is developing a generalised method of predicting ternary equilibrium data, and also kinetic data, from detailed studies of the $CO_2/C_2H_4/C_2H_6$ system on zeolite. This is a complex non-ideal mixture for which established models fail and is therefore a good system for analysis. This mixture is relevant to the use of adsorption for light hydrocarbon separations as an alternative to cryogenic distillation.

CNRS-LIMSI Orsay is developing a model to describe simultaneous heat and mass transfer for multi-component gas flowing in an adsorber. This model is being validated in collaboration with IFP and cross-checked with those from other tasks, particularly with respect to extracting thermo-Kinetic data from experimental measurements.

Rhone-Poulenc Recherches is providing adsorbents, characterising these and measuring equilibrium data using classical methods.

Process Design

Good process design methodology is based on a sound understanding of the fundamental scientific principles involved, but with an appreciation that detailed analysis on a micro-scale may not always be justified in the design of large process plants. These studies are intended to build on the basic data, but balancing this scientific knowledge with empirical relationships where appropriate.

The University of Bath is modelling the dynamics of pressure swing adsorption systems. Although PSA is established as a proven technology, much remains to be achieved in developing the process towards its optimum efficiency. A key parameter in optimising the process is the cycle time, and reduced cycle times can give rise to significant benefits in reducing plant size and operating costs. As the cycle time is reduced, the dynamic response of the system becomes more complex and current models are unable to take full account of this as cycle times become reduced. This task therefore aims to develop a rigorous model which can be used for the design of rapid cycle PSA plants and to study the controllability of the system.

Institut Francais du Petrole is using its facilities to validate models for hydrocarbon separations (normal/iso paraffins) and for fuel gas fractionation. These models will then be available for incorporation into in-house process design and simulation tools for testing industrial applications.

AEA Technology is consolidating the results from the studies described above into an outline selection and design procedure and is co-ordinating the project.

RESULTS

The 5A zeolite used in most of these studies was supplied by Rhone-Poulenc in the form of extrudates of mean pellet diameter 1.6mm, a belt density of 0.72g cm^{-3} and a pellet density of 1.23g cm^{-3}. The pellet composition is approximately 80% zeolite and 20% binder.

of the breakthrough curves in Figure 7 it can be concluded that diffusion is not limiting at 300°C. The presence of a roll-up phenomenon is characteristic of the displacement of the lightest normal paraffins by the heaviest. The adsorption capacity of the 5A sieve for each species in competition can be calculated from this data. Simulation of multicomponent breakthrough curves has also been carried out, using the commercial package, ADSYM. The results were very good for nC_6, fair for iC_5, but very poor for C_7s.

FURTHER WORK

The current JOULE project is continuing for a further year, comprising the same two lines of research: improving methods for obtaining reliable data and improving methods for process design. The extension will allow the partners to extend and enhance the programme of work and will therefore present greater opportunities for European Community companies to adopt or improve adsorption processes as part of a chemical process, or for environmental protection.

Further to this work, a JOULE II project has been proposed. This addresses two issues highlighted by European industry as of vital importance, namely coking and ageing of adsorbents and the use of thermal systems for regeneration and product recovery. These fields are seen as central to the energy efficiency of adsorption processes and a better understanding of these effects will enable a much wider range of adsorptive separations to be considered by industry. Process designers and operators will also gain a greater understanding of the factors affecting deterioration in performance and will therefore be better placed to minimise these.

Coking is a particular problem with zeolites when they are used to sweeten natural gas or separate hyrocarbons particularly where thermal regeneration is employed. Coke formation results in blocking of pores and adsorption sites, leading to a reduction in capacity or dramatically altered kinetics. Water also affects the capacity and life of adsorbents and any expansion and contraction of the intraporous water can generate mechanical stresses in the particles, leading to cracking or disintegration. The project aims to gain a better understanding of the mechanisms involved in coking and ageing, with a view to minimising the effects and providing better predictive models.

Thermal desorption methods are used when the adsorbed components have medium or low volatility and pressure swing is not sufficient to drive the desorption process. These TSA cycles have not been fully investigated, both with respect to the operation of classical operating modes and with respect to the design of advances cycles, therefore data will be generated for particular modes and models will be devised and validated. This will aid the accomplishment of efficient design and operation of processes involving thermal swing adsorption.

CONCLUSIONS

The mechanisms of adsorption of compounds onto solid surfaces has been extensively researched, particularly in the United States and Japan. More recent work in Europe, under the JOULE programme, has aimed at enhancing this understanding through the development of better methods to determine key physical and thermodynamic data such as multi-component equilibrium and kinetic data, diffusion coefficients and improved modelling of processes which can be applied in European industry. This project has fostered a good collaboration between key European research groups and industrial users and therefore should result in the wider and better use of adsorption processes.

Industry uses more energy separating and purifying products than in almost any other process operation. This is particularly so when separating products to high purity from very dilute mixtures, or where two products are present with similar physical properties but widely different market values. It is important that these separations are performed as efficiently as possible to maximise yield and selectivity at minimum energy cost.

Adsorption as a process for removing hydrocarbons from gas streams is obviously more environmentally benign than uncontrolled release and less energy intensive than incineration of waste gases.

ACKNOWLEDGEMENTS

We thank the CEC JOULE programme for part funding this project (Contract number JOUE-0052-C) and acknowledge the support of L'Air Liquide, BP Research, Domnick Hunter Filters and Shell-KSLA and the work of the partners: J. Granger, P. Remy, H. Kabir and D. Tondeur of CNRS-ENSIC, Nancy; A.E. Rodrigues, University of Porto; S. Swanton and J. Di Sanza, AEA Technology; G. Calleja, University of Madrid; P. Schweighart and A. Mersmann, Technical University of Munich; L.M. Sun, Ph. Grenier and F. Meunier, CNRS-LIMSI, Orsay; E. Garcin, Rhone-Poulenc, Aubervilliers; W.N. Ng, W.J. Thomas and B. Crittenden, University of Bath and A. Deschamps, C.L. Dezael and S. Julian, Institut Francais du Petrole.

REFERENCES

1. Reay, D. and Pilavachi, P.A., Needs for Strategic R&D in support of improved energy efficiency in the process industries, CEC Contract No. EN3E-0158-UK.

ADVANCED FLOW MODELLING FOR INDUSTRIAL APPLICATIONS - CFD MODELS OF ADSORBERS

G. Gouvalias[*], N. C. Markatos[*], J. Panagopoulos[*], M.J. Tierney[**], S.Huberson[***], G.Zhong [***]
* Department of Chemical Engineering
National Technical University of Athens
9, Iroon Polytechniou str., 157 73 Zografou, Athens, Greece
** Computational Fluid Dynamics Services
AEA Industrial Technology, B8, Harwell Laboratory
Oxfordshire OX11 ORA, UK
*** L.I.M.S.I. - C.N.R.S, BP 133
91403 Orsay cedex - France

ABSTRACT

The present study is concerned with an area of process engineering which has been identified as a priority area with respect to energy saving, namely the adsorption/regeneration cycles in packed - bed reactors. The objective of this work is to incorporate recognized fundamental description of adsorption into flow patterns measured and predicted by Computational Fluid Dynamics (CFD) techniques; the latter predictions being obtained by solving the full two-dimensional Navier - Stokes equations, energy and species conservation equations, for a fixed bed adsorber in steady-state and transient modes.

INTRODUCTION

The results of this study were made possible by the cooperation of the N.T.U of Athens (Computational Fluid Dynamics Unit), CFDS - AEA Industrial Technology (Harwell Laboratory) and CNRS - LIMSI institutes, in conjunction with the Advanced Flow Modelling, JOULE I research programme, Group II.

This paper describes the work performed on assessing the utility of pressent Computational Fluid Dynamics (CFD) approaches to predict the performance of packed-bed adsorbers, for improved energy efficiency.

Adsorber-regenerators are used extensively in the process industries for gas cleaning and separation. In the field of gaseous separations, adsorption is used to dehumidify air and other gases, to remove objectionable odours and impurities from industrial gases such as CO_2, to recover valuable solvent vapours from dilute mixtures with air and other gases, and to fractionate mixtures of hydrocarbon gases containing substances such as methane, ethylene, ethane, propylene and propane.

Adsorption of gases should not be considered as an energy-efficient process "per se", in some absolute sense. It certainly may be viewed as a low energy alternative to some other separation processes, such as cryogenic operations or azeotropic distillation, for certain applications. Of course, it is also a means of treating or recuperating energy vectors (such as hydrogen or hydrocarbons) in various mixtures, to improve chemical - reactor operation by adjusting the gaseous composition, to abate pollution, and in general to save energy by process improvement.

From the point of view of thermodynamic analysis, gas adsorption/ regeneration cycles imply a separation (of an impurity from a main stream, say), and therefore some minimal work of separation, thus implying in turn the destruction of some minimal amount of free energy. Owing to irreversible processes, the free energy destroyed is actually much larger than that thermodynamics minimum. However, adsorption/desorption processes sometimes operate relatively close to equilibrium, and irreversible dissipation is then smaller than in many other proceses. A source of free energy is thus needed to drive the separation and compensate for the irreversibilities, and this source strongly depends on the process and its context. It may be, for example, the enthalpy of a hot stream used for regeneration cycle, or the enthalpy of the pressurized treated stream itself; it may also be the free energy of a pure stream used as desorbent, or a combination of those elements.

CFD methods and tools can be used for the study of the above processes, [1,2]. This work is the first step in this direction. A lot of research remains to be done in order to improve and validate mathematical models of such physical processes as particles-wall interaction, porosity, etc.

Heat and mass transfer and flow fields were obtained with the model developed, by using a prescribed constant turbulence viscosity inside the fixed bed.

DESCRIPTION OF EXPERIMENTS

For the main objectives of this work, column dynamic response, such as gas pressure and temperature, needs to be measured. For this reason, an experimental system illustrated by Fig.1., has been built: a column filled with adsorbent grains is connected to a vapour generator at one end and is closed at the other end. A single - component vapour may enter or leave the column through the open end.

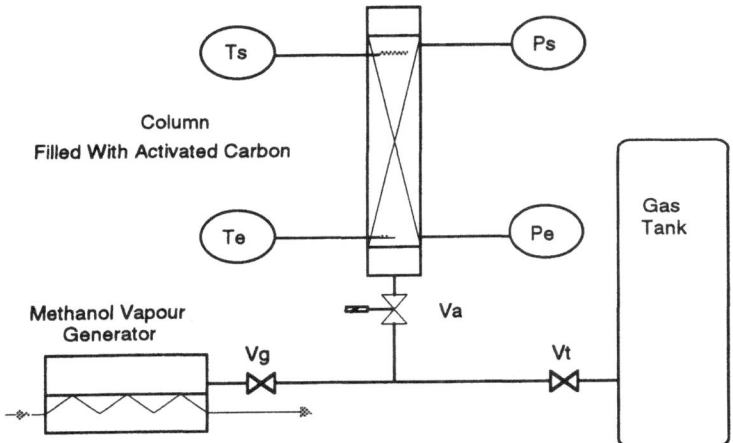

Figure 1. Dead-end column filled with adsorbent connected either to a vapour generator or to a gas tank. T : temperature sensors, P : Pressure sensors, V : Valves.

This experimental set up was used for different kinds of vapour and adsorbent grains. The purpose is to use the collected data to check the accuracy of the mathematical model of adsorbtion, and eventually to build up an identification process based on that model. This identification step is expected to provide the numerical values of the various parameters involved in the numerical model.

BASIC ASPECTS OF ADSORPTION

In this section observations that have been made in the laboratory, and the results of fundamental calculations are summarised.

Mechanisms of Adsorption.

Adsorption is a process whereby gas molecules are retained within a solid. Important material properties include surface shape, polarity, and specific surface area [3,4].

Adsorbents are generally porous pellets of ≈1mm diameter, themselves consisting of porous particles of ≈100 microns diameter. There is therefore an intricate system of micropores among the particles, and micropores within them. In exceptional circumstances the micropores are sufficiently small, perhaps 0.5 nm in diameter, acting as 'molecular sieves' and selectively retaining molecules.

More often, however, species are trapped by a strong electrostatic bonding. Moving from a strongly negative to a strongly positive charge, possible adsorbents include activated carbon, silica gel, alumina, and zeolites. The energy released during bonding is of the same order as the heat of vaporisation of the gas species.

Equilibrium Considerations

When adsorbents are in contact with a gas for sufficient time, equilibrium occurs between solid and gaseous phase concentrations. These quantities are generally measured in terms of Kmol of adsorbents per Kg in the bed, and partial pressure. Various correlations are available, of which perhaps the best known are the five classical Brunhauer classifications [5], that model:

- Strong adsorbent -adsorbate interactions (classification I);
- weak interactions (classification III);
- wide distributions in pore size (classifications II,IV,V).

The first classification, originally attributed to Langmuir in 1908 [6], is certainly the most widely reported, and is incorporated in the present model. It can be derived from simple theoretical considerations, and takes the form:

$$q = \frac{a \cdot b \cdot p}{1 + b \cdot p} \tag{1}$$

where, **q** is the bed loading (Kgmol/Kg), **a** is the asymptotic maximum solid-phase concentration, **b** is the equilibrium constant and **p** is the partial pressure.

The equilibrium is strongly dependent on temperature, and consequently the coefficient 'a' is normally fitted to an Arrhenius law.

When several species are adsorbed there will be competition for the available sites on the material surface, and pure component data will not necessarily be appropriate. Several procedures are available for multi-component equilibria, although in general most of these still require extensive experimental validation [5]. Options include the extended Langmuir method, and the ideal adsorbed solution theorem, or IAST [7]. IAST is regarded as more accurate, but it is iterative and therefore demanding on computer time.

MATHEMATICAL FORMULATION

The independent variables of the problem are the two components of a cylindrical polar coordinate system [z,r] and the time, t.

The dependent variables (time - averaged values) are the two velocity components w, v in the z and r directions, respectively, the pressure p, the enthalpy H and the concentration C.

The governing differential equations

The time - dependent equations for continuity, velocity components, temperature and chemical species (where density - fluctuation correlations are ignored) can be expressed in the following general form :

$$\frac{\partial}{\partial t}(\rho\Phi) + \text{div}\{(\rho v\Phi - \Gamma_\phi \ \text{grad}\Phi)\} = S_\phi \tag{2}$$

where ρ, v, Γ_ϕ and S_ϕ are density, velocity vector, " effective exchange coefficient of Φ " and source rate per unit volume, respectively. The source rate and effective exchange coefficient for the variables considered here are already well documented in the literature [1,2,8] and are not repeated here.

Auxiliary equations

The effective *exchange* coefficients are assumed constant, using a value for turbulence viscosity equal to approximately a thousand times the laminar viscosity. Fluid-to-wall friction is calculated using a logarithmic wall function [8], momentum loss at the perforated screen is calculated using a frictional drag formula and momentum loss in the packed bed is calculated using Ergun, Handley and Hegg or Hicks equation depending on the Reynolds number's magnitude, (by CMRS/LIMSI experiments).

Mass Transfer

Gas - phase concentration : Adsorption from a gas flowing through a packed- bed of solids is described by:

$$\rho_g \frac{\partial c_g}{\partial t} + \nabla \overline{v} c_g = \nabla D_{eff} \nabla c_g + k_{ad}(c_s - c_g)\rho_g \frac{1-\varepsilon}{\varepsilon} \tag{3}$$

where :

c_g $\Rightarrow$ adsorbate concentration in the fluid phase (Kg/Kg fluid)

c_s $\Rightarrow$ adsorbate concentration in the solid phase at the fluid - solid interface in equilibrium with c_g . This is obtained from the equilibrium relationship.

ρ_g $\Rightarrow$ density of the gas

ε $\Rightarrow$ void fraction in the bed.

v $\Rightarrow$ fluid velocity in the bed based on total cross - sectional area

D_{eff} $\Rightarrow$ diffusivity of gas phase.

k_{ad} $\Rightarrow$ mass transfer coefficient, given by a Colburn factor formula.

Bed-Phase concentration: The case is considered where adsorption rate is governed by the solid-phase mass-transfer processes. The rate of change of concentration within the bed is :

$$\frac{\partial c_p}{\partial t} = k_{ad}^p (c_s - c_p) \tag{4}$$

where c_p is the concentration of solute at any point inside the particle;

The coefficient k_{ad}^p is given by a formula recommended by de-Xin [9] and equilibrium adsorption is represented by the two - parameter Langmuir isotherm.

Heat transfer between gas and bed. : The convective heat transfer between the gas and the bed is given by:

$$S_{Tgas} = hA(T_s - T_g) \tag{5}$$

where A is the molecular bead surface area per unit bed volume and the local heat transfer coefficient is given by a J_H - Colburn factor formula

Heat transfer within the particle. : An unsteady-state, energy balance on the solid particle phase contained in the packed-bed leads to :

$$\frac{\partial T_s}{\partial t} = \underbrace{\frac{hA_p}{\rho_p c_p V_p}(T_g - T_s)}_{\textit{sensible heat}} + \underbrace{\frac{1}{\rho_p c_p} k_{ad}^p \lambda_{ad} (c_s - c_p)}_{\textit{latent heat}} \tag{6}$$

where , $V_p = \pi D_p^3$ and $A_p = \pi D_p^2$, λ_{ad} = heat of adsorption.

The finite - domain equations formulation.
Finite - domain equations (FDE) are derived by the integration of the above differential equations over finite control volumes, that taken together fully cover the entire domain of interest. These control volumes are called "cells" or "sub-domains". Within each cell is a "typical point" (called a 'grid node'), say P, for which the fluid property values, Φs, are regarded as representative of the whole cell. It is surrounded by neighbouring nodes which we shall denote by N(north), S(South), E(East), W(West), H(High), L(low) and T (grid node at earlier time). Cells and nodes for velocity components are 'stagerred' relative to those for all other variables. This practice is convenctional. The cells in the situation considered are strictly Cartesian, but in general they can be "Topologically Cartesian" (polar cylindrical or curvilinear) always having six sides and eight corners in the three-dimensional case, unlike the cells used by other techniques like the 'finite element' technique where they can be triangular, hexagonal etc.

The 'integration' involved is different to the usual Taylor series expansion used in the classical finite difference technique, and results in different coefficients of the algebraic equations that are finally obtained, in the general case. This integration allows for injection of physical considerations into the formal mathematical manipulations (e.g. the conservation principle that leads to the differential equations at the first place, is satisfied exactly) and permits a fully conservative formulation. Integration entails 'interpolation assumptions' about Φ values and values of Φ-gradients, prevailing at the cell boundaries, [10]. Integration leads to finite-domain equations (FDEs) having the following form, [10]:

$$\alpha_P \Phi_P = \alpha_N \Phi_N + \alpha_s \Phi_s + \alpha_E \Phi_E + \alpha_w \Phi_w + \alpha_H \Phi_H + \alpha_L \Phi_L + \alpha_T \Phi_T + b \tag{7}$$

where α_P, α_N, etc. are coefficients representing the influence (diffusion and convection) of the neighbouring cells (N,S,E,W,H,L) and time (T) to the balance of Φ_P; b is a representation of the source appropriate to Φ for the cell. Partial differential equations must satisfy 'boundary conditions' of the type:

$$(\Phi, \text{grad } \Phi) = 0 \qquad (8)$$

at specified points, lines, areas or volumes. These points, lines, etc. need not be at boundaries, but can be within the flow domain, when such additional information is given.

For a boundary cell the boundary condition is nothing more than a replacement of the unknown Φ_n value at the corresponding neighbouring cell (which is now missing) by the known value. Therefore, the treatment of boundary cells can be identical to that of any other cell; and the known boundary relations can be expressed again by integration over the cells containing the boundaries. In this manner the boundary (and internal) conditions simply make contributions to the b and ap of the FDEs (eq. 7)

Finally the set of the equations to be solved is closed by auxilliary relations, that refer in general to thermodynamics and transport expressions.

METHOD OF SOLUTION

The sets of FDEs for the various Φs are solved in an iterative manner by using CFD Computational Codes (NTUA-CFD code and the CFDS-ASTEC code [13]). The solution technique used to solve the above equations employs the SIMPLEST, [11], algorithm which is an improved version of the well-known SIMPLE algorithm. Details have already been discussed elsewhere and are not repeated here, [12].

DETAILS OF CALCULATIONS

Geometrical data, selection of solution domain and coordinate system.

Case 1. Two dimensional (NTUA Code): A vertical vessel of 4.200 mm height and 1.1304 m^2 cross-sectional area was considered.

Detailed geometrical data are given in Fig.2. The packed-bed itself is made up of molecular sieve beads approximately spherical in shape, and with a size distribution of between 0.0028 m and 0.0039 m in diameter. During the adsorption stage air enters the vessel through the lower inlet and leaves via the upper pipe. Baffles are indicated in Fig. 2 and, in addition, there are perforated screens at the flow distributor and below the packed-bed.

A two-dimensional cylidrical-polar grid was chosen to represent the vessel geometry, with its axis aligned with the vertical axis of the cylindrical vessel containing the packed bed. The extent of the solution domain and the choice of coordinate system are first described, together with details of the grid prescribed to partition the solution domain into subregions (cells) for integration purposes.

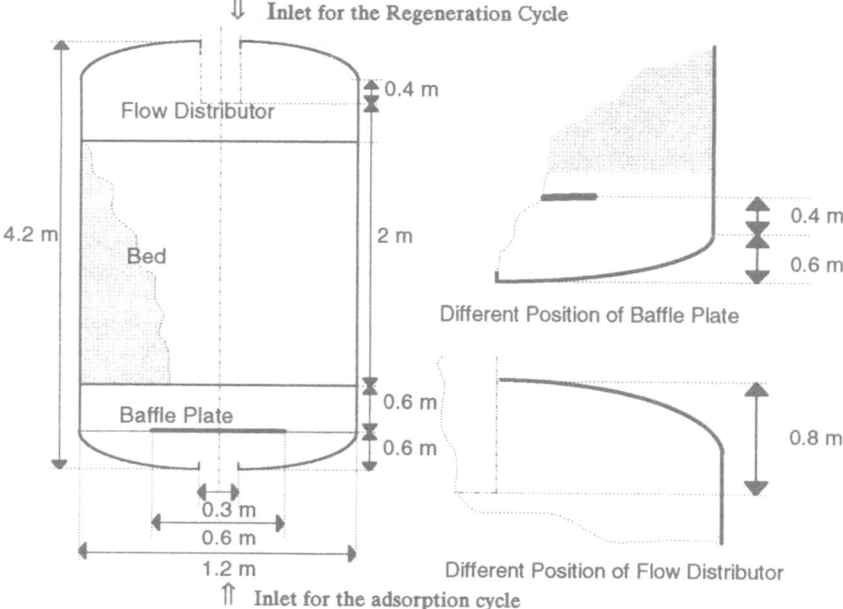

Figure 2. Geometry of the Adsorber / Regenerator.

Case 2. One dimensional system, (ASTEC Code): De Xin's experiments and predictions have been based on the passage of methane and carbon dioxide through a narrow, 45 cm long tube. They are very much on the laboratory bench scale.

The mesh is simply a length of 20 elements, of which the first and the last represent regions of void space. The velocity into the inlet header is low - only 0.0662 m/s.

Case 3. Two dimensional system, (ASTEC Code): These predictions refer to pilot studies on a shallow bed unit. The mesh is cartesian, and constructed from 1198 elements. (Fig 3.). The packed bed is 1m in height and 2m in diameter. An inlet velocity of 1m/s is assumed. The gas physical properties are those of air at 20 °C .

Grid specification, (NTUA Code)

Three cases were studied:

- **Case 1(a).** In the steady-state (adsorption cycle-hydrodynamic field) runs, half of the cross-sectional area was considered, divided into 40 cells. In the z-direction about 60 cells were used with the exact number depending on the geometry employed dyring the run.
- **Case 2(b).** In the transient runs (adsorption cycle-heat and mass transfer, regeneration cycle), half of the cross-sectional area was considered, divided into 40 cells, and 60 cells in the z-direction were used.
- **Case 3(c).** A finer grid was used: one eighth of the cross-sectional (y-direction) area was considered,divided into 50 cells. The axial distance z was divided into 70 cells (for steady-state and transient modes).

For the transient and steady-state runs 20 cells in the z-direction were located in the packed bed.

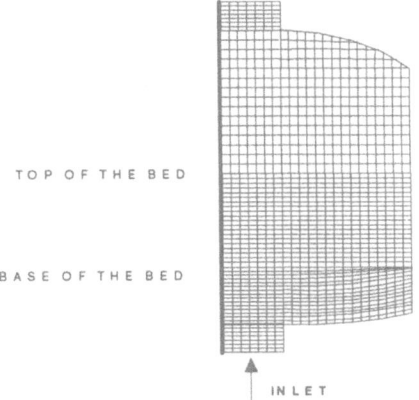

Figure 3. Mesh of an Adsorber (ASTEC Code)

Inlet conditions

For all runs there was one inlet boundary. This was located at the base of the vessel for the adsorption cycle (Fig.2 and Fig.3), and at the top for the regeneration cycle (Fig.2). To start the solution procedure for all runs it is necessary to provide initial values for the field values of the dependent variables. The initial values during the adsorption cycle prescribed in this instance were: v=0,w=2.88 m/s, P=6.55x10^5 Pa, T=298.15 K - Case I, and v=0,w=0.0662 m/s ,Inlet partial pressure = 5000 Pa, T=298.15 - Cases II and III. The basic inlet conditions during the regeneration cycle are : Air flow = 9000 Kg/h, P=1.2 x 10^5 Pa, and T= 596 K - Case I.

Outlet conditions

Fixed uniform exit pressure. As the fluid is assumed incompressible, this pressure is set equal to zero and the computed pressures are relative to this pressure.

RESULTS AND DISCUSSION

Due to limitations of space only a few results of this work will be shown. Fig 4. shows velocity vectors for the steady - state adsorption calculation. Fig. 5 shows pressure profile for the steady - state adsorption operation. Fig. 6 shows pressure profiles for the regenerations cycle and Fig. 7 shows bed exit temperature for the regeneration cycle.

From the results of this work it may be concluded that :

➡ **Flow distribution.** The predictions have shown that the high velocity of the inlet stream drives large scale recirculation zones in front of the packed bed. This effect was predicted for both regeneration and adsorption cycles. Baffle position and flow distributor position was shown to have a significant influence on the flow distribution and most importantly on the velocity of air impinging on the bed surface. For the regeneration phase the 600mm flow distributor height was shown to give a significantly more uniform flow distribution than the 800 mm height.

Velocity distribution in the bed itself was shown to be uniform in the case of the adsorption cycle but to be influenced by the effect of heat transfer in the regeneration cycle. Axial mass flux through the bed was shown to be practically uniform in each case. Pressure drop through the bed was linear for the adsorption cycle.

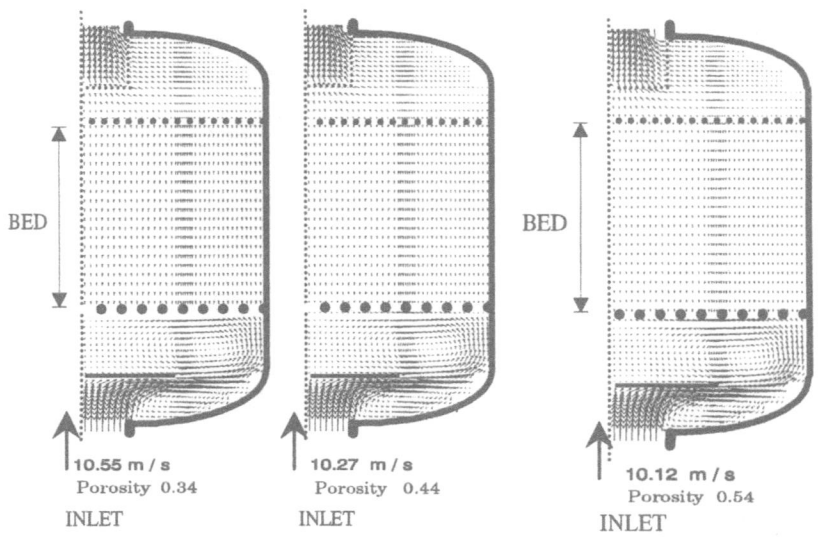

Figure 4. *Velocity Vectors For The Steady - State Adsorption Calculation Baffle Height 600 mm and Porosity 0.34, 0.44, 0.54*

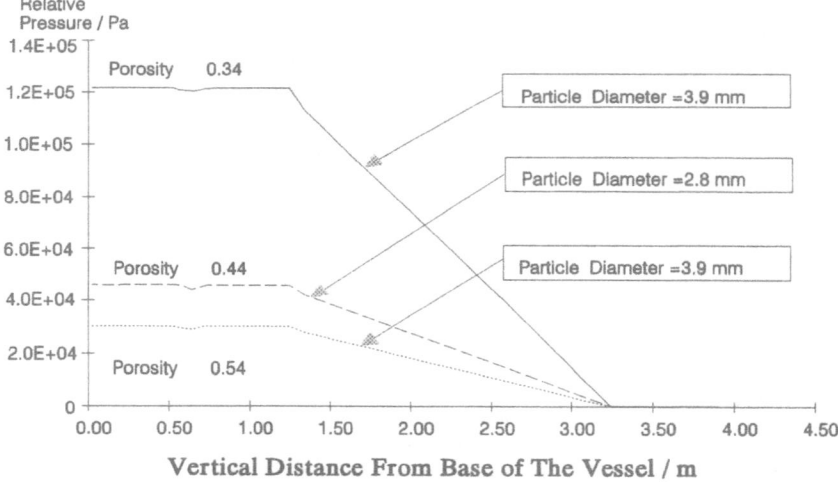

Figure 5. *Pressure Profiles For The Steady - State Adsorption Prediction Showing The Effect of Different Porosity (0.34, 0.44, 0.54)*

PRESSURE PROFILES THROUGH THE VESSEL
POROSITY 0.44, DISTRIBUTOR HEIGHT 800 mm

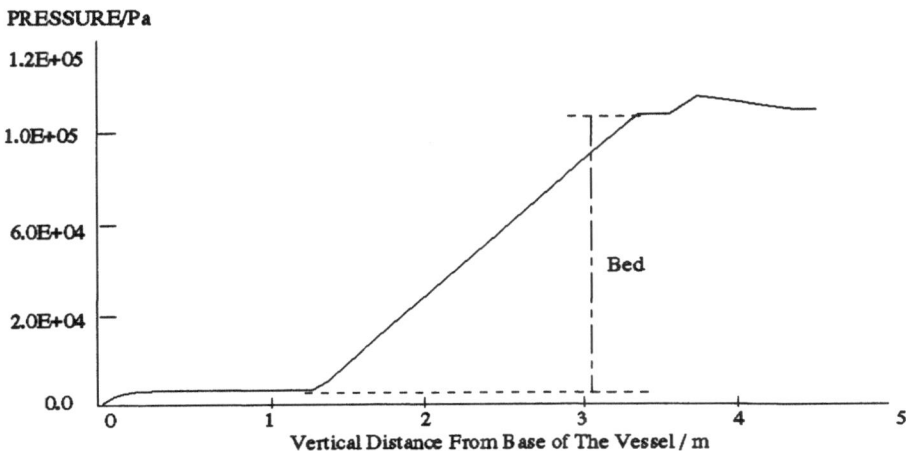

Figure 6. Pressure Profiles For The Regeneration Cycle Predictions Showing Change With Time.

AXIAL TEMPERATURE PROFILES DURING REGENERATION
POROSITY 0.44, DISTRIBUTOR HEIGHT 800 mm.

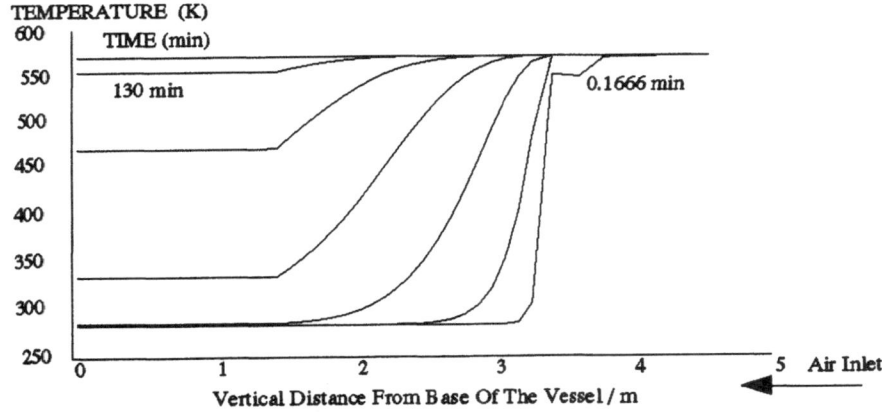

Figure 7.   Temperature Profiles Through The Vessel For The Regeneration Cycle Predictions.

Greater confidence could be placed on the predicted results by developing the model further. Parametric variations could examine the sensitivity of the results to changes in bead size and shape, properties of bed and air, variation in void fraction at the boundaries. etc.

➤➤ *Incorporation of the Adsorption Process into the Present Mathematical Model:*

1. One Dimensional system.: Figs. 8,9 and 10 show histories of both partial pressure and bed loading (the latter quantity has been scaled up by a factor of 10^6 to permit plotting.). The concentration histories at the bed exit are commonly referred to as " breakthrough curves". The lower plots represent temperature changes. The mass balance over this system has been checked, to ensure the consistency of units used. Table 1. presents the results - note that in this case the size of time step is 500s. Columns 2 to 4 list the levels of carbon dioxide that accumulate in the bed and the gas. These are calculated using straighforward arithmetic averages of the concentrations of all nodes within the bed. Columns 5 to 7 list cumulative net input terms.

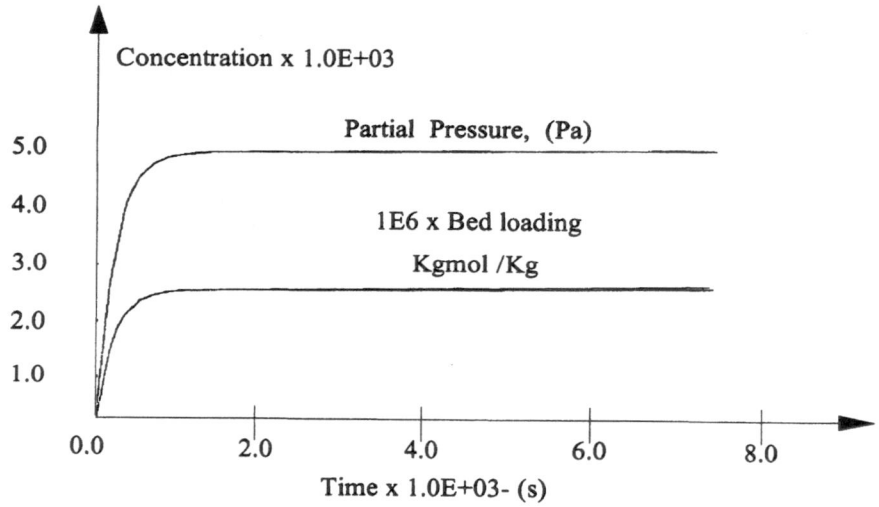

Figure 8. Concentration Histories Near the Inlet

TABLE 1
Mass balance

Time (s)	Bed (Kmol/m2)	Gas (Kmol/m2)	Total (kmol/m2)	Inlet (Kmol/m2)	Outlet (Kmol/m2)	Total (Kmol/m2)
0	0.0000	0.0000	0.0000	0.0000	-0.0000	0.0000
500	0.0709	0.0001	0.0710	0.0657	-0.0000	0.0657
1000	0.1406	0.0001	0.1407	0.1314	-0.0000	0.1311
1500	0.2090	0.0002	0.2092	0.1971	-0.0003	0.1948
2000	0.2700	0.0006	0.2706	0.2628	-0.0023	0.2539

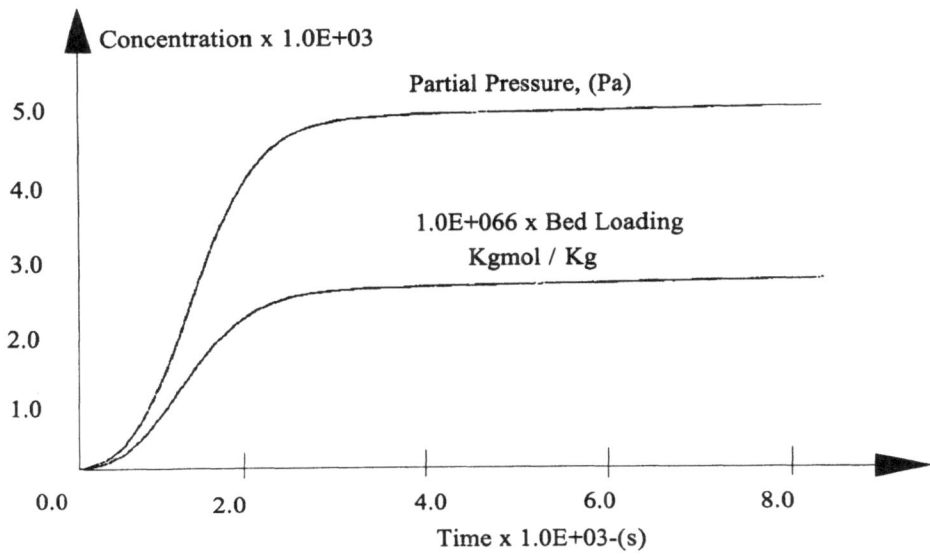

Figure 9. Concentration histories at the mid-point.

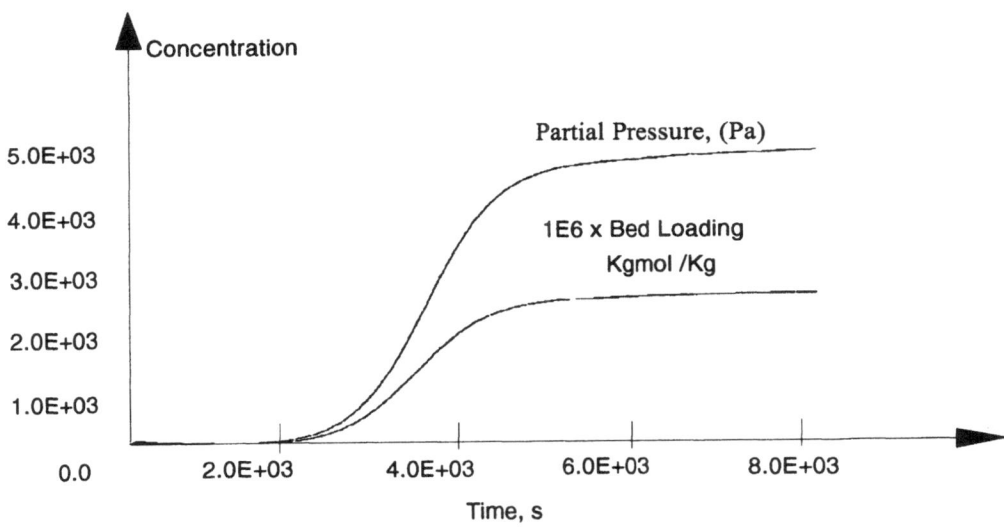

Figure 10. Breakthrough Curves.

There is a small discrepancy between the two columns of totals. Simple averaging does not account properly for nodes at the extreme ends of the bed, for which the control volumes are shared equally between the porous and non-porous regions. Given that, it is estimated that the observed discrepancy should be 5.6% on full saturation of the bed. In fact, it changes with time from 8.1% to 6.6%.

2. Two Dimensional System : Not surprisingly, there is no noticeable horizontal variation between breakthrough curves. A sample plot is provided, along with the temperature history (Figs 11,12). The time required for full saturation is far more a function of the zeolite properties than of fluid dynamics.

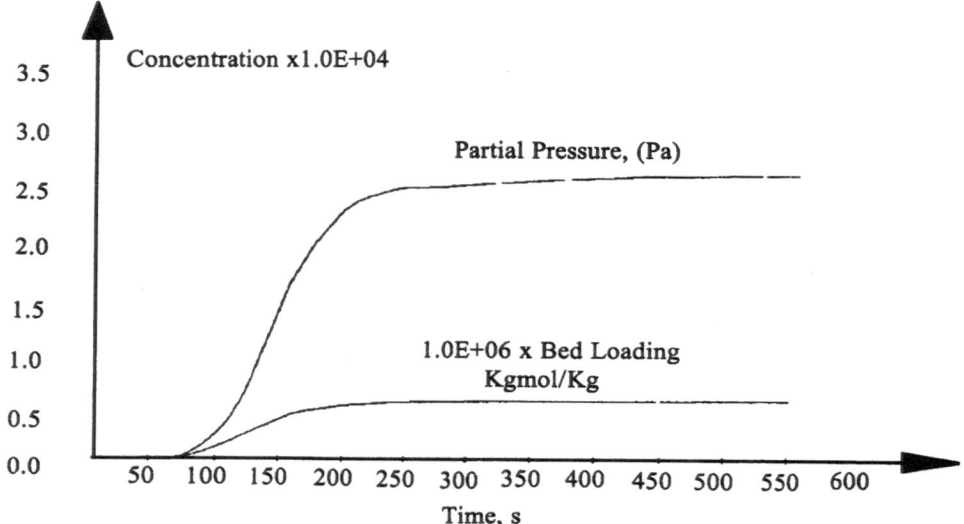

Figure 11. Breakthrough Curves.

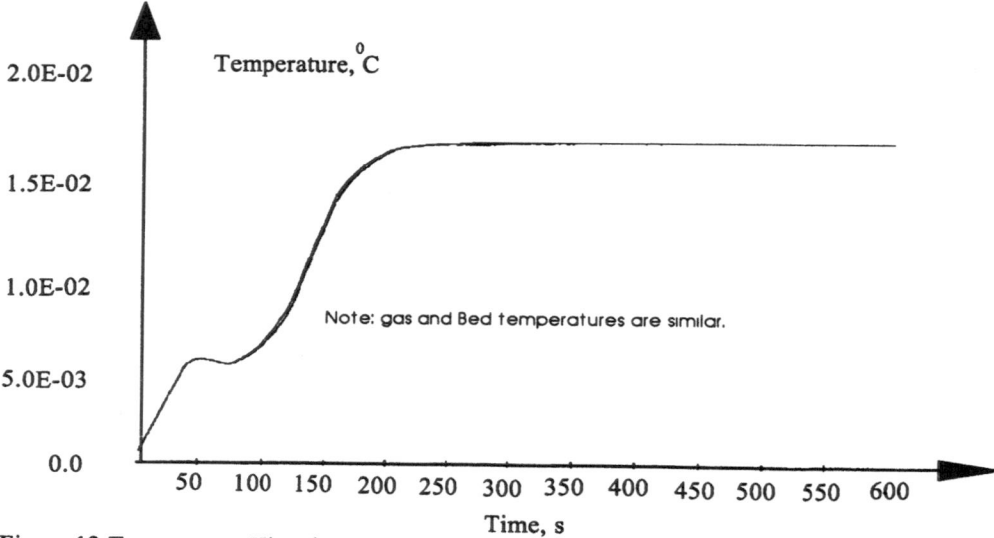

Figure 12 Temperature Histories.

ACKNOWLEGMENTS

The authors wish to thank CHAM Ltd. of London for permitting them to use their code, PHOENICS.
    They also thank the EEC, DGXII for providing partial funding for this work within the JOULE programme.

REFERENCES

1. Markatos, N.C and Spalding, D.B "Computer simulation of fluid flow and heat/mass transfer phenomena" The PHOENICS Code system, A lecture course, Thames Polytechnic, Scool of Mathematics., 1983

2. Markatos, N.C and Moult, A.   " The computation of steady and unsteady, turbulent, chemically - reacting flows in axisymmetrical domains". Trans. Inst. Chem. Engrs. 57, N_0=3, pp 156-162., 1979

3.Moreu, S. " Physical factors governing gas separation by adsorption processes", Proceedings of the International Symposium on Gas Separation Technology, Antwerp, Belgium, Semptember 1989.

4. Kast,W. " Adsorption from the gas phase - Fundamentals and Processes ", German Chemical Engineering, Vol. 5, 1981.

5. Yang, R.T., " Gas separation by adsorption processes", Published by Butterworths, Boston, USA, (Chapter 5), 1987.

6. Stone L.H, Siam J.Numer. Anal.,Vol 5, pp 530-558., 1958

7. Myers, A.L, and Prasnitz, J.M., A.I.Chemical Engineering, Volume 11, 1965, pp 121 onwards.

8. Launder, B.E and Spalding, D.B " The Numerical computation of turbulent flow ", Comp. Meths.Appl. Mech. Eng, pp 269-289., 1979

9. De Xin, Z. and Zhi-Jing, X. " Prediction of Breakthough Curves of Oxygen-Nitrogen Co-Adsorption Systems on Molecular Sieves", Gas separation and Purification. December 1988.

10.Markatos, N.C., " Computational fluid flow capabilities and software " Ironmaking and Steelmaking, Vol. 5, No 1, pp 32-38.

11.Serag - eldin, M.A and Spalding, D.B "A Computational procedure for three-dimensional recirculating flows inside can combustors". Numerical Methods in Heat Transfer., John Wiley and Sons Ltd, pp 445-466. 1981

12.Patankar, S.V and D.B.Spalding, D.B, " A calculation procedure for Heat, mass and momentum transfer in parabolic flows", Int J.Heat Mass Transfer,815, pp.1787-1806., 1972

13.Lansdate, R.D and Webster, R., The application of finite volume methods for modelling three-dimensional incompressible flow on an unstructured mesh. Proceedings of the 6th International Conference on Numerical Methods in Laminar and Turbulence Flow, University College of Swansed, U.K, 1989.

SOME ENERGETIC ASPECTS OF GAS PHASE ADSORPTION SYSTEMS

A. Mersmann, P. Schweighart, W. Sievers

Lehrstuhl B für Verfahrenstechnik, Technische Universität München,

Arcisstrasse 21, D-8000 München 2, Germany

ABSTRACT

A detailed model of an adsorption process using an LDF (Linear Driving Force) approach is employed to simulate a case system. Since in most literature studies dealing with the optimum operation of adsorption systems linear isotherms were assumed, this study employs a highly skewed isotherm. This type of isotherm is commonly observed for the adsorption on microporous solids. The desorption step which is the energy-intensive part of an adsorption process is optimized. The results lead to the recommendations to operate the desorption step of a temperature swing adsorption unit at high regeneration temperatures and at low purge gas velocities.

NOTATION

B	-	heat flux over column wall versus heat flux into the bed
C	-	sorbate-sorbate interaction parameter
c	-	dimensionless concentration in the gas void space
c_p	J/(kg K)	heat capacity at constant pressure
D_{ax}	m²/s	axial dispersion coefficient
E	J	exergy
d	m	diameter
H	J	energy input during desorption
H	m	height of adsorbent bed
K	1/Pa	Henry's Law adsorption constant
k_w	W/(m²K)	heat transfer coefficient over the column wall
L	-	dimensionless bed length
n	kmol/kg	equilibrium loading
p	Pa	pressure
q_0	J/kmol	enthalpy of adsorption for loading against zero

$\Re$	J/(kmol K)	general gas constant
R	J/(kg K)	specific gas constant
S,s	J/K, J/kgK	entropy / specific entropy
T	K	absolute temperature
t	s	time
V	m³	Volume
W	-	ratio of $(c_p \, \rho)_s$ versus $(c_p \, \rho)_g$
w_g	m/s	superficial gas velocity
β	m/s	mass transfer coefficient
ϵ	-	gas void fraction in fixed bed
λ	W/mK	conductivity of heat
μ_p	-	tortuosity factor
ρ	kg/m³	density
τ	-	dimensionless time $\tau = t \, w_g / H$
ξ	-	dimensionless loading of adsorbent
χ	-	loss factor
Θ	-	dimensionless temperature $\Theta = T/T_{ads}$
Θ_0	-	dimensionless temperature gradient over wall $\Theta = T/T_0$
Θ_{max}	-	dimensionless maximum bed temperature
ν	m²/s	kinematic viscosity
ψ	-	friction factor
K	-	average dimensionless slope of the isotherm
Pe	-	Peclet number for mass transport $Pe = Pe(H, w_g, D_{ax}, \epsilon)$
Pe_{ax}	-	Peclet number for heat transport
Re	-	Reynolds number
St	-	Stanton number $St = St(H, w_g, d_p, \beta, \epsilon)$

Indices		ads	adsorption	p	pellet
des	desorption	en	energetic	k	column
ex	exergetic	g	gas	s	solid
out	outlet	0	standard state		

INTRODUCTION

The energy requirement of separation processes in the chemical and allied industries can amount up to 40% of the manufactuoring costs. So beside the investment costs the energy costs gain more and more importance in the design of separation processes. With the increasing demand for environmental protection, the effort of eliminating polluting, toxic or hazardous components from gas streams which are discharged to the atmosphere, will even increase. For the purification of gas streams polluted by trace components or for the separation of supercritical gas mixtures, the use of adsorption processes is the preferable choice. Compared to the unit operations distillation or absorption, the technological maturity of gas adsorption processes is still underdeveloped [1] with a special lack of studies dealing with the optimization of process configurations. So far, only the studies of LeVan and coworkers [2-4], Fair and coworkers [5,6], Kumar and Dissinger [7] and Smith and Westerberg [8] discuss aspects of optimum desorption of adsorbent fixed beds. Only the studies of Banerjee et al. [9,10] employ an exergy analysis for the thermodynamic evaluation of an adsorption process.

Industrially popular adsorption processes in fixed beds are performed as follows: the adsorption step is carried out until the outlet concentration exceeds a specified value fixed by legislation or the maximum acceptable concentration of the adsorptive gas component in a subsequent step of a complex process. The desorption step can be operated either by reducing the pressure (Pressure Swing Adsorption, PSA) or increasing the temperature (Temperature Swing Adsorption, TSA) in the fixed bed. The choice which method to use depends on the question if the adsorption process aims at separating or purifying gas streams and on the degree of non-linearity of the adsorption isotherm (i.e. equilibrium loading of the adsorbent as a function of the concentration of adsorptive in the gas phase). In the case of purification tasks and in the presence of gas components which have highly curved convex isotherms, the temperature swing operation is preferable. For a TSA cycle the heating is with either a noncondensable hot purge gas or steam.

The total operation costs for a thermal swing process consist of the cost of hot regenerant, the cost of replacing spent adsorbent, and the costs associated with pressure drops [2]. The major expenses are the energy costs for heating the regenerant, so the regeneration will be the focus of attention in this paper. To shed light upon the influence of various operation variables, a parametric study for a case system will be presented and different possibilities to evaluate the energy losses of thermal regeneration processes will be discussed.

THEORY

Thermodynamic Evaluation

The engineering parameters valuating the desorption process are the energy input H, energetic and exergetic loss factors χ_{en} and χ_{ex}, desorption time t, volume V of purge gas required and pressure drop Δp. The energy input is the sum of the isothermal compressor and heater work (neglecting kinetic and potential energies). The energetic loss factor is defined as the ratio of the energy output H_{out} to the energy input H which is equal to

$$H = V R T \ln\left(\frac{p_0 + \Delta p}{p_0}\right) + V \rho_g c_p (T_{des} - T_0) \qquad (1)$$

The exergetic and the energetic loss factors are expressed in the same way

$$\chi_{ex} = \frac{E_{out}}{E} \qquad \chi_{en} = \frac{H_{out}}{H} \qquad (2)$$

with the exergy

$$E = (H - H_0) - T_0 (S - S_0) \qquad (3)$$

The specific entropy s is calculated for ideal gas behaviour with

$$s - s_0 = c_p \ln\frac{T}{T_0} - R\ln\frac{p}{p_0} \qquad and \qquad (4)$$

$$S - S_0 = V\rho (s - s_0) \qquad (5)$$

where the reference state (T_0 and p_0) is used at ambient temperature and pressure [9,10]. The pressure drop Δp is given by

$$\Delta p = H \psi \frac{\rho}{2} w_g^2 \frac{1-\epsilon}{\epsilon^3} \frac{1}{d_p} \qquad (6)$$

with the friction factor ψ according to the Ergun-equation [17]

$$\psi = \frac{150}{Re} + 1.75 \qquad Re = \frac{w_g d_p}{(1-\epsilon) v_g} \qquad (7)$$

Mathematical Model

The simulation of the desorption process is carried out using a Linear-Driving-Force (LDF) model. The concentration profile within the adsorbent particles is considered to be parabolic in shape, which has been shown to be a good approximation for adsorption processes [11,12]. In the LDF approach, the transient diffusion into the pellets is given by a simple rate expression similar to the film theory. This means that all resistances to mass transfer outside and inside the solid particles are placed into an imaginary mass transfer resistance within the film surrounding the adsorbent particles. Employing following assumptions:

- axially dispersed plug flow
- no gas velocity changes over the adsorbent bed
- ideal gas law is valid for the components

and from the component balance and energy balance over a differential slice of the fixed bed following dimensionless differential equations are obtained.

$$\frac{\partial c}{\partial \tau} = -\frac{1}{\epsilon} \frac{\partial c}{\partial L} + \frac{1}{Pe} \frac{\partial^2 c}{\partial L^2} - KSt \, (\xi_i^* - \xi_i) \tag{8}$$

$$W\frac{\partial \Theta}{\partial \tau} = -\frac{\partial \Theta}{\partial L} + \frac{1}{Pe_{ax}} \frac{\partial^2 \Theta}{\partial L^2} + \sum_i W \, \Theta_{maxi} + B \, (\Theta_\infty - \Theta) \tag{9}$$

$$Pe_{ax} = Pe_{ax}(\lambda_s, d_p) \qquad B = B(k_w, w_g d_k H) \tag{10}$$

These equations with the initial and boundary conditions for their solution, are given in much detail elsewhere [13]. The energy balance employs the so-called equivalent one-phase model which can be derived from the balances over the gas void space within the adsorption column and the balance over the solid adsorbent particles [14]. The equation system employs an equilibrium model based on Statistical Thermodynamics [15]; this study, however, is confined to single component calculations.

The equation system is solved by the method of finite differences (FD). Typical computing times for the desorption runs presented in this study range between 2 and 6 hours on a 386 CPU computer.

Calculations Performed

The calculations are carried out using the following assumptions:

- The bed is considered to be completely loaded with CO_2 before the desorption step. This assumption is justified by the observation that the regeneration efficiency is higher for fully saturated than for partially saturated beds [5]. For positively skewed isotherms, characteristic for adsorption systems using zeolitic adsorbents, the adsorption saturation front shows self-sharpening behaviour with a very steep breakthrough [16]. Thus the adsorption cycle is generally operated until the bed is nearly completely saturated.

- The desorption is considered to be completed if the concentration of the adsorptive gas component in the effluent gas stream dropped down below 1/1000 of the raw gas loading with adsorptive. This kind of definition is also used by Schork and Fair [5], whereas Kumar and Dissinger [7] defined desorption to be complete when the effluent gas reaches the regeneration temperature, a definition which is not applicable for non-adiabatic systems. Table 1 gives the parameters used for the computations discussed below.

TABLE 1

Parameters employed in the model calculations

Column diameter: $d_t = 0.4$ m $\qquad T_0 = 273$ K

Particle diameter: $d_p = 0.003$ m $\qquad p_0 = 1.E+5$ Pa

Height of fixed bed: $H = 1$ m

Tortuosity factor: $\mu_p = 4$

Heat conduction in particle: $\lambda_s = 0.58$ W/(m K)

Prior to desorption the bed is completely in equilibrium with

$c_{CO2} = 5$ Vol% and T=313 K

The isotherm equation is given by

$$n_i = a \frac{K_i p_i + (K_i p_i)^2 C}{1 + K_i p_i + (K_i p_i)^2 C/2} \quad with \quad K_i = K_{0i} \exp\left(-\frac{q_0}{\Re T}\right) \quad (11)$$

with the following parameters for CO_2 MS 5Å (case system):

$K_0 = 9.99E-11$ Pa^{-1}, $q_0 = 42.150$ kJ/(mol K), a = 1.45E-3, C = 0.57. Note that this case covers the most commonly observed shapes of isotherms on a microporous adsorbent.

RESULTS AND DISCUSSION

Simulations of the desorption process were carried out with the LDF-model described above depending on variations of

desorption temperature T_{des} in the range from 60 to 200°C

velocity of purge gas w_g in the range from 0.1 to 0.4 m/s

heat transfer coefficient k_w in the range from 0 to 100 W/(m^2 K).

The desorption half cycle starts and ends at the conditions described above. The purge gas is dry air. Figure 1 shows the typical shape of desorption breakthrough curves and their temperature dependence. The outlet gas concentration and temperature, normalized with respect to the raw gas concentration and temperature, show the strong influence of the gas

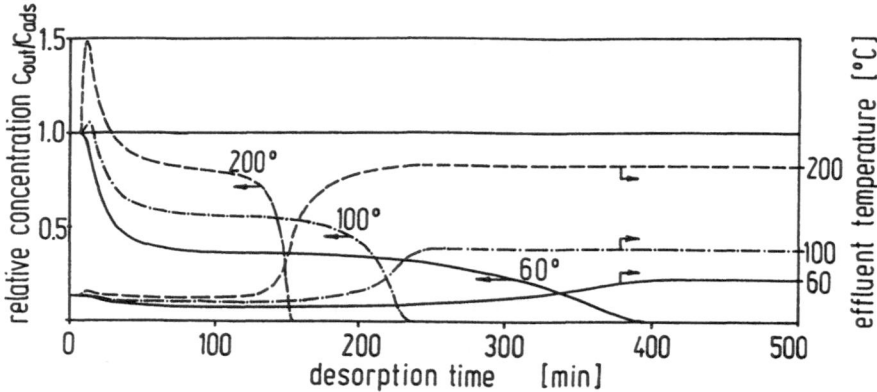

Figure 1. Desorption depletion curves for case system at various purge gas temperatures

temperature on the desorption time. It is interesting to note that both outlet concentration and temperature remain stable during most time of the desorption cycle. This so-called plateau is characterized by a temperature near to that during adsorption. This means that irrespective of purge gas temperature the outlet gas stream has about the same temperature. As soon as the bed is completely regenerated the concentration drops towards zero and the temperature rises sharply.

Figure 2 depicts for an adiabatic process both the energetic and the exergetic loss factors over the temperature and velocity of the purge gas stream. A value of $\chi_{en}=45\%$ indicates the existence of an energy consumption of 55%, mainly consumed by the heating of the adsorbent for

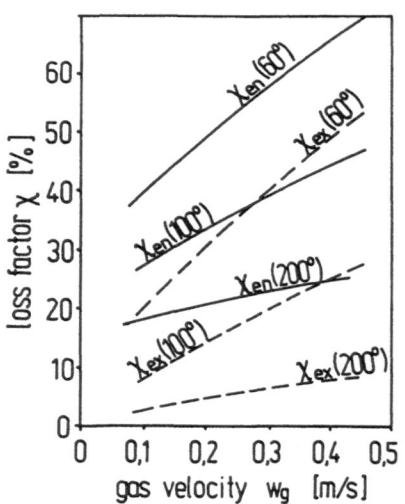

Fig.2 Case study: exergetic and energetic loss factors as a function of gas velocity and temperature

regeneration and, in the case of a diabatic process, by heat losses over the wall. The remaining 45% of the input energy is the physical energy loss which can only be recovered in heat regenerators, because high effluent temperatures are only to be expected in the final period of the desorption step, which leads to a non-permanent operation state of subsequent heat exchangers or heat pumps [18]. Only an exergetic consideration gives information concerning the maximum useful energy which might be obtained from the hot effluent stream, whose composition varies with the concentration of the adsorptive feed gas. The energy input ranges in this case study between 10 and 50 MJ; the work of the rotary blower calculated from the pressure drop only amounts to about 1‰ of the heating and might be neglected. As can be seen in figure 2, the exergetic loss factors range nearly independently of temperature a constant increment of about $\Delta\chi = 15\%$ below the energetic loss factors. This indicates that the desorption unit should be operated at a high temperature and a low purge gas velocity, where the loss factors of energy and exergy are low.

So far we considered only the efficiency of a desorption step, but neglected the purge gas requirement and the time consumption. In figure 3 the amount of purge gas required is plotted against the velocity and the temperature of the purge gas stream.

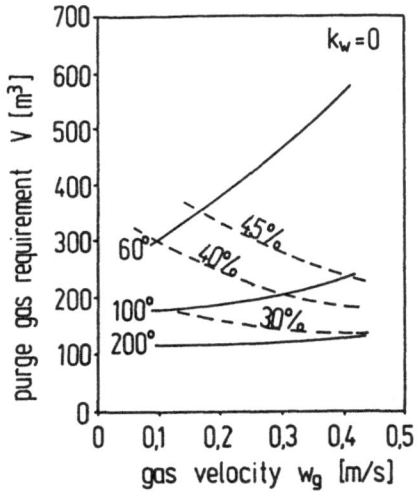 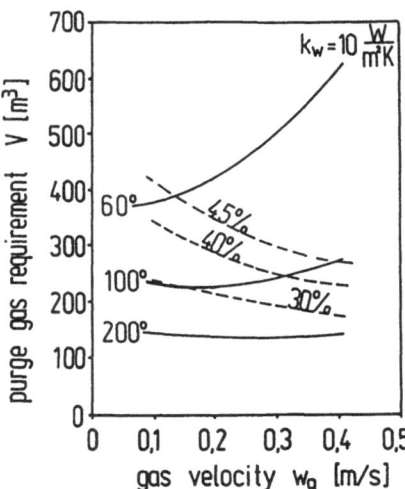

Fig. 3 Case study: purge gas requirement and energetic loss factor over velocity and temperature for adiabatic (left) and diabatic (right) desorption system,
solid lines: purge gas required, dashed lines: energetic loss factor

The dashed lines show the energetic loss factors of this desorption step.

In technical two- or multibed units the amount of product gas determines the theoretical upper limit for the required purge gas while a high amount of purge gas, as demonstrated in the left diagram of figure 3, leads to a higher energetic loss factor. For diabatic systems with heat losses over the wall, a minimum purge gas requirement can be found at a certain gas velocity as shown in the right part of figure 3. This minimum point moves from low gas velocities to high ones with increasing temperature. A longer mean residence time generally enhances the heat transport. At too low gas velocities, however, the effect is counterbalanced by the increasing heat losses over the column wall. Together with the information obtained by the energetic loss factor, this minimum in purge gas required offers a recommendation for the optimum operation range.

At high temperatures the purge gas requirement and the input energy H become nearly independent of the gas velocity, even for the diabatic case. This is demonstrated in figures 4 and 5 where the energy input, the desorption times, and the pressure drop are plotted against velocity and temperature.

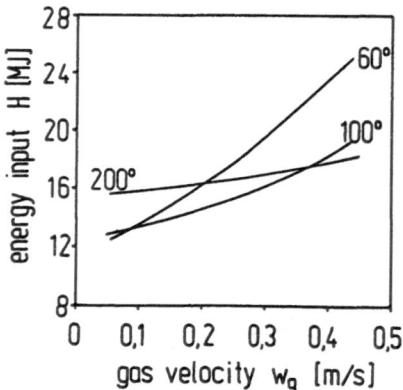

Fig.4 Case study: Energy input over velocity and temperature of purge gas, adiabatic bed

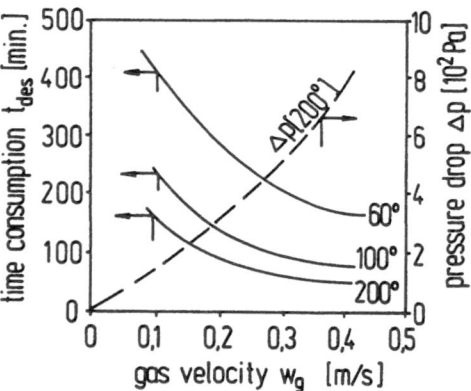

Fig. 5 Case study: Pressure drop and desorption time as a function of gas velocity and temperature, adiabatic bed

Recommendations for the avoidance of energy waste:

Desorption with a high temperature purge gas at a low superficial gas velocity (however, time consuming)

In the case of a low temperature purge gas and a high superficial gas velocity heat recovery in a regenerator is advantageous (however, increased investment)

ACKNOWLEDGMENT

Financial support by the Commission of the European Communities is gratefully acknowledged.

REFERENCES

[1]   Keller, G.E., Separations: New Directions for an Old Field, AIChE Monograph Series, 17, 83 (1987)

[2]   LeVan, M.D., McAvoy Jr., R.L., Davis, M.M., Dolan, W.B., Studies on the Optimal Thermal Regeneration of Adsorption Beds, Proc. 3rd Intern. Conf. Fund. of Adsorption, A. Mersmann, St. Scholl, Eds., Engineering Foundation, 1991

[3]   Davis, M.M., LeVan, M.D., Experiments on Optimization of Thermal Swing Adsorption, Ind. Eng. Chem. Res. 28 (1989), 778

[4]   Davis, M.M., McAvoy, R.L., LeVan, M.D., Periodic States for Thermal Swing Adsorption of Gas Mixtures, Ind. Eng. Chem. Res. 27 (1988), 1229

[5]   Schork, J.M., Fair, J.R., Parametric Analysis of Thermal Regeneration of Adsorption Beds, Ind. Eng. Chem. Res. 27 (1988), 457

[6]   Huang, C.-C., Fair, J.R., Parametric Analysis of Thermal Swing Cycles for Multicomponent Adsorption, AIChE J. 35 (1989), 1667

[7]   Kumar, R., Dissinger, G.R., Nonequilibrium, Nonisothermal Desorption of Single Adsorbate by Purge, Ind. Eng. Chem. Process Des. Dev. 25 (1988), 456

[8]   Smith IV, O.J., Westerberg A.W., The Optimal Design of Pressure Swing Adsorption Systems, Chem. Eng. Sci. 46 (1991), 2967

[9]   Banerjee, R., Narayankhedkar, K.G., Sukhatme, S.P., Exergy Analysis of Pressure Swing Adsorption Processes for Air Separation, Chem. Eng. Sci. 45 (1990), 467

[10]   Banerjee, R., Narayankhedkar, K.G., Sukhatme, S.P., Exergy Analysis of Kinetic Pressure Swing Adsorption Process: Comparison of Different Cycle Configurations, Chem. Eng. Sci. 47 (1992), 1307

[11]   Yang, R.T., Gas Separation by Adsorption Processes, Butterworths, 1987

[12]   Buzanowski, M.A., Yang, R.T., Approximations for Intraparticle Diffusion Rates in Cyclic Adsorption and Desorption, Chem. Eng. Sci. 46 (1991), 2589

[13]   Schweighart, P., Mersmann, A., Equilibrium and Kinetics of Gas Adsorption Process Design, Fourth Periodic Report to the CEC, Contract JOUE-0052-C

[14]   Vortmeyer, D., Packed Bed Thermal Dispersion Models and Consistent Sets of Coefficients, Chem. Eng. Process. 26 (1989), 263

[15]   Schweighart, P., Sievers, W., Mersmann, A., Nonideal Equilibrium Behaviour in Low and High Pressure Adsorption Systems, Recents Progres En Genie Des Procedes, 5 (1991), 17, 23

[16]   Mersmann, A., Adsorption, Ch. 9, Vol.B3, Ullmann's Encyclopedia of Industrial Chemistry, Verlag Chemie, Weinheim, 1989

[17]   Ergun, S., Fluid Flow Through Packed Columns, Chem. Eng. Progr. 48 (1952), 89

[18]   Fratzscher, W., Brodjanskij, V., Michalek, K., Exergie, VEB Verlag, Leipzig, 1986

DESIGN AND ANALYSIS OF SUPER-CRITICAL EXTRACTION PROCESSES

Glen Hytoft, Rafiqul Gani and Aage Fredenslund
Engineering Research Centre IVC-SEP, Department of Chemical Engineering, The Technical University of Denmark, DK-2800 Lyngby, Denmark.

ABSTRACT

Computer based design and analysis of super-critical extraction processes is performed within the context of choice and need of computational tools, consistency and accuracy of simulation results and importance of the need to study the conditions of operation of the process. Brief descriptions of the computational tools employed in this work are given. Numerical results demonstrating the correct and consistent use of the thermodynamic models in design, steady-state simulations and dynamic simulations are presented for two super-critical extraction processes whose details have been obtained from the open literature. The sensitivity of the conditions of operation on the phase stability of the streams which may cause operability problems are also illustrated for the two super-critical extraction processes. Finally, with a simple example, the scope for the reduction of energy utilization in super-critical extraction processes is demonstrated.

INTRODUCTION

Super-critical extraction (SCE) processes operate at temperatures and pressures where the solvent, usually a gas at normal temperature and pressure, becomes a liquid at or near its critical pressure. At these conditions the behaviour of the mixture of solutes together with the super-critical solvent becomes highly non-ideal. Properties such as density and phase equilibrium therefore become sensitive to process conditions.

It is well known that for accurate design and analysis of chemical processes, prediction of thermodynamic properties play an important role. Because of the nature of the mixture and its sensitivity to process conditions, accurate and consistent prediction of thermodynamic properties become particularly important for SCE processes. Thermodynamic properties influence the choice of operating conditions (for example, temperature and pressure of a liquid-liquid extraction column), the simulated steady-state condition, the simulated dynamic behaviour of the process, the sizing of equipments and many more.

If computational tools are to be employed in design and analysis of a process, it is essential that they are verified not only for accuracy and efficiency of calculations but also for consistency of thermodynamic principles. For example, it may be possible to achieve convergence in the simulation for a distillation column without checking for the identity of the phases. While the mass balance will not be effected if the phase identity is incorrect, the energy balance will be severely effected even though near the critical condition there is not much difference between a liquid and a vapour. Also, even if convergence of global mass and energy balance is achieved, they do not guarantee convergence for mass and energy balance for individual unit operations (unless explicitly tested for each unit operation). Since SCE processes are sensitive to operating conditions and they operate near the phase boundaries, phase identities of the streams must be consistent with respect to phase diagrams.

The objective of this paper is to demonstrate how computer-based design and analysis of SCE processes can be carried out in a consistent and efficient manner. This paper therefore, describes the methods of prediction of properties and process simulation tools that are needed for studies related to SCE processes. Also, brief descriptions of the computational tools employed in this work is described. These tools include thermodynamic models like the GCEOS-model (1) and the MHV2-model (2), a steady-state process simulator especially adapted for SCE processes (3), a dynamic process simulator (4), a dynamic liquid-liquid extractor model and various computer programs for computation of phase equilibria and phase diagrams.

Simulation results from earlier works ((5) and (6)) are analyzed in terms of consistency and accuracy. New, consistent and qualitatively correct simulation results are presented for the two SCE processes. Finally, a preliminary energy analysis is presented to demonstrate the potential of SCE processes for reduction of energy usage.

This work has been funded by the Commission of the European Communities (JOULE Programme). The main institutions involved in this project are the Technical University of Denmark (Denmark), University of Cadiz (Spain) and University of L'Aquila (Italy). The Technical University of Denmark have been responsible for the development of thermodynamic models (the MHV2-model) and process simulators to be used for design and analysis of SCE processes. The Universities of Cadiz and L'Aquila have been responsible for supplying the experimental phase equilibrium data.

SUPER-CRITICAL EXTRACTION PROCESSES

Super-critical extraction processes may be used to obtain pure alcohols and other oxy-organic compounds from aqueous mixtures. The SCE processes exploit the strong dependence of properties such as density and phase composition on pressure and temperature.

The concept of super-critical extraction is that in the critical region, the partial molar volumes of the extracted component in the solvent is highly dependent on the temperature and pressure. The fugacity coefficient of the solute in the super-critical solvent is given by

$$\left(\frac{\partial \ln \varphi_2}{\partial P}\right) = \frac{\overline{v_2}}{RT} - \frac{1}{P}$$

It may be noted that in the critical region it is possible to change the extracting

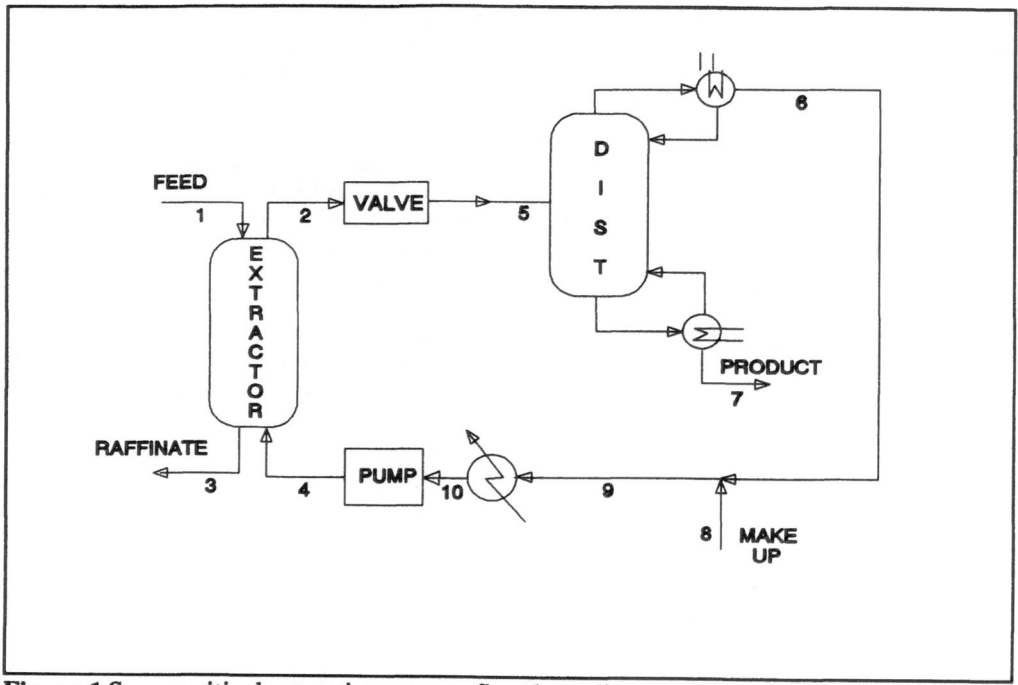

Figure 1 Super-critical extraction process flowsheet diagram. Dehydration of acetone with super-critical carbon dioxide.

super-critical compound has to satisfy several criteria. The solvent must have low solubility in water and therefore, probably should be non-polar. The solvent should not form binary azeotrope with the solute under conditions in the recovery column(s). The recovery of the solvent in this case would be difficult and add to the cost of separation. If VLLE is not considered in the recovery column(s) (as in earlier works (5)), then phase stability test for the liquid phase on each plate is also not needed. The critical temperature of the candidate solvent should be well above ambient conditions thereby keeping the process temperature low. The solvent must be a good entrainer so that it is able to separate the solute from water in the solvent recovery column(s).

A typical SCE process essentially includes a liquid-liquid extraction column (where the separation of solute from water takes place) operating at high pressure. A valve unit to reduce the pressure of the extract stream before it enters the solvent recovery column(s) is needed. It is cheaper (from an energy point-of-view) to operate the recovery column(s) at lower pressures. The temperature, however, should be higher than ambient (otherwise cooling costs will have to be added). The solvent recovery usually takes place in one or two distillation columns. Since the solvent is the lightest compound, the energy requirements for their columns are much less than processes where the solvent is the heaviest compound. The solvent stream from the recovery column(s) is sent back to the liquid-liquid extractor via a pump where the stream pressure is increased to that of the extractor pressure. The main energy costs therefore involve the solvent recovery column(s), the pump and the heat exchangers. A typical flowsheet for an SCE process is shown in Figure 1.

COMPUTATIONAL TOOLS

In order to design and analyze SCE processes it is necessary to have a number of reliable computational tools. It is important that the available tools can be employed to study the entire process and/or one or more units of the process separately. For example, after the entire process has been simulated it is important to know the stream properties with respect to the phase boundaries - are they too close to the boundaries? What will happen if the conditions are changed, will the boundaries be crossed? If yes, does it cause any operability problems? What about the effect on energy demands?

In this section, the need and importance of computational tools are highlighted followed by brief descriptions of the tools used in this work.

Thermodynamic Properties

Process simulations require thermodynamic routines for prediction of physical properties. These include equations of state (EOS), models for computations of activity coefficients in highly non-ideal solutions at moderate temperature and pressure (g^E-based models), and models for describing the phase behaviour of highly non-ideal mixtures over large temperature and pressure ranges (combined EOS and g^E-based model). Also a set of pure component correlations describing densities, enthalpies and heat capacities are needed.

Since in the critical region the extracting power of the solvent is enhanced, it is important that this behaviour is predicted correctly by the selected thermodynamic model for accurate simulation of super-critical extraction processes. In the case of dynamic simulation, the selected thermodynamic model should not only describe the phase equilibria correctly, it should also represent the PVT behaviour of the mixture accurately, since the volume of the mixture has significance in dynamic simulation.

Phase equilibrium related calculations such as determination of binary and ternary azeotropic points, binodal curves as a function of temperature, vapour-liquid-liquid equilibrium, mixture critical-point and the so-called distillation boundaries are also needed in design and analysis of SCE processes. These calculations have to be made at high pressures and temperatures because at these temperatures and pressures the operation of the liquid-liquid extractor and the solvent recovery columns are most effective. Also, the correct identity of the phase effects the energy demands (enthalpy of a vapour phase being higher than a liquid phase). The high non-ideality of the mixtures coupled with high temperatures and pressures makes it necessary to use thermodynamic models like GCEOS (1) and/or MHV2 (2) for accurate and consistent design and analysis of SCE processes (as will be shown later, at the cost of computing time).

Process Simulation

While the thermodynamic properties and individual process unit calculations help to design/analyze the operation of a single process unit or state of a stream, process simulations allow the study of an entire process. Design, analysis and optimisation problems can be solved through steady-state and dynamic simulations of SCE processes. Problem specific simulation tasks (leading to different simulation strategies) are needed. In this section, we outline some reasons why this is so.

In each of the two types of simulations, the mass and energy balance equations which are being solved for each process unit can be written in generic form as follows:

Rate of accumulation = input – output + net. production   (2)

There are differences in how Eq. (2) is interpreted for the two types of simulation tasks. For steady-state simulation the left hand side of (2) is equal to zero. Therefore all equations are algebraic, and for SCE processes and other complex chemical processes, the problem requires a solution of non-linear Algebraic Equations (AEs). For dynamic simulation, Eq. (2) are Ordinary Differential Equations (ODEs) and these are solved together with a subset of algebraic equations. The problem to be solved therefore involves the solution of a set of equations containing Differential and Algebraic Equations (DAEs).

In the case of optimisation problems yet another simulation strategy is required. The variables needed by the optimisation routines to calculate the objective function are limited to stream properties and process unit parameters. It is therefore sufficient for the optimisation routines to have exact information only of these variables. In both steady-state and dynamic simulation of chemical processes the complex thermodynamic models are spending up to 80% of the computational time. Since for any simulation task, the importance of the accuracy and efficiency of the prediction of thermodynamic properties is the same, therefore, these simulation tasks must also make sure that relevant thermodynamic properties are computed consistently, accurately and efficiently. It is however possible to simplify the thermodynamic model calculations as shown recently by Perregaard and Sørensen (7) and thereby reduce the computational time significantly.

Computational Tools Used in this Work
The computational tools that have been employed to obtain the results presented in this work are described briefly in this section. Most of these tools have been developed earlier. In this work, however, these tools have been tested for their suitability to studies related to SCE processes and have been modified where necessary.

Phase equilibria: 2-phase PT-flash, PH-flash and multi-phase PT-flash programs using GCEOS- and MHV2-models have been used. The flash routines are based on the algorithms proposed by Michelsen (8) and have been found to be robust and efficient. Phase diagram programs for calculating the PT-phase envelopes (vapour-liquid) and vapour-liquid-liquid phase boundaries (as a function of temperature) for ternary systems have been implemented and tested. The introduction of the MHV2-model into these programs have given the possibility of studying the effect of temperature and pressure on the "physical state" of the stream.

Steady-state simulation: The modular based SEPSIM steady-state simulator has been used in this work. SEPSIM has a variety of unit operation models that have robust individual solution methods. The simulator includes an algorithm for flowsheet decomposition and several methods for convergence of the tear-stream variables. The residuals of each partition are represented by the following equation

$$f(x) = x - g(x)$$ (3)

where x is the vector of present values of the tear-stream variables (the component flowrates, pressure, temperature and a number of design specifications) and $g(x)$ represent the values of the same variables resulting from a flowsheet evaluation. The flowsheet evaluation is performed modularly to complete an iteration for convergence of the tear-streams variables. The advantage of this strategy is that when one or more units contains highly non-linear equations (like flash-calculations in the critical-region), the robustness is unaffected since only one unit is being solved. The updating of x (i.e., tear-stream variables and design variables) is done by one of the following options: successive substitution, a Quasi-Newton method using Broyden update of the Jacobian, the two-tier approach and simultaneous solution of tear-stream and design variables.

SEPSIM offers 19 different process units and eight different thermodynamic models including GCEOS and MHV2. This makes SEPSIM especially suitable for studies related to high pressure processes such as the SCE process. It also includes the feature which allows the calculation of phase envelopes of any stream of the flowsheet. The steady-state simulation is fast (few minutes on a 80386-based personal computer). Consequently, a small increase in computing time is relatively unimportant, if a failure of the solution procedure can be avoided. One could say therefore, that **in steady-state simulation, robustness is more essential than speed**. This is the main reason why a modular approach has been used.

Dynamic Simulation: In the case of dynamic simulation, as we have mentioned earlier, the interpretation of Eq. (2) changes somewhat. Since the rate of change in accumulation is not zero, one has to take into account a new set of variables, describing the accumulation in the unit. For dynamic simulations, DYNSIM (4), has been used in this work. DYNSIM is an equation oriented based which can solve dynamic and steady-state simulation problems and employs problem specific simulation strategies for different types of simulation problems. DYNSIM is very flexible and allows various choices,

- Between dynamic and steady-state simulation modes.
- Between dense, sparse, sequential and simultaneous partitioned mode.
- Between "stiff" and "non-stiff" numerical methods.
- Simulations of flowsheets with dynamic and non-dynamic units.

The integration methods are able to solve ODEs, DAEs and AEs. The numerical integration methods available are the Adams-Moulton, the 4/5th order Runge-Kutta methods for "non-stiff" problems and the Backward Differentiation Formula (BDF) and the Diagonally Implicit Runge-Kutta (DIRK) for "stiff" problems. DYNSIM offers several different thermodynamic models including the GCEOS- and the MHV2-models.

Once a dynamic simulation is running, a good initial guess on the solution of the implicit algebraic equations is always available from the solution at the previous time step (4). However, in contrast to steady-state simulation, the total flowsheet problem needs to be solved several times. This is because each point on the transient path requires at least one solution of the total flowsheet problem. Therefore, it is quite crucial that the equations on each step are solved as fast as possible. One could therefore say that **in dynamic simulation speed of the solution procedure is more essential than robustness**. This is the main reason why an equation oriented approach is used.

Process simulation for optimisation: The optimisation simulations are done through SEPSIM. The Successive Quadratic Programming (SQP) algorithm of Biegler et al. (9) has been added to SEPSIM (10). It allows the solution of optimisation problems through simultaneous convergence of tear-stream variables and optimisation variables and through the "black box" mode where for each iteration in the optimisation loop the flowsheet simulation problem is solved.

RESULTS/DISCUSSION

Comparison of Thermodynamic Routines

In this section we will investigate the computational efficiency of three thermodynamic models for the purpose of phase equilibrium calculations in the critical region. As a test system we have considered a mixture of propane, ethanol and water. Computational times needed to calculate a representative set of physical properties using the three models are given in TABLE I.

TABLE I

Computing times (sec.) for three thermodynamic models.

Time in sec.	SRK	GCEOS	MHV2
PT-flash	0.2198	2.2527	1.2088
VT-flash	0.6593	5.7142	2.9670

The general impression from the calculations mentioned above is that the SRK equation is extremely fast compared to the two others while the MHV2-model is about twice as fast as the GCEOS-model. Note that the simulated results for all the three models in this case is approximately the same. The SRK-model however has been correlated to match the results from the other two models. Skjold-Jørgensen (1) and Dahl et al. (2) have verified the accuracy of prediction of properties with the GCEOS-model and the MHV2-model respectively.

Each calculation of the thermodynamic property (such as equilibrium constants) with the MHV2-model requires computations of activity coefficients (use of a g^E-model), MHV2-model parameters and fugacity coefficients (SRK-like computation). With the GCEOS-model, the parameters for the model need to be calculated from group-contribution principles at each new temperature before the fugacity coefficients (and the equilibrium constants) can be evaluated. It is therefore understandable that the two group-contribution equations of state are relatively slow compared to the SRK equation (11). Predictions with the SRK equation of state are however subject to the availability of interaction coefficients. Moreover, it is not always possible to correlate the data for highly non-ideal mixtures by fitting the interaction coefficients of cubic equations of state such as the SRK equation of state. We propose therefore, the use of the MHV2-model for predictions of phase equilibria at high pressures. Dahl et al. (12) have already verified the suitability of the MHV2-model for prediction of thermodynamic properties related to SCE processes.

Simulation of SCE Processes

Dehydration of Acetone with CO_2: Figure 1 illustrates the flowsheet of a super-critical extraction process with CO_2, acetone and water presented earlier by Cygnarowicz and Seider (6). We are using this process to show the importance of consistency of simulation results and the importance of correct prediction of phase equilibrium. The valve after the extractor in Figure 1 is usually simulated as an adiabatic PH-flash. This means that there should be no exchange of heat with the surroundings and the exit conditions (temperature, phase fraction and compositions) are to be determined for specified pressure and enthalpy (meaning no change in enthalpy). The Cygnarowicz and

Seider results were obtained with the GCEOS-model. In this work the same flowsheet has been simulated with the GCEOS-model and the MHV2-model. TABLE II lists the exit pressure from the valve unit and the deviation of enthalpy between inlet- and outlet-streams for the valve unit. The TABLE II results clearly indicate that the simulated results of Cygnarowicz and Seider are inconsistent since the energy balance for the valve did not converge. In order to achieve convergence with the GCEOS-model, the pressure needs to be increased from 65 atm to 68 atm (see TABLE II). However, with the MHV2-model, convergence for the energy balance in the valve unit can be achieved at 65 atm. This confirms our previous statement that the choice of the thermodynamic model is important. A comparison of the results of Dahl et al. (12) with those obtained in this work appears to indicate that the MHV2-model results are closer to reality than the GCEOS-model results.

TABLE II

Comparison of the simulation results of this work and Cygnarowicz and Seider (6) for the dehydration of acetone process.

	Cygnarowicz and Seider	This work	
	GCEOS	GCEOS	MHV2
Enthalpy deviation over the valve (J/mole)	1595.0	0.0	0.0
Pressure (atm.)	65	68	65

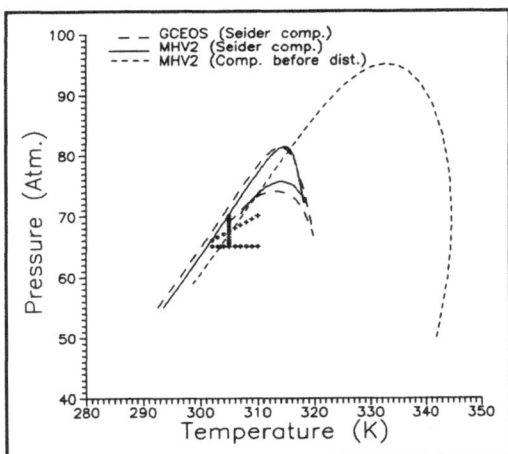

Figure 2 Phase envelope for CO_2-acetone-water with the composition from Cygnarowicz and Seider (0.9793, 0.0156, 0.0051), and the composition of stream 5 (0.9526, 0.0418, 0.0056) from this work. * and + marks three phase points.

We will now analyze the phase diagrams given earlier by Cygnarowicz and Seider (6) and that determined in this work for the composition of the stream going into the recovery column (stream 5). Note that the phase diagram of Cygnarowicz and Seider did not correspond to the composition of the stream going to the recovery column (there is a small difference). Figure 2 shows the phase envelopes for the two different compositions and also, a comparison of phase envelopes computed with the GCEOS- and MHV2-models. These phase envelopes indicate the regions where there are two phases (vapour-liquid). Simulations at various combinations of temperatures and pressures have pointed out that a part of the two-phase region has in fact three phases. That is, in this region, we have vapour-liquid-liquid equilibrium (in agreement with the results of Dahl et al. (12)). These 3-phase points are also indicated in Figure 2 (* and + for the composition of Cygnarowicz and Seider and this work respectively). Another interesting feature of the SCE process can be noted from the location of these points (all are very close to the phase boundary). Since the exit from the valve is the inlet stream for the distillation column (see Figure 1), it can be seen that small disturbances in the process conditions may cause a crossing of the phase boundary thereby causing operability problems (effecting energy demands) as well as numerical problems (related to convergence of mass and energy balances).

TABLE III
Results of multi-flash simulation for CO_2, acetone and water with the MHV2-model at P=65.0 atm and T=303.6 K.

Stream	Fraction	CO_2	Acetone	Water
Feed	1.000	0.95265	0.04179	0.00556
L_1 product	0.001	0.02699	0.01738	0.95562
L_2 product	0.966	0.95256	0.04302	0.00441
V product	0.033	0.99309	0.00544	0.00146

The 3-phase points in Figure 2 is made with simulations with a multi-phase flash program using the MHV2-model and the results from a single multi-phase flash are shown in TABLE III. These results confirm the need to check for liquid-phase stability (not considered in the earlier works) if consistent simulation results are to be obtained. It should be pointed out also that except for the valve unit results, the simulation results of Cygnarowicz and Seider (6) have been matched with the GCEOS-model in this work.

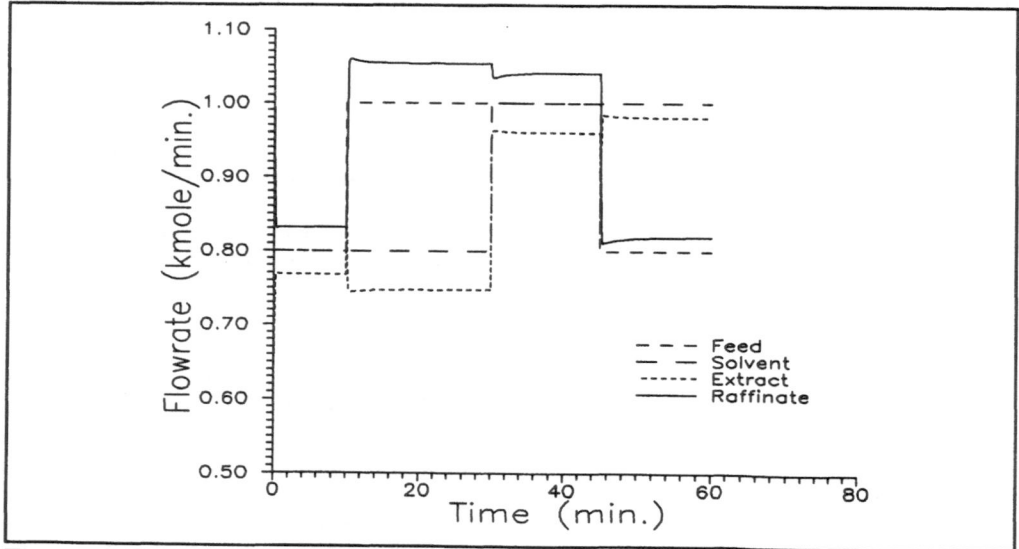

Figure 3 Simulated transient behaviour with dynamic simulation of an extractor with CO_2, ethanol and water at 134.22 atm.

Dynamic simulation of a liquid-liquid extractor

The effect of disturbances in a SCE process is analyzed through dynamic simulation of a liquid-liquid extractor unit. Super-critical CO_2 is used to extract ethanol from water at 134.2 atm pressure in a column with 3 plates. The simulated transient responses for the feed-, solvent-, extract-, and raffinate-flowrates are shown in Figure 3. Starting from an initial condition, t_0 (time zero), a steady-state is obtained at time, t_1. At t_1 the feed-flow rate is increased. This causes the extract-flow rate first to increase for a very short period of time and then to decrease (this is an example of inverse response) while the raffinate flow rate increases. Ultimately, the system comes to a new steady-state at time t_2. At t_2, the solvent rate is increased and this causes the extract-flow rate to increase and the raffinate-flow rate to decrease. The new steady-state is obtained at time t_3. At t_3 the feed-flow rate is now decreased which causes the raffinate-flow rate to decrease and the extract-flow rate to increase. The simulation results therefore confirm that the physical behaviour of the process is correctly represented.

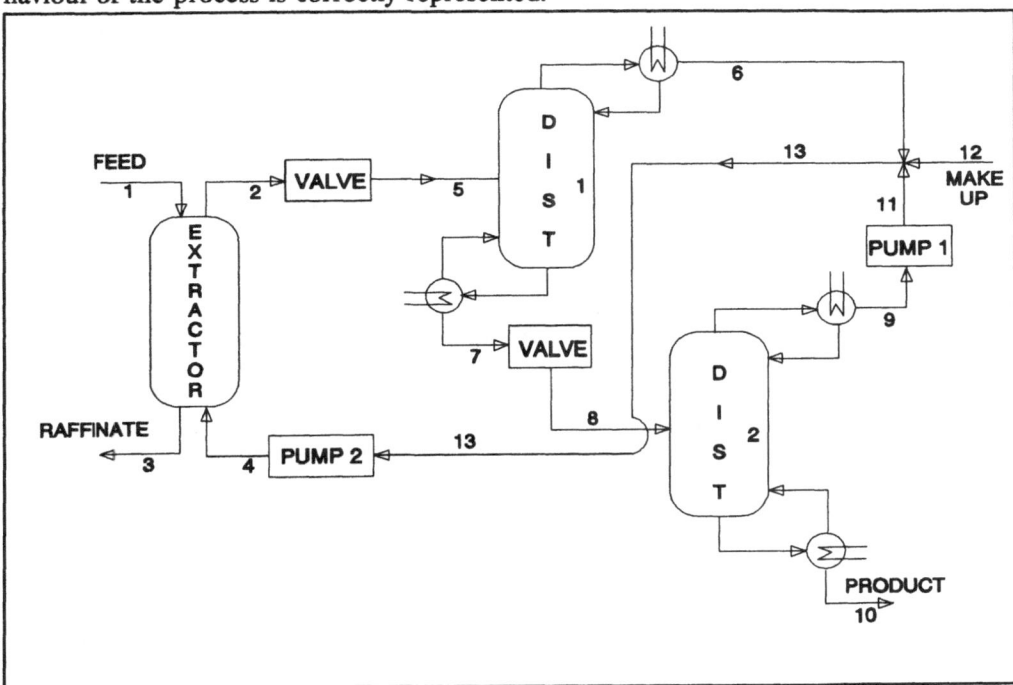

Figure 4 Flowsheet diagram of dehydration of ethanol using super-critical propane as solvent.

Steady-State Optimisation

In Figure 4 the flowsheet for dehydration of ethanol by super-critical propane is shown. This process has been studied earlier by Brignole et al. (5). In this work, we have first tried to reproduce the results of Brignole et al., then we have analyzed the results with respect to consistency of prediction of properties and finally, we have made an optimisation of the operating conditions with respect to the use of energy. Brignole et al. (5) used the GCEOS-model in their simulation. Since the MHV2-model has been shown

TABLE IV

Comparison of simulated stream data. Numbers of streams are given in Figure 1.
A : This work, **B** : Brignole et al. (5).

Flowrate	1		3		4	
mole/hr	**A**	**B**	**A**	**B**	**A**	**B**
Water	58.260	58.260	58.210	58.204	3.395	5.314
Ethanol	21.740	21.740	0.463	0.385	6.391	6.090
Propane	0.000	0.000	0.020	0.022	320.000	320.000
T in K	320.44	320.44	365.00	380.00	341.29	338.89
P in atm.	75.00	75.00	75.00	75.00	75.00	75.00

5		8		10		12	
A	**B**	**A**	**B**	**A**	**B**	**A**	**B**
3.401	5.323	0.010	0.014	0.007	0.009	0.000	0.000
26.969	26.765	26.969	26.765	20.579	20.676	0.000	0.000
319.980	319.979	3.501	3.851	0.019	0.021	0.039	0.042
347.83	345.12	387.45	382.44	433.48	433.45	340.00	340.00
26.00	25.00	12.50	12.50	12.50	12.50	26.00	25.00

to be more efficient and accurate, we have used in this work, the MHV2-model. We have noticed that the Brignole et al. results can only be matched if we relaxed the condition for convergence of energy balance for the valve unit. A consistent simulation therefore gives a different temperature (higher than that of Brignole et al.) for the exit stream for the valve. Lowering this temperature to that of Brignole et al. (5) gives two liquid phases. TABLE IV compares the stream values of the consistent Brignole et al. (5) results with the optimal simulation results from this work. Taking the "consistent" simulation results as the reference design, we have investigated the possibilities of reduction of energy demands. Since Brignole et al. (5) have already shown the advantage of SCE processes over non-SCE processes for the separation of ethanol from water with propane, in this work, we are only studying the possibilities of making even further reductions in terms of use of energy.

The operating temperatures and pressures for the different unit operations play a vital role in SCE processes. As the phase diagrams have shown, small deviations can cause a crossing of the phase boundary. Changes in the "right" direction may also cause

TABLE V

Energy analysis of the SCE process where ethanol is dehydrated with propane.

Energy in kJ/hr	Unit	Brignole	This work
Total heat energy	DIST 1	4637.1	4251.3
	DIST 2	139.4	125.2
Total work	PUMP 1	1.719	1.847
	PUMP 2	3.085	1.635

a reduction of the energy costs without any change in process specifications (negligible solvent loss, high purity of ethanol). TABLE V compares the energy requirements for the "consistent" simulation results at the condition of operation given by Brignole et al. (5) with a new condition of operation. Only the temperature of the liquid-liquid extractor (T_E) and the pressure of the first solvent recovery distillation column (P_{D1}) are different for the two sets of conditions of operation (see TABLE V). It can be noted that a saving of 8.4% (in terms of energy utilization) is achieved with the new operating condition.

Like the dehydration of acetone process, here also, we have made analysis of simulation results. Brignole et al. (5) have simulated the distillation columns for solvent recovery as 2-phase vapour-liquid equilibrium problem. We have calculated phase diagrams for the ternary propane-ethanol-water system at the operating condition of the distillation column (DIST1). This phase diagram, which shows the liquid-liquid-vapour equilibria in terms of the heterogeneous liquid boiling surface and the vapour line in equilibrium with the two liquid phases, is shown in Figure 5. Also shown in this figure are the liquid and vapour compositions from each plate of the distillation column. The vapour composition from the plates near the top of the column and the top plate liquid composition are clearly within the two-phase region. This indicates that when the vapour from the top of the column is condensed, it will split into two liquid phases. Bossen et al. (11) have shown that considerably different results can be obtained if the effect of liquid-liquid split is not considered. Also, from a separation point of view it is advantageous to consider the liquid-liquid split in the condenser.

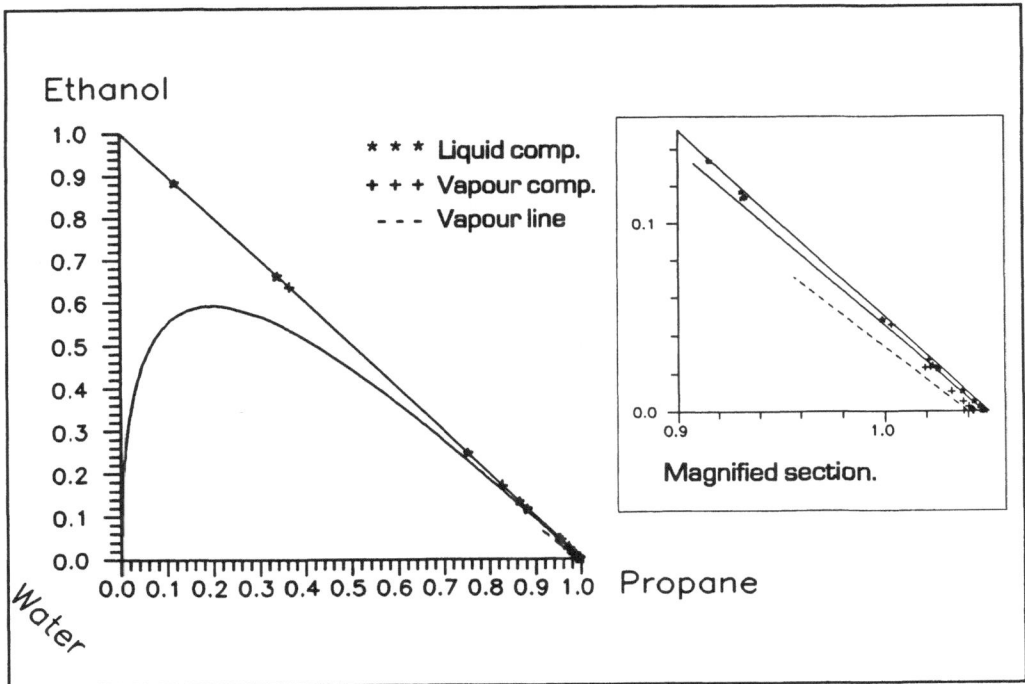

Figure 5 Phase diagram of propane-ethanol-water with liquid and vapour compositions profiles from the first distillation column. The vapour composition with the lowest amount of ethanol indicates the top plate.

CONCLUSIONS AND FUTURE WORK

Computational tools especially suited for studies related to super-critical extraction processes have been presented. Through numerical examples and analysis of the simulation
results, it has been demonstrated how these tools can be employed for design and analysis
studies of super-critical extraction processes. With the help of two examples from the
open literature the importance of efficiency and accuracy of prediction of thermodynamic
properties with respect to consistency of simulation results has been illustrated. Some of
the potential problems with respect to the design and operability of super-critical
extraction processes have been identified and the scope of process optimisation with
respect to reduction of energy utilization has been demonstrated.

Emphasis has been given in this paper to present the computational tools and how
they can be used efficiently and consistently to solve problems related to design and
analysis of super-critical extraction processes. Therefore, the design of earlier workers
((5) and (6)) have been analyzed and used in this work. As a result of this analysis, there
is reason to believe that it is necessary to take a closer look at the choice of the operating
conditions not only to reduce the cost of energy but also to avoid potential operability
problems (such as crossing of phase boundaries which may also effect the energy
demands). Also, the criteria for selection of the super-critical solvents need to be revised.
For example, is it necessary to have a super-critical component or can a near-critical
component do equally well? These questions need to be answered taking into account,
the cost of equipment. Current work is involved with answering these questions as well
as preparing a computer based system that can be used for synthesis, design and analysis
of super-critical extraction processes.

Finally it can be concluded that the results presented in this paper show clearly that
it is possible to perform computer based studies of super-critical extraction processes.
However, one must be aware of the dangers of inconsistent results even though the
simulated results satisfy the numerical conditions. Simulation programs, design methods
and use of phase diagrams have to be integrated in an efficient manner and be able to
avoid any violation of thermodynamic principles. This integrated technique will hopefully
lead to the generation of better process alternatives. Current and future research is
directed towards the development of an integrated method for design, simulation and
thermodynamic properties.

REFERENCES

1. Skjold-Jørgensen, S.: Group Contribution Equation of State (GC-EOS) : A
    Predictive Method for Phase Equilibrium Computations over Wide Ranges of
    Temperature and Pressures up to 30 MPa. Ind.Eng.Chem.Res., 1988, **27**, 1, 110-18.

2. Dahl, S., Fredenslund, Aa. and Rasmussen, P.: The MHV2 model - A UNIFAC
    based model for Prediction of gas solubility and Vapour-Liquid Equilibria at Low
    and High Pressures. I&EC Research, 1991, **30**, 1936-45.

3. Andersen, P.M.: Steady-State Simulation of Chemical Processes. Ph.D. Thesis. Dept.
    of Chem. Eng., Tech. Univ. of Denmark, 2800-Lyngby, Denmark, 1985.

4. Sørensen, E.L.: Dynamic Process Simulation using Accurate Thermodynamic Modelling. Ph.D. Thesis. Dept. of Chem. Eng., Technical University of Denmark, 2800-Lyngby, Denmark, 1991.

5. Brignole, E.A., Andersen, P.M. and Fredenslund, Aa.: Supercritical Fluid Extraction of Alcohols from water. Ind.Eng.Chem.Res., 1987, **26**, 254-61.

6. Cygnarowicz, M.L. and Seider, W.D.: Effect of Retrograde Solubility on the Design Optimisation of Supercritical Extraction Processes. Ind.Eng.Chem.Res., 1989, **28**, 1497-1503.

7. Perregaard, J. and Sørensen, E. L. : Simulation and optimisation of Chemical Processes : Numerical and Computational Aspects. Comp. Chem Engng.,1992, **16**, Suppl., S247-54.

8. Michelsen, M.: Phase Equilibrium Calculations. What is easy and what is difficult ? Computers Chem. Engng., 1992, **16**, Suppl. S19-S30.

9. Biegler, L.T. and Cuthrell, J.E.: Improved Infeasible Path Optimisation for Sequential Modular Simulators - II: The Optimisation Algorithm. Comp.chem.Eng., 1985, **9**, 257-67.

10. Perregaard, J.: A Steady-State Chemical Process Simulator with Optimisation. MAN8702. Dept. of Chem. Eng., Tech. Univ. of Denmark, 2800-Lyngby, Denmark, 1991.

11. Soave, G.: Equilibrium Constants From a Modified Redlich-Kwong Equation of State. Chem.Eng.Sci., 1972, **27**, 1197-1203.

12. Dahl, S., Dunalewics A. and Fredenslund, Aa.: The MHV2 Model : Prediction of Phase Equilibria at Sub- and Super-critical Conditions. Journal of Supercritical Fluids, 1992 (in press).

13. Bossen, B.S., Jørgensen, S.B. and Gani, R.: Simulation, Design and Analysis of Azeotropic Distillation Operations. AIChE Annual meeting, paper no. 60d, Los Angeles, Nov. 1991.

PHASE EQUILIBRIA AND PROCESS SIMULATION FOR HIGH-PRESSURE SUPERCRITICAL EXTRACTION PROCESSES: EXPERIMENTAL INVESTIGATION

G. Di Giacomo[*], V. Brandani[*], G. Del Re[*] and E. Martinez de la Ossa[**]
[*]Dipartimento di Chimica Ingegneria Chimica e Materiali, Università di L'Aquila
Monteluco di Roio, 67100 L'Aquila, Italy
[**]Departamento de Ingenieria Quimica
Universidad de Cadiz, Apdo. 40, 11510 Puerto Real, Cadiz, Spain

ABSTRACT

The purpose of this work, carried out within the European community Joule Program, is to collect new reliable experimental data of phase equilibria of synthetic mixtures a supercritical or near critical solvent, water and an organic product to be recovered from the aqueous solution. These data are required for the development and quantitative testing of predictive phase equilibria model, including generalized group contribution models, to be applied near the critical conditions of one or more component mixture. The experimental apparatuses used to perform the experimental measurements are described together with some results obtained.

INTRODUCTION

The work described in this paper is a part of a research carried out within the frame of the EEC JOULE R&D programme, contract n. JOUE 0053-C(MB). To this contract collaborate three institutions: Danmarks Tekniske Hojskole, Intitut for Kemiteknik (DK), as contractor and cordinator, which is mainly involved in modelling and process simulation; the University of L'Aquila, (IT), as a contractor, and University of Cadiz, (E), as a subcontractor of University of L'Aquila, which both contribute for the experimental measurements as well as for the validation and selection of phase equilibrium data.

Supercritical solvent extraction allows to separate a multicomponent mixture by combining the characteristics of distillation (differences in component volatilities) and of liquid extraction (differences in the specific interactions between solvent and the mixture components). In contrast to liquid-liquid extraction processes, where the solvent in the extracted phase is recovered by distillation, a supercritical or near critical solvent can be recovered by simply decreasing the system pressure. In addition the solubility in a solvent close to its critical point increases dramatically since the density of the goseous solvent approaches liquidlike densities.

The above mentioned features of supercritical or near critical fluids may turn out quite interesting in minimizing the energy requirements of several separation processes as discussed in detail by different authors, namely McHugh and Krukonis (1) and Brignole et al. (2).

Anyway a prerequisite to correctly design and evaluate a supercritical extraction process is the knowledge of the phase equilibria and in particular of the solute loading (amount of solute in the light phase) and of the solvent selectivity. This last property is defined as follows:

$$selectivity = \frac{\left[(x_1)_F\right] / \left[(x_1)_L\right]}{\left[(x_2)_F\right] / \left[(x_2)_L\right]}$$

where, for the system water-ethanol-solvent, x denotes composition, subscripts 1 and 2 denote ethanol and water respectively, subscript F indicates the light phase and L indicates the liquid phase.

To improve the experimental data base it is necessary to perform new measurements which are carried out using three different experimental apparatuses:

-a single pass flow apparatus, which allows the determination of product loading and solvent selectivity at pressures up to 100 MPa;

-an optical flow type apparatus which allows to measure the compositions of both supercritical fluid rich-phase and liquid phase;

-an optical static cell which can operate up to 30 MPa and is currently in the final stage of calibration at University of Cadiz.

Our single pass flow type apparatus has been extensively used for both fluid-fluid and solid-fluid systems (3). The optical flow type apparatus has been used for phase equilibria measurements of the system water-ethanol-propane, for which it was not possible to find experimental data in the literature. For this reason the apparatus was previously tested using the system water-ethanol-carbon dioxide, for which several sets of experimental data are available.

EXPERIMENTAL APPARATUSES AND RESULTS

A simplified flow sheet of the single pass flow apparatus is shown in figure 1, it has been described in detail in a previous paper (3). This apparatus allows to measure the composition of the fluid phase, but, although it is not possible to measure the liquid phase composition, it is possible to calculate the solvent selectivity, too, as was described in a paper dealing with the system carbon dioxide-water-ethanol (4).

Supercritical fluid from the gas cylinder is compressed by the pump to the operating pressure and flows through a first cell (200 cm^3) which dumps pulsations and acts as preheater. The solvent flows through the saturator, which is composed of two identical stainless steel cylinders. Each cylinder is 80 mm long and it has an internal volume of 100 cm^3. The first cylinder is packed with inert material while the second is filled with the solution to be extracted. The stream leaving the saturator is expanded to atmospheric pressure through a micrometering valve into cold glass traps where the solute condenses. The mass flow rate of the solvent is continuously monitored by a mass flow meter which also acts as totalizer. The flow rate is measured with an accuracy of $\pm 0.4\%$. The pressure is measured by a pressure transducer located at the top of the saturator with an accuracy of 0.15%. The temperature is measured by a type J thermocouple located inside the saturator. The whole saturator is kept at constant temperature by means of an aluminum mantle through which a thermostating fluid flows. Both micrometering valves V1 and V2 are used to adjust the pressure and flow rate at the desired value. At the end of each run all the collected solute is recovered from the glass traps using a suitable solvent and the resulting solution is analyzed. The flow rate of the supercritical solvent is set to a value (typically 0.5 g/min), depending on the solvent density at the operating temperature and pressure, that allows a sufficient contact time to attain equilibrium between liquid and fluid phases.

The single pass flow apparatus has been extensively used to measure the solubility

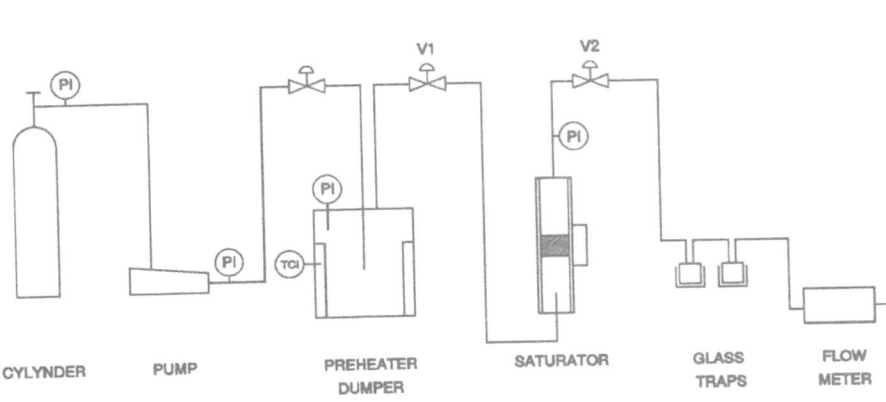

Figure 1 - Schematic diagram of the single pass flow experimental apparatus.

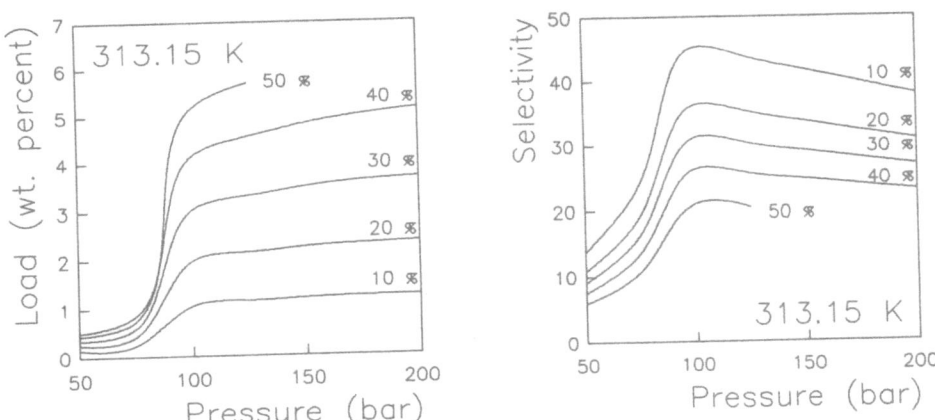

Figure 2 -Ethanol loading and carbon dioxide selectivity as function of system pressure and ethanol liquid phase composition at 313.15 K for the system ethanol-water-carbon dioxide.

of limonene and of citral in carbon dioxide (5) and the loading and selectivity for the system carbon dioxide-water-ethanol. For this last system the experimental data determined with this apparatus were used for validation of literature data as discussed detail in previous papers (3,4). In addition, figure 2 shows the behaviour of ethanol

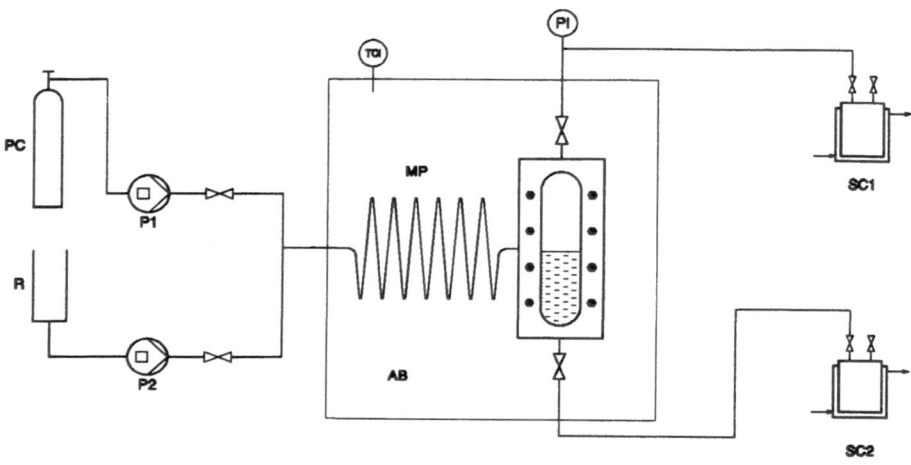

Figure 3 - Schematic diagram of the flow type apparatus.
SC1, SC2 : Sample collectors; PC: Propane cylinder; R: Reservoir;
P1, P2: HP Pumps; AB: Air bath; OC: Optical cell; MP: Mixer-preheater.

loading and carbon dioxide selectivity as function of system pressure.

In figure 3 is shown a schematic diagram of the optical flow type apparatus, which enables to measure the equilibrium composition of both liquid and light phase at pressure up to 30 MPa. Supercritical or near critical fluid and the solution to be extracted at fixed composition and constant rate are fed to an high pressure through window optical cell of about 30 cm^3 by two metering pumps (Milton Roy). Before entering the high pressure optical cell, the two streams pass through a preheater or cooler which also acts as a static mixer where equilibrium between the two coexisting phases is reached. The mixer is made up by a 5 meters long stainless steel coil of about 1.5 mm internal diameter which is inserted, together with the optical cell, inside a recirculating air bath equipped for precision automatic temperature control. The equilibrated phases separate by gravity in the optical cell which is equipped for precise pressure regulation and control. The liquid stream coming from the bottom of the cell is collected in a sample collector of about 0.2 liter equipped with a liquid circulating jacket for cooling. Similarly, the solvent rich phase, coming from the top of the cell is collected in a second sample collector as shown in figure 3. At the end of each run, the composition of the mixtures contained in the two collectors are obtained by volumetrically measuring the amount of supercritical solvent and by weighing and analyzing the residual liquid mixtures.

This apparatus besides the possibility of measuring both liquid and fluid phase composition, allows to check if more than two phases are present and, possibly, to measure the compositions of every existing phase.

The reliability of this apparatus was checked by comparing the results obtained with the ternary mixture ethanol-water-carbon dioxide with literature data and with the results obtained in our laboratory with the single pass flow apparatus. The apparatus was then used to collect experimental equilibrium data for the system water-ethanol-propane.

Some row experimental data for the system water-ethanol-propane at 300 and 350 K and at 13 and 50 bar are reported in Table 1; x and y indicate mole fractions in the water rich phase and in the propane rich phase respectively. Subscripts A, W and P are used to indicate ethanol water and propane respectively. In fig. 4 a comparison between experimental and calculated compositions of both phases is shown. The calculated compositions were obtained using a non cubic equation of state based model (6). Work in progress to collect further data at different pressure, temperature and feed composition.

TABLE 1

Raw experimental phase equilibrium results for the ternary system ethanol-water-propane.

T, K	P, bar	x_A	x_W	x_P	y_A	y_W	y_P
300.0	13.0	0.6940	0.2945	0.0115	0.1472	0.0677	0.7851
300.0	13.5	0.6706	0.3094	0.0200	0.2016	0.0122	0.7862
300.0	12.8	0.6499	0.3334	0.0167	0.2269	0.0211	0.7520
300.0	13.0	0.5911	0.3920	0.0166	0.2650	0.0127	0.7223
349.0	50.0	0.6718	0.3030	0.0252	0.5374	0.0924	0.3702
350.0	50.0	0.6707	0.2453	0.0840	0.5648	0.0777	0.3575
350.0	50.0	0.7071	0.2634	0.0295	0.5383	0.0945	0.3672

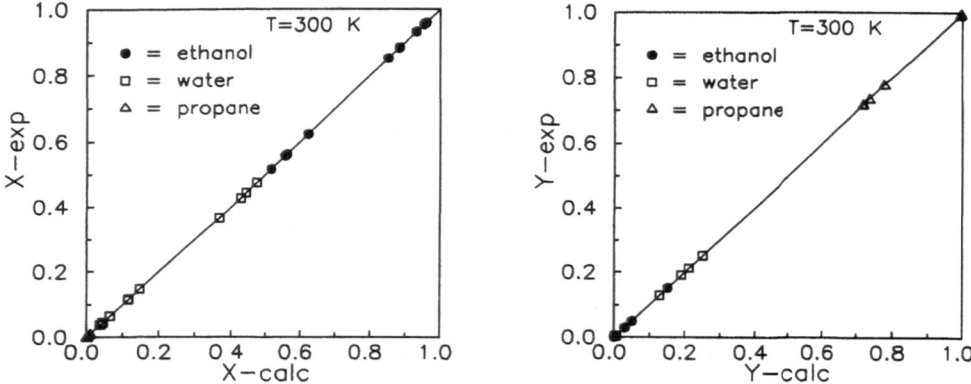

Figure 4 - Comparison between calculated and experimental composition of liquid and light phase for the system water-ethanol-propane.

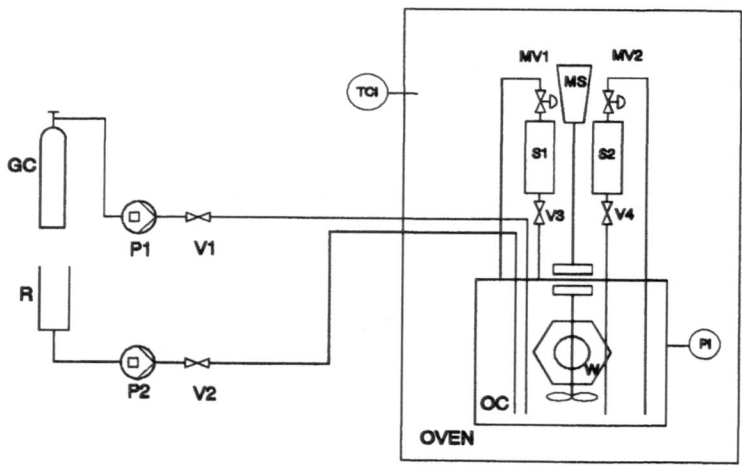

Figure 5 - Schematic diagram of the static optical apparatus for measuring high pressure phase equilibria data. GC: gas cylinder; R: liquid reservoir; P1,P2: High pressure pumps; V1,V2,V3,V4: on-off valves; MV1,MV2: metering valves; MS: magnetic stirrer; SC1,SC2: sample collectors; OC: optical cell; W: quartz glass window; OVEN: recirculating air bath.

In figure 5 is shown a schematic diagram of the optical static cell that is currently in final stage of calibration at University of Cadiz. Two high pressure metering pumps are used to feed the supercritical solvent and the solution into the equilibrium cell and to regulate the pressure. At the beginning of each test the two sample collectors S1 and S2 are filled with mercury. The solution loaded into the equilibrium cell is stirred using a magnetic-driven stirrer until equilibrium conditions are reached, then stirring is stopped and valves V3, V4, MV1 and MV2 are opened. The mercury contained in the sample collectors S1 and S2 drops into the equilibrium cell, at the same time, an equal amount of each phase come into the sample collectors. In this way the total volume of the cell is kept to a constant value, minimizing the pressure fluctuations. Then both valves are closed and the sample collectors are removed for sample analysis.

CONCLUSIONS

Different experimental apparatuses have been used to determine experimental equilibrium data for both the ternary systems water-ethanol-carbon dioxide and water-ethanol-propane under pressure and temperature conditions close to the critical point of the solvent. In both cases the behavior of ethanol loading and solvent selectivity show that low energy process for ethanol recovery from aqueous dilute solutions under pressure and temperature conditions close to the critical point of the solvent could be feasible.

The experimental data for the system water-ethanol-carbon dioxide have been compared with several sets of literature data and have been used to discriminate between contradictory results. In addition, by using the new measurements for the system water-ethanol-propane it is now possible the validation of the predictive model which are used for process simulation.

REFERENCES

1 McHugh M., Krukonis V., Supercritical Fluid Extraction, Butterworths, Boston, 1986, 69-78
2 Brignole E. A., Andersen P. M. and Fredenslund A., Supercritical Fluid Extraction of Alcohols from Water. Ind. & Eng. Chem. Res., 1987, 26, 254-261.
3 Martinez de la Ossa E., Brandani V., Di Giacomo G., Del Re G., Ferri E., Binary and ternary phase behaviour of the system water-ethanol-carbon dioxide. Fluid Phase

Equilibria, 1990, **56**, 325-340.
4 Di Giacomo G.,Brandani V. Del Re G., Martinez de la Ossa E.,Selectivity and loading behaviour in liquid carbon dioxide extraction of ethanol from dilute aqueous solutions. Chem. Biochem. Eng. Q., 1991, **5**, 141-144.
5 Di Giacomo G., Brandani V., Del Re G. and Mucciante V., Solubility of essential oil components in compressed carbon dioxide. Fluid Phase Equilibria, 1989, **52**, 405-411.
6 Masciulli L.,Estrazione di Composti Organici da Soluzioni Acquose con Solventi Supercritici. Thesis for the Degree in Chemical Engineering, University of L'Aquila, L'Aquila, Italy, 1992.

SESSION 5:

Separation Processes
Continued

Chairman: Prof. K.E. Porter

IMPROVEMENT OF MELT CRYSTALLIZATION'S EFFICIENCY FOR INDUSTRIAL APPLICATIONS

R. de Goede,
TNO Institute of Environmental and Energy Technology,
Laan van Westenenk 501, P.O. Box 342, 7300 AH Apeldoorn,
the Netherlands

ABSTRACT

From calculations based on replacing part of a conventional distillation process by single step melt crystallization in a conceptual separation train it appeared that an energy saving of about 30 % is likely to be attainable. This project focusses on comparing the two current melt crystallization technologies being (i) layer growth and (ii) suspension growth with respect to their ability to upgrade a 99% pure feed to 99.99%, the composition of the waste flow being 90%. This implies that the distribution coefficient, being defined as the ratio of the impurity concentrations in the crystals to that in the waste should have a value of 0.001. Experiments were carried out with two test systems. Crystal layers grown from stagnant melts yielded distribution coefficients in the range 0.4 - 0.8. This result could not be improved by sweating. In the case of crystal layers growing from moving melts distribution coefficients of 0.1 - 0.5 were found. A further reduction of the impurity content by a factor of almost ten could be realized with a washing treatment. Crystals grown in suspension yielded distribution coefficients in the range 0.005 - 0.025. These results indicate that single step suspension crystallization is likely to be sufficient to attain the preset separation duty provided that an efficient washing treatment is added.

INTRODUCTION

General

Melt crystallization is an energy efficient separation methodology capable of producing ultrapure materials. The power of the technique was recognized a long time ago as a laboratory standard for purification of organics. Industrial applications have been known for quite some time in cases where conventional technologies like distillation or extraction fail. An example is the separation of isomeric mixtures: distillation is hardly feasible because the boiling points of the components of the mixture are too close. Upon cooling however, one of the components will crystallize and thus can be separated from the mixture in pure form.

During the last decades it has been recognized that natural sources of energy are not inexhaustable. This has led to research and development attempts in the area of alternative separation technologies requiring less energy input than the conventional ones. Melt crystallization turns out to be an attractive alternative, because in most cases no thermodynamically stable miscibility occurs in the solid phase [1]. Consequently, very high separation efficiencies are feasible in single step operation using small to zero reflux flows.

Two principally different technologies are available for melt crystallization on an industrial scale:

1. Layer growth.
Crystallization takes place at externally cooled surfaces until the mother liquid is sufficiently expired. Then the mother liquid is drained off and the crystal layer is melted. Accomplishment of the required product purity in single step operation is not necessarily the most efficient way of processing because of low production capacity. As an alternative, higher crystal growth rates are applied leading to serious contamination of the solid layer due to entrapment of liquid entities. As a consequence, the process has to be repeated thereby reducing the energy benefits. Alternatively, end purification can be realized by sweating and washing with pure product which decreases the overall recovery. However, due to the simplicity of the method and easy scale up layer growth is already proven technology. The most widely known processes are Sulzer MWB [2] and Proabd [3].

2. Suspension growth
In this case, crystal growth occurs in suspension. Crystal-liquid separation is carried out either by centrifuges or wash columns, the latter being advantageous because due to the countercurrent washing operation the amount of contaminated mother liquid remaining attached to the crystal surface is reduced to a minimum. In suspension, much more crystal surface area is available for crystal growth allowing for lower growth rates yielding the same production per unit of equipment volume as layer growth. Contamination of the crystalline material occurs only to a minor extent so that high single step separation efficiencies are to be expected. However, the large specific surface area implies that the purity of the product is strongly determined by the slurry separation step. For a limited number of applications crystallization from suspension has been applied on a technical scale (paraxylene, paradichlorobenzene, naphthalene). For several other applications this technology has been proven on a laboratory and pilot plant scale. Due to a lack of knowledge in the area of scaling up, the process industry is rather reluctant to adopt this technology. Processes are being sold by Tsukishima [4] and Niro PT, the former Grenco PT [5]. Wash columns for slurry separation have been developed by Grenco, Phillips and TNO [5,6,7].

The aim of this project is to optimize and compare both crystallization technologies with respect to their separation efficiency and to promote application of melt crystallization on an industrial scale. In this paper some examples of attainable energy savings when incorporating melt crystallization into the separation train for some bulk organics are given. Secondly, the main results of the underlying project and their relevance for industrial application will be presented.

Scope of the project

The scope of this project can be defined as: (i) to investigate how incorporation of impurities in the crystal layer depends on process conditions such as crystal growth rate and initial impurity content in the feed, and (ii) to what extent removal of these impurities by operations like washing and sweating can be realized.

The purification duty is defined as to upgrade a feed containing 99% of the desired component to a 99.99% ultrapure product and a waste flow containing 90% of the desired component:

$$Feed\ (99\%) === > Product\ (99.99\%) + Waste\ (90\%)$$

Set-up of the project

The project is financially supported by JOULE, the Dutch Government and a number of industrial companies which are listed in the "Acknowledgements". Experimental work is being carried out at laboratory and pilot plant scale. The following contractors are cooperating in this project:

* TNO-IMET, the Institute of Environmental and Energy Technology of the Netherlands Organization for Applied Scientific Research, takes care of project coordination and laboratory and pilot scale experiments in the area of suspension crystallization;

* The University of Nijmegen provides fundamental support by (i) performing single crystal experiments with standardized measurement techniques under well defined experimental conditions and (ii) calculation of solubility of impurities in the solid phase with highly sophisticated theoretical models.

* The University of Bremen is doing laboratory scale experiments on layer growth under a variety of conditions.

* BASF, Ludwighafen, Germany, takes care of experiments on layer growth on laboratory and pilot plant scale.

Scientific support is provided by the following consultants:

* Prof.dr. J.N. Sherwood and Dr. K.J. Roberts, from the University of Glasgow, UK;

* Prof.ir. E.J. de Jong, Prof. em. of the University of Delft, the Netherlands;

* Dr. G.J. Arkenbout, after his retirement as workgroupleader of the Physical Separations Group of TNO-IMET.

Two industrially important test compounds were chosen i.e. caprolactam and naphthalene. In order to facilitate explanation of the results on a theoretical basis binary model systems were used instead of technical feedstocks. Caprolactam was contaminated with cyclohexanone and naphthalene with biphenyl.

Analytical procedures were provided by BASF. Samples with known concentrations of impurities were distributed by TNO and analyzed by all contractors to ensure comparable analytical data. The experimental results will be used for a technical study to be executed by BASF, the University of Bremen and TNO.

THE ENERGY SAVING POTENTIAL OF MELT CRYSTALLIZATION

The energy saving potential of melt crystallization will be illustrated by its incorporation into a conceptual separation train. We suppose 1400 kg of feed to be converted to 1000 kg product, 140 kg unreacted, 130 kg light components and 130 kg heavy components. When all separations are carried out by distillation the process scheme looks as follows:

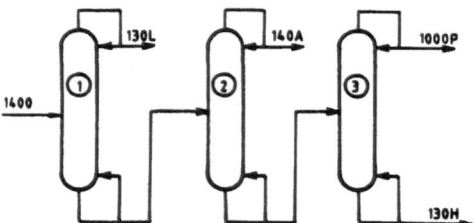

Fig.1. Simplified process flow diagram for separation of the conceptual reaction mixture by distillation.

It will be assumed that the total heat input equals the sum of the reboiler duties, which are related to the top product flows by:

$$Q = (R+1) \cdot D \cdot \Delta H_{evap}. \tag{1}$$

with D = the distillate flow, R = the reflux ratio, ΔH_{evap} = the heat of vaporization and Q = the heat input. In this example, the reflux ratio is taken to be 3 and for the heat of evaporation a value of 400 kJ/kg is chosen. This value is representative for most organics. Now if we assume that 50 % of the product can be recovered with melt crystallization then the revised flowscheme would be as shown in fig 2:

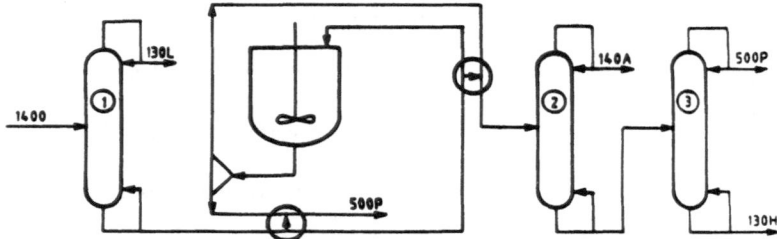

Fig.2. Flowscheme for the conceptual separation task, now recovering 50 % of the product by crystallization from the melt.

Further assumptions being made here are (i) the heat of crystallization amounts to 1/3 of the heat of vaporization, which is generally accepted as a rule of thumb, and (ii) crystallization occurs of a temperature trajectory of 100°C and the numerical value of the heat capacity amounts to 2% of the heat of crystallization.

The energy balance for the first flowscheme reads:

Tower 1: $Q = (3 + 1) * 130 * \Delta H_{evap.} = 520 * \Delta H_{evap.}$
Tower 2: $Q = (3 + 1) * 140 * \Delta H_{evap.} = 560 * \Delta H_{evap.}$
Tower 3: $Q = (3 + 1) *1000 * \Delta H_{evap.} = 4000 * \Delta H_{evap.}$

Total heat requirements: Tower $(1) + (2) + (3) = 5080 * \Delta H_{evap.}$

For the second flowscheme the following energy balance is obtained:

Tower 1: $Q = (3 + 1) * 130 * \Delta H_{evap.} = 520 * \Delta H_{evap.}$
Tower 2: $Q = (3 + 1) * 140 * \Delta H_{evap.} = 560 * \Delta H_{evap.}$
Tower 3: $Q = (3 + 1) * 500 * \Delta H_{evap.} = 2000 * \Delta H_{evap.}$

Heat requirements distillation: $\quad 3080 * \Delta H_{evap.}$

Crystallizer: $Q = 500 * 1/3 * \Delta H_{evap.}$ (latent heat) +
$\qquad\qquad 500 * 1/3 * \Delta H_{evap.} * 0.02 * 100$ (sensible heat)
$\qquad\qquad = 500 * \Delta H_{evap.}$

So the total heat requirement amounts to $3080 + 500 = 3580\ \Delta H_{evap.}$

From the calculations presented above it appears that the energy consumption when applying single step crystallization from the melt is 3580/5080 = 70 % of the total heat requirement in the case where product recovery is carried out with distillation alone. Hence, incorporation of melt crystallization would lead to an energy saving of about 30 %. However, when repetition of the crystallization process is required the energy advantage decreases rapidly [8].

Analogous calculations have been performed for a number of real bulk organics. Basic conventional process schemes were taken and melt crystallization was incorporated where appropriate [9]. The results are summarized in the following table:

Table 1.
Estimated energy savings due to incorporation of melt
crystallization into the conventional separation train.

Component	energy saving (%)
benzene	27
styrene	32
caprolactam	50
phenol	63
dimethylterephthalate	96

Regarding the total production capacities of bulk organics and organic fine chemicals in the Netherlands, a general acceptance of melt crystallization technology could lead to a total energy saving of $2.7 * 10^{15}$ J/year. This is equivalent to 85 million cubic meters of natural gas.

THEORETICAL BACKGROUNDS

Just as with vapor-liquid equilibria, solid-liquid equilibria occur in various types. However, as in vapor-liquid systems total miscibility and monotonous connection of the boiling points of both components is regarded as the most regular behaviour, total immiscibility in the solid phases is most commonly observed. This is not surprising when realizing that the crystal structures of both componenents have to be very much alike to allow for mutual solubility. The presence of a second component leads to depression of the freezing point of the first component which implies that two solid phases coexist at temperatures well below the freezing points of both pure components thus leading to phase diagrams of the eutectic type. More complicated phase behaviour occurs when solid compounds with a fixed composition (e.g. aqueous inorganic salt sytems, in which hydratation by fixed numbers of molecules of water frequently occurs) are formed or when mutual solubility in the solid phase exists. This being over a limited range or throughout the entire concentration range. Matsuoka [1] investigated the frequency of occurrence of the different types of phase diagrams for over 200 organic compounds. The results are summarized in fig. 3. It is apparent that most of the systems form simple eutectics. This observation directly illustrates the power of crystallization compared to conventional technologies like distillation and extraction: in principle, only one theoretical equilibrium stage is required to isolate a pure component. However, it has been shown that metastable miscibility may occur when high driving forces are applied [10].

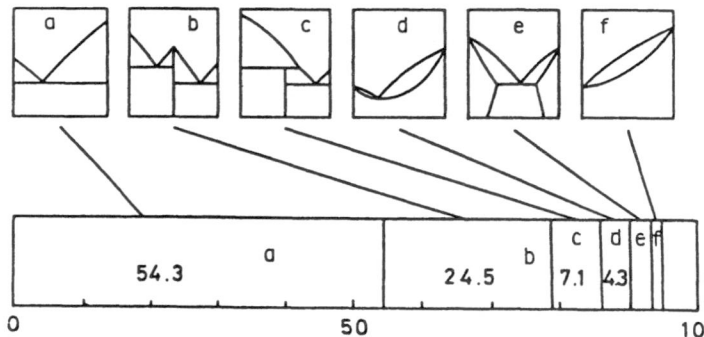

Fig.3. Occurrence of different types of solid - liquid equilibria in binary organic mixtures [1].

Although thermodynamic equilibrium considerations look very favourable the real situation is more complicated because kinetic effects may lead to contamination of the crystalline phase. Entrapment of liquid entities may occur especially when crystals grow irregularly by needles and dendrites. Such events are favoured by applying high driving forces, leading in their turn to high crystal growth rates.

The ratio of impurity concentrations in the solid and in the liquid phase is expressed in terms of a distribution coefficient:

$$k \equiv \frac{C_{i,S}}{C_{i,L}} \qquad (2)$$

in which i refers to the impurity, C denotes the concentration of impurity i in weight fraction, and S and L refer to the solid and the liquid phase respectively. The distribution coefficient is in principle a function of temperature, impurity concentration, crystal growth rate and mass transfer coefficient.

Removal of impurities can be realized by repeated crystallization, which requires an additional energy input, or by a sweating operation. Sweating is carried out by bringing the contaminated crystal phase to its melting temperature; this will induce melting of the contaminated regions in the crystal and diffusion of the impurities out of the solid phase through micropores [1]. The disadvantage of this method is the decreasing recovery due to consumption of product for sweating.

Another complication arises with phase separation: incomplete solid-liquid separation results in contamination of the product due to mother liquid remaining attached to the crystal surface. Washing with pure melt to replace the contaminated mother liquid is the most effective remedy.

Returning to the separation task as given in the scope of the investigations, an overall distribution coefficient of 0.001 is being aimed at. Arkenbout [11] showed already in a previous paper that in the case of single step operation such low values are only feasible when low crystal growth rates ($< 10^{-7}$m/s) are being applied. Because the production per unit equipment volume equals the product of crystal surface area per unit volume and linear growth rate, application of low growth rates is favoured by crystallization in suspension. In this case the crystal surface area is much larger than that of a crystal layer growing at a cooled wall. In suspension, crystal growth rates in the range 10^{-8}m/s - 10^{-7}m/s are common, while for layer growth linear growth rates in the order of 10^{-6}m/s are applied. The disadvantage of suspension crystallization is that because of the larger specific surface area more sophisticated slurry separation technologies are required to reduce contamination of the product due to adhering mother liquid to a minimum.

EXPERIMENTS, RESULTS AND DISCUSSION

Layer growth processes

The equipment used by the University of Bremen is schematically shown in Fig.4. In technical equipment the tubes are cooled at the outer side but for laboratory experiments the situation is reversed to allow for visual observation of the crystal layer. The diagram shows the configuration for dynamic experiments. For static experiments there is no feed being circulated but a heating medium which heats the vessel at the outer side (dashed region).

Fig.5 shows the distribution coefficients for the system caprolactam/cyclohexanone as a function of the crystal growth rate for different initial impurity concentrations under static and dynamic conditions measured at the University of Bremen. Experiments with moving melts resulted in distribution coefficients in the range 0.1 - 0.3. Static experiments in the same range of crystal growth rates yielded distribution coefficients of 0.4 - 0.8. A sweating treatment did not improve the results of the static experiments. Low crystal growth rate static experiments lead to distribution coefficients that lie in the same range as those from dynamic experiments.

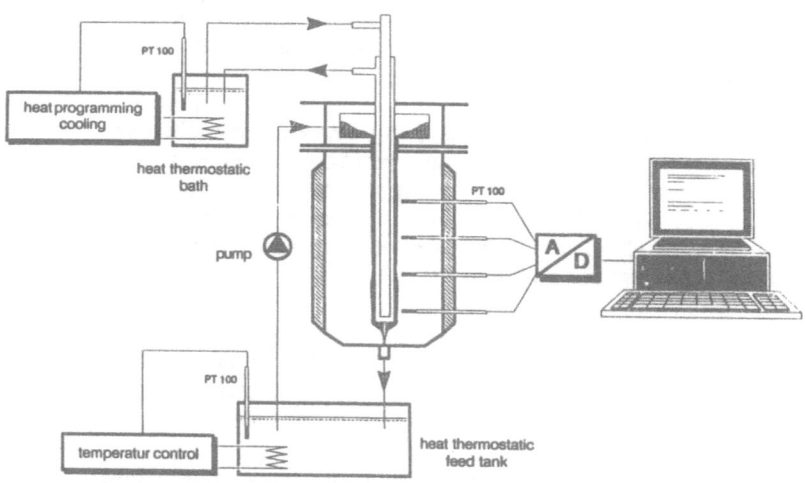

Fig.4. Test unit for layer growth [12]

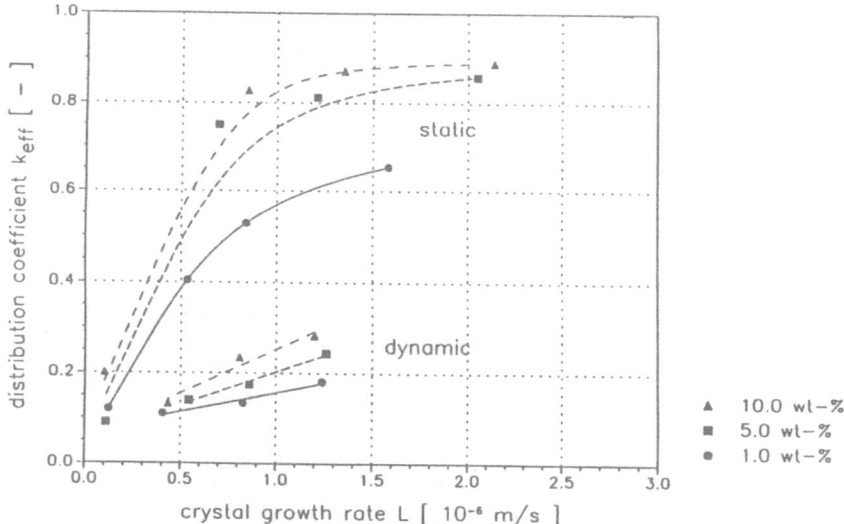

Fig.5. Comparison of distribution coefficients obtained for layers grown from static and dynamic melts [13].

For laboratory purposes BASF used a vessel in which a crystal layer was allowed to grow on the bottom. The pilot plant configuration is shown diagrammatically in fig.6:

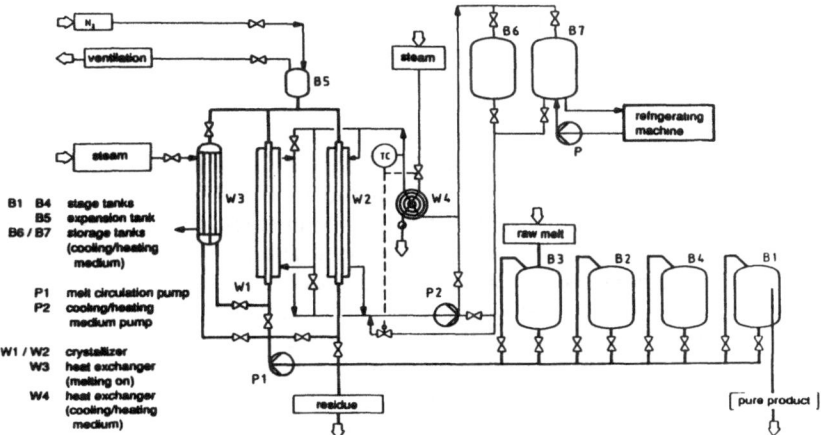

Fig.6. Pilot plant equipment used by BASF [14]

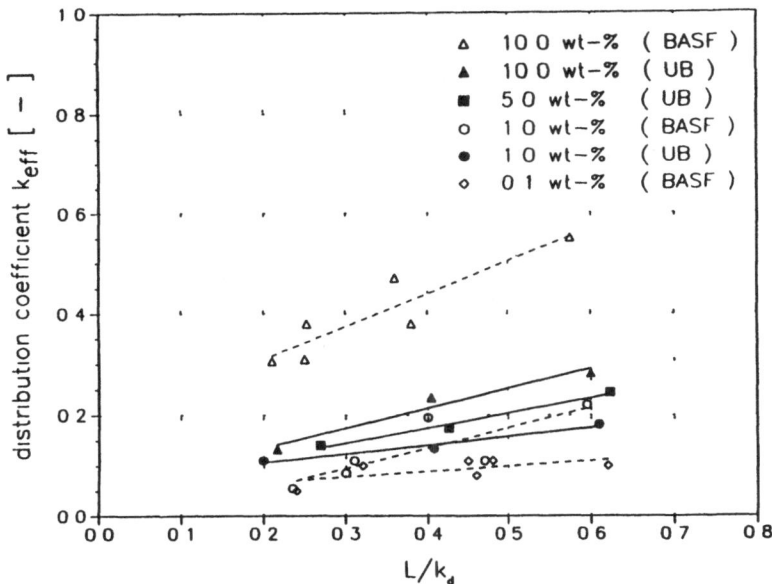

Fig.7. Comparison data UB-BASF. L = crystal growth rate in m/s, k_D = mass transfer coefficient [13].

The melt to be purified is pumped through the heat exchangers W1 and W2 (tube diameter 25 mm, length 5 m) arranged in series. Layers of crystals are frozen on the inner surfaces of the tubes by lowering the temperature in the heat carrier circuit. After the freezing step the residual melt is drained off. The crystal layer can then be washed with melt which is purified in a preceeding step. At the end of the washing treatment the crystal layer is melted and drained off.

A comparison of the caprolactam/cyclohexanone data from the University of Bremen and BASF is presented in fig.7. Both datasets are measured under dynamic conditions with laboratory scale equipment. All data lie in the range 0.1 - 0.5. Experiments with the system naphthalene/biphenyl carried out by BASF showed a comparable result although in the range of low growth rates and low initial impurity concentration ultrapure crystal layers were formed. Plotting distribution coefficients as a function of the ratio of growth rate and mass transfer coefficient turns out to be an adequate way of eliminating equipment characteristics. Further experiments by BASF showed that a further reduction of the impurity content by almost a factor ten was achieved by subjecting the crystal layer to a wash process. A theoretically justified way of correlating the results has proven to be a powerful tool to predict the separation effect on a larger scale.

Suspension growth

The experimental program on suspension growth consists of (i) zone melting experiments, a repeated normal freezing technique to measure distribution coefficients, (ii) experiments with a 100 ml batch crystallizer, (iii) experiments with a 3 ltr. continuous crystallizer, and (iv) experiments on 70 ltr. pilot plant scale. Pilot plant data is the closest to technical application but the results are not available yet. Therefore only 3 ltr. results will be presented in this section. A schematic diagram of the 3 ltr. crystallizer is presented in fig.8:

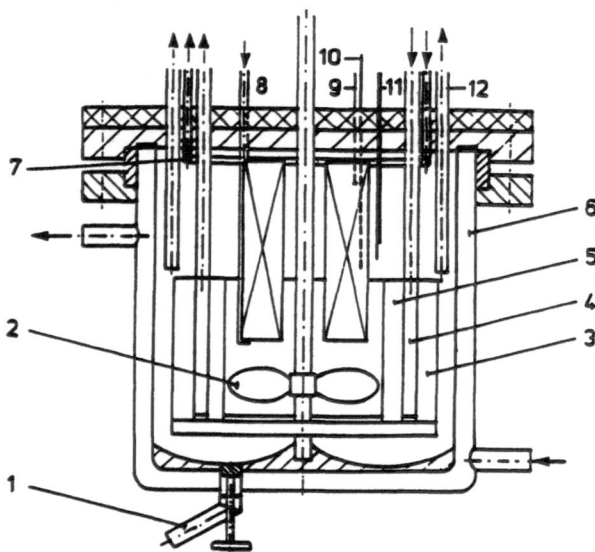

Fig.8. Diagram of 3 ltr. crystallizer [15]

Heating occurs via the double-walled glass cylinder (6). A separate heating coil (7) is installed in order to prevent crystallization above the liquid level. Cooling is performed with a cylindrical heat exchanger (4) which serves as a draft tube as well. In order to prevent incrustation, both sides of the heat exchanger are provided with scrapers (3 and 5). Propeller (2) takes care of proper mixing: streamline flow is approached by installation of baffles (9). Feed enters the crystallizer via the top port (8); intermittant slurry removal takes place through port (12). In this way, sufficiently high slurry outflow velocities are ensured in order to prevent classification during slurry drain-off. When the slurry outlet valve opens the slurry is pressed out by a slight overpressure of nitrogen gas. The frequency is controlled by level-controller (11). The temperature is measured with a Pt-100 element. Continous operation is supported by melting the slurry after which it is used as fresh feed again.

Crystallizer residence times of about 30 minutes were applied. It was assumed that a steady state was reached after 8 residence times. A slurry sample was taken then and separated in a bowl centrifuge until no liquid drain off was observable (about 5 minutes). The crystal size distribution was determined by sieve analyses. A sieving time of 15 minutes was applied. For both test systems, the crystallizer temperature and thus the slurry density was varied.

Crystal growth rate, nucleation rate and mean crystal size were estimated from the crystal size distribution using the well-known population balance theory [17]. Distribution coefficients are presented as a function of the crystal growth rate in fig.9:

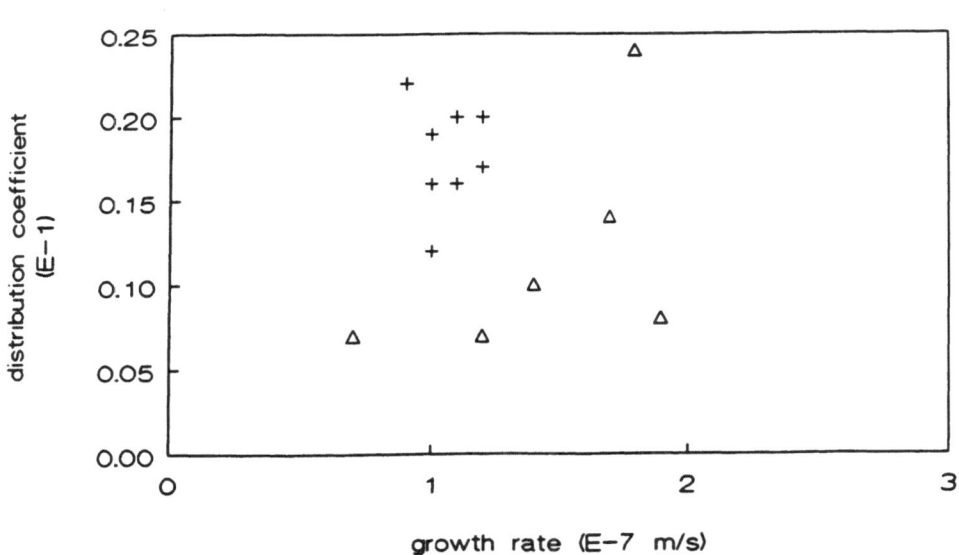

Fig.9. *Distribution coefficient vs. crystal growth rate from 3 ltr. suspension crystallization experiments [16].*

Distribution coefficients in the range 0.005 - 0.025 resulted here which is about an order of magnitude lower than those obtained from layer growth experiments. This means that a combination of crystallization in suspension and an effective washing operation is likely to achieve the preset separation target. Allthough the temperature was varied, the crystal growth remains almost constant at a value of 10^{-7} m/s. This observation is due to the fact that a lower temperature directly yields a higher slurry density thus keeping the driving force for crystal growth at a constant value. At the lowest measurement temperatures slurry densities up to 300 kg per cubic meter were observed. Production capacities of 360 and 660 kg per hour per cubic meter crystallizer volume were attained for caprolactam and naphthalene respectively. Mean crystal sizes were in the range 600 - 1000 microns which is coarse enough for further processing with a wash column.

Fundamentals

The University of Nijmegen is focussing on studying growth properties of single crystals in the presence of an impurity under well defined conditions. For these purposes well known crystal growth techniques like the Tammann and Bridgman method are being used but innovative designs of measuring techniques to control the temperature profile in the vicinity of a crystal are under development as well. The observed phenomena such as changes in morphology are being explained on a theoretical basis. Distribution coefficients are being estimated with computer models by calculating the energy and entropy changes involved in replacing host molecules in the crystal lattice by impurity molecules. The generated knowledge is valuable for the results of both layer and suspension growth processes.

ACKNOWLEDGEMENT

The author expresses his gratitude to Directorate General XII of the European Community, DSM, Enichem Anic, Amoco Belgium, NOVEM, Xytel Europe, Niro Proces Technology and the Goudsche Machinefabriek for sponsoring the project, to the contractors mentioned already in the introduction for the fruitful cooperation and to the consultants for their valuable scientific support.

CONCLUSIONS

From the results and discussion reported in the foregoing section it was concluded that:

* From rough calculations it is suggested that incorporation of a single step melt crystallization process in conventional separation trains could lead to substantial energy savings. A saving of $2.7 * 10^{15}$ J/year was estimated for the Netherlands alone. Therefore, further attempts to develop melt crystallization into a mature technology are justified.

* When growing crystal layers from moving melts distribution coefficients of 0.1 - 0.5 were obtained. A substantial improvement was achieved by a subsequent washing treatment. Crystal layers grown from stagnant melts yielded distribution coefficients in the range 0.4 - 0.8; only when extremely low (10^{-7}m/s) growth rates were applied to the growth of a crystal layer from a stagnant melt, were distribution coefficients comparable with dynamic experiments obtained. Crystal layers grown from stagnant melt could not be upgraded by sweating. The results obtained thus far by the University of Bremen and by BASF are in good agreement.

*   Crystallization in suspension yields distribution coefficients of 0.005 - 0.025 which is about one order of magnitude lower than the results of crystal layers grown at a cooled wall. On a 3 ltr. scale suspension crystallization yielded satisfactory production capacities for both test systems: 360 and 660 kg per hour per cubic meter crystallizer volume for caprolactam and naphthalene resp. The crystals produced in the 3 ltr. laboratory continuous crystallizer system are coarse enough for further purification with a hydraulic wash column. Based on the results obtained so far, combination of suspension crystallization and an effective washing procedure is likely to achieve the preset separation duty.

FUTURE OUTLOOK

The experimental work of the current project approaches its completion and the technological impact of the results will be evaluated thoroughly. The EC made extra money available for extension of the project for a period of 18 months starting at 1 October '92. In the extension part we will use the same test systems and the same experimental techniques but we will focus on separation rather than purification, the separation target being defined as:

$$\text{Feed (90\%)} ===> \text{Product (99.9)} + \text{Waste (10 - 20\%)}$$

For the extension, the project team remains unchanged. A majority of the sponsors have already agreed to continue their sponsorship.

REFERENCES

[1]   M. Matsuoka, Bunri Gijutsu (Separation Process Engineering 6(1977) 245.

[2]   O. Fischer, S.J. Jancic, K. Saxer,"Purification of compounds forming eutectics and solid solutions by fractional crystallization", in: Industrial Crystallization '84, ed. S.J. Jancic and E.J. de Jong, Elsevier 1984 pp. 153-157.

[3]   J.G.D. Molinari, "The Proabd refiner" Fractional Solidification, ed. M. Zief, R.W. Wilcox, Marcel Dekker 1, chapter 13, 1967

[4]   K. Takegami, N. Nakamura, M. Morita, "Industrial Molten Fractional Crystallization", in: Industrial Crystallization '84, ed. S.J. Jancic and E.J. de Jong, Elsevier 1984 pp. 143-146.

[5]   W.H.J.M. van Pelt, H.A. Jansen, "Freeze Concentretaion Economics and Application, in: Process Technology Proceedings 5, Preconcentration and Drying of Food Materials (ed S. Bruin), Elsevier 1988, pp. 77-86

[6]   D.L. Mc Kay, H.W. Goard, "Continuous Fractional Crystallization", Chem.Eng.Prog. 61(1965) pp. 99-104

[7]   G.J. Arkenbout, A. van Kuyk, L.H.J.M. Schneiders, "The TNO-Thijssen Crystallization Process and Wash Column", in: Industrial Crystallization '84, ed. S.J. Jancic and E.J. de Jong, Elsevier 1984 pp. 137-142

[8]   G. Wellinghoff, K. Wintermantel, "Schmelzkristallisation- theoretische Voraussetzungen und technische Grenzen", Chem.Ing.Techn. 63(1991) pp 881-891.

[9]   M.Nienoord, "Energy saving potential in the Netherlands when applying cystallization from the melt als a separation technique for organics", March 1989, TNO-report 89-114

[10]   R. de Goede, G. Hakvoort, G.M. van Rosmalen, "Solid-liquid equilibria in meta-paraxylene and ortho-paraxylene binary mixtures at atmospheric pressure", Thermochimica Acta (1990) pp.

[11]   G.J. Arkenbout, in: Proceedings "Improved energy efficiency in the proces industry, ed. P. Pilavachi, Brussels 1990.

[12]   M. Neumann, J. Ulrich in: "Improvement of melt crystallization's efficiency for industrial applications", first progress report, Appendix B, Apeldoorn 1991.

[13]   M. Neumann, J. Ulrich in: "Improvement of melt crystallization's efficiency for industrial applications", third progress report, Appendix B, Apeldoorn 1992.

[14]   G. Wellinghoff in: "Improvement of melt crystallization's efficiency for industrial applications", first progress report, Appendix C, Apeldoorn 1991.

[15]   R. de Goede in: "Improvement of melt crystallization's efficiency for industrial applications", third progress report, Appendix D, Apeldoorn 1992.

[16]   R. de Goede in: "Improvement of melt crystallization's efficiency for industrial applications", fourth progress report, Appendix D, Apeldoorn 1992.

[17]   A.D. Randolph, M.A. Larson, Theory of particulate processes, Academic Press, New York 1971.

ABSORPTION-DRIVEN MULTIPLE EFFECT EVAPORATORS
A STUDY OF THE ABSORBER-REGENERATOR COUPLE

S.YANNIOTIS
Agricultural University of Athens
11855 Athens, Greece

P.Le GOFF
Lab.des Sciences du Genie Chimique
Nancy, France

ABSTRACT

The principle of operation, the advantages and disadvantages of absorption-driven multiple effect evaporators are presented. Some basic thermodynamic relations for the regenerator-absorber are developed. The efficiency and the basic characteristics of an experimental four effect falling film evaporator coupled with a two effect regenerator operated with NaOH solutions are given.

INTRODUCTION

The use of hygroscopic solutions in evaporators as heating medium instead of steam has been initially proposed by Schwartzberg [1]. In these evaporators a concentrated hygroscopic solution i.e. lithium bromide or sodium hydroxide solution, is introduced in a falling film evaporator and flows down the outside surface of the evaporator tubes, while the liquid product flows down the inside surface of the tubes. If the product side is connected to the solution side through a vapour line, the water vapour generated from the evaporation of the water from the product is transferred and absorbed by the hygroscopic solution because its water vapour is less than that of the product. The heat of condensation and solution produced by the absorption of water vapour by the solution sustains the evaporation process. The diluted solution is recycled to a regeneration station.

Hygroscopic solutions can be concentrated in a "batch" mode in parabolic solar collectors [2] or using some low cost energy source (i.e. combustion of residues). Alternatively, conventional multiple effect evaporators can be used as regenerators of the hygroscopic solutions. In the last case the evaporator can be coupled to the regenerator and work in a "continuous" mode [4].

If the boiling point elevation of the solution is sufficiently high, multiple effect product evaporators can be operate using hygroscopic solutions instead of steam. In this case only the first effect has to be of the absorption type (absorber). A vapour return line connects the product side of the last effect with the absorbing solution side of the first effect and the hygroscopic solution absorbs the vapour generated in the last effect. The vapour produced in the other effects is used as in conventional multiple effect evaporators where the vapour produced in one effect is used as heating medium in the next effect.

A four effect falling film evaporator coupled with a two effect regenerator is shown in Figure 1. In such a system one might expect that the amount of steam required in the regenerator would be 50% of the steam required in the first effect of a conventional evaporator that would produce the same evaporation rate, because the evaporation of the absorbed water vapour takes place in two effects in the regenerator. In practice the energy requirements of an absorption-driven evaporator is higher than 50% because of inefficiencies of the system. In any case they have less energy requirements, less cooling water requirements in the condenser and they can operate with low boiling temperature in the first effect. The last characteristic is very important for heat sensitive products like liquid foods where the highest temperature can not exceed certain point i.e. 70-72°C in milk evaporators. Their disadvantage is the extra cost of the regenerator.

439

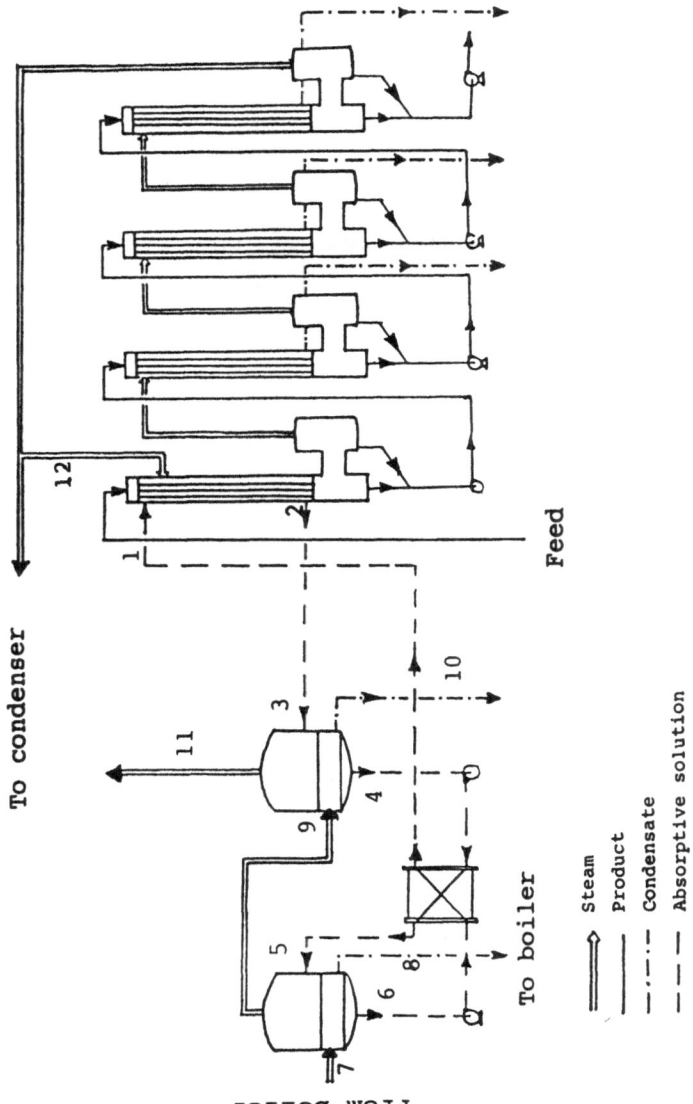

Figure 1. Four effect absorption-driven evaporator coupled with a two effect regenerator

THERMODYNAMIC CONSIDERATIONS

The absorption-driven evaporator is in fact an open-cycle absorption heat pump coupled to a multiple effect evaporator. In the regenerator (DESORBER-CONDENSER) thermal exergy of high value is converted to chemical exergy while in the absorber chemical exergy upgrades the low value heat (the vapour from the last effect of the evaporator) to high value heat (high exergy) which drives the multiple effect evaporator.

To get a better understanding of the system the mass, enthalpy and exergy balances of the absorber-regenerator couple for the system of Figure 1 are written.

Mass balance

$$m_{12} - m_{10} + m_{11} \qquad (1)$$

Enthalpy balance

$$m_7(h_7 - h_8) + m_{12}h_{12} - Q_{ev} + Q_{loss} + m_{11}h_{11} + m_{10}h_{10} \qquad (2)$$

from eqns (1) and (2) we get

$$m_7(h_7 - h_8) + m_{10}(h_{12} - h_{10}) + m_{11}(h_{12} - h_{11}) - Q_{ev} + Q_{loss} \qquad (3)$$

Also

$$Q_{ev} - m_1 h_1 + m_{12}h_{12} - m_2 h_2 \qquad (4)$$

where

m=mass flow rate (kg/s)
h=specific enthalpy (kJ/kg)
Q_{ev}=heat transferred to the multiple effect evaporator (kW)
Q_{loss}=heat losses (kW)

Exergy balance

$$m_7(e_7-e_8)+m_{10}(e_{12}-e_{10})+m_{11}(e_{12}-e_{11})-Ex_Q+Ex_D \tag{5}$$

where

   e=specific exergy (kJ/kg)

   Ex_Q=exergy transferred due to heat transfer (kW)

   Ex_D=exergy destroyed (kW)

In making the enthalpy balance it is helpful to represent the operation of the absorber/regenerator on the PONCHON diagram, where the specific enthalpy of the solution is plotted vs. the solution concentration. The isotherm lines and the equilibrium isobar lines provide the necessary information to represent the absorption/regeneration cycle on the plot. The enthalpy of the solution is read directly from the diagram. If one draws the tie lines on the diagram the temperature and concentration at the surface of the film can also be found. Details on the design of the absorber have already been presented [5]. A similar representation of the unit on an entropy-concentration diagram is helpful in calculations of the exergy balance.

Evaluation criteria

The performance of the system can be evaluated using the following criteria:

   The enthalpy-economy efficiency, which is a measure of the useful heat transferred to the high value heat supplied, is defined as:

$$\eta_{ee}=\frac{Q_{ev}}{m_7h_7+m_{12}h_{12}} \tag{6}$$

(if we consider that the condensates have not any economic value, while the vapour from the last effect of the evaporator has).

The exergy efficiency, which is a measure of the exergy transferred to the evaporator as compared to the exergy used in the regenerator/absorber, is defined as:

$$\eta_{ex} = \frac{Ex_Q}{m_7(e_7-e_8)+m_{10}(e_{12}-e_{10})+m_{11}(e_{12}-e_{11})} \qquad (7)$$

The exergy-economy efficiency which is defined as the exergy transferred to the evaporator over the high value exergy used:

$$\eta_{exe} = \frac{Ex_Q}{m_7 e_7 + m_{12} e_{12}} \qquad (8)$$

EXPERIMENTAL INSTALLATION

The system shown in Figure 2 has been built and operated within the frame of a JOULE project in which Hellas Energy, Greece and the Laboratoire des Sciences du Genie Chimique, France, participate.

The regenerator

The regenerat or is composed of two effects connected in series. It operates as a backward feed two effect evaporator. The dilute solution coming out of the absorber enters the second effect where it is partially concentrated. It is then pumped to the first effect where regeneration is completed. Energy is supplied to the first effect with an electrical resistance. The vapour produced in the first effect is used as heating medium in the second effect. A solution plate heat exchanger is used between the two effects to preheat the solution as it flows from the second effect to the first effect. A shell and tube heat exchanger is used as condenser.

The absorber

The absorber is composed of two columns in series each with four 1" tubes, 1.90 m long. The strong solution enters the column at the top and flows down on the outside surface of the tubes as a film. The product flows down the inside surface of the tubes as in a falling film evaporator. Thus the absorber is also the first effect of the evaporator.

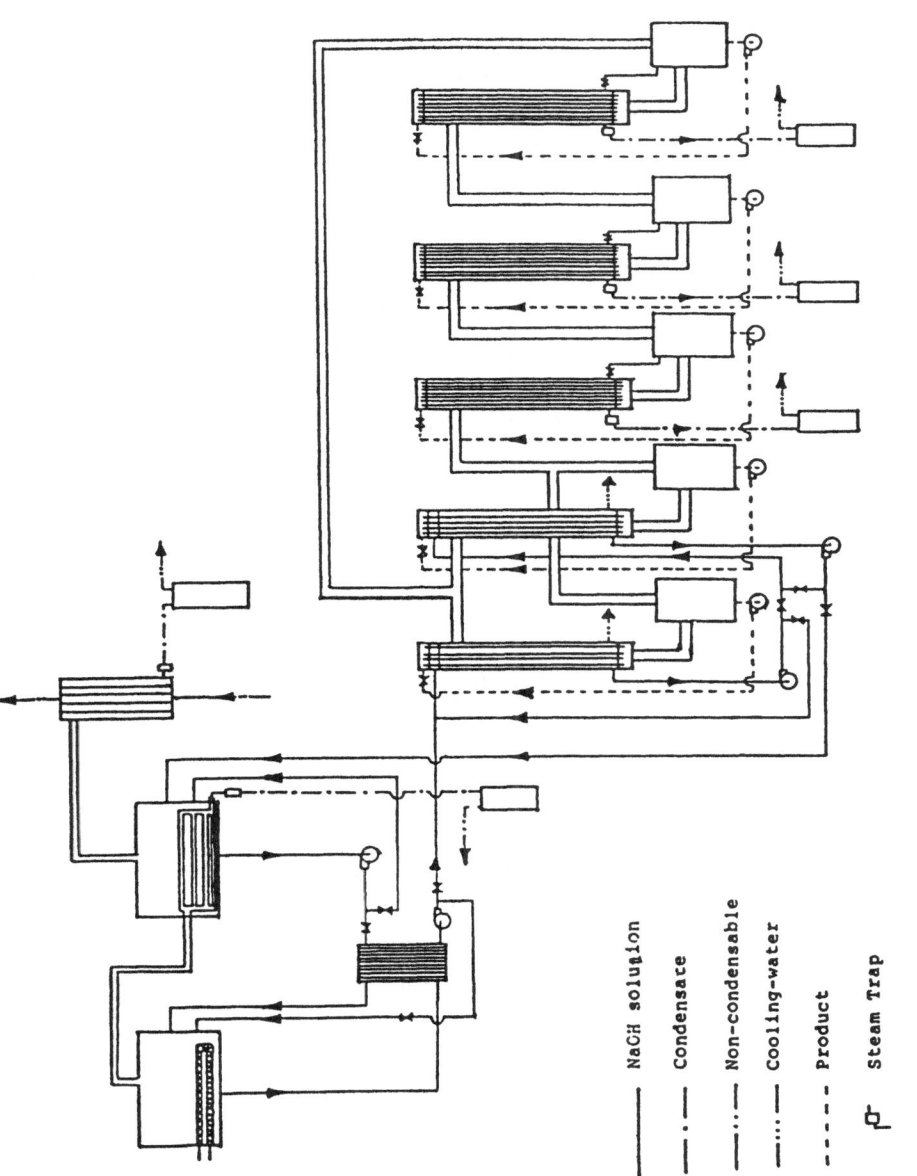

Figure 2. Experimental unit

—— NaOH solution

—·— Condensate

—··— Non-condensable

······ Cooling-water

------ Product

⌐ Steam Trap

The evaporator

Three more effects of the falling film type are connected in series with the absorber. Each effect has eight 1" tubes, 1.90 m long. The vapour produced in the absorber is used in the second effect. The vapour produced in the second effect is used in the third effect etc. The vapour produced in the fourth effect is returned to the absorber. A liquid ring vacuum pump is used to remove non-condensables from the system.

Instrumentation and control

A data acquisition system that is connected to a desktop computer is used to collect data for the temperature and the pressure at various points of the system every two minutes. The unit also controls the temperature of the first effect of the regenerator and the cooling water flow rate to the condenser.

Rotameters are used to measure the flow rate of the various streams. A kWh meter is used to measure the energy consumption of the system.

RESULTS AND DISCUSSION

The experimental unit was operated using sodium hydroxide solution as absorptive medium, but it is very flexible so that other hygroscopic solutions, i.e. lithium bromide solution, can be used. It was also operated as conventional evaporator using steam in the first effect as heating medium, instead of sodium hydroxide solution. It was proved that absorption-driven multiple effect evaporators are technically feasible.

The efficiency of the system, operated with 50 % NaOH solution as absorptive medium, was calculated according to Equations (6), (7) and (8) assuming that the heating medium is steam at $160°C$ and using as reference state $25°C$:

$n_{ee}=0.6-0.7$

$n_{ex}=0.6-0.7$

$n_{exe}=0.5-0.6$

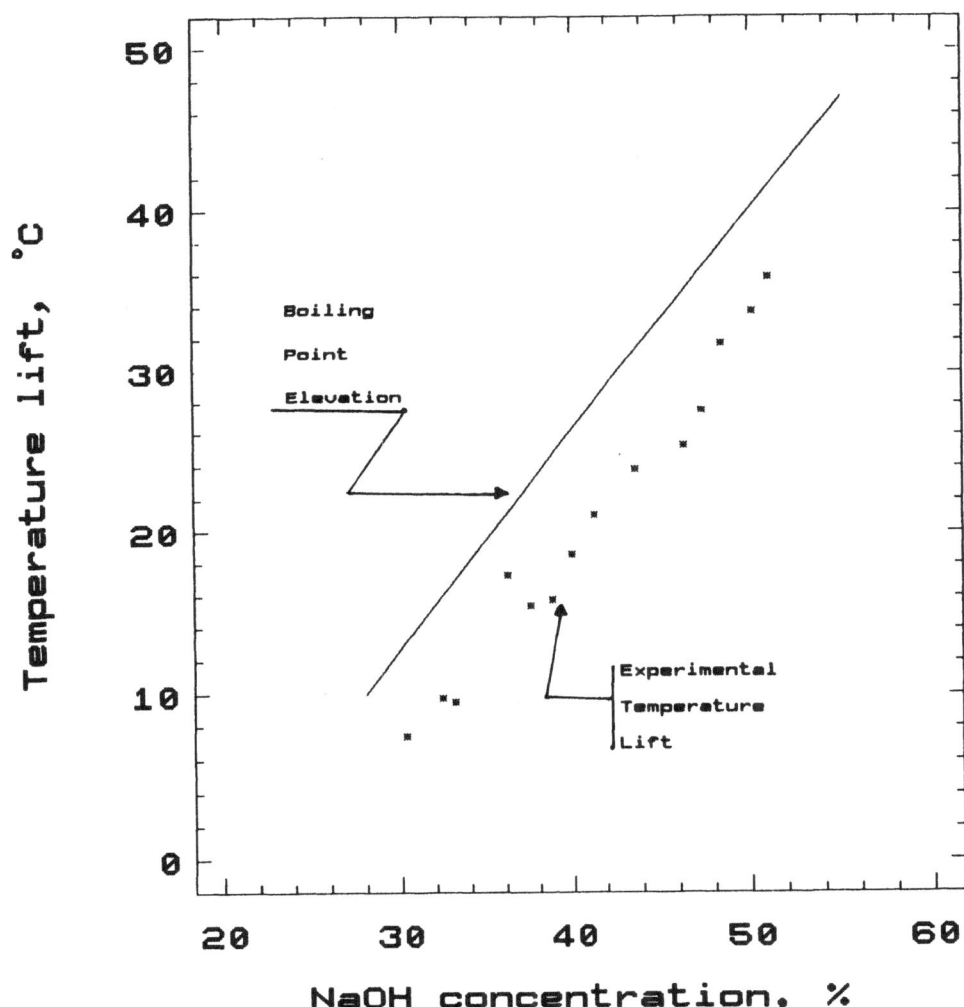

Figure 3. Experimental temperature lift in the evaporator
and theoretical boiling point elevation of the NaOH at the
pressure of the absorber (9.6 kPa)

The energy required to reconcentrate the solution (50% NaOH) was 2450 kJ/kg water evaporated from the solution. For every kJ of energy supplied in the regenerator 1.3-1.4 kJ were transferred to the multiple effect evaporator. In the absorber, for every kg of water vapour absorbed 1.5-1.6 kg of water were evaporated. Theoretical calculations have shown that the performance of the system could be further improoved with the proper optimization of the system.

The overall temperature difference (temperature lift) between the bulk temperature of the solution in the absorber and the fourth effect of the evaporator was less than the theoretical boiling point elevation of the absorptive solution, as shown in Figure 3. High temperature lift is necessary to drive multiple effect evaporators. In this respect sodium hydroxide has the advantage over other solutions i.e. lithium bromide, but on the other hand its high viscosity and its high solidification temperature are disadvantages. Mixtures of sodium and potassium hydroxide provide high temperature lifts and lower crystallization temperature than NaOH solution alone (Smith and Carey [3]). Such mixtures can operate at higher concentrations without the danger of solidification when the system is not in operation.

ACKNOWLEDGMENT

The authors are grateful to CEC and Hellas Energy for the financial support of this project.

REFERENCES

1. Schwartzberg H., Energy requirements for liquid food concentration. Food Technology, 1977, 31(3):67.

2. Schwarzberg H. Rosenau, K. Kim and S. Yanniotis, The use of solar concentrated brines for food processing. In Agricultural Energy, 1981, ASAE 5-81 v.3, p.386.

3. Smith I. and C. Carey, The alkali metal hydroxide/water absorption heat pump. Proceedings of the Congress on absorption heat pumps, CEC Paris 20-22 March 1985, p.165

4. Yanniotis S., Absorption-driven multiple effect evaporators. In **Industrial processes**, Proceedings of a contractors meeting, Brussels June 1988. Edited by P. Pilavachi, CEC puplic. EUR 12246 EN. p.121.

5. Yanniotis S., P. LeGoff, Absorption-driven multiple effect evaporators. In the proceedings of the European Seminar on Improved Energy Efficiency in the Process Industry, Brussels, July 1990, Edited by P. Pilavachi. p.129.

THE ENHANCEMENT OF HEAT TRANSFER IN FALLING FILM EVAPORATORS BY A NON-UNIFORM FLOW RATE

A. RAMADANE*, P. LE GOFF**, Bq. LIU**

*SNEA - DRDIE - PARIS (France).
**LSGC-CNRS-ENSIC-INPL BP 451 54001 NANCY-CEDEX

ABSTRACT

In an evaporator, with a falling film of uniform thickness and velocity, it is well known that the heat transfer coefficient h (w m^{-2} K^{-1}) is a decreasing function of the fluide velocity in laminar regime and an increasing function in turbulent regime.

However, if the liquid is flowing in the form of rapid rivulets on some parts of the wall, and in the form of thinner films on other parts, we show that such a non-uniform distribution of the fluid velocities, produces an enhancement of the overall heat transfer coefficient h, by a ratio of the order of 2. The addition of vertical fins on the wall may have a double effect : increasing the evaporating surface area, and also increasing the non-uniformity of the liquid flow distribution.

The present research programme is a part of a contract joue concerning an "Absorption-driven four-effect evaporator" coupled with a two-effect regenerator under the direction of Prof. S. Yanniotis and the company "hellas energy".

We will show that, in a falling film evaporator, a non-uniform distribution of the thickness of the falling film induces an enhancement of the heat transfert coefficient.

The non-uniformity of flow may be obtained either in a spatial geometric structure (corrugated wall surface, adjunction of metallic grids ...) or in a temporal structure (periodic waves).

We consider the case of a liquid film being evaporated as it flows down a vertical heated surface. First let us assume that the flow is perfectly uniform. In other words the film has the same thickness and velocity at all points at the same level. Theoretical and experimental results in the literature show that the heat transfer conductance h between the surface and the liquid-vapour interface versus flow rate decreases in the laminar regime, passes through a minimum and then rises in the turbulent regime. The decrease in h in the laminar regime is due the progressive increase in the film thickness. Then, at a certain critical flow rate waves start to form in the film (wavy laminar regime). As flow rate is increased these waves are replaced by eddies which become more and more intense as flow rate is increased. These eddies increase the value of h, first by compensating for, then by more than compensating for, the increase in film thickness.

THE ROLE OF THE SURFACE ROUGHNESS

Now let us assume that the flow down the surface is rendered non-uniform by the presence of vertical grooves of sinusoidal cross-section, or by rods fixed vertically on the surface (see schematic diagram given in figure 1). These two configurations can be schematized by the simplified representation shown in figure 1c. Here zones of length L1 are covered by liquid flows thicker than the average where there is a liquid flow faster than the average (flow rate V1), and by zones of length L2 which are covered by films which are thinner than the average and where the liquid flows less fast than the average (flow rate V2).

$$\text{Let } \beta = \frac{L_1}{L_1 + L_2} = \frac{L_1}{L} \qquad (1a)$$

$$\text{and } \gamma = \frac{V_1}{V_1 + V_2} = \frac{V_1}{V} \qquad (1b)$$

Here β is the fraction of the surface covered by fast flows.

and γ is the fraction of the total volume flow rate V flowing in these zones.

We can calculate an overall value of the Reynolds Number Re for the whole surface and also a separate value for each of the two zones :

$$Re_m = \frac{\rho\ V}{\mu\ L} \quad Re_1 = \frac{\rho\ V_1}{\mu\ L_1} = \frac{\gamma}{\beta} Re_m \quad Re_2 = \frac{\rho\ V_2}{\mu\ L_2} = \frac{(1-\gamma)}{(1-\beta)} Re_m \qquad (2)$$

To each of these three values of Reynolds number, Re_1 Re_2 Re_m there corresponds a value of conductance, h_1, h_2, h_m and an apparent overall conductance :

$$\bar{h} = \beta h_1 + (1 - \beta)\ h_2 \qquad (3)$$

For example point Q on figure 2 represents a non-uniform flow composed of a thin film (point F) having $Re_2 = 10$ and and which occupies a fraction $(1 - \beta) = 90\%$ of the surface and a thicker stream (point R) of $Re_1 = 600$ which occupies the complementary fraction of the surface (10%). The fraction of the total flow rate which flows in these streams is $\gamma = 87\%$. Under these conditions $Re_m = 69$ and heat transfer is increased by a factor of :

$$F = \frac{\bar{h}}{h_m} = 1.82$$

The standard correlations for the evaporation of a falling film are :

In the laminar regime   (Re < 100) $Nu_L = 0.9\ Re^{-0.33}$
In the turbulent regime   (Re > 400) $Nu_T = 0.0066\ Re^{0.4}\ Pr^{0.65}$

$$\text{with} \quad Nu = \frac{h}{g}\left(\frac{V2}{g}\right)^{1/3} \qquad\qquad Re = \frac{r\ V}{m\ B} \qquad (4)$$

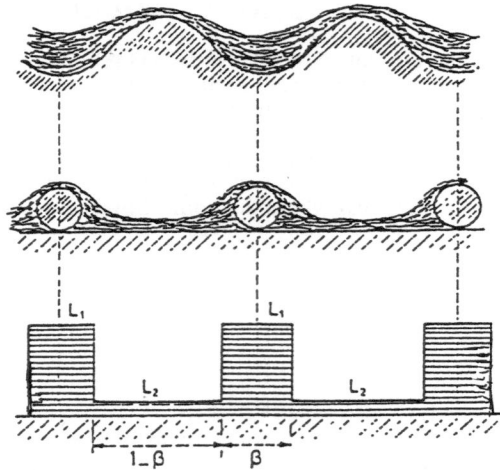

Figure 1 : Horizontal section through a vertical wall with
a flowing liquid film non-uniform thickness.

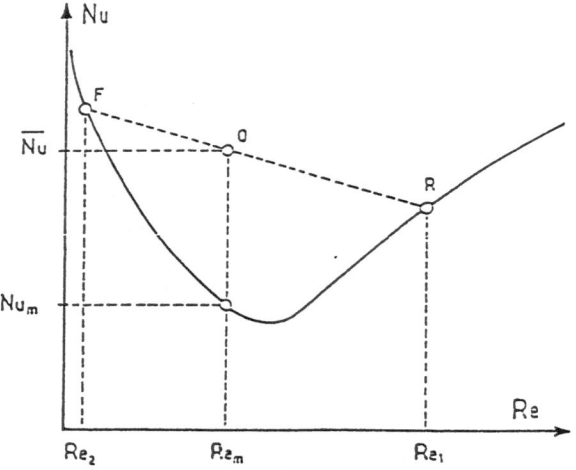

Figure 2 : Nusselt number as a function of the Reynolds number
for the evaporation of a flowing film. A non-uniform
flow (Nu) is more effective than a uniform flow (Nu_m).

Schnabel [1] proposed the following expression to cover both these regimes and include the intermediate regime :

$$Nu = (Nu_L^2 + Nu_T^2)^{1/2} \tag{5}$$

We have used these expressions to calculate the enhancement factor F, the factor by which heat transfer is increased, as a function of the coefficient β and γ which were varied between 0.1 and 0.9. Calculations were performed fot the three different flow regimes using the following values :

$Re_m = 50$ (laminar regime), $Re_m = 200$ (intermediate regime) and $Re_m = 600$ (turbulent regime).

The results are presented in figure 3 and suggest the following comments :

- At the point on each curve where $\beta = \gamma$, the enhancement factor is equal to 1. This means that if, for example 10 % of the flow rate is over a wetted width of 10 %, then the flow distribution is not disturbed.

- For $\gamma = 0.5$ each graph has an overall symetrical shape. This means that the same value of the factor is obtained for the couples (β, γ) and $(1-\beta, 1-\gamma)$.

- Finally, comparing the three graphs, shows that the enhancement factor F is only significantly greater than unity in the laminar and intermediate regimes and that the gain varies between 40 % and 90 %. We have in fact measured increases in heat transfer coefficient of this order of magnitude in our experiments.

CASE OF CONDENSATION ON A SMOOTH VERTICAL WALL

It is well known that vapour condensation as a film on a smooth wall gives results similar to those described above for evaporation. Here again the Nusselt number varies with Re-1/3 in the laminar regime and as Re+0.4 in the turbulent regime. The non-uniformity in flow would therefore have the same beneficial effect on condensation as it does for evaporation, and under the same flow conditions.

THE DOUBLE ROLE OF THE VERTICAL FINS

Figure 4 is a horizontal section through a flowing film evaporator fitted with vertical fins. Let β be the fraction of the total surface of the heated wall S due to the side and ends of the fins (index 1).

Let α be the fraction of the section Ω of the film which flows on the surface of the fins. It follows that $(1-\beta)$ is the fraction of the surface between the fins and that $(1-\alpha)$ is the fraction of the section through the film which flows on that surface (index 2).

Whatever the flow regime the action of capillary forces ensures that the film on the concave part of the wall (index 2) is thicker than on the convex part of the wall (index 1). The following inequality is therefore always valid :

$\alpha < \beta$ with $u_1 < u_2$

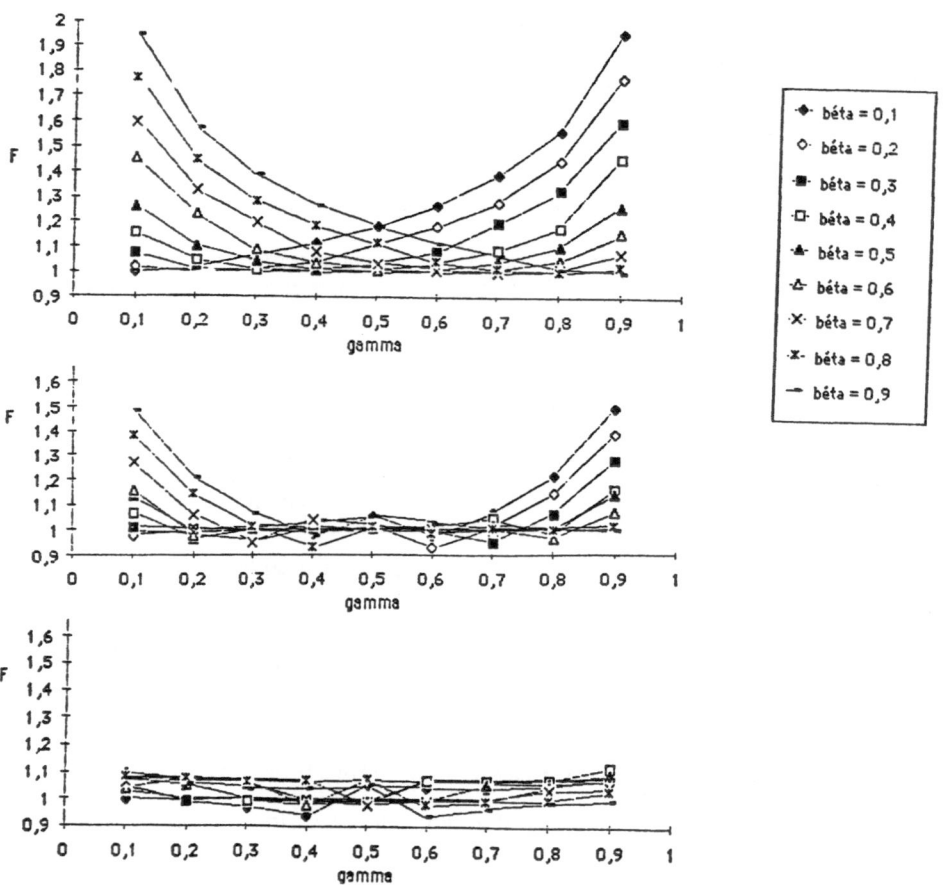

Figure 3 : The enhancement factor as a function of the coefficient γ
for three values of Reynolds number (50, 200, 600) and
various values of the coefficient β.

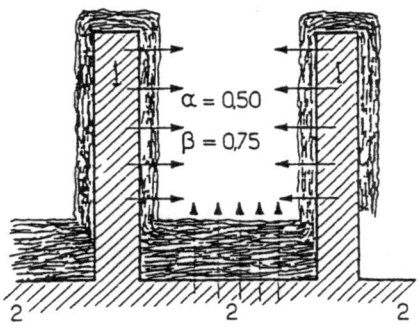

Figure 4 : Horizontal section through a falling film
evaporator fitted with vertical fins.

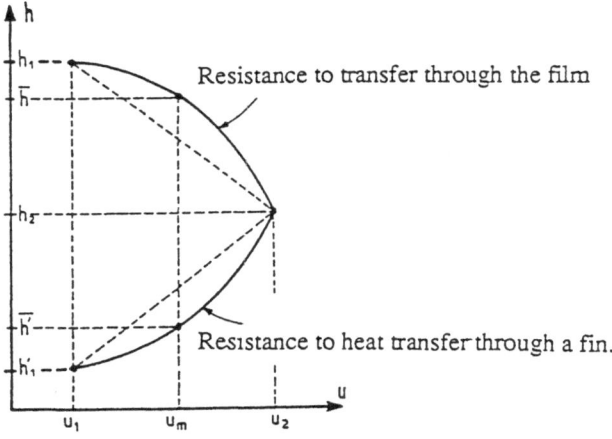

Figure 5 : Heat transfer as a function of the fluid velocity.

Two cases may then be considered, depending on whether the fin has a greater or lesser thermal resistance than the film.

First let us assume that the fin is made of a material which is an infinitely good conductor of heat such that the resistance to heat transfer is only due to conduction through the flowing film. In this case the transfer conductance h_1 on the faces of the fins would be greater than the conductance on the base surface. Therefore the apparent conductance h will be even greater than that which would result from a simple linear combination of h_1 and h_2. This is illustrated on figure 5 using the following numerical values :

$\alpha = 0.50$ and $\beta = 0.75$ (upper curve).

Let us now assume the opposite, that the fins are very poor conductors of heat. In this case we would have h'_1 less than h'_2. It is a simple matter to verify that under these conditions the value of the apparent heat transfer conductance will be smaller than the value given by a linear combination of h'1 and h'2. This is shown be the lower curve on figure 5.

CONCLUSION

In conclusion the use of vertical fins on the smooth wall of an evaporator increases the heat transfer flux firstly by the increase in the surface for evaporation. There is however a second effect due to the non-uniformity induced in fluid flow caused by the convexity and concavity created on the surface. This latter effect is only beneficial if the thermal conductance of the fins is sufficiently high.

The enhancement of heat and mass transfer coefficient obtained with the help of the present method, will be directly applied to the evaporators in operation in the laboratory of our greek partner within the Joule Project.

REFERENCES

[1]   G. SCHNABEL, Dr. Ing. Thesis, Karlsruhe (1980)
[2]   A. RAMADANE, M. BARKAOUI , H. LE GOFF, R. JURKOWSKI and
       P. LE GOFF
       World Congress III on Chemical Engineering
       Tokyo - Sept. 1986, Vol. 2, pp. 532-535

ENERGY-EFFICIENT EVAPORATORS IN THE DAIRY INDUSTRY

Sander Bouman, René Waalewijn, Peter de Jong, Joost H.J.L.J. van der Linden
Netherlands Institute for Dairy Research (NIZO),
P.O. Box 20, 6710 BA Ede, The Netherlands.

ABSTRACT

The possibilities for reducing the energy consumption of evaporator-spraydryer plants were investigated. Increasing the total solids content of the concentrate by reduced heat load offers the potentiality to reduce the energy consumption by 10 %.
Based on experimental results the computer program EvaDes (Evaporator Design) for the design of multistage falling-film evaporators in the dairy industry has been developed. It quantifies the heat-transfer area of the stages for different arrangements, the process conditions and the dimensions of the evaporator tubes.

NOTATION

			Subscripts	
D	inner diameter of evaporator tube	m		
E	specific energy consumption,			
	kg steam per kg water evaporated	-	c	concentrate
$4f$	friction factor	-	d	dryer
m	wetting rate	$kg.m^{-1}s^{-1}$	e	evaporator
p	pressure	$kg.m^{-1}s^{-2}$	g	gas
q	heat flux	$W.m^{-2}$	i,n	number of segment
r	correlation coefficient	-	m	milk
T	temperature	K	p	boiling product
TS	total solids content	%	s	condensing steam
v	vapour velocity	$m.s^{-1}$	V	vapour
X	moisture content of powder	%		
Re	Reynolds number, $\rho v D/\eta$	-		
α	heat transfer coefficient	$W.m^{-2}K^{-1}$		
η	viscosity	$kg.m^{-1}s^{-1}$		
ρ	specific mass	$kg.m^{-3}$		
Φ	mass flow rate	$kg.s^{-1}$		

INTRODUCTION

From the point of view of energy consumption, the falling-film-evaporation proces is a much more attractive process than spray drying. In addition, the efficiency of the evaporators has been greatly improved in the last ten years by increasing the number of stages (1). The experience gained with these large evaporators, however, has also revealed a serious drawback. It appears that in spite of the smaller differences in temperature between the stages, much product is left behind in these evaporators as a result of fouling on the very large areas. The value of this product loss and the cost of waste water purification contribute considerably to the total production costs (1). Minimizing fouling by optimizing the design and the process operation will lead to the use of energy-saving evaporators in the future. In this connection research projects have been initiated to describe the relations between the process conditions, energy consumption and the properties of the product.

The results of two research topics are presented here: the saving of energy by increasing the total solids of the final concentrate, and the methodological design of the stages of the falling-film evaporator. This type of evaporator is used in the dairy industry for milk and whey products but also in the sugar industry and for fruit juices.

The maximum total solids content is limited by the viscosity of the concentrate. Whole milk, for example, is usually concentrated not higher than about 48 % total solids. By decreasing the denaturation of whey proteins, resulting in lower viscosities, the total solids content of the concentrate can be increased.

For the design of falling-film evaporators it is necessary to know the factors which control the heat transfer and pressure drop in the evaporator tubes. These phenomena are complex, and it is therefore hard to find any correlations (2, 3). In a recent paper (4) the performances of falling-film evaporators used in the New Zealand dairy industry are reported. It appeared that there is a considerable scatter in the values for the heat transfer coefficients.

In this paper quantitative relations for the heat transfer and the pressure drop in evaporator tubes are given. For the knowledge acquired to become assessible and usable the EvaDes program has been developed.

MATERIALS AND METHODS

1. Energy Saving by Increasing Total Solids

The experiments were carried out with the four-stage falling-film evaporator and multistage dryer plant at NIZO. Whole milk was concentrated to total solids contents varying from 48 to 58 %. The flow rate of the milk was 2400 kg/h.

The milk was preheated in a plate heat exchanger at 70 °C and subsequently in another plate heat exchanger at 85 to 120 °C or in a DSI (direct steam injection) at 90, 110, 115 and 130 °C. The holding times were 2 and 10 s.

The total solids content of the concentrated milk was measured by the oven-drying method at 102 °C (5). The viscosity of the concentrated milk was measured with a Haake Rotovisco RV 2, System NV, MK 500 at various shear rates. The Whey Protein Nitrogen index (WPNi) was measured according to the American Dry Milk Institute (6).

2. Evaporator Design

The flow and boiling phenomena were investigated with a laboratory model evaporator. The milk flows along the outside of the tube so that flow and boiling behaviour are visible. Dimensions of the stainless steel tube: length 1 m, outside diameter 38 mm, wall thickness 1 mm. The tube was filled with copper and provided with an electrical heating wire.

The heat transfer and pressure drop were investigated with an experimental plant with one evaporator tube: length 7 m, diameter 38 mm. The experimental set up is illustrated in Fig. 1.

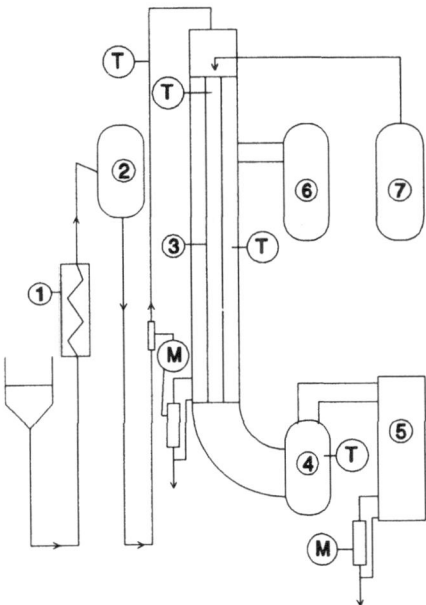

Fig. 1. Arrangement of the experimental evaporator plant. T = thermocouples; M = capacity measurement; 1 = plate heat exchanger; 2 = de-aerator; 3 = evaporator; 4 = separator; 5 = condenser; 6,7 = steam generator.

To reduce heat losses to a minimum the evaporator stage is insulated with mineral wool (thickness insulation 150 mm).

The milk is preheated in a plate heat exchanger at 90 °C, de-aerated by flash evaporation and concentrated in one pass. A spray condenser is used for the de-aerator and a surface condenser for the evaporator. Two vapour generators are installed, one for heating the tube and the other for additional vapour in the tube to manage the heat flux and the vapour velocity as independent variables.

During the experiments the following data were collected:

- The temperatures of the milk and the vapour were measured by thermocouples at the positions given in Fig. 1.

- The flow rate of milk was measured with a magnetic flow meter; the steam condensate and the condensate from the condenser were gathered discontinuously in a measuring device.

The data were collected with a data acquisition workstation (Keithly system 570) connected to an IBM personal computer (PS/2, model 30).

RESULTS AND DISCUSSION

1. Energy Saving by Increasing Total Solids

Intensifying the heat load causes denaturation of the whey proteins, measured by decreasing WPNi, and gives an increase in the volume fraction of the proteins, which results in an increase of the viscosity (7).

The effect of the preheat treatment on the viscosity of the concentrated milk was determined from the experimental results. Figure 2 shows the viscosity of the concentrated milk, measured with a shear rate of 392 s^{-1}, at different reductions of the WPNi. Decreasing the reduction of WPNi from 80 to 20 % allows for an increase in the total solids content of 2 to 4 %, depending on the actual level of the viscosity.

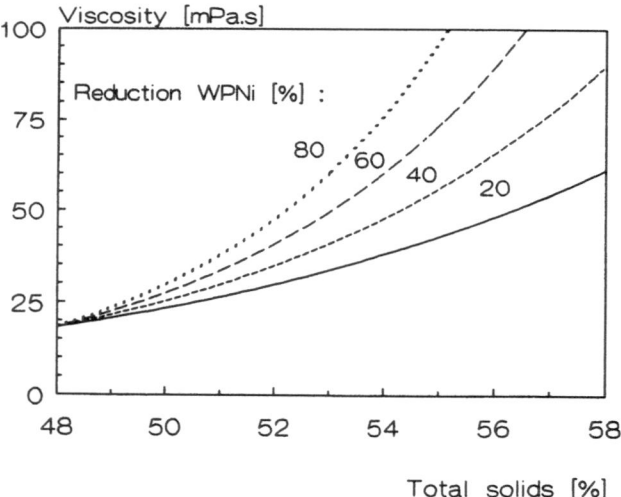

Fig. 2. Effect of preheat treatment on the viscosity of whole milk concentrate; shear rate 329 s^{-1}.

The process of water removal in an evaporator-spraydryer plant becomes cheaper by increasing the total solids content of the concentrate.

The conversion heat E, the number of kilograms of steam to convert fluid milk into one kilogram of milk powder, can be calculated from

$$E = \left[\frac{100 - X}{TS_m} - \frac{100 - X}{TS_c} \right] E_e + \left[\frac{100 - X}{TS_c} - 1 \right] E_d \qquad (1)$$

The results of experiments with a multistage dryer at different total solids content of the whole milk concentrate are given in Fig. 3.

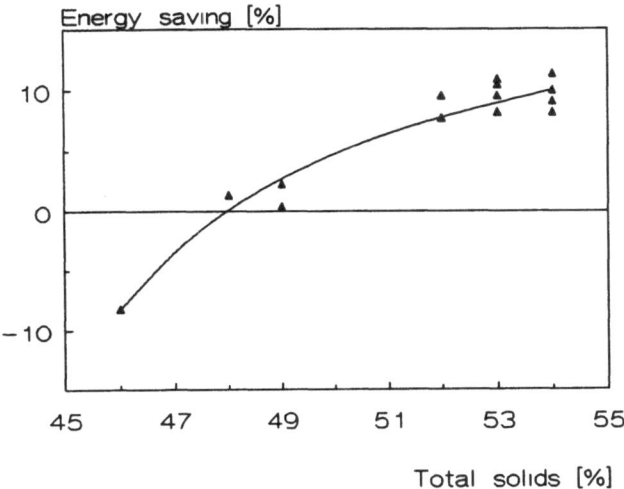

Fig. 3.   Saving energy by increasing the total solids content of the concentrate; the reference solids content is 48 %; ▲ = results with a multistage dryer.

The energy savings are related to the reference solids content of 48 %. The calculated reduction was based on a specific energy consumption of the evaporator $E_e = 0.12$ kg steam per kg water evaporated. It appears that the energy consumption of an evaporator-dryer plant can be reduced by 10 % by increasing the total solids content to about 54 %.

2. Evaporator Design
Boiling Phenomena
Figure 4 shows the evaporation behaviour of skim milk at (a) a low and (b) a high heat flux. In (a) convective boiling occurs and in (b) nucleate boiling. Both evaporation regimes are important in film evaporators for dairy products. Convective boiling occurs when the temperature differences across the film are small; there is only evaporation at the liquid-vapour interface of the film. Nucleate boiling occurs at larger temperature differences, when vapour bubbles are formed at the metal surface. Compared with water the transition from convective boiling into nucleate boiling with skim milk takes place at much lower heat fluxes (8).

(a)
(b)

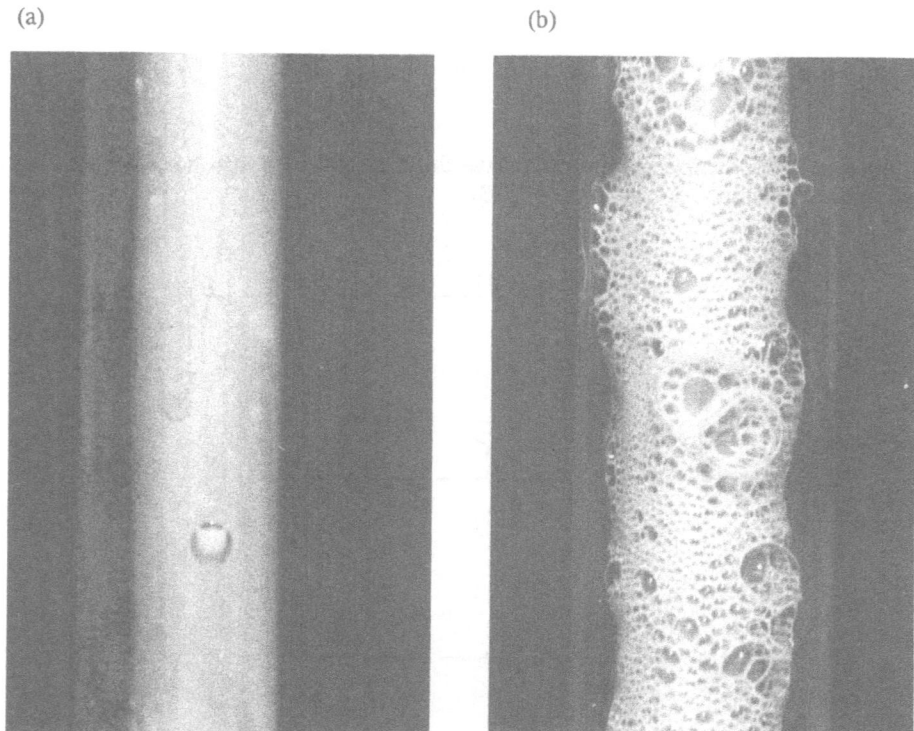

Fig. 4. Skim milk; boiling temperature 70 °C; wetting rate 400 kg.m⁻¹h⁻¹.
(a) Convective boiling, heat flux 0.8 kW.m⁻².
(b) Nucleate boiling, heat flux 6.2 kW.m⁻².

Heat Transfer and Pressure Drop
Experiments with whole and skim milk were carried out using different operating conditions. For whole milk the derived correlation for the heat transfer coefficient is

$$\alpha_p = 6.05 \cdot q^{0\,47} \cdot m^{0\,26} \cdot \eta^{-0\,44} \qquad (r = 0.912) \qquad (2)$$

and the correlation for the pressure drop is

$$4f = 84,500 \cdot q^{0\,24} \cdot m^{0\,88} \cdot Re_g^{-1\,47} \qquad (r = 0.911) \qquad (3)$$

The viscosity of the milk is obviously important: with increasing viscosity the heat transfer coefficient will decrease. The heat flux has a great influence on the heat transfer coefficient, which is in agreement with the boiling behaviour of milk. Nucleate boiling causes higher heat fluxes than convective boiling, which has been confirmed by many investigators (8). No significant effect of the vapour velocity on the heat transfer coefficient has been established.

Evaporator Design

The heat transfer and pressure drop correlations were used in a computer program for the simulation and design of multistage falling-film evaporators. The calculation procedure is given in Fig. 5.

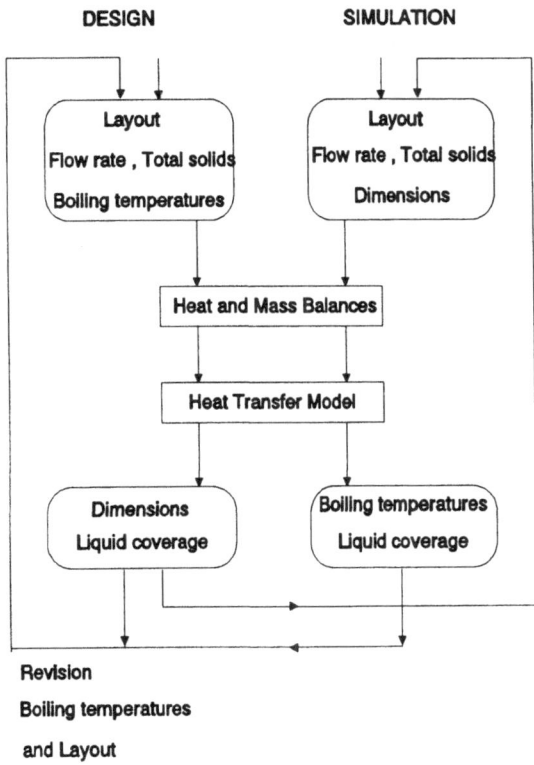

Fig. 5. Evaporator design and simulation procedure: EvaDes.

In this program the equations of the heat and mass balances of the various parts of the evaporator, e.g. preheaters, pasteurizer, evaporating stages and reheaters, and the heat transfer and pressure drop models of the evaporator tubes are solved.

For existing evaporators it may be investigated how the wetting rate of the evaporator tubes may be improved by plugging part of them and/or splitting the stages. For new evaporators it can be established to what extent the length and the diameter of the tubes, the arrangement of the stages, the efficiency of the thermocompressor and the boiling temperatures are determinative for the total heat-exchanging surface area required. This is of great importance in the optimization of the evaporator design in that it is aimed at low investments in combination with low energy consumption and small losses of the product.

The calculation procedure, which corresponds with the description of Perry and Green (9), is as follows:
1. Estimation of the boiling temperatures in the evaporator, taking into account the boiling-point elevations.
2. Determination of the total evaporation required and estimation of the steam consumption for the number of effects chosen.
3. From assumed feed temperature and assumed steam flow to the first stage, calculation of the evaporation of the succeeding stages.
   Steps 1 to 3 will be repeated until the concentrate flow from the last stage agrees with the actual requirements.
4. Calculation of the heating-surface area in each stage from the heat loads and temperature differences according to the heat transfer and pressure drop models.

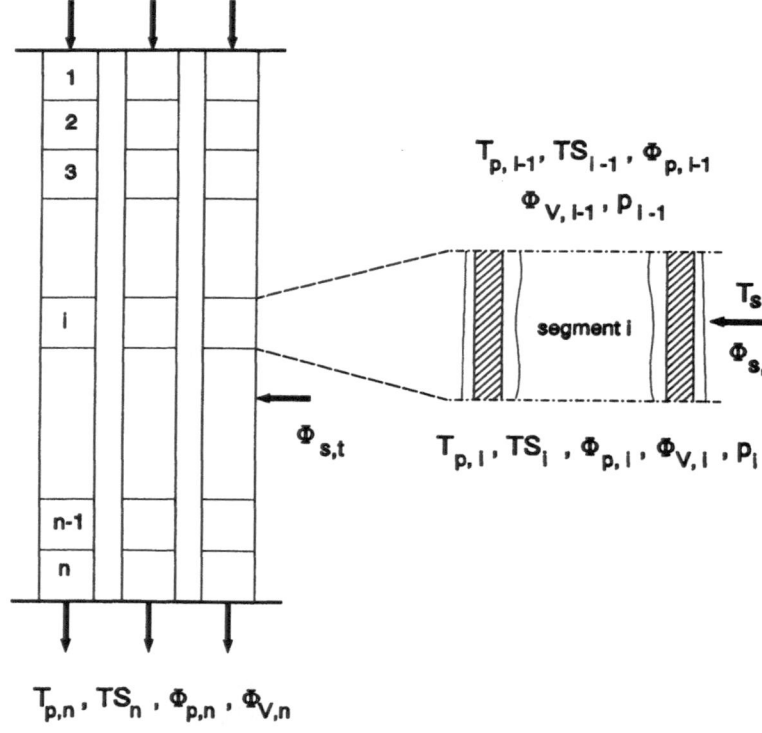

Fig. 6. Stepwise model of evaporator tubes.

The finite element method is applied for numerical solution of the heat transfer and pressure drop equations as indicated in Fig. 6. The length of the tubes is divided into a finite number of segments.

The calculations start with the segment at the bottom of the tubes where the boiling temperature T_p, the total solids content TS, the product flow Φ_p and the vapour flow Φ_v are the known variables. The calculations in a segment are repeated until the evaporating conditions in a segment have met the requirements of convergency. These calculations will be repeated with a revised number of tubes until the sum of the steam flows of all segments is equal to the steam flow of the stage, calculated from the heat and mass balances.

If the wetting rate in the stages is not as desired, the entire calculation, steps 1 to 4, needs to be repeated with revised boiling temperatures or arrangement of the plant.

Tube Length and Diameter

The designer has to determine the heating surface area in each stage of the evaporator, taking into account the desired wetting rate to prevent fouling. As described by Gray (10) the wetting rate can be increased by installing longer tubes or by dividing the stage into two or more passes in series. Recirculation, as a method to increase the wetting rate, cannot be considered for milk on bacteriological grounds.

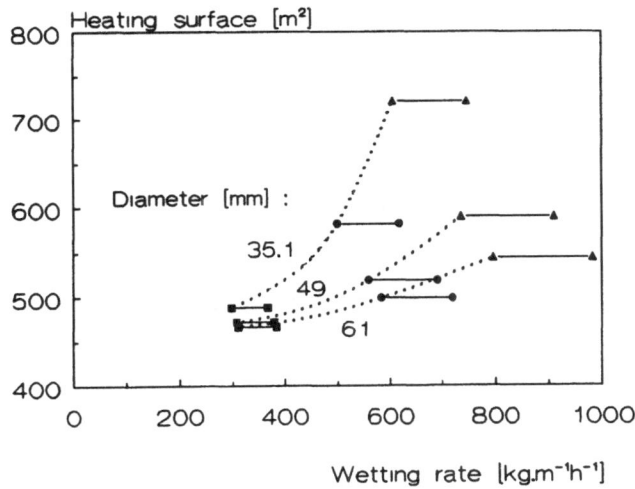

Fig. 7. Heating surface area and wetting rate in the first stage of a six-stage evaporator in relation to tube length and diameter.
Tube length: ■ 6 m; ● 12 m; ▲ 18 m.

The effect of tube length and diameter on the heating-surface area and the wetting rate for the first stage of a six-stage evaporator for whole milk is shown in Fig. 7.

The feed capacity is 30,000 kg/h, the steam temperature is 73 °C and the boiling temperature is 70 °C. The horizontal lines indicate the change of the wetting rate in the tubes from top to bottom. For a desired minimum wetting rate the tube length and diameter can be chosen. The surface area to be installed will decrease with increasing tube diameter.

The temperature difference between condensing steam and boiling product and the wetting rate are interdependent variables. A lower wetting rate accompanied by a smaller temperature difference will offer prospects for decreasing the energy consumption by increasing the number of stages.

The results of the calculations show that the selection of the tube diameter and length and the wetting rate are of great importance in designing falling-film evaporators.

CONCLUSIONS

The energy consumption of an evaporator-dryer plant can be reduced by 10 % by increasing the total solids content of the concentrate from 48 to about 54 %.

A design process has been developed to determine the dimensions of the evaporator stages, i.e. the diameter, the length and the number of the tubes, with regard to the process conditions and the physical properties of the product.

The computer program EvaDes, Evaporator Design, provides for a design method for equipment and process operation, based on first principles.

ACKNOWLEDGEMENT

The investigations formed part of a project realized in the framework of research into the reduction of energy consumption in the industry, and was co-financed by the Netherlands Society for Energy and Environment (NOVEM).

REFERENCES

1. Bouman S, Brinkman D W, de Jong P and Waalewijn R (1988) Multistage evaporation in the dairy industry: energy savings, product losses and cleaning. Preconcentration and Drying of Food Materials, pp 51-60, Bruin S, ed. Amsterdam: Elsevier Science Publishers B.V.

2. Munro P A (1988) Evaporation fundamentals. Evaporation and spray drying in the New Zealand dairy industry. pp 1-13, Jebson R S, ed. Palmerston North: Massey University.

3. Agarwala S P and Ojha T P (1976) Heat transfer studies in single tube effect falling film skim milk evaporator. Indian Journal of Dairy Research 29 231-233.

4. Jebson R S and Iyer M (1991) Performances of falling film evaporators. Journal of Dairy Research 58 29-38.

5. Determination of moisture content (1961) FIL-IDF 15.

6. American Dry Milk Institute (1961) Bulletin 912 rev.

7. Bloore C G and Boag I F (1981) Some Factors affecting the Viscosity of Concentrated Skim Milk. New Zealand Journal of Dairy Science and Technology 16 143-154.

8. Van Stralen S and Cole R (1979) Boiling phenomena. Washington: Hemisphere Publishing Corporation.

9. Perry R H and Green D (1985) Perry's Chemical Engineers' Handbook, p 11.40. 6th ed. New York: McGraw-Hill.

10. Gray R M (1981) Technology of skimmed milk evaporation. Journal of the Society of Dairy Technology 34 53-55.

AN EXPERIMENTAL STUDY OF DIRECT-CONTACT STEAM CONDENSERS AS NONCONDENSABLE GAS SEPARATORS

VASILIS BONTOZOGLOU* AND ANASTASIOS J. KARABELAS
Chemical Process Engineering Research Institute and
Department of Chemical Engineering, Aristotle University of Thessaloniki
P. O. Box 1517, GR 54006 Thessaloniki, Greece

ABSTRACT

Experimental results of direct-contact condensation of steam on water, in the presence of noncondensables, are reported. The performance characteristics of a column filled with structured packing are exploited, in an effort to enhance energy efficiency by achieving higher heat transfer rates at lower pressure drop. Local heat transfer coefficients as well as equivalent heights of a transfer unit (HTU) are reported for various conditions. The extent to which the noncondensable gas (CO_2) dissolves in the water/condensate and the factors which influence it are investigated. This process is of interest in the separation of noncondensables by condensation and reboiling of the steam.

INTRODUCTION

Direct contact heat transfer is already applied throughout the process and related industries in diverse areas such as physical separations, water desalination, steam condensation in power plants etc. The use of direct-contact versus surface transfer presents some distinct advantages. First, the exchange surfaces represent a major expense of the total system and are subject to corrosion and fouling. Furthermore, using a solid surface to transfer heat between two fluids requires a significant temperature difference, which results in loss of overall system efficiency. As the cost of energy grows, direct-contact devices are being given new consideration because they offer the possibility of improved performance with reduced capital expenses [1].

Retarding the development of direct-contact processes is the lack of reliable design techniques of general applicability, like the ones available for conventional configurations. Indeed, direct-contact devices are typically viewed as special situations, and empirical design procedures are often developed but without the underpinnings of a basic physical understanding of direct-contact phenomena [1]. The long-term goal of this research project is to contribute to a more sound design procedure, based on transfer rates at the local level. Carefully obtained

---

* *Present address:* Mechanical Engineering, University of Thessaly, 38334 Volos, Greece

experimental data, such as those presented here, will be used in the validation of this approach.

The present work deals with direct-contact condensation of steam on water , in the presence of noncondensable gases. This particular combination is encountered in various applications such as power-plant condensers, ocean thermal energy conversion systems, and geothermal installations. It is also of interest in the nuclear industry in certain safety evaluation scenarios [2].

The application motivating this study is the separation of noncondensable gases from high pressure geothermal steam upstream of the turbines. "Upstream removal" protects turbine components from corrosion, leads to higher conversion efficiency and permits more effective H_2S pollution abatement. A type of process, which has been proposed as suitable for this task [3,4], involves condensation of steam at high pressure, where the bulk of noncondensables escapes with the gas stream. Subsequent flashing of the condensate produces clean steam and a cooler liquid for recycle to the condenser. For this concept, however, to succeed it is essential to minimize the amount of noncondensables dissolved in the liquid. Preliminary calculations have indicated that countercurrent flow of steam and water leads to less gas absorbed. Therefore, a design involving a column filled with structured packing (Mellapak® 250.Y) was selected.

Literature in the area of direct-contact condensation is limited. Fair [5] summarizes available correlations for gas and liquid side heat transfer coefficients for various types of contactors including beds with random packing. His design methods, however, are limited to conditions where the sensible and latent heat loads are of comparable magnitude. Very few data are included for conditions where the latent load is the major heat duty of the contactor.

Sideman and Moalem-Maron [6] provide a thorough review of relevant studies until about 1980. Wilke et al. [7], Harriott and Wiegandt [8] and Jacobs et al. [9] report data on the condensation of pure vapors on different liquids, using columns filled with various random packings. In all these works, the major resistance to condensation resides in the liquid. An extensive study of various condenser configurations for steam condensation on water, in the presence of noncondensable gases, was undertaken by Bharathan and Althof [10]. Data are reported for low temperatures and pressures, relevant to open-cycle ocean thermal energy conversion systems.

In the present work two phenomena are given major consideration. First, the rate of heat transfer is studied and heat transfer coefficients are reported. The data complement relevant studies of condensation in packed beds with random packings ([7], [8], [9]) and commercial cooling-tower fills [10]. Further, the extent of gas absorption in the liquid is measured. These results test the suitability of the device proposed for the particular geothermal task. In addition they provide motivation for a study of the process of condensation with *noncondensable but slightly soluble* gases present. Gas dissolution, which takes place practically always, has been mostly neglected in heat transfer models [11].

In the next section, the apparatus built and the experimental procedure, followed during data collection, are described. Then, the results pertaining to heat and mass transfer are presented and discussed. Efforts, currently in progress, include modeling of the process of direct-contact condensation in packed beds, by a computational procedure based on the numerical integration of local heat and mass transfer rates. The algorithm will be validated by comparing results of the simulation with the present experimental data.

EXPERIMENTAL APPARATUS AND PROCEDURE

A schematic of the experimental system is shown in Figure 1. Tap water is demineralized and delivered to a spray manifold at the top of the column. The water flow rate is controlled by a PID controller (Shimaden, SR-24) and a flow meter (Signet, Rotor XLF), which electronically actuate a 1/2" Badger valve. Steam is provided by a central generator and is fed to the bottom of the column after going through a condensate separator and a pressure reducer. CO_2 is delivered to the steam line from supply cylinders and its flow rate is measured with rotameters and adjusted by precision valves.

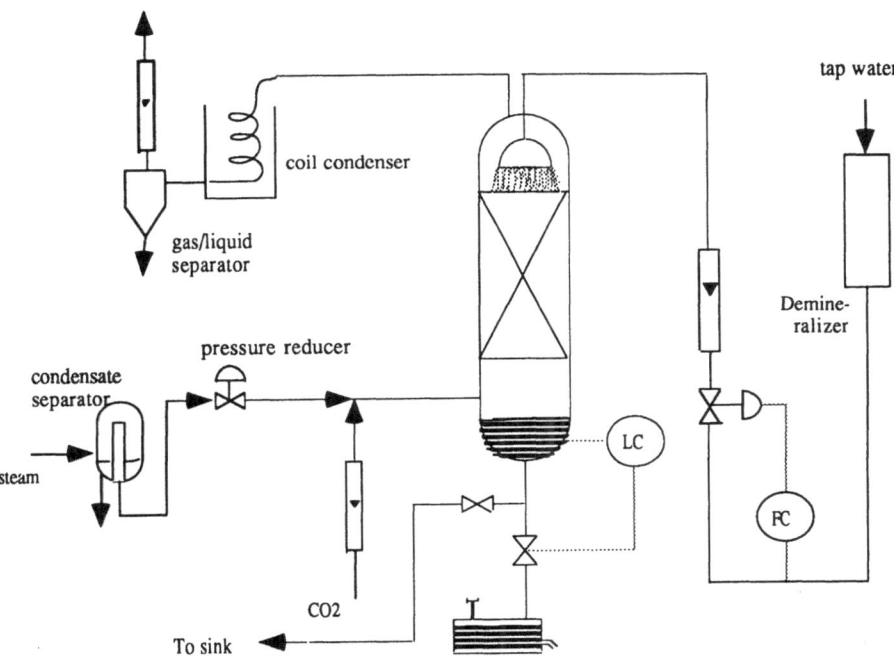

Figure 1. Schematic of the experimental set up.

A stainless steel tank with a volume of 25 liters is provided at the bottom exit of the column. Through a system of automatically operated pneumatic valves, the effluent liquid stream can be directed to the tank and the time needed to fill it is recorded. This serves as an indirect method of determining the steam flow rate (= total liquid flow - cold water inflow).

The vapor stream exiting from the top of the column passes through a coil condenser so that the steam vent condenses and separates from the noncondensable gas. The remaining (CO_2, saturated at ~20°C) passes through a rotameter and is then released to the atmosphere.

The condenser is a 1.05 m long, 0.15 m internal diameter column made of stainless steel. The body of the column consists of three sections 0.1, 0.2 and 0.4 m long, which can be combined to provide useful length ranging from 0.1 to 0.7 m. A hot-well, equipped with a magnetic floater, forms the bottom of the column. The liquid outlet valves are electrically actuated through this floater so as to prevent the passage of vapor in the drainage pipe. The entire condenser is externally covered by a layer of glass-wool to provide insulation.

The column is filled with Mellapak 250.Y® structured packing marketed by Sultzer. Interest in this type of packing stems from its favorable performance characteristics [12]. In particular, low pressure drop is highly desirable in the reboiling of geothermal steam, because any reduction in pressure carries with it a severe energy penalty. It is also noted that condensation data do not seem to be available for this kind of packing.

Temperatures are monitored with K-type thermocouples calibrated to ±0.2°C. The thermocouples are located at the water inlet, the water-condensate outlet and the vapor inlet lines. Four more thermocouples are embedded in the packing at 4-5 cm intervals from the bed support. Their tips are positioned close to the center of the column cross-section, in contact with the packing, so as to provide the temperature of the local liquid film. All temperature measurements are indicated on a central panel and are also recorded by a data acquisition system (AI13 interface on an Apple IIe) for later processing.

Pressure measurements are taken at various locations along the flow. Analog pressure indicators are installed on the steam/condensate separator, the steam line prior to the entrance in the hot-well and the vapor exit line at the top of the column. A mercury manometer is connected to the hot-well and measures the pressure difference with the ambient. Its indication is used to determine the liquid-vapor saturation point. Finally, a water manometer is installed to measure pressure drop along the packing. However, such measurements are not reported in the present study.

Determination of the amount of CO_2 dissolved in the liquid involves the following procedure. The indications of the entrance and exit CO_2 rotameters are recorded, after the entrance value is corrected by taking into account the pressure at the point where CO_2 joins the steam line. The difference between the two values is the amount of CO_2 dissolved in the water. This amount is also measured directly, to close the mass balance, by the following procedure: A sample of liquid is drawn from the hot-well using a 10 ml syringe. The syringe contains a small quantity of dilute NaOH solution with phenolphthalein indicator, to ensure that the CO_2 dissolved in the condensate is trapped. The amount of liquid withdrawn is such that the indicator does not turn colorless, which guarantees that the CO_2 is in the form of HCO_3^- and CO_3^{2-}. Determination of the total amount of CO_2 is accomplished by titration with HCl solution and recording the amount of acid needed to bring the liquid from pH = 9 (phenolphthalein indicator) to pH = 4.2 (methyl red indicator). A similar determination is done for the inlet water and the difference gives the amount of CO_2 absorbed during condensation.

RESULTS AND DISCUSSION

The results presently reported are the preliminary part of a more extensive investigation. The critical parameters varied in this set of experiments are the mass flow rates of water, steam and CO_2. The height of the packing is kept constant and equal to 72 cm and the steam vent rate from the top of the column is always zero. Temperatures at intermediate locations along the column are used to derive local heat transfer coefficients, which can also be viewed as average coefficients for the respective small length of packing.

Most data are taken under such conditions that the change in the liquid temperature takes place in the lower part of the packing. Indeed, the location along the column where most of the condensation occurs is subject to control by suitable choice of the steam-to-water ratio. This is demonstrated in figure 2, which shows the variation of water temperature upward from the hot-well when decreasing the water flow rate or increasing the steam load. In particular, the following is observed: For small steam rates, there is a considerable change in the liquid temperature from the end of the packing to the hot-well. This indicates that significant condensation may take place in the free-fall space at the bottom of the packing and the measured rates may reflect the performance characteristics of this region rather than of the packing itself.

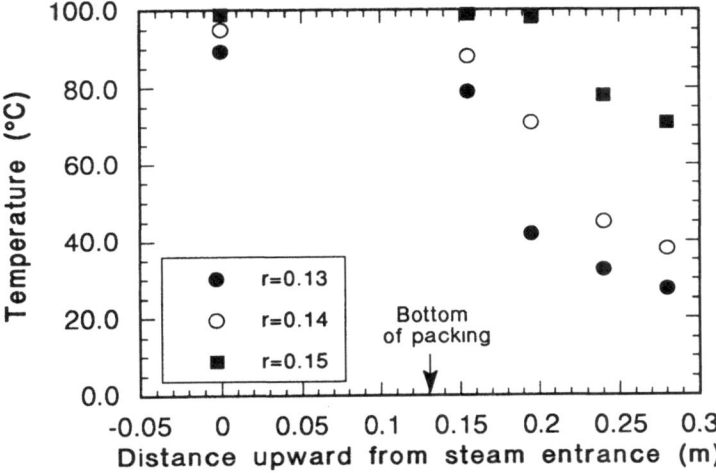

Figure 2. Variation of liquid temperature upstream from the hot-well for different inlet steam/water ratios.

As the steam flow rate increases for constant water rate, the reading of the first thermocouple (embedded 3.5 cm from the bottom of the packing) approaches that of the hot-well, indicating that the reduction in vapor loading at the bottom of the column becomes insignificant. Further increase in the steam flow rate moves the region of drastic change of the liquid temperature towards the upper part of the bed. Finally, when the steam supply is increased to the point that complete condensation is not possible, the liquid along most of the column length is practically at the saturation temperature and condensation takes place in the free space below the water spray manifold and in the first few centimeters of packing. The data in Figure 2, corresponding to steam/water ratio r = 0.15, are typical of the temperature distributions in the experiments reported.

Heat transfer measurements
Temperature measurements are made at various points along the column and volumetric coefficients are calculated from the total heat transferred and the log-mean driving force. Figure

3 shows representative data for four different liquid flow rates and steam loads, such that the water effectiveness is very close to one. The water effectiveness is defined as $\varepsilon_w = (t_{wo}-t_{wi})/(t_{si}-t_{wi})$ and provides a measure of the water temperature rise as it relates to the overall available temperature difference (subscripts s,w,i,o stand for steam, water, inlet and outlet).

It is readily observed from Figure 3 that local heat transfer coefficients change drastically from point to point. The higher values are measured close to the bottom of the packing and they drop almost exponentially with height. In these runs the amount of CO_2 in the entrance steam is 3% w/w.

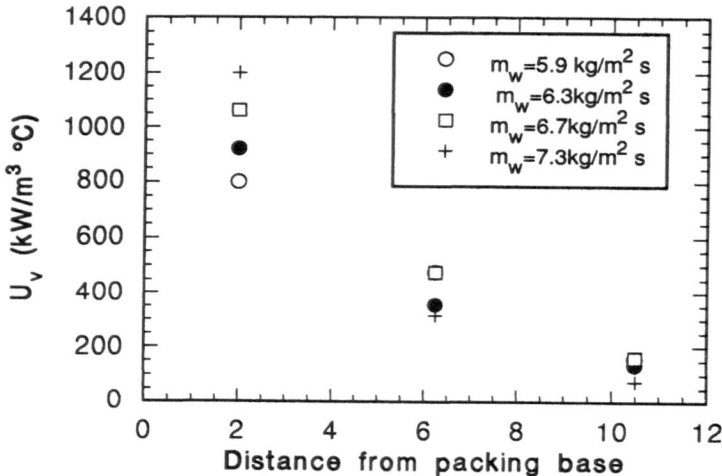

Figure 3. Local heat transfer coefficients at different locations along the packing, for constant steam/water retio r=0.15.

The strong variation of heat transfer rate along the packing is attributed to the existence of noncondensables in the steam. It is well known that inert gas builds up at the vapor/liquid interface, leading to a lower partial pressure of the steam, which, in turn, lowers the interface temperature at which condensation occurs and correspondingly decreases the thermal driving force. This phenomenon has been shown [13] to depend directly on the mass fraction of noncondensables. Therefore, as part of the steam condenses along the column, the concentration of CO_2 in the remaining vapor phase continuously increases, leading to significantly lower heat transfer rates.

The range of liquid and vapor flow rates covered in the present series of experiments is such that the hydrodynamic loading at the bottom of the column is around 40 to 60% of flooding. Given that this is the location with the highest loading, it is of interest to know what is the effect of the approach to flooding on the local heat transfer coefficient. Figure 4 shows such results for the first 4 cm of packing and the liquid flow rates and steam loads of Figure 3. Inspection of Figure 4 indicates that there is an almost linear dependence of the heat transfer coefficient on the liquid flow rate. This, however, is a preliminary conclusion and experiments closer to flooding are needed to check its validity.

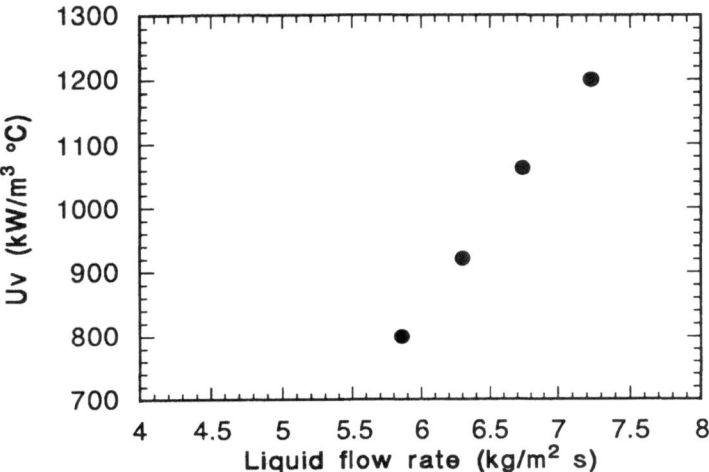

Figure 4. Heat transfer coefficients 2 cm above the base of the packing as a function of liquid flow rate, for constant steam/water ratio r=0.15

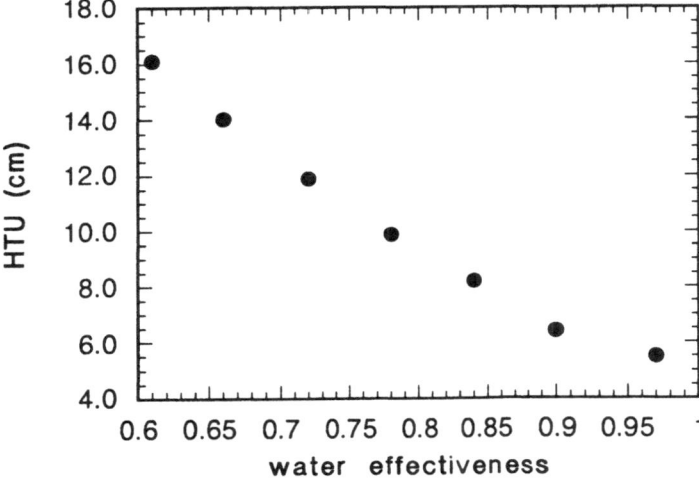

Figure 5. Height of a transfer unit (HTU) as a function of the water effectiveness, ε_w.

Overall volumetric heat transfer coefficients may be expressed in terms of the height of a transfer unit (HTU). Plotted in Figure 5 is the variation in HTU with the water effectiveness factor, ε_w. It is observed that the HTU decreases with increasing ε_w, being almost inversely proportional to ε_w for ε_w values less than one and approaching a minimum value for ε_w close to 1. These observations are in agreement with similar findings of Bharathan and Althof [7], except that the minimum value presently found is around 5 cm, whereas their value for the packed bed is 27 cm. The difference can be attributed to the lower ambient pressure in their experiments and to the superior performance characteristics of structured packings.

Mass transfer measurements

In the experiments reported, condensation takes place in the lower part of the packing and the undissolved CO_2 stream flows upward, countercurrently with the free falling water stream. It is found that, with this setup, there is always enough contact area for the cold water to get saturated in CO_2 at the upper part of the packing. Since this concentration is higher than any other measured at the exit of the column, the experiments reported represent the most unfavorable conditions for the separation of the noncondensable gas from the condensate.

The extent to which CO_2 remains dissolved in the water/condensate is found to depend on its concentration in the vapor phase and on the liquid temperature. Figure 6 shows the concentration of CO_2 in the condensate as a function of the exit water temperature, for a steam feed with 3% w/w CO_2. An increase of the water temperature by 30° C is seen to result in the reduction of the CO_2 concentration in the liquid to 1/4 its original value. This outcome can be qualitatively explained by the well-known reverse effect of water temperature on gas solubility.

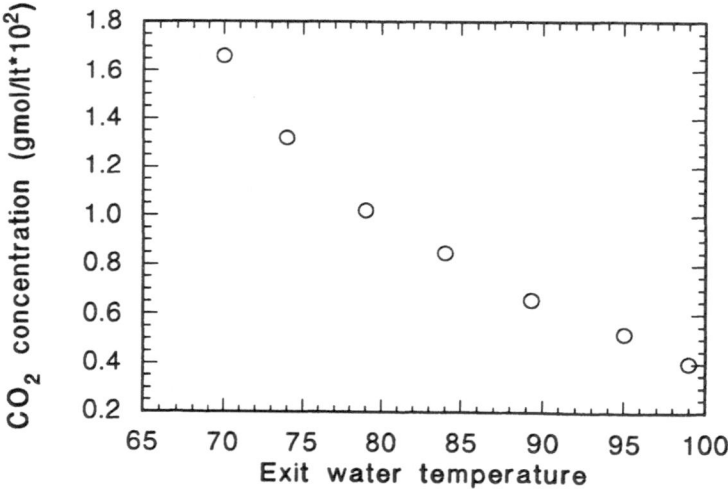

Figure 6. The concentration of CO_2 in the effluent liquid as a function of the exit water temperature.

The total amount of CO_2 dissolved in the effluent liquid decreases even more drastically. This is so because higher water exit temperatures are related to lower mass flow rates. For example, the amount of CO_2 remaining in the condensate for the experiments depicted in Figure 6, is found to decrease from ~27% to less than 4% of the CO_2 originally contained in the inlet steam.

Figure 7 shows the results of a series of experiments with inlet steam containing varying amounts of CO_2, and water effectiveness $\varepsilon_w \approx 1$. Inspection of the figure indicates that the concentration of CO_2 in the exit liquid scales linearly with the CO_2 content of the steam.

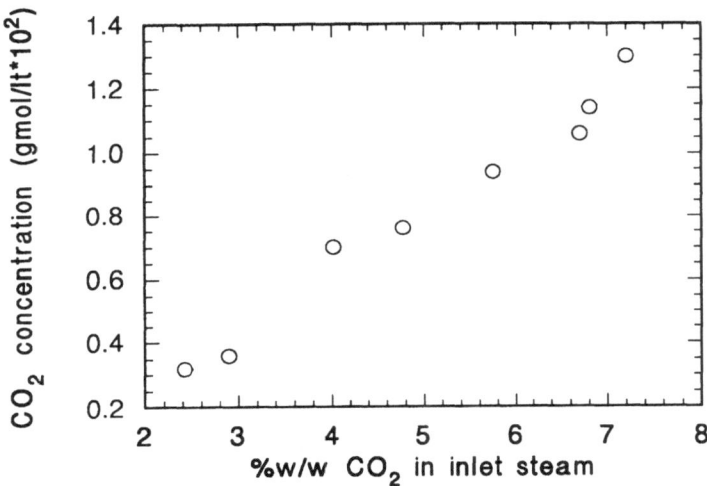

Figure 7. CO_2 concentration of the effluent liquid as a function of the CO_2 content of inlet steam.

CONCLUDING REMARKS

A column filled with structured packing is tested as a direct-contact condenser of steam in counterflow with water, in the presence of noncondensables. Local heat transfer coefficients are calculated and shown to change along the column by roughly an order of magnitude. Overall performance is expressed in terms of the height of a transfer unit. Minimum value of HTU for the present configuration is approximately 5 cm.

The extent of CO_2 retained in the water/condensate was measured for the highly unfavorable case when incoming cold water is saturated in CO_2. This was found to decrease with increasing water effectiveness, ε_w. For temperatures of the exit water close to that of inlet steam ($\varepsilon_w \approx 1$), the amount of noncondensables dissolved in the liquid is ~4% of the incoming feed.

As pointed out in the introduction, there is an ongoing modeling effort of the process of direct-contact condensation (to be validated by comparison with the present experimental data), based on the numerical integration of local heat and mass transfer rates. This procedure is expected to provide an alternative design approach, fundamentally different from the one based on average transfer coefficients and equivalent height of a transfer unit (HTU). Expected benefit is a more realistic modeling of the rapidly changing conditions (flow rates, temperatures, concentrations) along the condenser. In particular, the approach seems highly appropriate for the complex problem of condensation coupled with partial dissolution of noncondensables in the liquid.

ACKNOWLEDGEMENTS

This work has been supported by the General Secretariat for Research and Technology of Greece and by the Commission of European Communities (under the programme VALOREN). The help of Dr. S. Paras and Mrs. K. Mouza in the preparation of the computerized data acquisition system is gratefully acknowledged.

REFERENCES

[1] Kreith, F. and Boehm, R. F., Direct-contact heat transfer processes. In *Direct-Contact Heat Transfer*, ed. F. Kreith and R. F. Boehm, Hemisphere Publishing, New York, 1988, pp. 223-236.
[2] Tung, V. X. and Dhir, V. K., A hydrodynamic model for two-phase flow through porous media. *Int. J. Multiphase Flow*, 1988, **14**, pp.47-65.
[3] Coury, G., Geothermal Steam Purification by Evaporation to Improve Process Efficiency. *AIChE Annual Meeting*, Chicago, Nov. 1985.
[4] Awerbuch, L., Van der Mast, V. and Weckes, M., Geothermal Flash Evaporation Process Technology. *AIChE Annual Meeting*, Chicago, Nov. 1985.
[5] Fair J. R. Designing direct-contact coolers/condensers. *Chem. Eng.*, 1972, **2**, pp. 91-100.
[6] Sideman, S. and Moalem-Maron, D., Direct contact condensation. *Advances in Heat Transfer*, 1982, **15**, pp. 227-281.
[7] Wilke, C. R., Cheng, C. T., Ledesma, V. L. and Porter, J. W., Direct contact heat transfer for sea water evaporation. *Chem. Eng. Progr.*, 1963, **59**, pp. 69-75.
[8] Harriott, P. and Wiegandt, H., Countercurrent heat exchange with vaporizing immiscible transfer agent. *AIChE J.*, 1964, **10**, pp. 755-758.
[9] Jacobs, H. R., Thomas, K. D. and Boehm, R. F., Direct contact condensation of immiscible fluids in packed beds. In *Condensation Heat Transfer*, ASME, New York, 1979, pp. 103-110.
[10] Bharathan, D. and Althof, J., An experimental study of steam condensation on water in countercurrent flow in presence of inert gas. *Proceedings of the ASME*, paper 84-WA/Sol-25, 1984, New Orleans.
[11] Jacobs, H. R., Direct-contact condensation. In *Direct-Contact Heat Transfer*, ed. F. Kreith and R. F. Boehm, Hemisphere Publishing, New York, 1988, pp. 223-236.
[12] Fair J. R. and Bravo J. L. Distillation columns containing structured packing. *Chem. Eng. Progr.*, Jan. 1990, pp. 19-29.
[13] Sparrow, E. M., Minkowycz, W. J. and Saddy, M., Forced convection condensation in the presence of noncondensables and interfacial resistance. *Int. J. Heat Mass Transfer*, 1967, **10**, pp. 1829-1845.

THE HYDRODYNAMIC AND MASS TRANSFER CHARACTERISTICS OF A LARGE CENTRIFUGAL WATER DEOXYGENATOR

Dr K Al-Shaban, Dr V Balasundaram, Dr C R Howarth, Professor C Ramshaw, Mr J R A Peel, Department of Chemical and Process Engineering, University of Newcastle upon Tyne,Newcastle upon Tyne, NE1 7RU, UK

ABSTRACT

A rotary stripper is being developed to deoxygenate seawater to promote better oil recovery from marginal offshore oil fields. If the size of stripping equipment can be reduced and performance increased, there are benefits to be gained in terms of the reduced size and cost of the offshore platforms with consequential saving in energy. The characterisitics of rotating contactors make them ideally suited for this type of duty therefore, in order to improve the design, small scale laboratory bubble impact studies together with mass transfer work on a 1 m diameter deoxygenator have been undertaken. 4.3 mass transfer units were obtained in this large unit, but it is clear that this can be improved if the maldistribution of the bubbles and the larger bubbles found in the peripheral zone of the packing can be eliminated.

INTRODUCTION

The concept of Process Intensification whereby individual plant modules are reduced in size results in less civil engineering work and piping etc. with a consequential lowering of plant system costs and associated energy savings (1). A prime example is to be found in equipment employed to strip oxygen from seawater which is then used to promote better oil recovery from marginal offshore oil fields. Using small and technically advanced equipment saves the energy cost of the massive support structure and enhances oil reserves. This paper describes the development of a rotary deoygenator.

If centrifugal HiGee™ separators (2) are used to intensify a process, the lower energy consumption is primarily represented by the reduced system costs. Using centrifugal fields to enhance a process incurs an obvious energy penalty as a result of accelerating liquids to "g" values in excess of 100 and the associated linear velocities. However, as every chemical engineer knows, the design of a stripping column involves striking a balance between the number of stages (linked to column height) and energy requirements. This is a particular problem in offshore engineering where space is at a premium and column heights are generally minimised by using high reflux ratios. However if a small rotary oxygen stripper can be used no such compromise will be necessary and a net energy saving will result. Thus it is clear that centrifugal machines have a role in reducing energy consumption in some specialised applications (3), (4). Table 1 provides an estimation of the savings which can be made if rotary equipment is used in in offshore engineering. In Table 1 the weight and estimated cost of main plant items (MPI) do not include any savings which can also be made on structural steel work or installation. (At 1992 prices the cost of supporting 1 Tonne of offshore platform equipment is approximately \$45,000.) Running costs are similar because power requirements are comparable and the stripping gas for the rotary equipment will be well gas which can then be used as fuel.

TABLE 1
Space and Weight Reduction if Rotating Contactors are used in Offshore Duties

Seawater Deoxygenation(5) for 46 kg/s seawater down to 20 ppb residual O_2

Space/Weight		Packed Rotating Contactor(*)	Vacuum Column
Space:	Height	2 m	20 m
	Footprint area	4 m^2	4 m^2
Weights:	Unit	2.9 Tonnes	9.5 Tonnes
	Packing	0.2 T	1.6 T
	Hold up liquid	0.6 T	12.3 T
	Pretreatment liquid	1.2 T	10.0 T
Power:	Motor	0.6 T	
	Vacuum	_____	5.0 T
Total		5.5 T	38.4 T

Gas Dehydration Using Glycol(6) for 0.2 kg/s glycol, treated gas 140 kg/s at 50 bar g

Weight/Cost	Rotating Higee™(**)	Conventional Column
Weight (dry)	3.3 T	23 T
Cost (MPI)	£54,000	£97,000

* Based on a 1 m diameter rotor with 2 packed lengths of 0.2 m (0.8 m O.D; 0.4 m I.D.) and an axial length of 0.6 m/pass. Liquid in continuous phase.

** Based on a 0.5 m diameter rotor with a 1 pass packed length of 0.111 m (0.45 m O.D.; 0.225 m I.D.) and an axial length of 0.2 m. Gas is continuous phase.

It is clear from Table 1 that a significant size reduction is predicted for large throughput plants, but in order to accurately quantify energy running costs it will be necessary to commission higher throughput machines to allow optimisation of drive motor characteristics with power requirements. While the potential to save overall costs is clear, practical development work is needed and the following sections describe this research . Oxygen is removed from water by contacting an oxygen-free gas with the liquid in a rotating packing. Previous "proof-of-concept" work on a 0.5 m diameter seawater deoxygenator containing a 0.1 m radial depth of packing showed that 2.5 transfer units could be obtained to process saturated seawater flowing at 18 m^3/hr (7). In order to improve this mass transfer, a deeper radial depth will be necessary and a full understanding of the interaction between the gas bubbles and the flowing liquid is required to allow improvement of the design, scale up and performance characteristics of rotating packed contactors. Therefore small laboratory scale bubble impact studies were carried out to attempt to model bubble formation mechanisms. In addition, a 1 m diameter section of a large scale deoxygenator has been developed which allows the influence of the hydrodynamic flows on the overall residence time and mass transfer to be evaluated. This 1 m section will form the basis of a large scale unit which will contain a multiplicity of similar sections to allow 200 Tonnes per hour of seawater to be processed. Hitherto little work has visually demonstrated the interactions occurring within a rotating packing.

SMALL SCALE BUBBLE IMPACT STUDIES

Equipment

A shallow rotating pool of liquid (2 cm deep by 11 cm radial length) was used with a single nozzle gas jet "firing" into various packings (see Figure 1). This rotated from 300 to 1500 rpm with no net flow of the liquid which was enclosed by a transparent lid to allow bubble visualisation. Two liquids were used, distilled and sea waters, also two packings, Retimet™ and Declon™ of 1.7 pores/mm and 0.6 - 1 pore/mm respectively.

Results

The Effects of Speed on Bubble Size

The bubbles were observed by a photographic technique and produced Figure 2 showing that the average diameter falls with speed, but the packing had little effect for the two types used. The nozzle gas flowrate varied with speed because of the change in pressure at the nozzle, but further work (8) showed that the effect of a tenfold change in flowrate produced only a 10% decrease in diameter. The nature of the liquid did however have a significant effect and Figure 3 shows the fall in size with seawater.

The Effect of Packing Thickness

Figures 2 and 3 are for a 1 cm thick of packing, however the thickness was also varied from 1 to 5 cm as shown in Figure 4 shows that, apart from the slow speed, increasing the thickness has little effect on the diameter. However the packing does have the very important role of minimising coalescence and preventing the bubbles growing according to Boyles law expansion as they migrate to atmosphere at the centre line. Figure 4 shows the data points but the solid lines are derived by modelling the probability of bubbles colliding and splitting with the packing as they move to the centre (8). The bubbles experience 40 - 50 collisions in traversing 5 cm of packing.

Modelling and Discussion

Model comparisons are made using either a) mechanisms associated with bubble generation at nozzles or pores, or b) energy balance consideration. Figure 5 represents a comparison between the experimental results and these models. The dotted curve A is the data curve from Figure 2. D_p is the average diameter of the packing pores.

Curves B, C and D are based on models which predict the bubble volume at a single orifice extrapolated to high g* values by:

$$d_b = C (Q^2/g^*)^{0.2}$$ 1

where d_b = Bubble diameter, Q = gas flowrate per pore and g* is given by $\omega^2 r$ at a radius r and an angular velocity ω. C is a constant with a value of 1.13 suggested by Walter and Davidson (9), 1.27 from Wraith (10) and 1.38 by Davidson and Schuler (11).

A static equilibrium model balances the surface tension and buoyancy forces. This model was applied to the work of Voit et al (12) who empirically modelled of bubble measurements at a nozzle with no packing. Curve E best describes this work.

Bowander and Kumar (13) produced a model for bubble formation at a porous media under normal gravity which was based on a force balance between the buoyancy, viscous and inertial surface tension forces. This balance generated complex equations from which Curve F was derived for higher g* values.

By considering an energy balance before and after drop breakdown, Ramshaw and Thornton (14) modelled droplet diameters for liquid/liquid systems. This can be applied to bubble splitting and sizes are correlated with experimental data as:

$$0.15 \Delta\rho \, g \, d_e + \rho \, V_t^2 \, d_e/12 - 0.26 \, \sigma = 0$$ 2

where d_e = equilibrium diameter below which a bubble or drop cannot breakdown on impact with a baffle or knife edge, V_t = terminal velocity, $\Delta\rho$ = density difference, ρ = liquid density and σ = interfacial tension.
This model was extended to bubbles and high centrifugal fields and generated curve B'.

<u>Choice of Model</u>: It is clear that no model accurately predicts the average bubble size. The static equilibrium and porous distribution models are not accurate and single nozzle models over predict the sizes at speeds above 900 rpm. By contrast energy balance considerations over predict the size below 900 rpm, however this is probably still the best approach for modelling through packings at these lower "g" regions.

LARGE DIAMETER HYDRODYNAMIC AND MASS TRANSFER STUDIES

Equipment

Figure 6 shows a schematic cross sectional view through the rotor. Water is passed through a rotary union into the shaft of the rotor, and is flung into the packed section through 12 holes 4 mm diameter where it passes counter currently with the nitrogen gas, which has been sprayed out of the gas injector nozzles near the base of the packing. The water then passes through to the back of the rotor into the MAIN RETURN CHAMBER where it returns to near the centre line to reduce its kinetic energy and is then collected by four pick up tubes. There is also a second smaller chamber called the SAMPLE RETURN CHAMBER which collects water taken from internal sample tubes. The purpose of these tubes is to allow mass transfer determinations at various packed lengths. A metal shield with two nitrogen jets covers the area around the pick-up tubes, to prevent any oxygen re-dissolving in the de-oxygenated water.

One important aspect of the project was the visualisation of the bubble flows. For these large scale experiments a high intensity nano second flash unit was used to view through the front window. This 0.9 m diameter "viewing" window was 1 cm thick polycarbonate, which, despite its strength, could only allow speeds lower than 500 rpm. The speed was also further restricted to 300 rpm because the window "flexed" and leakages at the outlet seal occurred. Modifications are in hand to rectify this. The bubbles were photographed within "windows" cut out of the packing and Figure 7 shows the positions of these.

Results and Discussions

Bubble Distribution and Hydrodynamic Observations

Stroboscope observations showed that the bubbles are not uniformly distributed throughout the radial length. From high speed photographs non homogeneous zones of bubbles are observed which do not appear to be affected by speed or gas/liquid flowrates. Figure 7 schematically shows these zones. In ZONE A (up to window 3) 120 plumes of bubbles are seen emitting from the 120 nozzles. These plumes dispersed in ZONE B (window 3 to 5) and few bubbles are observed. The bubbles reappear in ZONE C (window 5 to 8) as a uniform high concentration of bubbles, but on reaching ZONE D (around window 9) only a few bubbles are seen. The final ZONE E up to the liquid/gas interface contains a high concentration of bubbles. It is clear from this description that there are complex gas/liquid/packing interactions and further measurements will be made on the bubble dynamics in each ZONE to relate this to the mass transfer occurring at differing radial depths.

Bubble Size Determinations

Bubbles were measured in windows 2, 4, 7, 9, 12 (Figure 7) for a variety of gas/liquid ratios and speeds. In windows 2 and 4 no bubbles were observed, but Table 2 could be obtained for the other windows (7 and 12, high concentration zones, and 9, low bubble concentration).

TABLE 2

Bubble Sizes Measured in Large 1 m Diameter Unit

Window	Radial Depth (m)	Speed (rpm)	Gas Flow (m³/hr)	g* Field (x 1g)	Measured Bubble Diam.(μm)	Bubble Di Lab. unit(μm) (See Figure 5)
7	0.39	250	0.5	27	800	470
7	0.39	250	1.0	27	800	470
7	0.39	250	2.0	27	900	470
7	0.39	250	3.0	27	1000	470
9	0.27	200	2.0	12	800	750
9	0.27	250	2.0	19	650	600
9	0.27	300	2.0	27	500	470
12	0.21	200	2.0	9	750	850
12	0.21	250	2.0	15	600	700
12	0.21	300	2.0	21	550	570

In Table 2 a comparison is made between the diameters obtained in this unit and those measured and predicted for the small unit (See Figure 5). It is clear that the bubbles formed at the periphery are not small enough and further work on the nozzle type and packing density in this zone will be undertaken to improve this.

Mass Transfer Measurement

Some data were taken at different radial positions but because of sampling errors in the SAMPLE RETURN CHAMBER, measurements reported here refer only to the full packed lengths. Figure 8, shows the effect of speed and flowrates. It is surprising to find a reduced performance at higher speeds which is obviously linked to the unusual hydrodynamic patterns and bubble sizes reported above. On the positive side the data confirms that an increased packed length produced a better performance than the "proof-of-concept" design (7).

481

CONCLUSION

Small scale work has developed bubble size data which may be used to model the mass transfer performance of larger machines. Hydrodynamic studies in a large 1 m diameter machine has revealed a non uniformity of bubble distribution. Nevertheless an increased performance is still obtained as the packed length increases. Further work will concentrate on improving the bubble distribution in order to further improve the mass transfer performance to allow the system energy saving potential of rotating machines to be realised.

REFERENCES

1. Ramshaw, C.R., "Process Intensification: A Game for n Players". Chem. Eng., 1985, July/Aug., pp. 30-33.
2. Fowler, R., "Potential Applications of HiGee Offshore". Symp. - "Offshore Separation Processes", May 1986, Teesside, UK, publish I.Ch.E., London.
3. French, M.J., "Centrifugal-Field-Effect Machines: A New Class of Convertors for Fluid Energy". C. Mech. Eng., 1983, Dec., pp. 25-28.
4. Mersmann, A., Voit, H. and Zeppenfeld, R., "Do We Need Mass Transfer Machines?". Inter. Chem. Eng., 1988, 28 (1), pp. 1-13.
5. Fowler, R., AP Technology, Amersham, UK, 1992, (private communication).
6. Starkey, P.E. and Dobson, B., "Potential Offshore Applications for ICI HiGee Technology". I.Ch.E. - Offshore Gas Tech. Seminar, 1983, publish I.Ch.E., London, Nov, pp. 4.2-4.10.
7. Balasundaram, V., Porter, J.E and Ramshaw, C., "Process Intensification - A rotary seawater deaerator", 5th BOC Priestley Conference, 1985, Birmingham, Oct., No 80. R.Soc.Chem, pp. 306-329.
8. Al-Shaban, K., "Bubble Sizes in Centrifugal Fields". PhD Dissertation, University of Newcastle upon Tyne, UK, 1990.
9. Walters, J.K. and Davidson, J.F., "The inital motion of a gas bubble formed in an inviscid liquid: Part II - the three dimensional bubble". 1963, J. of Fluid Mech., 17, pp. 321.
10. Wraith, A.E., "Two stage bubble growth at a submerged plate orifice". 1971, Chem. Eng. Sci., vol. 26, pp. 1659.
11. Davidson, J.F. and Schuler, B.O.G., "Bubble formation at an orifice in a viscous liquid". 1960, Tran. Instn. Chem. Engrs., 38, pp. 144.
12. Voit, H., Zeppenfeld, R. and Mersmann, A., "Calculation of primary bubble volume in gravitational and centrifugal fields". 1987, Chem. Eng. Tech., 10, pp. 99.
13. Bowander, B. and Kumar, K., "Studies in bubble formation IV bubble formation at a porous disc". 1970, Chem. Eng. Sci., 25, pp. 25.
14. Ramshaw, C. and Thornton, J.D., "Droplet breakdown in a packed extraction column. Part I: The concept of critical droplet size". 1967, I.Chem.E. Symp, No. 26,

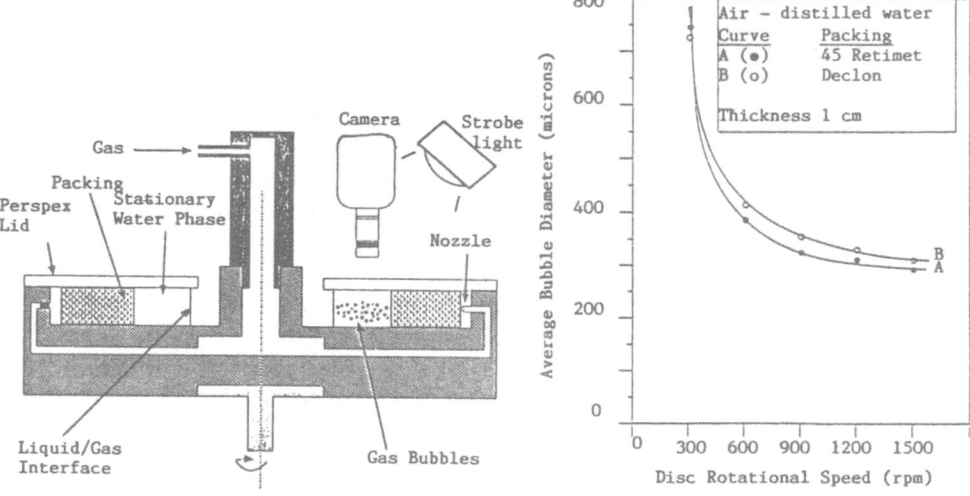

FIG. 1. LABORATORY SCALE BUBBLE SIZE APPARATUS

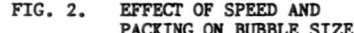

FIG. 2. EFFECT OF SPEED AND
PACKING ON BUBBLE SIZE

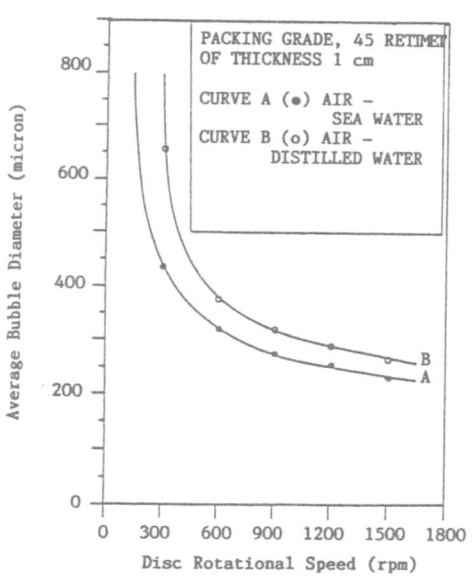

FIG. 3. BUBBLE SIZES IN DISTILLED AND
SEAWATER

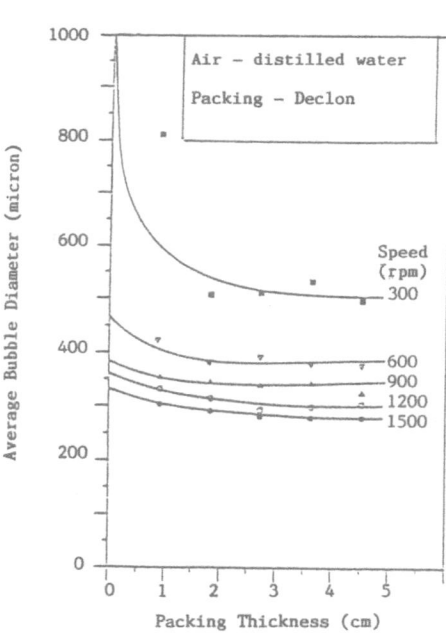

FIG. 4. EFFECT OF PACKING THICKNESS ON
BUBBLE SIZE

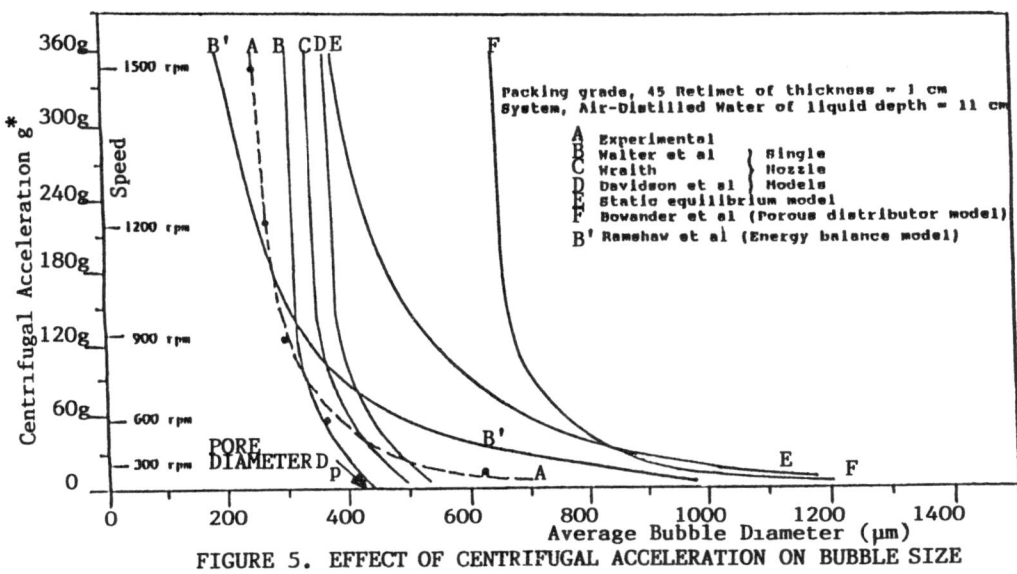

FIGURE 5. EFFECT OF CENTRIFUGAL ACCELERATION ON BUBBLE SIZE

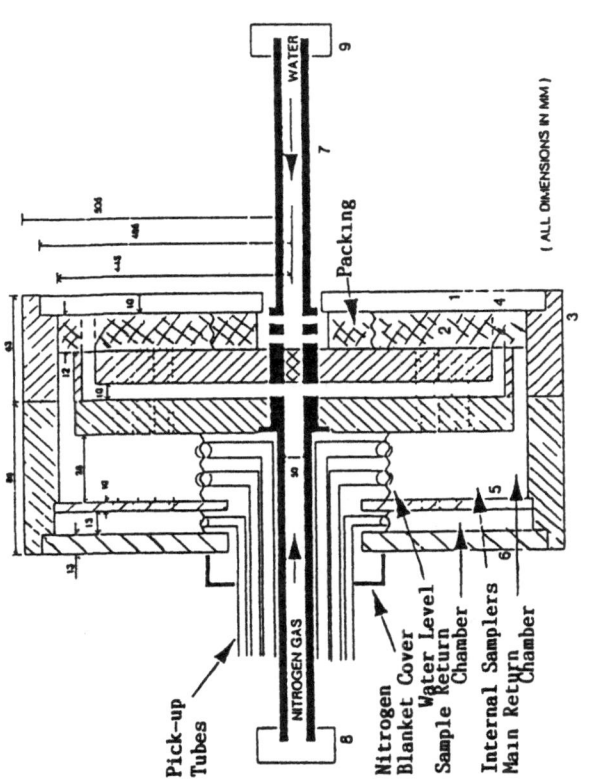

FIGURE 6. CROSS SECTION THROUGH ROTOR

MAJOR STRUCTURAL ITEMS

1 - POLYCARBONATE WINDOW 108 mm I D   970 mm O D
2 - DECLON PACKING PVC IMPREGNATED POLYURETHANE
     PORE SIZE 1 0 - 1 7 mm
3 - MAIN CHAMBER 60 mm THICK ALUMINIUM
4 - GAS INJECTOR NOZZLES
5 - CHAMBER SEPARATOR PLATE 300 mm I D   927 mm O D
6 - BACK PLATE 300 mm I D 970 mm O D
7 - STAINLESS STEEL SHAFT 50 mm I D 76 mm O D 1006 mm LONG
8 - GAS ROTARY UNION
9 - LIQUID ROTARY UNION

FIGURE 8. EFFECT OF SPEED AND GAS/LIQUID FLOWRATES ON MASS TRANSFER

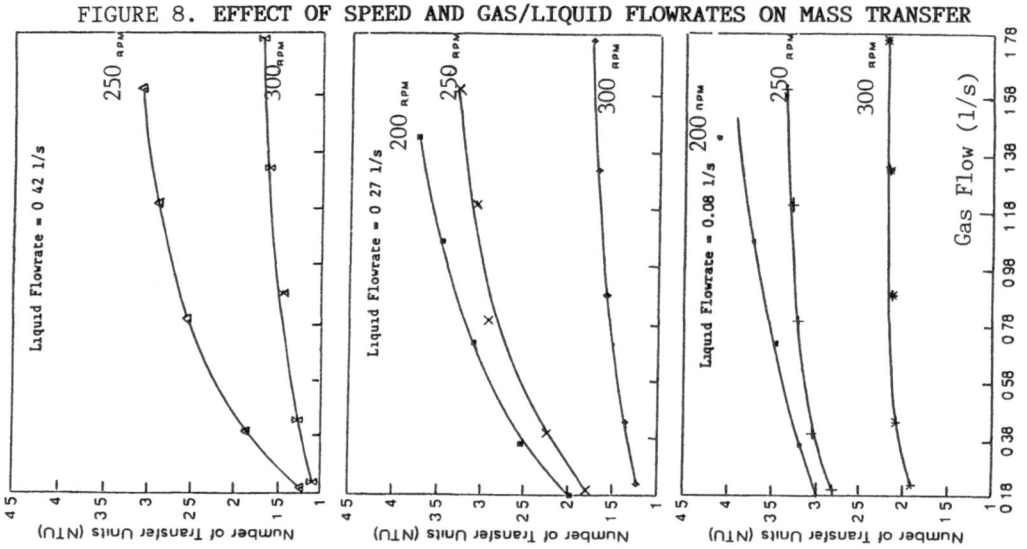

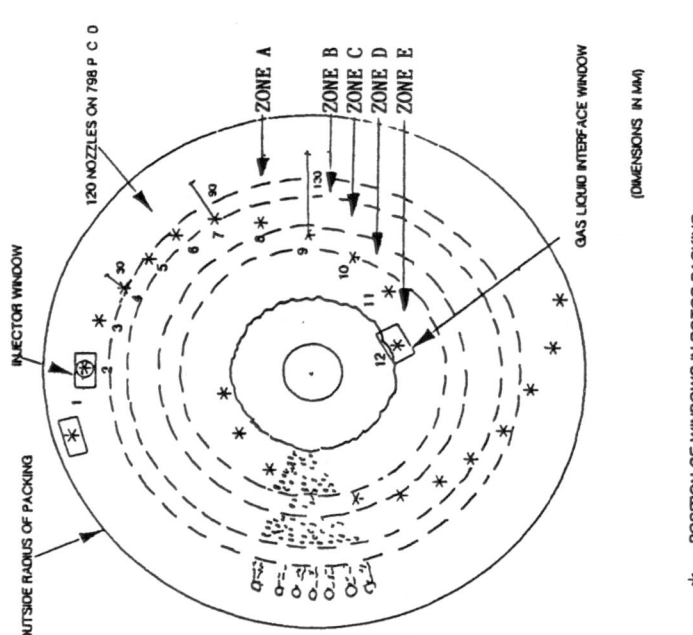

* - POSITION OF WINDOWS IN ROTOR PACKING
WINDOW NUMBERS 2 4 7 9 12 EXAMINED IN PHOTOGRAPHS

DIAGRAM TO SHOW POSITION OF WINDOWS

FIGURE 7. NON-UNIFORM BUBBLE ZONES

DEVELOPMENTS IN ELECTRICALLY ENHANCED SEPARATION PROCESSES

K.J. PTASINSKI and P.J.A.M. KERKHOF
Department of Chemical Engineering
Eindhoven University of Technology
P.O. Box 513, 5600 MB Eindhoven, The Netherlands

ABSTRACT

Superimposed electric fields are able to improve separation processes. The mechanism of the electrically aided processes involves some thermodynamical and hydrodynamical effects in separation systems, such as the change in a phase equilibrium and the electrical body forces of electrohydrodynamics. These effects can be used to form fine dispersions and to enhance the rates of heat and mass transfer. The efficiency of energy consumption by using electrical techniques is higher than that of traditional separation processes. The paper presents some recent engineering applications of the above aspects in separations, where a liquid is a continuous phase and the dispersed phase is either in the form of droplets of another immiscible liquid, gas bubbles or fine solid particles. Finally some results of our studies on gas-liquid processes in electric fields are also discussed.

INTRODUCTION

Separation processes can usually be enhanced by an energy input or by external forces. Often the energy to drive a separation process is supplied in the form of heat or mechanical work in combination with pressure or gravity forces. The ability of superimposed external fields, such as electric, magnetic, sonic etc, to improve separation processes has been well-known for many years. The familiar industrial applications of electric fields range from solid-solid separations in the beneficiation of ores in the mining industry, coalescence of "water-in-oil" emulsion in the petroleum industry, to cleaning of exhaust gases from solid particles in various technologies.

The direct utilization of electrical energy offers several advantages, particularly in multiphase systems:
- high rates of heat or mass transfer across the interface due to the selective interaction of an electrical energy with an interface
- the dispersed phase can be levitated or stabilized in a separator using electric body forces which contribute to the existing gravity force
- the energy efficiency of electrically driven processes is higher in terms of the second law compared to traditional mechanical or thermal proces-

ses. Due to the reversible nature of the electric stresses, the work done by electrical forces is nondissipative.

Recently an extensive effort has been directed to finding new applications of electrical energy for heat- and mass transfer operations and phase-separation processes. The most important examples are liquid-liquid extraction, dewatering of fine suspensions and electrophoretic separation of biological materials. It is the purpose of this paper to evaluate the major developments in this area and to discuss the engineering application of the phenomena. In particular the results of our own studies concerning an electrically driven gas-liquid absorption are also included. In order to understand the underlying principles of operations, the second section describes thermodynamical and hydrodynamical aspects of separation systems in electric fields, whereas section four discusses the influence of electric fields on mass transfer.

ELECTRIC FIELDS EFFECTS IN SEPARATION SYSTEMS

Phase equilibrium in polarized systems

An electric field penetrates into a dielectric fluid and the field energy contributes to the internal energy of a system. This interaction has a great effect on thermodynamic properties of a dielectric and it also leads to the change in a phase equilibrium. In the thermodynamics of systems in an electric field [1] the modified internal energy of a fluid in an electric field is defined as the internal energy of the total system (the fluid including the space) diminished by the energy needed to establish a field in the space of volume V:

$$U^* = U - \tfrac{1}{2}\,\epsilon_o E^2 V \tag{1}$$

Similarly the augmented pressure $p^*(\rho, T, E)$ of a fluid in an electric field is equal to the sum of the zero-field pressure $p_o(\rho, T)$ at the same density and temperature and the extra terms derived from the electric stresses:

$$p^* = p_o + p_p + p_s \tag{2}$$

where $\qquad p_p = \tfrac{1}{2}(\epsilon - \epsilon_o)E^2 \qquad$ is the fluid polarization pressure

$\qquad\qquad p_s = -\tfrac{1}{2}\,\epsilon\left(\frac{\partial \epsilon}{\partial \rho}\right)_T E^2 \qquad$ is the electrostrictive pressure

The modified fundamental equation for an open multicomponent system in an electric field is then given by:

$$dU^* = TdS - p^* dV + Ed(PV) + \Sigma\mu_i dn_i \tag{3}$$

whereas the modified Gibbs free energy is:

$$dG^* = -SdT + Vdp^* - PVdE + \Sigma\mu_i dn_i \tag{4}$$

The above relationship can be used to predict criteria of equilibrium in multiphase systems. The criteria for equilibrium for heat transfer and transfer of species between phases remain unchanged in the presence of a field. However, according to Eq. 4 the chemical potential in electromagnetic systems also depends on the electric field and thus the presence of a field is able to change the activity coefficients of some species [2]. The criterion of mechanical equilibrium in the presence of a field is altered from the usual equality of pressure in the phases and it depends on the orientation of the field and the geometric arrangement of the phases. If the electric field is oriented normal to the boundary of two phases α and β then the condition of mechanical equilibrium is

$$p_\alpha^* + \frac{P_\alpha^2}{2\varepsilon_o} = p_\beta^* + \frac{P_\beta^2}{2\varepsilon_o} \qquad (5)$$

whereas in the case of the tangential orientation the condition becomes (6)

$$p_\alpha^* = p_\beta^*$$

Electrohydrodynamics

Electrohydrodynamics deals with the fluid mechanics influenced by electrical forces which can arise either due to free charges present in a fluid or due to polarization of matter. The motion of a fluid under an applied electric field can be described by the usual principles of conservation of mass and of momentum. For incompressible fluids these principles are the equation of continuity

$$\nabla \cdot \mathbf{v} = 0 \qquad (7)$$

and the equation of motion [3]:

$$\rho \frac{D\mathbf{v}}{Dt} = -\nabla(p^*) + \mathbf{P} \cdot \nabla \mathbf{E} + \eta \nabla^2 \mathbf{v} + \rho \mathbf{g} \qquad (8)$$

The equation of motion can be represented in an alternative form using the concept of the body forces (per unit volume):

$$\rho \frac{D\mathbf{v}}{Dt} = f_p + f_e + f_v + f_g \qquad (9)$$

The terms familiar from fluid mechanics are the pressure force f_p, the viscous force f_v, and the gravitation force f_g.
The electrical body force f_e is given as [4]

$$f_e = \rho_e \mathbf{E} - \tfrac{1}{2}E^2\nabla\varepsilon + \nabla[\tfrac{1}{2}\rho\,(\frac{\partial\varepsilon}{\partial\rho})_T E^2] \qquad (10)$$

The first term is the ordinary electrostatic Coulombic force, the second term represents a force in an inhomogeneous dielectric in an electric field, and the last term (the electrostriction) gives a force on a dielectric in an inhomogeneous electric field. Electrical forces are in general quite small, but in situations where the other body forces are small, the forces associated with an electric field become significant.

Most of the electrically driven separation processes apply one or two of these driving forces. Electrophoretic separations rely on the application of the momentum exchange due to the charged particles (term 1 in Eq. 10) whereas the momentum transfer due to different phases, e.g. solid particles or gas bubbles in a liquid (term 2 in Eq. 10) is called "dielectrophoresis".

FORMATION OF DISPERSIONS USING ELECTRIC FIELDS

Dispersion systems are commonly used to increase the interfacial area and improve the rates of heat and mass transfer between one or more phases. The majority of studies on surface area generation has been devoted to systems composed of electrically conducting droplets in insulating liquids. Taylor [5] has shown that a neutral droplet placed in a dc electric field elongates due to polarization in a nearly prolate ellipsoidal shape. At the critical value of the electric field the disruptive electric stresses in excess can overcome the cohesive effect of interfacial tension and the original droplet disintegrates into a number of smaller daughter droplets.

Figure 1 shows two different electrical techniques that can be used to create large interfacial surface areas in liquid-liquid systems Such areas have been obtained in most designs [6] by droplet formation in charged nozzles or orifices. The second way of electrical dispersion is the direct use of strong electric stresses to rupture a liquid-liquid interface. The most efficient are high-intensity pulsed electric fields which allow the creation of a very large interfacial area by means of droplet rupture into extremely small droplets, even as small as 5 μm [7].

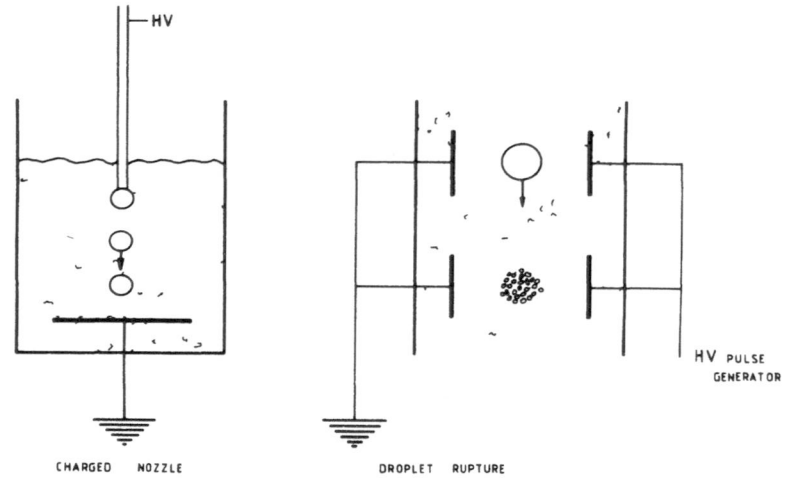

Figure 1. Formation of liquid dispersions in an electric field.

The application of electric size reduction of gas bubbles in a liquid has been much less investigated [8]. The studies have been restricted to the formation of small bubbles from charged needles in insulating liquids under a strong nonuniform electric field. Fig 2 shows that substantial bubble size reduction can be obtained, but the mechanism of this dispersion technique is not completely elucidated. It involves electrical pressure effects, the existence of corona discharge, and electrohydrodynamical flow [9].

Finally, it should be stressed that electrical techniques show a substantional energy saving compared to existing methods. In both investigated systems (liquid-liquid and gas-liquid), the energy input of an electrical technique is estimated to be about 1% of that required when mechanical agitation is used to produce droplets or bubbles of the same size.

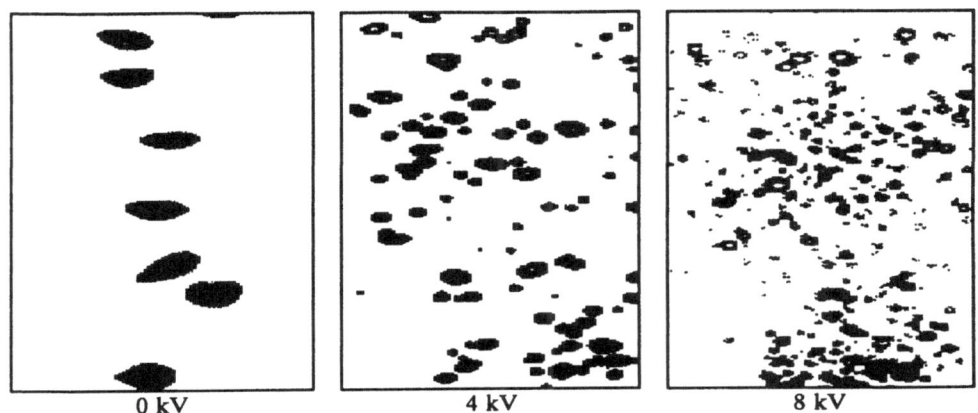

$$0 \text{ kV} \qquad 4 \text{ kV} \qquad 8 \text{ kV}$$

Figure 2. Shadow photographs of gas bubbles in an electric field
(system: nitrogen-ethanol, spacing of the electrodes 1 cm).

INFLUENCE OF ELECTRIC FIELDS ON MASS TRANSFER

Many separation processes such as solvent extraction, gas-liquid absorption, or liquid-vapor distillation involve a contact between a dispersed phase in the form of droplets or bubbles and a continuous fluid phase. The mass-transfer rate in these systems is proportional to the interfacial surface area and to the overall mass-transfer coefficient. This section concerns the electrical enhancement of the mass-transfer coefficient in dispersed systems whereas the application of electric fields to increase the interfacial contact area has been discussed in the preceding section.

A number of experimental and theoretical studies have been made on electrically augmented mass-transfer separations, in particular on electrical dropwise extraction. Most of the experimental results demonstrate that significant transfer rate enhancement can be achieved by using electric fields, typically the rise is by 2 to 6-10 factor compared to the no-field case [6]. The available theoretical analyses suggest that the mass transfer improvements are mainly obtained by producing a higher degree of fluid turbulence within and around the dispersed phase as the result of the interactions between the field and the interface. Different mechanisms promoting interfacial turbulence in a dispersion system in the presence of an electric field have been suggested. The most important mechanisms are higher terminal drop velocities, generation of the electrically driven circulating flow near the interface, alteration of the velocity profiles and interfacial-tension-induced Marangoni effects due to the presence of electric charges [10].

More recently attention has been paid to the possibility of mass transfer enhancement by electrical means in gas-liquid systems [9]. Typical test data plotted in Fig. 3 show that the volumetric mass-transfer coefficient can increase by a factor of nearly 8 using an applied field of approximately 8 kV/cm. Further investigation is in progress to elucidate the observed phenomena.

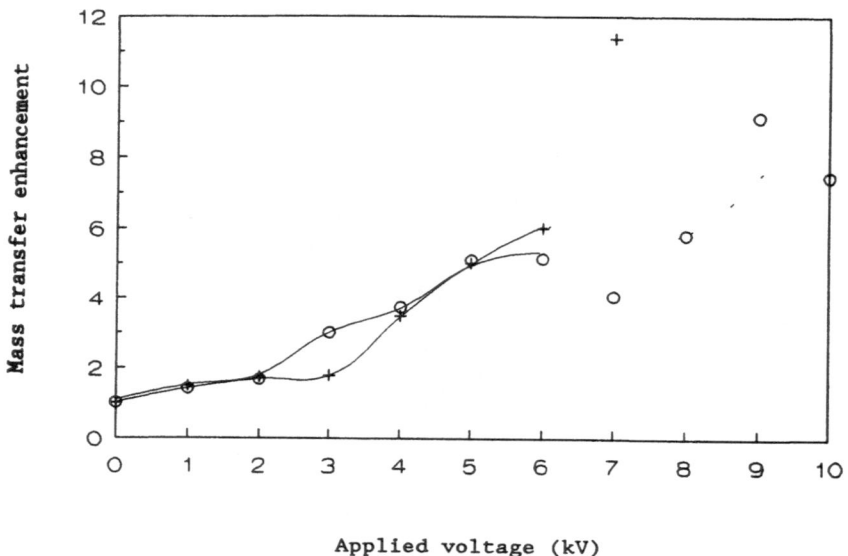

Applied voltage (kV)

Figure 3. Enhancement of the mass transfer rates by an electric field (+ N_2-ethanol, o Ar-ethanol, spacing of the electrodes 1 cm).

ELECTRIC FIELD DRIVEN SEPARATION PROCESSES

Electric field driven separations have been in commercial use for nearly 70 years. The applications which have the longest history and are still in use are the beneficiation of ores in the mining industry, electrostatic gas cleaning, electrical emulsion breaking, and dielectric heating and drying. Recent advances in electric field phenomena stimulate incentives for the improvement of existing separations and for the introduction of new processes. Electrical separation techniques nowadays receive wide attention both on a laboratory scale and in industrial practice, especially as fossil fuels become less accessible. Specific cases of electrical separations are considered below in some detail.

Electrofiltration of fine suspensions

A number of industrial processes deal with dewatering of fine or colloidal suspensions. Conventional mechanical dewatering such as cake filtration or centrifugation is restricted to particles larger than 10 μm. Suspensions containing smaller particles are usually dewatered by costly thermal drying. The application of a dc electric field provides a promising alternative in these cases.

Solid-liquid separations using electric fields are based on electro-kinetic effects, namely electrophoresis and electroosmosis. Most solids suspended in polar liquids (e.g. water) acquire a negative surface charge whereas a cloud of liquid containing positively charged ions (the diffuse

double layer) surrounds each solid particle. During electrophoresis the
charged solid particles migrate in an external electric field through a
relatively stationary liquid to an electrode. If solid particles are
immovable, as in a porous filter cake, the presence of an external electric
field will force liquid to flow through the porous solid, and this phenomenon is called "electroosmosis".

The majority of applications of electrokinetic effects relate to the
enhancement of conventional solid-liquid filtration known as electrofiltration. Electrofiltration techniques can be classified into two principal
categories:
- cake-forming methods, where a feed suspension is separated into a high
  solid cake and a filtrate [11]
- methods for limited cake growth, where the final products are thickened
  slurry and a filtrate [12]
Devices operating according to the first principle are commercially applied
on an industrial scale to dewater mineral slurries and to concentrate
polymer suspensions. Figure 4 shows the principle of cake-forming electrofiltration. The main advantage of this type of electrofiltration is a very
high solid content, up to 80%, in the cake produced and therefore the final
drying stage can be significantly shortened. The second electrofiltration
technique employs an electric field in a direction normal to the flow of a
suspension in order to keep solid particles away from the filtration
medium, and cake formation is prevented in this way.

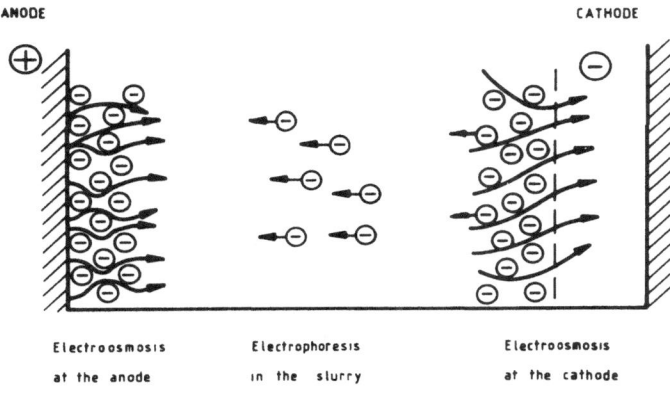

Figure 4. Schematic representation of cake-forming electrofiltration.

Continuous flow electrophoresis
Electrophoresis provides a useful tool for the purification and separation
of biological products such as cells and macromolecules. The majority of
analytical- and preparative-scale electrophoretic techniques is restricted
to batch separations. The commercialization of developments in biotechnology requires large-scale separation for downstream processing. In order to
separate small quantities of a product, a direct scale-up of an appropriate
batch analytical technique may be sufficient, whereas in many cases a
continuous separation may be needed.

Continuous flow electrophoresis (CFE) is a promising method for large-scale continuous separation, as this technique has been successfully used on a preparative scale for cell separations and isolation of biopolymers. A

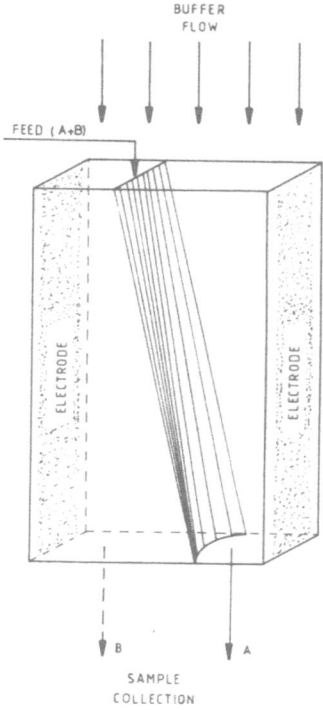

Figure 5. Principle of continuous flow electrophoresis.

schematic representation of CFE is presented in Fig. 5. In the CFE apparatus the buffered carrier fluid flows through a narrow slit and a uniform electric field is applied normal to the direction of flow. A multicomponent feed mixture is continuously injected into the fluid as a thin stream. The mobile species migrate laterally in the electric field with different specific velocities, and they are finally collected as separate fractions at the device exit. Scaling-up of CFE is hindered by electrokinetic and hydrodynamic factors such as the electroosmotic flow and the instabilities resulting from the thermal convection created by Joule-heating [13].

Electrical coalescence

The problem of effective removal of dispersed droplets from another continuous liquid phase is important in many industrial processes. The classical example of separating emulsified drops in liquid-liquid extraction is performed in stagewise mixer-settlers, where rapid phase separation is required after each mixing step. The fundamental drawbacks of existing gravity separators are their large size and high capital costs.

Of the many proposed improvements of phase separations (e.g. packings, centrifuges, additives), the application of an electric field seems to be the most promising. Electrical coalescers have been successfully used over many years for the removal of water from crude oil in the petroleum industry. More recently, electrical coalescence has been considered as a promising technique for the emulsion breaking in the liquid membrane extraction and for the direct liquid-liquid emulsion heat transfer (e.g. water desalination).

Efficient demulsification can be achieved both with dc fields and, more commonly, with ac fields. The entire process of liquid-liquid separation proceeds in three stages: coalescence of small drops into larger drops, settling of large drops and coalescence of large drops with the continuous phase. The presence of an electric field probably promotes all stages, but the way in which it occurs remains in doubt. A number of different mechanisms have been proposed, and they involved such effects as chain formation, electrophoresis, dielectrophoresis, dipole coalescence and random collision [14].

Electrically driven solvent extraction

The electrically augmented extraction has received considerable attention over the past two decades, but this process, contrary to the previously described coalescence, is not yet widely used in industry. However, recent developments achieved on a laboratory scale demonstrate the expected economic advantages of the direct use of electrical energy in this area. Extraction can be considered as an example of energy-augmented separation as the efficiency of this process usually depends upon the input of mechanical energy (e.g. agitation or pulsation) to the extraction equipment. The specific advantages of the direct utilization of an electrical energy is its selective interaction with the interface or with the dispersed phase only, whereas the mechanical energy influences all the phases present in the system, which is not generally required. It results in a high efficiency of energy utilization and, typically, the expenditure of a few watts can improve the efficiency of a laboratory extractor by a factor of 2 to 3.

The developed continuous flow extraction systems can be broadly classified as charged-nozzles or plates devices, spray columns with vertical electrodes, and emulsion-phase contactors. The charged-plates devices [6] are a modification of the existing sieve-plates extraction columns where the plates also serve as the electrodes. The second type of electrical contactor [15] uses a column with long vertically installed rod electrodes with reverse polarity. The most promising industrial application is the emulsion-phase contactor developed by the Oak Ridge National Laboratory [16], where a high-intensity ac electric field is used to create a high-surface-area emulsion.

CONCLUDING REMARKS

The place in separation technology currently occupied by electric field driven processes is modest but expanding. The results up to now demonstrate the potentialities of direct use of electrical energy in many fields as well as the higher energy efficiency of electrical processes. Much development work is still needed before new electrically augmented separations will be used to a significant extent in industry. Future research should include studies on electrical bubble columns and slurry reactors, emulsion coalescence, scale-up of the continuous flow electrophoresis and membrane fouling.

Acknowledgement
The autors wish to thank A. Staring and F. Geurts for their help in the experimental work on gas-liquid dispersions.

NOTATION

E	electric field (V/m)	v	velocity (m/s)
f	force density (N/m^3)	V	volume (m^3)
g	acceleration of gravity (m/s^2)	ϵ	permittivity (F/m)
G	Gibbs free energy of system (J)	ϵ_o	permittivity of free space
p	pressure in the presence		(8.854.10^{-12} F/m)
	of field (Pa)	η	viscosity (kg/m.s)
P	polarization (s.A/m^2)	μ	chemical potential (J/mol)
S	entropy of system (J/K)	ρ	density (kg/m^3)
t	time (s)	∇	nabla
T	temperature (K)	*	modified
U	internal energy of system (J)	i	species i

REFERENCES

1. Astarita, G., Thermodynamics, Plenum Press, New York, 1989.
2. Chandresekharan K., Padmanabhan K. and Mohan V., Effect of superimposed electric fields on transport processes. Proceedings of the 5th Int. Congr. Chem. Engng., Chem. Equip. Des. Autom., Vol. E, E 1.21, 1975.
3. Melcher J.R. and Taylor G.I., Electrohydrodynamics: a review of the role of interfacial shear stresses. Annu. Rev. Fluid. Mech., 1969, 1, 111.
4. Panofsky W.K.F. and Philips M., Classical Electricity and Magnetism, Addison-Wesley, Reading, Massachusetts, 1962.
5. Taylor G.I., Desintegration of water drops in an electric field. Proc. R. Soc. London, 1964, A280, 383.
6. Thorton J.D., The application of electrical energy to chemical and physical rate processes. Rev. Pure Appl. Chem., 1968, 18, 197-218.
7. Scott T.C., Surface area generation and droplet size control using pulsed electric fields. AIChE J., 1987, 33, 1557-1559.
8. Ogata S., Tan K., Nishijima K. and Chang J-S., Development of improved bubble disruption and dispersion technique by an applied electric field method. Ibid., 1985, 31, 62-69.
9. Ptasinski K.J., Staring A. and Kerkhof P.J.A.M. To be published.
10. Chang L.S. and Berg J.C., The effect of interfacial tension gradients on the flow structure of single drops or bubbles translating in an electric field. AIChE J., 1985, 31, 551-557.
11. Bollinger J.M. and Adams R.A., Electrofiltration of ultrafine dispersions. Chem. Eng. Prog., 1965, 61, 51-58.
12. Henry J.D., Lawler L.F. and Kuo C.H.A., A solid/liquid separation process based on cross flow and electrofiltration. AIChE J., 1977, 23, 851-859.
13. Reis J.F.G., Lightfoot E.N. and Lee H.L., Concentration profiles in free flow electrophoresis. Ibid., 1974, 20, 362-368.
14. Waterman L.C., Electrical coalescers. Chem. Eng. Progr., 1965, 61, 51-57.
15. Kowalski W. and Ziolkowski Z., Increase in rate of mass transfer in extraction columns by means of an electric field. Int. Chem. Eng., 1981, 21, 323-327.
16. Scott T.C. and Wham R.M., An electrically driven multistage countercurrent solvent extraction device: the emulsion phase contactor. Ind. Eng. Chem. Res., 1989, 28, 94-97.

NEW PROCESS FOR RECONCENTRATION OF ACID WASTES

GEORGES **LOUIS** - PATRICK **LE PEURIAN**
EDF DER Service AEE 6, Quai Watier - 78401 CHATOU CEDEX - France

FRANÇOIS **ROMBAUT**
SNPE Ingénierie 8, Cours Louis Lumière - 94306 VINCENNES CEDEX - FRANCE

ABSTRACT

This memorandum describes a new process for concentration of acid wastes. The heating technique here involves passing an electrical current directly through a metal Wall. The fluid to be concentraded in this specific case in sulfuric acid, diluted through a chemical operation to concentrate and recycle it into the process. The fluid passes through a tube, to which an electric current is applied directly. Heat is generated within the tube material itself due to the joule effect and transferred from the heated wall to the fluid by forced convection. An alternating current, as for example the 50 Hz of the electrical circuit, counteracts the electrolytic effects.

INTRODUCTION

The present text mainly concerns the development of an electric process to reconcentrate acid wastes. Partnership research, conducted in collaboration with the SNPE Ingénierie and EIVS (CORNING) Companies has made it possible on one hand to identify the potential market for reconcentration of waste sulfuric acids ($\approx$ 300,000/year) [1], and, on the other hand, to develop a demonstration prototype.

In order measure interest in a reconcentration stage, it is good to recall briefly the order main ways for treating sulfuric acid wastes :

- disposal : it involves chemical transformation of the acids and impurities in complex mixtures and of heat decomposition that makes it possible to obtain SO_2 and then sulfuric acid. These processes are complex and time-consuming ; to our knowledge, this method of treatment is

rarely used in FRANCE.
- Neutralization : this is a solution that is becoming less and less profitable, since it does not always resolve the problem of soil and water pollution and generates large amounts of salts.

Reconcentration thus is an interesting alternative to the treatment of acid wastes : it makes it possible to save in the following ways :
- reduction in purchases of concentrated sulfuric acid ;
- reduction in transport costs ;
- removal of neutralization or disposal costs.

On the other hand, new expenses are going to involve :
a) power consumption for reconcentrating the acid wastes (see figure 1).
b) maintenance and operating costs.

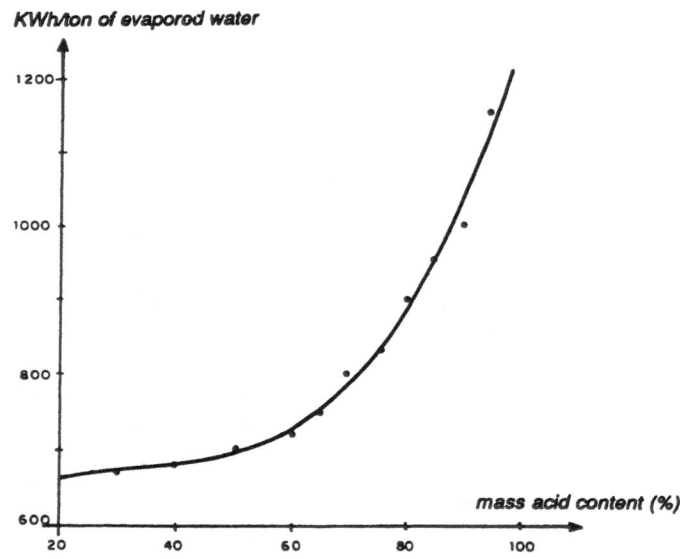

Figure 1. Evolution of specific consumption with the acid content.

MATERIALS AND TESTS

Reconcentration of acid wastes consists of eliminating the water by evaporation. The execution of this operation resides thus in developing an adequate and competitive evaporator called the "electric exchanger" from now on. The basic idea comes from the joule effect ; it is a question therefore of heating the wall of a tube directly by applying a current to the latter. The heat generated in this way causes a heat transfer between the hot wall and the fluid by convection.

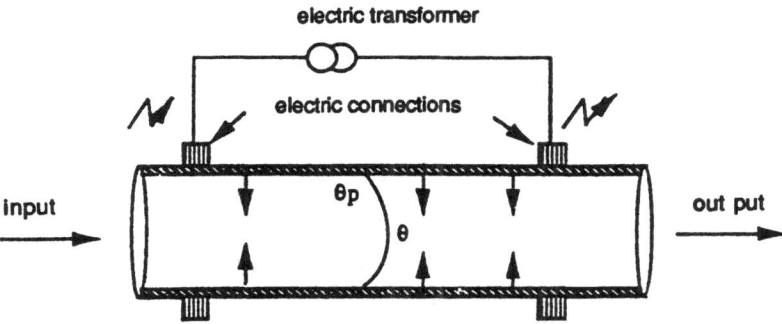

Figure 2 : Basic principle of the electric exchanger.

- the tube, the large hydraulics of the fluid being heated provides homogeneity of the flows and a uniform distribution of the heat exchanges, which because of this are perfectly controlled ;
- the heat fluw density φ is almost homogeneous along the tube ; it is strictly limited by :

a) the exchange cœfficient h between the tube and the fluid at a temperature (θ) ;

b) the maximum skin temperature of the tube (θ_p) authorized by the physicochemical characteristics of the fluid being heated.

The fundamental heat transfer equations are written.

$$\varphi = h \, (\theta_p - \theta) = \frac{U^2}{R\,S}$$

where

$$(\theta_p - \theta) = \frac{U^2}{R\,S\,h}$$

R : electrical resistance of the tube
S : heat exchange surface (m^2)
U : electric current applied (V)

The result of equation (1) is that for a controlled hydraulic process, the heat gradient is constant all along the tube ; this is not the case with a classic heat exchanger with a hot source (steam, thermal fluid). It involves a significant feature of the electric exchanger, expressing the fact that the temperature profile is perfectly controllable all along the exchange wall.

In the current case where the water is eliminated by evaporation, the fluid state advances all along the exchange surface [2] (see figure 3) ; the following can be perceived :

a) heating area (illegible) boiling. It assumes the introduction of the subcooled liquid.

b) two phase mixing area : it is characterized by a high heat transfer intensity (high exchange cœfficient).

c) Steam phase area : it is a question of a critical zone corresponding to relatively high transfer densities φ_c (critical heat flux) ; it is complety inadvisable to work here because there is a risk of transfer surface deterioration (tube breaking melting). In fact, the resistance to heat transfer is especially localized in the steam phase and as a result, the heat gradient (for steam) would be very high, thus causing a sharp increase in the skin temperature.

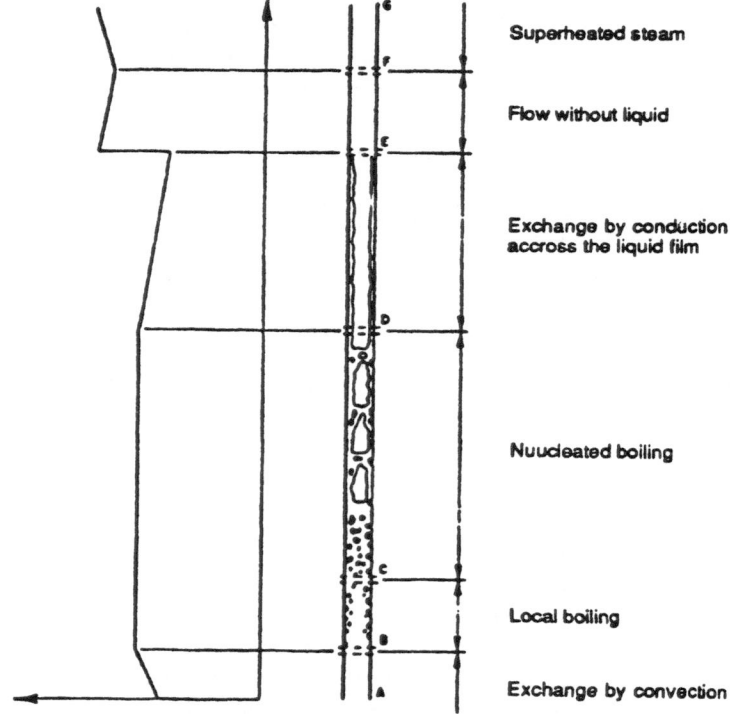

Figure 3 : Evolution of the fluid state and the skin temperature along a vertical tube

With a double purpose, to make the most of the advantages of electric exchanger and to demonstrate the technical feasibility of the recommended heating, a 20 kw prototype was conceived.

Figure 4 illustrates the installation diagram including :

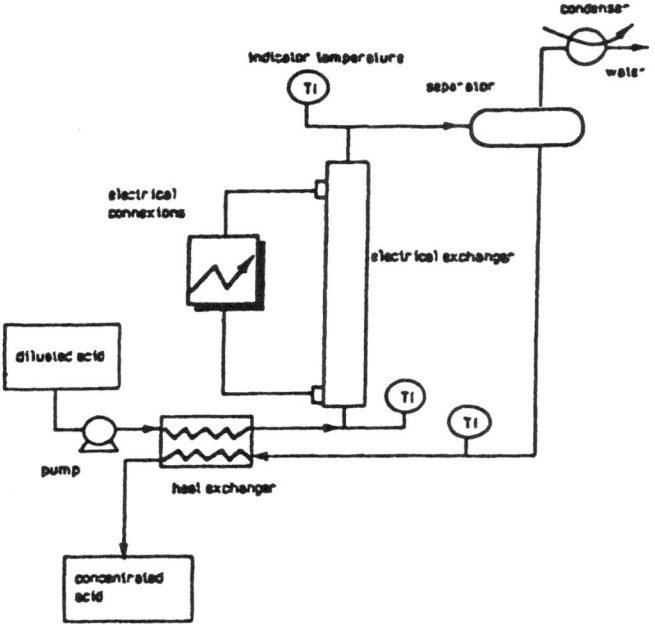

Figure 4 : General principle of the reconcentration process

- 2 storage tanks of diluted and concentrated acids ;
- a supply pump for the acid being treated ;
- a load pre-heater ;
- a tantalum tube under low current (6 - 12 V) with the following characteristics :
    + length = 0,8 m
    + diameter = 13 mm
    + thickness = 1 mm
- a glass separator balloon ;
- a glass head condenser.

ANALYSIS OF THE RESULTS

Tests at atmospheric pressure were conducted on site at the National Company of Gunnpowder and Explosives (SNPE) at Angoulème (France). They had a double objective :

1) to tally the performance data of installation and identify the operating zones (optimal flows)
2) validate the concept of direct heating in a metallic tantalum skin.

First, we identified the critical fluws on an experimental scale so that we could use some basic units that are indispensable to equipment rating studies. Figure 5 presents the comparison between the experimental values and those expected from the literature [3] (SHAH correlation) ; on the whole, the results were in good agreement.

CRITICAL HEAT FLUX CHF (kW/m²)

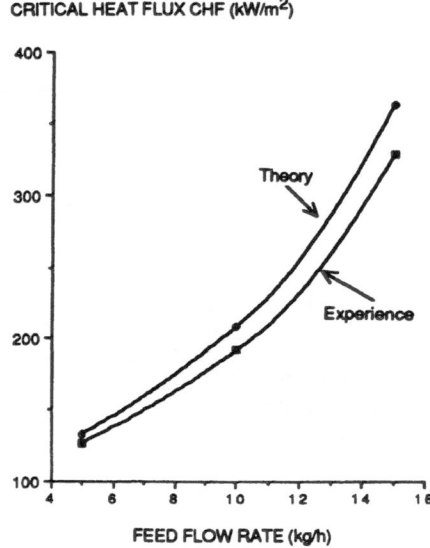

Figure 5 : Evolution of the critical flux with the feed rate

Secondly, we sought to identify better the performance data of the electric exchanger, i. e., to evaluate the maximum sulfuric acid concentration that could be expected at the equipment output.

The following table shows some of the results obtained.

input flow (kg/h)	input concentration	output concentration	heat yield %
13.6	53 %	69 %	73
20	53 %	71 %	83
42.6	53 %	57 %	82
52	53 %	78 %	89
68	53 %	55 %	84
40	70 %	79 %	80
30	70 %	77 %	85
25	70 %	81 %	88
15	70 %	79 %	80

Yield is defined by the ratio $\dfrac{\text{theoretical power}}{\text{experimental power}}$

Analysis of the set of tests reveals some important points :

a) The relatively high heat yield of the installation (80 - 90 %) is a question of an important element showing the advantage of electric heating compared to steam where the distribution and exchange yield scarcely exceeds 50 %.

b) Examination of the fluid temperature at the exchanger output as a function of the concentration obtained gives us completely satisfactory measurements, as are shown in Figure 6, the difference between the experimental values and theoretical ones is low.

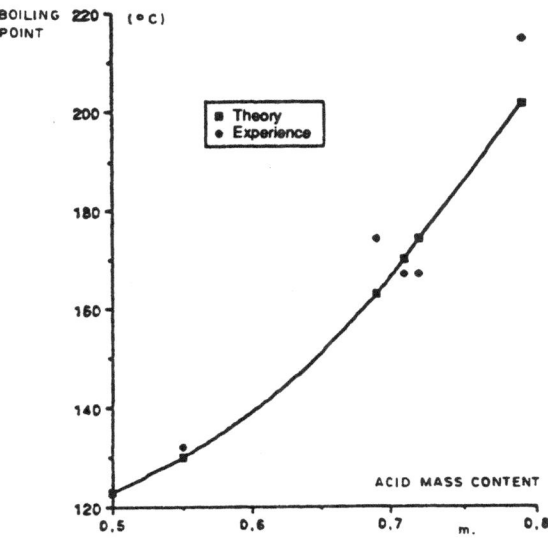

Figure 6 : Boiling point vs. mass content

On the whole the performance of the installation with tantalum as the heat exchange material is at about 80 % acid. This threshold can hardly be surpassed because of the chemical behavior 3of tantalum in a sulfuric acid medium ; on the other hand, ferro-silicon can be used to expect high concentrations (90 - 95 % acid) despite its heat shock that is widely alleviated by the fact that the skin temperature evolution would be well controlled by the recommended electric heating.

TECHNICAL AND ECONOMIC ANALYSIS

For better reliability of the sulfuric acid waste reconcentration, we examined the technical and economical feasibility of an industrial scale project with a capacity of 1,000 tons per year. The data of the problem are the following :
- continuous operation at a rate of 125 kg/h
- input concentration of 50 % ;
- output concentration of 78 %.

The study conducted included 3 sections :
- equipment rating ;
- evaluation of investment and operating costs ;
- examination of the profitability of the installation.

Beginning with the following economic data (furnished by an industrialist)
- cost of the purchase of 78 % unpolluted acid 79 $[**]/ton
- cost of 50 % acid treatment : 52 $/ton.

Assuming that the reconcentrated (polluted) acid is sold to other users at a rate of 50 % of its purchase price (unpolluted acid), the results described in the following table are obtained :

Battery limits ($)[**]	110,000
Cost of utilities ($/t)[*]	38[(1)] - 50[(2)]
Maintenance expense ($/t)	6[(1)] - 9[(2)]
Total cost ($/t)[*]	44[(1)] - 59[(2)]
Payback time (years)	2[(1)] - 2.5[(2)]

(1) cost of electric kwh : 0.045 $ cooling water = 0.07 $/m^3
(2) cost of electric kwh : 0.062 $ "   "
* at 100 % acid.

CONCLUSION

In light of the results obtained the "electric" sulfuric acid reconcentration process seems to be of some interest on the technical and economical level ; on the other hand, the cleaning of the wastes is not taken into account. Within this context, taking into consideration the nature of the impurities contained in the waste being treated, the following combination of electric techniques can be imagined :

- cleaning technique (dialysis, electrodialysis)
- reconcentration by the joule effect.

This combination of waste treatment is quite adapted to the mineral impurities contained in the fluid in question ; as far as organic imporities are concerned, different ways can be envisioned. Oxidation, extraction, heat decomposition, etc. can be mentioned.
The process presented seems to introduce new advances to the conception of high-performance evaporators ; it was validated on sulfuric acid in a concentration range from 50 to 80 %. Of the numerous existing advantages, the following can be mentioned :

- ease of operation and absence of inertia ;
- composite and modularity of equipment ;
- little inhibition and good process behavior.

The next step concerns reconcentration tests beyond 80 % and extrapolation to the industrial scale of the process. Within this context, Electricité de France (France Electricity) intends to continue collaboration with its partners on efforts to develop an industrial acid waste reconcentration process with the view of producing new installations of this type.

[**] 1 $= 5.6 French Franc

REFERENCES

[1]   G. LOUIS-V.CARON, utilisation du tube à passage de courant pour la reconcentration d'acide sulfurique EDF-DER-HES1/92.13

[2]   J.F. SACADURA, Initiation aux Tranferts thermiques, Lavoisier, Paris 1982

[3]   M. SHAH, Heat and Fluid flow vol. 8 n°4, Improved general for critical heat flux during upflow in uniformily heated vertical tubes, December 1987.

SESSION 6:

Refrigeration and Refrigerants

Chairman: Prof. K. Cornwell

REPLACEMENT OF R12 IN REFRIGERATION SYSTEMS

J.T. McMULLAN[*], H. KRUSE[†], R. CAMPORESE[‡] & R.J.M. VAN GERWEN[§]

[*]   Centre for Energy Research, University of Ulster, Cromore Road, Coleraine, Co Londonderry, BT52 1SA.

[†]   Universitat Hannover, Institut für Käletechnik und Angewandte Warmetechnik, Welfengarten 1A, Germany.

[‡]   Consiglio Nazionale delle Ricerche, Instituto per la Technica del Freddo, Corso Stati Uniti I-35100, Padova, Italy.

[§]   Institute of Environmental and Energy Technology, TNO, Laan van Westenenk 501, PO Box 342, 7300 AH Apeldoorn, The Netherlands.

ABSTRACT

Three aspects of work relating to the replacement of R12 refrigeration and heat pump applications. In the first, the use of non-azeotropic fluid mixtures in a mixture cascade system is discussed and both theoretical and experimental results are presented. In the second, analyses are presented of the performance of R134a-R125 mixtures in a number of configurations. Finally, the interaction of lubricants with the single fluid R134a is discussed, together with some aspects of control and the use of compact heat exchangers.

INTRODUCTION

This paper presents the incomplete results of three strands of work aimed at the problem of replacing R12 in refrigeration and heat pump applications.

The first strand is concerned with the use of zeotropic mixtures and a mixture cascade to permit the use of only one compressor in high temperature lift applications. The second strand concerns the use of zeotropic mixtures of refrigerants, notably R134a and R125 in industrial applications. The third strand is concerned with the now standard replacement R134a, and its interaction with the lubricants needed to protect the compressor and other moving parts. The work was carried out with the support of the CEC JOULE Energy R&D Programme.

MIXTURE CASCADES

The principle of a mixture cascade is described in Figure 1. The mixture leaves the compressor "A" at condensing pressure. After partial condensation in the condenser "B", separation occurs in vessel "D". Two fractions of refrigerant with different concentrations (defined as the mass fraction of the lower boiling component) leave the vessel "D". The liquid refrigerant (3') with the lower concentration is throttled in an expansion valve and mixed with the refrigerant (6) coming from the evaporator. The total flow evaporates in the cascade cooler "E" while the vapour from vessel "D" (3") is condensed in counterflow. The condenser liquid (4) with the higher concentration is throttled (5) and evaporates, producing the refrigerating effect at the lowest temperature in the cycle.

The temperature difference between the heat sink and the heat source can be increased by utilizing the temperature glide of refrigerant mixtures during the phase change. Therefore the mixture cascade is extended by two internal heat exchangers (Figure 2). The heat exchanger "C" reduces the quality of the total flow, while on the other side the mixture will be superheated in the suction line. In the heat exchanger "G" the partially condensed vapour is entirely condensed and subcooled before the throttling in the expansion valve.

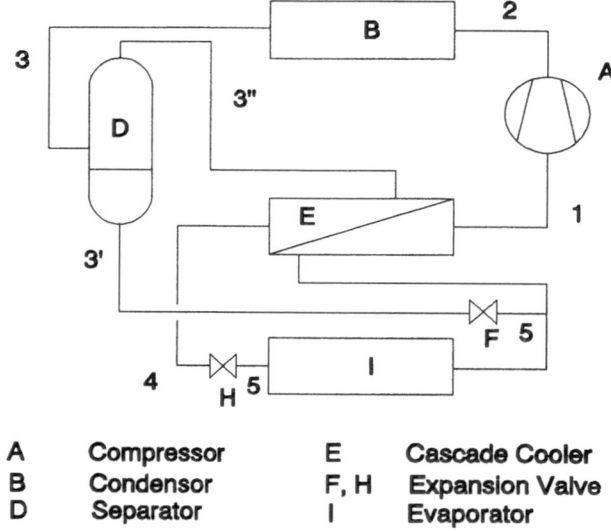

A	Compressor	E	Cascade Cooler
B	Condensor	F, H	Expansion Valve
D	Separator	I	Evaporator

Figure 1. Mixture Cascade System.

Theoretical Results

A computer model has been developed for the mixture cascade as shown in Figure 2. The simulation model has already been described elsewhere [1]. The compressor simulation has been improved by using parabolic equations for volumetric efficiency as a function of the pressure ratio p_c/p_o and for isentropic efficiency as a function of p_o. These equations were derived from measurements obtained with a reciprocating compressor "Bock F3", but the results under these assumptions differ not much from those which had been calculated using constant efficiencies.

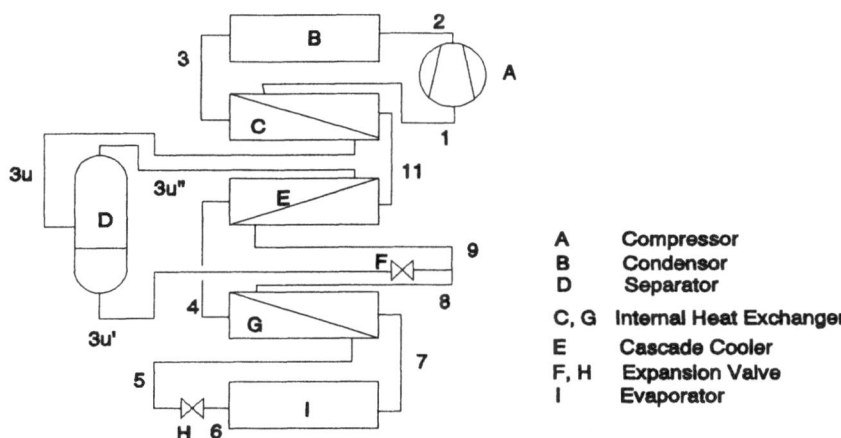

Figure 2. Mixture Cascade System with Internal Heat Exchangers.

The calculations have been performed assuming a minimum temperature difference of 10 K for heat exchanger "C" (Figure 2) and cascade cooler "E", and 5 K for heat exchanger "G". It is assumed that the temperature at the condenser outlet is 32 °C, the evaporator inlet temperature is -100 °C and the compressor displacement is 7.56 m³/h. Based on these assumptions, the concentrations of calculated mixtures in the mixture cascade were optimized for mixtures of R14/R134a, R14/R12, R14/R227/R123 and R23/R227.

The calculated COP for the mixture R14/R134a is 0.31 in comparison with the 0.34 of the FC-CFC mixture R14/R12. The reason for the higher COP with the mixture R14/R12 is the larger quality change in the evaporator, which results in a higher specific refrigeration capacity. The higher COP for the mixture R14/R12 occurs despite the lower isentropic compressor efficiency and the lower mass flow ratio through the evaporator.

The best result was obtained with the ternary mixture R14/R227/R123. The COP increases remarkably when the R123 content increases from 10% to 15% (wt). For the optimized concentration it is noticeable that there are very low values of the discharge temperature (67 °C), the discharge pressure (0.78 MPa) and the pressure ratio (4.00). With the ternary mixture a very large COP of 0.59 has been calculated.

Through the use of the binary mixture R14/R227 it is possible to get the relatively high COP of 0.51.

With increasing evaporation temperatures, up to -60 °C at evaporator inlet, and with decreasing condensation temperature from 37 °C down to 30 °C, the COP of the plant using the mixture R23/R227 increases significantly up to 0.91.

Experimental Results

A test plant for the refrigerant mixture R14/R134a (Figure 3) has been built using conventional components. Two parallel reciprocating compressors and an air-cooled refrigerant condenser form the hermetically sealed condensing unit (type Danfoss "SC 21/21 BX T2 Twin"). Between the compressors and the condenser, the mixture passes through a Danfoss OHB 1 oil separator. To allow for further investigations with another refrigerant mixture, the mixture cascade includes a small heat exchanger "C" of the type "Danfoss HE 1.0". The cascade cooler "E", is a double coaxial heat exchanger with plain tubes. Two "Wieland" heat exchangers of the type WKK4 and WKK5 are connected in line to form heat exchanger "G". The evaporator FWA20 made by "KÜBA" is placed in a modified form of a mini freezer cabinet. Pressure regulated expansion valves "AW3" from "Flica" are used. The separator of the type "Danfoss OHB4" is oversized to realise smaller pressures when the compressor is not operating. The lubricating oil (1.3 dm^3 "Fluisil S 55 K") is an ester of a poly-silicic-acid for low temperature applications and is manufactured by Bayer. A bypass line, with a solenoid valve, connects the separator inlet with evaporator inlet to protect the mixture cascade from excessively high discharge pressures.

Figure 3 shows the locations for pressure and temperature measurements in the cycle. The pressure transducers (made by Transamerica) are of the resistance gauge type and have been calibrated. The calibration curve is described by a polynom. Cu/CuNi thermocouples are attached with heat transfer paste on the surface of the tube. The temperatures corresponding to the voltages for the thermocouples were taken from /DIN IEC 584 Teil 1/. A 3497 Data Acquisition/Control Unit from Hewlett Packard as a converter links the computer with the pressure and temperature measuring transducers. At two locations (1 and 3u") provisions are made to take gas samples, which have to be analysed with the gas chromatograph in order to determine the concentrations of the refrigerant mixtures.

In order to reach the lowest evaporating temperatures, the refrigerant mass charge and the refrigerant concentration have to be optimized. With 0.72 kg R14 and 2.44 kg R134a ($\xi = 0.23$) the temperature in the evaporator glides from $t_6 = -80\ ^{\circ}$C to $t_7 = -72\ ^{\circ}$C. The experiment was conducted at the ambient temperature $t_2 = 21\ ^{\circ}$C, with 40 dm^3 air in the freezing cabinet as a heat source. With 35 km^3 petrolether in the freezing cabinet the temperature in the evaporator glides from $t_6 = -75\ ^{\circ}$C to $t_7 = -65\ ^{\circ}$C, which corresponds to a calculated heat flow rate of about 120 W from the environment into the freezing cabinet. With increasing refrigerant mass the temperature at the evaporator inlet decreases to below -100 $^{\circ}$C (air in the freezing cabinet). These experiments were conducted using up to 0.90 kg R14 and 2.64 kg R134a (concentration charged mixture, $\xi_c = 0.25$). At the lowest temperatures reached, the measured suction pressure was 0.17 MPa. The measured discharge pressure was 1.73 MPa.

During the test phase some very interesting effects occured. In the first minutes after the start of the mixture cascade, the direction of the heat flow in heat exchanger "G" alternates. At first, the temperature t_7 at the evaporator outlet (Figure 3) is higher than the temperature t_5 at expansion valve "H". This leads to a heat flow from the low pressure side to the high pressure side and therefore a decrease in temperature from point "7" to point "8". After about 15 minutes of operation, the heat flow decreases and later on reverses.

During the first 30 minutes, while the refrigerant mixture at point "8" is still superheated, the temperature t_9 behind the mixing point is 14 K less than the temperature at point "8". This effect appears in the h-ξ-diagram for the measured pressure $p_9 = 0.215$ MPa. This ultimately results in a large temperature difference between t_4 and t_9 and gives a larger heat flow rate in the cascade cooler (see Figure 3).

Noticeable is the high pressure of 1.35 MPa at stand idle also the high discharge pressure, which is reduced with a bypass to 2.5 MPa. A pressure drop of 0.07 MPa has been measured from the evaporator inlet to the compressor inlet (Figure 3).

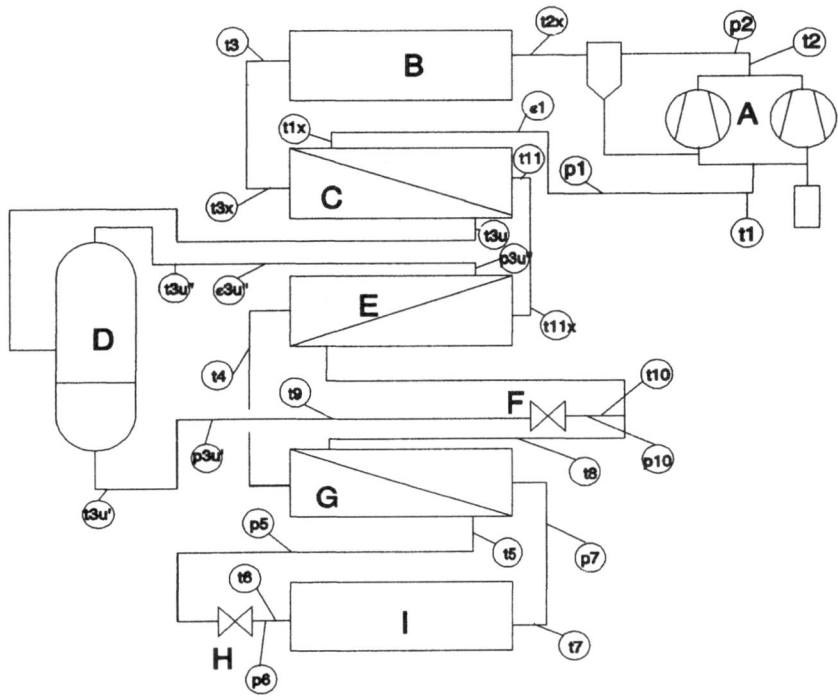

Figure 3. Experimental Test Rig.

NON-AZEOTROPIC MIXTURES (NARM) CYCLE SIMULATIONS

The objective of this strand was to establish applications and conditions with a high potential for achieving energy conservation through the use of NARMS. This paper reports some simulations of the performance of "industrial" systems based on mixtures of R134a and R125.

The simulation model used, CYCLE-11, was developed by the National Institute of Standards and Technology (NIST) in the USA and is described in detail in [2].

The calculations of the refrigerant properties (1 or 2 components) are based on the Carnahan-Starling-DeSantis (CSD) equation of state [3]. In this paper, it is concluded that the CSD equation of state accurately represents both vapour and liquid phases for HCFC refrigerants and their mixtures. An independent set of property subroutines [4] is linked with the cycle model code. Properties of 22 refrigerants and 20 binary interaction coefficients are incorporated in the model. Values for binary interaction coefficients may be put in manually as well as automatically.

For the mixture R134a/R125, the program does not deliver a binary interaction parameter. In general, a value of 1.011 is used (see later). In the CYCLE-11 program, the term mixing coefficient is used, where <mixing coefficient> = 1.0 - <binary interaction coefficient> [5].

Many different compressor models can be used in (isentropic or polytropic, with inclusion of volumetric efficiency and heat transfer to or from suction and discharge gas (hermetic compressors)) CYCLE-11. For this simple simulation, two compressor models are used: one for an isentropic compressor with ideal, 100% volumetric efficiency and an isentropic compressor with realistic volumetric efficiency for a reciprocating compressor (including an arbitrary leakage factor of 0.96 and a clearance volume factor of 0.04).

Counterflow heat exchange is assumed in both evaporator and condenser. The performance is specified in terms of average effective temperature difference δT, defined by $\delta T = Q/(U.A)$. Different sections for superheated vapour, two phase flow and subcooled liquid are considered. When using NARMS, the two phase region is split into a number of sub-sections, because of the non-linear character of the temperature profile.

The stationary simulation of a refrigerating cycle results in a table with refrigerant properties at several cycle conditions, (volumetric) refrigerating capacity (in kJ/kg and kJ/m^3), compressor work (in kJ/kg), pressure ratio and gliding temperatures in evaporator and condenser.

Simulation Conditions

The temperatures for the ingoing and outgoing secondary fluids (air, water etc.) of evaporator and condenser, as well as the average effective temperature differences, have to be chosen for each simulation run. In these simulations, zero refrigerant subcooling and superheating are assumed. No internal heat exchanger is used. For the simulation runs, 30 conditions have been selected:

- high, medium and low evaporator temperatures;
- high and low condenser temperatures;
- high and low temperature differences between secondary inlet and outlet (water, air) of evaporator and condenser;
- high and low average effective temperature differences in evaporator;

These selected conditions are realistic for practical industrial applications and are shown in Table 1.

For each of the 30 conditions, simulations have been carried out for the following mixing ratio's (mass percentages):

100 % R134a	0 % R125
80 % R134a	20 % R125
60 % R134a	40 % R125
40 % R134a	60 % R125
20 % R134a	80 % R125
0 % R134a	100 % R125

The simulations for 100% R134a and 100% R125 have been carried out with the program option for one component simulation. A check for the differences between the one and the two component simulation approach has been made by comparing the results of one component with results of the two component simulations (99% R134a, 1% R125 resp. 1% R134a, 99% R125). All the simulation results were exactly equal, except for the C.O.P., where a difference of 0.01% occurred. Therefore it may be concluded that it is valid to compare one and two component simulation results.

The simulations for the mixing ratio of 40% R134a and 60% R125 were also carried out for an alternative value of the binary interaction parameter, to investigate the influence of this value, the choice of which may be critical. Based on the Plöcker method [5], a value of 1.011 has been calculated by the University of Hanover. NIST has estimated a value of 0.995 for the binary interaction parameter [6]. For the mixture ratio 40/60%, simulations with both of the values have been carried out and comparison of the simulation results shows only a small influence of the two values. The maximum differences are:

- 3% increase of refrigerating capacity;
- 0.5% decrease of C.O.P.;
- 0.1 to 0.2 K increase of gliding temperature in condenser and evaporator;
- 2% decrease in pressure ratio.

TABLE 1
Selected Simulation Conditions

condition numbers:			1 - 10	11 - 20	21 - 30
inlet temperature condenser (°C)			20	30	20
outlet temperature condenser (°C)			25	35	30
avg. effective temperature diff. cond. (K)			10	10	10
evaporator:			inlet temp.	outlet temp.	avg. eff. temp. diff.
condition nrs.			(°C)	(°C)	(K)
1	11	21	5	0	5
2	12	22	5	0	10
3	13	23	-5	-10	5
4	14	24	-5	-10	10
5	15	25	-15	-20	5
6	16	26	-15	-20	10
7	17	27	10	0	5
8	18	28	10	0	10
9	19	29	-10	-20	5
10	20	30	-10	-20	10

Based on these small differences in simulation results, it seems to be legitimate to use a binary interaction parameter value of 1.011 for all the simulations.

Simulation Results
180 simulation runs have been carried out and, as an example, the C.O.P. (ideal refrigerating capacity divided by the compressor work), results are presented in Figure 4. These bar graphs for each of the 6 simulated mixture ratios are grouped in three series of ten simulation conditions with equal condenser conditions (see also Table 1).

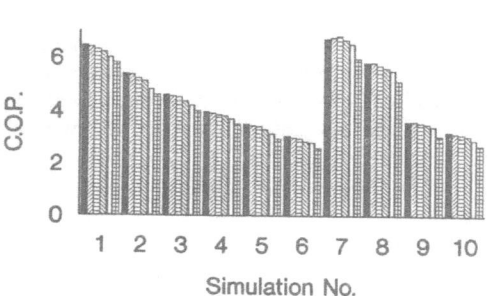

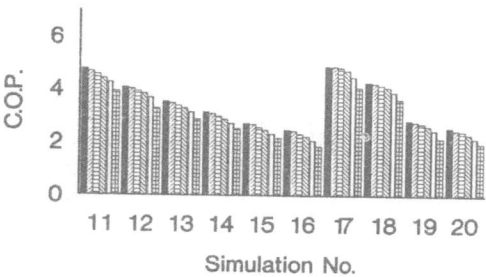

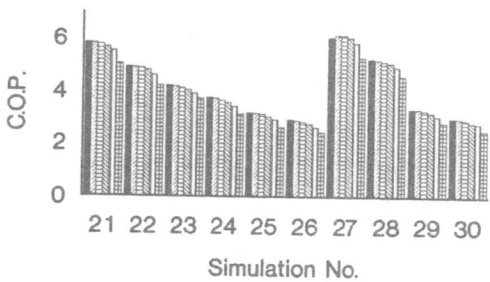

Figure 4. Sample Simulation Calculations (see Table 1).

The following general results have been obtained and are presented in [7].

Refrigerating Capacity:
-    The ideal volumetric refrigerating capacity increased linearly with the mixture ratio. For pure 125, this capacity is 1.6 to 1.7 times the capacity for pure R134a, using the same compressor displacement volume, for all of the simulation conditions.
-    Using the typical compressor model, (including volumetric efficiency), the results are equal to the results, obtained with the ideal compressor.

C.O.P.:
-    The C.O.P. for pure R125 is 10 to 20% lower than the C.O.P. for pure R134a, for all the simulation conditions. For high condenser temperatures, the decrease is larger than for low temperatures.
-    A significant increase of the C.O.P. of R125 may be obtained by mixing R125 with R134a; 20% R134a already leads to a significant increase of the C.O.P. (by 10%), specifically at low evaporator and high condenser temperatures (simulation conditions 15 and 16).
-    Using the mixtures, the influence of increasing percentage of R125 on C.O.P. is more negative for the high value of the effective temperature difference in the evaporator, probably due to the lower resulting evaporator temperature.
-    The largest C.O.P. increase using R134a-R125 mixtures, arises at high evaporator temperatures, low condenser temperatures and large secondary temperature differences (simulation conditions 7 and 27). In these cases, a slight increase of C.O.P. (1 - 2%) occurs for mixture ratio's of 20 and 40% R125.

Gliding Temperatures
-    The maximum gliding temperature difference in the condenser occurs at mixtures with 40 and 60% R125. The use of the mixtures leads to an extra gliding temperature in the condenser of 2 to 6 K, depending on the simulation conditions (in all cases, a temperature glide also occurs due to the superheated vapour at condenser inlet).
-    The maximum gliding temperature difference in the evaporator occurs at mixtures with 60% R125. The use of the 60% R125 mixture leads to a gliding temperature in the evaporator of 2.6 to 3.2 K, depending on the simulation conditions. The temperature glide at 20 % R125 is less than 80% R125.

Pressure Ratio
-    The pressure ratio using pure R134a is 10 to 30% higher than the ratio using pure R125.
-    The pressure ratio is linear dependent of the mixture ratio.

Conclusions and Recommendations
-    The C.O.P. for pure R125 is 10 to 20% lower than that for pure R134a. This is an important disadvantage of pure R125.
-    The largest advantages of the use of R134a-R125 mixtures, compared to pure R134a, arise at high evaporator temperatures, low condenser temperatures and large secondary temperature differences. In these cases, a slight increase of C.O.P. (1 - 2%) occurs for mixture ratio's of 20 to 40% R125. These results are obtained without using an internal heat exchanger; using this exchanger, an additional increase is expected.
-    A significant increase of the C.O.P. of pure R125 may be obtained by mixing R125 with R134a; 20% R134a already leads to a significant increase of the

C.O.P. by 10%.

- For low evaporator temperatures, mixtures of R134a and R125 may be a good, energy efficient alternative. These applications become more and more important, because of a lack of replacements for refrigerant R502, which has to be phased out soon. the only potential single-component replacement is R125, but this is not acceptable because of its poor energy efficiency. Mixtures of R125 and R134a raise the energy efficiency significantly, and thus will be a promising replacement for R502.

- Mixtures of R134a and R125 may also be a good, energy efficient alternative for R22. The phase-out of R22 (in Germany already started in 1995) will cause big problems, because of a lack of alternatives.

  The potential single-component replacement for R22 is R32, but his refrigerant is not yet acceptable because of its flammability. Mixtures of R125 and R134a will be a promising replacement for R22.

- Based on the previous conclusions, it is proposed to focus the experimental phase of the work and the detailed simulations on two fields of application:

  * low temperature industrial applications (industrial freezing processes, display cabinets), using mixture ratio's with 10 to 40% R134a. In these cases the mixtures may be an alternative for R502 and R22.

  * applications with high evaporator temperatures and low condenser temperatures (industrial process chilling, large air conditioning systems), using mixture ratio's with 20 to 40% R125. In these cases, the mixtures may be a more energy efficient alternative for pure R134a.

- Because of the discrepancy between the values of the binary interactions parameter of R134a-R125 mixtures, delivered by University of Hannover and by NIST, emphasis this topic requires further investigation.

R134a INVESTIGATIONS

This aspect centred on the investigation of the affect of the lubricating oil on system performance, but in the course of the work, two ancillary studies became important.

Expansion Devices

The first of these was the selection of a suitable expansion device. Conventional wisdom states that a conventional R12 cross charged valves are suitable for R134a systems, but this view seems to have developed through necessity because of the shortage of alternatives. The opportunity was taken to compare the performance on R12-cross-charged valve with a R134a charged alternative. The full results are discussed elsewhere [7], but in summary.

(a) both expansion valves can be made to work, but with limitations to their performance.

(b) Superheat settings higher than the usual factory preset values appear to be necessary to achieve stability in both cases.

(c) The R134a valve would appear to require a smaller orifice than the R12 valve for the same duty.

(d) The R12 valve appears to offer a fairly stable superheat value over quite a wide evaporating temperature range, while the R134a-valve induced a change in superheat value which is non-linear with evaporating temperature.

In general, it would appear that conventional wisdom is correct and that existing R12 cross-changed valves can be used in R134a plant, but probably with increased superheat control settings.

Compact Heat Exchangers

The second side-issue concerns the use of compact heat-exchangers on evaporators in refrigeration plant. This choice was made because of the restricted availability of R134a at the beginning of the programme, because of the reduced refrigerant change that would be required and because of the high refrigeration capacity. Control problems were encountered from the beginning of the experimental phase.

Before the experiments began, it was decided to use PI rather than PID control, setting the differential term to zero. It was believed that there would be a large difference in refrigerant residence time between the conventional shell-and-tube condenser and the compact plate heat exchanger acting as an evaporator. Therefore it could be said that the condenser added a considerable lag to the system and according to [8], control of systems with dominant time delays are not improved by the inclusion of derivative action.

Ramchandra and Durgaprasnada [9] believed that a plate heat exchanger conformed to a first order lay-delay model and although this is in conflict with the later work of [10], it suggests that PI control is adequate.

Approximate proportional and integral coefficients were obtained by the standard open loop Zeigler and Nichols step response method and subsequently manually tuned. Figure 5 shows that relative stability could be maintained providing that the evaporator proportional control was at least an order of magnitude less than that of the condenser.

Once it was established that a form of stability could be maintained, the superheat was raised and lowered and the effects on control observed. There are two main points that should be highlighted, the first is the common problem of hunting at low superheat. Additionally, however, at high superheat, a similar effect can be seen as under-utilisation of the expansion valve ultimately leads to instability. The combination of these processes provides the operational range of "sensible" superheat tests, within which it should have been a relatively simple task to maintain stability almost indefinitely.

However, there were times when the superheat took inexplicable dives. This situation was improved by increasing the distance between the thermostatic expansion valve sensing bulb and the evaporator exit, but it was never totally removed.

It was thought that the coalesced bubble/slug flow found in turbulent vertical flows was the origin of this effect, however the Reynolds number for typical evaporator inlet conditions was found to be so low that the necessary slug flow probably did not occur.

The test bed was then modified by the installation of sight-glasses on the inlet and outlet of the evaporator. On the evaporator inlet, it could be clearly seen that the refrigerant formed two distinct layers, but what was more interesting was that on occasion, short bursts of liquid could be seen in the evaporator outlet sight-glass. This caused the rapid drop in superheat which ultimately led to a period of instability.

Proposed Mechanism for Control Instability

The evaporator inlet sight-glass confirmed the separate layers of liquid and vapour entering the evaporator. Also, it should be noted that the cross-sectional area of each of the channel of the compact plate heat exchanger approached that of the inlet pipe. It can therefore be thought of as a phase separator as outlined by Mei [11] in which a series of vapour risers is used to segregate liquid and vapour.

If the compact plate heat exchanger is operating in this way in that the vapour rises first and then the liquid, then the "early" channels would contain vapour and the remainder liquid. As a result, the liquid/vapour interface in the heat exchanger will move under the influence of fluctuations in evaporator inlet quality. Thus, it is possible that a liquid may suddenly use a channel formerly occupied by vapour or vice-versa. When liquid displaces vapour, no problem is foreseen. However, when vapour displaces liquid, the vapour fills the channel and will push some liquid out ahead of it because the ratio of vapour velocity to liquid velocity is approximately 60 to 1. Thus, a control problem can be anticipated.

This problem can be alleviated by increasing the distance from the evaporator outlet to the superheat sensing device. However, it is believed that a turbulent flow regime in the evaporator inlet and subsequent well mixed two-phase fluid may be the best solution.

Alternatively, the problem may be reduced by mounting the heat exchanger horizontally.

Conclusions on the use of a Compact Plate Heat Exchanger as an Evaporator

Compact plate heat exchangers perform well and in general, PI control was found to be adequate. The effects of liquid surges on the thermostatic expansion valve sensing bulb can be alleviated by increasing the distance between the evaporator outlet and the sensing bulb. This also had the general effect of improving control.

Finally, a study of some of the refrigerants presently on the market or about to be marketed as alternatives to CFC's, showed in particular that ammonia, a potential rival to HFC 134a has a slip velocity ratio of approximately 130 at an evaporating temperature of 10 °C. This may imply that there could be difficulties in using compact plate heat exchangers with ammonia.

The Effects of Oil Solubility on Evaporator Performance

The presence of oil in the evaporator of a refrigeration system causes a capacity reduction due to the solubility of the refrigerant in the oil leading to a certain percentage of the refrigerant unavailable for heat transfer. At higher evaporator superheats, this reduction may be smaller because little of the refrigerant remains trapped in the oil. However at low superheats, the fraction of the refrigerant held in the oil increases dramatically causing a sudden drop in evaporator capacity.

As was previously reported, oil can be added into either the condenser or the evaporator. For the condenser, oil is added in to the vapour stream before the condenser by the aid of a fog-nozzle. This effectively turns the oil into a mist by mixing it with a portion of the refrigerant vapour and passing it through a nozzle into the main stream. Due the position of the oil recovery system, oil injection into the condenser automatically applies to the evaporator. For evaporator injection only, the oil is added in the liquid line via a turbulent mixer. The nozzles create a spray of oil while the ball bearings ensure a turbulent flow.

Oil recovery is achieved via a pair of oil separators placed in the compressor suction line. However, due to the solubility of the refrigerant in the oil, a certain amount of refrigerant is inadvertently removed from the system. This is returned to the main system by a combination of heating the oil-refrigerant mixture and reducing the pressure with the aid of a secondary compressor. The refrigerant is then condensed before injection into the compressor discharge of the primary circuit.

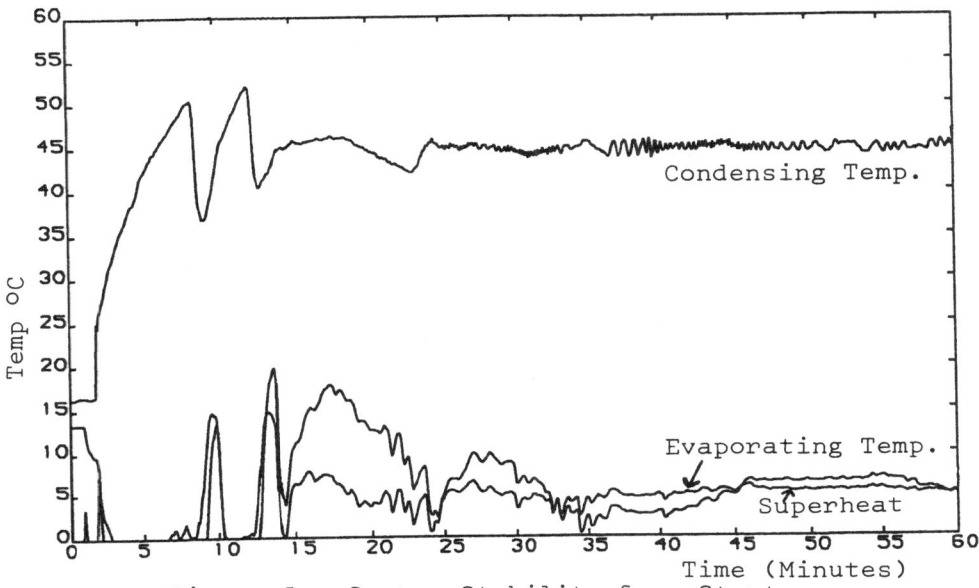

Figure 5.   System Stability from Startup.

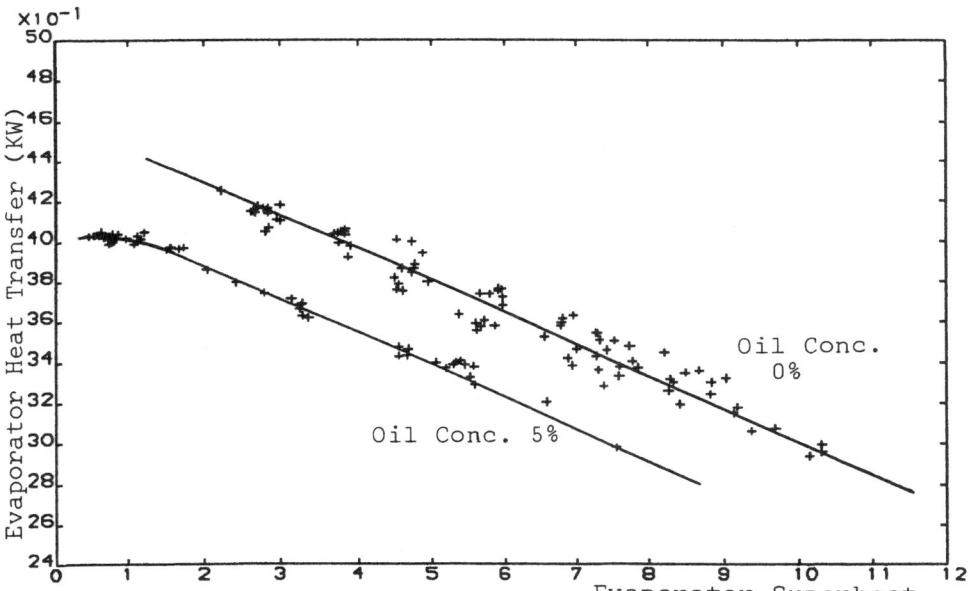

Figure 6.   Influence of Oil on Evaporator Heat
Transfer.

Experimental Procedure and Results

For HFC 134a, the initial experiments were carried out at an evaporator outlet temperature of 15 °C. This was to ensure that the mixture was operating as a single phase. Under these conditions, there is a noticeable peak in the superheat (Figure 6) for an oil content of 5% of the total mass-flow . However, it was pleasing to note that HFC 134a and this ester-based lubricant showed a less dramatic drop in performance than was expected.

Also, the presence of oil can have both positive and negative effects on control. The presence of oil can have a limiting effect on hunting at low "superheat" (the "superheat" term is that for the pure refrigerant only). However, in some instances, the presence of oil reduces the capacity leading to instability i.e. similar to the high superheat case discussed under the choice of expansion valve section.

By and large, the results reflect those noted earlier for R12 and R22 but the peak occurs at a lower superheat value. The improvement in control stability that results from the presence of the lubricating oil can clearly be seen in Figure 6. More detailed results will be presented elsewhere.

REFERENCES

1. McMullan, J.T., Kruse, H., van Gerwen, R.J.M. and Comporese, R., Replacement of R12 in Refrigeration Systems, Contract JOUF-0055 Interim Report No. 2. (1992) CEC, Bussels.
2. Domanski, P. et al., A simplified cycle simulation model for the performance rating of refrigerants and refrigerant mixtures. IIR-Purdue refrigeration conference (USA), July 1990.
3. Morrison, G. et al., Application of the Carnahan-Starling-DeSantis equation of state to mixtures of refrigerants. ASHRAE annual Meeting, Anaheim (USA), 1986.
4. National Institute of Standards and Technology. NIST Standard Reference Database 23: REFPROP version 2.0 NIST, Gaithersburg, (USA), 1990
5. Plöcker, U., Berechnung von Hochdruck Phasengleichwichten mit einer Korrespondenzmethode unter besonderer Berücksichtigung assymetrischer Gemische. Diss. Technische Universitat Berlin (Germany) 1977.
6. Domanski, P., Personal Communications. NIST, Gaithersburg, (USA) July 1991.
7. McMullan, J.T., Kruse, H., van Gerwen, R.J.M. and Camporese, R. Replacement of R12 in Refrigeration Systems, Contract JOUF-0055 Interim Report No. 3. (1992) CEC, Brussels.
8. Anstrom K.J. and Hagglund, T. "Automatic Tuning of PID Controllers". Pub. Instrument Society of America. (1988).
9. Ramachandra V.S. and Durgaprasnada, Ch., "Dynamics and control of plate heat exchangers" Indian Journal of Technology, Vol. 21, November, pp 456-460 (1983).
10. Khan A.R., Baker, N.S. and Wardle, A.P., "The dynamic characteristics of a countercurrent plate heat exchanger" International Journal of Heat and Mass Transfer, Vol. 31, No. 6, pp 1269-1278 (1988).
11. Mei, V.C., "Two phase flow measurement with a liquid and vapour separation technique." ASHRAE Transactions, (1988) Vol. 94, Part 2, pp 238-243.

DEVELOPMENT OF ADVANCED ABSORPTION SYSTEMS DRIVEN BY LOW TEMPERATURE HEAT SOURCES

I. BORDE,[1] M. JELINEK[2] AND N. C. DALTROPHE[2]

[1]Mechanical Engineering Department and [2]The Institutes for Applied Research, Ben-Gurion University of the Negev, P.O. Box 1025, Beer-Sheva, Israel

ABSTRACT

Two refrigerants 1,1,1,2-tetrafluoroethane (R134a) and 2-chloro-1,1,1,2, tetrafluoroethane (R124) that are alternatives to chlorofluorocarbons, in combination with two absorbents N,N'-dimethylacetamide (DMAC) and dimethyl ether tetraethylene glycol (DMETEG) were evaluated for possible utilization in absorption machines powered by low temperature heat sources. The thermodynamic properties of the pure refrigerants R134a and R124 were calculated by a predictive method based on the molecular structure of the substances, the boiling temperature, and the critical temperature and pressure. Enthalpy-concentration diagrams of the refrigerant-absorbent pairs were constructed on the basis of temperature-pressure-concentration relationships. A computerized simulation program was used to compare the different refrigerant-absorbent pairs. The program was based on an advanced complex single-stage cycle containing two generators and an ejector-type mixer and preabsorber for the refrigerant-absorbent pair. Equilibrium pressure-concentration relationships, activity, excess enthalpy, and enthalpy-concentration diagrams are presented for the pair R124-DMAC. Preliminary stability tests of the four investigated working fluids showed that stability of the components was not impared with time. In terms of coefficient of performance and circulation ratio the best refrigerant absorbent pair was found to be R124-DMAC, followed by R134a-DMAC.

INTRODUCTION

Since most energy recovery systems based on sorption cycles operate with working fluids that are not chlorofluorocarbons (CFCs), they are attracting ever increasing attention. The systems offer a substitute to CFC-based systems in both industrial and commercial applications.

Current developments in absorption technology include the search for new working pairs and new advanced cycles that would facilitate increased efficiency of absorption units and extend applicability to different temperature ranges. Our laboratory is engaged in an ongoing search for absorption fluid systems that can be used with low-temperature heat sources up to 150°C. Our investigation has focused on organic refrigerant-absorbent fluids. For refrigeration purposes monochlorofluoromethane (R22) was chosen for its thermal and chemical stability, nontoxicity and good solubility in a number of high-boiling organic absorbents. Even though R22 is considered to be a long-term alternative to the CFCs, we are presently studying two other alternative refrigerants, 1,1,1,2-tetrafluoroethane (R134a), which is considered to be an alternative to R12, and 2-chloro-1,1,1,2, tetrafluoroethane (R124).

The goals of our work are to investigate the possibility of combining R134a and R124 with different absorbents, to evaluate the performance of these refrigerant-absorbent pairs in absorption units, and to improve the performance of the absorption single-stage cycle. The data needed to evaluate working pairs for possible use in heat pumps comprise properties of the pure components, properties of mixing, transport properties, chemical and thermal stability, toxicity, and flammability.

PREDICTION OF THERMODYNAMIC PROPERTIES OF R124 AND R134a

In order to select suitable alternative refrigerants free of CFC for combination with absorbents for utilization in absorption machines the thermophysical properties of these refrigerants must be known. Even though our research may at times start in parallel with that of companies that are developing these refrigerants measured thermophysical properties of these compounds are either not available or very sparse. This is the reason that we have developed a prediction method for the thermodynamic properties of refrigerants.

The proposed predictive method for calculating the thermophysical properties of new working fluids is based on their molecular structure, their boiling points, and their critical temperatures and pressures [1,5]. On the basis of the above-mentioned data, the saturated vapor pressure was obtained as a function of temperature by the Riedel vapor pressure equation. Then, using an equation of state such as the Peng-Robinson equation, the specific volumes of the saturated liquid and gas phases were calculated. The enthalpy and entropy of the liquid and gas phases can be calculated if the departure functions based on this equation of state and the specific heat of the ideal gas are known. If these data are not available, appropriate estimation methods have to be used, such as the Joback modification of Lydersen's method for the estimation of the critical temperature and pressure and Benson's group contribution method for the estimation of the specific heat of the ideal gas.

The proposed prediction method was applied to predict the thermophysical properties of the alternative refrigerants. When our calculated data were compared with the data obtained by Kubota and Basu for R124 and by Basu and ICI for R134a [1], the proposed method was shown to have acceptable accuracy.

THERMOPHYSICAL PROPERTIES OF REFRIGERANT-ABSORBENT MIXTURES

Thermodynamic Properties

For measurements of equilibrium properties of refrigerant-absorbent mixtures an experimental set-up was designed, and a deduction methodology of vapor pressure-temperature concentration data was developed in our laboratory [2,3].

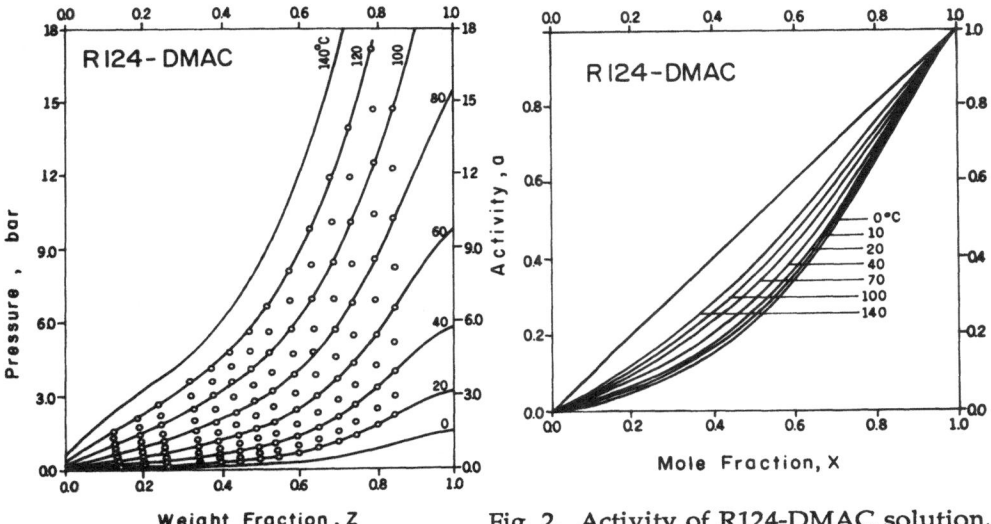

Fig. 2. Activity of R124-DMAC solution.

Fig. 1. Equilibrium pressure-concentration relationships of R124-DMAC. The dots indicate measured equilibrium points.

The pressure-temperature-weight fraction concentration relationship of the refrigerant R124 in combination with the absorbent N,N'-dimethylacetamide (DMAC) (b.p. 166.15°C) is presented in Fig. 1.

The fact that a large number of points was measured over a broad span of the temperature-concentration-pressure field is evident from the presented figure.

The excess thermodynamic properties of the investigated refrigerant-absorbent pairs were calculated by describing separately the vapor and the liquid phases and by comparing their fugacities at equilibrium [3]. In Fig. 2 the activity coefficient of the mixture of R124-DMAC calculated from the UNIFAC contribution model is presented as a function of temperature and mole fraction concentration of the refrigerant in solution. As can be seen from this figure the behavior of the activity coefficient is typical of the solution displaying a negative deviation from Raoult's law, and deviations from ideal behavior decrease with rising temperature.

The calculated excess enthalpy of the solution as a function of temperature and mol fraction concentration is presented in Fig. 3. From this figure we can see that the heat of mixing is negative for this temperature range and its absolute value is higher for lower temperatures.

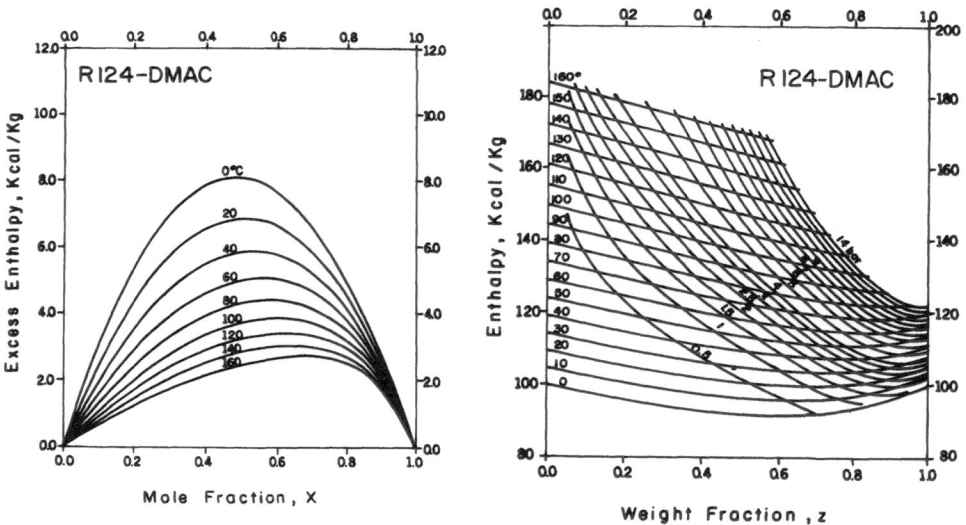

Fig. 3. Excess enthalpy of R124-DMAC   Fig. 4. Enthalpy-concentration diagram
         solution.   of the 134a-DMAC.

Enthalpy-concentration diagram

The evaluation and design of absorption systems requires a knowledge of the enthalpy of the mixtures in the liquid phase and of the enthalpy of the vapor phase as a function of temperature, pressure and concentration. To construct enthalpy-concentration-pressure-temperature diagrams for the solutions (refrigerant + absorbent in the liquid phase), the specific heat must be obtained as a function of the temperature of the pure components in the liquid phase. The calculation of the enthalpy of the pure components in the liquid phase as a function of the temperature is based on the obtained specific heat.

From the enthalpy of the pure components and the excess enthalpy of mixing, it is possible to calculate the enthalpy of the solution in the liquid phase as a function of the temperature and concentration, consisting of the sum of the weighed contributions of the two components.

Enthalpy-concentration diagrams depict the working fluid pair in equilibrium and describe the enthalpy of the mixture as a function of the concentration of the refrigerant based on the correlations between the pressure, temperature and concentration in a state of equilibrium. The enthalpy-concentration diagram for the pair R124-DMAC is given in Fig. 4. The diagram describes the enthalpy of the solutions in kcal per kg within the following ranges: weight concentration of refrigerant in the solution 0-1; temperature 0-160°C; pressure 0.5-16 bars.

Density and viscosity of refrigerant-absorbent mixtures

The design of the components of an absorption system requires in addition to the knowledge of thermodynamic properties information on the densities and viscosities of the working fluid mixtures. To measure simultaneously the densities and viscosities of the working fluid mixtures a measurement system was built in our laboratory for a temperature range of 20-120°C and pressure range of up to 20 bars. The system includes an Anton Paar density meter (type DMA-60) with a U-tube measuring external cell (type DMA-512), a SOFRASER S.A. viscosity meter (type Mivi 3003) for a viscosity range of 1-100 cP, a Micro Gear circulation pump, and an equilibrium vessel.

A calibration graph for viscosity as a function of the temperature that was based on five liquids with a wide range of viscosities was constructed to cover the lower range of the viscosity measurements. A computer routine was devised to convert the time period given by the density meter into density units.

Measurements of the densities and viscosities of the pure absorbents and absorbent-refrigerant solutions were carried out at different temperatures, weight fractions and pressures at equilibrium. The densities and viscosities of the mixtures were expressed in a polynomial form as a function of weight fractions (0-1) and temperatures of 20-120°C. The accuracy of the density measurements was 10^{-5} g/cm^3 and that of the viscosity measurements was 0.1 cP. The density and viscosity of the R134a-DMAC mixture as a function of temperature and weight fraction at equilibrium are presented in Fig. 5 and 6.

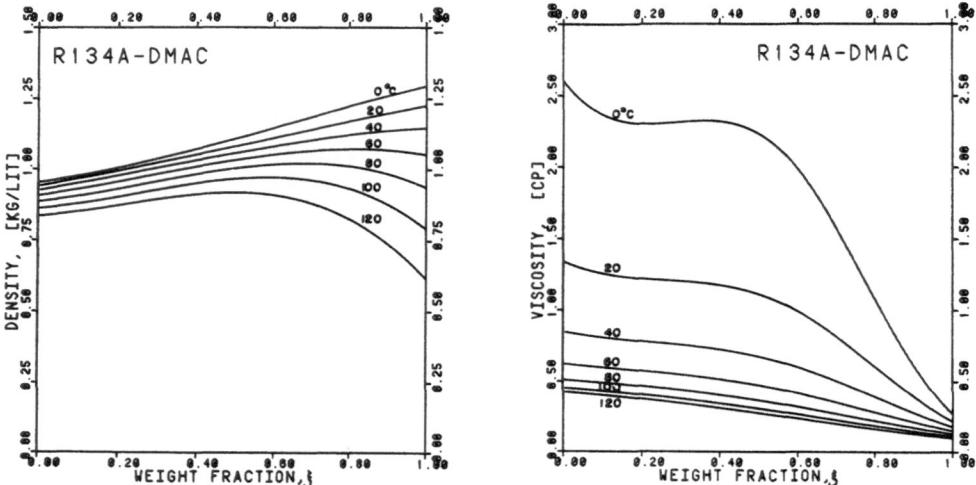

Fig. 5. Density of R134a-DMAC   Fig. 6. Viscosity of R134a-DMAC

CYCLE ANALYSIS

The performance characteristics of the investigated working fluids were based on a complex single-stage cycle represented schematically in Fig. 7.

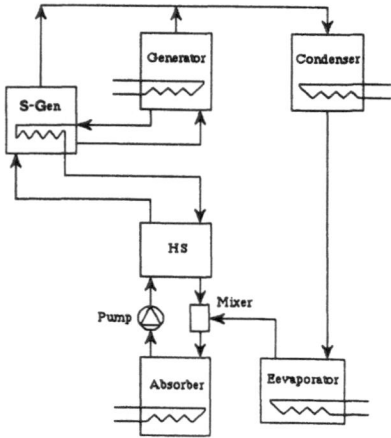

Fig. 7. Schematic diagram of the absorption refrigeration cycle.

This cycle contains two generators, the main generator receiving heat from an external heat source and the secondary generator (S-Gen) being heated by the poor solution emerging from the first generator in the direction of the absorber. This configuration reduced the amount of heat that had to be supplied from the external source for each unit of refrigeration produced, thereby increasing the coefficient of performance (COP) of the cycle. In addition to the generator, condenser, evaporator and absorber the cycle contains a solution heat exchanger (HS) and an ejector-type mixer and preabsorber for the refrigerant absorbent pair. The purpose of this latter component is to improve the mixing process and the preabsorption by the poor solution of the refrigerant coming from the evaporator. The process is very intensive as a result of spray generation of the liquid phase and of increased mass transfer area and as a result of extensive subcooling of the poor solution in the solution heat exchanger. The ejector-type mixer generally improves the whole absorption process.

In order to design such a jet mixer for absorption machines especially for the low flow rates of the poor solution in these devices, a numerical model of non-isothermal mass and heat transfer between the gas phase and the phase of liquid drops was developed. This model enables us to design ejector-type mixture devices for different working fluids to be used in absorption units. The absorption cycle was represented in terms of the heat and mass balance for each component, and the calculations were based on the thermophysical properties of the refrigerant-absorbent pairs measured and evaluated in our laboratory. The aim of the cycle analysis was to evaluate the highest COP and lowest circulation ratio (f) obtainable under different generator temperatures and for a constant evaporating temperature at different heat rejection (condensing and absorbing) temperatures for the investigated working fluids. The two-generator absorption cycle was evaluated under the following operating conditions: generator temperatures in the range of 80-150°C, evaporator temperature −5°C, condenser temperature 32°C, and absorber temperature 28°C.

The COP and f for the combinations R134a-DMAC, R134a-DMETEG (dimethylether tetraethylene glycol, b.p. 275.80°C) and R124-DMAC, R124-DMETEG are presented as a function of the generator temperature in Fig. 8 and 9. As can be seen from the figures the COP of the refrigerant-absorbent pair R124a-DMAC is up to 0.53 and that for R134a-DMAC up to 0.52. R124 and R134a with DMETEG exhibited lower COP, up to 0.46. The lowest circulation ratio was found for the mixture of R124-DMAC, followed by R124-DMETEG and R134a-DMAC, which had similar circulation ratios. The highest f was exhibited by R134a-DMETEG.

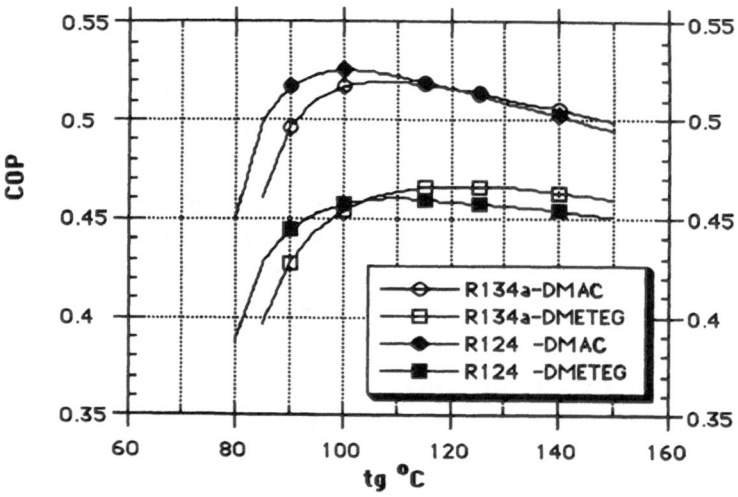

Fig. 8. Variation of COP with generator temperature

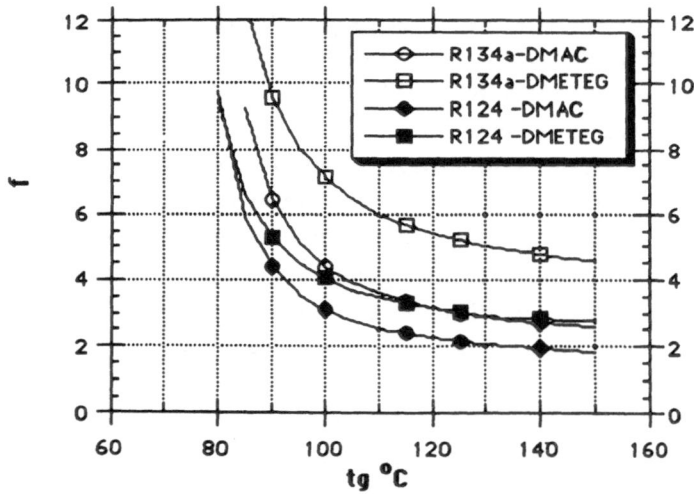

Fig. 9. Variation of f with generator temperature

STABILITY TESTS OF POTENTIAL WORKING FLUIDS

The stability of the four new fluid pairs, R134a-DMETEG, R134a-DMAC, R124-DMETEG and R124-DMAC was tested using the method presented at the conference of the I.I.R. [4]. The stability of working fluids may be defined as the ability to retain constant thermophysical properties under working conditions so as not to significantly affect the performance of the system over long periods of service. This is a broad definition that includes the various aspects of stability that are generally mentioned - and sometimes studied - separately, such as the thermal stability of the pure components, the chemical stability of the fluid, the corrosiveness of the fluid towards the equipment, or the catalytic action of the equipment on the fluid.

Preliminary stability tests of new working pairs are carried out in our laboratory as one of the first steps in the search for new fluid mixtures. The experimental set-up includes three reacting vessels, which are reusable stainless steel minireactors of about 8 cm^3 in volume with a net volume of the liquid sample of 3 cm^3 and a limiting temperature of 260°C imposed by the gasket. The reactors are gradually heated in a thermostated copper equalizer cylinder block for about two hours until a temperature of 200°C is reached. This temperature is then maintained for six hours, after which the reactors are cooled to room temperature. Before opening the minireactors, they are warmed moderately to elevate the pressure of the gas above the ambient pressure so that by connecting the tap to the sampling valve of gas chromatograph analyser we can see any evolution in the gas phase. Then, the minireactors are opened and the remaining liquid is poured into a glass tube and sealed. This liquid is also analyzed by gas chromatograph analysis. Changes in color, formation of precipitate or crystals, and turbidity are followed for several days in the glass tube. The results of the tests for the four new fluid pairs are given below:

	Gas GC analysis	Liq. GC analysis-color-precipitate-turbidity
R134a-DMETEG	Nothing	Nothing
R134a-DMAC	Nothing	Nothing
R124-DMETEG	Nothing	Weak color
R124-DMAC	Nothing	Weaker color than with DMETEG

The weak coloration does not preclude the fluid from further testing. As can be seen from the presented results, the four pairs did not show any degradation and are therefore acceptable candidates for further feasibility examination. We, in fact, tested a number of types of absorbent with R134a and R124, and some of them did show impairment of stability, for example, dimethyl methyl phosphonate with R134a.

The best refrigerant-absorbent pair in terms of the preliminary stability tests and the calculated COP and f will be tested in a long-term stability experiment in one apparatus made of stainless steel, and containing two other metals, steel and copper.

CONCLUSIONS

All four investigated working fluids—the refrigerants R124 and R134a in combination with the absorbents DMAC and DMETEG—for potential utilization in absorption machines showed no impairment of stability in the preliminary tests and are acceptable for further feasibility examination.

In terms of COP and f the best refrigerant absorbent pair is R124-DMAC, followed by R134a-DMAC. The absorbent DMETEG has a lower COP and higher f with the investigated refrigerants.

ACKNOWLEDGEMENTS

The research was supported by a grant from GIF, the German Israeli Foundation for Scientific Research and Development.

REFERENCES

1. Jelinek, M., Daltrophe, N.C. and Borde, I., Prediction of thermophysical properties of new alternative CFC-free refrigerants. Proceedings of the 24th Israel Conference of Mechanical Engineering, May 1992, 3.1.1., pp. 1-3.

2. Borde, I., Jelinek, M. and Daltrophe, N.C., Refrigerant-absorbent mixtures based on the refrigerant R134a. Proceedings of the XVIIIth International Conference of Refrigeration, Montreal, Aug. 1991, pp. 653-658

3. Borde, I., Jelinek, M. and Daltrophe, N.C., Research on thermophysical properties of absorption working fluids at the Energy Laboratory of Ben-Gurion University. In *Thermophysical Properties of Pure Substances and Mixtures for Refrigeration*, International Institute of Refrigeration, Commission B1, Herzlia (Israel), 1990, pp. 37-46.

4. Daltrophe, N.C., Jelinek, M. and Borde, I., Stability test for working fluids for absorption systems. In *Thermophysical Properties of Pure Substances and Mixtures for Refrigeration*, International Institute of Refrigeration, Commission B1, Herzlia (Israel), 1990, pp. 233-240.

5. Jelinek, M., Daltrophe, N.C. and Borde, I., Prediction and evaluation of the thermophysical properties of monomethylamine and dimethylamine, International Institute of Refrigeration, Commission B1, Herzlia (Israel), 1990, pp. 123-128.

A THERMODYNAMIC METHOD FOR A SECOND LAW ANALYSIS OF TWO-PHASE BINARY MIXTURE PROCESSES

Prof. Dr. D.A.KOUREMENOS, Dr. E.D.ROGDAKIS and G.E.HOUZOURIS
National Technical University of Athens
Mechanical Engineering Department, Thermal   Section
42 Patission Street, Athens 106 82, GREECE

ABSTRACT

The entropy production in two phase binary mixture parallel flow processes, due to mass fraction and temperature deviations from the corresponding equilibrium states, is examined. A physical-mathematical method is proposed, considering that such deviations are owing to the "heat and mass exchange delay". According to this method an actual state of the mixture along the process is bounded between two extreme states, the equilibrium one (zero delay) and the non-equilibrium of total delay and its location is determined by the "grade of delay". By the prediction of the "grade of delay", successive actual states of the process are located, the corresponding entropy production is evaluated and entropy quality indices are calculated for the process. As an illustration of this model, the evaporation and condensation of NH_3/H_2O liquid-vapor mixture are applied.

NOTATION

c	volatile component consistency (for mixture), kg/kg
d	vapour mass fraction, kg/kg
f	liquid mass fraction, kg/kg
h	enthalpy, J/kg
q	heat, J/kg
RD	delay ratio or grade of delay, kg/kg
s	entropy, J/kg,K
T	absolute temperature, K
t	temperature, C
x	volatile component consistency (for liquid), kg/kg
y	volatile component consistency (for vapour), kg/kg

Superscripts

e   equilibrium state
i   intermediate state
s   superheated or subcooled state

Subscripts

d   delay
e   equilibrium
l   liquid
m   mixture
v   vapour
p,q,r,n points P,Q,R,N in diagrams.

INTRODUCTION

For a system with heat exchange to its surroundings the entropy balance states that the entropy variation of the system ds consists of two terms, the entropy flow from the surroundings $d_a s$ and the entropy production inside the system $d_i s$:

$$ds = d_a s + d_i s \qquad (1)$$

In relation (1) ds designates the change of the state property s and $d_a s$ is related with the heat exchange amount dq at an absolute temperature of the system T, according to the Clausius - Carnot theorem:

$$d_a s = dq/T \qquad (2)$$

so that for $d_a s$ the following hold:

$d_a s > 0$, when heat is added to the system
$d_a s < 0$, when heat is extracted from the system
$d_a s = 0$, when there is no heat exchange

while according to the second law of Thermodynamics $d_i s$ is always a no negative quantity:

$d_i s \geqslant 0$, in general
$d_i s > 0$, for irreversible processes
$d_i s = 0$, for reversible processes.

IRREVERSIBILITIES OF THE TWO-PHASE FLOW OF A BINARY MIXTURE

In general, in a two-phase binary mixture process with heat exchange to the surroundings, the entropy production is due to temperature and mass fraction deviations from the corresponding equilibrium states and the pressure drop.

In the present work the flow is assumed to be inviscid and the temperature deviations between the liquid and vapour phases vanish. The entropy production $d_i s$ is provoked by an interior delay of heat and mass exchange between the two phases

in the binary mixture and as a result, the actual state of the
mixture is displaced from equilibrium.

    This delay is graded between a minimum value 0 (no delay)
and a maximum value 1 (total delay). The process of no delay
is a reversible one (successive equilibrium states) while the
process of a grade of delay greater than 0 is an irreversible
one followed by an entropy production.

IRREVERSIBLE EVAPORATION OF A BINARY MIXTURE

    According to the present model, an actual evaporation pro-
cess is bounded between two extreme processes, the reversible
one and the irreversible of the total delay. Respectively, the
successive states of the working medium are located between
two extreme states, the equilibrium one and the non-equilibrium
of total delay. Any intermediate state between them is a possible state.

    This can be shown schematically in an enthalpy (h) - mass
consistency (x or y) diagram of a binary mixture as those in
Fig. 1 and 2. The equilibrium point P is taken to be the starting point of the process (Fig.1). If the process is reversible

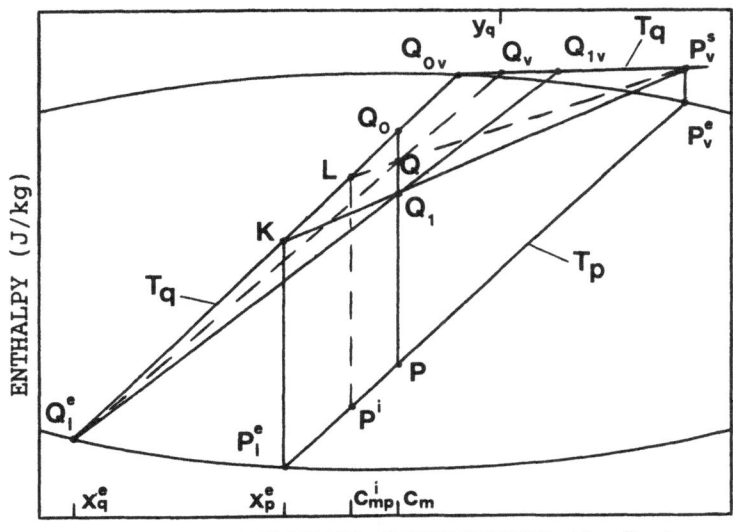

Figure 1. The initial elementary evaporation process

and the evaporation is expanded from temperature T_p to T_q, the final state of the mixture is an equilibrium one and designated by the point Q_0. In the case that a total delay is taking place for the variation from T_p to T_q, the saturated liquid P_1^e is evaporated to the equilibrium state K which consists of saturated liquid Q_1^e and saturated vapour Q_{0v}, while the saturated vapour P_v^e is superheated to the state P_v^s, obtaining the temperature T_q; the saturated vapour Q_{0v} and the superheated vapour P_v^s constitute the superheated vapour state Q_{1v}. In such a way the final non-equilibrium state Q_1 is created, consisting of the saturated liquid Q_1^e and the superheated vapour Q_{1v}, both being at the same temperature T_q.

When the delay of vapor is not total but partial, that is, when a part of the saturated vapor P_v^e is subjected to delay and the rest one obeys the saturated liquid P_1^e quasi-statically constituting an intermediate equilibrium state P^i, the evaporation of P^i leads to the equilibrium state L, consisting of the saturated liquid Q_1^e and the saturated vapour Q_{0v}, while the delaying vapour is superheated to the state P_v^s. The saturated vapour Q_{0v} and the superheated vapour P_v^s form the superheated vapour Q_v. The final non-equilibrium state Q is constituted by the saturated liquid Q_1^e and the superheated vapour Q_v. From an other point of view the final state Q is created by the equilibrium state L and the superheated vapour P_v^s.

In proportion to the grade of delay, the final state Q is located on the straight line of the concistency c_m of the mixture, between the states Q_0 and Q_1, consisting of saturated liquid Q_1^e and superheated vapour Q_v being located on the line of temperature T_q between the states Q_{0v} and Q_{1v}, or equivalently, is created by the superheated vapour P_v^s and the equilibrium state L being located on the straight line of temperature T_q between the states K and Q_0.

The grade of delay for the evaporation is defined as the ratio of the mass of delaying vapour to the mass of total vapour:

$$RD_v = (P^i P)/(P_1^e P) = (c_m - c_{mp}^i)/(c_m - x_p^e) \tag{3}$$

It is apparent that $0 \leqslant RD_v \leqslant 1$, $RD_v = 1$ when $c_{mp}^i = x_p^e$ (total delay) and $RD_v = 0$ when $c_{mp}^i = c_m$ (reversible change).

An intermediate stage of evaporation from the non-equilibrium state Q to the non-euilibrium state R is illustrated in Fig.2.

The liquid and vapour mass fractions f_q and d_q respectively in the mixture of non-quilibrium state Q, are given by the relations:

$$f_q = (Y_q - c_m)/(Y_q - x_q^e) \tag{4}$$

$$d_q = (c_m - x_q^e)/(Y_q - x_q^e) \tag{5}$$

while for the equilibrium state N, the corresponding quantities are:

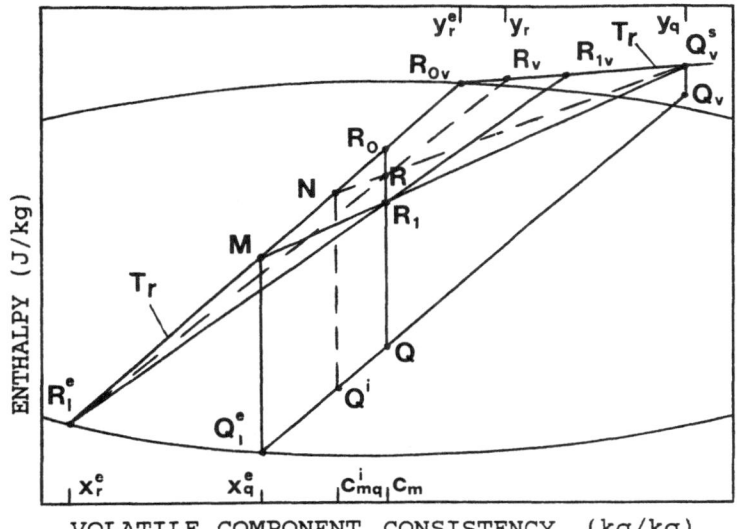

VOLATILE COMPONENT CONSISTENCY (kg/kg)

Figure 2. An intermediate elementary evaporation process

$$f_n = (y_r^e - c_{mq}^i)/(y_r^e - x_r^e) \qquad (6)$$

$$d_n = (c_{mq}^i - x_r^e)/(y_r^e - x_r^e) \qquad (7)$$

The mass fraction of the superheated vapour Q_v^s with reference to the total quantity of the mixture is:

$$d_{vd} = RD_v * d_q \qquad (8)$$

and the corresponding quantity of the saturated vapour R_{ov} is:

$$d_{ve} = (1 - RD_v * d_q) * d_n \qquad (9)$$

Thus the mass consistency y_r, of the superheated vapour R_v, is obtained from the equation:

$$y_r = (y_r^e * d_{ve} + y_q * d_{vd})/(d_{ve} + d_{vd}) \qquad (10)$$

Then the liquid and vapour mass fractions f_r and d_r respectively in the mixture of the final non-equilibrium state R are calculated from the relations:

$$f_r = (y_r - c_m)/(y_r - x_r^e) \tag{11}$$

$$d_r = (c_m - x_r^e)/(y_r - x_r^e) \tag{12}$$

IRREVERSIBLE CONDENSATION OF A BINARY MIXTURE

In a similar way, an actual condensation process is bounded between the reversible process and the irreversible one of the total delay. Correspondingly, the successive states of the mixture along the condensation are located between two extreme states, the equilibrium one and the non equilibrium of the total delay.

As the starting point of the process is considered to be the equilibrium state P (Fig.3). In proportion to the grade of delay, the final state Q is located on the straight line of

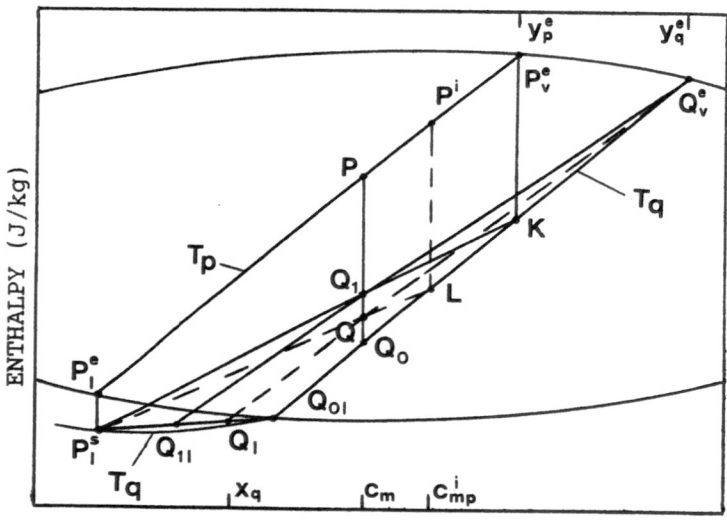

Figure 3. The initial elementary condensation process

the consistency c_m of the mixture between the states Q_0 and Q_1, consisting of saturated vapour Q_v^e and subcooled liquid Q_{1l} being located on the line of temperature T_q between the states Q_{0l} and Q_{1l}.

The grade of delay for the condensation process is defined from the ratio:

$$RD_1 = (P^i P)/(P^e_v P) = (c_m - c^i_{mp})/(c_m - y^e_P) \qquad (13)$$

which takes values in the region $0 \leqslant RD_1 \leqslant 1$, where $RD_1 = 1$ when $c^i_{mp} = y^e_P$ and $RD_1 = 0$ when $c^i_{mp} = c_m$.

An intermediate stage of condensation from the non-equilibrium state Q to the non-equilibrium state R is illustrated in Fig.4.

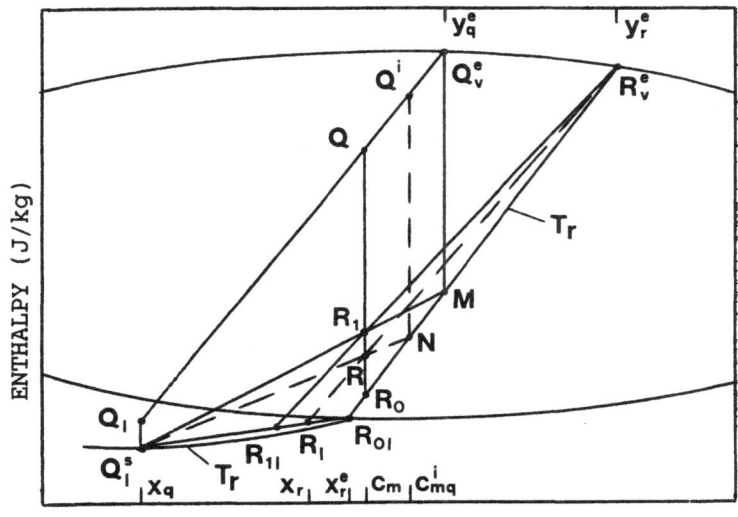

Figure 4. An intermediate elementary condensation process

The liquid and vapour mass fractions of the mixture at the state Q are respectively:

$$f_q = (y^e_q - c_m)/(y^e_q - x_q) \qquad (14)$$

$$d_q = (c_m - x_q)/(y^e_q - x_q) \qquad (15)$$

and the same quantities of the state N are:

$$f_n = (y^e_r - c^i_{mq})/(y^e_r - x^e_r) \qquad (6)$$

$$d_n = (c^i_{mq} - x^e_r)/(y^e_r - x^e_r) \qquad (7)$$

The mass fraction of the subcooled liquid Q^s_1 with reference to the total quantity of the mixture is:

$$f_{1d} = RD_1 * f_q \qquad (16)$$

and the corresponding quantity of the saturated liquid R_{01} is:

$$f_{1e} = (1 - RD_1 * f_q) * f_n \qquad (17)$$

Thus the mass consistency of the subcooled liquid R_1, is obtained from the equation:

$$x_r = (x_r^e * f_{1e} + x_q * f_{1d})/(f_{1e} + f_{1d}) \qquad (18)$$

Then the liquid and vapour mass fractions f_r and d_r respectively in the mixture of the final state R are calculated from the relations:

$$f_r = (y_r^e - c_m)/(y_r^e - x_r) \qquad (19)$$

$$d_r = (c_m - x_r)/(y_r^e - x_r) \qquad (20)$$

ENERGY AND ENTROPY BALANCE

From the foregoing it is apparent that the state of the mixture at any point of the process is completely determined, so the enthalpy h and the entropy s can be calculated from the relations:

$$h = f * h_1(T,p,x) + d * h_v(T,p,y) \qquad (21)$$

$$s = f * s_1(T,p,x) + d * s_v(T,p,y) \qquad (22)$$

For a differential change from Q to R (Fig.2, 4), the exchanged heat is:

$$dq = dh = (QR) \qquad (23)$$

and from eq. (1), (2) and (23) the corresponding entropy production is obtained from the relation:

$$d_is = ds - dq/T \qquad (24)$$

APPLICATION TO THE NH_3/H_2O MIXTURE

To obtain a quantitative demonstration of this model the NH_3/H_2O liquid-vapour mixture is applied to the evaporation and condensation processes.

The thermodynamic data and equation of state of the NH_3/H_2O mixture are taken from the known sources (Macriss et al, 1964), (Schulz, 1971), (Ziegler-Trepp, 1982). A detailed compilation of all available relevant information is given by Macriss

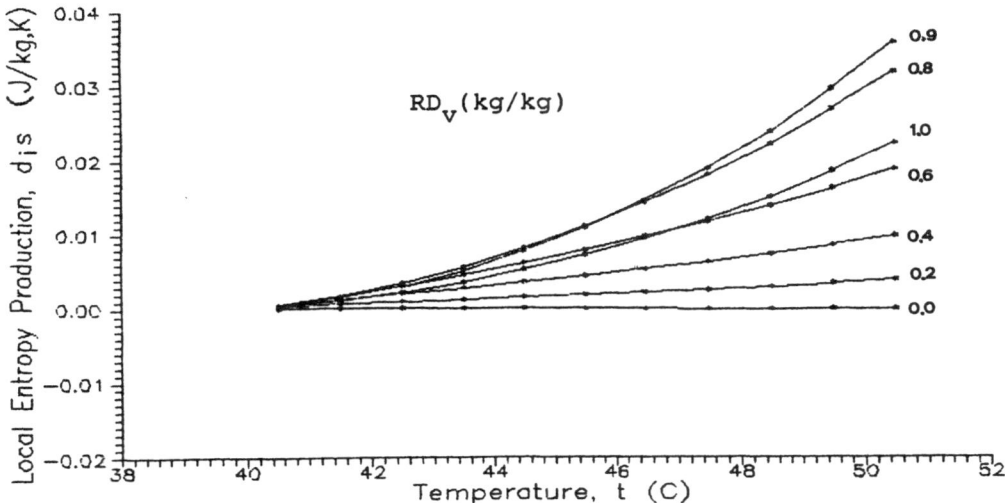

Figure 5 . The local entropy production $d_i S$ with temperature t and vapour grade of delay RD_v (Evaporation).

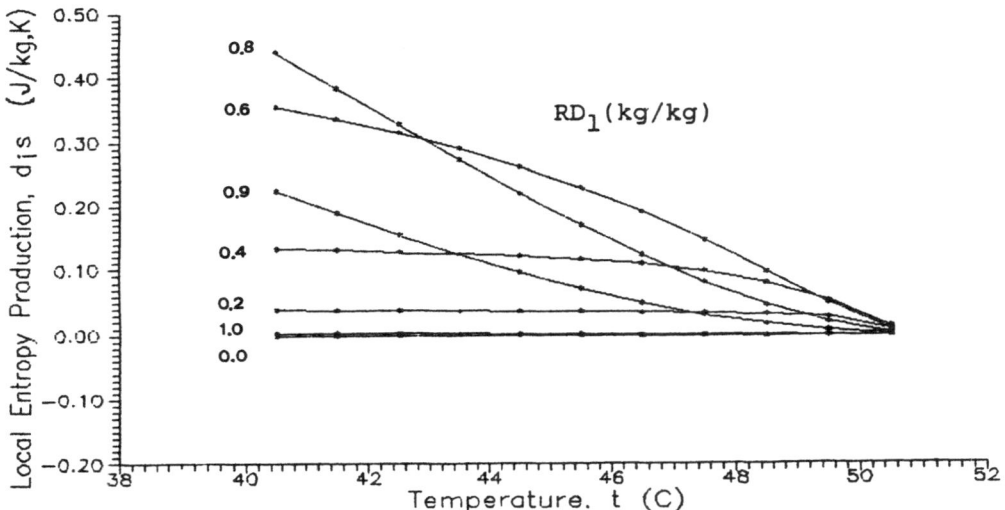

Figure 6. The local entropy production $d_i S$ with temperature t and liquid grade of delay RD_1 (Condensation)

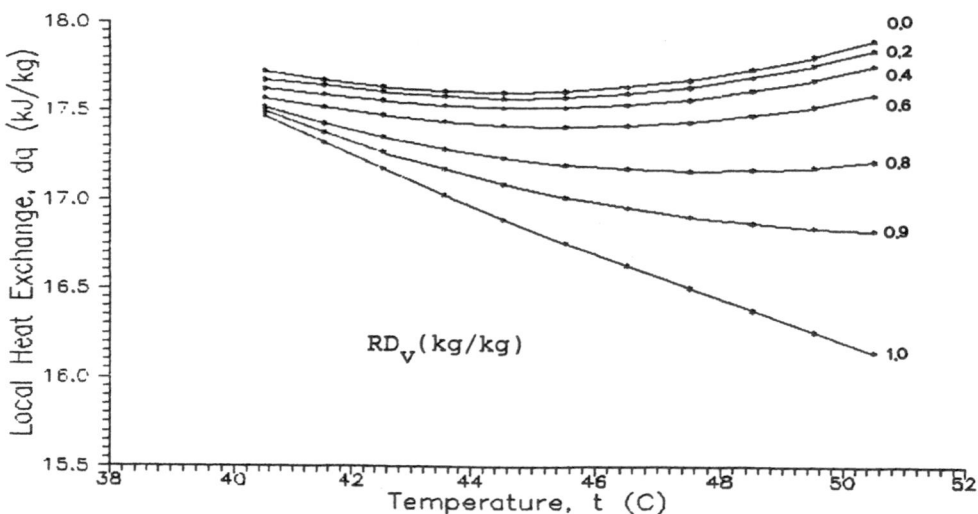

Figure 7. The local heat exchange dq with temperature t
and vapour grade of delay RD_v (Evaporation)

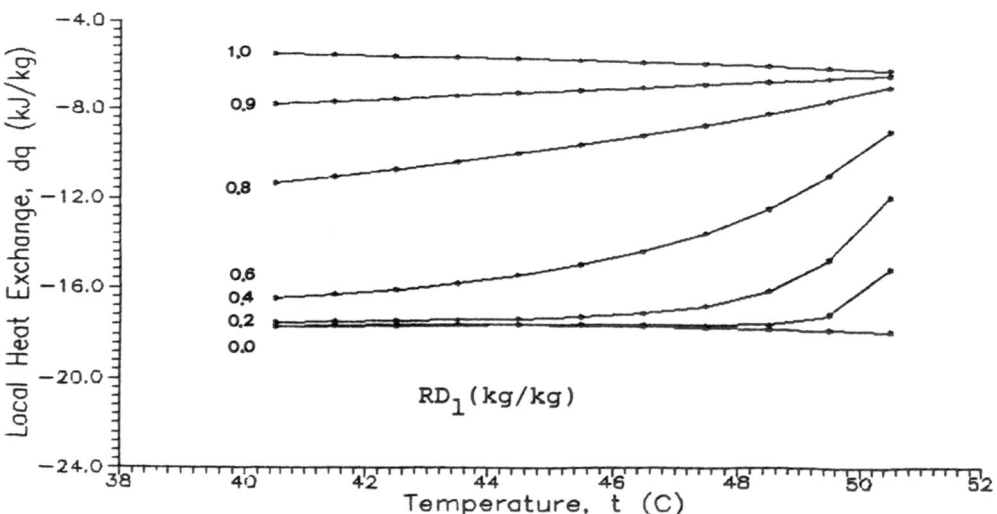

Figure 8. The local heat exchange dq with temperature t
and liquid grade of deday RD_l (Condensation)

and Zawacki (1984).

The evaporation and condensation processes of the NH_3/H_2O mixture are considered at a pressure p=1 bar, a mixture consistency c_m=0.25 kg NH_3/kg mixt., between the temperatures t_1= 40C and t_2=51 C, with a temperature step dt = 1 C, for different values of RD_v and RD_l in the region from 0 to 1.

A collection of numerical results are performed in Fig.5 to **8** presenting the entropy quality of the processes and making comparisons of the irreversible processes to each other and to the reversible ones.

CONCLUSIONS

A thermodynamic method for a second law analysis of two-phase binary mixture parallel flow processes is developed considering heat and mass exchange delay between the liquid and the vapour phases.

According to this method an actual process of evaporation or condensation is bounded between two extreme process, the reversible one and the irreversible of total delay.

The determination of the actual process is achieved by the prediction of the "grade of delay" which is correlated with the "entropy quality"of the process.

Numerous evaporation and condensation processes are developed for various values of grade of delay, the corresponding entropy quality indices are calculated and comparisons of the irreversible processes to each other and to the reversible ones are realized.

REFERENCES

1. Kouremenos,D.A., "Second Law Analysis of Non-Equilibrium Fluid Streams", Forschung im Ingenieurwesenss (1989) Nr1 pp 10-15.

2. Kouremenos,D.A., Rogdakis, E.D., "The irreversible NH_3/H_2O absorption cycle and its graphical representation", Proceedings of WAM ASME, San Francisco, California, 1989 H.T.D. Vol. 124(AES-Vol.6, pp 97-103.

3. Kouremenos, D.A., Rogdakis, E.D., "COP values of non-equilibrium NH_3/H_2O absorption refrigeration cycles", Proceedins of the second World Congress on Heating, Ventilating, Refrigerating and Air Conditioning-CLIMA 2000, Sarajevo, Yugoslavia, August, 1989, pp.68-74.

4. Kouremenos, D.A., Rogdakis E.D., "An Entropy Quality Index for Binary Mixture Parallel Flow Units". (To be published).

5. Macriss, R.A., Eakin, B.E., Ellington, R.T., Huebler, J., "Physical and Thermodynamic Properties of Ammonia-Water Mixtures", Research Bulletin No 34, Institute of Gas Technology, Chicago IL (1964).

6.   Macriss, R.A., Zawacki, T.S., "Absorption Fluids Data Survey", U.S. Dept. of Energy, ORNL/Sub/84-47989/1. Report prepared by Institute of Gas Technology, Chicago IL.

7.   Schulz, S.G., "Equations of State for the system Ammonia-Water for Use with Computers", Progr. Refrig.Sci.Technol., Proc. 13th Intern. Congr. of Refrigeration, 1971, Vol.2, pp.431.

8.   Ziegler, B., "Waermetransformation durch einstufige Sorptions-prozess mit dem stoffpaar Ammoniak-Waaser", Ph.D. Thesis, ETH Zurich, No 7070. (1982).

9.   Prigogine I, "Etude Thermodynamique des Phenomenes Irreversible", Thesis, Liege, Desoer, 1947.

HEAT TRANSFER COEFFICIENT OF A FILM CONDENSER IN A DUAL CYCLE REFRIGERATION SYSTEM

SAURO PIERUCCI, RENATO DEL ROSSO, ANGELO SOGARO,
CLAUDIO FERRARI, STEFANO GAUDENZI
Chemical Engineering Department,
Politecnico di Milano
P.za Leonardo da Vinci 32, 20133 Milano, Italy

ABSTRACT

The superiority of absorption dual cycle refrigeration systems over competitive configurations is widely accepted. The adoption of a sequence of film evaporations and condensations on parallel walls have been proposed and experimented for several years. The wall-to-wall geometry reduces the machine dimensions, increases the overall heat transfer efficiency, decreases the investiments. A short term project with the aim of a feasibility study on a dual cycle machine has been started two years ago in our Department with the cooperation of Industrial Partners and National Institutions. The pairs H2O-NH3 and H2O-LiBr have been preliminary chosen to drive the cycles. Condensing Propylene-glycole enables the primary heat flux to the generator.
The main component of the system is equiped by three coaxial tubes which provide for heat transfer from the high temperature cycle to the low temperature cycle, thus staging the heat input to the system in much the same manner as in a conventional, multiple-effect absorption heat pump.
The purpose of this work has been to investigate the change of heat transfer coefficient of film condensation on vertical surfaces. Sand blasted and Teflon-coated surfaces have been adopted. The experimental results show the superiority of Teflon-coated surfaces over the Sand blasted ones; possibly two different mechanisms occur,

drop condensation for Teflon-coated surfaces and film condensation for the sand blasted ones.

INTRODUCTION

The Absorption Dual Cycle (ADC) system basically links two simple single effect absorption heat pump cycles. One of the possible layouts of the system pairs in common the two desorbers and the two condensers in a single modular component which is mainly constituted by coaxial vertical tubes. This arrangement also provides the primary heat flux from the burner through the film condensation of an auxiliary fluid (ex: propylene gycole) and the heat discharge to the external utility. This promising module is the key to the simplicity of manufactured components for ADC; it stages the heat input to both the absorption cycles, it warrants modularity concepts, it is intended to also allow for low manufacturing cost through the use of high performance surfaces which enhance the heat transfer coefficients. The key, however, to the compact design of the module is a heat exchange surface for both desorber and condenser sections that allows for very effective heat and mass transfer.
Since 1981, significant advances have been made in exploring the mechanism of enhanced film condensation through the understanding the role that surface tension force plays in draining the condensate from the surface. Analytical models have been proposed since the early 80's by Adamek (1), Webb (2), and Rudy and Webb (3). Altough the works of the cited authors were mainly developed for understanding the mechanism of enhanced film condensation on finned surfaces for plates and horizontal tubes, it is reasonable assuming that the basic concepts may be adopted also for different geometries and surfaces. It was therefore reasonable and also "a priori" predictable that a Teflon coated surface might have an enhanced behaviour compared with a Sand blasted one. Moreover being the mechanism at a low heat flux values mainly of a "drop condensation" type, the enhancement between the two surfaces should reduce with the encrease of the heat flux .The experimental work was therefore based upon theoretical concepts which qualitatively proved and anticipated the obtained results so that the work was mainly conducted to identify the order of magnitude of the enhancement at different heat flux values.
This paper refers to the results obtained on the side of the primary heat flux which is given by condensing 1-2 propylene glycole on the inner surface of a boiling LiBr-H2O mixture.

MATERIALS AND METHODS

The main component of the system, the Desorber-Condenser (EC), consists of two coaxial vertical tubes and a jacket for heat removal. A sketch of the EC unit is reported in fig. 1. The primary heat flux is obtained by condensing propylene glycole on the inside of the inner tube. The desorbing fluid film flows down on the outside of the inner tube. The refrigerant is condensing on the internal surface of the outer tube and the heat is removed by water circulating inside the external jaket.

The EC length was taken 1 m in order to have fully developed flow and heat transfer before the bottom end of the evaporator (Chun and Seban (4) predict a shorter length =0.80 m). The evaporating film was maintained by a properly designed distributor on the head of the inner tube.

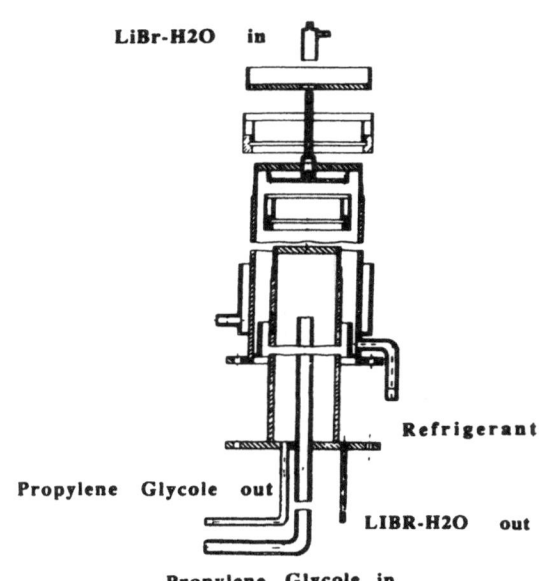

Figure 1 . Sketch of the EC unit.

In fig. 2 is shown the schematic diagram of apparatus.
Propylene glycole is vaporized in G and flows to the EC unit where it is condensed and fed back to the boiler. Three heat exchangers S1,S2, and S3 recover heat from the hot stream out of EC by warming the feed stream to EC. The heat exchanger S4 uses cooling water. The system was designed to operate at a maximum of 5 Atm pressure, with a boiling mixture of LiBr-H2O at 190 °C . The composition of the rich feed stream was maintained at 40% wt in LiBr. The Propylene glycole was

boiled at a vapor pressure of 3 Atm. The heat was provided by firing methane at a maximum value of 10 Kw.

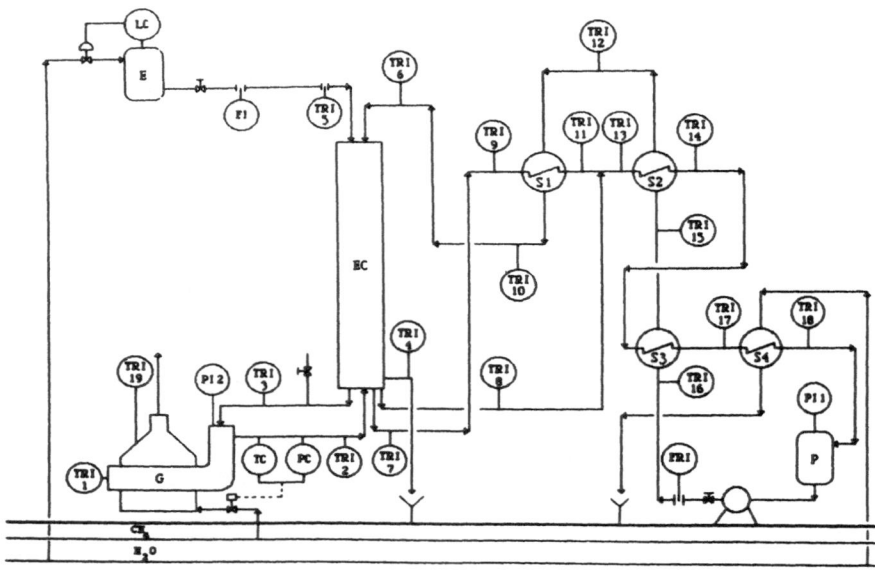

Figure 2. Schematic diagram of apparatus

Two sets of experiments were conducted with two types of surfaces at the propylene glycole condensing side: sand blasted and teflon coated. The heat transfer coefficient calculation was based on the assumption that the heat transfer coefficient of the boiling LiBr-H2O mixture might be estimated by the Chun-Seban (4) relation. This contribution was subtracted from the total measured heat flux so that an estimated coefficient value at the condensing side might be obtained. The total measured heat flux was corrected at the best by the duties required to heat the LiBr-H2O to its boiling point and to subcool the condensate.

RESULTS

Experiments were conducted by varying the total heat flux at the burners and the LiBr-H2O mixture flowrate. The estimated heat transfer coefficients for the condensing Propylene glycole vs Reynolds number are reported in Tables 1, 2 and 3 .

Table 1

Estimated heat transfer coefficients for sand blasted surface (hglic.non.tefl) and teflon coated surface (hglic.tefl.) versus Reynolds number (Re_{nt} and Re_{tef})

Total LiBr-H2O flowrate = 33 Kg/h $\qquad Re = 4\,\dfrac{\Gamma}{\mu}$

[Kw]	hglic. non tefl. [Kcal/h m^2 K]	Re_{nt}	hglic. tefl. [Kcal/h m^2 K]	Re_{tef}
5.12	1227.7	171.8	1514.0	122.4
6.69	1155.1	261.5	1418.0	220.3
7.43	1119.0	295.6	1295.0	266.4
8.54	1100.4	404.3	1352.6	335.6
9.93	1044.6	480.8	1250.1	435.1

The behaviour of the heat transfer coefficient for teflon coated surface is decreasing with the Reynolds number and the total heat from the burners. The difference between the behaviour of the two surfaces becomes negligible at an estimated Reynolds number greater than 750. Possibly at lower heat flux the mechanism of condensation is governed by drop-type condensation while at higher heat flux values the mechanism turns to a film condensation so that the two surfaces show a similar beahviour.

Table 2

Estimated heat transfer coefficients for sand blasted surface (hglic.non.tefl) and teflon coated surface (hglic.tefl.) versus Reynolds number (Re_{nt} and Re_{tef})

Total LiBr-H2O flowrate = 38 Kg/h $Re = 4 \dfrac{\Gamma}{\mu}$

[Kw]	hglic. non tefl. [Kcal/h m^2 K]	Re_{nt}	hglic. tefl. [Kcal/h m^2 K]	Re_{tef}
4.60	1268.4	162.6	2192.0	103.5
5.50	1224.0	205.3	1781.0	152.9
6.43	1178.5	260.0	1581.5	203.0
7.14	1156.5	318.4	1537.2	241.8
8.03	1129.5	393.6	1481.1	282.9
9.67	1043.0	492.2	1250.0	385.0

This hypothesis is partially validated by the fact that the heat transfer coefficient of the sand blasted surface may be easily interpolated by the Nusselt equation

$$h_{Nu} = 0.925 \, K_L \left(\frac{\rho^2 \, g}{\mu \, \Gamma} \right)^{1/3}$$

where:

ρ = mixture density [Kg/m³]

μ = mixture viscosity. [Kg/h m]

g = acceleration due to gravity [m/h²]

K_L = thermal conductivity [Kcal/h m K]

Γ = peripherical flowrate= $\dfrac{W}{\pi \, D}$ [Kg/h m]

D = diameter [m]

$$W = \frac{Q}{\Delta h_{ev}} = \text{condensing Propyl. glycole flowrate [Kg/h]}$$

Q = total duty [Kcal/h]

Δhev = heat of vaporization [Kcal/Kg]

Table 3

Estimated heat transfer coefficients for sand blasted surface (hglic.non.tefl) and teflon coated surface (hglic.tefl.) versus Reynolds number (Re_{nt} and Re_{tef})

Total LiBr-H2O flowrate = 43 Kg/h $Re = 4 \dfrac{\Gamma}{\mu}$

- -

[Kw]	hglic. non tefl. [Kcal/h m² K]	Re_{nt}	hglic. tefl. [Kcal/h m² K]	Re_{tef}
4.40	1286.8	145.0	1763.7	112.9
5.06	1234.7	171.3	1498.6	149.0
5.98	1191.5	225.6	1402.5	200.0
6.93	1145.3	275.7	1294.0	251.5
7.76	1139.3	373.9	1237.1	297.0
8.41	1061.0	395.7	1117.5	332.6

The three sets of data are drawn in Fig. 3

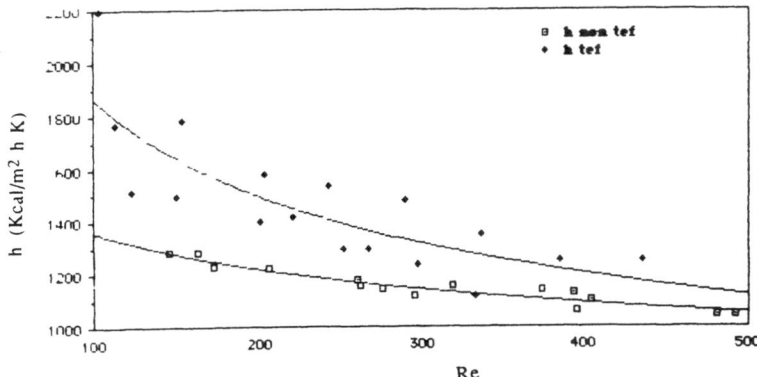

Figure 3. Estimated heat transfer coefficients for teflon coated and sand blasted surfaces versus Reynolds number

The heat transfer coefficients values reported in Fig 3 have been interpolated by an empirical equation of the type:

$$\frac{h_{prev}}{h_{Nu}} = a\,Re^{\beta}$$

where the coefficient h_{Nu} is evaluated according to the Nusselt equation and h_{prev} comes from the measures.
The resulting empirical equation is :

$$h_{prev} = (3.08\,Re^{-0.16})\,h_{Nu} \quad (1)$$

The next Table 4 summarises the comparison between calculated and observed values:

Table 4

Comparison between h_{prev} values by equation (1)
and the observed values hsperim.

Re	hNu	h sperim.	h prev.	E%
115,00	1260,0	1783,6	1816	1,84%
134,30	1178,0	1698,0	1657	2,44%
154,00	1166,6	1626,1	1605	1,30%
176,00	1138,0	1559,3	1533	1,72%
185,00	1146,0	1534,4	1531	0,22%
214,70	1137,4	1463,7	1484	1,37%
218,00	1134,0	1456,7	1476	1,31%
250,00	1083,5	1395,0	1379	1,11%
269,30	1087,4	1362,4	1368	0,41%
279,00	1100,0	1347,2	1376	2,14%
311,00	1056,0	1301,7	1298	0,26%
348,00	1050,3	1256,2	1268	0,96%
385,00	1025,0	1216,0	1218	0,15%
435,10	1033,0	1170,4	1204	2,84%

CONCLUSIONS

An enhancement of the heat transfer coefficient for a Teflon coated surface has been observed for condensing Propylene glycole. The enhancement decreases with encreasing the total transfered duty so that the effect of coating versus a sand blasted surface diminishes.

An empyrical equation valid for the analysed system has been proposed.

ACKNOWLEDGMENT

The authors wish to thank ENEA instituiton for having supported the work and P.i P. Giacometti for his valuable suggestions.

REFERENCES

1. Adamek, T., Bestimmung der Kondesationgros-sen auf feingewellten Oberflachen zur Ausle-gun aptimaler Wandprofile. <u>Warme-und Stoffubertragung,</u> Vol. **15,** pp. 255-70

2. Webb, R.L., Shell-side Condensation in Refrigerant Condensers. <u>ASHRAE,</u> 1984, **90,** Part 1, pp. 5-25

3. Webb, R.L., Rudy, T.M. and Kedzierski, M.A., Prediction of the Condensation Coefficient on Horizontal integral-fin tube.
<u>J. Heat Transfer</u>, 1985, **107,** 369-76

4. Chun,K.R., and Seban, R.A., Heat transfer to evaporating liquid films. <u>ASME J. Heat Transfer</u>,1971, **4,** 391-96

WATER VAPOUR COMPRESSION AND ADIABATIC COOLING IN THE PROCESS INDUSTRY

HANS MADSBØLL/FRANK ELEFSEN
Danish Technological Institute (DTI)
Dept. of Energy Technology/Refrigeration and Heat Pumps
Teknologiparken, DK-8000 Aarhus C

ABSTRACT

A description of a 4.6 MW process cooling plant based on Water Vapour Compression is given. The cooling plant is a two stage plant with two water vapour compressors, a low pressure evaporator, an intermediate cooler, and a high pressure condenser. The cooling unit meets the cooling load at a process cooling water temperature of 10°C. The condenser is cooled by means of cooling towers. The evaporator and condenser are direct contact types of heat exchange, i.e without any heat exchanging surfaces.

A dynamic computer model of the system is made.

The annual energy consumption is calculated to 1.8 TJ/year. This is an energy saving of 50% compared to a traditional ammonia cooling plant. Calculations of temperatures, power consumption and COP are presented as a function of the wet-bulb temperature. Dynamic and steady-state simulations of the cooling plant is performed. The results of the calculations clearly show that there are several areas where mechanical water vapour compression has many advantages in respect of energy saving and environmental protection.

INTRODUCTION

The demand for research in the field of energy saving processes and environmentally safe refrigerants has caused that DTI in cooperation with several large Danish companies has started a research and development programme. The involved processes are among other things sewage treatment, separation of sludge and liquid manure, drying of different articles of foods. The common process is to separate these aqueous solutions in a solid part and a distillate. This programme will in the near future result in process equipment which in quite a new way uses a very energy saving method of treating the above-mentioned substances.

The thermodynamic process for separation is to recycle the energy for evaporation of the water by means of a mechanical water vapour compressor and heat exchanger.

The above-mentioned method has required development of several new technologies, new types of water vapour compressors, new heat exchanger designs etc. This development work has also made the use of mechanical water vapour compressors possible in connection with refrigeration processes, for example process cooling in the industry, vacuum ice production and heat pumps. It appears that water with respect to the environment and the energy consumption is a very appropriate refrigerant - [1] and [2].

The present research and development work has included the following issues:

- **Evaporation** or separation of process sewage, sludge from local sewage plants and liquid manure

- **Drying** of aqueous solutions

- **Process Cooling** - vacuum ice - and cooling plants

- **Heat recovery** by means of heat pumps

The advantage of using water as refrigerant in connection with process cooling plants, vacuum ice production plants and heat pump plants is that the heat exchange in evaporator and condenser can be performed in direct contact with the surroundings without any physical separation of the heat exchanging fluids.

Four advantages can be obtained in this way:

- No temperature difference between fluids in the heat exchanger. The suction pressure and the discharge pressure of the compressor therefore correspond to the outlet saturation temperature of the water from both the evaporator and the condenser.

- The energy consumption of the cooling unit can in many cases be reduced by 50%. The emission of CO_2 is reduced correspondingly to the reduced energy consumption, thereby causing a significant smaller contribution to the Global Warming Potential.

- No fouling in the heat exchangers.

- Water is absolute harmless to the environment, unlike some of the traditional refrigerants which have a high Ozone

Depletion Potential and a Global Warming Potential too.

The Coefficient of Performance (COP) of a refrigeration plant is defined as:

$$COP_r = \frac{\dot{Q}_e}{P_{comp}} \tag{1}$$

Where $\dot{Q}_e$ is the cooling demand and P_{comp} is the power consumption of the compressor.

In general, the COP is better for cooling plants based on water vapour compression (WVC) due to the lack of any temperature differences in the evaporator and the condenser. A comparison between cooling plants based on water vapour compression and traditional NH_3 cooling plants is shown in Figure 1. The figure shows thermodynamic calculations using a temperature dif-

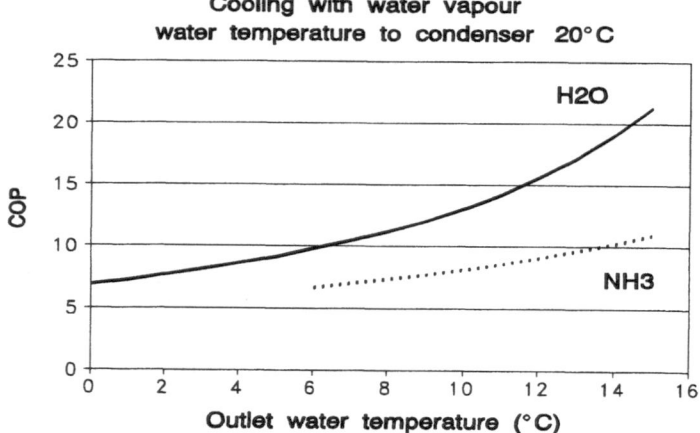

Figure 1 **Comparison of COP for cooling plants based on water vapour compression and NH₃**

ference in the evaporator and the condenser of 4°C for the NH_3 plant, and in both cases a compressor efficiency of 0.7.

THE DESCRIPTION OF THE COOLING PLANT

The cooling demand can be met by a cooling water flow of max. 350 m³/h at a temperature of 10°C. The water is heated in the process to a temperature of approx. 24°C.

In the period from November to March, the necessary cooling demand can be met by cooling towers. In the remaining part of the year, the cooling demand is met by water from a local bo-ring. The annual water consumption amounts to approx. 1 million m³ fresh drinking water, which corresponds to the water consumption in a town with approx. 20,000 inhabitants.

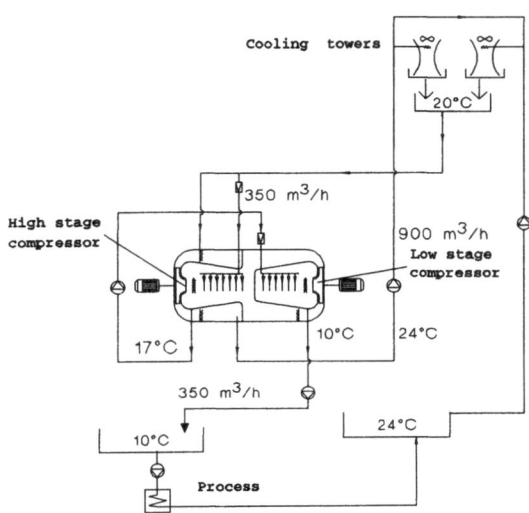

Figure 2   **Flow diagram of a two stage process cooling plant, using mechanical water vapour compression.**

The plant is a two-stage unit with a low and a high pressure compressor. If the wet-bulb temperature is below 14°C, only the low stage compressor is operating. If the wet-bulb temperature exceeds 14°C, the high pressure compressor will start.
The cooling unit is a stand alone unit including compressors, evaporators and condensers. The unit is operating in vacuum.

The pressures in the shells correspond to the saturation pressures of the vapours at the outlet of the evaporators and the condensers. Special precautions have to be taken concerning the air extraction.

A flow diagram of the process cooling plant operating in a two stage mode is depicted in Figure 2. The process cooling water is pumped to the cooling towers where it is cooled to a temperature of 3°C above the actual wet-bulb temperature. From the reservoir below the cooling towers a constant flow of 350 m³/h is led to the high pressure evaporator. In the evaporator, less than 1% of the water evaporates by the expansion, and in this way is causing the rest of the water to be cooled. The water to the high stage condenser is also taken from the reservoir below the cooling towers, and at a constant flow of 900 m³/h. The water is heated approx. 4°C in the condenser and led back to the cooling towers. The precooled water from the high pressure evaporator is led to the low pressure evaporator where further evaporation takes place by expansion. The remaining cooled water liquid is led to the process. The water vapour is compressed in the compressors and led to the condenser.

ENERGY CONSUMPTION

In order to estimate the annual energy consumption, theoretical calculations are made. The calculations are based on the following assumptions:

- 350 m³/h process water is cooled from a process outlet temperature of 24°C to an outlet temperature from the cooling towers of 3°C higher than the actual wet-bulb temperature. If the water temperature at the outlet of the cooling towers exceeds 10°C, the refrigeration plant starts in order to keep the process inlet temperature below 10°C.

- The maximum capacity of the refrigeration plant is approx. 4.6 MW, assuming a wet-bulb temperature of 18°C, which is the design value in Denmark.

- The temperature rise in the condenser at maximum capacity is 4°C.

- The flow of water through the condenser is constant.

- The value of the wet-bulb temperature is taken month by month as the average of the last 30 years, as depicted in fig. 3.

Assuming these conditions for the cooling plant using WVC, the annual energy consumption is calculated to approx. 1.8 TJ/year.

Dependent on the detailed assumptions on the temperature difference over the heat exchangers, the corresponding energy consumption of a traditional ammonia plant is at least the double. Assuming a temperature difference of 8°C across the evaporator and 6°C over the condenser, the annual energy consumption is 10 TJ/year for a NH_3 plant. An optimized system with plate heat exchangers and a 4°C temperature difference on both sides would lead to an annual energy consumption of 3.6 TJ/year.

CONTROL APPROACHES

The control of the cooling plant has to consider two main factors:

- The variations in wet-bulb temperature as a function of time.

- The variations in the process water flow as a function of time.

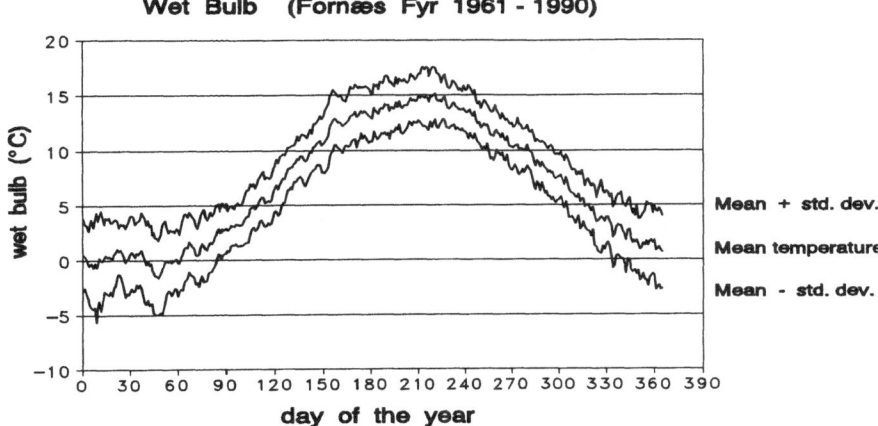

Figure 3   **Measured annual variations of wet bulb temperature, showing mean values and standard deviation.**

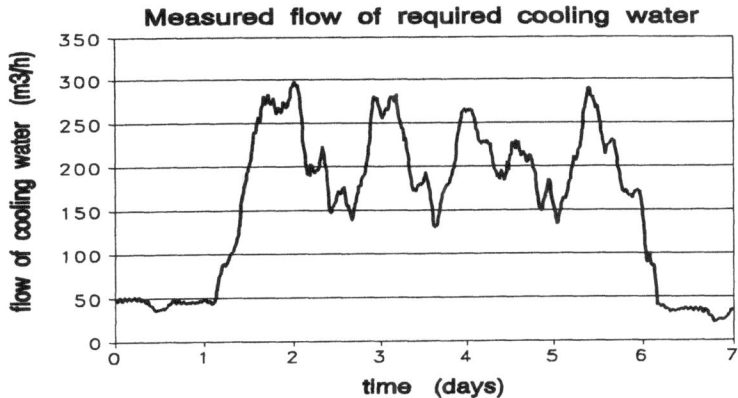

Figure 4   **Measured values of the flow of process cooling water, during one week.**

Because of the restricted range of water vapour flow through the compressor, the flow of water through the cooling unit must be constant. Therefore, the variations in the flow of process cooling water must be managed by an on/off control of the compressors.

Varying wet-bulb temperatures result in variations in the necessary pressure ratio over the compressors. This pressure ratio can be changed by controlling revolution number of the compressors. If the pressure ratio exceeds the available pressure ratio for the one stage operation, the second stage compressor starts.

Measured values of wet-bulb temperature and flow of process
cooling water are shown in figure 3 and figure 4.

DESCRIPTION OF COMPUTER MODEL

In order to model the total system, and thereby investigate the
possible operating conditions of the cooling unit, a dynamic
computer model was made. The model is based on the equations of
mass and energy. A flow diagram of the model is shown in figure
5. The model variables and some of the parameters are indicated.

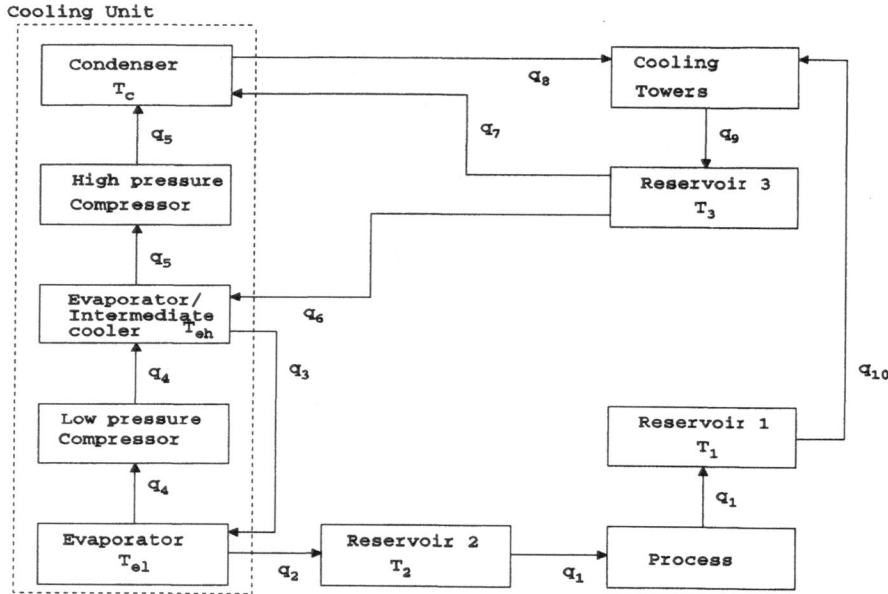

Figure 5   **Flowdiagram of the computer model.**
         q: Mass flow [kg/s], T: Temperature [°C]

The characteristics of the compressor are expressed in terms of
the dimensionless volume flow ϕ, the dimensionless pressure
ratio ψ and the efficiency η by the following equations:

$$\phi = \frac{q_v}{U_t \, A_1} \, C \qquad\qquad (2)$$

where
q_v is the compressor volume flow (m^3/s)
U_t is the peripheral speed of the compressor wheel (m/s)
A_1 is the area of the compressor wheel (m^2)

C   is a geometrical factor

and

$$\psi = \frac{2\ Y_t}{U_t^2} \tag{3}$$

where Y_t is the specific work of compression (Nm/kg) done by the compressor wheel, assuming $\eta = 1$.

The two functions expressing ψ as a function of ϕ and η as a function of ϕ must be determined by experiments. For each of the reservoirs equations for the time derivatives of temperature and mass using mass and energy balances are part of the model.

The energy balance for the low pressure evaporator gives:

$$0 = q_3 h_{s,1}(T_{eh}) - q_4 h_{s,g}(T_{el}) - q_2 h_{s,1}(T_{el}) \tag{4}$$

where
$h_{s,1}(T_{eh})$
    is saturated enthalpy of liquid at T_{eh}
$h_{s,g}(T_{el})$
    is saturated enthalpy of gas at T_{el}
$h_{s,1}(T_{el})$
    is saturated enthalpy of liquid at T_{el}

A similar equation for the high pressure evaporator/intermediate cooler is given.

The compressor power of the low stage compressor is given according to (2) and (3):

$$Q_{comp} = \frac{Y_t}{\eta(\phi)}\ q_4 = \frac{\psi(\phi)\ U_t^3\ \phi\ A_1}{2\ \eta(\phi)\ v_4\ C} \tag{5}$$

where v_4 is the specific volume of water vapour at the inlet of the compressor.

COMPUTER SIMULATION RESULTS

For the dynamic simulation, a typical variation in the process water flow and the wet-bulb temperature during one day (24 hours) is used as input to the model.

The dynamic equations for the reservoirs together with the steady state equations are used to determine the size of the reservoirs and to investigate different types of control approaches.

Results of such simulations show that the numbers of start/stop during a typical week will be 30 - 40, assuming a reservoir volumen of 400 m³. The variations in evaporator temperature are in most cases unimportant, less than 0.5°C. The cooling unit will be on for 2 - 4 hours and off for 1 - 2 hours, assuming flow rates as shown in figure 4.

In the following, two figures steady state simulations are shown. Due to the fact that the wet-bulb temperature is determinative for the power consumption of the system, the results are given as a function of the wet-bulb temperature. It is assumed that the number of revolutions can be controlled on one of the compressors, and this control is used to optimize the working point on the compressor characteristics if possible.

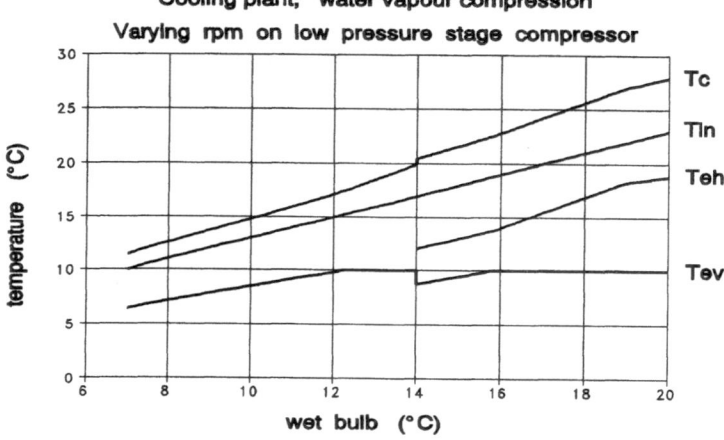

Figure 6   **Results of computer simulations, showing T_{ev}, T_{eh} and T_c as function of wet-bulb temperature.**

The outlet temperature T_{ev} of the low pressure evaporator is depicted in figure 6 along with the temperature T_{eh} in the high pressure evaporator/intermediate cooler and the condenser temperature T_c. T_{in} is the outlet temperature from the cooling towers, which is the inlet temperature to both evaporator and condenser. T_{ev} will in some cases be at a temperature below the necessary 10°C because there is a lower limit for pressure ratio obtainable due to the compressor characteristic.

In figure 7, the calculated average COP is shown as a function of the wet-bulb temperature. The values for the water vapour compression cooling unit is compared to a fully optimized traditional ammonia cooling plant with screw compressors and plate heat exchangers. On the figure it is indicated how many percent of the running time for the cooling unit the wet bulb temperature is within the shown interval. 50% of the running time the wet-bulb temperature will be in the interval 7°C to 12°C. So at

Cooling plant

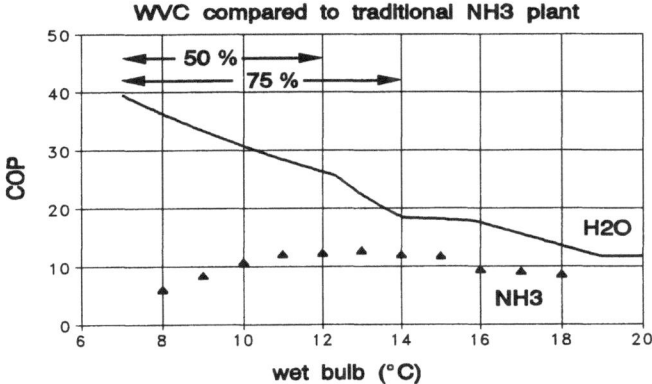

Figure 7 COP for the water cooling unit

low wet bulb temperatures, the big difference in energy consumption is important because of the time spent at these temperatures. The interval 7°C to 14°C corresponds to the interval, where the cooling unit is in the one-stage mode, so in 75% of the running time, it will only run on one compressor.

At higher wet-bulb values, the COP of the water vapour compression unit is still 50 - 85% higher than the NH_3 plant, partly because of the lack of a temperature difference in heat exchangers, and partly because it is run in two stages with open intermidiate cooler.

The theoretical calculations will be compared with measurements when the process cooling plant will be installed in 1993.

REFERENCES

[1] Ophir A., Paul J., The Ecochiller - A Mechanical Vapour Compression Cycle Using Water as Refrigerant, XVIIIth International Congress of Refrigeration, Montreal 1991.

[2] Paul J., Alternative, Environmentally Safe Refrigerants and Processes for Cooling, Heat Pump, and Power Generating (ORC) Systems, XVIIIth International Congress of Refrigeration, Montreal 1991.

SESSION 7:

Heat Exchangers

Chairman: Prof. D.A. Reay

COMPACT HEAT EXCHANGERS FOR THE PROCESS INDUSTRY

R.K. Shah

Harrison Division, General Motors Corporation, Lockport, NY 14094, USA

J.M. Robertson

Consultant, AEA Petroleum Services, AEA Technology, Harwell, Oxon OX11 0RA, UK

ABSTRACT

Compact heat exchangers (CHEs), having light weight, minimum volume and high effectiveness, have played a major technological role in the advancement of the transportation industry. CHEs will also play a similar important role in the advancement of the process industry. However, the operating environment and applications for the process industry are significantly different. After providing basic background information for the CHEs, the present barriers to the general application of CHEs have been described, and the considerable energy-savings potential with the use of CHEs has been outlined. Finally, future developments for the process plant heat exchangers are summarized. Users and manufacturers of process industry exchangers must now seriously consider the use of CHEs in future applications to make substantial savings in cost and energy.

INTRODUCTION

Compact heat exchangers (CHE) have historically played a major technological role in the development of aerospace, vehicular and marine transportation systems as a result of their light weight, minimum volume and/or high effectiveness. However, the major limitations of compact exchangers have been on maximum operating pressure and temperature, minimum allowed fouling, and these have prevented their widespread adoption and use in the process industry. However, significant advances have taken place in manufacturing technology as well as in the design theory and our knowledge of fundamentals of flow phenomena in heat exchangers. These new developments together with new and improved types of CHEs present an excellent opportunity to introduce compact heat exchangers in the process industry, particularly when energy saving is also an important objective.

In this paper, starting with the definition, types of CHEs used in process plants will be described. Characteristics of CHEs of importance to process plants will then be presented along with benefits of their use and limitations. Subsequently, technical and commercial issues that inhibit the broader use of CHEs will be discussed. Some details will also be provided on energy savings using CHEs. Finally, future potential uses and applications of CHEs will be summarized.

DEFINITION OF A COMPACT HEAT EXCHANGER

A gas-to-fluid heat exchanger is referred to as a compact heat exchanger if it incorporates heat transfer surface having a surface area density above about 700 m^2/m^3 (213 ft^2/ft^3) on at least one of the fluid sides which usually has gas flow. It is referred to as a laminar flow heat exchanger if the surface area density is above about 3000 m^2/m^3 (914 ft^2/ft^3). It is referred to as a micro heat exchanger if the surface area density is above about 10000 m^2/m^3 (3050 ft^2/ft^3) [1]. A liquid/two-phase to liquid/two-phase heat exchanger is referred to as a compact heat exchanger if the surface area density on any one fluid side is above about 400 m^2/m^3 (122 ft^2/ft^3). A typical process industry shell-and tube exchanger has a surface area density of less than 100 m^2/m^3 on one fluid side with plain tubes, and 2-3 times that with the high fin density low finned tubing.

As will be shown later, extremely high heat transfer coefficients h are achievable with small hydraulic diameter flow passages with gases, liquids, and two-phase flows. A typical PHE has also about two times h or U that of a shell-and- tube exchanger for water/water applications. Basic constructions for gas-to-gas compact heat exchangers are plate-fin, tube-fin, and all prime surface recuperators and compact regenerators; basic flow arrangements of two fluids are single-pass crossflow, counterflow, and multipass cross-counterflow. Basic constructions for liquid-to-two-phase compact exchangers are: gasketed and welded plate-and-frame heat exchangers, welded stacked plate heat exchangers (without frames), and printed circuit heat exchangers. Schematic examples of gas-to-fluid and liquid/two-phase to liquid/two-phase compact heat exchangers are shown in Figs. 1 and 2.

WHY COMPACT HEAT EXCHANGERS?

A common understanding is that turbulent flows (having Reynolds numbers greater than

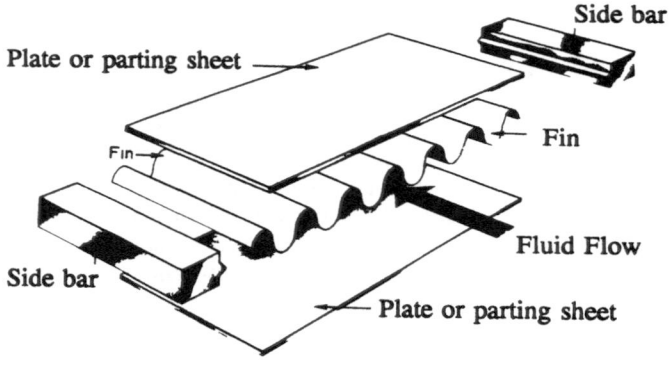

FIGURE 1. (i) A plate-fin assembly.

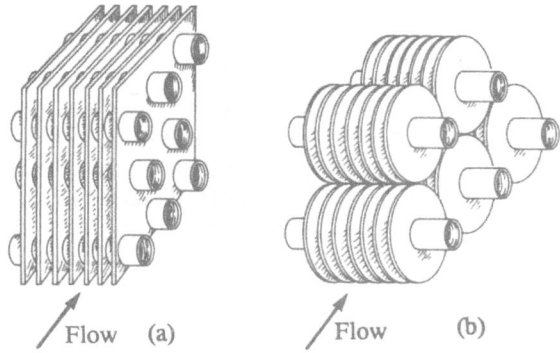

FIGURE 1. (ii) Tube-fin exchanger: (a) Flat (continuous) fins on an array of tubes; the fin is shown as plain, but can be wavy, louvered, etc.; (b) Individually finned tubes.

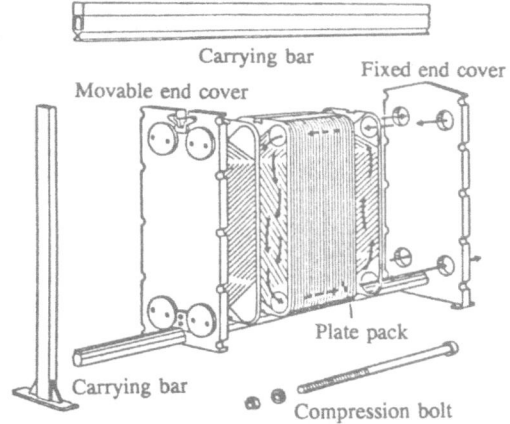

FIGURE 2. (i) A plate heat exchanger assembly.

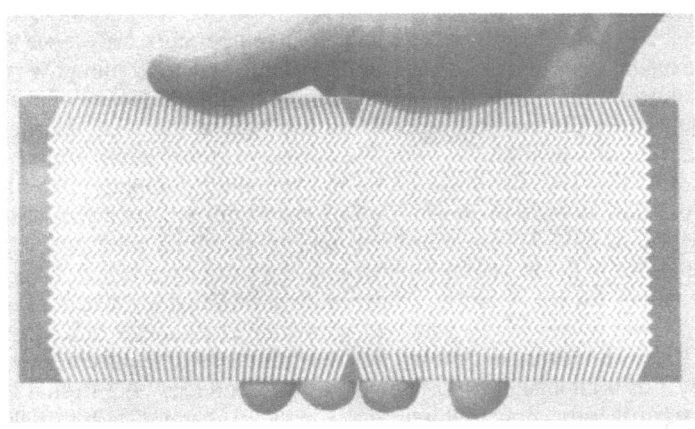

FIGURE 2. (ii) A chemically milled printed circuit heat exchanger plate [5].

10000) provide high heat transfer coefficients (although with high pressure drops) and hence are desirable in heat exchanger applications. However, laminar flows with small hydraulic diameter flow passages in a heat exchanger can provide high heat transfer coefficients with reasonable pressure drops. Consider water flow in a circular tube at 310 K. Using correlations for laminar flow (Nu=3.657) and turbulent flow (Gnielinski correlation, see [2]), the heat transfer coefficient h is computed as a function of the tube inside diameter D_i and is shown in Fig. 3. It can be seen that h for a 20 mm diameter tube at Re=10^4 is the same as that for a 1 mm diameter tube in laminar flow! Similarly, the h's are the same for a 20 mm diameter tube at Re=5×10^4 and 0.3 mm diameter tube in laminar flow!

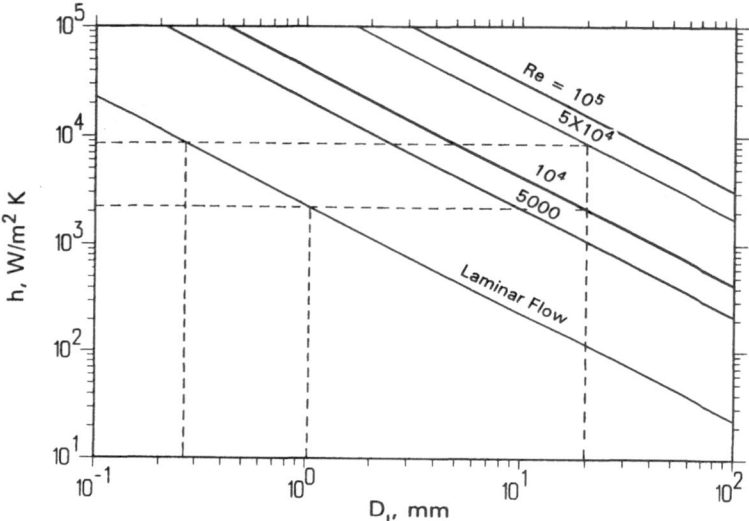

FIGURE 3. Heat transfer coefficient as a function of tube diameter for a long circular tube with water flow at 310 K.

For a given heat duty, fluid flow rates and mean temperature difference in the exchanger, laminar flow exchangers can offer a substantial volume and fluid pumping power reductions over the conventional shell-and-tube exchangers. Consider the case of thermal resistances distributed equally on the hot and cold sides, and where the wall thermal resistance and fouling resistances are negligible. For flow over a tube bank with X_T/D_o=1.25 and X_L/D_o=2.00 and Re_L=100, the volume (V_L/V_T) and fluid pumping power (P_L/P_T) ratios as a function of the tube outside diameter (D_L/D_T) ratio with Re_T as a parameter are shown in Fig. 4 [1]. Here X_T and X_L are transverse and longitudinal tube pitches, D_o is the tube outside diameter, Re is the Reynolds number and the subscripts L and T denote laminar and turbulent flow values for Re, V and P. This figure shows that a substantial reduction in both volume and pumping power can be achieved with smaller diameter tubes in the range of 10^{-2} < D_L/D_T < 10^{-1}. However, not shown are the frontal area requirement and the length of the exchanger. For a laminar flow exchanger, as D_L decreases, the frontal area continues to increase very significantly compared with that for the exchanger with turbulent flow and the length continues to decrease signficantly. Hence, Fig. 4 cannot be used arbitrarily without consideration of the frontal area and flow length requirements. Also, note that Fig. 4 is not

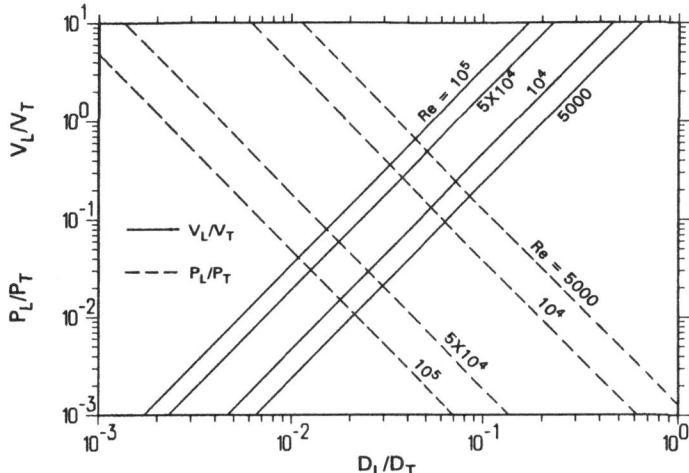

FIGURE 4. Ratios of heat exchanger volumes and pumping powers for laminar to turbulent flows as a function of a ratio of tube diameters for laminar to turbulent flows.

general; it is specific for the selected geometry of a tube bank and for selected operating conditions.

CHARACTERISTICS OF COMPACT HEAT EXCHANGERS

Gas-to-Fluid Exchangers

The unique characteristics of compact extended (plate-fin and tube-fin) surface exchangers, as compared with the conventional shell-and-tube exchangers, are: (1) many surfaces available having different orders of magnitude of surface area density; (2) flexibility in distributing surface area on the hot and cold sides as warranted by design considerations; and (3) generally substantial cost, weight, or volume savings.

The important design and operating considerations for compact extended surface exchangers are: (1) usually at least one of the fluids is a gas to compensate for its low h; (2) fluids must be clean and relatively non-corrosive because of small hydraulic diameter (D_h) flow passages and no easy techniques for cleaning; (3) the fluid pumping power (i.e., pressure drop) design constraint is often as equally important as the heat transfer rate; (4) operating pressures and temperatures are somewhat limited compared to shell-and-tube exchangers due to joining of the fins to plates or tubes such as brazing, mechanical expansion, etc.; (5) with the use of highly compact surfaces, the resultant shape of the exchanger is one having a large frontal area and a short flow length; the header design of a compact heat exchanger is thus important for a uniform flow distribution among the very large number of small flow passages; (6) the market potential must be large enough to warrant the sizeable manufacturing research and tooling costs for new forms to be developed.

Some of the advantages of plate-fin exchangers over conventional shell-and-tube exchangers are as follows. Compact heat exchangers are generally fabricated from thin metallic plates thus forming many convoluted and narrow flow passages and giving a large heat transfer surface area per unit volume, typically up to ten times greater than the 50 to 100 m^2/m^3 provided by a shell-and-tube exchanger for general process application. Compact gas side surfaces have high to ultra high ratios of heat transfer surface area to volume (β) ranging from 1000 to 6000 m^2/m^3. Compact liquid or two-phase side surfaces have a heat transfer surface area to volume (β) ratio ranging from 400 to 600 m^2/m^3. This high surface area density in turn reduces substantially the exchanger volume, mass and capital cost for the same surface area or the same duty. This will also reduce considerably the structural support requirement and associated installation cost. The heat exchanger can be configured easily for counterflow or multipass cross-counterflow to yield very high exchanger effectiveness or designed for very small temperature differences between fluid streams, and for very small pressure drops compared to shell-and-tube exchangers. This saves energy and produces lower thermal pollution thus reducing impact on the environment. A compact exchanger provides a tighter temperature control, thus is useful for heat sensitive materials, improves the product (e.g. refining fats from edible oil) and its quality (such as a catalyst bed). Also, a compact exchanger could provide rapid heating or cooling of a process stream, thus improving the product quality. It requires a low fluid inventory (hold up) in the exchanger; this is important for valuable or hazardous fluids. The plate-fin exchangers can accommodate multiple (up to 12 or more) fluid streams in one exchanger unit [3].

The combination of its special operational features, its high heat transfer coefficients and large heat transfer surface area, makes the compact exchanger very attractive for broader use in the oil refinery, gas processing, petrochemical, chemical and associated process industries. The compact heat exchangers which are at present available and the new types which are evolving, offer process flowsheet designers the opportunity and stimulus, not only to make savings by replacing conventional equipment, but also to rethink the approach to process design.

The major limitations of plate-fin and other compact heat exchangers are as follows. Plate-fin and other compact heat exchangers have been and can be designed for high temperature applications (up to about 850°C), high pressure applications (over 200 bars), and moderate fouling applications. However, applications usually do not involve both high temperature and high pressure simultaneously. Highly viscous liquids can be accommodated in the plate-fin exchangers with a proper fin height; fibrous or heavy fouling fluids are not used in the plate-fin exchangers because mechanical cleaning in general is not possible. However, these liquids can be readily accommodated in plate heat exchangers. Most of the plate-fin heat exchangers are brazed. At the current state-of-the-art, the largest size exchanger that can be brazed is about 1.2 x 1.2 x 6 m. While plate-fin exchangers are brazed in a variety of metals including aluminum, copper, stainless steels, nickel and cobalt based superalloys, the brazing process is generally of proprietary nature and it is quite expensive to set up and develop specific brazing techniques. Note that due to environmental concerns, dip brazing is being eliminated by the industry, and is replaced by vacuum brazing and neutral environmental atmospheric brazing. Thus, the problem of flux removal (to avoid corrosion) with dip brazing for brazed heat exchangers will no more be a problem. The plate-fin exchanger is readily repairable, if leaks occur at the external border seams.

Fouling is one of the potential major problems in compact heat exchangers (except for plate-and-frame heat exchangers), particularly having a variety of fin geometries or very fine circular or noncircular flow passages that cannot be cleaned mechanically. Hence, extended surface compact heat exchangers may not be applicable in high fouling applications. However, with the understanding of the problem and applying innovative means to prevent/minimize fouling, compact extended surface heat exchangers may be used in at least low to moderate fouling applications. In order to reduce fouling, nonfouling fluids should be used where permissible such as clean air or gases, light hydrocarbons and refrigerants.

Other important limitations of compact heat exchangers are as follows: With a high effectiveness heat exchanger and/or large frontal area, flow maldistribution becomes important. More accurate thermal design is required and a heat exchanger must be considered a part of a system. Due to short transient times, a careful design of controls is required for startup for compact heat exchangers compared with shell-and-tube exchangers. Flow oscillation could be a problem for compact heat exchangers. No industry standards or recognized practice for compact heat exchangers are yet available, particularly for power and process industry (note that this is not a problem for aircraft, vehicular and marine transportation industries). Structural integrity is required to be examined on a case-by-case basis utilizing standard pressure vessel codes.

Liquid-to-Liquid Exchangers

Since plate heat exchangers are mainly used for liquid-to-liquid heat exchange applications, the characteristics of these exchangers will be briefly summarized here. The most significant characteristic of a PHE is that it can easily be taken apart into its individual components for cleaning, inspection, and maintenance. The heat transfer surface area can be readily changed or rearranged for a different task or anticipated changing loads, through the flexibility of the number of plates, plate type, and pass arrangements. The high turbulence due to plates reduces fouling to about 10 to 25% that of a shell-and-tube excahnger. Because of the high heat transfer coefficients, reduced fouling, absence of bypass and leakage streams, and pure counterflow arrangements, the surface area required for a plate exchanger is 1/2 to 1/3 that of a shell-and-tube exchanger for a given heat duty. This would reduce the cost, overall volume, and maintenance space for the exchanger. Also, the gross weight of a plate exchanger is about 1/6 that of an equivalent shell-and-tube exchanger. Leakage from one fluid to the other cannot take place unless a plate develops a hole. Since the gasket is between the plates, any leakage from the gaskets is to the outside of the exchanger. The residence time for fluid particles on a given side is approximately the same for uniformity of heat treatment in applications such as sterilizing, pasteurizing, and cooking. There are no significant hot or cold spots in the exchanger which could lead to the deterioration of heat sensitive fluids. The volumes of fluids held up in the exchanger are small. This is important with expensive fluids, for faster transient response, and for a better process control. Finally and importantly, high thermal performance can be achieved in plate exchangers. The high degree of counterflow in PHEs makes temperature approaches of up to 1°C possible. The high thermal effectiveness (up to about 93%) facilitates low grade heat recovery economical. Flow-induced vibration, noise, thermal stresses, and entry impingement problems of shell-and-tube heat exchangers do not exist for plate heat exchangers.

Plate heat exchangers are most suitable for liquid-liquid heat transfer duties requiring uniform and rapid heating or cooling as is often the case when treating thermally sensitive fluids. Special plates capable of handling two-phase fluids (e.g., steam condensation) are available. PHEs are not suitable for erosive duties or for fluids containing fibrous materials. In certain cases, suspensions can be handled but to avoid clogging the largest suspended particle should be at most, one-third size of the average channel gap. Viscous fluids can be handled but extremely viscous fluids lead to flow maldistribution problems, especially on cooling.

Some other inherent limitations of the plate heat exchangers are due to the plates and gaskets as follows. The plate exchanger is used for a maximum pressure of about 2.5 MPa gage (360 psig), but usually below 1.0 MPa gage (150 psig). The gasket materials (except for the recent Teflon coated type) restrict the use of PHEs in highly corrosive applications; they also limit the maximum operating temperature to 260°C (500 °F), but usually below 150°C (300 °F) to avoid the use of expensive gasket materials. Gasket life is sometimes limited. Frequent gasket replacement may be needed in some applications. Pin-hole leaks are hard to detect. For equivalent flow velocities, pressure drop in a plate exchanger is very high compared to a shell-and-tube exchanger. However, the flow velocities are usually low and plate lengths are "short", so the resulting pressure drops are generally acceptable. Some of the largest units have a total surface area of about 2500 m² (27000 ft²) per frame. Large differences in fluid flow rates of two streams cannot be handled in a PHE.

TYPES OF COMPACT HEAT EXCHANGERS

Compact heat exchangers as shown in Figs. 1 and 2 are used in specialized process industry applications. Some of these exchangers and their applications are now briefly summarized.

Plate-Fin Exchangers

Unique characteristics, advantages and limitations of plate-fin exchangers are described in the preceding section and hence will not be repeated here.

These exchangers are quite commonly used in cryogenic processing area in air separation plants to separate nitrogen and oxygen from air; and in large plants, even minute constituents of air, such as argon, neon, krypton, xenon, etc., are separated. Multiple streams in such a heat exchanger are common. This exchanger has brazed aluminum construction with, for example, a choice of plain, offset strip fins (serrated fins) or perforated fins. Operating pressure for the compressed incoming air is about 6 bar. While extremely high exchanger effectiveness (98-99%) are achievable, flow maldistribution is a major problem in this application.

Plate-fin exchangers are also used in separation and purification of light hydrocarbons such as methane, ethane, propane and butane by liquefaction. Operating conditions for these applications also include high pressure and low temperatures. Other applications of plate-fin exchangers include: production of liquified natural gas; separation, purification and liquefaction of helium; purification and liquefaction of hydrogen and the products of olefin plants.

Tube-Fin Heat Exchangers

These exchangers are of two major types as shown in Fig. 1(ii).

In most of the applications, any operating pressure on the tube side can be accommodated. However, the operating pressure on the airside is usually of the order of magnitude of one atmosphere since atmospheric air is used as a cooling medium. In most applications, the operating temperatures are lower than 260°C (500°F). Fouling is generally not a problem on the tube side since it can be cleaned by the usual techniques. However, cleaning may not be easy on the fin side. The most common materials for tubes and fins are copper, aluminum and steel.

Tube-fin exchangers with flat plain, wavy, louvered or slit fins are most commonly used in evaporators and condensers for air-conditioning and refrigeration industry when air is one of the two fluids.

Air-cooled heat exchangers (ACE) have generally round tubes and flat (continuous) fins or individually finned tubes with a number of plain, wavy and interrupted fin geometries. Generally they are not compact heat exchangers. They are extensively used in process industry for rejecting heat directly to the atmosphere. They compete with water cooling.

Plate Heat Exchangers

Gasketed plate-and-frame heat exchangers were developed for food industry (milk, juices,etc.) in 1920's. They are most common in the dairy, beverage, general food processing, and pharmaceutical industries where their ease of cleaning and the thermal control required for sterilization/pasteurization makes them ideal. They are used in synthetic rubber industry, paper mills and petrochemical plants. In addition, they are also used in process industry for water-water duties (heating, cooling and temperature control) with stainless steel construction when rather high pressure drops are available; in such applications, they compete well with shell-and-tube exchangers by cost, live time and close temperature approach. The above applications in food and process industries account for about 90% of the PHE production. PHEs are generally not used in boiling and condensation applications due to their high pressure drop characteristics. Specially designed plates are now available for condensing high density vapors such as ammonia and propylene. PHEs have been recently used as evaporators for sugar solution concentration [4].

The major advantages and limitations of the PHEs are described in the preceding major section. One major limiting consideration for the PHEs in the process industry is the gaskets due to potential fire and/or environmental pollution associated with gasket leaks with many of the fluids used in the process industry. Also, vacuum operation is excluded for the PHEs due to the length of the gaskets and difficulty to tighten them properly plus the cost of changing them. In specific applications, gaskets are replaced by the welded plates (welded at the periphery) or graphite PHEs.

Stacked Plate Heat Exchangers

This exchanger differs from the PHE in that it does not have any gaskets, and the rectangular plates are stacked and welded at the edges. The physical size limitations of PHEs (1.2 m wide x 4 m long max) are considerably extended to 1.5 m wide x 20 m long. The potential maximum operating temperature is 815°C (1500°F) with an operating pressure of 20 MPa (3000 psig) when the stacked plate assembly is placed in a cylindrical pressure vessel and 2 MPa (300 psig) when not contained in a pressure vessel. This exchanger is a new construction with limited current applications. One application of this exchanger is a feed effluent heat exchanger in manufacturing of unleaded petrol.

Printed Circuit Heat Exchangers

This exchanger has only primary heat transfer surface as PHEs. Fine grooves are made in the plate by using the same techniques as those employed for printed electrical circuits. Very high surface area densities (1000-5000 m^2/m^3 or 300-1520 ft^2/ft^3) are achievable. A variety of materials such as 316 SS, 316L SS, 304 SS, 904L SS, cupronickel, monel, nickel and superalloys can be used. It has been successfully used with relatively clean gases, liquids and phase-change fluids in chemical processing, fuel processing, waste heat recovery and refrigeration industries [5]. Again, this exchanger is a new construction with limited current special applications.

DESIGN AND SELECTION OF CHEs

Two-Fluid CHEs

Figure 5 illustrates a design methodology for an optimum compact heat exchanger and consists of four major phases: thermal and hydraulic design, mechanical design, and manufacturing considerations and cost estimates, and trade-off factors and system based optimization. There could be strong interactions and feedback among the aforementioned phases as indicated by double-sided arrows shown in Fig. 5. This methodology is discussed in some details in [6]. It may be best characterized as a case study method based on a particular *surface selection*.

Two most common heat exchanger thermal design procedures concern with rating and sizing. Determination of heat transfer and pressure drop performance of either an existing exchanger or an already sized exchanger is referred to as a rating problem. The design of a new exchanger to meet the specified heat transfer and pressure drops within all specified contsraints is referred to as a sizing problem. The design procedure for two-fluid CHEs is straightforward since there are no leakage and bypass streams in the exchanger as is found in a shell-and-tube exchanger. Detailed step-by-step solution procedures are available for rating and sizing of compact plate-fin and tube-fin heat exchangers [2,7-9], rotary regenerators [10], and plate-and-frame heat exchangers [11]. One can readily develop a computer program for rating and sizing of these exchangers. Hence, no such commercial programs are available in the USA, except for the plate heat exchangers. Mathematical optimization techniques may be employed after initial sizing to optimize heat exchanger size [2].

A number of proprietary computer programs are available from Heat Transfer and Fluid Flow Service of UK for thermal sizing and rating of plate-fin exchangers (MULE, MUSE, MUSC, PFIN), plate heat exchanger (APLE) and air-cooled heat exchangers (ACOL). For the last two exchangers, PHE and ACE programs are available from Heat Transfer Research, Inc. of USA.

Multistream CHEs

The thermal design theory for exchangers having more than two fluid streams is rather complex. Hence the detailed solution procedure for thermal rating is available only through proprietary computer programs such as MULE from HTFS. Prasad and coworkers [12-14] have developed differential methods for rating and sizing of multistream exchangers.

Guide to CHE Specification, Use and Selection

In order to help overcome the barriers to the broader use of aluminum plate-fin heat exchangers, a Guide to the Use and Specification was produced by the UK Study Group [9]. This Guide was aimed at filling the gap in information available about these exchangers which manufacturers felt reluctant to provide, mainly as a result of their proprietorial nature and their original use in a narrow range of applications. In this context, the manufacturers of

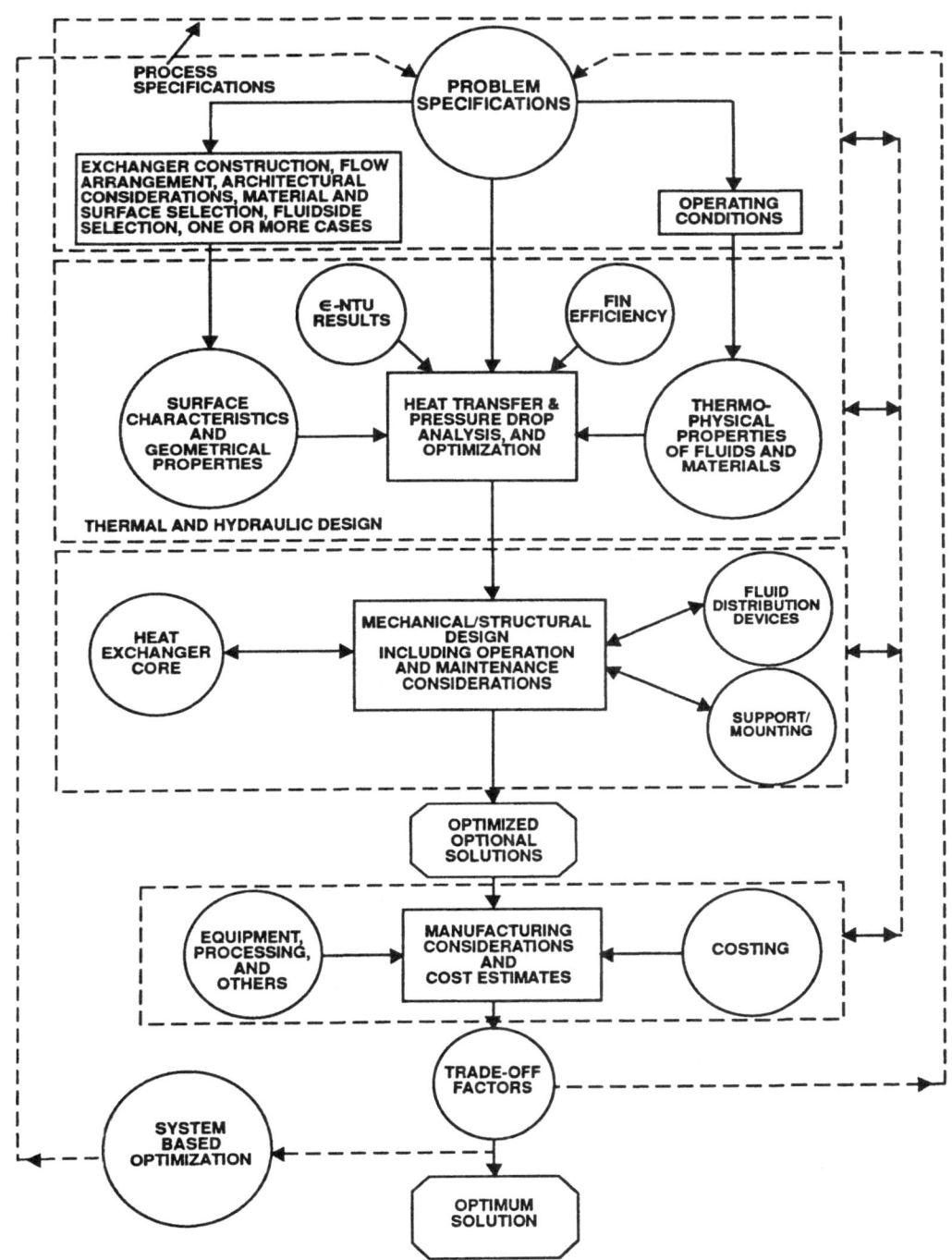

FIGURE 5. Heat exchanger design methodology.

brazed aluminum plate-fin heat exchanger have now formed an Association[1] and one of its aims is to promote the broader use of these exchangers. The Guide is useful as an introduction for the novice user and is currently being followed by the production of more formal documentation for allowing that user to specify and purchase many different forms of compact heat exchangers without having to treat each order as a "special".

Useful information on how to select CHEs is presented in [15].

BARRIERS TO THE GENERAL APPLICATION OF CHEs

Where the cost of the heat exchanger is a small fraction of the total cost of a process plant, there can be strong resistance to using new and untried compact heat exchangers: its failure may result in an expense much larger than the saivngs from its use. Moveover, fouling and flow maldistribtuion are perceived to be major uncertainties with many forms of compact heat exchangers and pose insurmountable difficulties to some companies.

The barriers to a more general application of these exchangers in the process industries and to their incorporation at process flowsheet design stage are as follows.

- The absence of industry standards for the construction and use of compact heat exchangers, and of detailed information about construction and use, etc.
- The absence of user companies' internal documentation and procedure for ensuring that the specification, purchasing, installation and use of these exchangers is as easy and straightforward as with more conventional exchangers.
- The absence of information about costs of installation, operation, commissioning, etc. prevents economic justifications being made for their use.
- The absence of rational methods for process flowsheet design using compact heat exchangers to their fullest potential, and not merely to replace a shell-and-tube exchanger.
- The absence of clear information on use with potential fouling duties.
- The absence of previous operational experience of these exchangers in a company requires "a change of company culture" to ensure that they have a chance of being adopted within that company.
- The absence of a real choice of compact heat exchangers to compete successfully with conventional exchangers over their full operating range capabilities. The current forms of compact exchangers available are capable of covering only segments of this full potential range. This points to the requirement for new and improved forms of compact exchangers to be developed and constructed in the future.

It is of interest to note that there are at least two industry-funded groups of companies, which use heat exchangers in their process plants, have formed Study Groups to overcome the barriers to broadening the use of these exchangers to save cost and energy in new applications [16,17]. Currently, these groups are studying improved methods for process flowsheet design incorporating multistream compact heat exchangers and producing the documentation and standards for user companies to specify and purchase compact heat exchangers as easily as shell-and-tube exchangers. Manufacturers will contribute to this latter activity.

---

[1] This Association is made up of members of six manufacturing companies, chaired by the co-author of this paper, J.M. Robertson.

ENERGY SAVINGS WITH COMPACT HEAT EXCHANGERS

Compared to shell-and-tube exchangers, compact heat exchangers offer either more throughput at the same energy expenditure or energy savings for the same throughput through improved process flowsheet design and in-plant operation. Here we will consider the case of energy savings for the same throughput for the following discussion. A similar discussion can be provided for the case of increased throughput at the same energy expense.
- Their special technical capabilities: for example to operate with very small temperature differences between streams and thus save compressor power on liquefaction plants.
- Their reduced capital and installation costs: for example by making better use of the capital funds available to provide improved heat exchanger networks to recover more energy in process plants.

The combination of their special technical capabilities and their reduced costs give them additional advantages. For example, their ability to accommodate more than two streams in one exchanger reduces the number of heat exchangers required in a heat recovery network thus further reducing the capital cost of the system.

Cost savings can be obtained as a result of their significantly lower capital costs and improved thermal performance. Additional cost savings can be obtained from installation, plot area, structural support, as a result of the reduced size, volume and weight. Since these additional costs can be several times the capital cost of the exchangers, they form a significant feature of the reduced overall costs.

The energy-saving capabilities of these exchangers arise as a result of the following.
- The ability to operate with very small temperature differences between streams which offers a greater flexibility in choosing the optimum balance between the capital cost of the heat exchanger network and the capital running costs of other major features of the plant, for example a compressor. Another example of this would be by improving the performance of a feed-effluent heat exchanger where a feed stream entering a reactor or some process heating step is preheated by the stream leaving. Energy can be saved as the temperature of the stream entering can be raised as close as possible to the temperature of the stream leaving.
- The ability to reduce the pressure drop in a gas compressor circuit will contribute significantly to the power required from the compressor, and of course, further savings can be obtained because of the reduced compressor size required. Compact heat exchangers with very short passage length and very large heat transfer surface areas associated with improved heat transfer coefficients offer solutions here, particularly with primary surface only. In another example, a large bank of air-cooled heat exchangers used as compressor intercoolers could, in principle, be replaced by a few compact exchangers using sea or surface water at much lower temperatures. The increased compression ratio could be used, for example, to achieve greater throughput in a given system.
- The ability to reduce fouling potential with certain types of compact exchangers has a direct effect on energy savings. For example, plate-and-frame exchangers which are constructed with convoluted (stamped or ribbed) surfaces and flow passages are able to reduce certain types of fouling, e.g. the particulates as a result of the high rates of shear within flow passages. The additional ability to permit dismantling of these exchangers permits reduced operating levels of fouling. An example with these convoluted passages is operation with very viscous streams which exhibit non-Newtonian characteristics; these passages produce considerably increased shear rates which reduce the viscosity of these fluids and significantly reduce pumping power and possibly improve product quality.

- The ability to improve heat exchanger networks for recovering energy results from the opportunity to use with very small temperature differences at the "pinch point". This, of course, will reduce the associated heat rejection to the environment. Multistream heat exchangers also have the opportunity of reducing the number of heat exchangers required.
- The ability of compact exchangers to provide combined functions such as simultaneous mass and heat transfer in one unit offers enormous opportunities for future energy saving which, as yet, are untapped.

The full potential of energy savings with compact exchangers can only be realized through properly conducted and thorough energy-savings studies. Studies which look at compact exchangers as replacing conventional exchangers will be inadequate for this purpose as the full range of opportunities and potential capabilities of compact exchangers must be considered if these studies are to be realistic.

In short, compact heat exchangers offer improved energy savings through their lower costs and unusual capabilities. An example of lower cost can be shown for hydrocarbon liquids cooled by water using estimates from Purohit [18] and Saunders [19] for six exchangers; stainless steel plate-and-frame exchangers had one fifth of the capital cost of carbon steel shell-and-tube units. The scope of energy savings through lower equipment cost is therefore very significant!

FUTURE DEVELOPMENTS FOR PROCESS PLANTS

New Applications

Compact Heat Exchangers are already successful in many industrial process applications, particularly where shell-and-tube heat exchangers are technically unsuitable, for example where there are requirements for regular dismantling for cleaning, true counterflow, expensive materials of construction, small temperature differences between streams, multistream exchangers, etc. In the drive to reduce cost, energy use and waste, improve products and increase output, compact heat exchangers are becoming more and more a viable alternative over a wide variety of standard process plant applications primarily because they can be cheaper. The requirement for plant owners and operators to make significant reductions in plant costs, including plot area, and to be able to "shoe horn" skid-mounted process equipment into a restricted area, makes the possible space and weight savings with these exchangers very attractive.

The potential benefits they offer in current and new applications are
- lower capital and installation costs, and improved safety
- reduced weight and support structure, space and footprint
- improved energy recovery networks, power savings
- improved process design and plant layout
- improved processing conditions (more throughput or yield of higher value product).

New applications which can be expected to be created are
- evaporators, boilers, thermosiphon reboilers (internal and external to columns)
- condensers, overhead condensers, with viscous and immiscible condensates under vacuum conditions.

Improved Forms

These will use
- new materials for construction, e.g. titanium alloys, high temperature plastics, ceramics, combination of metal and ceramics
- new methods of construction, e.g. diffusion bonding, laser stake welding
- new geometries of construction designed for specific plant duties.
- new methods of operation, e.g. pulsating flows, reversing flows, spasmodic flows (like in a human lung)!

Extended Capabilities

Compact heat exchange equipment is able to combine plant functions such as simultaneous mass and heat transfer. Some possible examples are: heat transfer integral with distillation, catalyst coated on the heat transfer surface for catalytic reactor, fractionation with heating and cooling. Combining functions will reduce the number of equipment components required in process plants, reduce cost, save energy, reduce the inventory of process fluids, etc. However, the most important feature is the stimulus this provides for a radical general reassessment of the opportunities for savings through the improved design of process equipment.

CONCLUDING REMARKS

In this paper, starting with the definition of a compact heat exchanger, it is shown that high heat transfer coefficients are achievable with the CHEs as well as a significant reduction in the exchanger volume and fluid pumping requirement. Characteristics and types of CHEs are then described together with design and selection methodology. There has been a significant resistance by the process industry to use CHEs because they are unconventional; some of the major barriers to the general application are outlined. CHEs can make a major impact on the energy use and/or throughput in many process applications; these energy savings capabilities are enumerated. Finally, future developments to fully utilize the potential of CHEs are briefly summarized.

In summary, we have made an attempt to outline a number of options available for process exchangers, to discuss the advancement of CHEs, and to review the lessions from other industries. It is hoped that the process engineers will become more knowledgeable to try the CHEs for applications with a minimum risk. It is anticipated that CHEs will make a major impact in the process industry in the future and revolutionize present day views of heat exchange and process flowsheet design.

REFERENCES

1. Shah, R.K., Compact heat exchanger technology and applications. In Heat Exchange Engineering, Vol. 2: Compact Heat Exchangers: Techniques for Size Reduction, eds. E.A. Foumeny and P.J. Heggs, Ellis Horwood Ltd., London, 1991, pp. 1-29.
2. Shah, R.K., Compact heat exchangers. In Handbook of Heat Transfer Applications,

Second Edition, eds. W.M. Rohsenow, J.P. Hartnett and E.N. Ganić, McGraw-Hill, New York, 1985, Chapter 4, Part III, pp. 4-174 to 4-311.

3. Bell, K.J., Applications of plate-fin heat exchangers in the process industries. In Compact Heat Exchangers, eds. R.K. Shah, A.D. Kraus, and D.E. Metzger, Hemisphere Publishing Corp., New York, 1990, pp. 591-602.

4. Patel, N., Plate heat exchangers for process evaporation and condensation. In Heat Exchange Engineering, Vol. 2: Compact Heat Exchangers: Techniques for Size Reduction, eds. E.A. Foumeny and P.J. Heggs, Ellis Horwood Ltd., London, 1991, pp. 385-397.

5. Johnston, T., Miniaturized heat exchangers for chemical processing. The Chemical Engineer, No. 431, December 1986, pp. 36-38.

6. Shah, R.K., Multidisciplinary approach to heat exchanger design. In Industrial Heat Exchangers, ed. J-M. Buchlin, von Kármán Institute, Belgium, 1992.

7. Shah, R.K. Compact heat exchanger design procedures. In Heat Exchangers: Thermal-Hydraulic Fundamentals and Design, eds. S. Kakaç, A.E. Bergles and F. Mayinger, Hemisphere Publishing Corp., Washington, DC, 1981, pp. 495-536.

8. Shah, R.K., Plate-fin and tube-fin heat exchanger design procedures. In Heat Transfer Equipment Design, eds. R.K. Shah, E.C. Subbarao and R.A. Mashelkar, Hemisphere Publishing Corp., Washington, DC, 1988, pp. 255-266.

9. Taylor, M.A., Editor, Plate-Fin Heat Exchangers: Guide to Their Specification and Use, Heat Transfer and Fluid Flow Service, Harwell, UK, Amended Edition-Oct. 1990.

10. Shah, R.K., Counterflow rotary regenerator thermal design procedures. In Heat Transfer Equipment Design, eds. R.K. Shah, E.C. Subbarao and R.A. Mashelkar, Hemisphere Publishing Corp., Washington, DC, 1988, pp. 267-296.

11. Shah, R.K. and Wanniarachchi, A.S., Plate heat exchanger design theory. In Industrial Heat Exchangers, ed. J-M. Buchlin, von Kármán Institute, Belgium, 1992.

12. Prasad, B.S.V., and Gurukul, S.M.K.A., Differential method for sizing multistream plate fin heat exchanger. Cryogenics, Vol. 27, 1987, pp. 257-262.

13. Prasad, B.S.V., and Gurukul, S.M.K.A., Differential methods for the performance prediction of multi-stream plate-fin heat exchangers. J. Heat Transfer, Vol. 114, 1992, pp. 41-49.

14. Prasad, B.S.V., The performance prediction of multistream plate-fin heat exchangers based on stacking pattern. Heat Transfer Engineering, Vol. 12, No. 4, 1991, pp. 58-70.

15. User Guide to Process Integration for the Efficient Use of Energy, Institution of Chemical Engineers, London, 1982.

16. The UK Compact Heat Exchanger Study Group: AEA Petroleum Services, Harwell Laboratory, Bldg. 392, Didcot OX11, 0RA, UK.

17. The Netherlands Advanced Heat Exchanger Users' Group: A.R. Braun, Chairman, Braun Consultants, Markt 306, Postbus 765, 7751 at Hengelo, The Netherlands.

18. Purohit, G.P., Estimating cost of shell and tube exchangers. Chem. Eng., August 1983, pp. 56-57.

19. Saunders, E.A.D., Heat Exchangers: Selection, Design and Construction, John Wiley, New York, 1989.

PROCESS INTENSIFICATION - THE UK PROGRAMMES TO ENCOURAGE THE DEVELOPMENT AND USE OF INTENSIFIED HEAT EXCHANGE EQUIPMENT AND TECHNOLOGY

A C MERCER
Energy Technology Support Unit
(on behalf of the Energy Efficiency Office of the UK Department of the Environment)
Harwell
Oxfordshire OX11 0RA

ABSTRACT

The Energy Efficiency Office of the UK Department of the Environment have ongoing programmes to encourage and stimulate the efficient use of energy. These programmes provide technical and financial support for energy efficiency R&D, for the use of new technology and for the further application of proven energy efficiency measures. Since 1989, the Energy Efficiency Office have had an increasing programme to encourage the development and use of intensified equipment and technology. The aim of this paper is to report a review of the scope for process intensification, to review the activities in the UK in key areas for intensification and to discuss activities on advanced heat exchanger technology.

INTRODUCTION

Within the UK, the Energy Efficiency Office (EEO) of the Department of the Environment have ongoing programmes to encourage the efficient use of energy within industry and commerce. The industrial programmes are managed on behalf of the EEO by ETSU at Harwell.

An important element of the current programme is financial support for research and development into a range of energy efficiency measures. Under the programme, ETSU have identified process intensification as an area offering the potential for significant national energy savings. As a result, the EEO have supported an increasing package of research and development initiatives in the area of process intensification.

This paper provides information on a selection of the EEO initiatives in this area. In particular, the background to the energy aspects of the technology is presented, along with details of the development areas, the findings to date and the impact the technology is having within UK industry.

BACKGROUND

Process intensification is a philosophy of design of chemical process plant which aims to achieve radical reductions in the physical size of the plant. Intensified process plant is where the same level of output is achieved from manufacturing equipment which is significantly smaller than the conventional. It is the established convention within the industry that process intensification relates to reductions of at least 3-4 fold in magnitude.

Why Intensification Now?

The chemicals industry utilises equipment that has become well established over many years of operation. The industry has invested heavily into R&D of new products but, in general, has not sought to develop the basic manufacturing equipment. The industry has been sufficiently profitable that it did not need to be concerned over the cost of manufacturing facilities. The objective has always been to ensure that market demands for the products were being met.

However, the profit margins on chemical products are being increasingly reduced. This is in part due to environmental pressures leading to additional capital and operating costs. It is also due to increased competition world-wide, particularly from areas such as Central Europe and the Middle East, where there are cheap supplies of primary natural gas or oil feedstocks. As a result, the industry is looking strategically at its future manufacturing, and one option being considered is the radical redesign of equipment in order to intensify the operations.

A major factor leading the industry in this direction is the success which has been achieved in other sectors of manufacturing industry through radical size reductions. Industrial sectors such as aerospace, defence and nuclear have developed fabrication technology which enables radically new, intensified equipment design concepts such as narrow channels to be realised. There is therefore both a pull from the industry looking to improve the plant economics, and push from a number of industrial sectors such as defence looking to diversify their activities.

INTENSIFIED HEAT EXCHANGER TECHNOLOGY

Within the process industries, the energy efficiency of a process is dictated by the ability of the process heat exchangers to transfer heat in a cost effective manner. Conventionally, the majority of heat exchangers in the process industries are of the shell and tube type. These have been developed for a wide range of possible applications to be functional, rather than high performance units. Typically they consist of 25-50 mm diameter tubes contained within a shell.

However, the performance of a heat exchanger in terms of heat transfer is directly related to the characteristic flow diameter. The smaller the diameter, the higher the heat transfer. Traditionally, impracticalities in manufacture and concern over actual operation have prevented developments to improve performance by reducing the diameter. Now, the use of new materials and fabrication techniques have shown that heat exchangers consisting of a large number of narrow channels can be manufactured in a highly cost effective manner. These new types of exchanger have been shown to be very competitive with conventional in terms of capital cost, installation cost and operating performance.(1)

The compactness of a heat exchanger is measured in terms of the heat transfer surface area per unit volume (m^2/m^3). In principle, a compact heat exchanger is any exchanger which is smaller than another designed for the same duty. It is generally accepted that a heat exchanger with a surface area per unit volume in excess of 700 m^2/m^3 can be classed as a compact exchanger.

Industry is justifiably cautious in implementing a radically innovative technology such as compact heat exchangers. This conservative nature of industry will inevitably mean that it will take a considerable time for the technology to become accepted. However, there are a number of companies actively considering and investigating the use of the technology. They have identified the scale of the potential benefits and are looking to reduce the risks associated with the new technology. In particular, in the UK, ICI are at the forefront of the use of compact heat exchangers. There is no doubt that their active involvement has stimulated many companies to take an interest.

Barriers to Exploitation

Within UK industry, there are a number of barriers to the exploitation of compact heat exchangers. These are discussed below:

Lack of awareness

Many companies in the process industries are simply not aware of the technology and the potential benefits from its application.

Lack of industrial experience

Where companies consider the use of the technology, they will always seek information and experiences from a similar application. In general, companies, in especially the chemicals sector, do not like to be first to use a new technology. Compact heat exchangers are novel and to date there is limited experience available. Apart from Government programmes such as the existing EEO initiatives, the only other source of information is from the suppliers, and industry will tend to sceptical of the manufacturers claims.

Lack of suitable equipment

Compact heat exchangers will only be installed where they offer capital, installation, or operating cost savings. For many potential situations, the currently available types of heat exchanger are not cost effective. For example, existing types of compact heat exchangers are unlikely to be attractive for fouling or low pressure, non corrosive duties where they cannot compete with plate heat exchanger units. To address these markets requires the development of some form of self cleaning, or non fouling unit, and a simple construction, low cost exchanger.

Lack of design information

The major companies have in house resources to carry out design and specification of plant items. They are, in general, reluctant to be dependent upon suppliers information and will seek to specify, or at least be able to confirm, the required design. For compact heat exchangers, the detailed design information is solely with the suppliers. This restricts the ability of companies to compare the use of compact exchangers with conventional technology.

Energy Saving Potential

The quantification of the national potential market and associated energy cost benefits for compact heat exchangers is a complex task. This is due to the diversity of heat transfer operations within the process industries. In principle, a compact exchanger can be used as a replacement for any existing heat exchanger, but there is limited information available within the UK on how many heat exchangers there are in industry, or on how much heat is actually transferred.

An analysis of a 'typical' process to assess the impact of high performance, low cost heat exchangers has shown that with compact exchangers, the optimum energy required was reduced by 5%. Extrapolating across the whole of the UK chemicals industry leads to energy savings of 16 PJ/year, worth £40m/year. Obviously, the technology is equally applicable across oil refining and the other process industries.

This simple assessment shows the significant potential for energy cost savings from the more widespread use of compact heat exchanger technology. However, it still does not reflect the full significance of the technology in terms of overall process intensification. The use of high performance heat exchangers can have a marked effect upon the whole process. For example, the ability to rapidly heat or cool a process stream, could have a dramatic effect on product quality. In addition, there are believed to be opportunities for combined operations of, for example, heat transfer integral with distillation, and the incorporation of catalyst inside narrow channels to produce catalytic plate reactors.

It is against this background of major potential energy benefits that the Energy Efficiency Office have initiated a programme of activities to address the technical and market barriers. The overall objective of the programme being to encourage the more widespread industrial use of the technology.

UK EEO HEAT EXCHANGER PROGRAMME

Within the UK, the framework for the current Energy Efficiency Office activities is the Best Practice programme. This is a collaborative programme with industry, commerce and the public sector to meet energy and environmental challenges in an efficient manner. Its main aims include information dissemination and support for research and development into innovative energy saving technologies.(2) The support available for selected collaborative R&D projects, involving at least two organisations, can be up to 49% of total costs. The ongoing package of projects on process intensification and advanced heat exchanger represents just one part of the overall EEO R&D programme.

Performance R&D

The aim of this area of activity has been to produce information for dissemination to industry on the performance and on the range of potential applications of compact exchangers. The availability of this information will help to offset the lack of awareness barrier to the use of the technology.

The existing activities are on the fouling characteristics of compact exchangers, on the mechanical integrity and the potential application of the technology. These are shown in Table 1.

At Heriot Watt University a facility has been built to compare the fouling propensity of two heat exchangers operating with cooling water from an open cooling tower. Tests were carried out with a Printed Circuit Heat Exchanger (PCHE) and a heat exchanger having tube dimensions representative of a typical industrial shell-and-tube unit. The results of the tests showed that a PCHE can operate satisfactorily with cooling water providing that the chemical treatment and filtration of the water is adequate. It also showed that facilities to permit backwash of the heat exchanger should be incorporated in any practical application of a PCHE with cooling water.(3)

Table 1
Performance R&D Activities

Project	Contractor	Status
PCHE fouling	Heriot Watt University	Completed
PCHE mechanical integrity	British Gas	Completed
Investigation of applications for compact exchangers	AEA Technology	Ongoing
PCHE water scaling	Heriot Watt University	Ongoing

At British Gas, a PCHE has been subjected to an extreme pressure cycle fatigue test equivalent to 100 years service life. Stress analysis showed that the test had no detrimental effect upon the unit with the high integrity of the diffusion bonding being maintained. The liquid side of the PCHE was subjected to controlled contamination tests with various levels and sizes of solid particulate contaminants (silting). With a 50 micron filter in the line, there was no increase in pressure drop over extended periods of operation, indicating no significant silting up of the PCHE passages. In these tests, the PCHE has withstood extreme conditions of operation, giving confidence for its use in normal duties and practical situations.(4)

AEA Technology, in collaboration with a number of major companies, have investigated the range of applications for compact exchangers within petrochemical processes. This work clearly showed the wide applicability of the existing types of compact heat exchangers. There were very few heat recovery duties where an advanced heat exchanger offering energy and cost benefits could not be used. Further work to detail specific applications is planned by AEA Technology.(5)

New Types of Compact Exchanger

There is a very wide range of potential duties for heat exchangers in the process industries. The currently available compact heat exchangers will only be cost effective for particular duties. EEO initiatives to date have led to the development of new types of compact exchangers. The subsequent availability of these types of exchanger will have a significant impact on the market. The programme activities are shown in Table 2.

Table 2
New Types of Exchanger R&D Activities

Project	Contractor	Status
Porous Matrix Heat Exchanger	NEL	Completed
Polymer Feasibility	Newcastle University	Completed
New types of heat exchanger	Various	Ongoing

The project at the National Engineering Laboratory (NEL) concerned the development of the porous matrix heat exchanger (PMHE). This new surface comprises a stack of thin sheets of expanded or punched metal sheet brazed or diffusion bonded between separating plates. The sheets are arranged so that adjacent fins are out of phase causing a three dimensional flow field. The experimental results have been very encouraging. There has been a direct measure of improvement over an equivalent serrated plate fin design.(6) The principal advantages are the following.

- It is more compact than existing plate fin surfaces in terms of overall volume for a given duty.
- Crossflow units could be cost effectively manufactured using punched or expanded metal plates with no edging bars.
- The surface can be constructed using diffusion bonding techniques.

The work at Newcastle University is concerned with the acquisition of heat transfer and flow friction data for a polymer film compact heat exchanger (PFCHE), when operating with a water/water system. A 100μ corrugated PEEK (poly ether ether ketone) film, was the chosen material for the heat exchanger under study, as it has an estimated continuous working temperature of about 250°C, combined with excellent chemical resistance.

The results (7) have shown that the heat transfer intensity per unit matrix volume and the power loss incurred in achieving a given heat transfer rate, are much more attractive for the PFCHE than for alternative compact units. Provided matrix fabrication and potential fouling problems can be overcome, PFCHE's could eventually play a major role in heating/ventilating, domestic condensing boilers and low temperature (<250°C) process heat exchangers.

In addition there are a number of further collaborative initiatives underway on alternative types of advanced heat exchanger. The nature of these collaborations are such that further information is not currently available. Details of the developments will be published on satisfactory completion of the various research initiatives.

Design

A barrier to the use of compact heat exchangers is a lack of available design information. Chemical engineers are taught how to design conventional shell and tube exchangers, and there is also a large number of commercially available software packages for the design of this type of exchanger. There is considerable familiarity, and hence a natural tendency, towards this type of exchanger.

To increase the use of compact heat exchangers, there is a need to increase the amount of information available to the engineer on the design and use of the technology in his process. EEO activity in this area has to date concentrated on the use of multistream heat exchangers, in a project with AEA Technology and UMIST to investigate exchanger design and integration into heat recovery networks. (8)

Industrial Experience

A major barrier to the use of advanced heat exchangers is the lack of experience of the industrial application of the technology. The Energy Efficiency Office programme aims to overcome this barrier by providing information on the use of the technology in industrial plant.

In this paper two applications of a printed circuit heat exchanger (PCHE) are briefly discussed.

The first industrial application in the UK was at Associated Octel, where the PCHE is used on a caustic soda evaporation plant. The PCHE is constructed of nickel, and transfers 220kW of heat to a caustic feed solution at 60°C from intermediate stage caustic at 200°C. The unit was installed in July 1988 and has operated well with no evidence of significant fouling. The PCHE has eliminated leakage difficulties resulting in increased heat recovery and plant utilisation.(9)

A further application is at ICI Fibres at Wilton, where four PCHE's are installed on the nitric acid plants. The PCHE's are a part of an environmental project to reduce, by almost 90%, the emission of NO_x in the tail gases from two 350 tonnes/day nitric acid plants. The largest duty involves pre-heating the acid plant tail gas to 200°C, before it enters the catalytic NO_x reactor, by condensing steam at a pressure of 18 barg. Using PCHE's installed in line with the 600mm diameter tail gas piping at an elevated level has resulted in significant benefits in terms of plant layout and installed cost. The exchangers were installed in 1989 and are performing to design specification without any operational difficulties. There has been no evidence of fouling of the narrow channels.(10)

Impact on UK Industry

The Energy Efficiency Office programme on advanced heat exchangers has been underway for 2 years. To date a total of 15 projects have been initiated.

Within the UK, there is an increasing level of industrial interest in the concepts of advanced heat exchangers. Particularly in the chemicals sector, companies are looking at the potential benefits of the technology. The EEO through its targeted programme of dissemination of information is increasing awareness of the technology. This has the effect of creating a pull for improvement from within industry, which then stimulates the manufacturers to develop appropriate technology. In due course, as developments of new types of exchanger come to completion, there will be an increased range of applications where advanced exchangers are competitive, and increasingly alternative supply options. These factors will lead to an increased market for the compact type of heat exchanger.

CONCLUSIONS

Process intensification is a highly innovative technical development which offers industry, major capital, energy, environmental and safety benefits. As a result of environmental and operating cost pressures, sectors of the process industries are actively investigating radically new initiatives such as intensification. In particular, the environmental benefits of intensified plant are stimulating considerable interest.

Intensified heat exchanger technology offers particular capital and energy cost saving benefits. There are the following major barriers to the more widespread industrial use of the technology.

- Lack of industrial awareness.
- Lack of industrial experience.
- Lack of suitable equipment.
- Lack of design information.

The Energy Efficiency Office of the UK Department of the Environment have an ongoing package of Case Studies and R&D projects on advanced heat exchangers. These projects currently cover the performance of advanced exchangers, the design and development of a range of new types of advanced heat exchanger and industrial experience in the use of the technology.

The technology of advanced heat exchangers offers significant capital and energy cost benefits. The increasing industrial use is making some types of advanced exchanger a `proven technology`. Research initiatives under the JOULE programme and the EEO initiatives are increasing the amount of information available, leading to new technical developments and new types of advanced heat exchanger for industrial use. Advanced heat exchangers will be increasingly used across a wide range of industrial sectors, and are poised to have a major impact on the energy efficiency of those sectors.

REFERENCES

1. Cross, W. T. and Ramshaw, C, Process Intensification: Laminar Flow Heat Transfer. Chem Eng Res Des, Vol 64, July 1986.

2. EEO, The Best Practice programme, a Guide to Participants.

3. EEO, An Investigation into fouling of a PCHE, Future Practice Report 13,

4. EEO, Testing of PCHE, R&D Profile 12

5. EEO, Opportunities for Compact Heat Exchangers, R&D Profile to be published.

6. EEO, Investigation of a Novel Compact Heat Exchanger Surface, R&D Profile 29.

7. EEO, Compact polymer film heat exchanger, R&D Profile to be published.

8. EEO, Plant Design using Multistream Heat Exchangers, R&D Profile 30.

9. EEO, PCHE in a Highly Corrosive Application, Good Practice Case Study 22.

10. EEO, Printed Circuit Heat Exchangers on Steam Condensing Duties, Good Practice Case Study 109.

COMPOUND ENHANCEMENT OF HEAT TRANSFER
IN A GAS-TO-GAS HEAT EXCHANGER

M.M. Ohadi, Associate Professor
S.S. Li, Doctoral Student
A.I. Ansari, Research Associate
Department of Mechanical Engineering
University of Maryland
College Park, MD   20742-3035

R.L. Whipple, Research Engineer/Scientist
Mechanical Engineering-Engineering Mechanics Department
Michigan Technological University
Houghton, MI   49931

ABSTRACT

The applicability of the electrohydrodynamic (EHD) enhancement technique when used in conjunction with finned tubes for compound heat transfer enhancement in a gas-to-gas heat exchanger was experimentally investigated. Three series of experiments were performed. The first series involved evaluation of heat transfer characteristics when the electric field was applied to the tube side of the heat exchanger only. The second and third series of experiments were concerned with excitation of the shell side only and simultaneous excitation of the tube and the shell sides, respectively. In each case, enhancements were studied as a function of the Reynolds number and the applied electric field voltage in the tube and shell sides. It is demonstrated that when used in conjunction with finned tubes, the EHD technique yields impressive enhancements over-and-above what is achievable with the already enhanced performance of these tubes.

NOMENCLATURE

A	Heat transfer surface area
b	Mobility

$D_{s,i}$	Inner diameter of the shell
$D_{t,i}$	Inner diameter of the tube
$D_{t,o}$	Outer diameter of tube
l	Effective length of the heat exchanger
LMTD	Logarithmic mean temperature difference
$\dot{m}_t$	Tube-side mass flow rate
$\dot{m}_s$	Shell-side mass flow rate
$\dot{Q}$	Heat transfer rate between cold and hot fluids
Re_t	Reynolds number based on inner diameter of the tube
Re_s	Reynolds number based on hydraulic diameter for the shell side
$T_{t,i}$	Bulk temperature of the tube-side air at the inlet
$T_{t,e}$	Bulk temperature of the tube-side air at the exit
$T_{s,i}$	Bulk temperature of the shell-side air at the inlet
$T_{s,e}$	Bulk temperature of the shell-side air at the exit
U	Overall heat transfer coefficient

Greek Letters

ρ	mass density
μ	viscosity
ω	humidity ratio

Subscripts

b	bulk, also ion mobility
i	inlet, also inner
o	outer
e	exit, also outlet
s	shell

INTRODUCTION

Compound enhancement methodology refers to simultaneously applying two or more enhancement techniques to the heat transfer surface to achieve better overall heat transfer performance. The objective of the present study was to investigate the extent by which heat transfer coefficients can be further enhanced when the electrohydrodynamic (EHD) technique is applied to a low-finned tube geometry. Among the various augmentation techniques, the EHD is a new and promising technique which has demonstrated impressive potential for heat transfer enhancement in heat exchangers involving single-phase (gas or liquid) or phase-change (boiling, evaporation, melting, and freezing processes). The technique utilizes the effect of electrically-induced secondary motions to destabilize the thermal boundary layer adjacent to the heat transfer surface, leading to substantial increase in the heat transfer coefficients. Previous studies [1 and 2] on certain refrigerants has demonstrated that significantly higher enhancements can be obtained when the EHD method is used in conjunction with low-fin tubes. However, our literature search failed to identify any previous work dealing with compound enhancement of single-phase region, forced convection of a gas for conditions investigated here.

The experiments reported here were performed in a double-pipe, air-to-air heat exchanger in which the inner tube was a commercially available tube that was enhanced both in the inner and outer surfaces. Heat transfer and pressure drop measurements were performed as a function of parametric values of Reynolds number and applied electric field potential in the tube side, shell side, and simultaneous excitation of the tube and shell sides. However, due to space limitations only heat transfer results are presented in this paper. Brief description of the apparatus and experimental procedure is given in the following with details available in [3-5].

EXPERIMENTAL APPARATUS

The schematic of the double-pipe heat exchanger configuration used in the present experiments is shown in Fig. 1. To facilitate execution of parametric studies on both the tube and shell sides, a single-tube, single-pass, shell-and-tube heat exchanger was used. The physical dimensions and electrode spacing in the tube and shell sides are depicted in Fig. 1. The inner tube was a commercially available low finned tube (Wolverine Tube, Inc., Decatur, Alabama, Catalog No. 64-1112065) with the physical configuration shown in Fig. 2. It had 12 equally spaced longitudinal fins on the inside and 4.33 circumferential fins per cm (11 fins/in.) on the outside. The outside diameter of the circumferential fins was 3.7 cm (1.46 in.). The nominal tube root diameter was 16 mm (0.75 in.) and the fin material was 3003 aluminum. The shell side of the heat exchanger had an i.d. of 12.7 cm (5.0 in.) and o.d. of 14.0 cm (5.5 in.). The inner tube and the annulus formed between the inner and outer tubes will hereafter be referred to as the tube and shell sides, respectively.

Measurement of temperatures at the inlet and exit ports of the heat exchanger was performed by type E gage 30 thermocouples. All thermocouple voltages were read and recorded to 1 μV by a 5 1/2 digit multimeter. To evaluate the extent by which the rise in heat transfer coefficients resulted in higher pressure drops, both heat transfer and pressure drop measurements were performed during the course of the experiments. Pressure taps were installed at the inlet and exit of both the tube and shell sides. Application of the high-voltage electric field to the tube side was through a 0.25 mm (0.010 in.) stainless steel (AISI 304) wire which was placed concentric with the tube as shown in Fig. 1. Supply of high-voltage power to the electrodes was via a high- voltage (0-30 kV), low-current (0-15 mA) lab grade DC power supply.

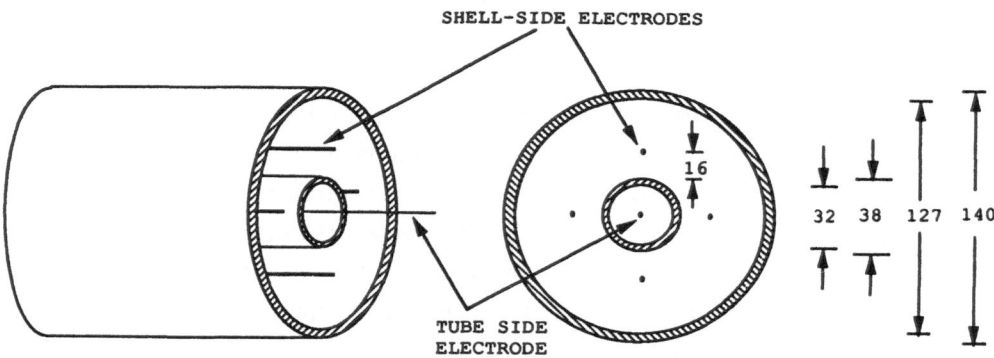

All dimensions in mm

Figure 1. Electrode Configuration in Tube and Shell Sides

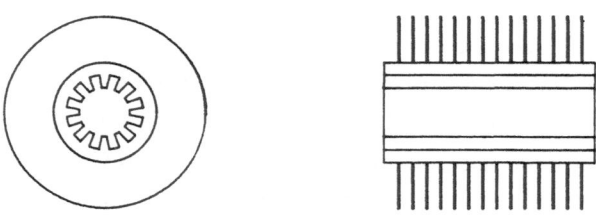

Figure 2. Schematic of the Finned Tube Geometry used in the Experiments

The electrodes on the shell side of the heat exchanger were four stainless steel wire which were oriented in 90 degree intervals as shown in Fig. 1. These electrodes were connected in parallel to a second high-voltage source. Each wire electrode had a diameter of 0.8 mm (0.031 in.) and a length of 127 cm (50 in.).

Room air was used as the working fluid for the shell side of the heat exchanger. Heating of the air for the hot side of the heat exchanger (the tube side) was provided through a series of circumferential resistive heating elements which were attached to a ceramic base plate. Care was taken to ensure that uniform heating of air in the tube was accomplished and that proper insulation for the tube wall was in place. The heater assembly was powered with a regulated low-voltage DC power supply. Mixing of the air before it entered the heat exchanger was accomplished with a mixing box downstream of the heater. The mixing box also housed the electrode tensioning mechanism for the tube side. Additional details of the experimental apparatus can be found in [3 and 4].

DATA REDUCTION

The average overall heat transfer coefficient U for the heat exchanger was determined by introducing the experimental data into the defining equation

$$U = \frac{\dot{Q}}{[(A)(LMTD)]} \tag{1}$$

where $\dot{Q}$ is the net rate of heat transfer between the hot and cold fluids and A is the heat transfer surface area based on the tube side outer surface given by

$$(2)$$

$$A = \pi D_{t,o} l$$

in which l is the effective length of the heat exchanger (the length over which heat transfer takes place) and $D_{t,o}$ is the tube outer diameter. The quantity LMTD is the log mean temperature difference for the heat exchanger defined as:

$$LMTD = \frac{(T_{t,i} - T_{s,e}) - (T_{t,e} - T_{s,i})}{\ln\left[\frac{(T_{t,i} - T_{s,e})}{(T_{t,e} - T_{s,i})}\right]} \tag{3}$$

in which $T_{t,i}$ and $T_{t,e}$ represent tube-side fluid bulk temperatures at the inlet and exit of the heat exchanger. The corresponding temperatures for the shell side of the heat exchanger are $T_{s,i}$ and $T_{s,e}$, respectively.

The LMTD was determined by measuring the fluid bulk temperatures at the inlet and exit of the heat exchanger for the hot and cold sides. At each position, the average fluid bulk temperature was determined by taking the average of measurements at several positions across the tube cross section. This practice was necessary to account for the cross stream temperature variations which are particularly important in the case of laminar flow. The rate of heat transfer $\dot{Q}$ was obtained by writing an energy balance equation for the tube side, details of which can be found in [3].

The Reynolds number used in parameterization of the results was evaluated for the tube side from

$$Re_t = \frac{4\dot{m}_t}{\mu_t \pi D_{t,i}} \qquad (4)$$

Similarly, the shell-side Reynolds number was evaluated from

$$Re_s = \frac{4\dot{m}_s}{\mu_s \pi (D_{s,i} - D_{t,o})} \qquad (5)$$

The quantities $\dot{m}_t$ and $\dot{m}_s$ in Eqs. (4) and (5) are the measured air mass flow rates in the tube and shell sides of the heat exchanger, respectively. The viscosities μ_t and μ_s were evaluated for air at the tube- and shell-side average fluid bulk temperatures, respectively. The quantities $D_{t,i}$ and $D_{t,o}$ refer to the inner and outer diameters of the tube side, while $D_{s,i}$ represents the inner diameter of the shell side.

RESULTS AND DISCUSSION

Figure 3 shows variation of the EHD heat transfer enhancements at various Reynolds numbers when electrical field was applied only on the tube side of the heat exchanger. The corresponding corona current is plotted in Fig. 4. The best fitted lines in Figs. 3 and 4 are based on a linear regression analysis of the experimental data.

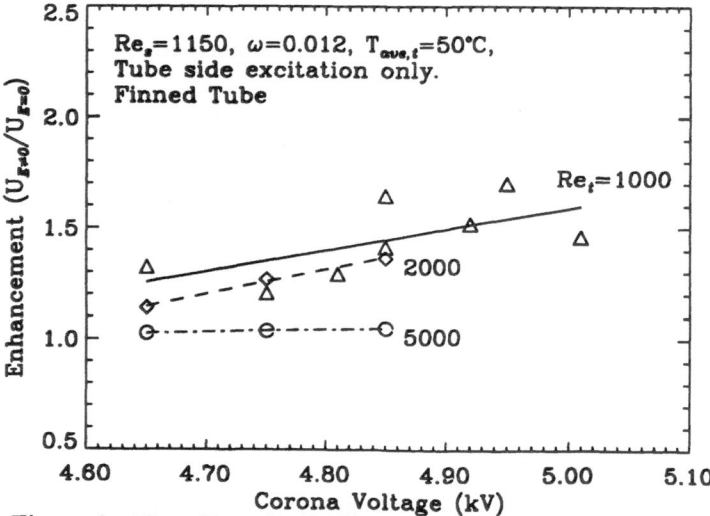

Figure 3. Heat Transfer Enhancement vs. Corona Voltage

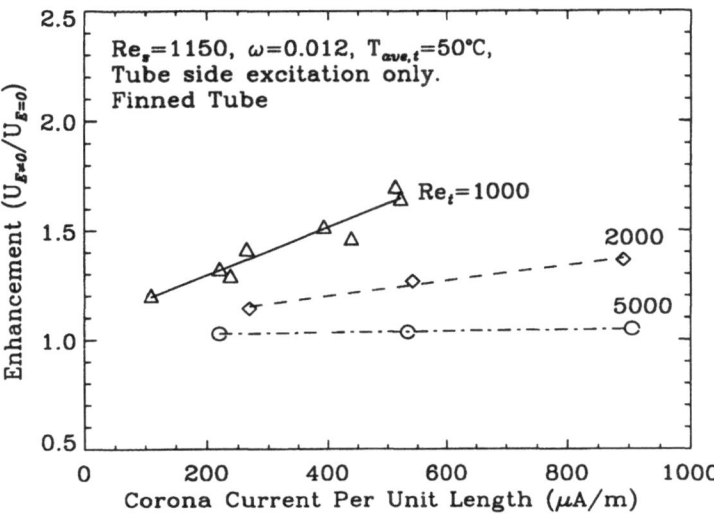

Figure 4. Heat Transfer Enchancement vs. Corona Current Per Unit Length

The $Re_t = 1000$ test was repeated several times and extra care was taken to minimize fluctuations in the experimental data points. Such fluctuations are typical at low Reynolds numbers due to the sensitivity of the flow to small disturbances in the laboratory conditions. Note that for $Re_t = 1000$ the data appears more scattered when plotted against corona voltage (Fig. 3) than that of the corona discharge current (Fig. 4). This difference is due to the higher sensitivity of the high voltage to non uniformities in the electrode or the test section configurations. In contrast, the corona current is a more robust variable and is not influenced significantly by minor changes in the test section conditions. In fact, it is the corona current that gives rise to the electrically-induced secondary motions which in turn result in enhancement of heat transfer rates [4].

The $Re_t = 1000$ was the lowest Reynolds number tested in the present experiments. As expected, it exhibits highest enhancement in the overall heat transfer coefficient. The EHD enhancement reduces to a negligible amount at $Re_t = 5000$. Because at high Reynolds number the turbulence momentum eddy diffusivity effects overwhelm those of the corona discharge. The $Re_t = 2000$ exhibits a characteristic intermediate to the 1000 and 5000 Reynolds numbers.

To investigate if the data for different mean flow velocity and corona current can be combined into a single variable, we attempted to plot the heat transfer data reported above against various combinations of independent parameters. It was found the $x^2 = ((j/b)/(\varrho.u_t^2))$ correlated well with the heat transfer enhancement data (Fig. 5).

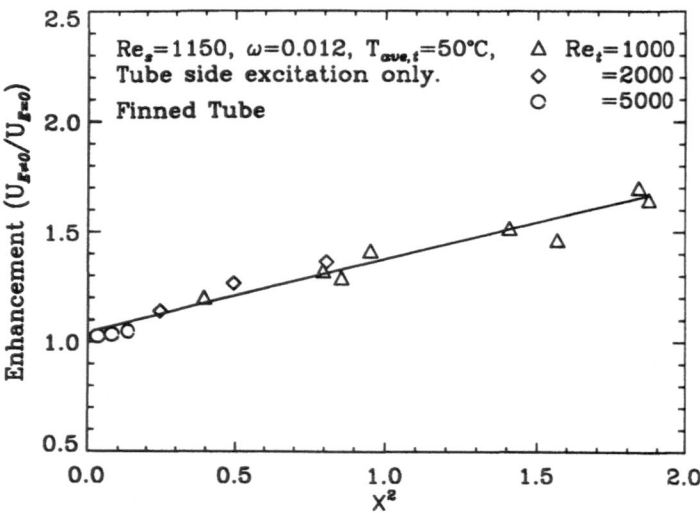

Figure 5. Heat Transfer Enhancement vx. X^2. (The U comes from LMTD method)

Figure 6 shows the EHD heat transfer enhancement at several tube side Reynolds numbers when the electric field was applied only on the shell side. Notice that in this case, for all tests, the shell side Reynolds number was kept constant at 1150. The data in this plot suggest that an increase in either the shell side corona current or the tube side Reynolds number results in further increase in the heat transfer rates. The reason for this behavior can be easily explained on the basis of convective thermal resistances on the inside and outside surfaces of the tube. An increase in the tube side Reynolds number results in a higher convective heat transfer coefficient and hence a corresponding increase in the overall heat transfer coefficient. Similarly, an increase in the electric field intensity on the shell side causes a stronger disruption to the boundary layer on the outside surface of the tube resulting in decreased thermal resistance and hence an improved overall heat transfer coefficient.

Figure 7 shows the heat transfer enhancement at Re_t = 2000 for three cases: (a) electric field applied on the tube side only, (b) electric field applied on the shell side only, and (c) simultaneous application of electric field on the tube and shell sides. In all cases four electrodes were used on the shell side, and a single concentric electrode was used in the tube side. It is seen that simultaneous tube and shell side excitation boosts the enhancement in heat transfer coefficient by as much as 80% in contrast to a negligible amount when only the tube side corona was utilized.

Although not shown here the pressure drop data indicated that in all cases the increase in heat transfer coefficient was consistently higher than the corresponding increase in the pressure drop coefficients.

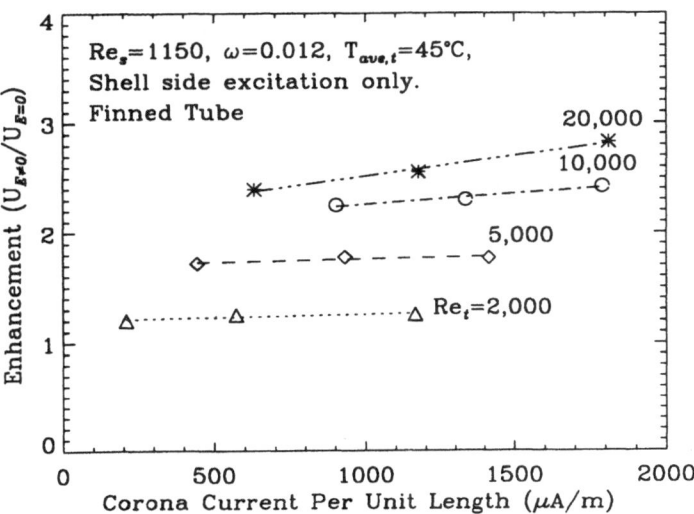

Figure 6. Heat Transfer Enhancement vs. Corona Current per Unit Length

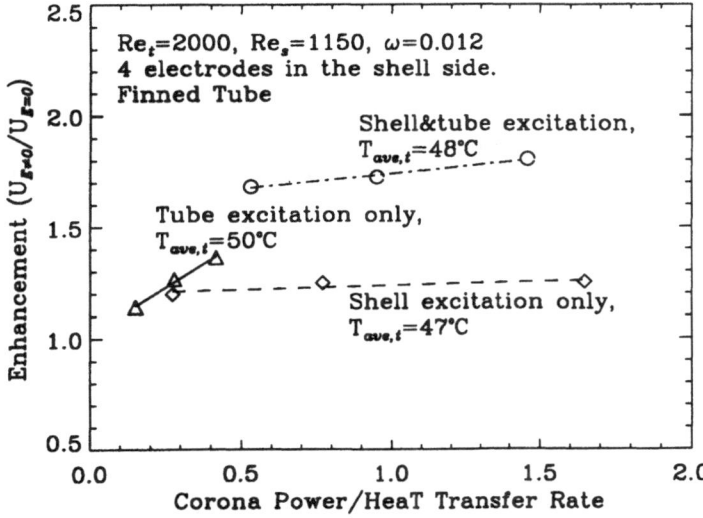

Figure 7. Heat Transfer Enhancement vs. Ratio of Corona Power to Total Heat Transfer Rate, $RE_t = 2000$

CONCLUSIONS

The applicability of the electrohydrodynamic (EHD) technique for compound air-side heat transfer augmentation in a gas-to-gas heat exchanger was studied experimentally. It was demonstrated that when used in conjunction with finned tubes the EHD technique can significantly add to the magnitude of the enhancements already achievable with these tubes. Up to 80% additional enhancements were found in the present experiments. It was found that enhancements are highest at the lowest Reynolds numbers and at the highest applied electric field voltage. The simultaneous excitation of the tube and shell sides led to maximum enhancement in the overall heat transfer coefficient. The results further suggested that for a given shell-side Reynolds number an optimum Reynolds number in the tube side can be identified under which maximum enhancements can be achieved. The overall findings of this study suggest that the compound EHD-enhanced technique can be particularly useful for air-side enhancement in situations where a low temperature difference between the two working fluids in the heat exchanger prevails.

ACKNOWLEDGEMENTS

The funding of this project by the Gas Research Institute under contract GRI contract number 5091-260-2185 is gratefully acknowledged. The finned tubes used in the present experiments were donated by Wolverine Tubes, Inc (Decatur, Alabama). The authors wish to thank Petur Thors and James Bogart (both of Wolverine Tubes, Inc.) for their efforts and assistance in this respect.

REFERENCES

1. Allen, P.H.G. and Cooper, P., "The potential of electrically enhanced evaporators." Proceedings, 3rd Int. Symp. on the Large Scale Application of Heat Pumps, pp. 221-229, 1987.

2. Damiannidis, Karayiannis, T.G., Al-Dadah, R.K., James, R.W., Collins, M.W., and Allen, Ph.G., "EHD Boiling Enhancement in Shell-and-Tube Evaporators and its Application in Refrigeration Plants," ASHRAE Transactions, Vol. 98, part 2, 1992.

3. Ohadi, M.M., Electrostatic Enhancement of Heat Transfer in Heat Exchangers," Final Report, GRI Contract No. 408-260-1528, GRI-91/0202, Gas Research Institute, Chicago, IL, 1991.

4. Ohadi, M.M. and Ansari, A.I., Electrostatic Enhancement of Heat Transfer in a Gas-to-Gas Heat Exchanger, GRI Final Report, GRI-92/0325, Gas research Institute, Chicago, IL, 1992.

5. Ohadi, M.M., Nelson, D.A., and Zia, S., Heat Transfer enhancement of laminar and turbulent pipe flows via corona discharge" Int. J. of Heat and Mass Transfer, Vol. 34, No. 4-5, pp. 1174-1187, 1991.

ENHANCED EVAPORATION HEAT TRANSFER SURFACES

MANFRED GROLL & STEFAN RÖSLER
Institut für Kernenergetik & Energiesysteme (IKE), Universität Stuttgart
Pfaffenwaldring 31 , W - 7000 Stuttgart 80

CHRISTOPHE MARVILLET
GRETh-Centre d'Etudes Nuléaires de Grenoble (CENG), F - 38041 Grenoble

JOHN E. HESSELGREAVES
National Engineering Laboratory (NEL), East Kilbride, GB - Glasgow G75 OQU

KEITH CORNWELL & PETER A. KEW
Energy Technology Unit (ETU), Heriot-Watt University, GB - Edinburgh EH14 4AS

ABSTRACT

Within the frame of the "JOULE" r & d programme of the Commission of the European Communities a project is carried out to develop and investigate novel enhanced evaporation heat transfer surfaces with the aim to employ them in improved industrial two-phase heat exchangers. Main surfaces which are investigated are structured and covered planar and tubular surfaces and planar surfaces with narrow channels. Typical results obtained so far are presented which demonstrate the excellent evaporation heat transfer characteristics of these surfaces.

INTRODUCTION

The development of improved, highly efficient heat exchangers has been found to be an important task in order to reduce the consumption of primary energy in industrial processes. In the past few years, also the aspect of pollution abatement has pushed forward the research and development work in this field. Especially heat transfer processes with phase change (mainly boiling and condensation) offer a great potential for application of different enhancement techniques.

In general the possible techniques can be subdivided into "active" and "passive" measures. The active techniques mainly comprise a movement of the heat exchange surface or the working fluid, e.g. a rotating or vibrating surface. Passive techniques are almost exclusively devoted to changing the properties of the heat exchange surface. This changing of surface properties is termed surface enhancement and can again be subdivided into surface treatment, surface structuring, surface extension. Enhanced surfaces have been comprehensively reviewed [1-3]. They comprise mainly coatings of porous material and mechanically machined or formed cavities to ensure continuous vapour trapping. These surfaces provide for continuous renewal of vapour at the nucleation sites. However, their high production costs are an obstacle to their widespread use. Cheaper alternatives are therefore searched for.

Comprehensive reviews of pool boiling are available [4,5] which indicate that good evaporators should have widespread nucleation sites (surface cavities of the order 5 to 10 microns), along with a geometry which enables the continuous wetting of the surface. As the thermal conductivity of a liquid is greater than that of a vapour, heat transfer coefficients from evaporators are increased the longer a liquid stays in contact with the surface.

Within the frame of the "JOULE" programme of the Commission of the European Communities a multi-national r & d project has been set up to develop and investigate novel enhanced evaporation heat transfer surfaces.

The project comprises the following partners and their respective tasks. Institut für Kernenergetik & Energiesysteme (IKE), Universität Stuttgart (M. Groll, S. Rösler) is project coordinator and investigating planar and tubular structured surfaces. The Groupe de Recherche sur les Echangeurs Thermiques (GRETh) of the Centre d'Etudes Nucléaires de Grenoble (CENG) (Ch. Marvillet) focusses on planar covered surfaces, whereas the National Engineering Laboratory (NEL), Glasgow (J.E. Hesselgreaves) is concentrating on tubular covered surfaces. The Engineering Technology Unit (ETU) of Heriot-Watt University, Edinburgh (K. Cornwell, P.A. Kew) is investigating planar heat exchanger surfaces with narrow vertical grooves. Both the Centre for Renewable Energy Sources (CRES) in collaboration with the Agricultural University of Athens (S. Yanniotis) and the University of Newcastle upon Tyne, Department of Chemical and Process Engineering (C. Ramshaw) are engaged in the study of rotating evaporation and condensation surfaces. The Department of Chemical Engineering of the National Technical University of Athens (N.C. Markatos, A.N. Karayannis) employs numerical methods to calculate the evaporation (condensation) thin film characteristics on stationary and rotating surfaces. Three industrial partners, TREFIMETAUX (M. Messant), CIAT (A. Bailly), both of France, and CAL GAVIN (J.V. Rogers) of England act as advisors to the project with respect to the evaluation of the industrial feasibility of the enhanced surfaces, and their technical and economical advantages over existing enhanced surfaces.

In the present paper only the passive enhancement techniques are covered and typical experimental results obtained so far are presented.

ENHANCED EVAPORATION HEAT TRANSFER SURFACES

Three different types of enhanced surfaces are investigated, viz. structured surfaces, covered surfaces, vertical narrow channels.

Structured Surfaces

The planar structured surface specimen (Fig. 1) comprise a circular base plate (30 mm diameter) with grooves 1.5 mm deep and 1.5 mm or 3 mm wide. They are covered by a perforated copper sheet (0.7 mm thick) which is diffusion-welded to the base plate. The perforations are conical, their diameters (in mm) are varied, viz. $d_1/d_2 = 0.6/0.45$, $0.4/0.3$, $0.25/0.1$; the number of holes is $50/cm^2$. Water and freon R113 are used as working fluids for the pool boiling experiments. The saturation temperatures are varied, for water: 80 °C, 100 °C, 120 °C. Tests with vertical and horizontal orientation (heated from below) are carried out.

The tubular structured surface specimen (Fig. 2) are copper tubes (length 130 mm, outer diameter 25 mm, wall thickness 2 mm) into which axial or helical grooves are machined. In the case of axial grooves, there are 24 grooves of rectangular cross section, 1.1 mm wide and 1 mm deep. A copper wire is helically wound around the tube and fixed by spot-welding. The wire diameter is varied between 0.4 and 1.2 mm. Again water and freon R 113 are used as working fluids for the pool boiling experiments. Tests are carried out with horizontal tubes only.

Covered Surfaces

The term covered surfaces stands for a surface from which no "free" evaporation is possible. Instead an adiabatic sheet or foil is placed in front of the tested surface thus generating a confined space. From this gap the generated vapour bubbles have to be ejected into the surrounding pool (see Fig. 3).

The planar covered surface specimen (Fig. 3) are square blocks made from copper-aluminium (30 mm x 30 mm). Separated by a gap (widths 5, 4, 3, 2, 1 mm) an adiabatic plate is installed in front of the heated plate. There are two different gap situations: in one case the periphery of the confined space is open all around, in the other case it is closed at the sides. The experiments are carried out for horizontal, 45° inclined and vertical orientation; in the first two cases heating is from below. Water and R113 are employed as working fluids at about 100 °C and about 50 °C, respectively.

The tubular covered surface specimen (Fig. 4) are brass tubes (102 mm long, 25.4 mm diameter) which are surrounded by different sleeves. Four different sleeves were tested, only one of which was perforated (0.8 mm holes on a 3 mm square pitch). This sleeve was wrapped around the surface of the tube with as small an annular gap as possible. The remaining sleeves had annular gaps of 1, 0.4 and 0.2 mm. These sleeves had a row of 4 mm diameter holes, on an 8 mm pitch, along the bottom to allow the entry of saturated liquid from the bath, and a row of 8 mm holes on a 12 mm pitch along the top to allow exit of the vapour and superheated liquid. The perforated sleeve was made of stainless steel, 0.25 mm thick, and the unperforated sleeves were made of 0.7 mm thick copper. R 113 was used for the first set of tests and water for the second set. The surface of the plain tube was roughened using fine emery, to check if this affected the performance of the sleeves. The experiments were carried out on horizontal tubes.

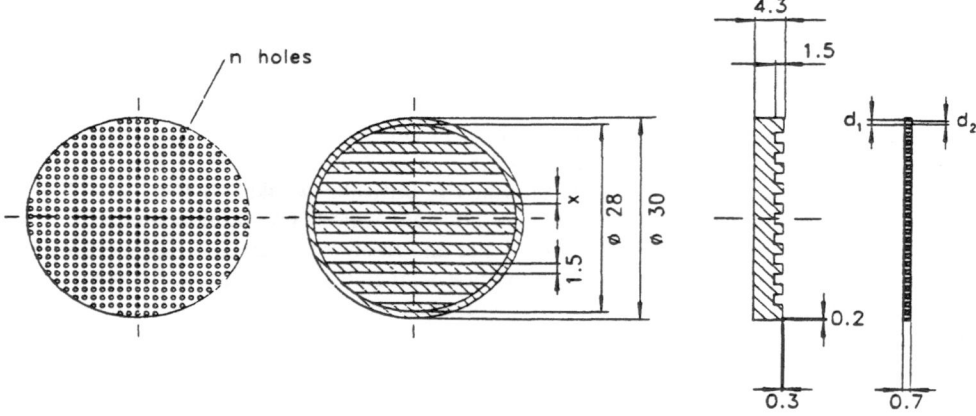

FIGURE 1: Planar test specimen with structured surface.

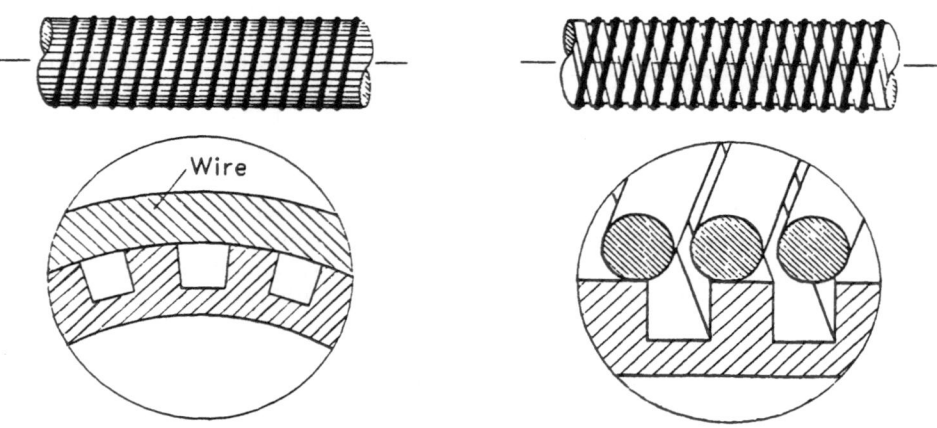

FIGURE 2: Tubular structured surface specimen. a) Axial grooves; b) circumferential grooves.

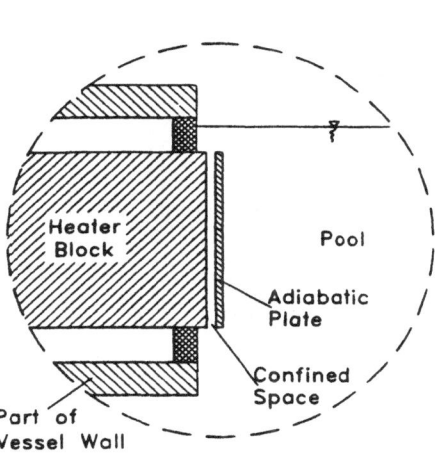

FIGURE 3: Schematic of test set-up for planar test specimen with covered surfaces.

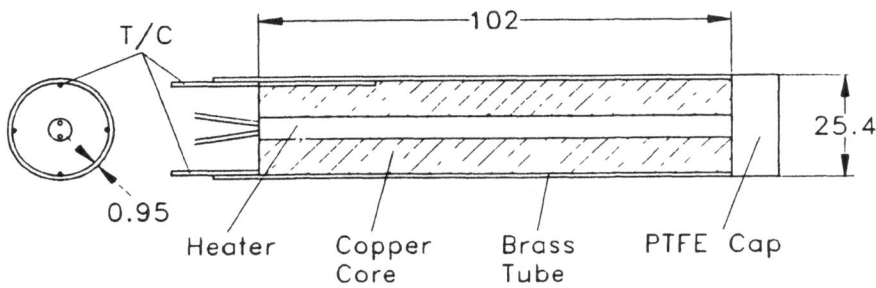

FIGURE 4: Tubular test specimen with covered surface.

Narrow Vertical Channels

The experimental work comprises investigations with a flow visualization test rig and a Printed Circuit Heat Exchanger (PCHE) test rig.

The test section of the flow visualization rig (Fig. 5) consists of an aluminium plate (180 mm long, 420 mm wide, 300 mm heated width) with 75 rectangular channels, each 1.4 mm wide by 1 mm deep. In particular images of the flow are obtained using flash photography and high-speed video camera. R 113 is used as the working fluid employing different flow rates up to about 2 l/min (including stagnant flow). The test section is always operated in a vertical orientation.

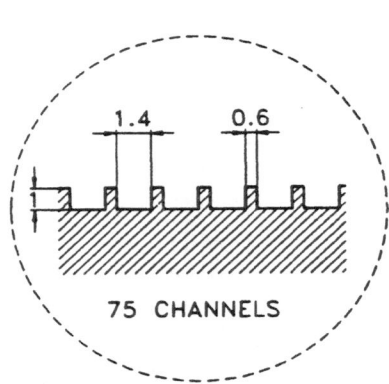

FIGURE 5: Channel geometry of flow visualization test section.

The <u>PCHE rig</u> employs a commercial PCHE (150 mm wide, 300 mm long, 25 mm thick), made of stainless steel, which is operated in the vertical orientation. The PCHE is heated by water in 5 cross-flow sections. The heat transfer fluid on the secondary side is R113, flowing vertically in small channels through the PCHE. The R113 mass flow rate can be measured up to 1.8 l/min. Typical water temperatures are around 40 to 50°C.

RESULTS AND DISCUSSION

Structured Surfaces

Experiments have been carried so far with <u>planar surfaces</u> using water as the working fluid. Typical results are shown in Figs. 6 a) and b), for vertical orientation of the test specimen. The test conditions are characterised by three characters. The first character stands for the channel width: O for 3 mm, II for 1.5 mm. The second character stands for the smaller diameter d_2 of the conical perforations: A for 0.45 mm, B for 0.3 mm, C for 0.1 mm. The third character stands for the orientation of the conical perforations: 1 for the wider openings (d_1) facing the pool, 2 for the narrower openings (d_2) facing the pool.

All enhanced surfaces have a significantly superior performance compared to the plain reference surface. This holds for the initiation of nucleate boiling at low heat fluxes, where only very small wall superheats of 1 to 2 K are required (OA1, OB1), and also for high heat fluxes. Derived heat transfer coefficients up to 20 to 30 kW/m²K are obtained for wall superheats between about 1 and 15 K and corresponding heat fluxes between about 30 to 500 kW/m² (OA1, OB1). Similar results are obtained for horizontal orientation.

The perforation diameter and the orientation of the conical perforations have a strong influence on heat transfer performance. The groove width is less important. In all cases with vertical surfaces OC1 and OC2 perform poorest. This may be due to the small perforation openings of the cover plates (0.25 mm and 0.1 mm, respectively), which can impede both

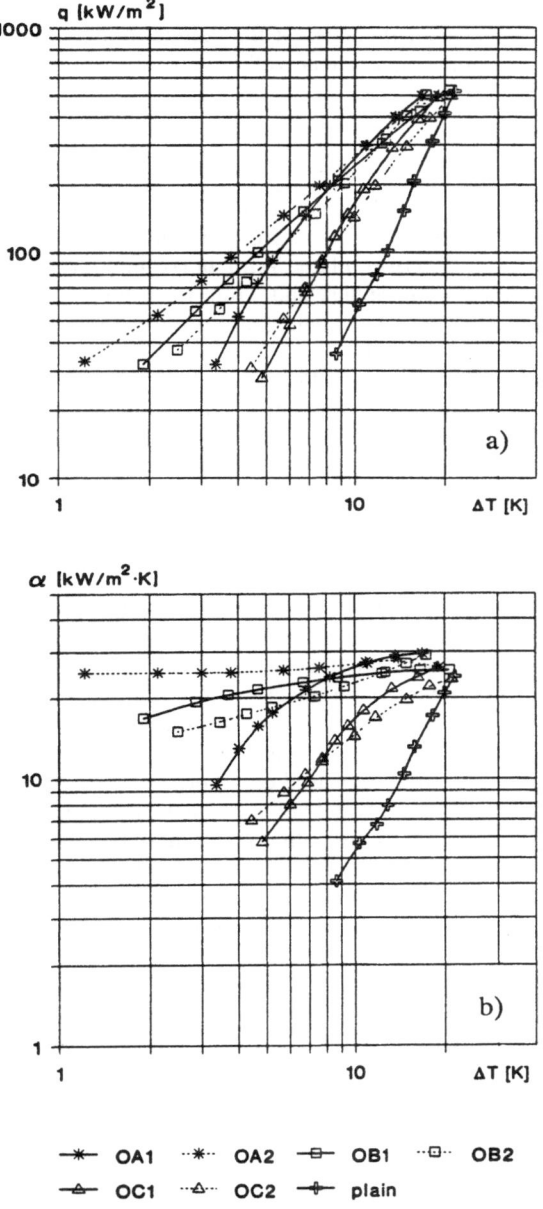

FIGURE 6: Planar structured surface specimen; vertical orientation; working fluid water at 100°C. a) Heat flux vs. wall superheat; b) heat transfer coefficient vs. wall superheat.

the liquid flow into and the vapour flow out of the re-entrant channels. For perforation orientation 1 the optimum outer diameter of the conical holes is between 0.4 and 0.6 mm, for orientation 2 the optimum diameter is well beyond 0.45 mm.

Covered Surfaces

Experiments have been carried out with planar covered surfaces in different orientations using water as the working fluid. Typical results are shown in Fig. 7 for sidewise closed peripheries. For all gap widths the heat transfer performance is superior to that of the plain (unconfined) surface without cover plate. With horizontal orientation the minimum gap widths were 2 mm for all-open periphery and 3 mm for sidewise closed periphery. For these gap widths dry heated areas have been detected during the experiments, associated with an unstable vapour and liquid flow in the confined space. Simultaneously, very unstable temperature measurements were noted. For vertical and 45° inclined orientation the (heat flux vs. wall superheat) - curves are very similar. The smallest gap width (1 mm) results in by far the lowest wall superheat (less than 2 K) for initiation of nucleate boiling (at around 20 kW/m^2). For higher heat fluxes the curves converge, and there are about the same wall superheats of 12 to 15 K at about 250 kW/m^2 for the 1 mm and 2 mm gaps (45°) and for the 1, 2, 3 and 4 mm gaps (vertical).

Visualization showed bubbly flow in the confined space for all gap widths down to 2 mm. For open periphery the vapour leaves the confined space by the top edge and sometimes by the top and side edges. For sidewise closed periphery the vapour leaves the confined space by the top edge and sometimes by the top and bottom edges. For 1 mm gap width a two-phase flow similar to slug flow is seen in the confined space; vapour leaves the confined space by all open edges in an unstable flow regime.

For the tubular covered surfaces experiments have been carried out using R113 and water as the working fluids. The experiments clearly show that when a plain cover sleeve, with apertures at the top and bottom, is used on a horizontal tube in evaporation, substantial increases in heat flux result, for a given wall superheat. In fact as the radial gap is reduced, the performance improves to a limiting value, which may change for different fluids. Since coalesced bubble boiling in the form described by Ishibashi and Nishikawa [6] was not observed it can be said that the optimum radial gap is a function of surface tension as well as latent heat, liquid/vapour density ratio, and other thermophysical properties. The optimum gap is probably of the order 0.2 mm for R113 and 0.4 mm for water but may also be dependent on tube diameter.

Best results are shown in Fig. 8 for R113. The 0.2 mm annulus system has a significantly higher performance than an uncovered tube (with the exception of very high heat fluxes), and compares well with available enhanced tubes. The wall superheats for low heat fluxes (~ 0.6 kW/m^2) are very low (~ 0.2 K), whereas at higher fluxes (~ 10 to 30 kW/m^2) the needed wall superheats increase beyond that of other enhanced tubes (~ 3 to 10 K). On the basis of a wall superheat of 4 K the obtained heat fluxes have been compared. For the 0.2 mm gap an enhancement of 1800% over the bare tube is observed for R113, the corresponding enhancement for water being 900%. The enhancing effect of the sleeve is unaffected by the surface finish which is in agreement with the observations of Ishibashi and Nishikawa [6].

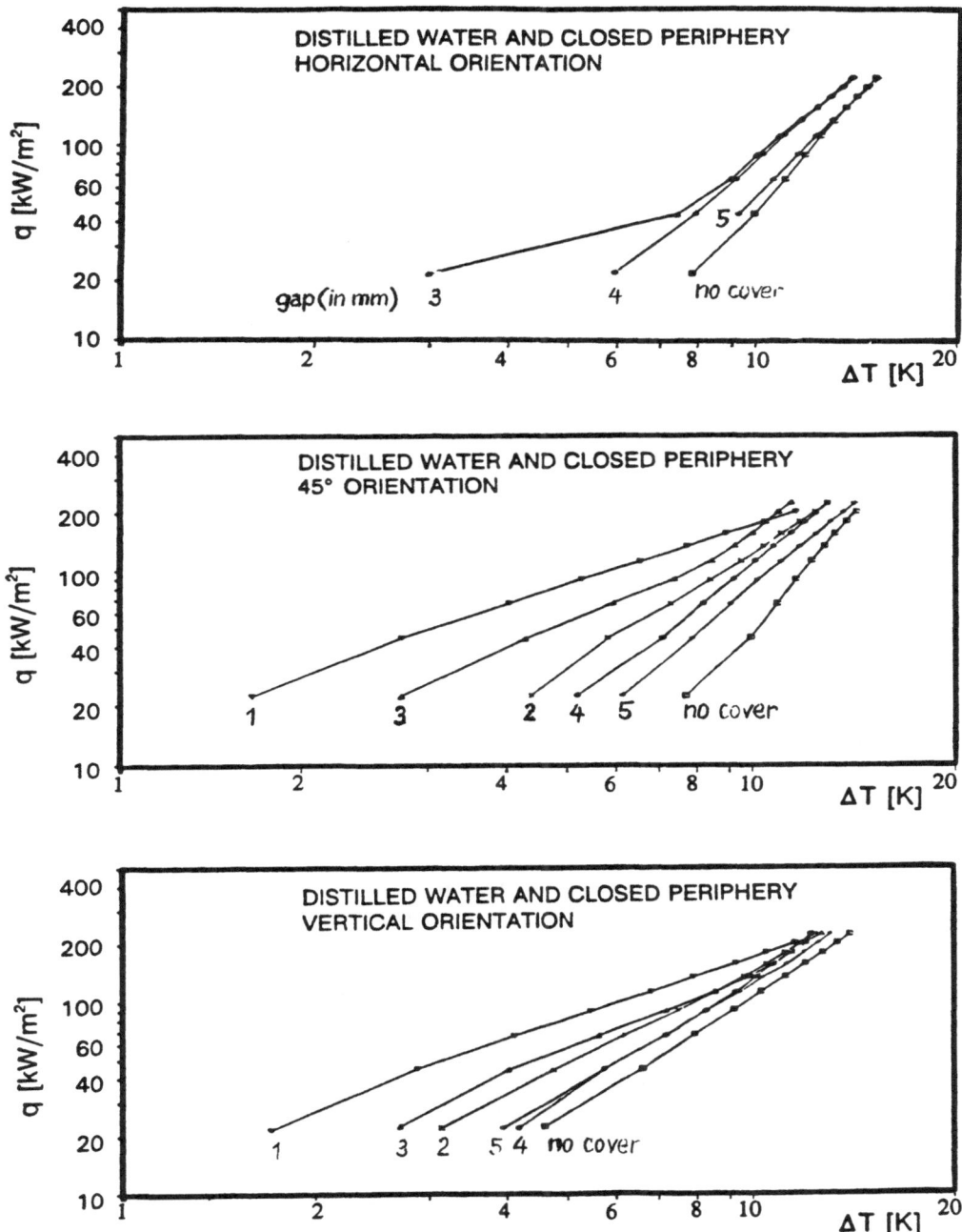

FIGURE 7: Heat flux vs. wall superheat for planar covered surfaces; working fluid water.

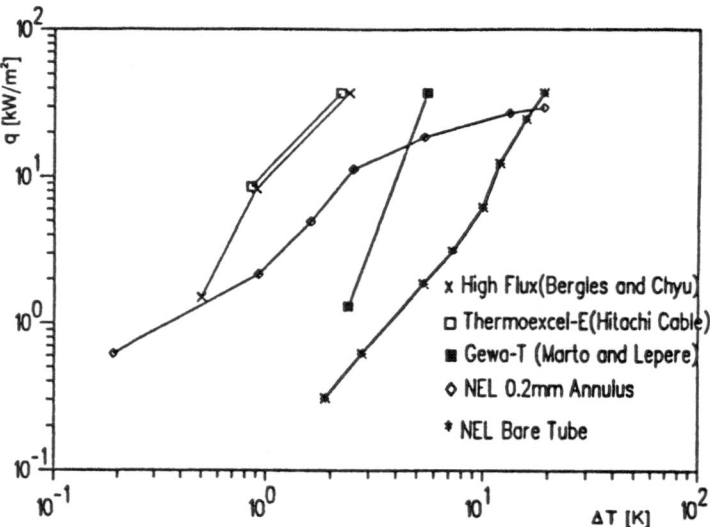

FIGURE 8: Heat flux vs. wall superheat for tubular specimen; working fluid R113. Comparison with other enhanced tubes.

Narrow Vertical Channels

Experiments with the visualization test rig showed that there are at least two flow regimes of interest in the study of evaporation in narrow channels. The flow regime at low flow rates is of unstable nature. Sections of the channels are alternately occupied almost entirely by liquid followed by the generation or movement of vapour bubbles into these sections which then occupy most of the channel width. Some nucleation has been observed in the predominantly liquid regions between the large vapour bubbles, but it appears that the majority of vapour generation (and hence heat transfer) occurs when the vapour region rapidly expands. During this phase a liquid layer remains on the wall. This flow regime is referred to as pulsed channel boiling. At higher rates the flow is steady and the transition from bubbly to slug or churn flow can be observed. The heat transfer in this regime is likely to be due to both nucleate boiling and convection. The results to date, carried out over a limited range and with a single geometry, suggest that the transition from pulsed channel boiling to conventional channel flow may be associated with the transition from laminar to turbulent flow in the liquid region.

The performance of the PCHE has been measured over a range of flow rates and water inlet temperatures. The majority of the tests falls in the region of pulsed channel boiling. Results of a series of tests with R113 are shown in Fig. 9. The boiling heat transfer coefficients appear to be independent of the dryness fraction for exit vapour qualities between about 0.2 and 0.8.

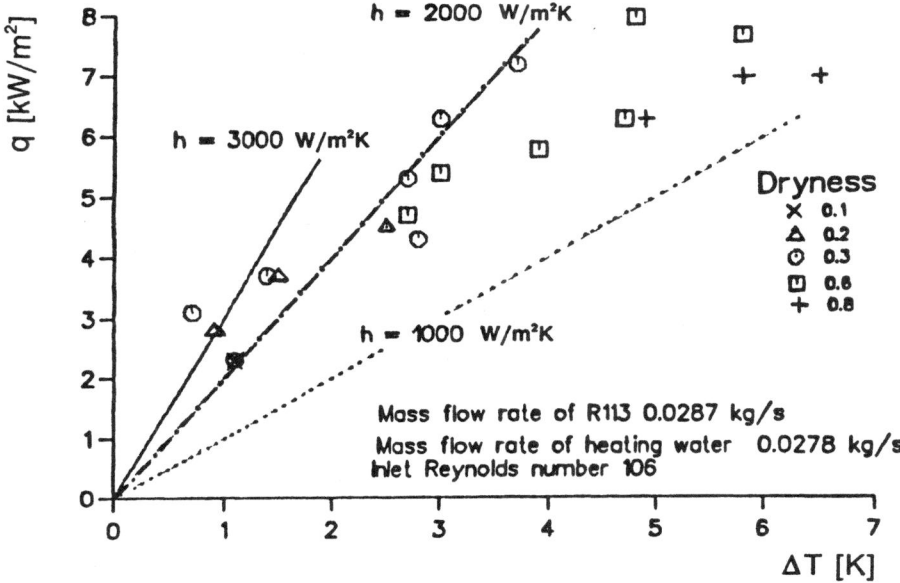

FIGURE 9: Heat flux vs. wall superheat for Printed Circuit Heat Exchanger.

SUMMARY AND CONCLUSIONS

Various enhanced planar and tubular evaporation heat transfer surfaces have been developed and tested with water and R113 as the heat transfer fluids. The experiments carried out so far reveal their excellent heat transfer characteristics which are comparable to or even better than those of commercially available enhanced surfaces.

The most likely short-term application of these surfaces is in the refrigeration industry (for example for flooded evaporators), where high performance surfaces are already applied. Here, the two problems that need to be addressed are those of production costing and the effects of oil contamination on performance. Cost is an extremely important criterion for commercial viability, and the various types of surface need to be examined by industry to establish whether they can compete with existing surfaces. The presence of lubricating oil has two potential effects: it can either enhance or reduce the surface performance (Thome [5]) and there is a possibility that an oil layer could contaminate (foul) the pores progressively, leading to gradual deterioration of the performance.

With the progressive replacement of CFC's by HFC's in refrigeration equipment, reduction of refrigerant inventory is becoming of increasing importance (Hesselgreaves [7]), and flooded evaporators equipped with tube bundles may give way rapidly to more compact types. It is likely, however, that the present research in confined channels will become increasingly applicable to the design of these compact types, as the mechanisms of enhanced boiling deviate from those on the outside of tubes.

For both refrigeration and the potential use of the new surface types in petrochemical and related industries, a greater understanding of the boiling mechanisms is necessary, as it is clear from the R113 and water studies that optimum geometry selection is dependent on a large number of fluid properties. Studies of these mechanisms should be undertaken in a systematic way as an inherent part of further surface development.

REFERENCES

1. Webb, R.L.: The evolution of enhanced surface geometries for nucleate boiling. HT Engng, Vol 2, Nos 3-4, pp 46-49, 1981.

2. Webb, R.L.: Nucleate boiling on porous coated surfaces. HT Engng, Vol 4, No 3-4, pp 71-81, 1983.

3. Pais, C., Webb, R.L.: Literature survey of pool boiling on enhanced surfaces. ASHRAE Semi-Annual Meeting, New York, January 1990.

4. Collier, J.G.: Enhanced Boiling and Condensation. McGraw-Hill, London, 1972.

5. Thome, J.R.: Enhanced Boiling Heat Transfer. Hemisphere, London, 1990.

6. Ishibashi, E., Nishikawa, K.: Saturated pool boiling in narrow spaces. Int. J. Heat Mass Transfer, Vol 112, pp 863-394, 1969.

7. Hesselgreaves, J.E.: The impact of compact heat exchangers on refrigeration technology and CFC replacement. Purdue/ASHRAE CFC Conference, July 1990.

A ROTATING DISC HEAT EXCHANGER

S. Yanniotis[*], D. Kolokotsa[**] and P. Valachis[**]

[*] Agricultural University of Athens
Iera Odos 75, 118 55 Athens, Greece.
[**] Centre for Renewable Energy Sources
Frati 6, Koropi, Athens, Greece.

ABSTRACT

An experimental unit to study the boiling heat transfer characteristics of spinning discs has been built and operated. The heat flux in a spinning aluminum disc was measured as a function of the rotational speed and the liquid flow rate. The first experimental results show that the heat flux increases substantially with the rotational speed, while the flow rate does not play a significant role.

INTRODUCTION

There are several passive and dynamic techniques to improve the performance of heat exchangers or to enable reductions in the size and the weight of heat exchangers of a given performance. The use of centrifugal fields for enhancement of heat transfer is one of the most promising active techniques.

Centrifugal heat exchangers have the following advantages: high heat transfer coefficients due to the fact that the film velocity is high and the film thickness very small; reduced fluid residence time on the heating zone; reduced equipment volume due to the high heat transfer coefficient and the high temperature difference that can be used even in heat sensitive materials; reduced fluid pressure drop; ability to handle viscous products; reduced exergy loss due to the small temperature difference that can be realised.

Devices in which liquid films are created due to the centrifugal force on a rotating disc or other surfaces of revolution, are technically suitable for numerous applications. Many unit operations in Chemical Engineering ,involving heat and mass transfer are susceptible to exploitation of centrifugal fields like distillation, absorption, extraction, boiling, condensation, etc.

Convection heat transfer of power low fluids in rotating systems is important for the thermal design of industrial equipment dealing with molten plastics, polymeric liquids, food stuffs, or slurries.

In vehicular applications where both compactness and low weight of heat exchangers are needed, rotation is one way to achieve heat transfer enhancement without suffering power losses in the ducts or weight and volume increase.

In space applications where a falling film cannot be achieved in a zero g environment, the artificial gravity created by the centrifugal force on a rotating disc to generate a thin film can be used.

The study of falling films has been performed by many researchers in the past due to the high heat transfer rates that can be achieved. Evaporation from thin liquid films on spinning surfaces has been studied to a limited extend, but very few reports are in the literature which deal with evaporation from treated spinning surfaces. Among other researchers Sparrow and Gregg [11], Bromley [2], Butuzof and Riffert [3], Bell [1], Kaplon et. al.[6], Khan [7], Mudawwar and El-Masri [8] can be mentioned.

Certain applications have also been studied. Hickman [5], and Rees [10] described and gave operating data on a centrifugal boiler still in which liquid to be evaporated flows in a thin film outward along the inside of a rapidly rotating cone. Gray [4] studied the performance of a wickless, rotating heat pipe with applications on cooling of motor rotors, cooling of jet engine turbine rotor blades and air conditioning with a compact unit having one moving part. Tleimat [12] studied the desalination of sea water in a rotating flat-disc wiped film evaporator. Ramshaw and Winnington [9] studied a compact rotary absorption heat pump with 10 kW duty. Finally the Alpha-Laval "ROTOTHERM" evaporator is in commercial use mainly in applications dealing with heat sensitive products.

To study the heat transfer characteristics of spinning discs made from enhanced heat transfer surfaces, an experimental project has been initiated within the frame of JOULE I program of CEC, where passive and dynamic techniques for heat transfer enhancement are studied. The experimental investigation of heat transfer on a spinning disc is carried out by CRES. Professor Markatos of the National Technical University of Athens is investigating the theoretical analysis of the heat transfer and fluid dynamics of a film on a spinning disc with the PHENIX software. In this article the experimental unit, the instrumentation and the first experimental results are presented.

EXPERIMENTAL WORK

Description of the apparatus
The experimental installation shown in Figure 1 has been constructed and operated. The main unit consists of an aluminum disc, 30 cm in diameter and 9 mm thickness, rotating on a central vertical shaft. The feed is introduced in the centre of the disc, through the rotating hollow shaft which has 6 holes 1 cm above the point where the disc is attached. An inverted cup attached on the shaft forms the liquid distributor. The liquid passes through a small gap (0.5 mm), which is formed between the disc and the cup and flows across the rotating disc as a thin film. Steam is introduced on the other side of the disc through the other end of the hollow shaft. The condensate is collected and removed by two stationary take-off tubes.

A DC electric motor with variable speed is used to rotate

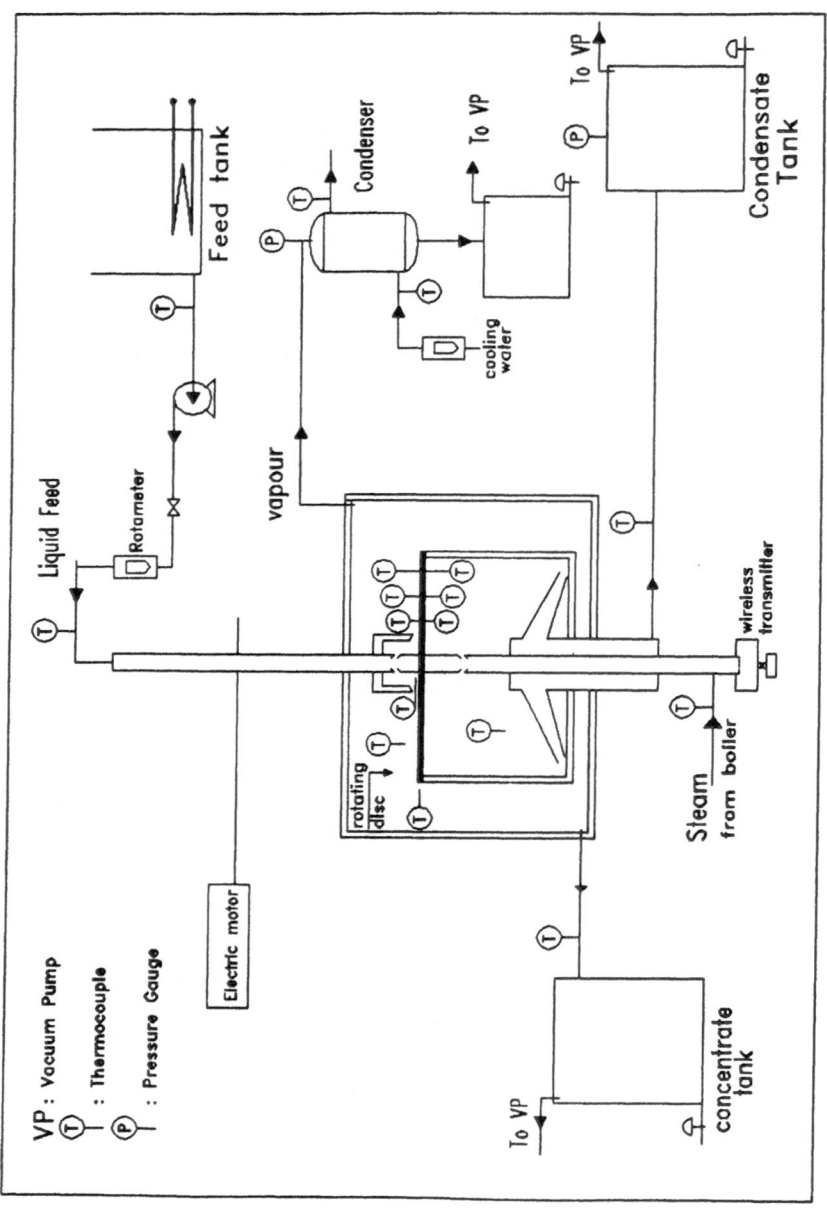

Figure 1.Experimental Set up

the disc. Mechanical seals are placed between the rotating and stationary parts to permit operation of the system under vacuum. Two liquid ring vacuum pumps remove the non-condensables from the system.

Steam and vapour condensate holding tanks, a feed tank and a concentrate holding tank are connected to the main unit. The vapour generated from the evaporation of the feed is condensed in a shell and tube heat exchanger. The experimental set-up is constructed from stainless steel. The main unit has certain parts made of plexiglass to allow visual observation on both sides of the disc.

The data acquisition system consists of two peripheral stations connected to a PC for central control and storage of the data. Station 1 equipped with a wireless optical transmitter is attached at the bottom end of the rotating shaft and is used for the measurement of the local disc surface temperatures on both sides of the disc using 12 thermocouples (type K) embedded just beneath the surface of the disc. Other thermocouples measure the steam and condensate temperature underneath the disc. The accuracy of the thermocoules and transmitter is $\pm 0.1^{\circ}$C. Station 2, which is stationary, is used for the measurement of the inlet and outlet temperatures of the fluids (with an accuracy of $\pm 0.5^{\circ}$C), the flow rate of the liquid and the rotational speed of the disc(± 10 rpm).

The thermocouples for the disc surface temperature measurement are placed in 6, 10 and 14 cm from the centre so that the middle point temperature in three zones of the disc of equal area are measured (the total useful heat transfer area of the disc is 0.0596 m^2). To place the thermocouples in the top side of the disc (liquid film side) a fine circumferential groove was opened for each thermocouple on the surface of the disc (1 cm long by 1 mm wide and 2 mm deep). At the end of each groove a hole 1 mm in diameter was opened through which the thermocouple wires with their insulation were emerged from underneath. The bare wire end was then bent and forced into the groove. The grooves were closed by pressing the aluminum and the surface was smoothed in a lathe so that the tip of each thermocouple was 0.5 mm from the surface.

Experimental procedure

Deionized tap water was used as evaporant. It was preheated with an electric heater to the evaporation temperature ($45\pm5^{\circ}$C) before feeding it to the disc. The pressure of the system and thus the boiling temperature, was adjusted by varying the cooling water temperature to the condenser. The amount of steam used and water evaporated were determined from the steam and vapour condensate that were collected in the holding tanks. The system was operated until steady state was reached (all the temperatures were stabilised). From that point on the data collected were utilised in the analysis. Each experiment lasted about 20 min in steady state operation.

Analysis of data

The heat flux was calculated knowing the disc thickness, the thermal conductivity of the aluminum (140 W/m$^{\circ}$C), as given by the supplier of the aluminum and the average temperature difference between the top and the bottom surface of the disc. To check the accuracy of the measurements the heat flux was also calculated from: a) the amount of steam condensate and b) the enthalpy balance on the condenser. In most of the experiments the agreement was within 10-20 %.

RESULTS AND DISCUSSION

The heat flux vs. the wall superheat for 200 and 600 rpm is shown in Figure 2 and Figure 3 for 1, 2 and 5 l/min flow rate. It can be observed that the effect of flow rate, if any, is smaller than the experimental error and can not be detected.

In Figure 4, the effect of rotational speed on heat flux is shown for a constant flow rate. It is clear that the heat flux increases with rotational speed. On the same figure results are included which were obtained with the disc stationary (0 rpm).

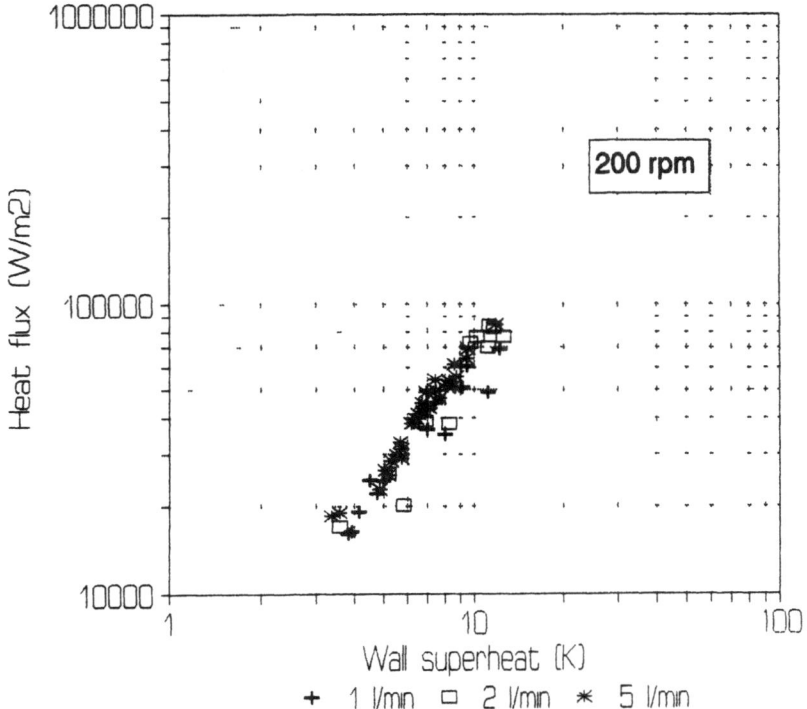

Figure 2. Heat flux vs. temperature difference between the surface of the disc and the vapour space temperature at 200 rpm and various fow rates

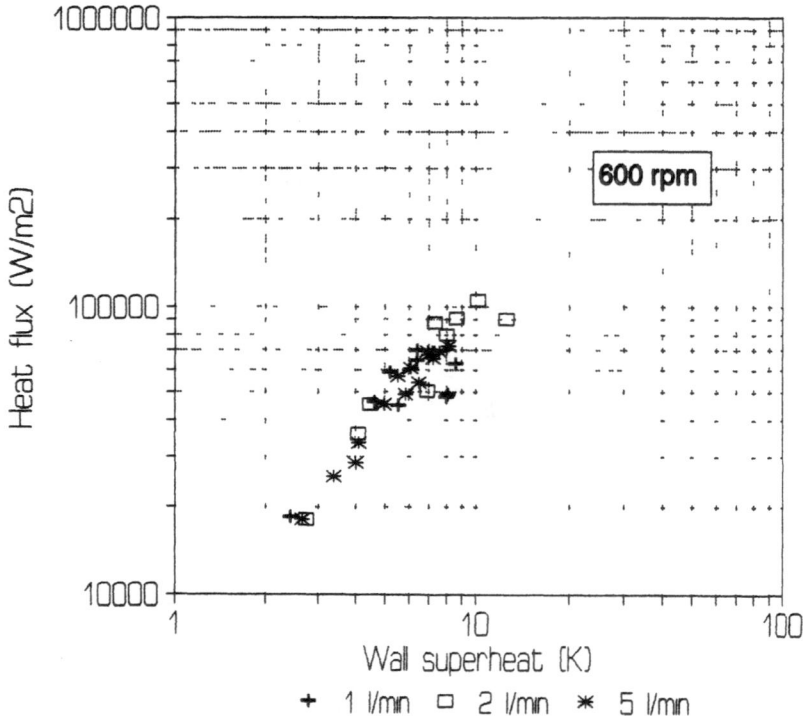

Figure 3. Heat flux vs. temperature difference between the surface of the disc and the vapour space temperature at 600 rpm and varius flow rates

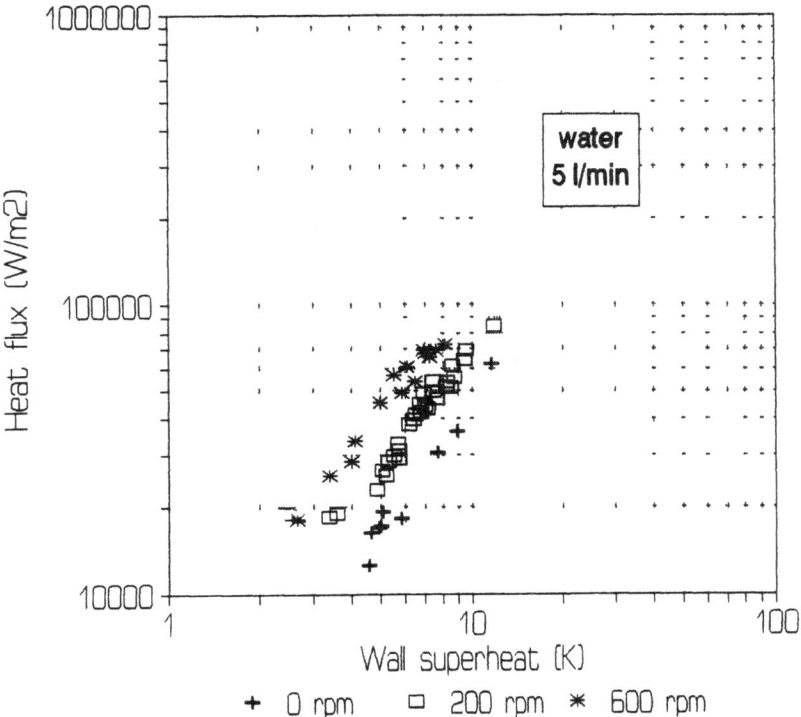

Figure 4. Heat flux vs. temperature difference between the
surface of the disc and the vapour space temperature at
5 l/min flow rate and various rotational speeds

Results in the literature suggest that the heat flux is a function of the rotational speed to 2/3 power. The same trend is observed in our data.

CONCLUSIONS

Centrifugal fields can be used in various unit operations in the food and chemical industry as well as in other applications. The experimental results support the idea that a substantial increase in the heat flux is attained with rotation. More work is needed to see the effect of the rotating surface treatment on the heat transfer.

REFERENCES

1. Bell, C., The hydrodynamic and heat transfer characteristics of liquid films on a rotating disc. Ph.D. Thesis, University of Newcastle upon Tyne, 1975.

2. Bromley, L.A., Humphrey, R.F. and Murray, G.W., Condensation on and Evaporation from Radially Grooved Rotating Discs. Journal of Heat Transfer, TRANS. ASME, 1966, 80-86.

3. Butuzof, A.I., and Riffert, V.F., An experimental study of heat transfer during condensation of steam at a rotating disc. Heat Transfer-Soviet Research, 1972, 4(6), 150-153.

4. Gray, V.H. : The Rotating Heat Pipe- A Wickless,Hollow Shaft for Transferring High Heat Fluxes. ASME paper 69-HT-19.

5. Hickman, K.C.D., Centrifugal Boiler Compression Still. Ind. Eng. Chem.,1957, 49, 786-789.

6. Kaplon, J., Kawala, Z. and Skoczylas, A. Evaporation Rate of a Liquid from the Surface of a Rotating Disc in High Vacuum. Chemical Engineering Science,1986, 41, 519-522.

7. Khan, J.R., "Heat Transfer on a Rotating Surface with and without Phase Change", Ph.D. Thesis, Univ. of Newcastle Upon Tyne, 1986.

8. Mudawwar and M. A. El-Masri, Boiling Incipience in Plane Rotating Water Films. Journal of Heat Transfer, TRANS. ASME, 1988, 110, 532-535.

9. Ramshaw, C. and Winnington, T.L., An Intensified Absorption Heat Pump. Proc. Inst. Refrigeration, November 1989, London.

10. Rees, G.J., Centrifugal Molecular Distillation-1. Fluid Dynamics, Heat Transfer and Surface Evaporation. Chemical Engineering Science,1980, 35, 837-840.

11. Sparrow, E.M. and Gregg, J.L., Heat Transfer from a Rotating Disc to Fluids of any Prandtl number. _Journal of Heat Transfer , Trans. ASME_ 1959, **81**, 249-251.

12. Tleimat, B.W., Performance of a Rotating Flat-Disc Wiped-Film Evaporator. ASME paper, 71-HT-37.

BOILING IN SMALL PARALLEL CHANNELS

Keith Cornwell and Peter A Kew

Department of Mechanical Engineering,
Heriot-Watt University, Edinburgh, EH14 4AS

ABSTRACT

The need for intensification of process heat exchangers has led to the development of several types of compact heat exchangers suitable for evaporation heat transfer. These heat exchangers are characterised by small multi-channel passages operating in parallel. The various regimes of two-phase flow which occur within these passages determine the heat transfer and hence the heat exchanger effectiveness. Currently there are limited data available in the literature of use in the design and selection of compact heat exchangers for evaporation.

New experimental work involving visualisation of the flow and measurement of the heat transfer suggests division into three types of flow boiling; isolated bubble, confined bubble and annular-slug. The mechanism of heat transfer in each regime is examined and guidance is given on the grouping of parameters for correlations which describe the heat transfer in a form suitable for design purposes.

NOTATION

d_e	hydraulic diameter m	C	Constant
F	Chen F-Factor	Bo	Boiling number
g	acceleration due to gravity, m/s²	Co	Confinement number
G	mass flux, kg/m²s	Nu	Nusselt number
h_{fg}	latent heat, kJ/kg	Pr	Prandtl number
q	heat flux, kW/m²	Re	Reynolds number
x	dryness fraction	Subscripts	
X_{tt}	Lockhart-Martinelli parameter	f	liquid
α	heat transfer coefficient, kW/m²K	g	vapour
σ	surface tension , N/m	ℓo	liquid only
μ	dynamic viscosity, kg/ms		
ρ	density, kg/m³		

INTRODUCTION

The work described here has emanated in part from collaborative research on a CEC JOULE Contract. Other partners included IKE, Stuttgart University; GRETh, Grenoble; National Engineering Laboratory, Glasgow; CRES Athens and NTUA, Athens.

The development of compact heat exchangers for evaporation is an essential part of process intensification. They allow a reduction in the quantity of the fluid in the evaporator with a subsequent increase in safety and, in the case of refrigeration systems, an increase in environmental acceptability. Their reduced physical size leads to less material being used in their manufacture and in the construction of associated equipment and buildings. Energy savings result from the reduction in heat losses and the greater heat exchanger effectiveness obtainable.

Compact heat exchangers may be of plate or plate-fin construction or laminated as in Printed Circuit Heat Exchangers (PCHE). Gap widths or channels are typically in the size range 0.5 to 2mm and lead to laminar flow in single phase applications. In the PCHE, channels are chemically etched into layers of metal rather like printed circuits which are then diffusion bonded together to form a heat exchanger matrix with large surface area/volume ratios. When used as evaporators they may exhibit flow instabilities at low flow rates due to bubble nucleation in the small channels and pressure fluctuations between the parallel channels. At higher flow rates these instabilities may subside, but compact heat exchangers will always tend to be more susceptible to pressure effects than conventional equipment.

The type of two-phase flow occurring in compact evaporators is equally as important as in larger equipment. It is necessary for the designer to predict the regime of flow occurring in each part of the evaporator and to use a correlation appropriate for that regime. In compact equipment the correlations are complicated by the strong dependence on channel size for some flow regimes. Currently the data for small channel evaporation under flow boiling conditions (with low mass flow rate) are so limited that general correlations of use to the designer do not exist.

References [1] to [10] report experimental work which has influenced this study. Unfortunately much of the data relate to experiments involving natural circulation of the boiling liquid in which no attempt was made to estimate mass flow rate; a parameter essential for heat exchanger design. Exceptions are [1,3]. Research using closely spaced plates or annuli is generally assumed applicable by using a mean hydraulic diameter, but some doubts arise owing to the increased confinement of bubbles growing in a channel over those between infinite plates. The concepts involved in the correlations developed for particular geometries by Bankoff and Rehm [1], Bar-Cohen and Schweitzer [2] and Lazarek and Black [3] were influential in our development of the dimensionless equations. the work of Nishikawa and Fujita [4] provides the only systematic study of channel size effects on heat transfer.

The purpose of this work is twofold. Firstly it reports experimental work on the flow regimes exhibited by flow in a multi-channel heat exchanger in which Refrigerant 113

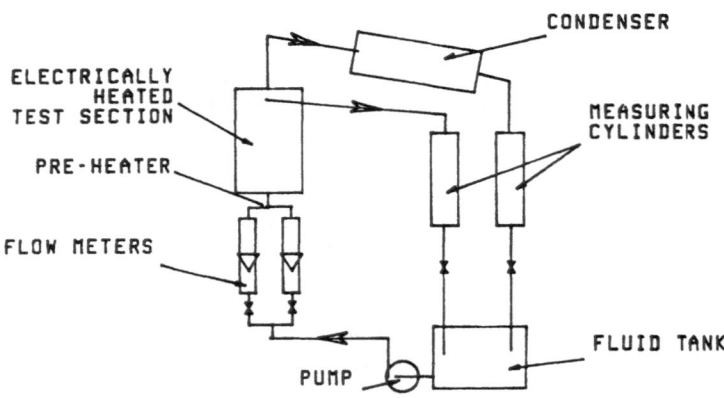

Figure 1. Schematic of Test Facility.

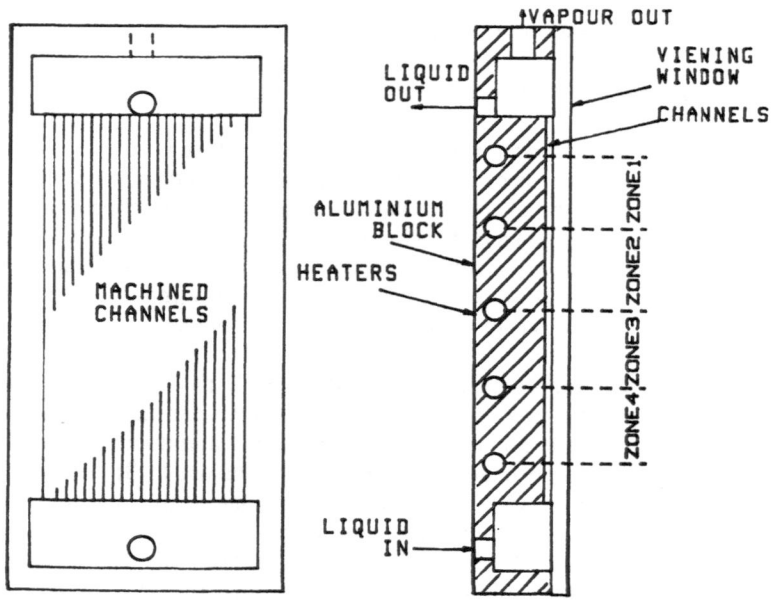

Figure 2. Schematic of Test Section.

(which well simulates organic process fluids) is evaporated. The mass flow rate and the mean heat flux are controlled and mean heat transfer coefficient is recorded together with the flow regime by use of photographic and high speed video techniques. Secondly some guidance is given on the dimensionless parameters appropriate for each regime and the type of correlation which will be of most use to the designer. A first attempt is made to estimate the constants in these equations by conceptual reasoning and comparison with the limited data which are available.

EXPERIMENTAL WORK

A rig has been constructed primarily for use in flow visualisation studies but including instrumentation for the measurement of heat transfer coefficients and pressure drops in test sections representative of a single plate within a compact evaporator. The test rig and test section are shown schematically in Figures 1 and 2. The test section comprised an aluminium block with inlet and outlet chambers connected by flow channels machined into the block, five cartridge heaters, each having a rating of 1kW and a 6mm thick glass plate clamped securely over the front surface. The channels were 320mm long with 300mm of the length being heated and the width of the flow area of the test section was 150mm.

A finite difference analysis of the temperature distribution within the block was carried out. The analysis showed that the temperature would, assuming constant heat transfer coefficient between the surface and the fluid and constant fluid temperature, be uniform along the length of the test section at a depth of 24mm below the channels to within ± .5K and, at the surface constant to within ± .005K. This justified the assumption of one-dimensional heat conduction used in calculating the surface temperature from a knowledge of the heat flux and measurement of the temperature of the block 21mm behind the surface.

The heat input to the cartridge heaters was controlled by a variable auto-transformer (Variac) and measured using a digital wattmeter. In order to account for heat losses from the block a relationship between block temperature and heat loss was determined: with the rig drained of fluid and the control valves closed to prevent circulation of air through the test section, the heaters were switched on at low power. The power input to the heaters, the temperature at which the block was in equilibrium with the surroundings and the ambient temperature were measured. Thus the rate of heat loss was correlated against the difference between block and ambient temperatures. The pressure drop across the test section was measured using pressure transducers connected to the inlet and outlet manifolds of the test section.

Tests have been carried out with two surface geometries:

> Geometry 1 - 75 channels 1.2mm wide x 0.9mm deep
> Geometry 2 - 36 channels 3.25mm wide x 1.1mm deep.

The working fluid was R113. The results of heat transfer tests are shown in Figures 3 and 4. During the tests the flow regimes were observed visually and recorded using flash photography and high-speed video. Figures 5(a)-(c) are samples showing the flow regimes described in the following section.

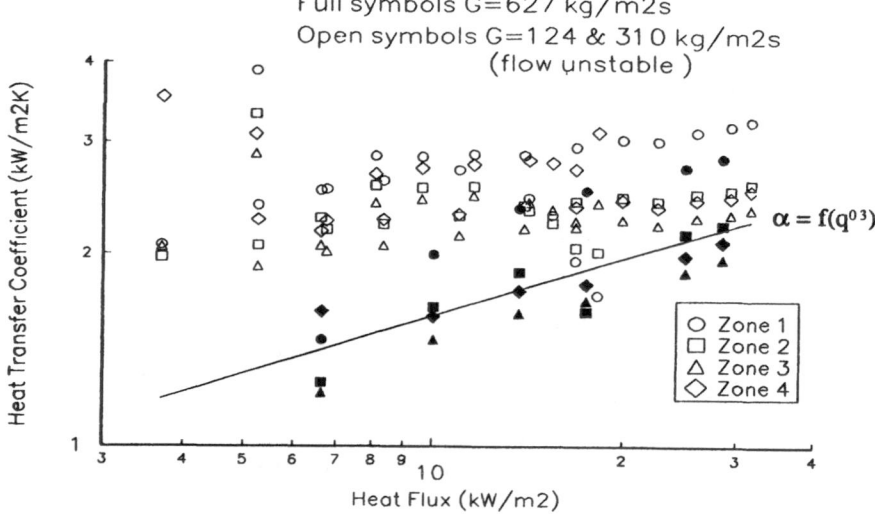

Figure 3. Results of Heat Transfer Tests - Geometry 1.

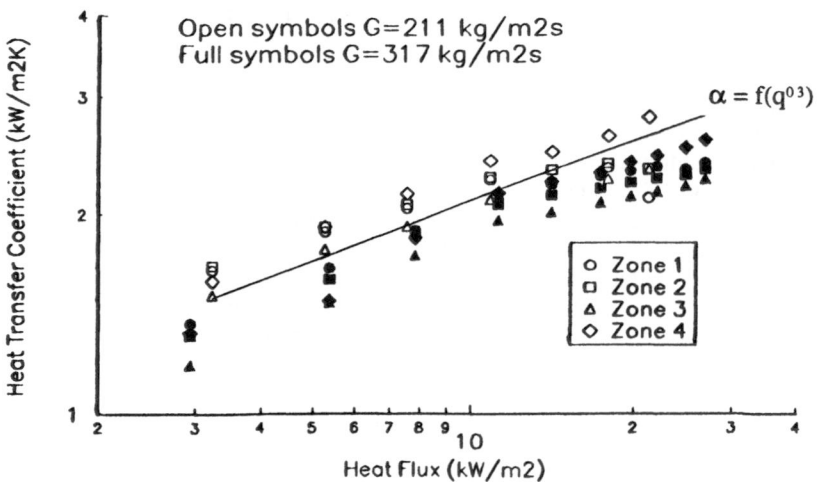

Figure 4. Results of Heat Transfer Tests - Geometry 2.

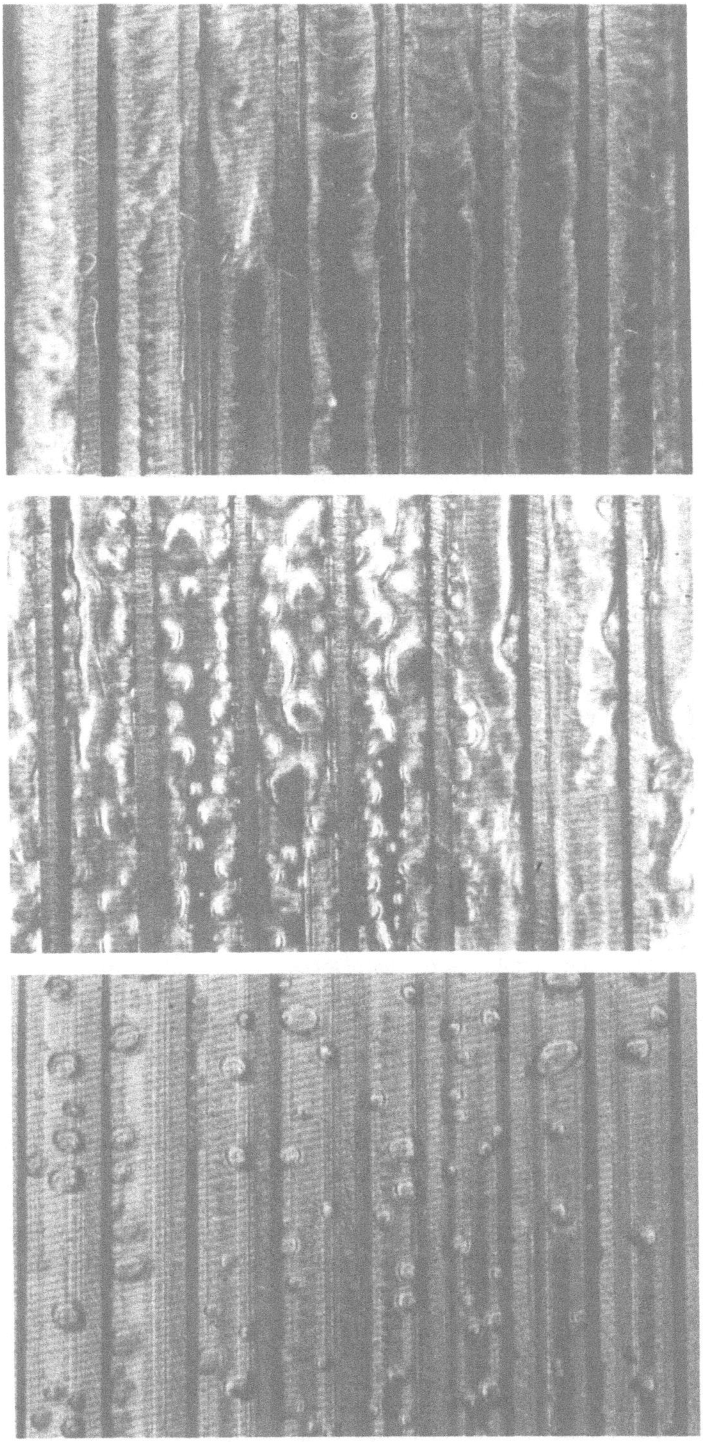

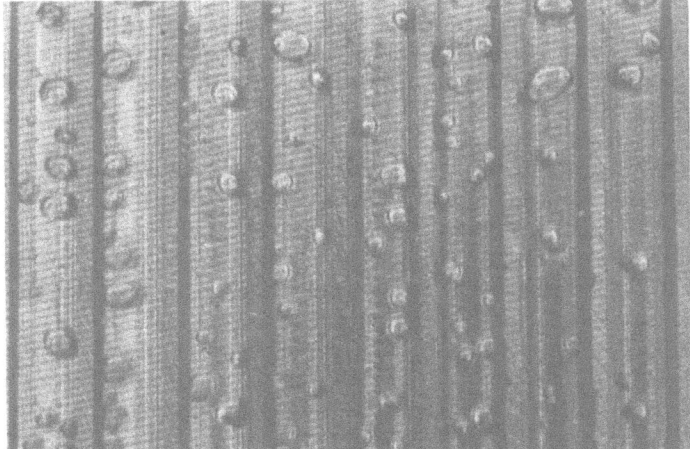

(c)
Geometry 2
Zone 1
3.25mm wide
G = 117kgm²
q = 20kW/m²

(b)
Geometry 2
Zone 4
3.25mm wide
G = 117kg/m²
q = 20kW/m²

(a)
Geometry 2
Zone 4
3.25mm wide
G = 316kg/m²
q = 4.3kW/m²

Figure 5. Flow Regimes Viewed in Test Section

At low flow-rates the flow was observed to be unstable. Note that Figure 5(b) shows that under such conditions several phenomena can be seen to occur simultaneously in adjacent channels: Isolated bubble, Confined bubbles and Annular flow appear to be present.

DEVELOPMENT OF CORRELATIONS FOR FLOW BOILING IN NARROW CHANNELS

Correlations for flow boiling in conventionally-sized heat exchangers have usually been based on a boiling regime and an evaporative convection regime. Typical evidence for the existence of these two regimes is shown in Figure 6 which shows $\alpha = f(x)$ for water in a 9.6mm tube and R113 flowing up a vertical channel of minimum width 6.2mm between the tubes of a tube bundle. The boiling regime includes nucleate boiling and the effects of sliding bubbles (in bubbly flow and slug flow) and is characterised by a predominant dependence on the heat flux q. The convective evaporation regime, on the other hand, is mainly influenced by heat flow through a thin liquid film on the surface and is therefore mainly dependent on the mass flow rate G and virtually independent of q. Correlations either add together the contributions to heat transfer from these two regimes with suitable suppression of boiling (for example Chen [12]), take the largest contribution alone, as proposed by Shah [13], or effectively offer a compromise in which the terms are squared before addition as in Kutateladze [14] and Gungor and Winterton [15].

There is no reason why correlations for narrow channels should not be subdivided in a similar way if some special features are recognised. As described earlier there is considerable instability at low flow rate which means that any design correlation must be taken as time-averaged and position-averaged. At low flow rates and low quality confined bubbles (CB) which straddle the channel may occur while at higher flow rates, isolated bubbles (IB) may occur at the same position and quality. Determination of the regime of flow is therefore a necessary pre-requisite to estimation of the heat transfer. At high qualities the annular flow characteristics may differ from those in large tubes owing to the cross-sectional geometry (corners) and the tendency of surface tension to thicken the liquid layer in small channels. Additionally, if flow in the entry region is unsteady due to the irregular formation of confined bubbles, the film thickness may vary with time.

From our studies of the literature [1,4,7,8,9] and the observations of our own work we have divided the regimes into 3 for the purposes of applying design correlations as shown in Figure 7. There will of course be transition regions between those regimes and not every system will exhibit all the regimes. At low flow rates for example bubbles may grow to fill the channel at the nucleation site, thus precluding the IB regime. Similarly to larger channels these regimes may be represented on the α/x plane as shown in Figure 8. The correlations are based on a physical dimension which is taken as the mean hydraulic diameter. Note that this is equal to twice the gap dimension for parallel plates. The following dimensionless numbers are used:

1. The liquid-only Nusselt Number is based on the Reynolds Number for flow of the liquid proportion alone. For turbulent flow,

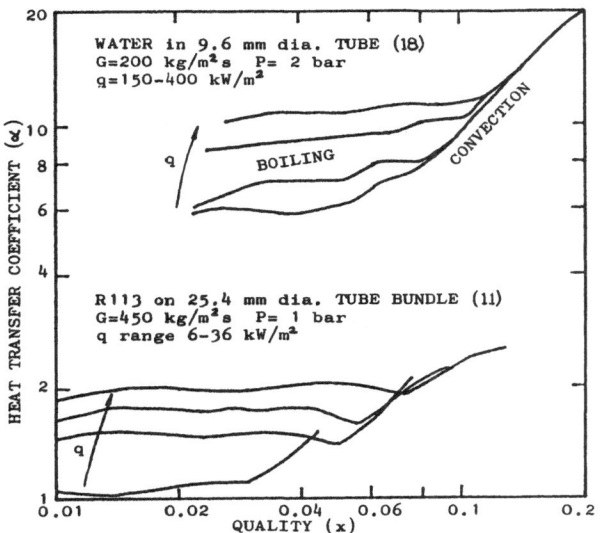

Figure 6. Comparison on In-Tube Boiling of Water with Tube Bundle Boiling of R113.

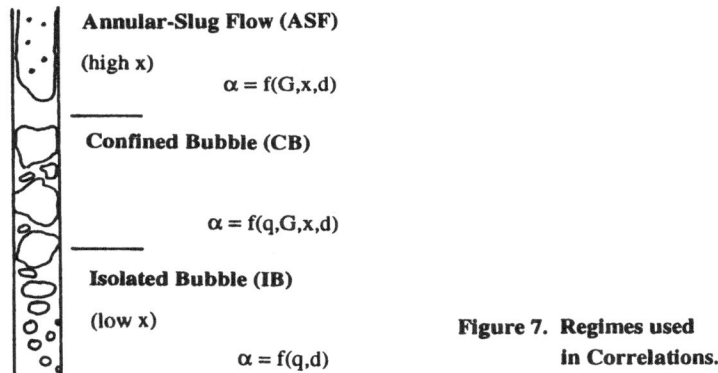

Annular-Slug Flow (ASF)

(high x)

$$\alpha = f(G,x,d)$$

Confined Bubble (CB)

$$\alpha = f(q,G,x,d)$$

Isolated Bubble (IB)

(low x)

$$\alpha = f(q,d)$$

Figure 7. Regimes used in Correlations.

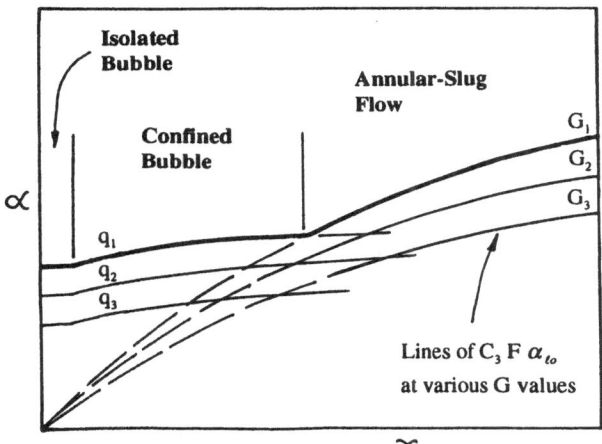

Figure 8. Flow Regime Map.

$$Nu_{to} = 0.023 \, Re_{to}^{0.8} Pr^{0.4} \qquad (1)$$

where $\qquad Re_{to} = \dfrac{Gd_e}{\mu_f}(1-x)$

For steady laminar flow (which may occur in narrow channels) Nu_{to} is not a function of G or x.

2. The (Shah) Boiling Number is essentially the ratio of the vapour production mass flow rate to the total mass flow rate, G,

$$Bo = \dfrac{q}{Gh_{fg}} \qquad (2)$$

3. A group termed the Confinement Number, Co in this study is used to represent the restriction to the normal flow regimes caused by the small size of the channel. Some previous researchers [1,2,5,16] have used gap dimension to characteristic bubble departure dimension ratio. We here arrange for the Number to increase with confinement

$$Co = \dfrac{\left[\sigma / (g(\rho_f - \rho_g))\right]^{\frac{1}{2}}}{d_e} \qquad (3)$$

Co may be usedf to find the transition from isolated to confined bubble regimes and to a first approximation confined boiling occurs when $Co > 0.5$ as indicated in Kew [16].

4. Finally for convective evaporation the enhancement factor F developed by Chen [12] is used. While more sophisticated expressions are available, the Chen-F is adequate for our purposes here.

$$F = 2.35(0.213 + 1/X_{tt})^{0.736} \qquad (4)$$

where $\qquad \dfrac{1}{X_{tt}} = \left(\dfrac{x}{1-x}\right)^{0.9}\left(\dfrac{\rho_f}{\rho_g}\right)^{0.5}\left(\dfrac{\mu_g}{\mu_f}\right)^{0.1}$

If $1/X_{tt} < 0.1$, $F = 1$.

Since we are primarily interested in developing correlations for design purposes, the normal practice in two-phase flow of giving heat transfer enhancement over the liquid-only flow is followed. Appropriate correlation forms for the three regimes in Figure 5 are therefore as follows:

IB
$$Nu = C_1 Bo^{0.7} Nu_{t_o} \qquad (5)$$

CB
$$Nu = C_2 Bo^m Co^n Nu_{t_o} \qquad (6)$$

ASF
$$Nu = C_3 F \, Nu_{t_o} \qquad (7)$$

The normal boiling exponent (e.g. Mostinski [17]) is used for the isolated bubble regime, but it does not follow that $m = 0.7$ in equation 2 where boiling may be a mixture of thin film and nucleation effects.

EXPERIMENTAL CORROBORATION OF CORRELATIONS

The purpose of this Section is to determine the extent to which the limited data in the literature and our own experiments fit the dimensionless equations 5 to 7 and where possible to give some indication of the power indices applicable in these equations.

Comparison with results of other workers is complicated in many cases by the absence of direct measurement of the mass flux G. Furthermore at low G local dryout is likely to occur even at low heat fluxes due to liquid starvation. This yields a reduction in the mean value of α and it is this mean which is normally measured. Bearing these points in mind the regimes are now examined in order:

Isolated Bubble Regime

Substitution of equation 1 into 5 yields for any particular fluid at low quality $(x \to 0)$

$$Nu = C_4 Re^{0.8} Bo^{0.7} \qquad (8)$$

The work of Lazarek and Black [3] is one of the few to include G measurements and for flow boiling of Refrigerant 113 in a 3.15mm diameter tube $(Co \sim 0.3)$ in this regime they obtained

$$Nu = 30 \, Re^{0.857} Bo^{0.714} \qquad (9)$$

The comparison is gratifying in view of the reported failure of other more complex equations. It was confirmed that heat transfer was independent of vapour quality thus implying nucleate or sliding bubble effects predominate.

On re-arrangement equation 8 may be written in a different form;

$$Nu = C_4 \left(\frac{qd_e}{h_{fg}\mu_f} \right)^{0.7} \left(\frac{Gd_e}{\mu_f} \right)^{0.1} \tag{10}$$

The first group is a two-phase Reynolds number where the mass flow rate is essentially the vapour flow away from the surface, q/h_{fg}. This group (to the power 0.67) has been used successfully for correlating pool boiling on tubes by Cornwell et al [11]. The second group indicates a slight but positive effect of G on the heat transfer. While in pool boiling there is no effect of G it is to be expected that this may not hold entirely for confined spaces. Figure 9 from the work of Nishikawa and Fujita [4] indicates that $\alpha = f(d_e^{-0.13})$ in this regime, which may be compared to the same function to a power of -0.2 (at constant q) indicated by equation 10.

Confined Bubble Regime

There are a number of papers available [2,4,7,8,9] which report heat transfer results for narrow channels in this regime under pool boiling conditions. In these cases the upward flow between plates or in channels is induced by the bubble action and the flow rate G is not recorded. However these data provide useful qualitative comparison as the dependence of α on G is usually fairly low and dependent on index m as, from equation 6;

$$\alpha = f(G^{0.8-m}).$$

This index m is the slope of the evaporation curve on the α/q plane and has a value of about 0.7 for nucleate boiling as mentioned earlier, but may be lower for confined spaces where thin film evaporation is important.

The results of a number of researchers [5,6,8,9] working with vertical channels immersed in a liquid pool indicate a variation of m with both confinement and heat flux. For example, Figure 10 shows some results of Fujita and Uchida [8] for water at 1 atmosphere boiling between plates at various spacings. The confinement numbers are given on the curves and it can be seen that m (the slope) varies from ~0.7 for the unconfined state to ~0.3 for $Co \sim 0.6$. For very narrow gaps ($Co \sim 2$) the characteristic is grossly affected by restriction to the liquid flow and the downward turn presumably indicates partial dryout. In flow situations where G is held constant the variation of m is likely to be less and from our work the slopes in Figures 3 and 4 indicate a value of 0.3 at high flows. At low flows, when instability was observed, the measured, time averaged, heat transfer coefficient was a weak function of q.

The value of n is difficult to estimate from the very limited data available. The work of Nishikawa and Fujita [4] (Figure 9) indicates that for this regime

$$\alpha = f[d_e^{-2/3}]$$

and from equation 6

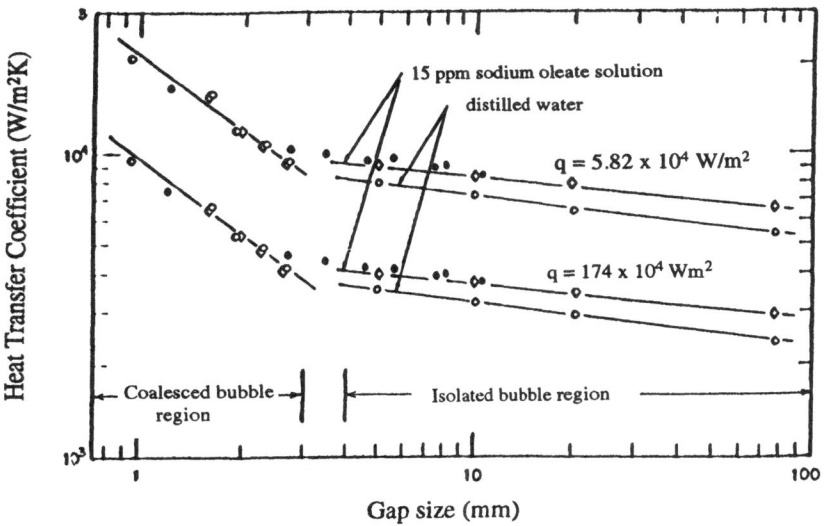

Figure 9. Variation of Heat Transfer Coefficient with Gap Size. (from Ref. [4]). Including Data of Ishibashi and Nishikawa [7] and Chernobyl'skii and Tananaiko [6].

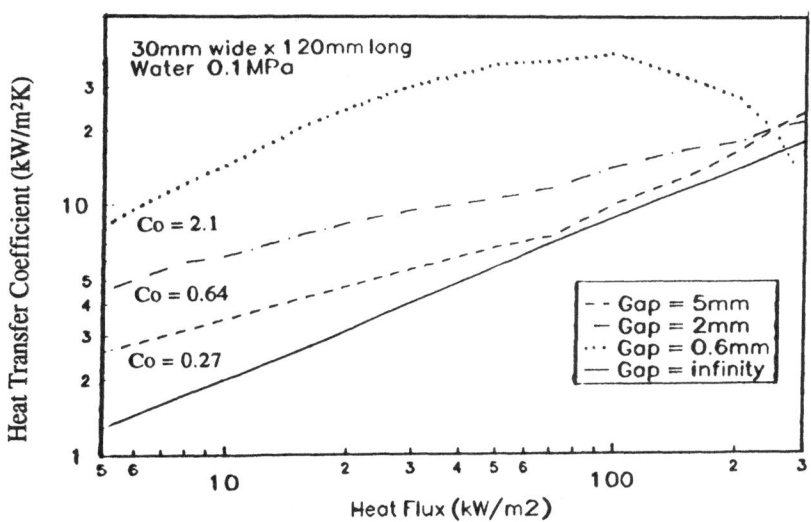

Figure 10. Variation of Heat Transfer Coefficient with Heat Flux (Adapted from data of Fujita and Uchida [8]).

$$\alpha = f[d_e^{-(n+0.2)}]$$

thus yielding $n = 0.47$. This in turn implies from equation 3 that approximately (with $n = 0.5$)

$$\alpha = f\left[\frac{\sigma}{g(\rho_f - \rho_g)}\right]^{1/4}$$

which is the dependence of this group in many boiling and dryout correlations.

Annular Flow and Dryout.

Equation 7 indicates strong dependence of α on x and G and low dependence on d_e. Bar Cohen and Schweitzer [2] examined the flow of water between plate spaced 2-12mm apart and found that for this regime there was no effect of the spacing. To a first approximation normal annular two-phase flow theory may be applicable at least up to Co values of around 0.5. In more confined situations the strong influence of local geometry and likelihood of dryout make steady annular flow more difficult. Bankoff and Rehm [1] in their study of flow in annuli at high Co values found that the Chen F factor needed to be modified to considerably higher values to describe the convective evaporation. More work is required on this important regime.

In general it would appear [2,8,9,10] as expected that the critical heat flux CHF is reached at lower levels in confined spaces. Fujita and Uchida [8] suggests that CHF occurs when the quality of the fluid leaving the channel is very close to unity. However Katsuta and Nagata [9] have shown that, in a partially submerged channel, the heat transfer coefficient increases with void fraction to a maximum at a void fraction between 0.8 and 0.9 and then decreases rapidly with further increase in void fraction. This indicates that the maximum heat transfer coefficient in a small channel must occur at a dryness fraction of the order of 0.1. At higher dryness fractions areas of the heat transfer surface dry out causing a progressive reduction in average heat transfer coefficient.

CONCLUSIONS

Experimental work on a compact multi-channel plate has led to the clear identification of three types of flow. The form of correlations for each type of flow is developed as follows

Isolated Bubble Regime	$Nu = C_1 Bo^{0.7} Nu_{to}$
Confined Bubble Regime	$Nu = C_2 Bo^m Co^n Nu_{to}$
Annular-Slug Flow	$Nu = C_3 F\, Nu_{to}$

The new Confinement Number Co represents the influence of the channel or gap size and indicates that the confined bubble regime occurs for $Co > 0.5$. The very limited data available suggest values of about 0.3 and 0.5 for m and n but is insufficient to estimate the other constants. These correlations yield a framework on which data from experiments on compact evaporators can be based. Their general form will allow them to be used in the design and evaluation of a wide range of new equipment.

REFERENCES

1. Bankoff, S.G. and Rehm, T.E., Convective boiling in narrow concentric annuli. Journal of Engineering for Gas Turbines and Power, Oct. 1990, **112**, 607-613.

2. Bar-Cohen, A. and Schweitzer, H., Thermosyphon boiling in vertical channels. Journal of Heat Transfer, Nov. 1985, **107**, 773-778.

3. Lazarek, G.M. and Black, S H., Evaporative heat transfer, pressure drop and critical heat flux in a small vertical tube with R113. Int. Journal Heat and Mass Transfer, 1982, **25**, No.7, 945-960.

4. Nishikawa, K. and Fujita, Y., Nucleate boiling heat transfer and its augmentation. Advances in Heat Transfer, 1990, **20**, 1-82.

5. Yao, S-C. and Chang, Y., Pool boiling heat transfer in a confined space. Int. Journal Heat and Mass Transfer, 1983, **26**, No.6, 841-848.

6. Chernobyl'skii, I.I., and Tananaiko, I.M., Heat exchange during boiling of liquids in narrow annular tubes. Soviet Physics - Tech Physi, 1956, **1**, 2244-49.

7. Ishibashi, E., and Nishikawa, K., Saturated boiling heat transfer in narrow spaces. Int. Journal Heat and Mass Transfer, 1969, **12**, 863-

8. Fujita, Y. and Uchida, S., Boiling heat transfer and critical heat flux in a confined space effects of gap, size, inclination angle and peripheral conditions at the space edge. Proc. 9th International Heat Transfer Conference, 1990, **2**, 153-158.

9. Katsuta, M. and Nagata, K., Boiling induced heat transfer enhancement using a narrow space, Engineering Foundation Conf., Pool and External Flow Boiling, Santa Barbara, March 1992, Publication pending.

10. Monde, M., Kusuda, H. and Uehara, H., Critical heat flux during natural convective boiling in vertical rectangular channels submerged in saturated liquid. Journal of Heat Transfer, 1982, **104**, 300-303.

11. Cornwell, K., The role of sliding bubbles in boiling on tube bundles. 9th Int. Heat Transfer Conf., Jerusalem, 1990, ('Heat Transfer 1990', Hemisphere), **4**, 455-460.

12. Chen, J.C., Correlation for boiling heat transfer to saturated fluids in convective flow, I & E.C. Process Design and Development, 1966, **5**, No.3, 332-339.

13. Shah, M.M., A new correlation for heat transfer during boiling flow through pipes. Transitions ASHRAE, 1976, **82**, 66-86.

14. Kutateladze, S.S., Boiling Heat Transfer. Int. Journal Heat Mass Transfer, 1961, **4**, 31-45.

15. Gungor, R.E., Winterton, R.H.S., A general correlation for flow boiling in tubes and annuli. Int. Journal of Heat Mass Transfer, 1986, **29**, No.3, 351-358.

16. Kew, P.A., Enhanced evaporation heat transfer surfaces - boiling in vertical narrow channels, Periodic Report for Period August 1990 - January 1991, CEC Contract JOUE-0041-C.

17. Mostinski, I.L., Application of the role of corresponding states for the calculation of heat transfer and critical heat flux. Teploenergetika, 1963, **4**, 66-76.

18. Cooper, M.G., Flow boiling - the 'Apparently Nucleate' Regime. Int.J.Heat Mass Transfer, 1989, **32**, 459-464.

SESSION 7:

Heat Exchangers
Continued

Chairman: Prof. M. Groll

NEW CONCEPTS IN LONGITUDINAL FLOW SHELL AND TUBE HEAT EXCHANGERS

J E HESSELGREAVES
NEL, East Kilbride, Glasgow, G75 0QU, Scotland
P MERCIER
CENG-GRETh, 38041 GRENOBLE, FRANCE
T MOROS
CRES, Koropi Attikis, 19400, Greece
S S MANSUR
CENG-GRETh, 38041 GRENOBLE, FRANCE
M McCOURT
NEL, East Kilbride, Glasgow, G75 0QU, Scotland

ABSTRACT

The work described in this paper has the object of seeking improvements to the performance of shell-and-tube heat exchangers. Traditional segment-ally baffled exchangers used widely in the process industries have significant shell-side leakage losses and stagnation zones which reduce their effectiveness and cause parasitic pressure drop.

Lower pressure losses are experienced in the fully longitudinal-flow types such as rod baffle and grid baffle. These allow higher fluid velocities and smaller exchangers, provided that erosion and tube vibration criteria are met. These types, together with a novel type utilising twisted tapes between tubes, are investigated in this paper.

The results indicate distinct promise for the new baffle types. Grid baffles, applicable to triangular tube arrangements and hence at lower spacings than rod baffles, have lower pressure drop and lower heat transfer augmentation than the latter. Twisted tapes offer increasing advantage over both as pitch to diameter ratio increases, (eg above 1:33), since the rod and grid types would involve both excessive material and high pressure drop.

The overall conclusions are that the new baffle types may well repay further investigation of engineering practicalities and costs.

NOTATION

A_c	Flow area	m^2
A_s	Surface area requirement	m^2
A_s*	Dimensionless surface area, defined by eg: 7	–
d_h	Hydraulic diameter	m
D	Tube diameter	m
f	Fanning friction factor	–
j	Colburn heat transfer factor	–
k	Enhancement factor	–
L	Flow length	m
$\dot{m}$	Mass flow rate	kg/s
N	Number of thermal units	–
Nu	Nusselt number	–
Δp	Pressure drop	Pa
P	Tube pitch	m
Pr	Prandtl number	–
Re	Reynolds number based on hydraulic diameter and mean throughflow velocity (longitudinal flow) or maximum velocity (transverse flow)	–
Re*	Operational parameter	/mm
u	Mean throughflow velocity	m/s
σ	Porosity	–
ρ	Fluid density	kg/m^3
ν	Kinematic viscosity	m^2/s

INTRODUCTION

Baffles in shell-and-tube heat exchangers serve two basic functions. Firstly, they provide tube supports, thereby preventing or reducing mechanical problems such as sagging or vibration. Secondly, they direct the flow over the tubes so as to introduce a cross-flow component, thereby increasing the heat transfer.

Segmented baffles have several sources of performance loss, some through various leakage flows and others caused by stagnation zones. Disc and doughnut, and disc and ring systems eliminate some of the losses, with the disc and ring system showing the higher performance. Longitudinal baffle forms offer further advantages for some applications, because of their elimination of stagnation zones. The basic heat transfer and pressure drop chracteristics, in the form of friction (f) factor and Colburn (j) factor are also advantageous for longitudinal flow, because of the relatively lower magnitude of parasitic pressure loss of the tube supports in longitudinal flow compared with that of the tubes in transverse flow. The overall benefits are discussed by Taborek [1].

In this paper studies are presented of the performance of the twisted tape, grid, and rod types of longitudinal flow exchanger. The studies, supported by the CEC JOULE programme, consisted of experimental measurements together with Computational Fluid Dynamics (CFD) calculations intended to provide an understanding of the flow structure in the baffle regions. To facilitate the overall three-dimensional modelling of flow and heat transfer throughout the exchanger, detailed measurements were made of the longitudinal and transverse f and j factors: the presentation and interpretation of these data form the main body of the paper. The longitudinal baffle types are then compared with each other on a common basis, and a preliminary discussion is given of the relative merits of these types with segmentally baffled types.

The National Engineering Laboratory Executive Agency (UK) is undertaking the longitudinal flow investigations, whilst transverse flow studies are being undertaken by Groupement pour la Recherche sur les Echangers Thermique - GRETh (FR). Numerical (CFD) aspects are being studied by GRETh (Global heat exchanger modelling), and Centre for Renewable Energy Sources - CRES (GR) (detailed flow calculations near baffles). Two industrial partners are assisting with the investigations: COVRAD Heat Transfer (UK), in the supply of heat exchangers for test, together with manufacturing experience, and ICI (UK), in advice on user aspects.

STUDIES OF LONGITUDINAL FLOWS

Scope of Investigations

Longitudinal flow forms of tube support offering enhanced shell-side heat transfer and increased pressure drop include rods (as in the commercial RODbaffle heat exchanger), grids and twisted tapes. The degree of support varies with the system used, but with all systems it increases as enhancement increases. All three types have been tested in this study, but only rod baffle and twisted tape arrangements are reported here. The range of pitch to diameter ratios was determined by practical considerations.

Twisted Tapes between Tubes

Arrangements with six twisted tapes with 360° twist length to diameter ratios, of 5, 7, 10 and 15 were tested in the triangular arrangement, for pitch ratios of 1.25, 1.33, 1.50 and 1.73. An opposite hand of twist was used in adjacent sub-channels (see Fig. 1), to reduce shear at the inter-sub-channel boundaries. The tapes were dimensioned to touch each of the three surrounding tubes, thus providing periodic support. A limited series of tests was also conducted with three tapes, in alternate rows (Fig. 2), giving half the number of support points but at the same longitudinal spacing.

The results, expressed in terms of Colburn j factor and Fanning friction factor versus Reynolds number, are shown in Figs 3 and 4, compared with plain duct values. For comparison with rod and grid baffles types, only pitch ratios of 1.25 and 1.33 are shown. For clarity, only the extreme curves, corresponding to twist ratios of 5 and 15 are shown: curves for the remaining values were intermediate between these. For each pitch ratio, they demonstrate increasing j factors as twist ratio increases. The gradients of the curves are also higher than those of the plain ducts, giving relatively greater enhancement at lower Reynolds numbers, a factor of about two being typical for the highest twist ratio (5) at an Re of 6000. This feature is more pronounced at the higher pitch ratios.

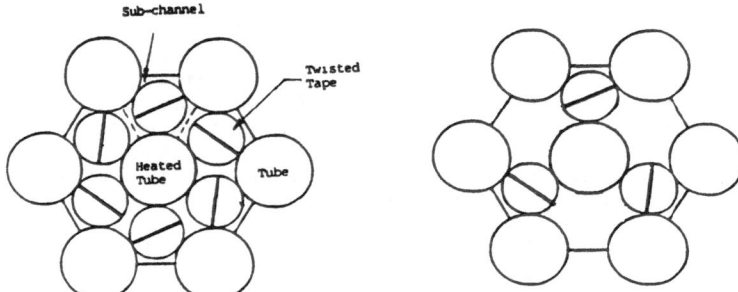

Figure 1 Twisted Tape Arrangement: Six Tapes   **Figure 2** Twisted Tape Arrangement: Three Tapes

Friction factor increases are considerably higher than those of j factor, penalties being about six for the same conditions. Of considerable interest are the results for three tapes, Figs 5 and 6. The loss in enhancement is remarkably small, suggesting that swirling flow is induced in the empty sub-channels by the adjacent tapes; the corresponding friction factors are reduced as expected by the reduction in exposed surface area. The design consequences are discussed further in Section 4.

It was not possible within the scope of the current investigation to analyse twisted tape flows by CFD techniques, since the boundary condition on the curved tape surface could not be adequately simulated.

Rod Baffles

Rod baffles were tested in the normal 'alternate' form recommended by Gentry et al [3], for a square pitch ratio of 1.25. Spacings of 125 mm and 250 mm were tested.

The results, shown on Fig. 7, indicate that except for Reynolds numbers below about 10 000, both j and f factor enhancements are rather larger than the equivalent enhancements for grid baffles. The correlations by Gentry [3] and Hesselgreaves [2] are also shown for comparison. Good agreement is observed for heat transfer for both correlations, but pressure drop is under-estimated, particularly by Gentry.

Computational Fluid Dynamics (CFD) studies were undertaken using a rod baffle geometry with rods in each row instead of in alternate rows, to simplify boundary conditions. Initial calculations, with the FLUENT code, were made with a rectilinear coordinate system, which enabled only an approximate representation of the tube wall. More recent calculations have used a body-fitted coordinate (BFC) system, giving the correct tube wall boundary condition. The heat transfer results, for three Reynolds numbers, are shown in Fig. 8. In an attempt to allow for the transitional region between fully turbulent and fully laminar conditions, the Nusselt numbers for each limiting condition were evaluated and combined on a quadratic basis. A reduced Nusselt number is shown, using an augmentation factor 'k' from Hesselgreaves [2].

Figure 9 shows the friction factor results, again showing comparison with Gentry's [3] correlation.

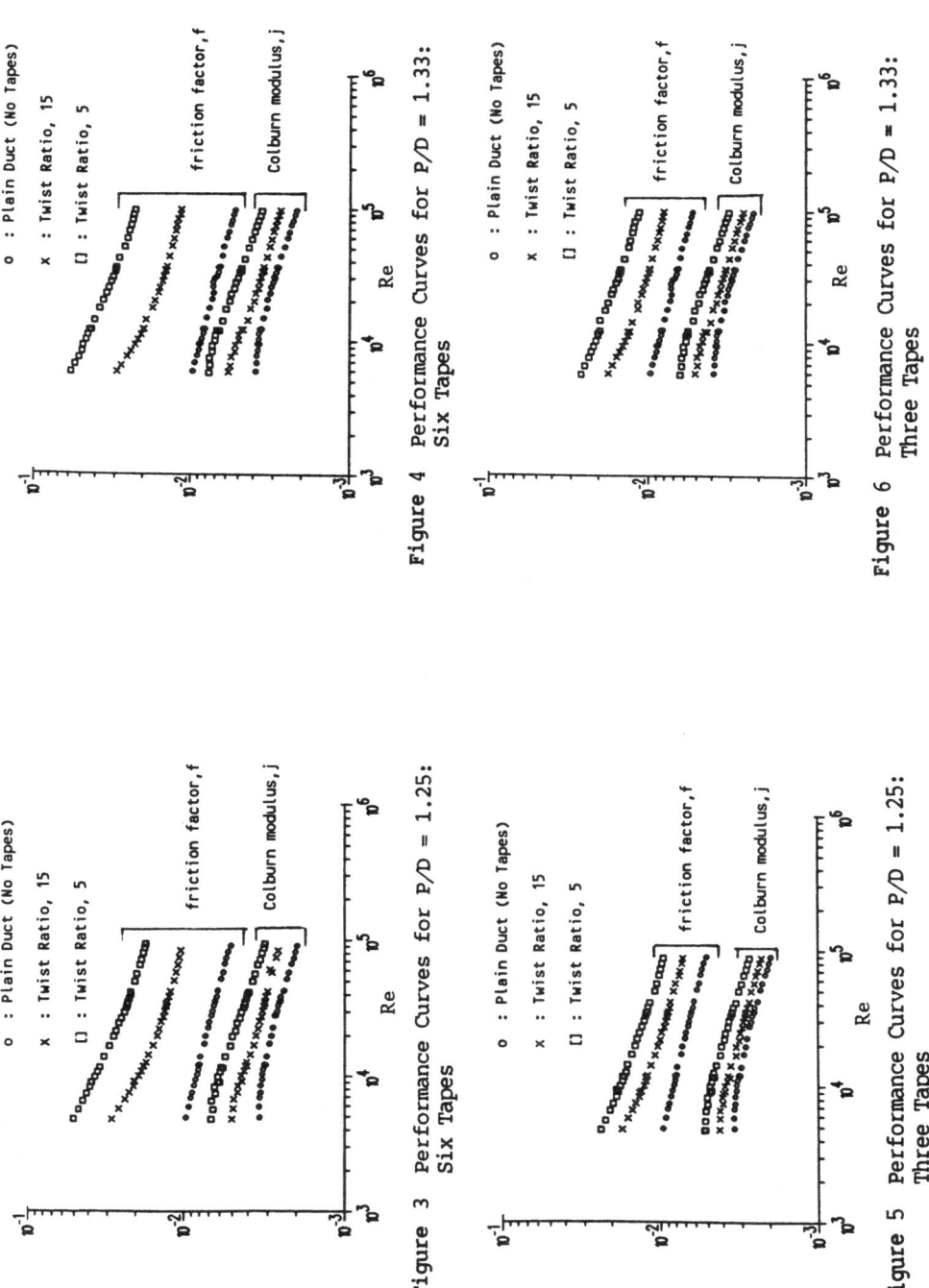

Figure 3 Performance Curves for P/D = 1.25: Six Tapes

Figure 4 Performance Curves for P/D = 1.33: Six Tapes

Figure 5 Performance Curves for P/D = 1.25: Three Tapes

Figure 6 Performance Curves for P/D = 1.33: Three Tapes

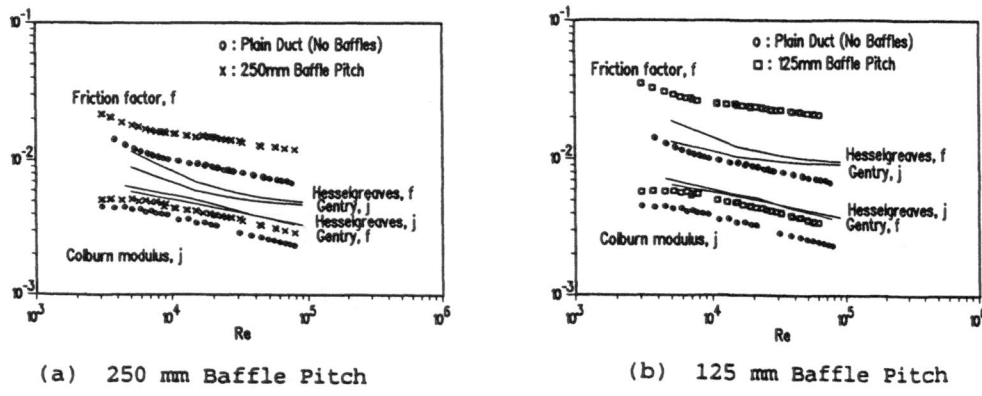

(a) 250 mm Baffle Pitch   (b) 125 mm Baffle Pitch

Figure 7 Performance Curves for P/D = 1.25; Rods

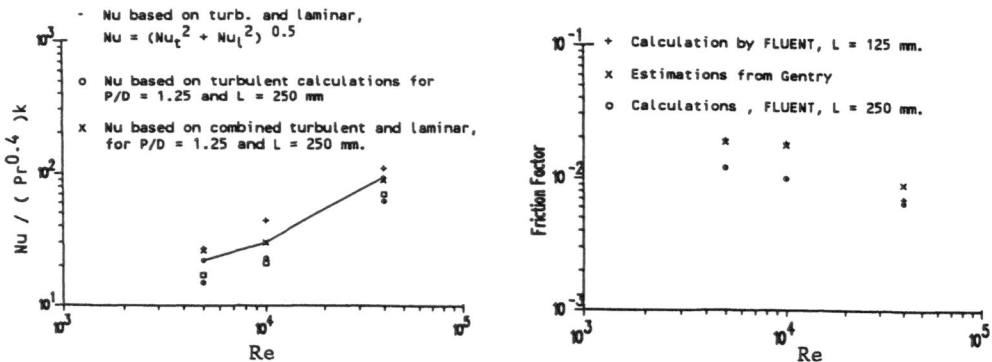

Figure 8 Numerical Prediction for   Figure 9 Numerical Prediction for
Rod Baffles; j factor   Rod Baffles; friction factor

Although the agreement between the CFD code and the correlations is quite good, significant qualitative and quantitative differences are evident. These are most likely to be the consequence of the basic k - ε turbulence model used in the code, which is only valid for fully turbulent flows. The present comparisons are in the transitional Reynolds number range, in which the turbulence structure and the consequent heat and momentum transport properties are changing markedly.

Friction Factor in Transverse Flows

The friction factor results, shown in Figs 10 and 11, demonstrate two striking features. Firstly, the friction factor of the randomly arranged tapes was significantly higher than that of the orientated tapes, a factor of 10-20 over plain bundle values being typical. This would be a logical consequence of the more ordered flow in the orientated arrangement. Secondly, the 'random' results were strongly dependent on twist ratio, (friction factor decreasing with decreasing twist ratio), whereas the 'orientated' results were almost independent of twist ratio. The former is thought to reflect the longer mean path length of the fluid with tapes of low twist ratio, whilst the latter suggests that the more ordered flow structure is little affected by the twist ratio.

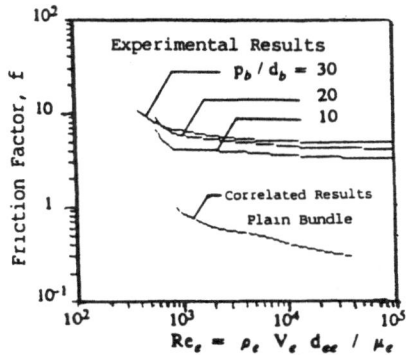

Figure 10 Friction Factor for Tapes in Transverse Flow: Effect of Twist Ratio for Random Orientation

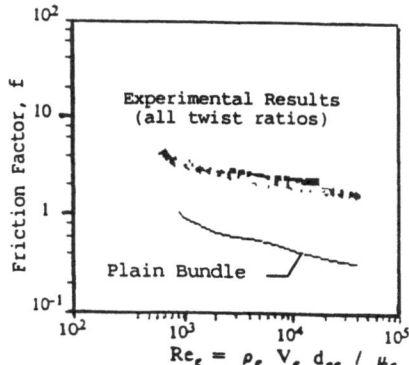

Figure 11 Friction Factor for Orientated Tapes in Transverse Flow

Heat Transfer in Transverse Flows

The heat transfer results, expressed in terms of reduced Nusselt number, are shown in Figs 12 and 13. The effect of orientation is much greater than that of twist ratio, the 'random' arrangements showing rather higher performance than the 'orientated' ones, yielding enhancement factors over the plain bundle case of about 1.7 compared with about 1.3.

Although transverse flow tests were not undertaken with rod or grid baffles in this investigation, it is safe to assume that plain bundle friction and heat transfer data are applicable in the end zones: it is unlikely that baffles would be located in these zones.

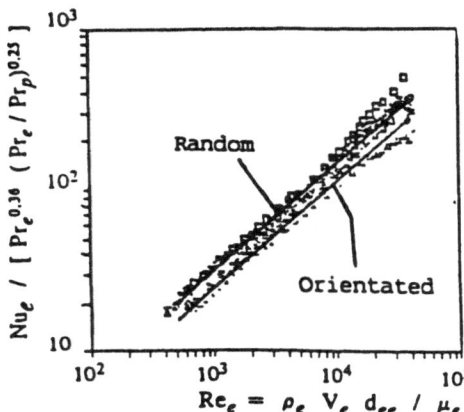

Figure 12 Heat Transfer in Transverse Flow: Experimental Results

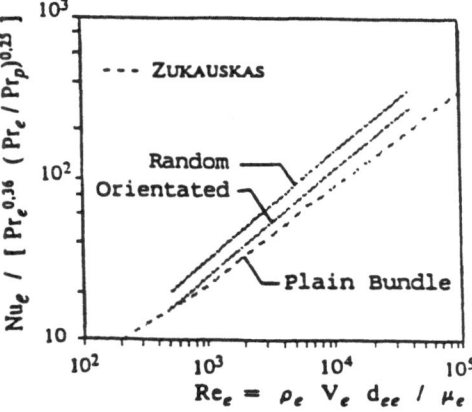

Figure 13 Heat Transfer in Transverse Flow: Correlations

GENERAL DISCUSSION: CONSEQUENCES FOR SHELL–AND–TUBE DESIGN

A shell-and-tube heat exchanger design for longitudinal flow has three basic flow regimes: an entry region of mixed crossflow and longitudinal flow, a nearly 'pure' longitudinal flow section, and a discharge region similar to the entry region. The central 'pure' longitudinal section is the critical one for realising the advantages of this flow, and is discussed first. Entry and exit regions are then treated together, and finally, the inter-relationship of shell-side and tube-side performance is discussed. It is assumed that the shell-side mass flow, Number of Thermal Units (NTU) and pressure drop are specified.

Longitudinal Flow Zone

The geometric design features of the longitudinal flow section are basically two-fold; that of surface area requirement (which directly affects the cost), and that of shape, which affects the installation requirements, and this can also affect the cost. A comparative assessment of the surface area rquirement of the various longitudinal forms can be made by combining the flow area requirement (via the core velocity equation, London [5] with the length requirement (from the specified NTU), to derive a volume comparison (Taylor, [6]). This can be transformed into a surface area parameter $A_s\star$

$$A_s\star = \frac{A_s}{\dot{m}Pr} \cdot \left(\frac{2\,\rho\Delta p}{N^3}\right)^{\frac{1}{2}} = \left(\frac{f}{j^3}\right)^{\frac{1}{2}}, \tag{1}$$

which is a function, for any given surface (or baffle type in the present investigation), of an operational parameter Re$\star$ given by:

$$Re\star = \frac{Re}{d_h}\left(\frac{f}{j}\right)^{\frac{1}{2}} = \frac{1}{\nu} \cdot \left(\frac{2\,\Delta p}{\rho NPr^{2/3}}\right)^{\frac{1}{2}} \tag{2}$$

The right hand side of equation (2) is purely a function of heat exchanger requirements and thus defines the operational Reynolds number of any given surface of specified hydraulic diameter and j, f versus Re characteristics. Thus a high hydraulic diameter requires a high working Re, and vice-versa.

From these equations it is clear that a plot of $A_s\star$ versus Re$\star$, both expressed in terms of f, j and d_h from equations (1) and (2), yields the comparative surface area requirements for the surfaces for any given operating condition expressed as a fixed point on the abscissa (Re$\star$).

Curve fitted values of this area parameter are shown on Figs 14 (for rod baffles) and 15 (for twisted tapes), for a pitch to diameter ratio of 1.25, together with plain bundle values. Figure 16 shows twisted tape values for P/D = 1.33. Figure 14 shows that for a pitch ratio of 1.25, the area for a rod baffle decreases as baffle pitch increases. This reflects the persistence of turbulence combined with reduced overall form drag.

For the pitch to diameter ratio of 1.25 (Fig. 15), twisted tapes perform similarly to rod baffles for the high twist ratio (15) but show a progressive advantage up to about 20%, as the twist ratio decreases to its lower limit (5). This applies over the whole operational range of Re$\star$.

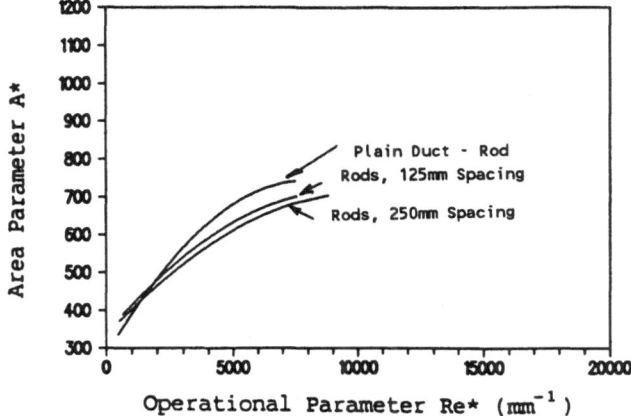

Figure 14 Area Parameter for Rod Baffles, P/D = 1.25

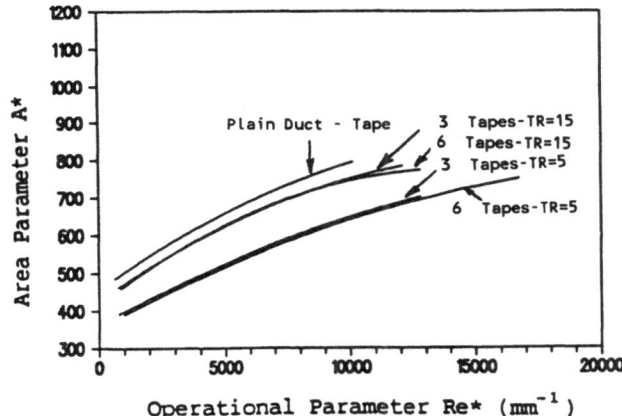

Figure 15 Area Parameter for Twisted Tapes, P/D = 1.25

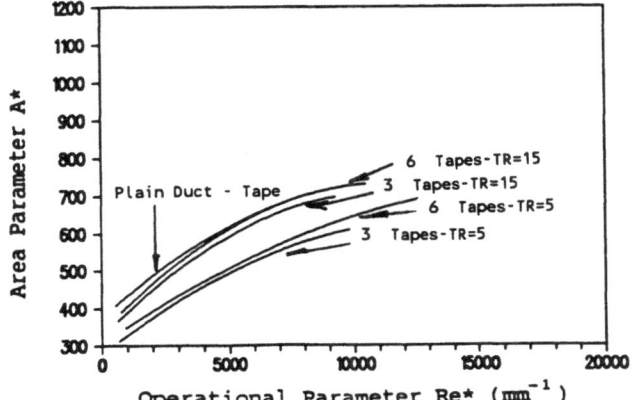

Figure 16 Area Parameter for Twisted Tapes, P/D = 1.33

It is noteworthy that the performance of 3 and 6 tape baffle systems is almost identical, indicating that a 3 tape system would be preferred on cost grounds. Similar results are observed for the higher pitch to diameter ratio of 1.33, shown in Fig. 16. In this case the 3 tape system is marginally superior to the 6 tape system. Rod baffle data are not available for this pitch ratio.

The mechanism of augmentation of the twisted tapes is that of mixing in the longitudinal vortex system induced by the tapes. The advantage over the other types is thought to be because the high twist gives strong mixing without separated flows (the tapes only contributing a skin friction component to the pressure drop), whereas the rod baffles generate increasing bluff-body drag without a relative increase in turbulent mixing. This parasitic drag component increases substantially as pitch ratio rises, and rods could be impractical for this reason at pitch ratios above about 1.5. It is also noteworthy that the heat transfer augmentation for twisted tapes at these higher pitch ratios is high, particularly at lower Reynolds numbers.

The shape of the longitudinal flow zone of the exchanger dominates the overall shape, and is characterised by the flow area and length, which are described by the parameters $(f/j)^{\frac{1}{2}}$ and $1/j$ respectively. The limiting case, of a long, thin bundle, is provided by the (impractical) plain bundle case, which has the lowest value of f/j of typically about two, and the lowest j value. For the same tube geometry, flow areas would typically be 25-40% higher, and flow lengths 40-50% lower for twisted tape configurations, depending on Reynolds number and twist ratio. For rod baffles the comparison is not clear because the necessarily square tube layout gives rise to different hydraulic diameters and porosities, but flow area increases are of the order of 20-40%, whilst flow length reductions are about 20-30%.

Entry and Exit Zones

These zones comprise the portion of the exchanger from each tubesheet, where the flow over the tubes is largely transverse (from the entry and exit ports), to the point along the bundle at which the flow becomes essentially longitudinal. The basic requirement for longitudinal flow exchangers is that the flow can penetrate to the core of the bundle in the shortest practical flow length. This is achieved in three possible ways: by having a small bundle diameter, by a low resistance to transverse flow, and by incorporating an annular port construction instead of the usual pipe entry. The bundle diameter is usually determined by design considerations of the longitudinal zone as above. The resistance to transverse flow becomes more important with increasing bundle diameter, and is highest at low pitch ratios. As mentioned in Section 3.2 above, the transverse resistance can be taken as the plain bundle resistance for rod baffles. Twisted tapes have considerable resistance to crossflow, which would be detrimental to flow distribution; this effect could be alleviated or eliminated by orientating the tapes or leaving a tape-free length respectively.

Matching Tube-side and Shell-side Performance

In a longitudinal flow exchanger there is much less scope for varying the shell-side performance than with a segmentally baffled one. It is therefore more important to match the shell-side and tube-side resistances. The tube side options for achieving such matching include: varying the pitch ratio, which controls the relative velocities and hydraulic diameters on the two sides; internal low fins (usually longitudinal), and inserts. These features in appropriate combination give a factor of about five or six in potential tube-side enhancement over plain tube values based on a typical pitch ratio of 1.25, provided that pressure drop constraints are met. This figure is to be taken in the context of a shell-side augmentation of up to two with plain tubes, or perhaps up to four with low fin tubes.

Tube Support

In conventional segmentally baffled exchangers the maximum baffle pitch specified by TEMA for avoidance of flow-induced vibration is about 1.5 m for a tube diameter of 18 mm, and is proportionally less for smaller diameters. The support for rod baffled exchangers is different in nature, being in the form of point support at intervals, rotated at 90° in turn. Typical spacings are 250 mm, giving an effective 0.5 m between supports in each plane, and clearly much greater rigidity than with most segmentally baffled exchangers.

The support for twisted tape systems is either six-point or three-point, depending on whether full 'six tape' or half 'three tape' support is used. The support length varies with both pitch ratio and twist ratio, being 20 mm for P/D = 1.25 and TR = 5, and 134 mm for P/D = 1.73 and TR = 15. It is clear from these figures that the twisted tape system is considerably more rigid than any other tube support system, thus being much less prone to flow induced vibration, and allowing considerably higher flow velocities consistent with pressure drop and erosion requirements. The main constraint would arise if tape-free zones were specified at inlet and outlet for flow penetration.

Overall Comparison of Longitudinal with Segmented Baffled Exchangers

In the longitudinal flow heat exchanger, provided that the length to diameter ratio of the longitudinal-flow zone is high (greater than 5), then there is a low loss of 'ideality' caused by the entry and exit zones. The only other main loss is that of leakage between the tube bundle and the shell. This would in practice be reduced by the use of sealing rings at appropriate intervals; in the case of rod baffle types these rings would form the baffle supports. For twisted tape systems, the sealing ring spacing would need to be determined for each application, and would be related to the shell diameter. It is expected that the overall loss of idelility from pure longitudinal flow would be of the order of 20-25%.

In segmentally-baffled exchangers, on the other hand, the many forms of loss from parasitic and turn-around pressure losses, bypass and leakage flows, result in a total loss of ideality of about 40% (Mueller [7]). The consequence of this difference, together with surface performance differences, has been evaluated by Tuma [8], with the result that longitudinal flow exchangers become progressively advantageous in terms of surface area requirement as operating Reynolds number increases beyond about 7000. Further discussion on this aspect is beyond the scope of this report.

CONCLUSIONS

1 Twisted tape inserts between tubes in a longitudinal flow heat exchanger have been shown to give superior shell-side performance to rod baffles, and may well repay further investigation of engineering practicalities and costs.

2   Longitudinal exchangers, particularly if provided with twisted tape 'baffles', offer the potential for lower surface area requirements than segmentally-baffled exchangers when the shell length to diameter ratio is high.

3   The best performance is obtained with the lowest twist ratio and with tapes in alternate passages.   The advantage increases if process conditions call for higher tube pitch to diameter ratio and high Reynolds number.

4   The advantage is dependent on achieving a low resistance to the transverse flows near to the entry and exit ports.   This can be assisted by having tape-free zones in these regions, consistent with tube support requirements, and by incorporating annular port arrangements.

5   The high degree of tube support offered by twisted tape systems gives the potential for significantly increased flow velocities without the risk of flow-induced tube vibration: this could in turn give lower surface area requirements if erosion criteria were met.

ACKNOWLEDGEMENT

This work was supported by the CEC JOULE Programme, under contract JOUE-0016-C.   The paper is published by permission of the Chief Executive, National Engineering Laboratory Executive Agency.   It is Crown Copyright.

REFERENCES

1   TABOREK, J., (1989) Longitudinal flow in tube bundles with grid baffles. AIChE Symposium, Heat Transfer, Philadelphia.

2   HESSELGREAVES, J.E., (1988) A mechanistic-model for heat transfer and pressure drop in rod baffled heat exchangers.   Second UK National Heat Transfer Conference, University of Strathclyde, Glasgow.

3   GENTRY, C.C., and SMALL, W.M.   (1985)   Rod Baffle Exchanger Thermal-Hydraulic Prediction Models over Expanded Baffle Spacing and Reynolds Number Ranges, 1985 National Heat Transfer Conference, Denver, Colorado, USA.

4   ZUKAUSKAS, A., Heat Transfer from Tubes in Crossflow. Advances in Heat Transfer, 18, pp 87-159, 1987.

5   LONDON, A.L., Compact Heat Exchangers - Design Methodology, in Low Reynolds Number Flow Heat Exchangers, Ed. Kakac, Shah and Bergles, Hemisphere, 1983.

6   TAYLOR, M.A., (1987)   Plate-fin Heat Exchangers: Guide to their Specification and Use. HTFS, Harwell, UK.

7   MUELLER, A.C.   Shell-and-Tube Exchanger Design, in Low Reynolds Number Flow Heat Exchangers, Ed. Kakac, Shah and Bergles, Hemisphere, 1983.

8   TUMA, J.   (1990)   The Enhancement of Heat Transfer to Gases in Longitudinal Flow in Tube Bundles.   PhD. Thesis, Brighton Polytechnic, UK.

ETUDE DE LA DISTRIBUTION DE MELANGE LIQUIDE-VAPEUR COTE TUBE DANS LES ECHANGEURS A TUBES ET CALANDRE

J. GARCIN, M. IDRISS, F. LAURO
Commissariat à l'Energie Atomique
Groupement pour la Recherche sur les Echangeurs Thermiques
GRENOBLE, France

RESUME

La redistribution d'un écoulement diphasique entre passes d'un échangeur à tubes et calandre est étudiée expérimentalement. L'expérimentation est réalisée sur une maquette en similitude eau/air à une pression de 20 bar et en condition adiabatique.

Une distribution homogène d'eau et d'air est mise en place dans les tubes de la passe amont de la maquette. A la sortie des tubes de la passe aval, des mesures de débit de chaque phase sont faites pour chacun des tubes ainsi que des mesures de pression.

L'influence de la géométrie de la boîte de retournement, du titre et de la vitesse massique sur la distribution du mélange diphasique dans les tubes de la passe aval est étudiée.

INTRODUCTION

L'efficacité d'un échangeur de chaleur dans lequel survient un changement de phase dépend fortement de la distribution spatiale du mélange liquide-vapeur. Cela vaut particulièrement pour les condenseurs et pour les évaporateurs à tubes et calandre où ce changement survient côté intérieur tube. En raison du large domaine d'emploi de ces échangeurs, la réduction des causes de maldistribution (injection, retournements) du fluide changeant de phase devrait conduire à des économies d'énergie substantielles. C'est la justification de cette recherche supportée par la C.E.E. [1].

Pour un échangeur tubulaire multipasse, le coefficient global d'échange thermique mesuré est toujours nettement inférieur au coefficient d'échange local calculé sur un tube présentant les caractéristiques moyennes spatiales de l'échangeur : titre, vitesse massique.

Une étude théorique paramétrique simplifiée d'un évaporateur tubulaire multipasse côté du changement de phase, conduite avec le logiciel de simulation CETUC [2] a donné les effets estimés sur le coefficient d'échange thermique global dus à une mauvaise distribution. La fiabilité de ce logiciel incorporant entre autres les lois d'échange diphasique de SHAH, de CHEN et de GUNGOR, a été vérifiée sur des essais d'évaporateur eau-R22. Dans cette étude

[3], l'échangeur évaporateur était à une seule passe. On a trouvé que :

- A répartition uniforme de débit diphasique total de R22, augmenter la maldistribution du titre entre les tubes diminue régulièrement le coefficient d'échange global et ce, d'autant plus qu'il y a plus de vapeur à l'entrée de l'appareil (titre moyen plus élevé). Pour citer un exemple, la réduction est maximale lorsqu'à l'extrême tout le liquide emprunte la moitié des tubes et la vapeur l'autre moitié : - 30 % par rapport au cas de répartition uniforme de titre et débit.

- A titre uniforme à l'entrée de tous les tubes, augmenter la maldistribution de débit diphasique diminue aussi régulièrement le coefficient d'échange global, lorsque la vitesse massique moyenne est en-dessous d'un certain seuil ; au-delà, le coefficient peut parfois être augmenté dans la zone des très bas titres mais il reste globalement décroissant lorsque la maldistribution de débit augmente.

Enfin, une étude expérimentale des effets hydrodynamiques d'une maldistribution imposée de titre et/ou de débit dans une maquette adiabatique d'échangeur évaporateur à plaque [4] a montré qu'une maldistribution originelle tendait à se résorber quand le titre moyen s'élève, par le jeu du retournement en fin de passe, c'est-à-dire le passage d'une passe à la suivante dans un échangeur multipasse.

Il y avait donc toutes chances d'obtenir, dans un évaporateur tubulaire multipasse, des gains de performance par optimisation géométrique au moins des premiers retournements et/ou de l'injection (introduction du mélange liquide-vapeur dans l'évaporateur).

Le programme de recherche objet du contrat Joule précité est exécuté en association de trois partenaires.

- Etude sur une machine frigorifique de puissance d'un évaporateur de 250 kW à H.W.U. (Heriot Watt University).
- Réalisation de la machine frigorifique par CIAT (Compagnie Internationale Aerolique et Thermique) équipé d'un évaporateur spécial pour essais à H.W.U.
- Etude en similitude eau/air sur une maquette BEATRICE 2 au GRETh (Groupement pour la Recherche sur les Echangeurs Thermiques).

Le présent article présente les travaux du GRETh sur une maquette en condition adiabatique.

ETUDE EXPERIMENTALE DU GRETh

1. Objectif de l'étude du GRETh

On veut mesurer sur une maquette (photo 1) la redistribution d'un mélange diphasique réalisée par une boîte de retournement de forme variable, à partir d'une distribution connue de débit diphasique sortant d'une passe amont, vers une passe aval. La simulation du retournement du mélange liquide-vapeur de fluide frigorigène aux conditions usuelles d'évaporation en unité frigorifique est faite en s'appuyant sur le respect du rapport des densités des phases gazeuse et liquide comme critère principal de similitude. Le couple eau-air maintenu à 20 bar absolus et à 30°C a été retenu pour cette étude expérimentale.

Un évaporateur comportant approximativement le même nombre de tubes que la maquette (machine CIAT) est étudié en conditions réelles dans le même temps par H.W.U. et les programmes d'essais sont coordonnés.

2. Equipement expérimental

Le moyen expérimental utilisé est la boucle BEATRICE 2 (photo 2) adaptée, pour la partie échangeur, à la géométrie tubulaire. Le faisceau de tubes et la boîte de retournement sont conçus pour représenter le retournement entre 2 passes successives quelconques de l'évaporateur industriel à quatre passes de diamètre intérieur de calandre 355 mm (fig. 1 et 2). La symétrie verticale a été mise à profit pour l'équipement expérimental. Le faisceau (fig. 3) comporte 76 tubes (Ø 14,85 x 15,5 mm) arrangés au pas triangulaire de 19,5 mm. Sur cette figure, est représenté la manière utilisée pour simuler les trois boîtes de retournement de l'évaporateur. L'alimentation du faisceau en fluides, le relevé de la répartition de débit en 2ème passe et la mesure du taux de présence de la phase gazeuse dans la boîte de retournement sont inchangés par rapport à BEATRICE 1 [4], [5]. Les aspects nouveaux sont :
- Vérification de la répartition de débit imposée en 1ère passe (passe amont).
- Mesures de pression sur la plaque tubulaire en entrée de la passe aval.
- Mesure de la perte de pression en fonction du titre, de la vitesse massique et de la longueur sur l'un des tubes du faisceau. Cette mesure sert à établir un ensemble de courbes d'étalonnage permettant à partir d'une mesure de pression sur les tubes de la 2ème passe de déduire, à partir du débit massique et du titre mesurés pour chacun des tubes, les pressions au droit de la plaque tubulaire.

La figure 4 montre le schéma de principe de l'installation expérimentale.

Le débit total d'eau est ajusté à l'aide de la vanne VRE2 et de la vanne du circuit de by-pass VRE1. ce débit est mesuré par le débitmètre DME1. De la même façon, le débit total d'air est ajusté à l'aide des vannes VRA2 et et VRA1 (by-pass), et mesuré par le débitmètre DMA1.

Deux systèmes de distribution DE et DA inclus dans GLFI permettent de fractionner les débits totaux d'eau et d'air, en autant de parties égales qu'il y a de tubes à alimenter. La répartition en débits égaux est assurée par une perte de pression locale importante dans GLFI par rapport à la perte de pression dans la maquette d'essais.

Une distribution non uniforme du mélange peut être imposée à l'entrée de la maquette par modification des injecteurs et diaphragmes contenus dans GLFI.

Un jeu de vannes de réglage VRM/44 a été placé en aval de la maquette, ce qui permet d'imposer la pression dans les tubes de la passe aval à une distance donnée de la plaque tubulaire. La différence de pression entre chaque tube aval et un tube amont pris comme référence est mesurée à l'aide d'un transmetteur de pression différentielle TPD2, et la voie de mesure est sélectionnée à l'aide des électrovannes EV/20.

Avant mesure de toute pression différentielle, une injection d'eau pendant quelques secondes assure le remplissage en eau de tous les piquages des prises de pression.

Les mesures de débit à la sortie de la passe aval se font selon une séquence complètement automatisée. Le mélange sortant d'un tube donné de la maquette est dévié, à l'aide des vannes trois voies V3VP/22, vers les systèmes de séparation des phases SM1 et FA1. Ensuite, les débits d'eau et d'air sont mesurés par les débitmètres DME2 et DMA2, respectivement.

Afin de ne pas perturber la répartition des fluides dans la maquette d'essais par la mise en ligne des systèmes de séparation et de mesure, la vanne de régulation VRG ajuste automatiquement la pression différentielle pour la rendre égale à celle mesurée avant la déviation, c'est-à-dire avant la manoeuvre des vannes trois voies.

Les circuits de sortie de la maquette sont reliés à un collecteur de mélange général (CMG) et à un système de séparation des phases SM2 et FA2. La pompe PE reprend l'eau à partir du réservoir R, tandis que le compresseur CA recomprime l'air.

L'échangeur RE sur le circuit eau et l'échangeur RA sur le circuit air assurent la constance de la température de l'eau et de l'air à l'entrée de la maquette.

3. Paramètres étudiés et résultats expérimentaux

Les paramètres qui influent sur la mauvaise distribution par retournement sont d'une part des paramètres géométriques (profondeur, courbures de la boîte de retournement, répartition des tubes) et d'autre part, des paramètres de fonctionnement : vitesse massique et titre du mélange en entrée. Ces paramètres sont étudiés expérimentalement dans ce projet.

Les résultats expérimentaux présentés ci-après ont été obtenus à partir de mesures de répartition du débit de chaque phase, réalisées sur la sortie des tubes de la passe aval. Des essais ont été faits pour différentes valeurs du débit total et du titre du mélange, et pour divers profils de la boîte de retournement. Les essais ont été réalisés pour une seule orientation de l'écoulement, sens ascendant dans la boîte de retournement.

Le tableau I indique les différents essais réalisés.

h	G	0,5 Gn			0,66 Gn			Gn			1,2 Gn	
32,5	X %	35	40	50	30	40	50	30	40	50	30	40
16	X %		40			40		30	40	50		40

Tableau I : essais réalisés pour la boîte de retournement (zone 1) (cf. figure 1)

    h   profondeur de la boîte de retournement (mm)
    G   vitesse massique (kg/m²/s)
    G_n vitesse massique nominale (225 kg/m²/s)
    X   titre massique (%)

Les hublots positionnés sur la boîte de retournement ont permis d'observer que dans les cas présentés, l'eau sortant des tubes est projetée sur fond de la boîte avec formation d'un film liquide entraîné vers le haut par l'air. L'effet tridimensionnel peut également être observé par l'apparition de tourbillons dans la partie haute de la boîte dans deux zones de tubes référencés n° 18-21-24 et n° 7-4-3 (voire figure 5a).

L'ordre des tubes indiqués ci-dessus donne le sens de rotation des tourbillons observés. Le tourbillon 18-21-24 est beaucoup plus important que le tourbillon 7-4-3. L'observation montre des zones à écoulement à bulles et des zones à écoulement à gouttes.

Les résultats des mesures de la distribution du débit d'eau et d'air dans les tubes de la passe aval de la maquette sont présentés sous la forme de graphiques. La position des tubes a été rapportée sur l'axe des abscisses selon une numérotation de 1 à 24, représentant le repérage des tubes tel qu'indiqué sur la figure 5.

Les figures 6, 7 et 8 présentent les résultats obtenus. Sur l'axe des ordonnées sont portés les débits massiques d'eau ou d'air recueillis en sortie de chaque tube de la passe aval de la maquette. Sur les figures 6 (relative à un titre de 40 %) et 8 (relative à des titres de 30, 40 et 50 %), on peut remarquer que, pour l'eau, plus la vitesse massique est élevée plus les tubes du haut sont suralimentés en eau (tubes 4, 11, 18). Les tubes 21 et 24 du tourbillon 18-21-24 ainsi que les tubes 3, 7 du tourbillon 3-7-4 sont également suralimentés en liquide. L'effet de la profondeur de la boîte de retournement (Fig. 6 et 7) pour des valeurs de 16 mm et 32,5 mm n'est pas significatif sauf pour le plus fort débit massique de (G = 272 kg/m²/s). La figure 7 relative au débit d'air recueilli pour un titre de 40 %, deux valeurs de profondeur de boite et trois vitesses massiques montre dans ces cas une suralimentation en air des tubes 1, 2, 5, 6, 8, 9, 12, 13, 15, 16, 19, 23. Les tubes 4, 11 et 18 sont par contre toujours très sous-alimentés en air.

4. Simulation numérique

Les résultats expérimentaux seront comparés à des calculs réalisés avec le code TRIO développé à Grenoble. Il est apparu lors de l'interprétation des résultats de redistribution en échangeurs à plaques [4] que la loi de frottement interfacial dû au cisaillement choisie arbitrairement faute de travaux scientifiques appropriés en 2D ou en 3D dans le domaine des taux de vide intermédiaires, $(0,3 \leq \alpha \leq 0,7)$ c'est-à-dire entre écoulement à bulles et écoulement à gouttes était insatisfaisante. L'ajustement du coefficient de frottement pour $0 \leq \alpha \leq 1$ dépendait des conditions limites qui n'étaient pas mesurées sur l'équipement expérimental. Dans le cadre de ce projet JOULE, une loi de coefficient de frottement interfacial fonction de la vitesse relative des 2 phases et du taux de vide est ajustée entre ces 2 limites d'écoulement tel qu'indiqué sur la figure 5b. Le principe de ce raccordement est la continuité de ce coefficient et de sa dérivée première en α aux courbes connues aux limites $\alpha = 0,3$ et $\alpha = 0,7$. Le paramètre d'ajustement dans la comparaison calcul-essais est la position du point d'inflexion sur la courbe dans la région $0,3 \leq \alpha \leq 0,7$.

CONCLUSION

Les premières observations de mauvaise distribution diphasique en écoulement adiabatique ascendant, par retournement en fin de passe d'évaporateur tubulaire révèlent un comportement déjà observé dans les études de redistribution sur échangeur à plaques : la phase liquide tend à se plaquer sur le fond de la boîte de retournement pour gagner préférentiellement les tubes supérieurs de retour, (l'air passant préférentiellement par les tubes inférieurs).

Un caractère nettement tridimensionnel a été observé dont il faudra tenir compte pour l'optimisation de la géométrie de la boîte de retournement.

Cette optimisation pourra être facilitée par le calcul si l'ajustement de la loi de frottement interfaciale s'avère réalisable à partir des résultats expérimentaux.

REFERENCES

[1]   JOULE contrat -0039 -C (programme JOULE 1 : Mars 1990).

[2]   G. RATEL, P. MERCIER : Présentation de la version 4 du logiciel CETUC, note interne GRETh 90/218 - Grenoble.

[3]   A. ANTARIKSAWAN, Influence de la maldistribution d'un échangeur de chaleur sur ses performances thermiques, note interne STT/LPML/89/14/A - Grenoble (Juin 1989).

[4]   MENDES DE MOURA, L.F. : Etude de la distribution d'un écoulement diphasique entre passes d'un échangeur à plaques, Thèse INPG Grenoble 1988.

[5]   MENDES DE MOURA, L.F., Experimental and numerical study on the 2 phase flow distribution between 2 passes of a heat exchanger, Proc. ASME Winter 1990 annual meeting, Dallas, Nov. 1990.

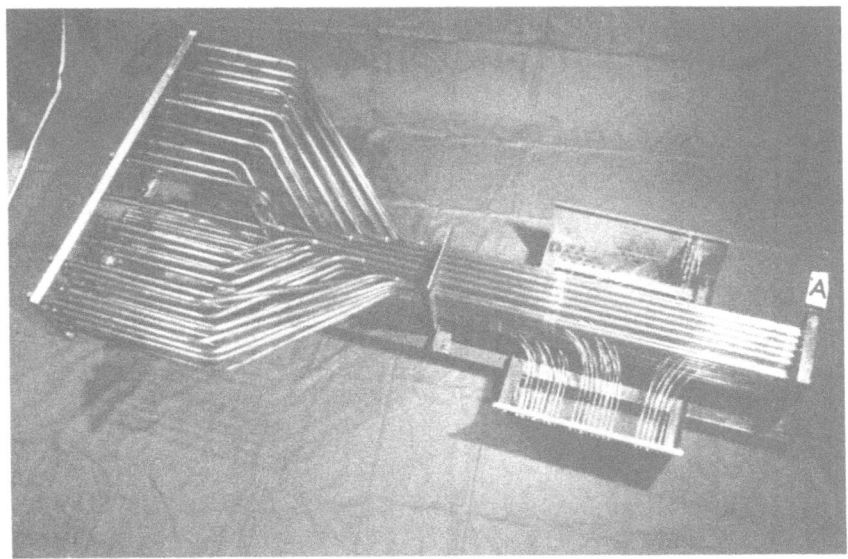

Photo 1 : BEATRICE-2 : Maquette d'échangeur: passe aller,
passe retour et plaque tubulaire (A) pour boîte de
retournement

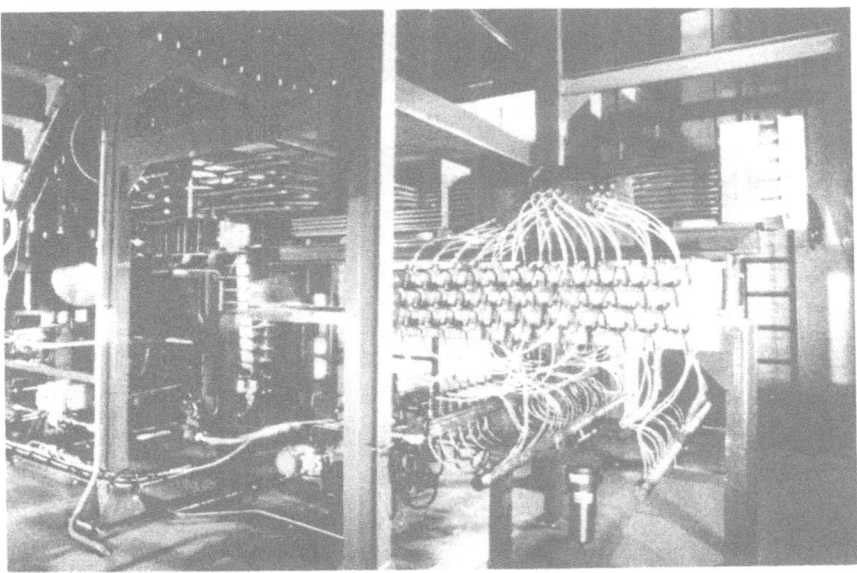

Photo 2 : Installation expérimentale BEATRICE-2

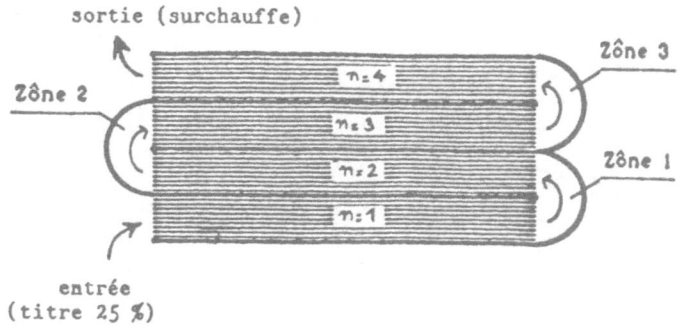

Figure 1: Evaporateur CIAT 4 passes testé par HWU

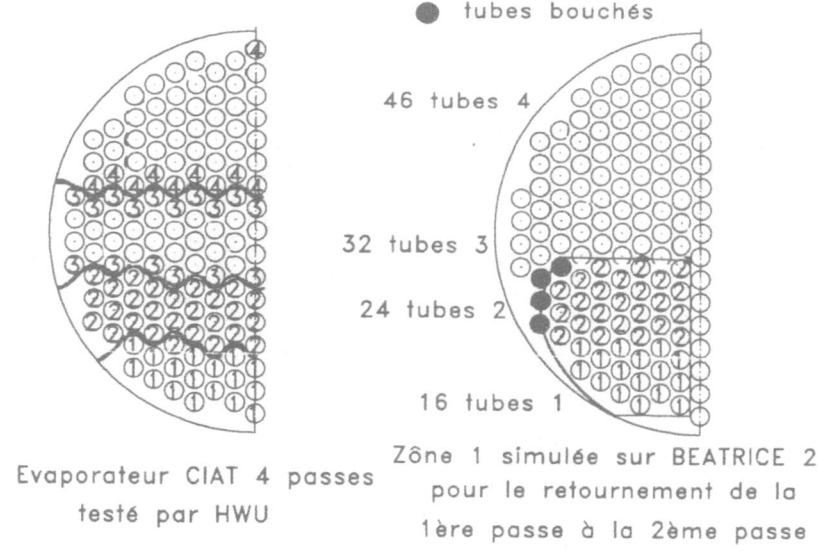

Evaporateur CIAT 4 passes
testé par HWU

Zône 1 simulée sur BEATRICE 2
pour le retournement de la
1ère passe à la 2ème passe

Figure 2: Simulation BEATRICE 2

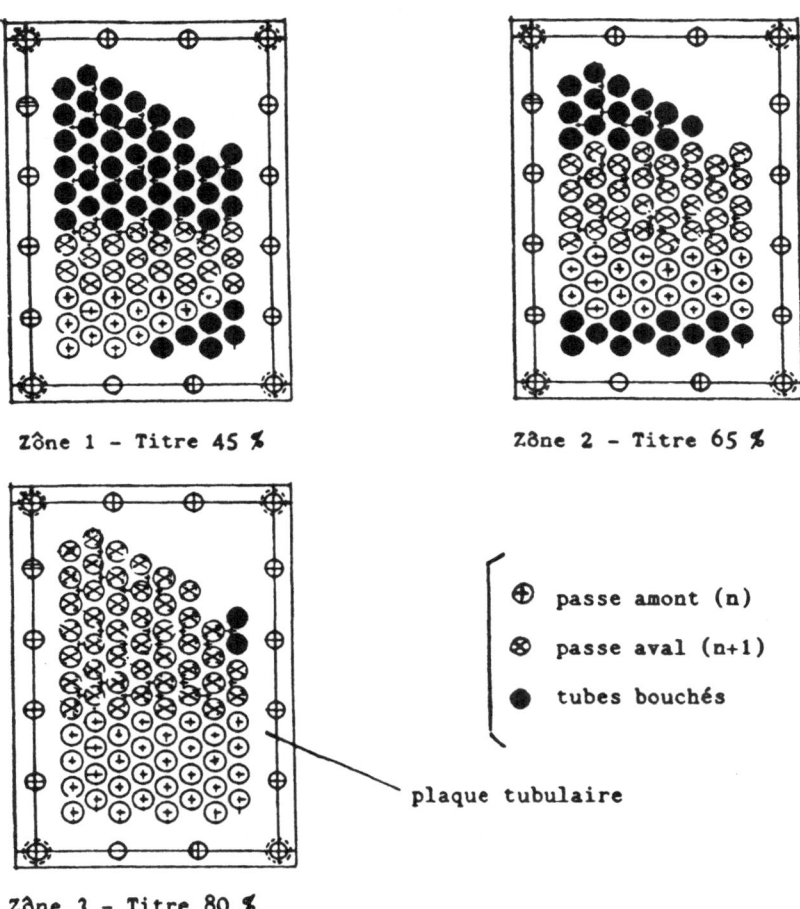

Zône 1 - Titre 45 %

Zône 2 - Titre 65 %

Zône 3 - Titre 80 %

⊕ passe amont (n)

⊗ passe aval (n+1)

● tubes bouchés

plaque tubulaire

Figure 3: Principe de simulation des trois zones de retour sur la maquette d'échangeur par bouchages successifs de tubes

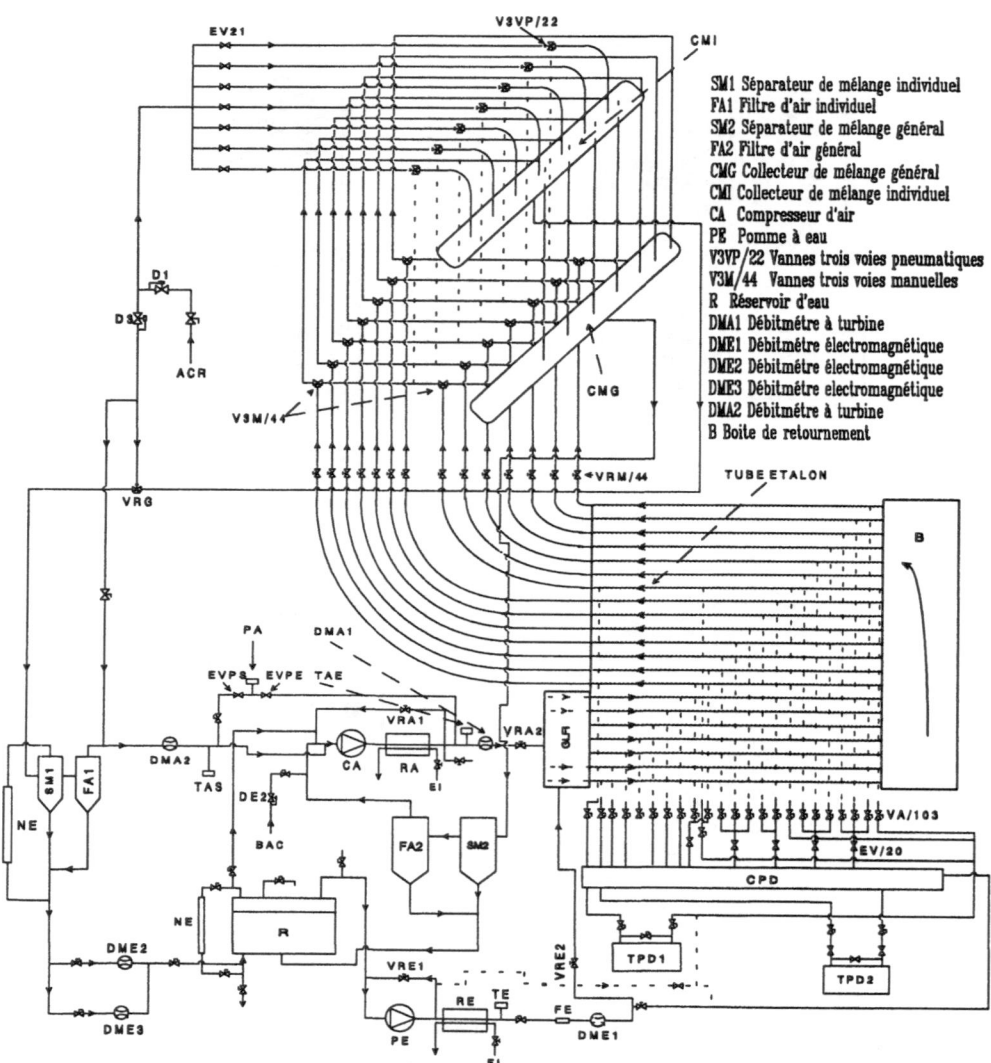

Figure 4. SCHEMA DE PRINCIPE DE L'INSTALLATION BEATRICE2

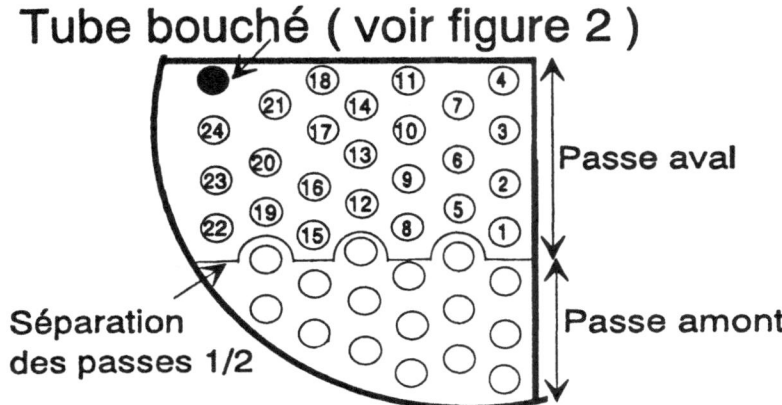

Figure 5a. Numérotation des tubes

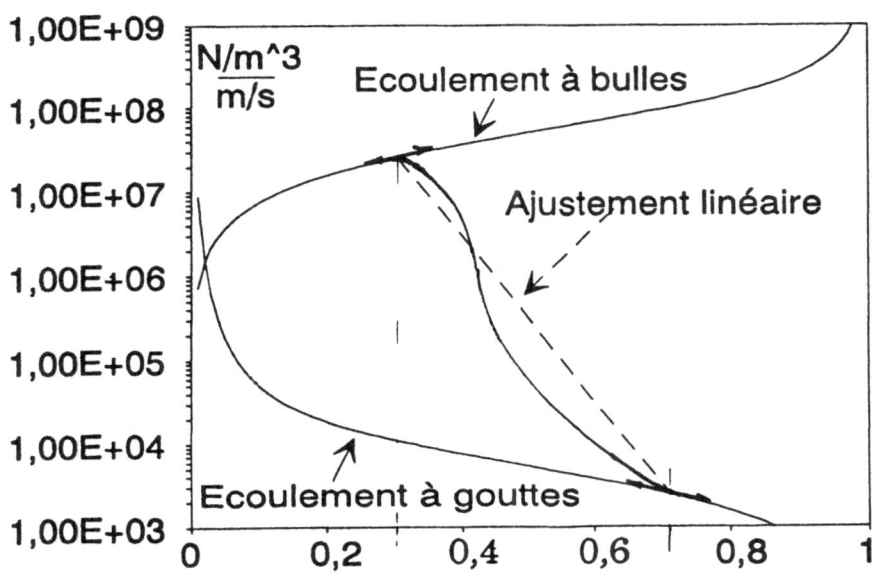

Figure 5b. Variation du coefficient de frottement interfacial en fonction du taux de vide

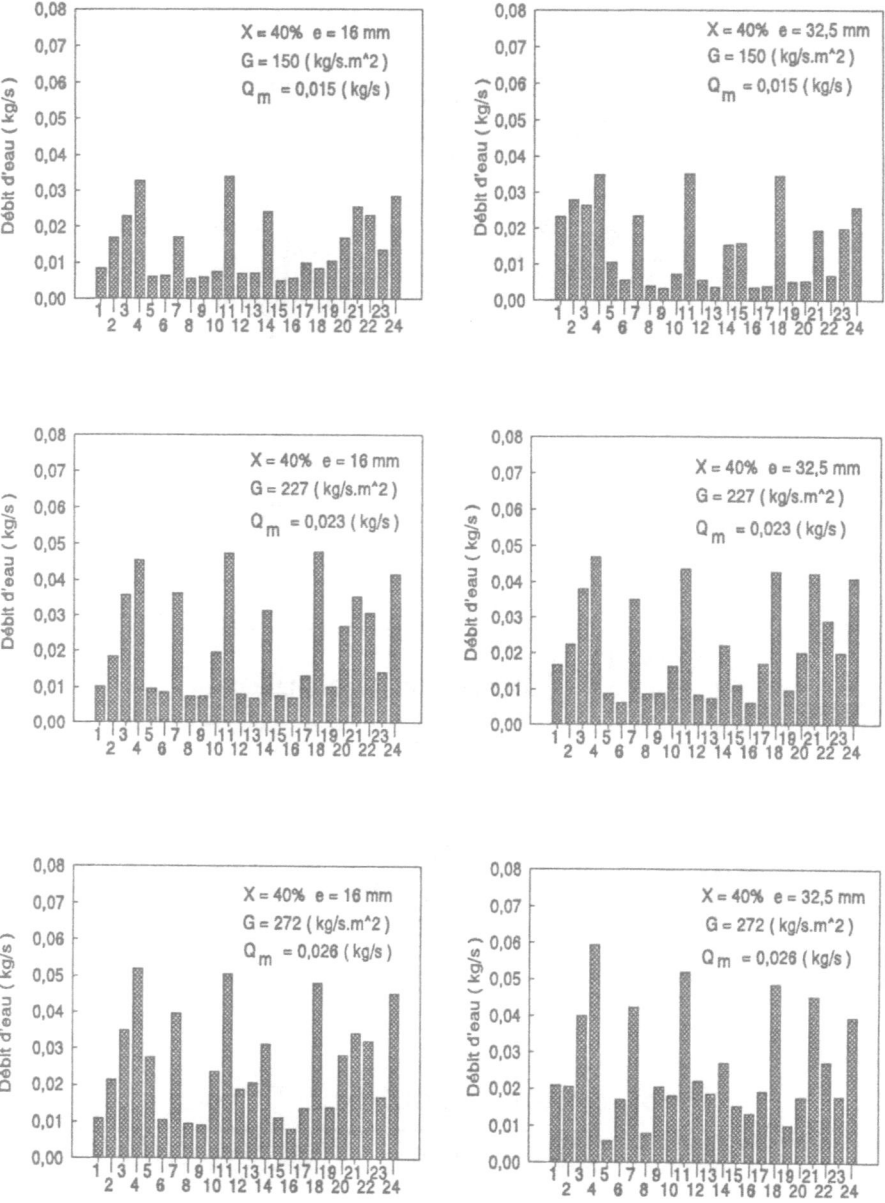

Figure 6. Redistribution de l'eau dans les tubes de la passe aval dans la zone 1 pour deux profondeurs de boîte (16mm et 32,5 mm), pour trois vitesses massiques G (150, 227 et 272 kg/s.m^2) et pour un titre de 40%

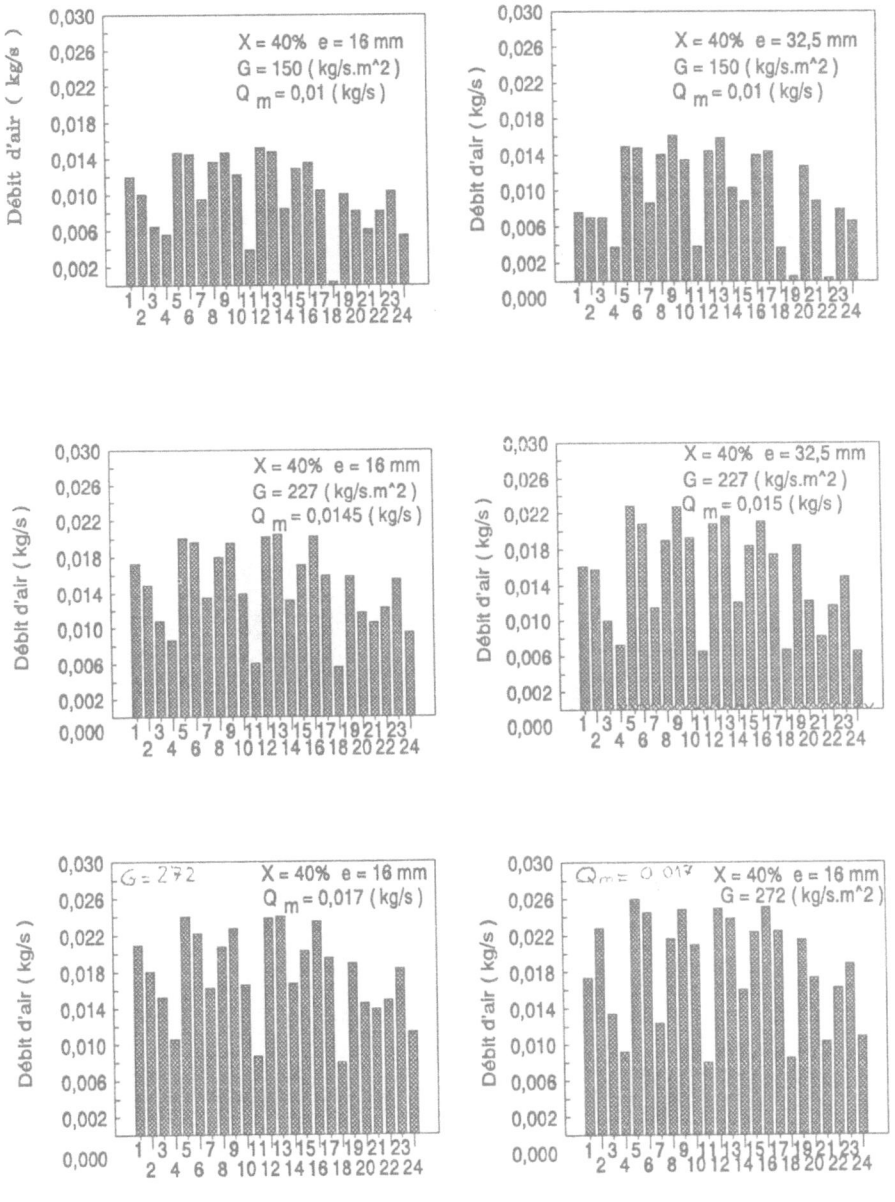

Figure 7 . Redistribution de l'air dans les tubes de la passe aval dans la
zone 1 pour deux profondeurs de boîte (16 mm et 32,5 mm) pour
trois vitesses massiques G (150, 227 et 272 kg/s.m^2) et pour
un titre de 40%

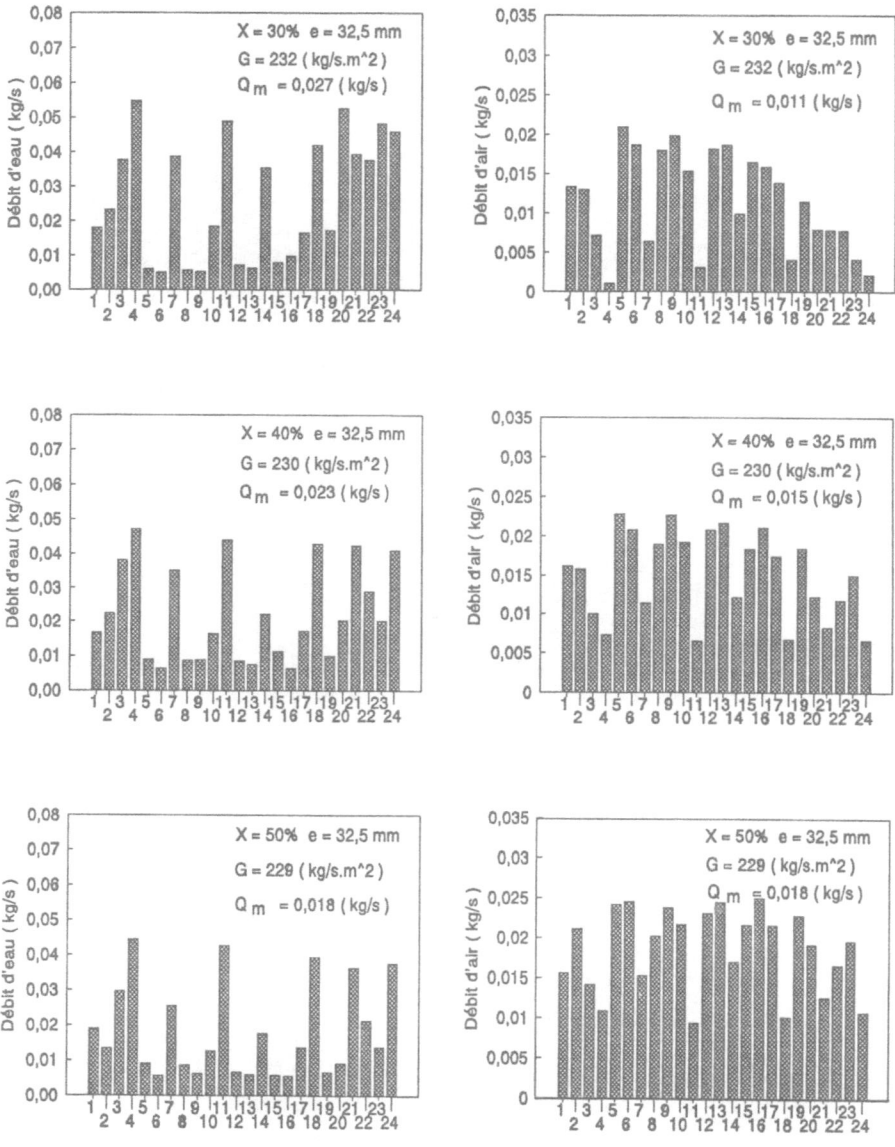

Figure 8. Redistribution de l'eau et de l'air dans les tubes de la passe
aval dans la zone 1 pour une profondeur de boite (32,5 mm),
pour une vitesse massique G (230 kg/s.m^2) et pour trois
titres massiques (30%,40%,50%)

HEAT PIPES HEAT EXCHANGERS AND ENERGY EFFICIENCY IN PROCESS TECHNOLOGY

L.L.Vasiliev
Luikov Heat & Mass Transfer Institute Byelorussian Academy
of Sciences,220728 Minsk,P.Brovka str.15, Republic Byelorus

ABSTRACT

The Luikov Heat & Mass Transfer Institute (LHMTI) together with a
number of other institutions of former USSR and Czeskoslovakia
have designed and fabricated a number of new types heat pipe heat
exchangers (HPHE) for different applications in industry and
agriculture (to preheat fresh air,to heat water,for steam
producing , for waste heat recovery in farms).
   For example HPHE (UTF-12) is fabricating in Republic
Byeloruss during 5 years and is used in the air-condition systems
for the agricultural establishments .
Air preheaters based on carbon steel HPs with naphthalene or water
had been tested during more then one year in one of the plants in
Minsk with the efficiency of 50 % and was successfully used for
heat recovery from the flue gases.

INTRODUCTION

HPHE could be considered as a new and very attractive heat
transfer devices for heat recovery. Usually HPHE are made as a
tube bundle of HPs,with the evaporator and condenser isolated,so
as not to mix the high temperature gas flow from the low
temperature one.
   Such a device is suitable for heat transfer between different
fluids of any kinds,including gas,liquid,steam or one containing
solid particles. The developed heat transfer surfaces on both
sides of HP, which are usually finned,ensure a large heat exchange
area and make this system most suitable for gas-gas heat
exchangers.
   The most important features of HPHE are:
- high thermal efficiency; large design flexibility ; excellent
mechanical isolation between the two fluids,as HP has a double

wall between fluids, simple realization of counter flows, small labor consumption in assembling and dismounting of a unit; effective and rapid cleaning of the heat transfer surface; possibility to adjust the heat exchange surface temperature according to the choice of the heat exchange areas on the hot and cold side; the hot and cold exchange surfaces are independant.

HPs guarantee constant temperature over all its surface, avoiding hot spots , local condensation and corrosion on cold points; high reliability and small maintenance costs; simple combination of HPHE with heat pumps and heat storage systems ; minimum capital costs of production due to wide unifications of the dimentions.

At this time there are a certain number of publications on HPHE (1-4).

A very good review of gas-gas heat recovery systems was made by D.Reay (1) , where some of advantages of HPHE were stressed .

Some of new types of HPHE designed in the USSR were described by L.Vasiliev in (2) .

HPHE activity in the East European Countries was presented by F. Polasek (3). M. Groll gave a detailed analysis on HP research and development in Western Europe (4) , and F.Dobran presented the review on HP activity in the Americas (5) .

A current survey on the HPHE activity was presented at the 5-7 IHPC . These surveys took into account the outcome of investigations in Europe, Japan and USA.

ANALYSIS.

HPHE analysis consists of breaking down of an exchanger into elementary evaporators and condensers and using a procedure of a calculation by NTU/Thermal Efficiency method. The final efficiency of the HPHE is obtained by coupling one evaporator with one condenser (one heat pipe) then by associating some of the HPs to a HPHE. Then HPHE could be analyzed from the point of view of the global efficiency and it is used to determine the fluid temperature and heat flow .

The efficiency of a one-row HPHE may be defined:

$$\eta_o = \frac{2 \pi N d L_m}{2 R_{ox} N w_{min} + \pi d L (N + 1) m} \qquad (1)$$

where N is the ratio of water equivalents of heat transfer media, L is the HP length, m is the number of HPs in row, w - the water equivalent of the heat transfer medium, d - the diameter of the finned tube, R_{ox} - the total thermal resistance of HP.

The efficiency of the HPHE consisting of n rows is equal to:

$$\eta = 1 - \left[N^{n-1} (1 - \eta_o)^n \right] / \left[(N - \eta_o)^{n-2} (N - \eta_o^2) + \right.$$

$$+ \, \eta_o \sum_{i=1}^{n-2} N^i \, (\, N - \eta_o \,)^{n-i-2} \, (\, 1- \eta_o \,)^{i+1} \, \Bigg] \quad . \tag{2}$$

Thus HP HE efficiency is :

$$\eta = \frac{W_{max} \, (\, T_1 - T_{2n+1})}{W_{min} \, (\, T_1 - t_{2n+1})} \quad . \tag{3}$$

The total heat flux is :

$$Q = G_1 c_{p1} \, (\, T_1 - T_{2n+1} \,) \quad . \tag{4}$$

The heat flux for each HP row is :

$$Q_k = G_1 c_{p1} \, (\, T_{2k-1} - T_{2k+1}), \tag{5}$$

where T , t - the temperature of a hot and cold gas, passing through the kth HP row; k -the row number and n is the number of rows in HP HE.

Now there is a number of computer programmes for such a modelization which can be used for HPHE design .

INDUSTRIAL HPHE APPLICATION FOR HEAT RECOVERY.

1. AIR CONDITIONING SYSTEMS ON HPHE .

In 1985 the Luikov Institute together with a Central Special Design Office,Brest put into operation a HPHE - (UTF-12) for the agricultural and industrial establishments. UTF-12 was made up of radially extruded aluminum finned pipes with the length 2 m (Fig. 1), working liquid - ammonia .

Now it is clear that besides the main goal to save energy for instance weight increments in calf and pig rearing have grown by as much as 10 - 12 % as a result of better ventilation system using heat recovery. The same results were pointed out for the poultry farms.

Fig. 1 Heat pipe heat exchanger UTF-12 for the air-condition
system in agriculture.

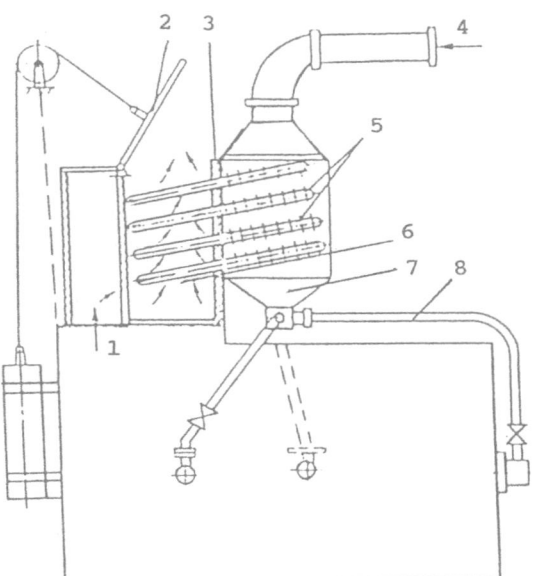

Fig. 2 Heat pipe heat exchanger to recover the energy of flue
gases from a gas furnace :
1 - flue gas; 2 - flue gas oven-door ; 3 - splitter
plate; 4 - cold air ; 5 - heat pipes ; 6 - by pass
flue gases; 7 - hot air collector ; 8 - combustion
hot air in .

Using UTF-12 there is a possibility to regulate the temperature and humidity of space all the year automatically and prevent frost generation on heat transfer surfaces even at low temperatures (till - 14°C).

The main parameters of UTF-12 :

1. Fresh air flow rate ,m^3/ h   12 000
2. Humid air flow rate ,m^3/ h   12 000
3. Heat flow recovered , kW   65
   ($\Delta T = 40^{o}$C)
4. Thermal efficiency , %   55
5. Mass of device , kg   2150
6. Dimensions   m   3 x 2,5 x 1,4

UTF -12 for the climate of Byelorussia gives the economy of 15 tons of petrol per year. Pay back period is 1.5 year .
Now there is a modification of UTF-12 for the air flow rate up to 18 000 m^3/ h .

2. HEAT RECOVERY ON FLUE GAS FOR THE COMBUSTION AIR PREHEATING.

To improve the effectiveness of thermal kilns and stimulate heat recovery a new type HPHE designed in the Luikov Institute was used on one of the plants in Minsk (Fig.2).

This HPHE recovers on flue gas coming from a gas furnace and preheats fresh air going to the gas burner of the furnace.
In such a type of HPHE as a working liquid could be used water,diphenil and naphthalene. Their operating temperature is below 400°C. To extend this limit of temperature some HPs using liquid metals could be used .
For our case 112 finned carbon steel thermosyphons with water and naphthalene as a working liquid were chosen.
Their length was 2 m , d$_{in}$ = 25 mm , d$_{fin}$ = 56 mm .
The first 3 rows of thermosyphons from the hot side of flue gases were filled by naphthalene, the other 3 rows of thermosyphons were filled by distilled water .
The HPHE was installed on the top of a furnace and has an angle of inclination 15° to the hot side of the heat exchanger .
There was ensured the possibility to change the flow rate of flue gases and air through the finned surface on heat exchanger during its exploitation .
The main parameters of gas and air flows on such a heat exchanger are:

- hot gas flow rate   - 2300 m^3/ h
- hot gas temperature   - 600 - 800°C
- fresh air flow rate   - 2030 m^3/ h
- fresh air inlet temperature   - 30° C
     fresh air outlet temperature   200 - 230 ^{o}C.

To know what is the influence of dust flow on the heat transfer surface in the hot side of HPHE some experiments were performed during which the cleaning of the finned surface of HPHE was deliberately omitted .

On the aluminum fins of HPs evaporators in the flue gases chamber with dust particles after 2 month of use there was a reduction of flue gas rate through the fins to 15 % of the original value and reduction of thermal efficiency from 45 % to 20 % .

What is essential as the result of these experiments that there were no signs of corrosion of the aluminum fins and steel surfaces of the thermosyphons.

It means, that the problem of fouling of the HPHE in such a situation is more important than the problem of external corrosion.

The life of the HPHE mostly depends on the life of the HP itself, and the period of working time of the carbon-steel naphthalene and carbon steel - water thermosyphons during 1 year was checked without any important changes of parameters.

Therefore we have an opportunity to use such type of HPHE for 5 years of exploitation,but it is necessary to provide the possibility of HPs repair and regeneration.

CONCLUSIONS.

Some of the institutions of the CIS Countries including the Luikov Heat & Mass Transfer Institute Byelorussian Academy of Sciences,Minsk are involved in the R & D on Hps and HPHEs . The main application of these devices are heat recovery in air conditions systems , flue gases of furnaces and drying chambers .

Since 1985 a number of HPHE were designed and installed on different plants ,agricultural farms and successfully used till now.

REFERENCES

1.   D.A. Reay A Review of Gas-Gas Heat Recovery Systems, J.Heat Recovery Systems, 1 , 3-42,(1981).
2.   L.L.Vasiliev Heat Pipe Heat Exchangers , Minsk , Nauka i Technika (in russian) ,180 p. ,(1981).
3.   J.Ptacnik,F.Polasek Present State of Heat Pipe Technology in the Countries of Mutual Economic Assistance, 3 IHPC - Tsukuba , Preprints,September 12-14 , p. 13 - 41 ,(1988).
4.   M. Groll Heat Pipe Research and Development in Western Europe J.Heat Recovery Systems and CHP , 9 ,19 - 66 , (1989).
5.   F.Dobran Heat Pipe Research and Development in the Americas , J.Heat Recovery Systems and CHP ,9 , 67 - 100 , (1989).

THE PERFORMANCE CHARACTERISTICS OF A WATER/WATER "POLYMER FILM COMPACT HEAT EXCHANGER" (PFCHE)

Dr C.R. Howarth, Dr R.J.J. Jachuck[*] and Prof. C. Ramshaw
Department of Chemical & Process Engineering
University of Newcastle upon Tyne
Newcastle upon Tyne, NE1 7RU, U.K
(* For communication)

ABSTRACT

Heat transfer and flow friction data has been obtained for a polymer film compact heat exchanger (PFCHE), when operating with a water/water system. The PFCHE was constructed using a 100μm thick film of PEEK (poly ether ether ketone) with 1mm high corrugations. PEEK withstands a temperature of 523 K and has excellent chemical resistance.
Design information in the form of "j_h" and "f" factor charts is presented. Both the heat transfer intensity / unit matrix volume and the power loss incurred in achieving a given heat transfer rate, were much more attractive for the PFCHE than for alternative compact units. Provided matrix fabrication and potential fouling problems can be overcome, PFCHE's could eventually play a major role in heating/ventilating, domestic condensing boilers and low temperature (<523 K) process heat exchangers.

NOMENCLATURE

A = Real heat transfer area (m^2)
C = m C_p (W/K)
C_h= Constant
C_p= Specific heat at constant pressure (J/kg K)
D_h= Hydraulic diameter of the channel (m)
E = Thermal effectiveness
  = Actual heat transfer / maximum possible heat transfer
f = Fanning friction factor
h = Heat transfer coefficient (W/m^2 K)
j_h= Colburn factor
k = Thermal conductivity (W/m K)
L = Length of the uninterrupted flow path (m)

l = Total exchanger length (m)
m = Mass flow rate (kg/s)
N = Number of transfer units(NTU) = $(UA)/C_{min}$
ΔP= Pressure drop (Pa)
ΔT= Temperature difference (K)
U = Overall heat transfer coefficient (W/m^2 K)
u = Linear fluid velocity (m/s)
w = C_{min} / C_{max}
X = Thickness of the polymer film (μm)
y = Constant
Z = Thermal concentration parameter
Nu= Nusselt number (h d / k)
Pr= Prandtl number (C_p ρ/ k)
Re= Reynolds number (D_h u ρ/ μ
St= Stanton number(Nu / Re Pr)
ρ = Density of the fluid
τ = Shear stress at the wall
Φ = Viscosity ratio
θ and α= Constants

INTRODUCTION

In the last few years the topic of intensified heat transfer has made gratifying progress, from university/industrial research laboratory to full scale application in process plant. Much of the initial interest in the general process applications of compact heat exchangers was on laminar flow between closely spaced chemically machined metal plates(1). It was shown that under these conditions, high performance could be achieved even at modest fluid velocities. Extensive work has also been carried out on diffusion bonded printed circuit heat exchanger (PCHE) (2). This consists of a diffusion bonded stack of metal plates 1-2mm thick which has been etched with zig-zag channels in a potentially complex array. The configuration gives excellent heat transfer and has great structural strength in view of the high quality of the diffusion bond (over 90% of the parent metal strength).

Having recognised the cost saving implications of the laminar approach to heat exchanger design, industrial interest is developing in the market opportunities presented by exploiting the latest range of high performance polymers in new heat transfer matrix configurations. For example, poly ether ether ketone (PEEK) and poly imide (Upilex) polymers both have long term working temperatures of at least 523 K, coupled with attractive corrosion resistances. Film thicknesses of 100μm appear to be appropriate on cost grounds. Fortunately a matrix constructed in corrugated 100μm PEEK film is remarkably robust and can withstand a differential pressure of about 10 bar at ambient temperatures. The fouling characteristics of polymer film should be intrinsically superior to those of metal in view of the smooth hydrophobic surface which can be generated.

PEEK is the chosen material for the heat exchanger under study. In addition to its high resistance to chemical attack, it can be used at high temperatures(<523 K) in steam or high pressure water environments without significant property degradation. The only common materials that attack PEEK, it is claimed, are concentrated nitric and sulphuric acids. The material is resistant to 50% H_2SO_4 and 50% NaOH at room temperature and has excellent chemical, thermal and mechanical properties.

THE TEST FACILITY AND CALCULATION TECHNIQUE

Tests were carried out on a polymer matrix compact heat exchanger, which is schematically shown in Figure 1. The heat exchanger consisted of 44 corrugated PEEK sheets (13.5 cm x 13.5 cm x 100μm), stacked vertically in cross corrugation and housed in a PERSPEX block.

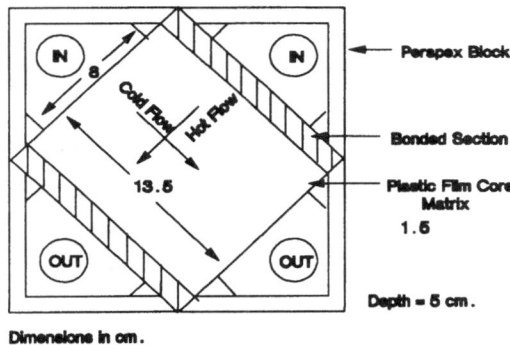

Dimensions in cm.

Figure 1. Schematic diagram of the test facility.

As shown in Figure 2 the polymer sheets were arranged such that a perfect cross flow was obtained and there was negligible lateral mixing of each stream as it flowed along the channel. This was achieved by carefully sealing the edges of the sheets, such that successive layers formed a cross corrugation (i.e corrugations cut one another at right angles as they run along the length of the heat exchanger). The compressed sheets were placed in a square PERSPEX block so that four isolated compartments one each for the inlet and outlet of the hot and cold streams respectively were formed. The heat loss through the walls of the PERSPEX block was negligible.

In order to accurately measure the inlet and the outlet temperatures, k-type thermocouples were placed mid way down the exchanger stack, in the inlet and outlet chambers as shown in Figure 1 , and the temperature recorded by a digital unit. In the outlet chambers of the hot and the cold streams, paddle mixers were used, as recommended by Kays et al(3), so that a uniform temperature distribution existed both laterally and vertically in the chambers. To measure the pressure drop, tappings in the form of glass tubes were taken from the inlet and the outlet chambers, which were connected to an inverted manometer, by means of flexible tubes. The inverted manometers had a sensitivity of ± 9.8 Pa. The characteristic dimensions of the heat exchanger matrix are presented in Table 1.

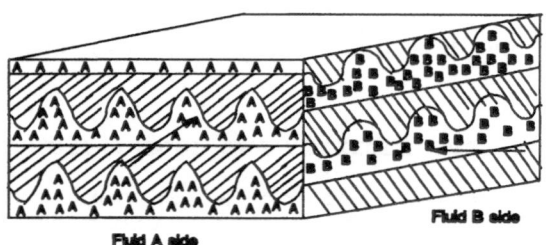

Fluid A side   **Fluid B side**

A – Channel for flow of fluid A.
B – Channel for flow of fluid B.
//// – Sealed channel

Figure 2. Flow distribution in the exchanger matrix.

TABLE 1
Details of the Heat Exchanger

Superficial heat transfer area	$= 0.624 \ m^2$
Real heat transfer area (A)	$= 1.21 \ m^2$
Hydraulic diameter D_H	$= 6.6 \times 10^{-4} \ m$
L/D_H	$= 3.03$

Determination of the Overall Heat Transfer Coefficient

The terminal temperatures, measured when the system had achieved a steady state, were used to calculate the overall heat transfer coefficient. The unit film conductance of one side was calculated by assuming equal conductance on both sides, as the flow rates of the cold and the hot water streams were equal and their channel geometries were identical.

$$1/U = 2/h + X/k$$

To calculate the overall coefficient(U), the simplified formula for cross flow heat exchanger effectiveness, presented by Baclic[4], using the modified Bessel functions(I_n) has been used.

$$E = 1 - \exp[-(w + 1)N][I_0(2N(w)^{1/2}) + (w)^{1/2}I_1 (2N(w)^{1/2}) - (1-w)/w \ \Sigma \ w^{n/2} \ I_n(2N(w^{1/2})] \qquad (1)$$

For a well balanced flow(w≪1), as is in the present study (i.e $w = C_{min}/C_{max} = 1$), the above relation reduces to:

$$E = 1 - \exp (-2 N [I_0 (2 N) + I_1 (2 N)]) \qquad (2)$$

From equation 2, the thermal effectiveness (E) was calculated and plotted for corresponding values of N. The plot was then used to estimate the overall heat transfer coefficient U. This technique is most accurate when NTU < 5.

RESULTS

The water flow rate was varied from 0.036 l/s to 0.416l/s, corresponding to a Reynolds number range of 27 - 311 and the mass flow rates of both the hot and the cold streams were approximately equal.

The heat balance for all the runs was well within ± 3% and the the thermal effectiveness of the exchanger varied from 0.59 to 0.42 corresponding to Re(Reynolds number) values of 27 and 313 respectively and was well within the sensitive region of the E / NTU relationship.

The pressure drop of the two streams, measured during the study, was used to calculate the friction factor (f) by using the following expression(5):

$$\Delta P = 2 \ f \ \rho \ u^2 \ l \ / \ D_h \ldots\ldots\ldots\ldots (3)$$

DISCUSSION

Effect of the film thickness

The corrugated film was produced from flat film having a nominal thickness of $100\mu m$ by pressing in a heated die. While the superficial film area was held constant, the pressing procedure caused local film thinning. Since the actual film area is readily determined by stretching the corrugations until they are flat, the average film thickness is reduced by a factor F which is the ratio of the corrugated surface area to the superficial (projected) area.

Since the resultant film resistance is approximately $1/4500$ (m^2 K/W), it exerts a significant limitation on the overall heat transfer performance when the fluid film coefficients exceed about 2 kW/m² K. This corresponds to a Reynolds number with the present matrix of 500. Therefore for heat transfer duties where both the fluids involved are liquids, there is little incentive for operating with Reynolds numbers much beyond 1000. When one of the fluids is a gas then its film coefficient will be limiting and so higher Reynolds numbers can be exploited (pressure drop permitting).

Correlation for the j_h factor(laminar condition)

Using the experimental results an attempt has been made to present a relation between the j_h factor and the Reynolds number for a polymer film compact heat exchanger. The j_h factor, usually called the Colburn factor for heat transfer is defined as follows:

$$j_h = St.Pr^{2/3} \ \text{or} \ j_h = \theta \ Re^{\alpha}$$

Figure 3 shows a plot between $\log(j_h)$ and $\log(Re)$, and is used to derive:

$$j_h = 0.0213 \ Re^{-0.263}$$

The coefficient of regression is 0.9627 and the standard error in calculating the exponent of the Reynolds number is 0.0204.

The j_h factors given in Figure 3, have been quoted on the basis of the total (i.e the corrugated) film area. When the resultant h and U values are computed they should be multiplied by the area factor F, defined above to convert them to the equivalent values based on the superficial area. This corresponds to an enhancement of approximately 80%.

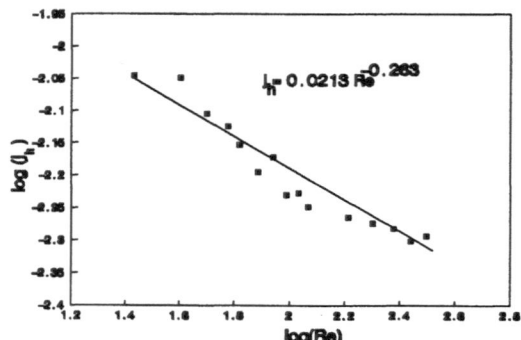

Figure 3. Heat transfer results for water in the polymer matrix
(Based on total film area)

The heat transfer characteristics of the polymer film compact heat exchanger may be compared with that of a typical plate frame and shell-tube exchanger. The details of the plate frame and the shell-tube exchangers, which have been used for a comparative study are presented in Ref(6), and the correlations which have been used to calculate the film coefficients are as follows:

(i) Plate Frame Exchanger (7)

$$Nu = h \, D_h / k = j_h \, Pr^{0.33} \, \Phi^{0.17} \qquad Re < 500$$
$$h = j_h \, k \, Pr^{0.33} \, \Phi / D_h \qquad\qquad \text{for} \quad 20 < Re < 500$$

(ii) Tube side coefficient in a shell-tube exchanger(7)

$$Nu = h \, D_h / k = 1.86 \, (x/D_h)^{-1/3} / Pe^{-1/3} \, f^{0.14} \quad \text{for } Re < 2000$$

(iii) Polymer film compact heat exchanger

$$h = 0.0213 \, Re^{-0.263} \, Pr^{-0.66} \, C_p \, \rho \, u \quad \text{for} \quad Re < 1000$$

Figure 4 shows the laminar flow film coefficients for a typical plate frame, shell-tube and polymer film compact heat exchanger. It is apparent that the film coefficient achieved by the PFCHE in this flow regime is much higher than both the plate frame and the shell-tube exchangers.

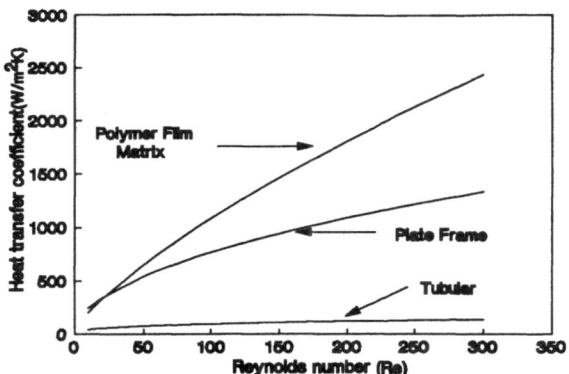

Figure 4. Study of the laminar flow film coefficients.

However, plate frame and shell-tube exchangers would seldom be operated at such low Reynolds number and hence a direct comparision of the above nature would be misleading. A more significant comparison is to evaluate the respective <u>volumetric</u> heat transfer performances for each unit while operating at its typical Reynolds number and at a fixed temperature difference of 5 K.

The ratio between the heat transferred and the volume of the heat exchanger may be defined as follows:

Z = Heat transferred / volume of the exchanger matrix
  = U A T / Effective Matrix volume

A temperature difference(ΔT) of 5 K has been considered for calculating the heat transferred per unit matrix volume and the U values for both the plate frame and shell-tube exchangers, have been taken from Ref(6). The overall heat transfer coefficient (U), which has been used in calculating the Z parameters for plate frame, shell-tube and the polymer film exchanger respectively are as follows:

(a) Plate frame at Re = 3000,   U = 6000   (W/m^2 K)
(b) Shell-tube   at Re = 50000, U = 1000   (W/m^2 K)
(c) Polymer film at Re = 1000,   U = 5000 (Based on the superficial area & taking the film resistance into account)
Re of 3000, 50000 and 1000 have been used for the plate frame, shell-tube and PFCHE exchangers respectively in order to simulate actual operating flow conditions.

From Table 2 it is apparent that Z_{PFCHE} / Z_{PF} = 3.0 and Z_{PFCHE} / Z_{ST} = 68.5, suggesting a significant increase in heat transferred per unit matrix volume for a polymer film compact heat exchanger. This is a reflection of the high value of the F factor, and highlights the advantage of the extreme compactness of the the PFCHE.

F(factor) = Real area / Superficial area

TABLE 2
Heat transferred per unit exchanger volume (Z parameters)

TYPE	Q / Effective Matrix vol. (For temp. difference of 5 K)(MW / m^3)	Re
PFCHE	24.0	1000
Plate frame	7.83	3000
Shell-tube	0.35	50000

Details of the Plate frame (PF) & the S-T exchangers have been taken from Ref(7).

So far only the heat transfer performance of the PFCHE has been studied and so to get an overall view of its performance the resistance offered to the fluid flow by the polymer matrix is now considered.

Correlation for the friction factor (f)
The pressure drop needed to overcome the wall shear stress generated when a fluid flows along a channel is given by equation (3) and in order to relate the friction factor (f) and the Reynolds number, Figure 5 was generated from the pressure drop data obtained for PFCHE, and a friction factor design correlation given by

$$f = 14.12 \ Re^{-1.06}$$

was obtained.

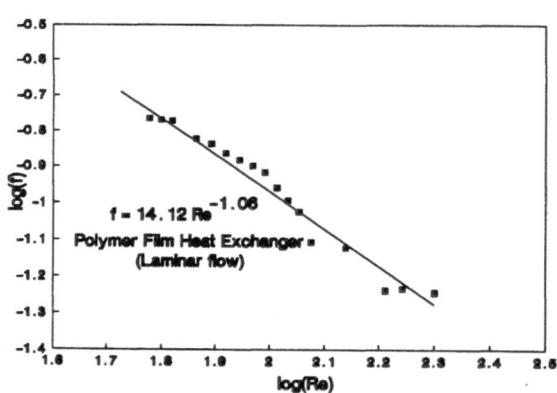

Figure 5. Friction factor results for water in the polymer matrix.

In order to get a complete picture of the heat exchanger performance, the pumping power necessary to achieve a given film coefficient, which will be of significant design interest, may be evaluated.

Heat transfer coefficient vs Pumping power loss
In Figure 6, the heat transfer coefficient based on the superficial (projected) area is plotted against the the pumping power = shear stress x velocity, for the PFCHE and also for the plate frame and the shell-tube exchangers.

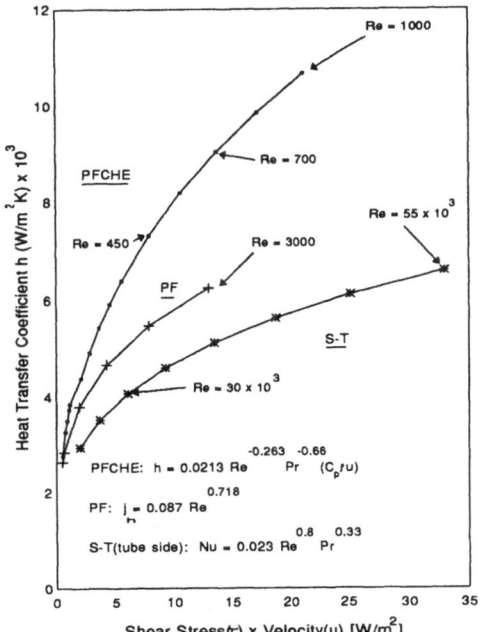

Figure 6. Study of the Pumping Power Loss.

It is apparent from Figure 6, that for a typical value of heat transfer coefficient of about 6000 W/m² K, the energies lost due to the fluid flow resistance are as follows:
(a) PFCHE   5 W/m²
(b) Plate frame   10 W/m²
(c) Shell tube   25 W/m²

It may therefore be concluded that there could be a significant energy saving if a plate frame exchanger is replaced by a polymer film compact heat exchanger(PFCHE). However the plate frame and the polymer film exchangers have their own merits and demerits, which should be taken into account before deciding on the type of exchanger to be used.

CONCLUSIONS

The following conclusions can be drawn from this investigation:
(1) Heat transfer and pressure drop data in the form of j_h and f factors have been obtained for a corrugated polymer film (PEEK) heat exchanger, using a water/water fluid system and this is hoped will allow preliminary estimates of the configuration and cost of polymer film heat exchangers to be made, with a view to the subsequent exploitation of this technology.

(2) Experimental limitations of the apparatus restricted the permissible pressure drop and hence the Reynolds number(Re) to 300.

(3) There is little point in extending Re beyond 1000 since the film thermal resistance then becomes limiting for the water/water system.

(4) For Re=1000 and assuming a mean temperature difference of 5 K, the matrix performance is equivalent to 24 MW heat transferred per m^3 for a total pumping power loss of 20 W/m^2 as compared to 8 MW/m^3 and 15 W/m^2 for a plate frame exchanger operated at Re=3000. Shell and tube produced values well below these.

(5) The PFCHE producing this duty was constructed from PEEK which has excellent corrosion resistance and therefore it will be an excellent choice for cooling flue gas. It is also possible that polymer film compact heat exchangers may eventually replace cooling towers, or in less aggressive areas, such as air/air heat exchangers for heating and ventilating, where cheaper plastics such as PVC or poly propylene may possibly be used.

(6) The heat transfer resistance of a 100μm PEEK film corresponds to a film coefficient of about 2.5 kW/m^2 K. While this is unlikely to be limiting in heat transfer situations involving gases, some performance penalty may be incurred if both fluids are liquids.

(7) As the coefficient of linear expansion of PEEK is quite high (10^{-4} K^{-1}) compared with metals (10^{-5} K^{-1}) it may be able to shed any scale deposit, by repeated expansion and contraction. However the topic of fouling in polymer film matrices is sufficiently important and poorly understood, to warrant an extensive research programme in its own right.

ACKNOWLEDGEMENT - The authors wish to express their appreciation for the financial support from the Energy Efficiency Office (U.K).

REFERENCES

1. Cross, W.T, and Ramshaw, C., Chem Eng Res Des, July 1986, vol.64,pp.293.

2. Johnstone, A., The Chemical Engineer, Dec 1986, pp. 36.

3. Kays, W.M., and London, A.L., Trans. ASME, November, 1950, pp 1075-1080.

4. Baclic, B.S., J.H.T. ASME, 1978, Vol 100, pp 746-747.

5. Kern, D.Q., Process Heat Transfer, Mc Graw-Hill, 1988.

6. Saunders, E.A.D., Heat Exchangers- Selection, Design & Construction, Longman Group, U.K, 1988.

SYSTEMS CONTAINING CONTACT ECONOMIZERS FOR FLUE GAS HEAT UTILIZATION

Prof. NIKOLAI KOLEV, DSc, Assoc.Prof. RUMEN DARAKCHIEV[*], DSc, Assoc.Prof. KRUM SEMKOV, PhD

Institute of Chemical Engineering, Bulgarian Academy of Sciences, Acad. G. Bonchev, Bl. 103, Sofia 1113, Bulgaria

[*]Institute of Heat Power Engineering, Sofia, Bulgaria

ABSTRACT

A reference is made to the fact that a significant part of the heat generated by combustion, is lost in the atmosphere with exhausted flue gases. Contact economizer systems are considered which help to utilize this heat by heating up the water before entering the boiler or by heating up urban heat supply water of the central heat supply network, as recycled back to the heating plant. Attention is drawn to some features of the new contact economizers which allow pressure drop values as low as 10-20 mm water head at high gas superficial velocities, i.e. up to 3 m/s. New process flow charts with contact economizers which helping to utilize heat at higher temperature levelsby simultaneous reduction of nitric oxides emissions down to 33 %, are given. The method developed for designing such systems is described, and data obtained in experiments with such industrial systems developed by the authors, are presented.

INTRODUCTION

It is well known that flue gases carry away in the atmosphere a huge part of the heat generated by combustion. Besides, the higher the temperature and steam content of the flue gases, the higher the losses. (One kg natural gas is equivalent to more than two kg steam production). Apart from the significant economic effect, heat utilization in this case, leads to considerable reduction of the waste gases exhausted into the atmosphere. In addition, the reduction of carbon dioxide emission in the atmosphere is proportional to reducing fuel consumption per unit produced heat.

The flue gas purification processes for removal of sulphur dioxide, in particular when the SO_2 is subject to utilization as concentrated gas, require preliminary cooling of the flue gases in order to improve the conditions for further absorption. The combination of this cooling with waste heat utilization in gases, presents large ecological and economic advantages which can make out of flue gas purification, a profitable process. In the present lecture, the results ar given of our studies in developing novel highly effective flue gas heat utilization plantscarried out in the Institute of Chemical Engineering by the Bulgarian Academy of Sciences in cooperation with the Institute of Heat and Energy Engineering in Sofia.

FIRST GENERATION CONTACT ECONOMIZER SYSTEMS

Waste heat utilization in conventional heat exchangers has two main disadvantages:

1. Very active corrosion in the area of condensation and high temperature;

2. Low gas-side heat transfer coefficient.

Both problems are solved simply by turning from conventional heat exchangers to contact economizers. Figure 1 contains the scheme of such a system [1]. The flue gas leaving the boiler enters contact economizer 2 where it gets scrubbed by the recycle water. The cooled gas goes out the chimney trough a blower The economizer-heated-up water is sent by means of pump 3, into heat exchanger unit 4, where it gives the heat recovered in the economizer to warm up pure water. The excess gas bypasses the contact economizer along line 1, provided with a regulating valve, by heating up

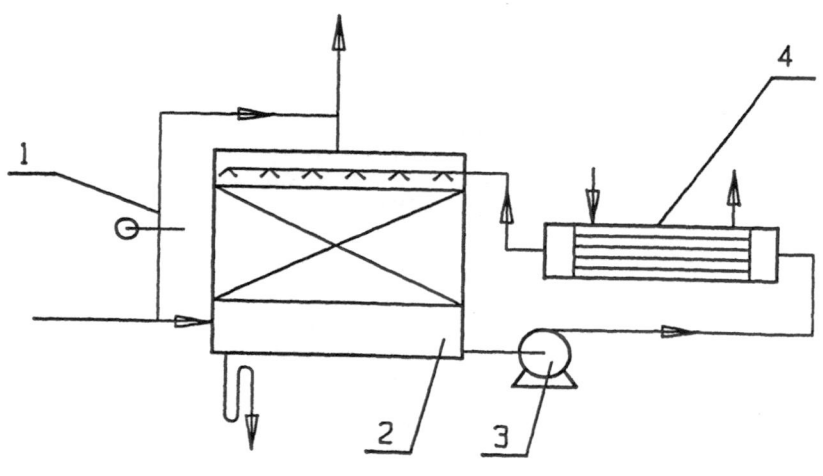

Figure 1.

the cooled flue gas. With regard to the conventional heat exchanger, this more complicated system has the following rather essential advantages:

    1. Cheap and highly effective contacting area is employed in gas phase heat and mass transfer;

    2. The pressure drop is low;

    3. Heat transfer between hot and cold water, especially in a plate heat exchanger, is much more efficient than indirect heating of water with gases. 4. There are no plant sections subject to strong corrosion influence.

    The only drawback of the system is that gas temperature in the initial economizer section is strongly reduced by partial evaporation of recycle water which leads to partial reduction of unit mean temperature difference.

SOME DESIGN FEATURES OF THE CONTACT ECONOMIZER

    As contact economizers, all types of heat and mass transfer units with direct interfacial contact may be used. However, packed columns are most suitable, since among all contactors, these apparatuses operate most near to complete countercurrent flow, and consequently, at the highest rate of heat utilization. Apart from the packed columns [1, 2], contact economizers may be also columns with shelf trays [3]. Wide-spread in the countries of the former Soviet Union, there are also contact economizers [1] loaded with Raschig rings, a packing patented by Dr. Raschig already in 1914.

    The contact economizers, mounted before the boiler flue gas fan in order to reduce the exhaust gas volume, require low gas pressure drop. The comparison of various designs of highly effective packings [4, 5] shows that, with respect to pressure drop per mass transfer unit, the honeycomb-type packings with vertical walls are the most effective ones. Based on these packings, two contact economizer systems with heat utilization capacities of 2,5 and 8 Gcal/h, respectively, were designed, erected and commissioned. The most important problem which was overcome, was to create uniformity of liquid distribution across the column cross section. Uniformity not only increases the process driving force, but also presents an opportunity for a relatively simple and extremely reliable sizing of the columns of that type. Table 1 contains some data of contact economizers developed by our team.

DESIGN OF THE DEVICE

    The heat exchanger unit design is a trivial problem. The design of contact economizers is carried out by solving the set of heat and mass balance non-linear differential equations representing the unit performance. By appropriate selection of a column type, and in particular of a spray liquid distributor, in order to ensure uniform liquid and gas distribution across the column cross section, one may achieve a considerable simplification of the system. A mathematical model for the units, to answer the above

TABLE 1

Performance Characteristics of the New Contact Economizer

Gas velocity relative to the unit cross
section taken for the conditions at the inlet .... 3 m/s

Heated pure water temperature after the
heat exchanger .... up to 55 $^{\circ}$C

Pressure drop low limit .... 9 mm w.h.

Cool flue gas temperature low limit .... 35° C

Waste heat utilized .... up to 0,6 Gcal/(m^2h)

Allowable loading for one economizer............. unlimited

requirements, by eliminating the axial mixing effect as negligible, is given, as follows:

$$G \frac{dX}{dh} = K_G a_e (Y - Y^*); \qquad (1)$$

$$W_0 \rho_G c_G \frac{dt_G}{dh} = K a_e (t_G - t_L); \qquad (2)$$

$$L \rho_L c_L \frac{dt_L}{dh} = K a_e (t_G - t_L) + rG \frac{dX}{dh}; \qquad (3)$$

$$\rho_L \frac{dL}{dh} = G \frac{dX}{dh}. \qquad (4)$$

The following relationships exist between the parameters in these equations:

$$W_0 = G (1 + X)/\rho_G; \qquad (5)$$

$$Y = X\rho_G/(1 + X). \qquad (6)$$

where G is the dry gas mass flow rate in kg/$(m^2.s)$; X is steam mass concentration relevant for the dry gas, in kg/kg; h is the packed bed height (as a current coordinate) in m; K_G is the gas phase mass transfer coefficient in m/s; a_e is packing effective area in m^2/m^3; Y and Y^* are steam performance and equilibrium concentration of gas in kg/m^3; W_0 is the gas superficial velocity in m/s; ρ_G and ρ_L are gas and liquid density in kg/m^3, respectively.; c_G and c_L are gas and liquid heat capacities in J/(kg.K), respectively; t_G and t_L are gas and liquid phase temperatures in $^\circ$C, respectively.; K is the overall heat transfer coefficient in W/(m^2.K); L is the superficial liquid velocity in m^3/(m^2.s); r is the heat of evaporation in J/kg.

The boundary conditions are:

$$\text{at } h = 0: \quad t_G = t_{G1}; \quad X = X_1 \text{ and}$$
$$\text{at } h = H: \quad L = L_1; \quad t_L = t_{L1}.$$

where t_{G1} is the inlet gas temperature in $^\circ$C; X_1 is the vapour concentration at the column inlet in kg/kg dry gas; t_{L1} is inlet liquid temperature in $^\circ$C, and H is the packing height in m.

By using a stacked packing with vertical walls, and with regard to the superficial liquid velocity employed in the contact economizer and in the column, the packing effective area is equal to its specific area [6]. In this case, the mass transfer coefficient can be determined from the Shavoronkov, Gildenblat and Ramm equation [7]. As our studies have shown [8], the overall heat transfer coefficient at these conditions can be determined from the mass transfer coefficient based on the analogy of heat and mass transfer.

THE NEW GENERATION OF CONTACT ECONOMIZERS

Another significant disadvantage of the system with contact economizer, as illustrated in Fig. 1, is that the temperature of the water heated in the system, is limited thermodynamically by the wet bulb temperature value of the flue gas which is much lower than its own temperature.

In order to increase the water final temperature by flue gas waste heat utilization, one may:

1. Combine the system containing contact economizers with a system of corrosion-resistant heat exchangers, in order to carry out an initial gas cooling, or

2. Improve the contact economizer system in a way such as to ensure a flue gas wet bulb temperature rise which leads to a heated water temperature rise. Since the wet bulb temperature, apart from the gas temperature which cannot be changed, depends also on its steam content, the only reasonable way to increase this temperature is to increase the humidity of the flue gases. This can be done easily by preliminary direct heating of air with warm water. Besides, together with the increase of the water heating potential in the contact economizer, one obtains also the following positive effects:

1. Reduction of nitric oxides concentration in the gases emitted, due to a flame temperature reduction, and

2. Reduction of the necessary excess air, because of the oxidative function of steam at the high flame temperatures obtained.

Our experiments with a 220 ton steam natural gas burning boiler showed that a 10 per cent increase of the steam content in the air leads to a nitric oxides reduction of 33 %, without any adverse effect on combustion.

Figures 2 and 3 present two process diagrams of second generation contact economizer systems.

The plant illustrated in Fig. 2 [9] operates as follows:

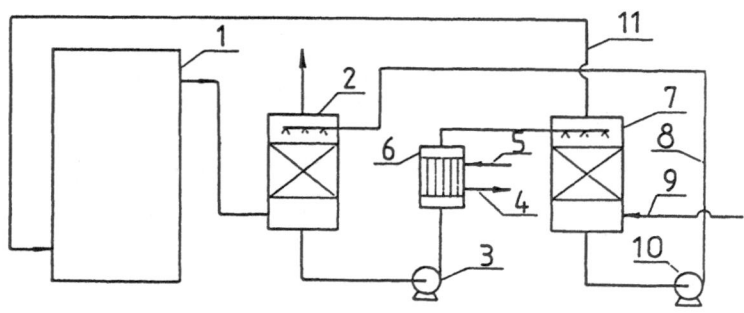

Figure 2.

Flue gas from boiler 1 enter contact economizer 2, where they get cooled by recycle water and then enter into the chimney, by means of a blower. The water heated up in the contact economizer is introduced into heat exchanger 6, by means of pump 3, where the main part of the heat is utilized by heating central heat supply water or other water entering the heat exchanger unit through line 5 and leaving it through line 4. The water, previously cooled in the heat exchanger, and additionally cooled in column 7, by a direct contact with atmospheric air, passes along pipeline 8 by means of pump 10, and enters the distributor of the contact economizer 2. The air, thus heated and humidiefied, enters the boiler burner.
A disadvantage of the process flow chart presented in Fig. 2, is that the liquid-to-gas ratios of contact economizer 2 and in column 7 are practically the same. Therefore no optimum performance regime can be maintained in any unit. Here we consider the "optimum" as the performance which allows the highest possible water temperature to be achieved

by heating in heat exchanger 6, at given flue gas wet bulb temperature following the boiler. This disadvantage has been avoided by some complication of the process flow chart, as presented in Fig. 3 [9]. Contact economizer 2 in this figure is divided into two parts. The flue gases coming from boiler 1, enter the lower section of the contact economizer where they are coold whih water coming from heat exchanger 3. The latter is cooled down by central heat supply water entering through line 4 and leaving through line 5. Circulation is carried out by the pump 6. The partially cooled flue gases

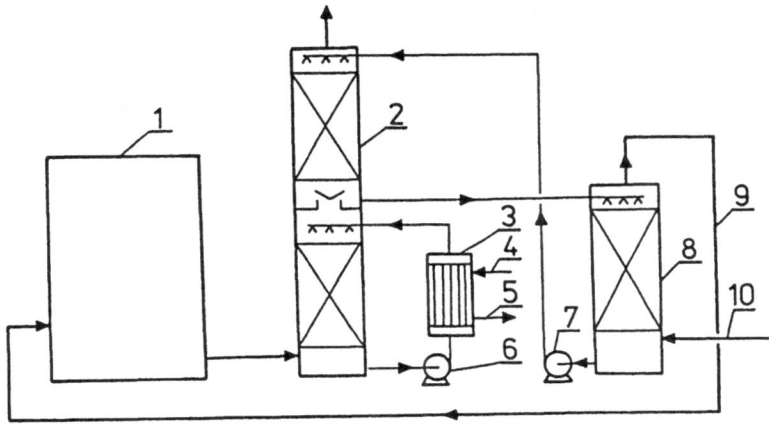

Figure 3.

are scrubbed additionallyin the contact economizer upper section by means of recycle water introduced by pump 7. The water heated in the economizer upper section goes into the distributor of column 8, where air is introduced through line 10, by a blower. The heated and humidified air enters the boiler along line 9, for fuel combustion.

On the basis of the arrangement shown in Fig. 3, a process flow chart for direct heating of heat supply water in contact economizer 2 lower section and subsequent carbon dioxide desorption and pH correction has been developed.

Figures 4 and 5 present the data on heat utilization Q for the systems shown in Figs 2 and 3 per 1 kg dry gas versus temperature t_{L1} of the water entering the contact economizer lower section. The water temperature t_{L2}, achieved in the economizer lower section, is plotted in the same figures. The lines are valid for flue gas temperature of 115 °C obtained by burning natural gas in the boiler at excess air coefficient of 1,1. By comparing the curves obtained in both contactors, it is seen that, given the temperature regime, the second system exhibits higher heat utilization.

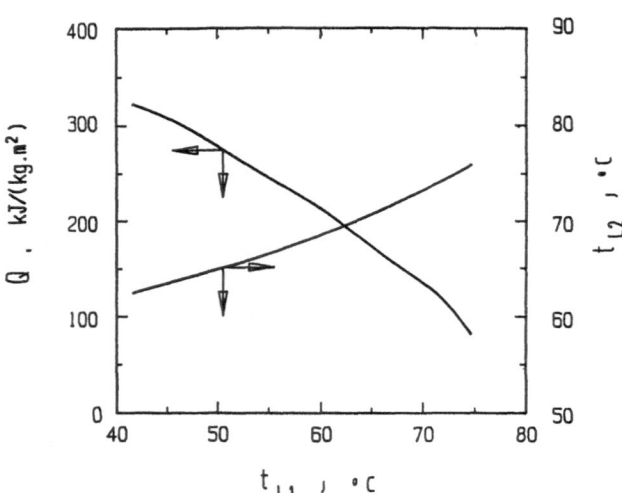

Figure 4.

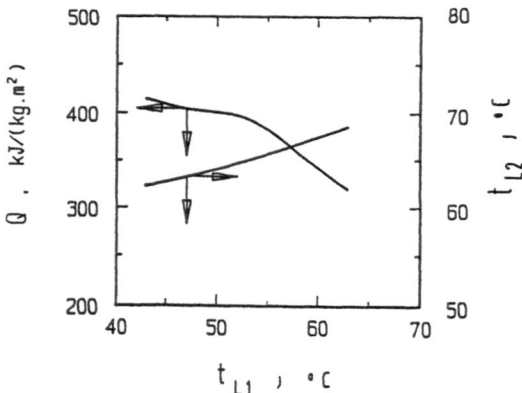

Figure 5.

ECONOMIC CONSIDERATIONS

The pay-off time of the contact economizer systems now in operation, as those shown in Fig. 1, is less then one year. The tentative calculations for the sytems presented in figures 2 and 3 showed that this time is less than 2 years. The ecological benefit due to reduction of both the concentration of nitric oxides in the flue gas and emission of carbon dioxide is not accounted in these economic considerations.

REFERENCES

1. Aronov, I.S., Direct heating of water by combustion products of natural gas, Nedra, Leningrad, 1978 (in Russ.).

2. Grigorov, G.W., Neiman, W.K. and Churakov, C.D., Utilization of secondary low-temperature sources of energy in chemical industry, Chimia, Moskwa, 1987 (in Russ.).

3. Levy, C., Chauffage, ventilation, conditionment, La recuperation de chaleur sur les fumees de chaudieres,1974, 50, 3, 11-20.

4. Kolev, N., Winkler, K, Daraktschiev, R. and Brosh Z., Development of highly effective packing applying mass transfer theory, Khim.Prom., 1986, 8, pp. 43-5 (in Russ.).

5. Kolev, N., Billet, R. and Semkov, Kr., Über die optimale Form von Füllkörpern und Kolonnen, FAT Science Technology, 1990, 92, 7, pp. 291-94.

6. Shaworonkov, N.M., Gildenblat, I.A. and Ramm, W.M., Analysis of the effective surface area in absorption packed columns, Trudy MKHTI, 1963, 40, 5-18 (in Russ.).

7. Shaworonkov, N.M., Gildenblat, I.A. and Ramm, W.M., Investigation of one-phase gas flow mass transfer in packed columns, Shurnal Prikladnoi Khimii, 1960, 33, 8, pp. 1790-800 (in Russ.).

8. Kolev, N., Filipova, N., Daraktsciev, R. and Winkler, K., Zur Auslegung von Packungskolonnen für die direkte Wärmeübertragung von heissen Gasen an Flüssigkeiten, Chem. Techn., 1986, 38, 7, pp. 287-9.

9. Kolev, N., Mirtscev, A, Semkov, Kr. and Daraktschiev, R., A method for burning process, Bulg.Patent Nr.77784/28.01. 1987.

ACKNOWLEDGMENT

The study has been supported financially by the National Science Fondation of Bulgeria under Contract No TH 95/91.

PREDICTION OF THE THERMAL PERFORMANCE OF
INTEGRATED HEAT RECOVERY AND FILTERING MATRICES FOR FLUE GASES

MARIO GAIA MATTEO PEROTTI
Dipartimento di Energetica
Politecnico di Milano
P.za L. da Vinci 32, 20133 Milano, Italy

ABSTRACT

The exhaust gases from a large number of high temperature industrial processes have to be filtered before the discharge to the atmosphere. Often they also have to be cooled in order to recover thermal energy and to allow the adoption of fabric filters which are damaged at high temperature. Cooling is normally accomplished in large and expensive heat exchangers. In the paper a different approach is considered, i.e. the utilization in the filter of a metal filtering matrix in good thermal contact with a coil of small diameter tubes in which the cooling fluid flows. Hence the two functions of cooling and filtering the flue gas are obtained in the same metal surface. In order to estimate the thermal performances of such device, a specific 2D heat transfer model is also developed.

INTRODUCTION

In industrial processes involving the discharge of gaseous effluents to the atmosphere it is often necessary to both cool and filter the gases to be discharged. The cooling of exhaust gases can be aimed to recover the heat content of the gases themselves, thus reducing the energy loss to the atmosphere, but in most cases the temperature of the gases has to be reduced, independently from energy gains, in order to increase the life of the fabric filters. Fabric filters, generally in the form of tubular filters arranged in a matrix inside a bag-house, are the most effective filters for the separation of small particles. More stringent environment protection rules will probably induce an increase of the use of fabric filters. Filtering by fabric filters follows a specific pattern: firsthand, the clean filter is effective in stopping only the particles larger than the cloth openings, in a second stage the build-up of particles on the cloth increases the filtering efficiency. As the particle layer increases, the pressure loss also increases until some cleaning action eliminates most of the particles. The cleaning action is obtained sometimes by mechanical devices like hammers or vibrators, but more often by a sudden jet of compressed air, crossing the cloth countercurrent to the gas. In order to obtain a good compromise between filtering effectiveness and pressure loss, the gas velocity through the particle layer has to be very low. Typical values of gas speed with reference to filtering of exhaust gases from ovens, are in the range $V_o = 0.01$ m/s to 0.025 m/s.

As a consequence, the frontal area of filtering fabric has to be quite large.

The temperature control is of primary importance to obtain an acceptable filter life: according to the kind of fiber used the accepted temperatures range from 120 to 250°C while the temperature must not descend below the dew point of the gas, to avoid packing of particles. The heat exchangers used to cool the exhaust down to a temperature acceptable for the filter are exposed to the unfiltered gas, hence they must have large tube spacing. Accordingly they exhibit low heat exchange coefficients. It is not uncommon to use ambient air as cooling medium, without any form of heat recovery. In the present paper the concept of integrating the heat exchange function in the filter is analyzed. For this purpose the filtering surface made of high conductivity material must be provided with a set of tubes, evenly distributed on the filtering surface and in good thermal contact with the filtering surface itself. Several arrangements of filtering surface and tubes are shown in fig. 1.

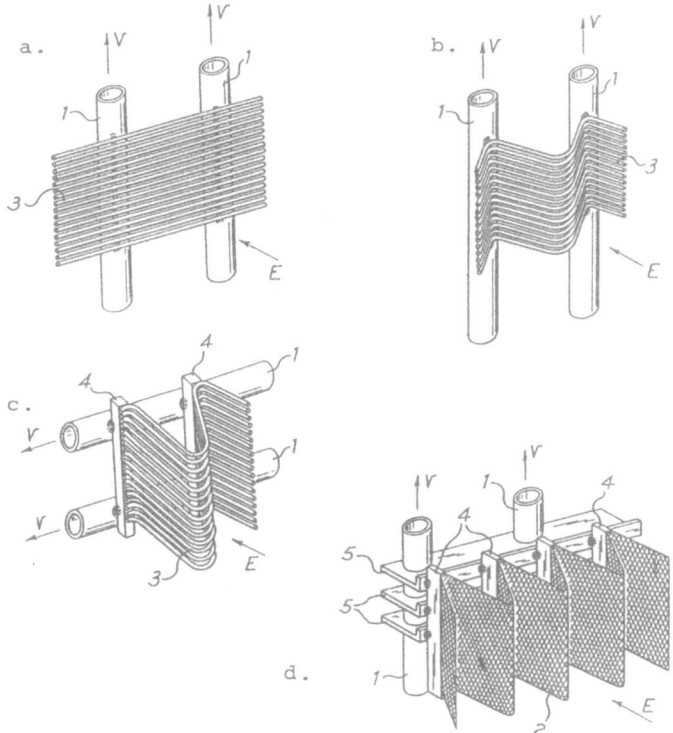

Figure 1. Typology of filtering surface, with heat drain tubes and rods.

In the straight wire-and-tube solution (Fig. 1a) the heat collected by the wire is directly transferred to the tubes. By this configuration the spacing of the tubes equals the wire length between the tubes. Hence many tubes are required. The wire can be vee shaped (Fig. 1b) to increase the filtering surface per unit of frontal area.

The wire length and tube spacing can be made independent by adding connecting rods as in Fig. 1c and in Fig. 1d. The wires can be made with non circular cross section. In particular they can be (for most arrangements) of rectangular shape. In the following pages reference is made to such a rectangular fin. From a thermal point of view a fabric filter is similar to a very compact cross-flow heat exchanger. Several models describing the behaviour of such devices have been developed [1,2]. Previous tests over compact heat exchangers [3]showed that the "classical" monodimensional approach overestimated local heat transfer particularly for geometries where bidimensionality of heat fluxes couldn't be neglected. In order to overtake this discrepancy a model was developed to describe the coupled phenomena of fin heat conduction and flowing gas temperature variation.

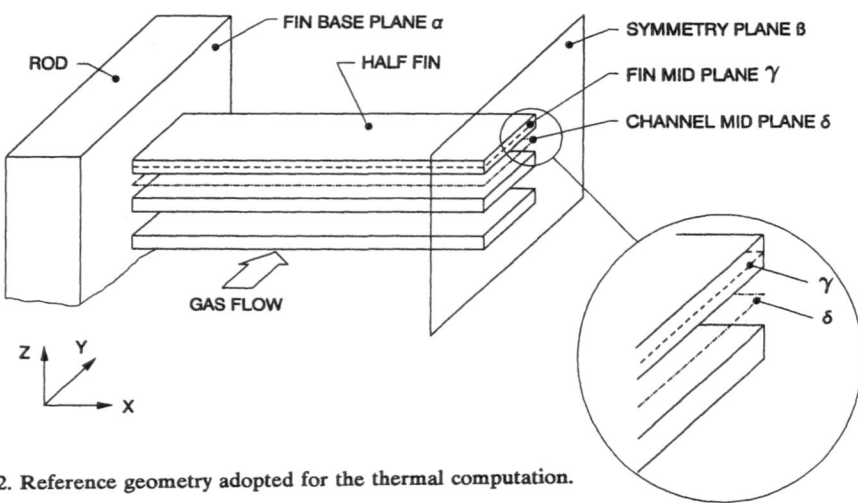

Figure 2. Reference geometry adopted for the thermal computation.

MODEL DESCRIPTION

The domain considered is represented in Fig. 2. Problem symmetry allows to consider a volume limited by one rod (plane α), by the mid plane between two adjacent rods (plane β), by the mid plane between two fins (plane δ) and the simmetry plane of a fin (plane γ). Planes β, γ and δ may be assumed adiabatic while the fin base temperature (plane α) could be assumed constant. Other assumptions made are: gas and fin temperature are functions of x and y coordinates only (average values are assumed in z-direction); fin material is homogeneous and isotropic; thermodynamical properties of flowing gas are equal to those of pure air; gas density and specific heat are constant and equal to the mean value between inlet and outlet; gas speed V is equal to the mean value within the core: $V = V_o/\sigma$; gas conductivity is negligible with respect to fin conductivity.

The convective heat transfer coefficient (h) is calculated through the equation $(N_u)_{air} = 7.54$ thus assuming the air flow field as laminar and fully developed. Hence local heat fluxes are slightly underestimated. With this assumption the solving system becomes:

$$\begin{cases} \lambda \left(\frac{e}{2}\right) \left[\frac{\partial^2 T_{fin}}{\partial x^2} + \frac{\partial^2 T_{fin}}{\partial y^2}\right] = h\left(T_{fin} - T_{gas}\right) & (1) \\[3mm] \rho c_p v \left(\frac{a}{2}\right) \frac{\partial T_{gas}}{\partial y} = h\left(T_{fin} - T_{gas}\right) & (2) \end{cases}$$

Eq. (1) expresses the identity between heat transmitted in the fin and heat exchanged by convection with the gas. Eq. (2) represents energy conservation for the flowing gas. Putting $\dfrac{e}{2}\dfrac{\lambda}{h} = A$ and $\dfrac{a}{2}\dfrac{\rho c_p v}{h} = B$ we can write:

$$\begin{cases} A\frac{\partial^2 T_{fin}}{\partial x^2} + A\frac{\partial^2 T_{fin}}{\partial y^2} = T_{fin} - T_{gas} & (3) \\[3mm] B\frac{\partial T_{gas}}{\partial y} = T_{fin} - T_{gas} & (4) \end{cases}$$

The numerical method

The numerical method employed to solve the above system is based on the finite volumes theory explained in [4]: the domain is decomposed in simple elements within which eqs (3) and (4) are integrated, the integral being expressed as function of grid point temperature. A centred differences approach was chosen for the fin and an upwind formulation was preferred for the gas. With reference to Fig. 3 for grid points layout and assuming that $\dfrac{\partial T_{gas}}{\partial y} >> \dfrac{\partial T_{fin}}{\partial y}$, hypothesis verified by the authors to be relevant only near the leading edge of the fin (integration of eq. (4)) within one element gives:

$$T_{gas_{out}} - T_{fin} = \left(T_{gas_{in}} - T_{fin}\right) e^{-\frac{\Delta y}{B}} \qquad (5)$$

$$e^{-\frac{\Delta y}{B}} T_{gas_{in}} - T_{gas_{out}} = \left(1 - e^{-\frac{\Delta y}{B}}\right) T_{fin}$$

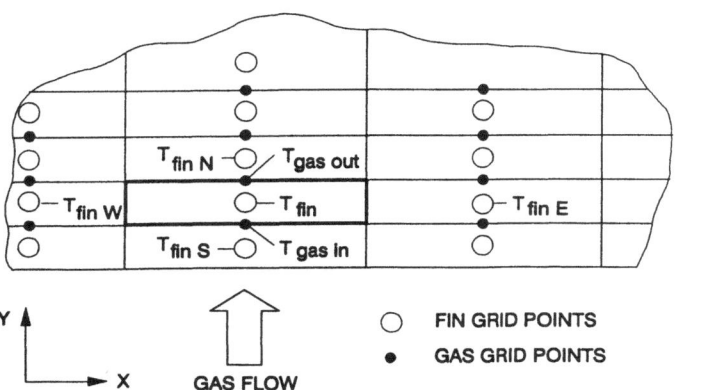

Figure 3. Layout of grid points for gas and fin in the x-y plane

Eq. (3) becomes after discretization (Fig. 3):

$$A\left(\frac{T_{fin\,W}+T_{fin\,E}-2T_{fin}}{\Delta x^2}\right)+A\left(\frac{T_{fin\,N}+T_{fin\,S}-2T_{fin}}{\Delta y^2}\right)=(T_{fin}-T_{gas})_{av.\,in\,elem.} \quad (6)$$

From eq. (4) we obtain

$$(T_{fin}-T_{gas})_{av.\,in\,the\,element} = (T_{fin}-T_{gas_{in}})\frac{1-e^{-\frac{\Delta y}{B}}}{\frac{\Delta y}{B}} \quad (7)$$

Introducing eq. (7) in eq. (6) and rearranging, it results:

$$\frac{A}{\Delta x^2}T_{fin\,W}+\frac{A}{\Delta x^2}T_{fin\,E}-\frac{2A}{\Delta x^2}T_{fin}=G(T_{fin}-T_{gas_{in}})-\frac{A}{\Delta y^2}(T_{fin\,N}-2T_{fin}+T_{fin\,S}) \quad (8)$$

with $G = \dfrac{1-e^{-\frac{\Delta y}{B}}}{\frac{\Delta y}{B}}$

Eqs. (5) and (8) can be written for all grid elements in the domain, with the conditions for the boundary elements:

$T_{fin} = T_{fin\,base}$ at fin base; $T_{gas} = T_{gas\,inlet}$ at filter inlet.

Starting from an initial temperature distribution, the following interative solving procedure was adopted: for each fin element, eq. (8) was solved for the left side and a better approximated fin temperature distribution was calculated through an underrelaxed method; eq. (5) was solved for the left side. Iteration continued until convergence was reached. The coating of surfaces due to fouling was not considered in the present thermal model, it can be easily accounted for through an equivalent heat transfer coefficient, without altering the structure of the calculation.

RESULTS

The thermal performance of the filtering surface was calculated as the thermal power drawn from the gas stream per degree of temperature difference per unit of frontal area of filter, for a given velocity of the gas upstream of the filter. Two different sizes of fin were considered: a thick geometry, having a fin pitch of 0.400 mm and a thin geometry, having a fin pitch of 0.200 mm. For both sizes different values of fin thickness to fin spacing were considered. Copper and stainless steel were assumed as suitable materials. Though copper exhibits a scarce corrosion resistance it can be protected by many metal and organic coating and is attractive for its high conductivity. The exchanged thermal power for a fin width (dimension in the y direction) of 1 mm is reported in Fig. 4. The gas velocity upstream of the filtering surface was assumed 1 m min^{-1} (0.0167 m s^{-1}). The performance for a copper fin is constant, regardless of fin thickness and fin spacing because of the high conductivity of the fin material. The temperature distribution in the fin is characterized by a very small region of strong gradients in the y direction near the gas inlet edge. The remaining portion of the fin is at near constant temperature.

The temperature distribution for stainless steel is quite different (Fig. 5) with evenly distributed gradients in both x and y directions, due to the lower thermal conduction coefficient. The figure refers to a fin base temperature of 100°C and an upstream gas temperature of 150°C.

As an effect of these gradients the thermal performance (Fig. 4) is larger for small values of a/e, that is for a geometry having a small gas channel and a thick fin. However, in the range of a/e considered, the variation is moderate (about 10% for a tenfold variation a/e). The performance reduction for larger fin spacing is due only to the metal gradient, while the variation of the liminar coefficient value, though large, is not relevant to the thermal performance (Fig. 6). This fact is confirmed by the near coincidence of the points of both the thick and thin geometry (a + e = 0.200 mm)

The mean velocity inside the gas channels increases with the fin thickness for the given velocity upstream of the filter surface. Though this fact is relevant to the pressure loss, it does not affect the heat transfer coefficient (laminar flow). The effect of variation of the upstream velocity for the same geometries, for stainless steel only, is described in Fig. 7. In Fig. 9 the performance is reported in terms of ratio of obtained power versus maximum available power.

For the low velocity the effect of fin thickness decreases, as the fin tends to become nearly isothermal. The power drawn from the gas at higher velocity is obviously larger, while the gas temperature decrease is only moderately affected. The modification of the temperature distribution can be obtained by comparing Figures 5 and 8. The increase of gas velocity draws the isothermal curves in the mid section of the fin towards the gas exit side, and a section of the fin on the inlet side is nearly isothermal, and hence substantially ineffective to heat transfer.

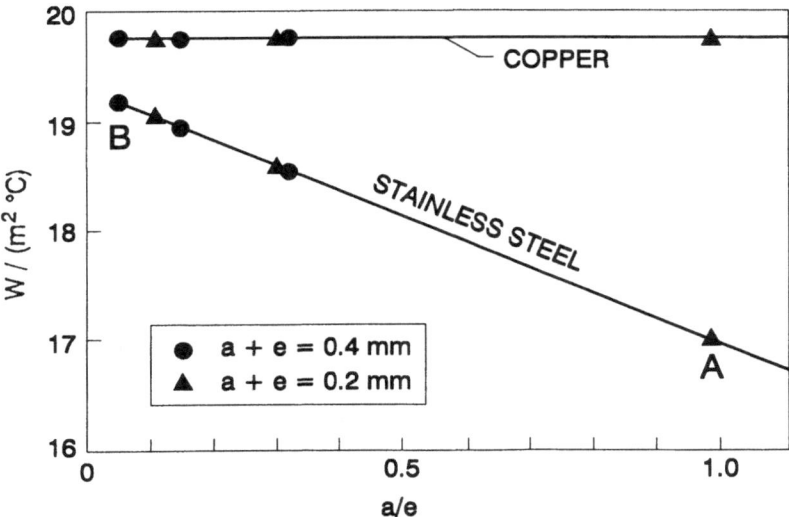

Figure 4. Exchanged thermal power per unit of frontal area copper and stainelss steel fin. Upstream air speed 0.0167 m/s, fin length 40 mm, fin width 1 mm.

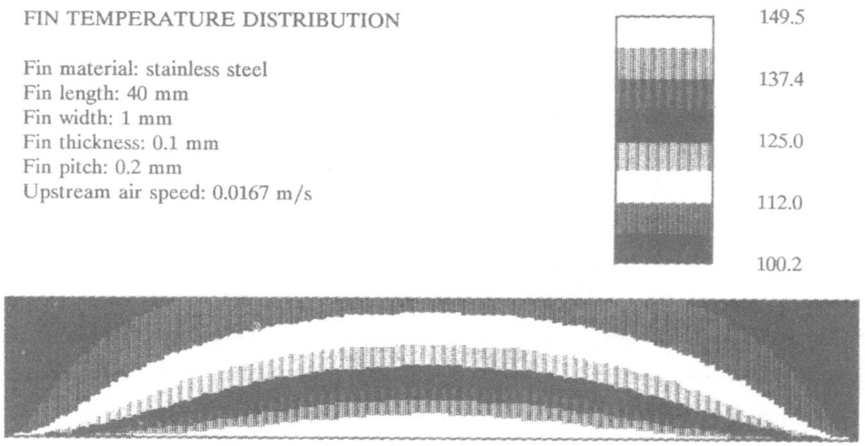

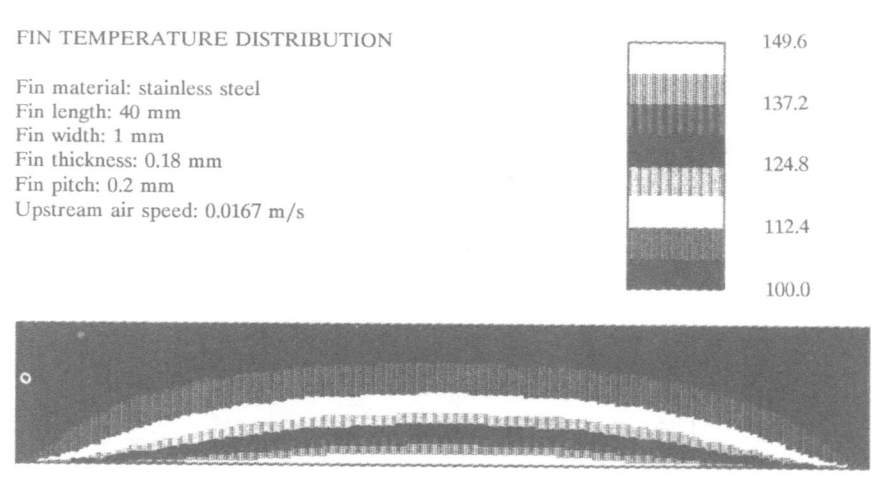

FIN TEMPERATURE DISTRIBUTION

Fin material: stainless steel
Fin length: 40 mm
Fin width: 1 mm
Fin thickness: 0.1 mm
Fin pitch: 0.2 mm
Upstream air speed: 0.0167 m/s

149.5

137.4

125.0

112.0

100.2

Figure 5. Fin temperature distribution for the stainless steel fin corresponding to point A in Fig. 4. (The y dimension is enlarged, gas flows from bottom to top of figure)

FIN TEMPERATURE DISTRIBUTION

Fin material: stainless steel
Fin length: 40 mm
Fin width: 1 mm
Fin thickness: 0.18 mm
Fin pitch: 0.2 mm
Upstream air speed: 0.0167 m/s

149.6

137.2

124.8

112.4

100.0

Figure 6. Fin temperature distribution for the stainless steel fin corresponding to point B in Fig. 4. (The y dimension is enlarged, gas flows from bottom to top of figure).

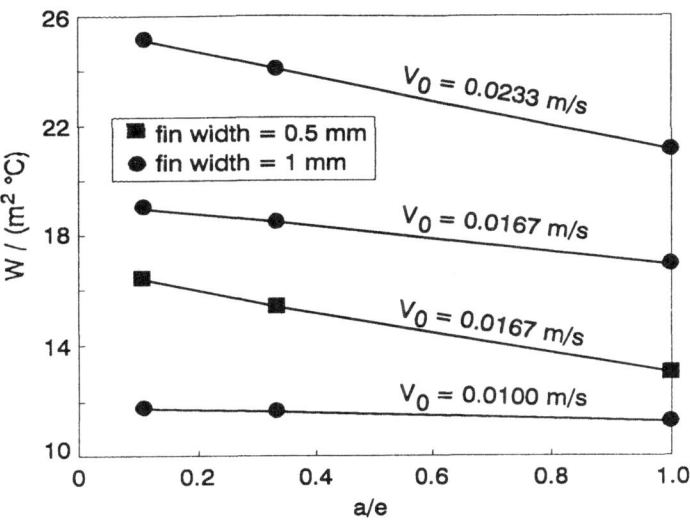

Figure 7. Exchanged thermal power per unit frontal area for stainless steel fin only, for three values of air speed V_0 and two fin width values as a function of air passage height (a) to fin thickness (e) ratio.

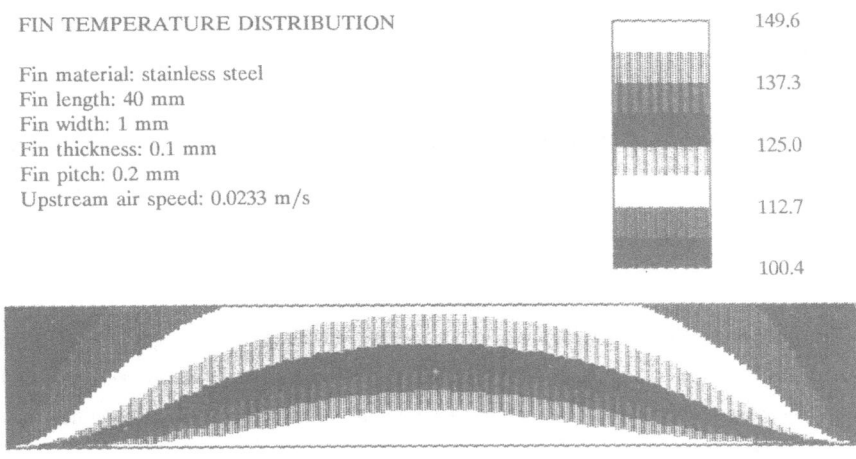

Figure 8. Fin temperature distribution for the stainless steel fin with the same conditions as Fig. 5 except for the higher upstream velocity. (The y dimension is enlarged, gas flows from bottom to top of figure).

700

CONCLUSIONS

The viability of the concept of integrating in the same surface both the function of filter and the function of heat exchanger has to be verified from many different points of view (filtering efficiency of metal surface, useful life expectancy, cost effectiveness, pressure drop, etc.). The results discussed above indicate that, for low gas speed applications, and for the geometry proposed, the thermal performance is very good. In fact in the considered cases, the thermal gradients in the wires are small while the gas/solid heat transfer coefficient is very high (typically in the range 1000-4000 W/m^2K). As a consequence the gas stream releases most of its heat content to the surface, in spite of the shortness of the gas path across the filtering surface (0.5 to 1 mm). The absolute value of the obtained thermal power per square meter is not large e.g. for a ceramic oven application having a gas temperature of 150°C, with a rod temperature of 100°C, (which could be consistent with a water temperature in the tubes around 80°C) the thermal power obtained with the conditions of point B Fig. 4 is 0.85 kW/m^2, as was to be expected for given the low gas velocity.

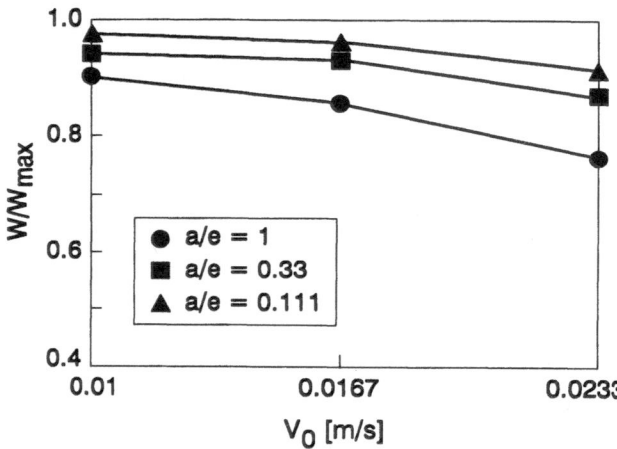

Figure 9. Ratio of exchanged thermal power to maximum thermal power available in the air flow (W/W_{max}) for different values of a/e.

Besides the heat recovery, the evenly distributed tube pattern throughout the filter can give other advantages. In particular it allows to heat and maintain the filter at a temperature high enough to avoid condensation and the following occlusion of the filter, when the conditions occur.

The use of an all-metal filter reduces the risk of electrostatic discharge in the opposite case of filtration of dry combustible particles. The paper deals only with the thermal aspect, with reference to a simplified geometry. In fact the fabrication aspect is fundamental and the actual geometry will be determined by the manufacturing process.

The shapes a and d of Fig. 1 seem to be less difficult to construct, as they avoid the joining of elements having too large a size difference. The manufacturing process should refer to the techniques used for the fabrication of thin mesh including flash and laser welding (or brazing); the authors will investigate the production aspect as a next step.

NOMENCLATURE

a = gas passage height = fin pitch - fin thickness
A_f = filter frontal area
c_p = average gas specific heat at constant pressure
d = gas passage hydraulic diameter
e = fin thickness
h = convective heat transfer coefficient
$Nu = h\, d/\lambda_{gas}$ = Nusselt number for gas
$T_{fin\ base}$ = fin base temperature
$T_{gas\ inlet}$ = gas temperature at filter inlet
$T_{gas\ outlet}$ = gas temperature at filter outlet
V = average gas speed in filter core
V_0 = upstream gas speed
$W = c_p\, V_0\, A_f\ (T_{gas\ inlet} - T_{gas\ outlet})\, \rho$
$W_{max} = c_p\, V_0\, A_f\ (T_{gas\ inlet} - T_{fin\ base})\, \rho$
x = grid element dimension in x-direction
y = grid element dimension in y-direction
λ = fin metal conductivity
λ_{gas} = flowing gas conductivity
ρ = average gas density
σ = (free filter frontal area)/A_f

REFERENCES

1. Kays, W.M. and London, A.L., Compact heat exchangers, 2nd edition, McGraw-Hill Book Company, New York, USA.

2. Rohsenow, W.M., Hartnett, J.P., Ganic, E.N., Handbook of heat transfer applications, 2nd edition, McGraw-Hill, Inc. New York, USA, 1985, pp. 4-274, 4-279.

3. Perotti, M., Sviluppo di un convettore a bassa temperatura in regime laminare, Politecnico di Milano, Ph.D. Thesis, Milan, Italy, 1990.

4. Patankar, S.V., Numerical heat transfer and fluid flow, McGraw-Hill Inc. New York, USA, 1980.

ENERGY EFFICIENT PROCESSING USING SPRAY EVAPORATION COOLING TECHNIQUES

Dr C.R. Howarth, Dr R.J.J. Jachuck[*] and Prof. C. Ramshaw
Department of Chemical & Process Engineering
University of Newcastle upon Tyne
Newcastle upon Tyne, NE1 7RU, U.K
(* For communication)

ABSTRACT

Experiments have been carried out to determine heat transfer coefficients for evaporative cooling due to water jet impingement onto a continuously moving sheet of polymeric material. Operating conditions were selected so as to simulate conditions encountered in industry wherever possible. The initial sheet temperature was varied between 373 and 423K. The maximum average air jet velocity was 40 m/s and the separation distance between the nozzle and the sheet surface was between 10 mm and 18 mm. The sheet velocity was kept constant during the course of the tests at 0.0025 m/s, and the temperature of the water used for the liquid spray was more or less constant at 291K. It has been found that an average heat transfer coefficient of 5000 W/m^2 K can be achieved for an nozzle exit Reynolds number of 30000 and for L/D, the ratio between the distance between the tip of the nozzle and the surface of the sheet, and the diameter of the air nozzle, of 13.3. Cost analysis has also been included in order to demonstrate that spray evaporation technique can be both economical and energy efficient.

NOMENCLATURE

D= diameter of the nozzle (m)
fi= factors of cost for piping, instruments,etc.
fi'= factors for indirect cost.
h= average heat transfer coefficient (W/m^2 K)
I_F= fixed investment cost of a complete plant
I_E= cost of major items of processing equipment
k= thermal conductivity of air (W/m K)
L= distance between the sheet surface and the nozzle (m)
q= tonnes of refrigeration
Nu= Nusselt number (-)
Re = Reynolds number (-)

INTRODUCTION

Many sectors of the process industry require rapid and energy efficient cooling after processing. The pasteurisation of packaged foods for the chilled food market and polymer extrusions are two growth areas which typify these operations. Techniques available for the cooling of extruded plastic profiles, packaged foods and other hot sheets include immersion cooling, spray cooling, contact cooling and simple refrigeration. In general heat transfer rates achieved in immersion cooling are low, as water baths are usually used and the mode of heat transfer is largely by conduction and partly by forced convection. Spray cooling could be effective if as in the case of metals the hot sheets leave the processing operation at very high temperatures. Contact cooling can be very effective where stainless steel and aluminium are used as the contact surfaces, though this technique has several limitations and disadvantages when applied to the cooling of polymeric sheets. Refrigeration requires bulky equipment and has low heat transfer rates.

In the extrusion of plastic sheets/profiles or in food pasteurisation the product leaves the processing unit at temperatures between 273K and 403K, hence heat transfer rates by the above "conventional" techniques (1-3) will not be high. Therefore a relatively new approach to the problem, spray evaporation, has been investigated and it is suggested that this technique may be used for the successful rapid cooling of extruded polymer profiles or sheets and for the pasteurisation of packaged foods. To demonstrate the potential of this technique work has been carried out on moving polymer sheets(4) and the effect of the system geometry, velocity of the impinging air jet, and the initial temperature of the sheet on the rate of cooling has also been studied. The heat transfer rates achieved by applying this technique to polymer sheets are very high compared to convective heat transfer achieved by the impingement of hot or cold air and cooling water bath, as can be seen from Table 1.

TABLE 1
Approximate values of heat transfer coefficient for different cooling media.

Medium	h (W/m^2 K)
Still air (5)	10
Air at 5 m s^{-1} (5)	50
Water Spray (5)	1500
Water at 5 ^{0}C (5)	1000
Spray Evaporation (present investigation)	300 - 6000
Refrigeration	(Same as still air)
Polymer to stirred water(6)	1500 - 3000

EXPERIMENTAL APPARATUS AND PROCEDURE

A schematic diagram of the test facility is shown in Figure 1. Description of the test facility and the operational procedure has been discussed in (4) and will therefore not be discussed in detail.

Water and then air were sprayed sequentially onto both the top and the bottom of a moving polymer the sheet. The nozzles were clamped onto a frame, which ran along the whole length of the cooling system, to keep them secure in position. The system was designed so that the nozzles could be moved up and down and also along the horizontal axis. Four air nozzles were fed from one supply pipe, therefore each air spray zone had four nozzles directing air from above the sheet and four from below. Thermocouples were located in matrix arrangement and measured temperatures at both surfaces and also within the polymer thickness. Using rapid data acquisition system cooling curves were obtained for each evaporation zone. The heat transfer coefficient was then obtained by modelling these curves, using the technique used by previous workers(7,8).

This involved a numerical technique(Finite difference implict method) to calculate the local and average heat transfer coefficients, from the experimental data. Once the heat transfer coefficients had been determined for each of the thermocouples, the local heat transfer coefficients were plotted against their respective positions on the sheet. Figure 2 shows the variation of the heat transfer coefficents along the width of the sheet for L/D ratios of 1.6, 5, 10, 13.3, 15, 16.6, 20, 25, and 30 respectively.

After calculating the average heat transfer coefficients, the Nusselt number for each of the experimental conditions was calculated by using the following expression:
$$Nu = (h\ d)/k = (h \times 0.006)/0.025 = 0.23\ h$$
In order to study the the variation of the heat transfer coefficient along the width of the sheet, coefficient of variation was calculated and studied.

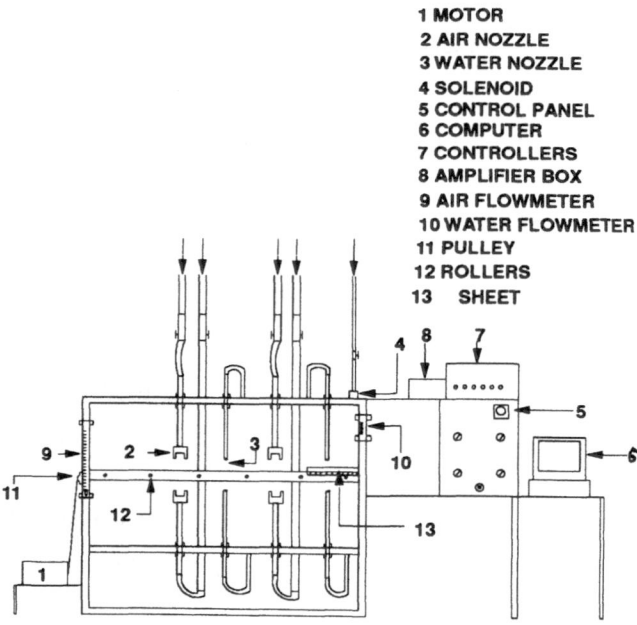

1 MOTOR
2 AIR NOZZLE
3 WATER NOZZLE
4 SOLENOID
5 CONTROL PANEL
6 COMPUTER
7 CONTROLLERS
8 AMPLIFIER BOX
9 AIR FLOWMETER
10 WATER FLOWMETER
11 PULLEY
12 ROLLERS
13 SHEET

Figure 1. Schematic diagram of the test facility.

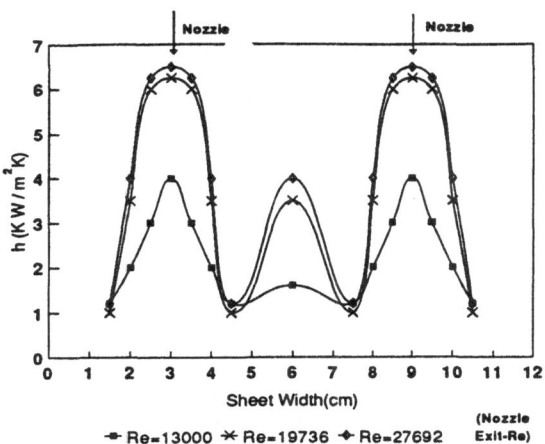

Figure 2. Variation of the heat transfer coefficient along the width of the sheet for L/D = 13.3

RESULTS AND DISCUSSION

(a) Effect of the impinging air velocity and the L/D ratio on the evaporative coefficient.
The Reynolds number(Re) of the impinging air jet can be classified into three distinct zones on the basis of Nusselt number:
(1) Laminar zone   Re<9000
(2) Transition zone 9000< Re <13000
(3) Turbulent zone 13000< Re <30000
   From Figure 3, it can be seen that in the laminar zone the Nusselt number(Nu) is very low and is more or less constant. There is a sharp increase in the Nusselt number in the transition zone and in the turbulent zone the Nusselt number increases up to Re=24000 and then the profile becomes more or less flat.

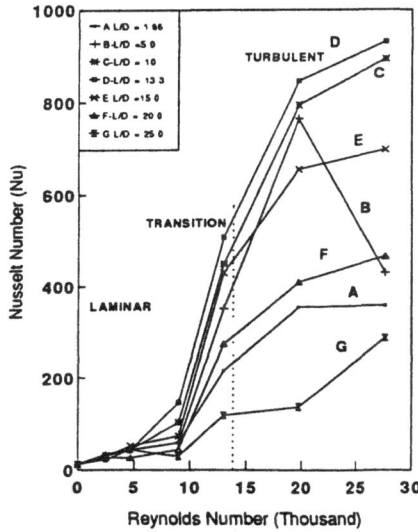

Figure 3. Heat Transfer Characteristics for Spray Evaporation Cooling Technique.

For low impinging air jet velocities, heat from the sheet is transferred by conduction to the water film, sprayed on to the sheet during the liquid spray section. The surface heat transfer coefficient is therefore low due to the poor thermal conductivity of water (0.597 W/m K). The effect of the impinging air is not very significant as it only succeeds in cooling the liquid film by convection, thereby increasing the temperature gradient between the surface of the sheet and the surface of the liquid film in contact with the sheet.

As the velocity of the impinging air increases, evaporation of the liquid film starts because the partial pressure of the water vapour in the impinging air is less than that of the water vapour in the water film covering the sheet. The latent heat required for the film to evaporate is taken up from the sheet, and as a direct consequence of which a high surface heat transfer rate is experienced.

If the velocity of the impinging air is further increased, there is a significant increase in the Nusselt number up to a certain point, after which the heat transfer coefficient drops for lower values of L/D ratios (L/D<5.5), and remains unaffected for higher values of L/D ratios (6<L/D<30)

This drop in heat transfer rate is due to the fact that higher velocities of the impinging air jet disrupts the liquid film and stops the process of evaporation. Forced convection is then the mode of heat transfer and therefore there is a significant drop in the heat transfer rate, which can be seen in Figure 3, for L/D=5.

For an L/D ratio of 1.66, it is seen from Figure 3 that the heat transfer coefficient achieved is not very high. This is because the velocity of the impinging air at the surface of the sheet is very high and as a result of which forced convection plays a more dominant role than evaporation.

As the L/D ratio is increased to 5, impinging velocity at the surface of the sheet is not high enough to disrupt the film and therefore evaporation takes place, and the heat transfer rates experienced are better than the ones achieved for L/D=1.66. However even for L/D=5, at higher Reynolds number that is Re>22000, the velocity at the surface of the sheet gets high enough to disrupt the film and therefore the rate of heat transfer falls sharply as forced convection comes into play.

It is seen that as the L/D ratio is increased towards 13.3, the impinging velocity at the surface of the sheet gives a very high heat transfer rate, due to evaporation of the water film. Beyond 13.3, the rate of evaporation drops steadily as L/D ratio is increased to 20 and then drops sharply as L/D approaches 30. This is because at higher values of L/D ratios (that is L/D=30) the velocity at the surface of the sheet is not high enough to evaporate the water film and hence conduction plays a significant role in determining the mode of heat transfer.

Figure 4. shows the relationship between the Nusselt number and the L/D ratio, for different values of Reynolds number.

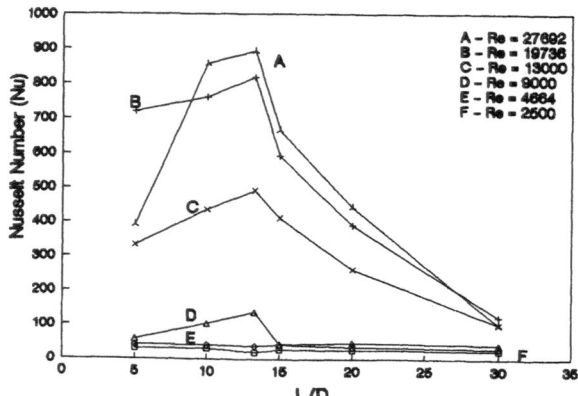

Figure 4. Heat transfer characteristics of Spray evaporation cooling (Nu vs L/D).

707

(b)   Effect of the initial sheet temperature on the evaporative heat transfer coefficient
In Figure 5, the calculated average heat transfer coefficient for Re=19736, and L/D=13.3(the optimum condition), have been plotted as a function of the initial sheet temperatures and it is evident that the rate of heat transfer achieved in the evaporation zone, decreases at temperatures above 373K. This decrease could be due to the fact that at higher temperatures, partial evaporation of water takes place in the liquid spray zone, which is the first step in the cooling process. This happens because the surface temperature is above 373K and results in film boiling heat transfer followed by evaporation. The film boiling reduces the amount of water left on the surface of the sheet to be evaporated in the evaporation zone resulting in a lower heat transfer coefficient.

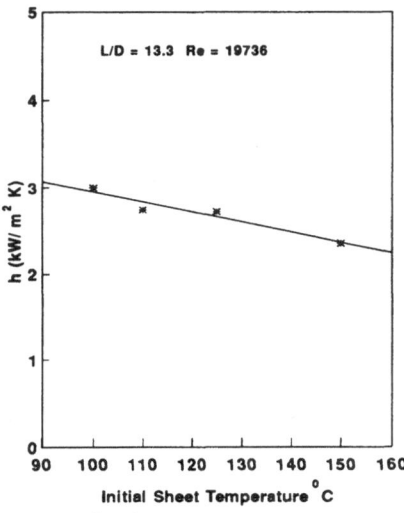

Figure 5. Effect of initial sheet temperature on the surface heat transfer coefficient.

(c)   Effect of the nozzle diameter on the evaporative coefficient.
Table 2 shows the effect of the diameter of the air nozzle on both the local and average evaporative heat transfer coefficient. Increasing the nozzle diameter from 0.006m to 0.008m decreases the heat transfer coefficient at the location directly under the nozzle, but increases the overall average heat transfer coefficient. This is because increasing the diameter of the nozzle, gives a more uniform distribution of the air on the surface of the sheet and this results in a lower variation of the heat transfer coefficient over the width of the sheet.

Table 2
Effect of the Nozzle Diameter

Distance from the the nozzle tip to the surface of the sheet. (m) L	Nozzle Diameter (m) D	Average heat transfer coefficient (W/m² K)
0.8	0.006	3553
0.8	0.008	3797

(d) Correlation of Nusselt and Reynolds numbers

Using the experimental results, correlations between the Nusselt and Reynolds number have been presented as Ref(4):

$$Nu = f(L/D, Re)$$

These correlations suggest that the optimum L/D ratio is 13.3 for a Reynolds number of about 20000, which corresponds to an average heat transfer coefficient of about 5000 W/m^2 K. It is this value which is used to simulate the effectiveness of cooling using this technique and provide the cost data below.

COST ANALYSIS

Choice of Technique

Assuming that, in a particular application, the cooling of a polymeric material by means of natural convection, forced convection and jet impingement is not desirable, and there is a need for rapid and efficient cooling, then there remains three alternatives: refrigeration, spray evaporation or a combination of both. If the final product is to be cooled below 278K then spray evaporation technique will not be suitable, as it is largely dependent on the liquid spray temperature.

Hence if it is desired to cool a packaged food sachet from about 323K to 268K, there remains two alternatives: refrigeration or a combination of spray evaporation and refrigeration.

It seems useful therefore to estimate the capital costs and power consumption of equipment to cool polymeric materials using the following:

Technique 1.   Refrigeration(323K-268K)
Technique 2.   Combination of spray evaporation(323K-278K) & Refrigeration(278K-268K).

Assumptions and Data

(a)   All the costs in this analysis are based on the assumption that a polypropylene food sachet (10 x 10 x 10 cm) is to be cooled from 323K to 268K and that each sachet is ready in 1.2 minutes and also the thermal properties of the contents of the sachet are similar to that of water.

(b)   The following expressions have been used to calculate the installed major equipment cost and the installed plant cost(9,10):

$$I_E = 4630 \, (q)^{0.73} \tag{1}$$

and $\qquad I_F = I_E \, (1 + \Sigma f_I) \, (1 + \Sigma f_{I'})$ $\qquad\qquad$ (2)

where I_E is the installed equipment cost, q is the capacity (tonnes of refrigeration) and I_F is the installed plant cost.

(c)   Heat capacity of refrigerant = 3.51 kW / Tonne

(d)   Cooling coefficient for spray evaporation is 5kW/m^2 K, and 10W/m^{2}K for refrigeration.

Technique 1.

i) Refrigeration system

In this estimation, the equipment cost is based on the assumption that 850 sachets can be cooled at any particular time and it is noted that rapid rates of cooling are promoted in refrigeration by high temperature gradients with regrettably low coefficients whilst in spray evaporation, high coefficients are achieved although the temperature gradients are somewhat more modest.

(a) Major Equipment Cost

Assuming that 50m^2 of material is to be cooled (equivalent to 850 sachets) from 323 K, using a refrigerant at 268K, then the rate of heat removal is:10 X 50 (323 - 268) = 27.5 kW = h A ΔT (A refers to 50m^2 = 850 sachets)

If the efficiency is = 70%, then the design power rating = 39.28 kW

Equivalent tons. of refrigeration =(39.28/3.5) = 11.22 tons, (1 T.of refrigeration = 3.51 kW Ref(7))

Therefore I$_E$ = 4630 (11.22)$^{0.73}$ = $ 27052 (Based on 1982 figures)

Assuming an inflation rate of = 50%, then

I$_E$ = $ 27052 =(27052 x 1.5) = **£ 40578**

(b) Total plant cost

Installed plant cost I$_F$ = I$_E$ (1 + Σf$_i$) (1 + Σf$_{i'}$)

Taking Σf$_i$ = 1.05 and Σf$_{i'}$ = 0.35 as recommended in Ref(3), I$_F$ can be calculated as:

$$I_F = 40578 (1 + 1.05) (1 + 0.35) = £137965$$

Fixed cost I$_F$ = **£ 137965** (3)

Auxilliary Building cost = 5% of I$_F$
= £5460

Product Storage cost = 1.8% of I$_F$
= £1965

Hence total auxilliary cost = **£7426** (4)

Total cost of the Refrigeration
section of the plant (3) + (4) = **£145391** (5)

(c) Cooling Cost (for 850 sachet's)

The cooling cost can be estimated by assuming that this is proportional to the production rate. The heat transfer coefficient for refrigeration in still air has been improved to account for refrigeration with air circulation and this was used to model Ref(4), the time taken to cool a sachet from 323K to 268K, to give approximately 13 hours.

If the plant is operated for 17 hours, then (50 x 17) = 850 sachet's (i.e 50m^2) will have to be cooled. Power as estimated previously is seen to be consumed at 27.5 kW for 50m^2 and therefore during 13 hours 357.5 kWh will be consumed.

Hence electricity costs @10p per kWh = **£35.75 / day**

For the repayment of the capital, the sum to be set aside each year can be calculated by using the expression[10]:

R = P i /((1 + i)^n - 1) £/year (6)

where P is the total capital invested in the plant, the money is invested at i per cent compound interest, and R is the amount to be invested each year to produce P in n years.

Assuming that a compound interest of 15 per cent can be obtained on the money invested and that the interest on the loan must be paid for 5 years, then the sum to be set aside each year is:

R = 145391 x 0.15/((1 + 0.15)^5 - 1) = 21564 (£/year)

or 21564/330 = 65 £/day

FINAL COST = (Power cost + Investment Cost) / Number of sachet's =((3575 + 6500)/850)

= **11.85p/sachet** (7)

TECHNIQUE 2

ii) Combination of Spray Evaporation & Refrigeration
(A) Spray Evaporation System
Fuller details of this cooling technique are found in Ref(4).

(a) Major Equipment Cost
The essential equipment for spray evaporation cooling includes:
(a) Rotary Screw Compressor for supplying air
(b) Pump for for the water spray
(c) Water recovery unit
(d) Air nozzles
(e) Water spray nozzles
(f) sheet conveying system

The following costs for this equipment are based on quotations from revelant suppliers.

Compressor	= £9000
Pump cost	= £400
Cost of 12 air nozzles @ £10	= £120
Cost of 8 water spray nozzles @£15	= £120
Conveying unit cost	= £2350
Hence the total cost of the major equipment I_E	= **£11990**

(b) Total plant cost
From equation (2), the total plant cost can now be estimated as follows:
Installed plant cost $(I_F) = I_E (1 + \Sigma f_i)(1 + \Sigma f_{i'})$
Taking $\Sigma f_i = 1.05$ and $\Sigma f_{i'} = 0.35$

$$\text{Fixed Cost } I_F = 11990 (1 + 1.05)(1 + 0.35) = £33182 \qquad (8)$$

Auxillary Building cost	= 5% of I_F
	= £1659
No Product Storage cost (not required)	
Hence total auxillary cost	= £1695 (9)

Total cost of the Spray Evaporation
section of the plant ((8) + (9)) for 850 sachets = **£34877** (10)

(c) Cooling cost
It is suggested that the food sachets be first cooled by spray evaporation technique and then once the temperature of the contents inside the sachet have dropped to about 278K then refrigeration technique should take over to cool the sachets down to 268K. Using the same basic cooling model as for refrigeration but employing a film coefficient of 5000 W/m^2 K for spray evaporation and 10 W/m^2 K for refrigeration, it was estimated that 2 hours of spray evaporation and 8 hours of refrigeration will be needed to cool the food sachet's to 268K, thereby reducing the cooling time by about 3 hours.

The power consumed in spray evaporation is 35 kW (as the compressor rating is 31kW and it is assumed that a pump of 4kW would be sufficient to deliver the water spray) and therefore after 2 hours 70 kWh will be consumed.

Hence electricity costs @10p per kWh = 700p
Using expression (3), the sum to be set aside each year is:
$$R = 34877 \times 0.15/((1 + 0.15)^5 - 1) = 5173 \text{ £/year}$$
$$\text{or} = 5173/330 = 15.5 \text{ £/day}$$

FINAL COST(S. Evaporation) = (Power cost + Investment Cost) / Number of sachet's
=((700 + 1550)/850) = **2.65p/sachet** (11)

The 850 sachet's (50m²) are to be cooled from 278K to 268K in the refrigeration unit and therefore the rate of heat removal is: 10 X 50 (278 - 268) = 5 kW = h A ΔT
For 17 hours 40 kWh will be consumed and hence electricity cost @ 10p per kWh = 400p .

Cost per sheet = 400/850 = **0.47p**

FINAL COST(Refrigeration) = (Power cost + Investment Cost) / Number of sachets = (400 + 6500) / 850 = 8.11p **(12)**

FINAL COST (Spray Evaporation + Refrigeration) = (11) + (12)
= **10.76p/ sachet** **(13)**

Hence it can be seen that during the first 5 years about 1.67p/sachet((13) - (7)) will be saved. After 5 years approximately 2.9p/sachet (power cost spray evaporation + refrigeration(Tech.2)) - (power cost refrigeration (Tech.1)) can be saved. This corresponds to a total saving of about 30%

CONCLUSION

1. Spray evaporation rates of heat transfer have been obtained and an average value of about 5000W/m^2 has been used for cost estimation.

2. A comparison between spray evaporation and refrigeration with respect to cost, cooling times and power consumption has shown that when spray evaporation technique is used, there is a significant reduction in plant and overall costs. Therefore this technique merits further investigation in order to explore the advantages to the full. However a combination of spray evaporation and refrigeration gave the best results and this technique is expected to improve the overall energy efficiency of processing operations where rapid cooling is necessary.

REFERENCES

1. Coulson, J. M., Richardson J. F., Backhurst, J. R., & Harker J. H., Chemical Engineering Volume 1., Permagon, Oxford, 1985.
2. Martin, H., heat and mass transfer between impinging gas jets and solid surfaces, Advances in Heat Transfer, 1977, vol. 13, pp. 1-60.
3. Bolle, L., and Moureau J. C., Spray Cooling of Hot Surfaces, Int. Multiphase Science and Technology, Hemisphere, 1982, pp. 1-97.
4. Jachuck, R. J. J., Rapid cooling of polymers using spray evaporation techniques, Ph.D Thesis University of Newcastle upon Tyne. 1992.
5. Mills, N. J., Plastics Microstructure. Properties and Applications, Edward Arnold Ltd, London, 1986.
6. Powell, P. C., Engineering with Polymers, J. W. Arrowsmith Ltd., Bristol, 1983.
7. Bambezha, A. V., Gimbutis, G. I., and Shvenchyanas, P. P., Heat transfer at a vertical flat surface with the combined effect of forced and free convection in the same direction, Int. Chem. Eng., 1981, vol. 21, No.1, pp. 135-138.
8. Romanenko, P. N. and Davidzon, M. I., Int.Chem. Eng., 1970, vol.10, No.2, pp.223-228.
9. Perry, R. H., Green D. W., and Maloney J. O., Chemical Engineer's Handbook, 6th. edition, McGraw-Hill, New York, 1984.
10. Backhurst J.R & Harker J. H., Process Plant Design, Heinemann Educational, London, 1973.

SESSION 7:

Heat Exchangers
Continued

Chairman: Prof. A. Karabelas

GENERIC STUDIES FOR INDUSTRIAL HEAT EXCHANGER FOULING

J D ISDALE
National Engineering Laboratory Executive Agency
East Kilbride
Glasgow G75 0QU, UK

ABSTRACT

The aims of the work are to develop a model for gas-side fouling which can be used for prediction and a fouling monitor, and, for liquid-side fouling to identify a viable basis for generic modelling and provide data. The project combines the work of ten partners from five EC member states. A summary of the work in progress is provided together with some preliminary results. For gas-side fouling a method has been identified based on an equation framework for analysis, combined with experimental data from both laboratory and industrial sites. The fouling monitor is based on an optical technique to measure surface temperature using a multi-wavelength infra-red scanning technique. This will be combined with heat flux measurement to allow determination of the thermal fouling resistance through the probe surface.

The liquid-side projects are similarly based on a combination of fundamental and practical studies. These include theoretical and experimental studies of fouling in industrial sugar beet evaporator streams (extended surfaces and plain surfaces with and without turbulence enhancement), laboratory studies of simulated process liquors and a study of the influence of colloidal particles on precipitation fouling.

NOTATION

A	area	m^2
d	diameter	m
M	mass deposition rate	$kg.s^{-1}$
P	probability	
U	velocity	$m.s^{-1}$
δ	density	$kg.m^{-3}$

σ   mass fraction
ε   collection efficiency

Subscripts

g	gas	r	retention
i	ith mechanism	tot	total theoretical
p	particle	x	experimental
proj	projected / frontal		

INTRODUCTION

This paper presents information on research activities aimed at controlling fouling behaviour of heat exchangers in industry. Since the work is still in progress the aim here is to present a summary of current progress and provide an indication of preliminary results. The research forms part of the JOULE Non-Nuclear Energy R & D Programme of the European Community [1].

For gas-side fouling the aims of the work are to develop a predictive model and a fouling monitor. For liquid systems, the studies will attempt to identify a generic modelling method, with particular emphasis on industrial process fluids. Both studies will provide data which will be of direct use in industrial systems.

The project combines the work of ten partners from five EC member states. The principal partners are: National Engineering Laboratory Executive Agency (NEL); Commissariat a l'Energie Atomique, (GRETh); Constructions Industrielles de la Mediterranee, (CNIM); Chemical Process Engineering Research Institute, (CPERI); TNO Division of Technology for Society, (TNO); Hamon-Sobelco SA, (H-S); and the University of Birmingham, (UB). Additional contributions are provided by Sogelberg (SO), CalGavin (CG) and British Sugar plc (BS). The gas-side work is nearing completion but that on liquid-side has started only recently.

The contributions from the partners fall broadly into three categories: model development, experimental measurements and monitor development. For gas-side work model development is carried out mainly by NEL with assistance from GRETh, but for liquid-side modelling the roles are reversed and assistance is also provided by CPERI and UB. A wide range of experimental measurements are clearly required including those in industrial process streams and in the laboratory. The industrial partners, CNIM, H-S, SO, CG and BS provide access to real process streams or facilities and, where needed, support and plant data for the site measurement teams. For the incinerator and cast steel furnace tests CNIM and H-S have also provided extensive chemical analyses of deposits and gas-streams. Laboratory measurements are carried out by GRETh, CPERI and UB with assistance from NEL. Development of a gas-side fouling monitor is being carried out by TNO.

The method developed involves theoretical calculation of particle transport to the heat transfer surface taking all active mechanisms into account to give the overall theoretical collection efficiency. Mass deposition rates measured experimentally are also used, in combination with particulate concentration data, to calculate experimental collection efficiencies. The ratio of these two collection efficiencies provides overall residence probabilities P for particulate fouling. These residence probabilities combine the effects of instantaneous particle retention, and of deposit removal. The P values so obtained are then used in a predictive model of the deposition process.

The above approach relies however, on quantifying the deposit fractions attributable to different mechanisms or groups of mechanisms. Clearly, if the chemical composition of different size fractions of gas-borne particulates is known, along with the composition of deposits, any differences between the two can be ascribed to either disproportionate deposition from particulate species, or to non-particulate deposition. This approach to understanding the deposition process has been reported elsewhere [2].

GAS-SIDE FOULING

Model development
An analytical framework has been developed [3] by extending modelling and analysis techniques and by utilising data from both laboratory and industrial plant measurements. This framework (or model) should therefore be capable of describing the observed data and of predicting fouling for similar environments. The main steps here are the identification and assessment of a suitable set of equations; the development of FORTRAN 77 analysis programs; the calculation of equation parameters such as particle sticking or retention probability from the experimental data; the development of suitable functions to describe these; and finally the incorporation of any adjustments found necessary by validation studies. Retention probability, the fraction of impacting particles retained by a surface, cannot at present be determined by calculation and is the key unknown in most fouling processes.

For particulate deposition, mass deposition rates can be calculated in terms of a collection efficiency, ϵ, defined as the fraction of the free stream particle mass flow (through the projected area of the body) which actually strikes the body. For each mechanism ϵ_i is a function of a number variables such as particle diameter, density and gas velocity. Thus:

$$\epsilon_i = \Sigma \, f(d_p \, , \, \delta_p \, , \, U_g \, , \, . \, .) \tag{1}$$

where the summation is over the full particle size range to be modelled, assuming all the mechanisms are independent. Industrial gas streams frequently contain polydisperse particulate material covering a range of particle sizes from 0.01 to over 100 μm, with the result that a number of deposition mechanisms are active as shown in Fig. 1. The total collection efficiency, ϵ_{tot}, is then given by:

$$\epsilon_{tot} = 1 - \pi \, (1 - \epsilon_i \,) \tag{2}$$

where the product is over all the active mechanisms.

The mass deposition rate can then be calculated from:

$$M = \epsilon_{tot} \, (\, \delta_g \, U_g \, \sigma \, A_{proj} \,) \tag{3}$$

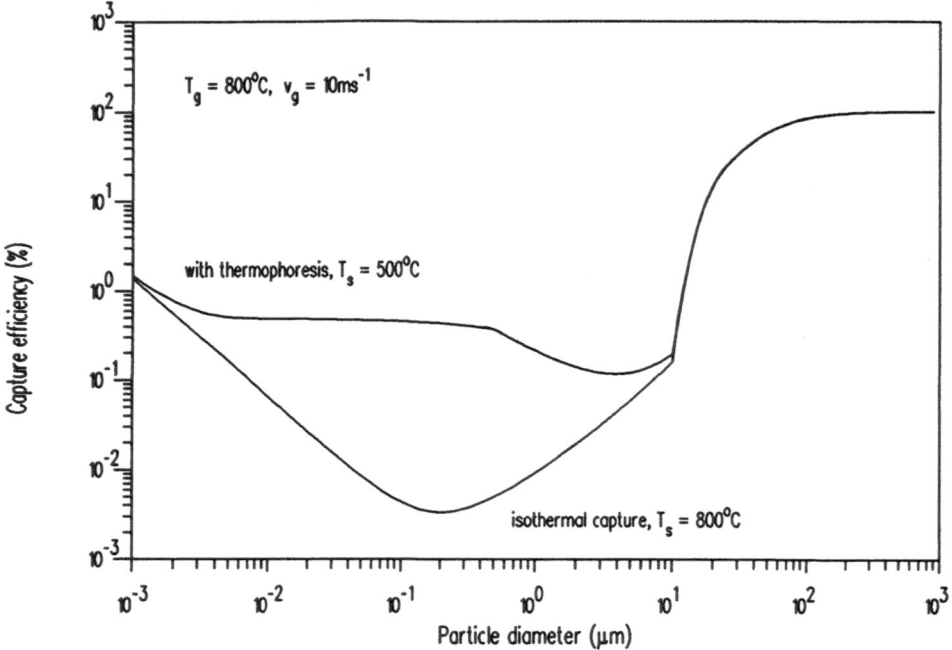

Figure 1. Variation of capture efficiency with particle diameter

A previous model [4,5] based on a correlation [6] for inertial impaction has been extended to incorporate a similar style of mass transfer correlation [5] for small particle diffusion and thermophoresis.

The theoretical expressions for collection efficiency for various mechanisms all assume unit sticking probability, ie. all material which reaches a surface remains there. In practice this is not the case, since particles may bounce, or deposit may be removed by incoming particles or fluid shear. At present there are no a priori methods for calculating instantaneous sticking probabilities and it is necessary to use a deposit retention probability, P_r. Equation 3 can be rearranged and used to calculate a collection efficiency ϵ_x from experimental data, which is the net result of all the deposition mechanisms and incorporates the effects of deposit removal. The deposit retention probability is thus given by:

$$P_r = \epsilon_x / \epsilon_{tot} \tag{4}$$

Values of P_r can then be correlated with relevant system parameters such as gas and surface temperatures, gas velocity or material properties. The correlations can then be used to modify theoretical collection efficiency calculations for other conditions.

An applications program will also be developed for convenient implementation of the results.

Experimental tests

The experimental work required includes measurements both under laboratory and full scale operating conditions. The aims of the former are to provide direct information on the validity of specific particulate fouling mechanisms, and on temperature and velocity distribution for selected finned tube configurations. The full scale industrial tests have provided data which will enable retention probabilities to be calculated from the model and related to appropriate system parameters. The parameters measured during the course of the work are listed in Table 1.

TABLE 1
Parameters to be measured

Quantity	Measured Parameter	Technique
Mass accumulation.	Deposit mass. Sample piece temperatures.	NEL probe.
Gas stream properties.	Temperature. Velocity. Pressure.	Thermocouple. Pitot-tube. Manometer.
Particulate.	Concentration. Size distribution.	British Coal Utilisation Research Association sampler. Cascade Impactor. Laser system.
Deposit chemistry.	All elements (other than C,H,O). Crystal structure. Deposit inspection. Ash fusion, sublimation and ionisation of salts. Deposit homogeneity.	X-ray microprobe. X-ray diffractometry. Scanning electron and optical microscopy. Differential thermal and thermogravimetric analyses. Visual observation.
Gas analysis.	Oxygen, carbon dioxide water, sulphur dioxide NO_x and HCl.	On-line analysers.
Fin tests.	Temperature distribution and temperature gradient. Air velocity. Alignment.	Infrared thermometry. TRIO software.
Steam quality.	Mass flowrate, temperature and pressure.	Plant operating data.

Laboratory tests were carried out to determine the temperature distribution on exchanger fins. The work was carried out in a rig designed for infrared measurements of temperature on heat exchanger surfaces. Aligned and staggered tube configurations of ten rows were measured at three different positions within the tube array, at several temperatures. The results have provided data which will be used for model validation.

Laboratory tests have also been carried out in diesel exhaust environments with the aim of providing data on the influence of surface temperature and velocity on deposition. For different conditions, the rig was used to measure the change of heat transfer coefficient and pressure drop with time as deposition progressed. Aligned and staggered configurations of finned tubes were tested to provide data for a range of velocities and surface temperatures. Particle size distribution and concentration in the exhaust were measured and deposit analysis was also carried out.

A similar series of measurements has been completed using an ultrasonic aerosol generator to produce solid monodisperse particles (in the size range 0.5 to 5.0 μm) in a heated stream of air, so providing a simulated fouling environment. For these experiments the exchangers consisted of a row of ten tubes, with two adjacent columns of uncooled half section tubes to simulate a complete array. Measurements of the change of heat transfer coefficient and pressure drop with time were carried out as the deposition proceeded. Fig. 2 shows the measured change of fouling resistance with time. These results have also allowed the influence of particle size on fouling mechanisms to be examined in detail [8]. This is illustrated in Fig. 3 which shows the variation of heat transfer coefficient with time for different particle sizes.

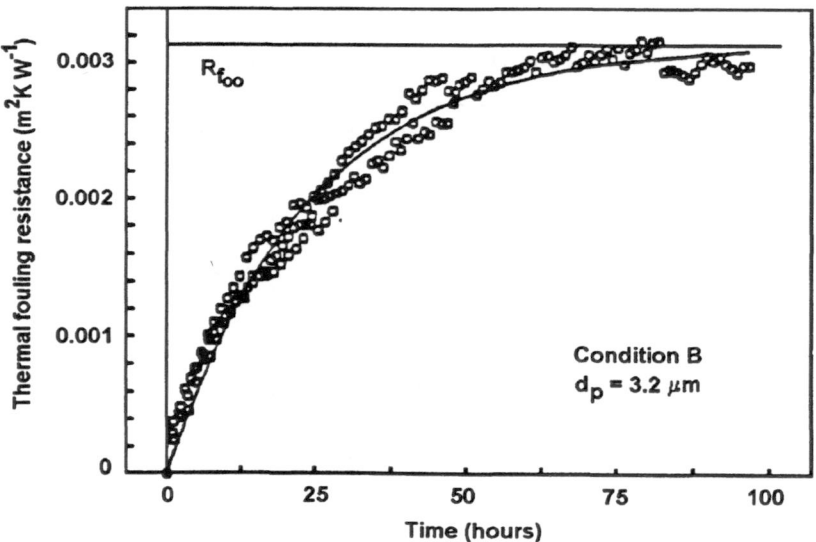

Figure 2. Variation of thermal fouling resistance with time

For the larger particle tests a significant fraction impacts on the surface thus forming a more compacted deposit. For the smaller size particles, deposition is mainly by diffusion and thermophoresis, resulting in a lower deposit density and conductivity with a higher fouling resistance.

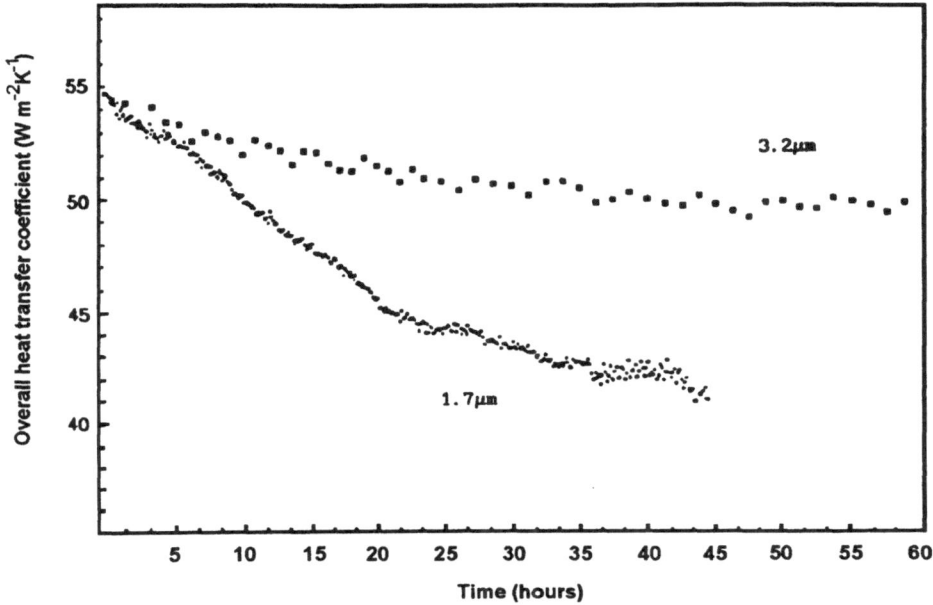

Figure 3. Variation of heat transfer coefficient with time and particle size

The industrial tests were carried out in a domestic refuse incinerator and in a cast steel plant. In each case the measurements included mass accumulation rates [9] and other relevant parameters over the range of conditions available in each plant. These tests provided data on the influence of gas temperature, gas velocity at a given temperature, time, plant operating conditions and the chemistry of the environment, on the deposition. Chemical analyses of the deposits and the gas streams were required to provide input information for model interpretation. A variety of advanced chemical analytical methods were used [2]. In each case consistent variations of the deposit composition were detected which will be related to the deposit formation mechanisms through the retention probability functions. An example of the data from the incineration fouling work is given in Fig's 4 and 5. Fig. 4 is a typical differential thermal analysis (DTA) trace from deposit samples, showing the presence of several endotherms indicating phase changes in the region of 700°C. Fig. 5 shows a corresponding increase in collection efficiency, measured using the mass accumulation probe, at a similar temperature.

Monitor development

The aim of the monitor development was to produce a device which gives some indication of the progress of fouling within a plant. Translation of the reading indicated by the monitor to the fouling of the actual heat exchanger may be achieved using the applications program combined with knowledge of the plant heat exchanger operating conditions. A multi-wavelength infra-red sensor has been developed which will scan a controlled heat transfer surface to determine deposit surface temperatures. The use of three or more wavelengths permits the equations to be solved for both the temperature

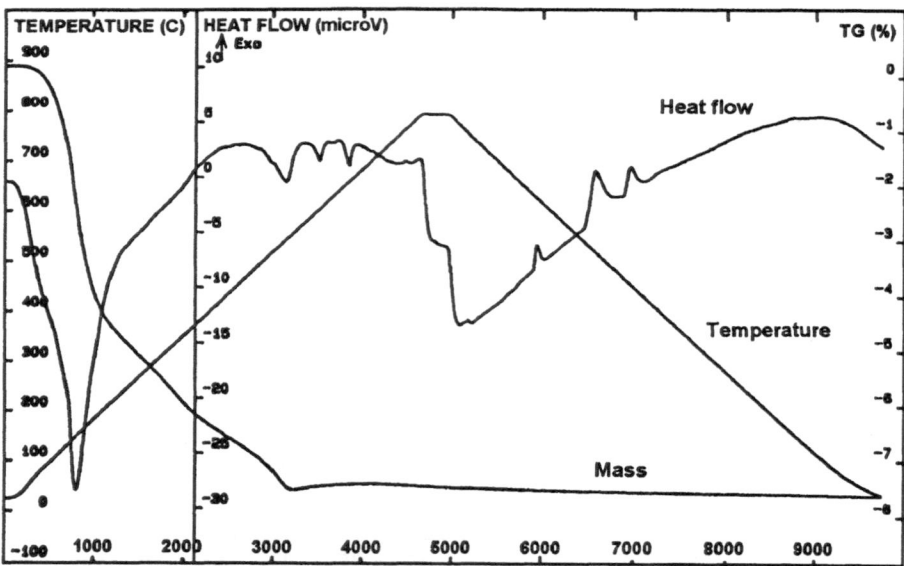

Figure 4. DTA trace for incinerator deposits

and emittance of the test surface. Measurements of heat flux through the surface will permit the thermal fouling resistance to be obtained by calculation. These data will thus provide a direct indication of local fouling effects and may be used in combination with other data to calculate fouling elsewhere in the system.

LIQUID-SIDE FOULING

The aim of this part of the project is to investigate the possibilities for generic modelling of fouling in liquid systems, with particular emphasis on industrial process fluids in the sugar industry. This work contains model development studies combined with measurements of process liquids in the laboratory and in industry.

The liquid-side studies aim to identify a viable basis for modelling in liquid systems and provide useful data on the thermal fouling resistance. Since there is no sufficiently general theoretical framework available at present, the problem will be studied from two directions. Firstly, for particulate fouling, theoretical and laboratory studies will be undertaken to identify suitable analytical methods for systems where this type of fouling is dominant. If appropriate, parallels will be drawn with the current gas-side modelling. Secondly, existing liquid-side models will be tested with data obtained using industrial fluids that give rise to fouling problems. These models are normally very highly specific to a particular environment or range of physico-chemical conditions, so the aim here is to determine if and how it will be possible to extend them to other areas. Data will also be provided on fouling in conventional and extended surface configurations

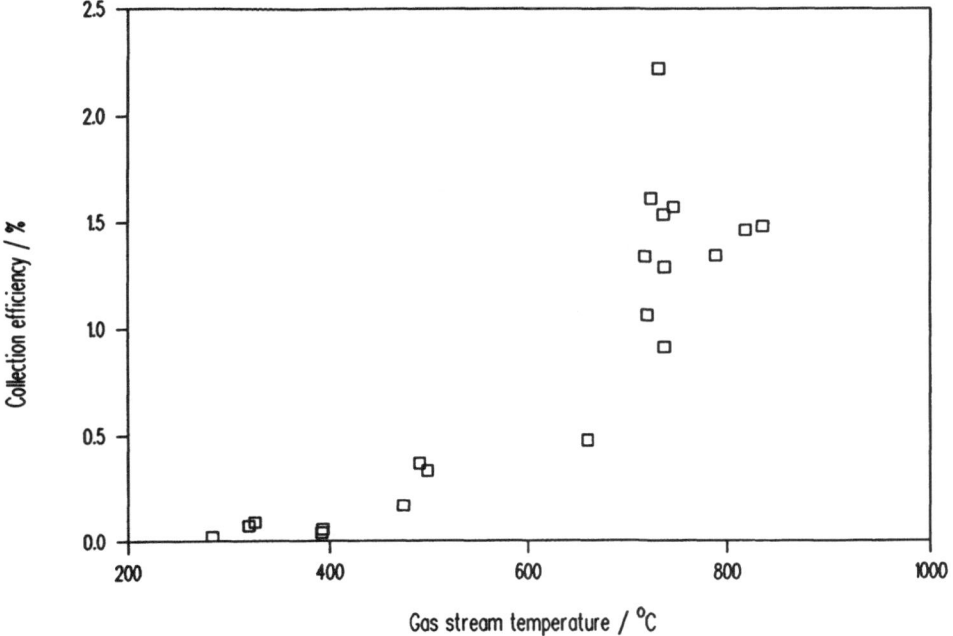

Figure 5. Measured collection efficiency at different incinerator exhaust gas temperatures

and on the influence of turbulence promoters, the importance of which should be stressed.

The technical content is divided into three parts, generic modelling, laboratory studies and site measurements.

Generic modelling

The objective here is to assess (and where necessary extend) particulate fouling models for liquid systems to allow their application to extended surface configurations. It thus concerns the more fundamental aspects of fouling mechanisms in the liquid phase, from which it will be possible to develop methods for a wide range of industrial environments including sugar beet extracts and others. Data from other parts of the project will be used as appropriate, and the modelling results will in turn be applied to the industrial environments studied for the prediction of thermal fouling resistance.

Laboratory studies

The aim of the work in this section is to extend the database for process liquid fouling under carefully controlled laboratory conditions. The data will be used to develop and test models of deposition processes in simulated sugar beet liquors which will be used to improve mitigation techniques in real streams. The advantage of this approach is that it will be possible, for example, to vary the means of pH control (and determine the fouling consequences) without impairing the quality of expensive real process streams.

Specifically to aid the generic model studies, measurements will be carried out of particulate fouling under controlled laboratory conditions on some commercial enhanced surfaces (ribs and plates). Experiments will initially be carried out on kaolin suspensions in water and subsequently in simulated sugar beet liquors. Existing models from the literature will be modified as necessary to suit extended surface applications, based on experimental data from these tests.

In addition, measurements of thermal fouling resistance will be carried out, in conjunction with chemical analyses of deposits, using synthetic liquids designed to simulate key aspects of sugar beet liquor streams. The measurements will be undertaken using an existing laboratory rig which will be modified to produce suitable process conditions. Tests will be carried out with and without turbulators using, initially, buffered magnesium and calcium solutions and their mixtures (used for pH control).

In many heat exchangers in the process industries the rate of precipitation fouling and the heat transfer characteristics of the deposits so formed are strongly affected by the concentration of colloidal particles. These can act as nuclei for the formation of crystals and influence the number and size of the crystallites. Very small colloidal particles in the size range of 10 to 100 nm will be used, under controlled conditions which relate to laboratory and industrial conditions studied in other parts of the project. Specific theoretical and experimental work will therefore be undertaken to provide the framework for data interpretation and modelling. The characteristics of the particles, the pipe surfaces and the fluid properties will be taken into account in laboratory tests using tubular test sections to simulate heat exchanger elements. The density and other physical and chemical characteristics of the resulting fouling deposits will be measured and related to the properties of the colloidal particles.

Site measurements

The concentration of sugar beet extract is normally carried out by stepwise evaporation prior to crystallisation, using steam heated short vertical tube evaporators. This process has high energy costs which are increased by the formation of deposit layers from the extract on the heat transfer surfaces. In this process, enhancement of heat transfer by means of extended surfaces or turbulators could potentially provide a significant saving. However the influence of fouling on the performance of such devices is an unknown which has tended to inhibit their implementation. In addition, chemicals added to the beet extract for pH control, along with natural components of the extract, such as calcium oxalate, contribute to deposit formation during the concentration process.

Measurements will be carried out to determine the rate of fouling on plain heat transfer surfaces in an industrial plant. A Portable Fouling Assessment Unit (PFAU) will be used to simulate conditions in evaporator tubes by passing a side- stream of the extract through fouling monitors: the effect of turbulence promoters will also be examined. Two series of tests are planned in successive seasons.

Measurements will be carried out to determine the effectiveness of an extended surface evaporator for improving the efficiency of heat transfer from beet extracts under fouling conditions. The performance of enhanced surface (fluted) evaporator tubes will be measured in two beet processing factories, using test runs of up to 1000 hours duration. The results will therefore provide important data on the performance and fouling of fluted surfaces under real conditions. These data will be used to develop and test fouling models.

REFERENCES

1. Pilavachi, P.A. and Isdale J.D. European Community R & D strategy in the field of heat exchanger fouling: projects. In proceedings of Fouling Mechanisms: Theoretical and Practical Aspects. Eurotherm Seminar No. 23, Centre d'Etudes Nucléaires de Grenoble, 8 - 9 April 1992. Editions Européennes et Thermique, Paris 1992.

2. Howarth, J.H., Tabaries, F. and Durand, J.P. Characterisation of incinerator fouling. In proceedings of Fouling Mechanisms: Theoretical and Practical Aspects. Eurotherm Seminar No. 23, Centre d'Etudes Nucléaires de Grenoble, 8 - 9 April 1992. Editions Européennes et Thermique, Paris 1992.

3. Glen, N.F., Flynn S., Grillot, J.M. and Mercier, P. Measurement and modelling of gas-side fouling in finned tubes. In proceedings of Fouling Mechanisms: Theoretical and Practical Aspects. Eurotherm Seminar No. 23, Centre d'Etudes Nucléaires de Grenoble, 8 - 9 April 1992. Editions Européennes et Thermique, Paris 1992.

4. Glen, N.F. and Howarth, J.H. Modelling refuse incineration fouling. 2nd UK National Conference on Heat Transfer, 1, Sept. 1988, pp. 401-420.

5. Glen, N.F. and Howarth, J.H. A deposition model for pulverised coal firing. In Deposition from Combustion Gases. I.O.P. Short Meetings Series No. 23. Institute of Physics, 1989, pp 33-48.

6. Israel R. and Rosner, D.E. Use of a generalized Stokes number to determine the aerodynamic capture efficiency of non-Stokesian particles from a compressible gas flow. Aerosol Sci. and Technol. 1983, 2, 45-51.

7. Gökoğlu, S.A. and Rosner, D.E. Correlation of thermophoretically-modified small particle diffusional deposition rates in forced convective systems with variable properties, transpirational cooling and/or viscous dissipation. Int. J. Heat and Mass Transfer, 1984, 27(5), pp. 639-646.

8. Grillot, J.M. Fouling of a cylindrical probe in crossflow in a diesel exhaust environment. In proceedings of Fouling Mechanisms: Theoretical and Practical Aspects. Eurotherm Seminar No. 23, Centre d'Etudes Nucléaires de Grenoble, 8 - 9 April 1992. Editions Européennes et Thermique, Paris 1992.

9. Ewart, W.R. Obtaining valid fouling data from industrial gas streams. 2nd UK National Conference on Heat Transfer, 1, 1988, pp. 421-430. The Institution of Mechanical Engineers, 1988.

AN ASSESSMENT OF DATA AND PREDICTIVE TOOLS FOR COOLING WATER FOULING OF HEAT EXCHANGERS

S. VALIAMBAS, N. ANDRITSOS and A.J. KARABELAS

Chemical Process Engineering Research Institute and Chemical Engineering Department,

Aristotle University of Thessaloniki, P.O. Box 1517, GR 540 06 Thessaloniki, GREECE

ABSTRACT

Results of a literature review on cooling water fouling are summarized in this paper. To model this complex phenomenon, reliable data sets characterized by completeness are necessary. the scarcity of such sets in the open literature as well as other data shortcomings are evident. Available data on initial fouling rates are identified as the most appropriate for evaluating models of crystallization fouling. An ionic diffusion model proposed by Hasson and coworkers is compared against data from the literature and from recently conducted tests in this Laboratory. This model appears to perform satisfactorily and may serve as a building block for a more general predictive algorithm. Research priorities are placed on studies on combined precipitation/particulate deposition, and of solids detachment phenomena.

NOTATION

D	diffusivity of $CaCO_3$	$m^2 \cdot s^{-1}$
d_t	tube diameter	m
E	activation energy	$J \cdot kg^{-1}$
k_D	convective mass transfer coefficient	$m \cdot s^{-1}$
k_f	deposit thermal conductivity	$W \cdot m^{-1} \cdot K^{-1}$
k_R	rate coefficient for surface crystallization	$m^4 \cdot s^{-1} \cdot mol^{-1}$
K_{SP}	Solubility product of $CaCO_3$	$(mol \cdot m^{-3})^2$
P_d	probability of scale deposition	
R_f	thermal fouling resistance	$m^2 \cdot K \cdot W^{-1}$
R_f^*	asymptotic thermal fouling resistance	$m^2 \cdot K \cdot W^{-1}$
R	universal gas constant	$J \cdot kg^{-1} \cdot K^{-1}$
Re	Reynolds number ($=U \, d_t/\nu$)	

Sc	Schmidt number ($=\nu/D$)	
t	time	s
T_s	temperature of fouling deposit surface	K
U	bulk flow velocity	$m \cdot s^{-1}$
w	wall deposition rate	$kg \cdot s^{-1} \cdot m^{-2}$
W	mass flow rate	$kg \cdot s^{-1}$
x	fouling film thickness	m
ν	dynamic viscosity	$m^2 \cdot s^{-1}$
τ	shear stress	$N \cdot m^{-2}$
φ_d	deposition rate in terms of resistance	$m^2 \cdot K \cdot W^{-1} \cdot s^{-1}$
φ_r	removal rate in terms of resistance	$m^2 \cdot K \cdot W^{-1} \cdot s^{-1}$
ψ	deposit strength factor	-
Ω	water characterization factor	ppm $CaCO_3$ (?)

Subscripts

i	interfacial conditions

INTRODUCTION

Heat exchanger fouling is a long-standing problem with very serious technical and economic consequences. Several comprehensive papers [1-3] review the technical difficulties and economic penalties resulting from heat exchanger fouling.

Despite the recognition of its significance and a great deal of work that has been done on this problem in recent years, the technology is in a truly sorry state especially as regards predictive tools or reliable methodology readily applicable for design purposes. Standards recommended by TEMA in the form of tabulated values of thermal fouling resistance R_f [1,4] appear to be the accepted design practice to account for fouling due to cooling water. The drawbacks of the TEMA standards are serious; e.g. largely unfounded and conservative R_f values, insufficient instruction for applications, neglect of heat exchanger operating variables, such as the flow velocity, the concentration of the fouling species, the bulk temperature and the surface temperature. The motivation of the work reported here is to improve this situation by taking systematic steps towards the development of reliable predictive tools for design.

Results of an extensive literature review are summarized in this paper. The main goal is to assess available experimental data and models or correlations, proposed for predicting cooling water fouling, thus setting priorities on research needs, and identifying possibly fruitful modeling approaches. A set of premises is adopted to narrow down somewhat the breadth of this review. Emphasis is placed on fouling (precipitation and particulate) in turbulent pipe flow of cooling water, due mainly to hardness salts (e.g. $CaCO_3$, $CaSO_4$). Some data, recently obtained in this Laboratory [5] with the $CaCO_3$ system, aid the clarification of fouling mechanisms and the assessment of models.

The paper presents a summary of work carried out at the Chemical Process Engineering Research Institute (CPERI), as part of a project funded by the JOULE Non-Nuclear R&D Programme of the European Community (contract No JOUE-0040-C). The aim of the project is to develop better methods to measure and to predict fouling in a generic way. The principal contributors are: National Engineering Laboratory (NEL); Groupment pour la Recherche sur les Echangeurs Thermiques Grenoble (GRETh); Construction Industrielles de la Mediterranée Toulon (CNIM); Hamon-Sobelco SA Brussels (H-S); TNO Division of Technology for Society Apeldoorn (TNO); University of Birmingham (UB); and Chemical Process Engineering Research Institute Thessaloniki (CPERI). Additional contributions are provided by Sogelberg (SO), CalGalvin (CG) and British Sugar plc (BS).

REVIEW OF DATA

It is generally recognized [4,6] that hard deposits are due to precipitation (or crystallization) right *onto* the solid surface. Initially, when the pipe surface is clean, the rate of mass deposition is constant. There is already sufficient evidence [6,7] that during this period the dominant mechanism is convective diffusion of lattice ions to the wall. The "depositing" crystalline material usually adheres quite strongly onto the pipe wall, in which case particle (crystal) detachment by the fluid is considered negligible.

Most data sets show that, after a sufficiently long time period, the *net* rate of deposition (and the resistance R_f) tends to drastically decrease and to reach an asymptotic value; e.g. Figure 1. This reduction is attributed to the effect of solids detachment by the flow. Factors contributing to, and conditions favoring, this mechanism (especially for crystalline deposits) are largely unexplored.

The assessment of experimental data from the open literature summarized here is made on the basis of a list of system parameters and properties (Table 1), which are considered necessary to characterize a data set as *complete*. There is generally a lack of "complete" data sets, even though a few sets reviewed appear to be close to this definition. Such "completeness" is considered essential for clarifying the dominant mechanisms and for model development. The main deficiencies in the reported data sets are as follows:

- There is a lack of measurements of deposit properties. Directly measured density and conductivity are non-existent. Moreover, simultaneous measurements of thermal resistance *and* of the deposit mass or thickness are scarce, while the description of deposit bonding or deposit strength seems to be quite subjective.
- The range of reported flow velocities and wall surface temperatures is relatively narrow.

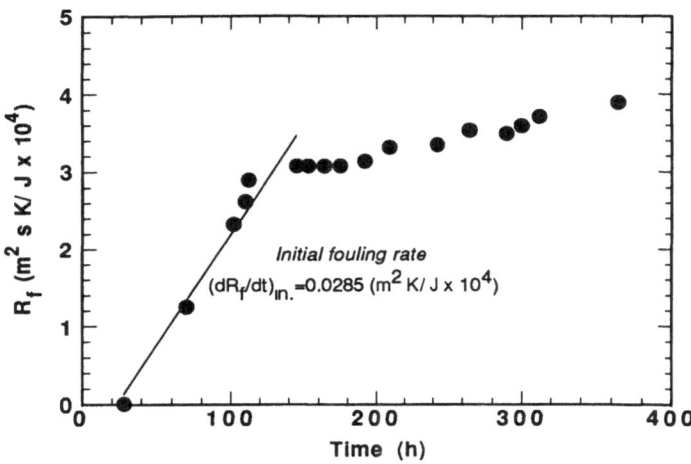

Figure 1. Definition of the initial fouling rate from experimental results (Morse & Knudsen [18], Run 18A).

- The possible variation of deposition rate and thermal resistance R_f along the flow path is ignored.
- In most data sets there is no detailed characterization of the fluid (in contact with the solid surface) especially as regards the presence of colloidal particles.

It is evident that a systematic experimental effort is required to collect *complete* sets of data, useful for model development and validation.

New computer software has been developed to treat all the useful data sets, collected from the open literature. This software is specially design to accomplish the following main tasks:

- store sets of experimental data with all the related information;
- perform statistical analysis of single data sets or combinations thereof;
- make comparisons with (selected) stored models;
- store the results of the aforementioned analyses and evaluations for possible future use in research and applications (comparisons, selection of reliable data sets, guidance in design calculations, etc.).

REVIEW OF MODELS

The first study dealing with fouling of heat exchangers appears to be that of McCabe and Robinson, who assume that the mass deposited on a heated surface is proportional to the heat

TABLE 1. Information required for completeness of a fouling data set

A. Scale Properties	B. Water Properties
- Chemical composition - Crystalline structure and phases - Density, ϱ_f - Thermal conductivity, k_f - Thickness (along the flow path), x	- Chemical composition - pH - Total suspended solids - Conductivity, I
C. Flow Characteristics	D. Temperatures, Heat Flow
- Flow field geometry, or test section - Velocity, U	- Wall temperature(s), T_s - Liquid inlet-outlet temperatures - Heat flux, $\dot{q}$
E. Solid Wall or Pipe	F. Experimental Technique
- Chemical composition - Roughness - ζ-potential	- Adequate description - Error estimates

G. Fouling Rates and Resistances
  - Temporal evolution of deposited mass (mass versus time curve) in order to evaluate
      • Delay period
      • Initial deposition rate m and resistance R_f
      • Asymptotic values (R_f^* and deposited mass)	

flux. The often quoted papers of Kern and Seaton [8,9] put modeling on the right track. They consider that the net fouling rate (expressed on the basis of deposit thickness, x) is the difference between a deposition rate and a removal rate,

$$\frac{dx}{dt} = \frac{d(R_f k_f)}{dt} = K_1 c' W - K_2 \tau x \tag{1}$$

where K_1 and K_2 are constants, W is the mass flow rate, c' is the concentration of active species (consistent with K_1) and k_f is the deposit thermal conductivity.

In 1972, Taborek *et al* [10], associated with the Heat Transfer Research Institute (HTRI), presented the results of a large scale investigation of the fouling problem. Their model is based on the deposition-removal approach of Kern and Seaton. The two rate terms, φ_d and φ_r are given by

$$\varphi_d = C_1 P_d \Omega^n \exp\left(-\frac{E}{T_s R}\right) \tag{2}$$

$$\varphi_r = C_2 \frac{\tau}{\psi} x^m \tag{3}$$

where, C_1 and C_2 are constants; P_d is a deposition probability (a function of fluid velocity); Ω^n is a water characterization factor; E is an activation energy; T_s is the surface temperature; τ is the shear stress; and ψ is the strength of the deposits. A close look at this model reveals that

- There is a rather strong dependence of deposition rate φ_d on solid surface temperature T_s. There is no confidence, however, in the determination of T_s, in view of the lack of physical properties of the deposits.

- The probability function P_d and, consequently, the deposition rate φ_d decrease with increasing flow velocity, U. However, this appears to be unacceptable on physical grounds. In fact, for crystallization (precipitation) fouling, either there is no velocity effect (surface control), or the deposition rate increases with velocity (diffusion controlled process) [11]. The deposition of particulate matter also increases with increasing velocity [12].

- There are very serious reservations about the physical significance of the "water characterization factor" Ω and the function ψ of the deposit structure.

- Despite the fact that the HTRI model for the first time considered the basic mechanisms involved in a fouling process, it seems to be of no great value to the design engineer, because of so many parameters which must be determined using data banks.

Overall, although the HTRI model is the most frequently quoted one in the literature, it is impossible at present to make *a priori* predictions of R_f with some degree of confidence. It appears that a basic shortcoming of this model is the proposed correlation of fouling rates, due to at least two different types of fouling (precipitation and particulate fouling), with the same form of equations.

Fouling models based on the mechanisms of wall crystallization appear to have an adequate theoretical foundation and to hold promise for possible correlation of experimental data. As mentioned previously, calcium carbonate is a major component of the deposits encountered in a cooling system. Thus measured fouling rates mainly due to $CaCO_3$ (>90% w.) appear to be suitable for model assessment.

Hasson and coworkers [11,13] have developed a rather successful model for predicting the deposition of calcium carbonate in pipes. The rate of wall crystallization can be described by combining the convective diffusional resistance and the surface reaction resistance. The rate of surface reaction for a wide range of pH conditions can be given by an ionic product model

$$w = k_R \{[Ca^{++}]_i [CO_3^=]_i - K_{SP}'\} \tag{4}$$

where i denotes the interfacial conditions and k_R is the rate coefficient for the surface reaction. The diffusion of the carbonic species can be simply described by assuming "high" and "low" pH conditions. At high pH conditions ($pH > 10$) only the carbonates have to be considered, while at low pH (pH<8) the distribution of CO_2 and HCO_3^- is sufficient to describe the diffusion rate. Thus the diffusion rate at *high pH* is given by

$$w = k_D \{[Ca^{++}] - [Ca^{++}]_i\}$$
$$= k_D \{[CO_3^=] - [CO_3^=]_i\} \tag{5}$$

and at *low pH*, by

$$w = k_D \{[Ca^{++}] - [Ca^{++}]_i\}$$
$$= \frac{k_D}{2} \{[HCO_3^-] - [HCO_3^-]_i\}$$
$$= k_D \{[CO_2]_i - [CO_2]\} \tag{6}$$

The interfacial concentrations in eqn (4) can be eliminated using expressions (5) and (6) and an expression for the carbonic species equilibria. Complete solutions can be found in the literature [13,14].

COMPARISON OF DEPOSITION MODELS WITH EXPERIMENTAL DATA

In this presentation measured *initial* fouling rates are compared with predictions based on the Hasson model as well as on correlations for *particulate* deposition. As pointed out previously, initial rates are considered representative of conditions whereby deposit removal is insignificant and measured rates are essentially due to the deposition process alone. Initial rates are, therefore, the most appropriate for evaluating models and correlations for deposition rate φ_d.

Measured initial fouling rates have been retrieved from several publications [14-19]. The initial deposition rate, $(dR_f/dt)_{in}$, if not reported, is either taken from the slope of the linear increase of R_f with time, or (in most cases) estimated from the asymptotic fouling curves, as shown in Figure 1. The same procedure is applied to fouling curves with or without a delay period. This procedure may cause a small error in the estimation of the rate, especially for cases with a rather pronounced delay period. The latter probably reflects the enhancement of heat transfer due to roughening of the heat transfer surface from the initial deposits [2]. An amount of $CaCO_3$ in the scale greater than 90% by weight is taken as a criterion for selecting data sets. It is noted that several sets obtained by Knudsen and his coworkers do not meet this criterion.

The mass transfer coefficient, k_D, used in the Hasson model is the well known Linton-Sherwood equation,

$$k_D = 0.023 \ U \ Sc^{-2/3} \ Re^{-0.17}, \tag{7}$$

although more recent correlations yield somewhat smaller values of the mass transfer coefficient. A value of 0.85×10^{-9} m²/s is used for the diffusivity of $CaCO_3$ at 25°C [20], which is corrected for temperature differences according to

$$D \sim D \ (\text{at } 25°C) \ \text{Temp. } / \ \text{Viscosity.} \tag{8}$$

A value of the product $\varrho_f k_f$ of 5000 kgW/m⁴K is used to convert mass deposition rates to initial fouling rates, $(dR_f/dt)_{in}$.

Comparisons between measured initial fouling rates and predictions are shown in Figure 2. It appears that the greatest deviation (in some cases more than one order of magnitude), occurs in certain data sets of Watkinson [14], where dispersed particles are present in the liquid phase and for high fouling rates, uncommon in real situations. However, it is difficult to establish whether this discrepancy can be attributed to the model or to data inaccuracies. Reasonable agreement is observed with some data sets obtained by Knudsen and his coworkers. Their low fouling rates, which the models underpredict, correspond to scales with significant non-$CaCO_3$ contents. In general, it appears that the Hasson model can afford fair (order of magnitude) predictions of the initial fouling rates, in cases where $CaCO_3$ is the dominant scale species.

Figure 3 presents a plot of measured fouling rates by Watkinson versus predicted values using correlations for *particulate* deposition recommended by Papavergos and Hedley [12]. In these calculations it is assumed that all available calcium carbonate is converted to spherical 0.5 and 5 µm particles. As anticipated, and also noted by Watkinson [14], particulate deposition rates alone cannot account for the high fouling rates observed at high bulk temperatures.

Scale formation data in *isothermal flows* are also considered appropriate in assessing the deposition term φ_d of the various fouling models. This is because isothermal deposition curves, especially those from once-through runs and with a single scaling species, do not exhibit simultaneous removal of the deposits, and it can be safely assumed that particulate

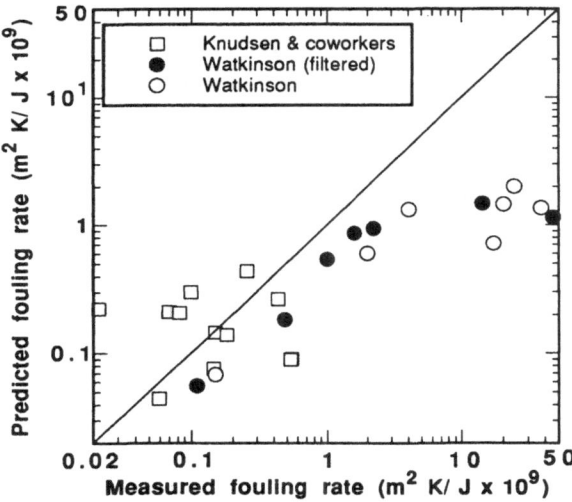

Figure 2. Comparison of measured initial fouling rates with predicted values from Hasson's model (D=0.85x10^{-9} m^2/s at 25°C).

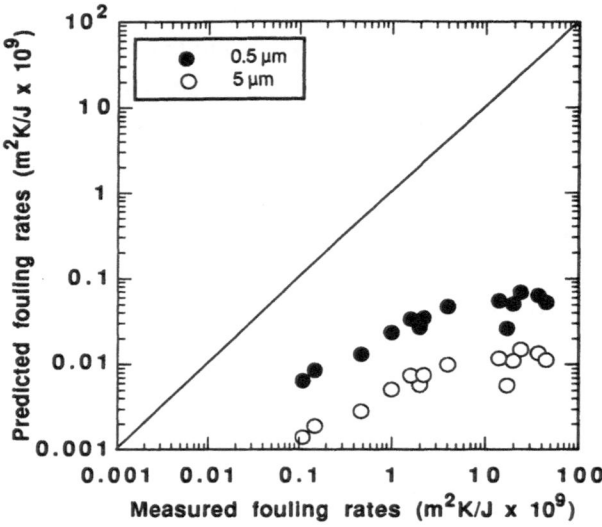

Figure 3. Comparison of measured initial fouling rates by Watkinson [14] with predictions for particulate deposition.

deposition is negligible. Figure 4 compares measured $CaCO_3$ deposition rates [5,11,21] with predictions from the "high pH" ionic wall crystallization model of Hasson and his coworkers. All data points correspond to once-through runs and represent initial deposition rates. Most of the data by Hasson and his coworkers exhibit a brief delay period. For such data, the initial deposition rates are obtained from the slope of the fouling curve after the delay period. These data are in fair agreement with model predictions, despite several assumptions made for the experimental conditions. New data from this laboratory [4] for pH>9.5 are systematically overpredicted by the model. The apparent discrepancy between the two seemingly similar experimental sets may lie in the definition of the initial deposition rate. The experiments of Kontopoulou [4] show no induction period and the rates are calculated from the linear deposit growth during short 2-hr to 4 hr runs. However, in some longer runs, the slope of the deposition curve tends to increase somewhat. Thus, better agreement with the model can be obtained, if the slope of the curve for long times is used.

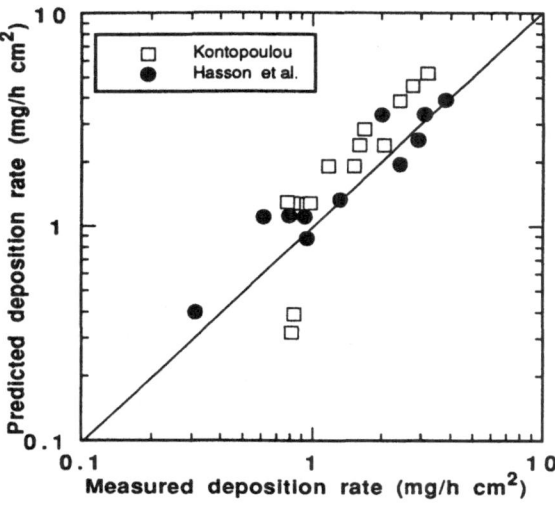

Figure 4. Comparison of measured deposition rates with predicted values. Precipitation under isothermal conditions.

CONCLUDING REMARKS

The best approach for modeling appears to be the framework originally proposed by Kern and Seaton, whereby the net fouling rate is the difference between deposition (φ_d) and removal (φ_r) rates. For systematic model development, it is necessary to treat each process separately. Moreover, to model φ_d it is necessary to study separately the mechanisms of *precipitation* (crystallization) and *particulate deposition*. The former is considered more important for cooling water systems, leading to hard scale. The latter can significantly influence the deposition rate and the scale properties.

With regard to precipitation fouling due to $CaCO_3$ (which is usually the main fouling species in cooling water), a model proposed by Hasson and coworkers [11,13] appears to be the best presently available. Thus, it can be used (with some improvements) as a building block in a generalized predictive algorithm. Towards the development of such a general tool, the following research priorities are identified

- Careful precipitation experiments in a once-through pipe flow system, with inverse solubility salts (e.g. $CaCO_3$), in order to collect *complete* data sets as defined in Table 1. These data will be essential for clarifying the influence of various mechanisms and for model validation. Two types of experiments are required, i.e. with no suspended particles (precipitation only) and with well-identified colloidal size particles (combined precipitation/ particulate fouling).

- A study of solids detachment from deposited crystalline scale and from layers of colloidal size particles. There is a lack of basic understanding of these phenomena. Careful experiments and modeling are required to develop reliable expressions for φ_r, thus completing the aforementioned general algorithms.

Research on these fronts is in progress in our Laboratory.

REFERENCES

1. BOTT, T.R. *Fouling Notebook*. Instn Chemical Engineers, London, 1990.

2. EPSTEIN, N. Fouling in heat exchangers. Proceedings of the *Sixth Int. Heat Transfer Conference*, Toronto. Hemisphere Publ. Co., 1978, pp. 235-253.

3. PRITCHARD, A.M. The economics of fouling. In *Fouling Science and Technology*, Melo, L.F., Bott, T.R. and Bernardo, C.A. Eds., Kluwer Academic Publishers, Dordrech, 1988.

4. KNUDSEN, J.G. Conquer cooling-water fouling. *Chem. Eng. Progr.*, **87**(4), 42-48, 1991.

5. KONTOPOULOU, M. *Calcium carbonate scaling in pipes*. Diploma Thesis, Dept. Chem. Engng, Aristotle University of Thessaloniki, May 1992 (in Greek).

6. ANDRITSOS, N. and KARABELAS, A.J. Crystallization Foulin: the effect of flow velocity on the deposition rate. Presented at *Fouling Mechanisms: Theoretical and Practical Aspects*, Grenoble, 8-9 April, 1992.

7. BOHNET, M. Fouling of heat transfer surfaces. *IChem. Eng. Technol.*, **10**, 113-125, 1987.

8. KERN, D.Q. and SEATON, R.E. A theoretical analysis of thermal surface fouling. *Brit. Chem. Eng.*, 1959, **4**, 258-262.

9. KERN, D.Q. and SEATON, R.E. Surface fouling - how to calculate limits. *Chem. Eng. Progress*, 1959, **55**, 71-73.

10. TABOREK, J., AOKI T., RITTER, R.B., PALEN, J.W. and KNUDSEN, J.G. Predictive Methods for Fouling Behavior, *Chem. Eng. Prog.*, 1972, **68**(2), 59 - 67, and **68**(7), 69 - 78.

11. HASSON, D. Precipitation Fouling. In *Fouling of Heat Transfer Equipment* (E.F.C. Somerscales and J.G. Kundsen, eds). Hemisphere Publ. Co., Washington, 1981, pp. 527-568.

12. PAPAVERGOS, P.G. and HEDLEY, A.B. Particle deposition behaviour from turbulent flows. *Chem. Eng. Res. Dev.*, 1984, **62**, 275-295.

13. HASSON, D., SHERMAN, H. and BITON, M. Prediction of CaCO3 scaling rates. Presented in the *6th Int. Symposium Fresh Water from the Sea*, 1978, pp. 193-199.

14. WATKINSON, A.P. Water quality effects on fouling from hard waters. In *Heat Exchangers Theory and Practice*, (ed. Taborek, J., *et al.*), Hemisphere Publ. Co., Washington, 1983, pp. 853-861.

15. COATES, K.E. and KNUDSEN, J.G. Calcium Carbonate Scaling Characteristics of Cooling Tower Water. *ASHRAE Transactions*, 1980, **86** (Part II), 68 - 91.

16. KNUDSEN, J.G. and ROY, B.V. Studies on the Scaling of Cooling Tower Water. In *Fouling of Heat Exchanger Surfaces*, Bryers R.W. (Ed.), Engineering Foundation, New York, 1983, pp. 517 - 530.

17. LEE, S.H. and KNUDSEN, J.G. Scaling Characteristics of Cooling Tower Water. *ASHRAE Transactions*, 1979, **85** (Part I), 281 - 302.

18. MORSE, R.W. and KNUDSEN, J.G. Effect of Alkalinity on the Scaling of Simulated Cooling Tower Water. *Can. J. Chem. Eng.*, 1977, **55**, 272 - 278.

19. DUNQI, X. and KNUDSEN, J.G. Functional correlation of surface temperature and flow velocity on fouling of cooling tower water. *Heat Transfer Eng.*, 1986, **7**, 63 - 70.

20. CUSSLER, E.L. *Diffusion - Mass Transfer in Fluid Systems*. Cambridge University Press, New York, 1984.

21. HASSON, D. and BRAMSON, D. Effectiveness of magnetic water treatment in suppressing $CaCO_3$ scale deposition. *Ind. Eng. Chem. Process Des. Dev.*, 1985, **24**, 588-592.

STUDIES OF PHOSPHATE SCALE DEPOSITION IN HEATED TUBES

D. HAWTHORN M. Phil.
Formerly of The Technology and Research Centre, PowerGen Division,
C.E.G.B., Ratcliffe-on-Soar, Nottingham, UK.
Now of *min*-DEP Research, 88 Clifton Road, Ruddington, Notts. NG11 6DE, UK.

ABSTRACT

An account is given of the experiments which lead to the development of
quantitative kinetic equations that describe the precipitation of calcium
phosphate from solution and the subsequent deposition of some of the
amorphous precipitated material within heated condenser tubes. The scaling
process was concluded to consist of two stages.
Initiation occurred when sub-micron particles were formed by the
reaction of the influent phosphate with calcium in the bulk of the warm,
high pH circulating water. Deposition then occured by particulate transport
and adhesion, restricted by the turbulent shear stress at the heated tube
surface.
Details are given of the experimental regime pursued and the
mathematical treatment of the results to produce a quantitative model. The
implications of this approach to the future study of calcium carbonate
deposition are discussed.
The scaling phenomenon cannot be described simply by relating it to
an index of the supersaturation of the circulating or influent waters. Like
any other production process it depends upon the rate at which raw
materials are taken into the system. The efficiency of production depends
upon chemical conditions within the plant whereas the rate of capture of
particles at a given surface is governed by physics and hydrodynamics.

INTRODUCTION

In the summers of 1975/6 the combined performance losses at nine power
stations on the river Trent were estimated to be £4M, [1]. These losses
were shown to be due to the presence of less than 50 microns of scale on
the water side of the steam condensers. The scale was mainly an amorphous
form of calcium/magnesium phosphate with some carbonate. TABLE 1 shows
typical analyses for river and circulating waters and scale.

Units and Symbols

CF_{Na}	concentration factor, Na_{cw} / Na_{in}	
S	supersaturation index for tricalcium phosphate	
R	rate of deposition as calculated by	
	Kingerley, Rantell and Willett [5]	$mgPO_4/m^2/hr$
log(R)	K-R-W Scaling Index	
$(PO_4)_{oppt}$	observed rate of precipitation	
	from equation (3)	$mgPO_4/hr$
$(PO_4)_{cppt}$	calculated rate of precipitation	
	from equation (4)	$mgPO_4/hr$
$(PO_4)_{dep}$	calculated rate of deposition	
	in heated tube	$mgPO_4/m^2/hr$
Mu	water make-up rate	kg/hr
$[PO_4]$	concentration of PO_4	$mgPO_4/kg$
[Ca]	concentration of Ca	mgCa/kg
T	temperature in Celsius	
V	mass of water in system	kg
P	purge rate	kg/hr
D	condenser tube diameter	metres
d	density of water	kg/m^3
v	linear velocity of water in tube	m/s
$\cap$	viscosity of water at mid-tube temp.	Ns/m^2
$t_\frac{1}{2}$	mean residence time of water,	
	circulating in the system,	
	$= -\ln\frac{1}{2}$ x V / P	hours
N_{Re}	Reynolds number	
	$= Ddv/\cap$	unitless
F	friction factor, for brass	
	$= 0.046$ / $(N_{Re})^{0.2}$	unitless
Γ	turbulent shear stress at tube wall	
	$= \frac{1}{2}Fdv^2$	Nm^{-2}
MU & CW	refer to make-up & circulating water (concentrations)	
in & out	refer to the inlet & outlet of condenser tube (temps.)	

The sources of phosphate were treated domestic sewage, industrial effluent and land drainage into the river Trent.

FIGURE 1, which is drawn so that the relative flow rates are to scale, shows a recirculating water system such as used in power stations, which incorporates condensers, where the bulk temperature of the water is raised about 10 C, and cooling towers, where the temperature is lowered by a similar amount; the resultant evaporative loss is replaced by fresh river water make-up. The excessive concentration of salts is prevented by a purge from the ponds with additional make-up to maintain a constant level in the system. The flow rates are typically in the ratio of 3:2:1, make-up, purge, evaporation whilst circulating water flow rate is 80 units. Thus it can be seen that the purge water is an aliquot or sample of the circulating water; a point that is vital to the later argument.

U.K. power stations were designed to operate at a concentration factor (CF_{Na}) in the region of 1.5, i.e. the concentration of a readily soluble ion (Na) in the circulating water (CW) would rise to 1.5 times the influent concentration. If the phosphate concentration in the CW rises above about $2mgPO_4kg-1$ then precipitation will cause the apparent phosphate concentration factor (CF_{PO4}) to be less than that of sodium, (CF_{Na}).

TABLE 1

Typical River, Circulating Water and Scale Analyses

Component	as	River Water mg/l	Circulating Water, at CF_{Na}=1.5 mg/l	Scale %
Sodium	Na	76	114	
Calcium	Ca	120	175	35
Magnesium	Mg	30	44	3
Manganese	Mn	0.02	0.03	3
Phosphate	PO4	7	8	50
Sulphate	SO4	230	340	
Fluoride	F	0.3	0.45	
Alkalinity(MO)	CaCO3	180	196	
Alkalinity(PP)	CaCO3	0	16	
pH		7.5	8.5	
Carbonate	CO3	0	5	7
Organic Carbon	C	5	7	2

The cost to the industry were so high that the problem was tackled by a number of approaches. The first of these was the collection of information on plant operating conditions and water chemistry. This showed that different designers built in a different allowance for fouling [2]. Thus given the same scaling rate, one station will report efficiency losses earlier than an adjacent one. This means that plant efficiency records do not give a reliable indication of the rate of scale deposition. The second approach was the rig testing of the efficacy of commercial scale inhibitors and pH adjustment, and the third was an investigation of the feasibility of using mechanical methods of scale removal eg. the Taprogge Process.

Most affected power stations adopted the Taprogge Process, only one used pH adjustment. With the introduction of mechanical cleaning, research into the fundamentals of scaling ceased. In 1989 the CEGB released its rig data base to the author. It has since been supplemented by further work at City University (London) [3], and is continuing at Nottingham Polytechnic.

The EDF observation [4] that copper is released to rivers receiving water from power stations fitted with similar ball cleaning systems suggests that deeper understanding of the fundamental processes of scaling is essential if mechanical systems are not environmentally acceptable.

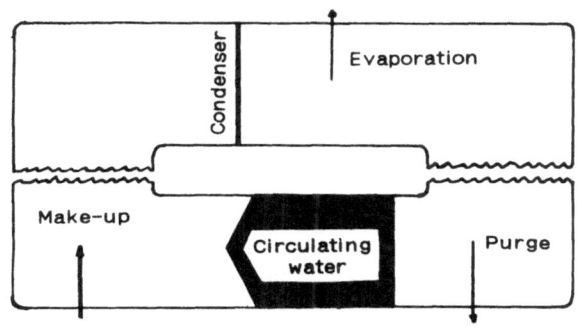

FIGURE 1.   Relative water flow rates in a power station cooling system.

EXPERIMENTAL WORK

During the investigation of scaling and its inhibition three rigs were built, each comprising twin parallel, low pressure steam heated tubes, separate forced draught evaporative cooling towers and ponds. Each was supplied with river water make-up and a controlled purge system.

The design of the rigs followed that of the station shown schematically in FIGURE 1. The aim was to attain the same temperatures, heating, cooling, storage and recycle times of water within the rigs, as seen on plant. The rig design parameters are shown in TABLE 2 compared with those of a 500MW(e) generating set. The low flow rig was designed to have heated tubes that were one tenth of the length but the same diameter and cut from the same stock as those on typical modern plant whereas the high velocity rig had the same length of tube as used on plant but with reduced diameter. These compromises were matched by a reduction of water mass flow to one tenth of that on plant whilst attempting to keep its time in the heated tube constant. The object of such compromise was to reduce the maximum power requirements from 120kW required for a full scale simulation to 12kW for the one tenth scale models.

TABLE 2
Comparison of Plant and Rig Parameters

Parameter	Units	Plant	Low Vel.	Medium Vel.	High Vel.
Tube length	m	18.3	1.83	7.32	18.3
Surface Area of Tube	m^2	1.46	0.146	1.30	0.64
Tube O.D.	mm	25.4	25.4	15.9	11.1
Tube Volume	dm^3	7.50	0.75	1.03	0.91
Water Velocity	ms^{-1}	1.8-2.8	0.24	1.37	2.12
Time of water in tube	s	6.6-9.7	7.6	5.35	8.63
Steam Temperature	$^\circ C$	29-43	45-70	40-70	40-70
Temperature of pond	$^\circ C$	5-27	5-27	5-27	5-27
Temp. rise of water	$^\circ C$	9-11	3-13	3-13	3-13
System vol/tube area	$m^3 m^{-2}$	0.9-2.6	2.5	0.9	0.43
Heat transfer/unit area	kWm^{-2}	20-29	11-44	10-40	2-10
Reynolds Number(approx)		55,000	7,500	24,000	22,000
Turbulent Shear Stress	Nm^{-2}	8.4	0.2	5.7	13.9

The controlled variations in temperature rise that were available facilitated an investigation of scale deposition within high, medium and low heat transfer regions of a condenser; the variation in water velocity between rigs promoted the investigation of hydrodynamic effects under conditions of near constant mass flow. The control available via the rigs' cooling towers provided constant temperature conditions day and night and broke the link between normally confounded variables; it provided the facility for investigating summer temperatures with winter-type water. It was also possible to study the separate effects of other normally confounded variables such as temperature, concentration factor and the influent and circulating water concentrations of phosphate and calcium.

THE SUPERSATURATION INDEX MODEL

The parameters governing scale deposition were thought initially to be solely those that control supersaturation locally , namely the temperature to which the water was raised (calcium phosphates are less soluble at higher temperatures), the influence of phosphate and calcium in the CW (elevated by evaporative concentration) and its pH (increased by the passage of the circulating water down the cooling towers causing loss of carbon dioxide from bicarbonates).

A semi-empirical equation had been derived previously [5], from limited plant monitoring, which correlated the observed rate of scale deposition with the concentration of calcium and phosphate in solution in the circulating water, its pH and the condenser outlet temperature. The equation was:-

$$R = 1.0 \times 10-3 \times S \times \exp [50 \times 103 \times (1/298 - 1/(T+273))] \qquad ....(1)$$

and on the basis of that equation the following Index was proposed

$$\log(R) = F_1(T)_{out} + F_2[PO_4]_{cw} + F_3[Ca]_{cw} + F_4(pH)_{cw} \qquad ....(2)$$

with the interpretation that positive values of $\log(R)$ suggested that scaling was occurring and that "their relative magnitude indicated the relative extent of deposition". This approach derived from work on supersaturation by de Boice and Thomas [6], Reddy and Nancollas [7], Ryznar [8], and others. Although Langelier [9], who pioneered the Scaling Index approach, had earlier stated that "...the Saturation Index is an indication of directional tendency and driving force *but it is in no way a measure of capacity*" (to deposit or corrode).

INITIAL RESULTS FROM THE MODEL RIGS

The object of building the low flow twin circuit rig was to test commercial scale inhibitors under controlled conditions. However, it was also used to study the process of scaling on the undosed "control" circuit at the same time. It was planned that these studies would be used to refine the coefficients for equation (2).

FIGURE 2 shows the plot of log(observed scale deposition rate) for data obtained from the low flow rig against the Scaling Index (log R) using the plant derived coefficients. The multiple correlation coefficient was only 0.43 and further, the effect of the phosphate concentration in the circulating water was eliminated as not statistically significant. Allowing the regression process to optimise the coefficients of the parameters improved the overall correlation coefficient slightly to 0.59 but again the circulating water phosphate concentration was eliminated. The relationship was therefore of log(deposition rate) against the effects temperature and pH only.

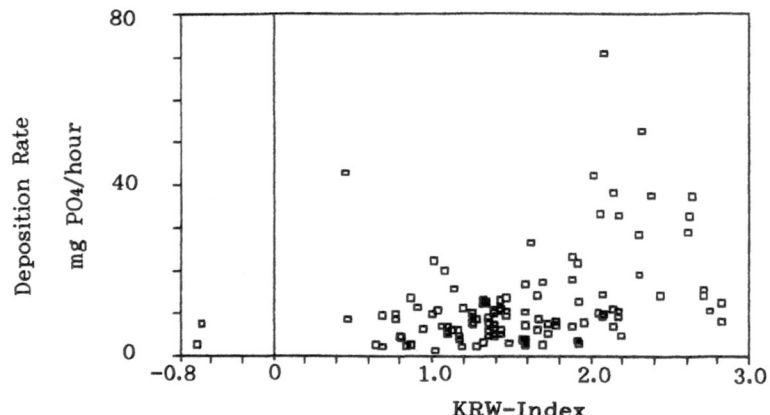

FIGURE 2.   Kingerley-Rantell-Willett Index versus observed
phosphate deposition rate.

Further, it was seen that for experiments with similar temperatures and
circulating water pH, concentrations of phosphate and calcium, which should
have produced identical scaling rates if equation (2) was valid, the rate of
deposition varied by an order of magnitude. However it was also observed
that this variation was associated with differences in the influent mass of
phosphate (see FIGURE 3) which   prompted a re-evaluation of the results
and a renewed search for the parameters which controlled phosphate
scaling.

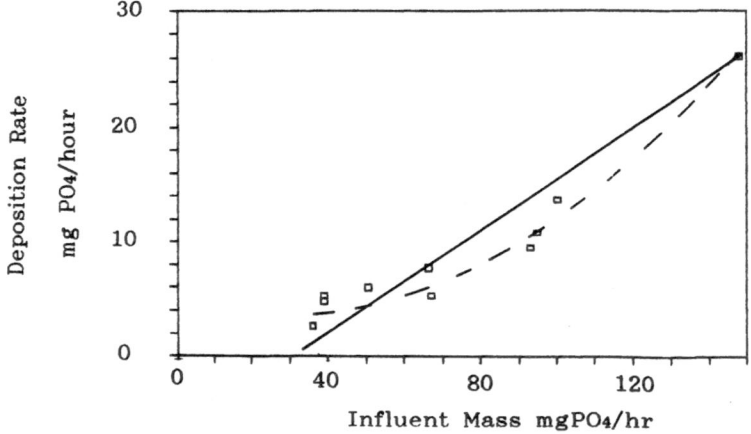

FIGURE 3.   Mass make-up rate versus phosphate deposition rate
(for similar circulating water analyses).

The apparent failure of the rig to provide data to validate the then accepted supersaturation index approach to scale deposition prompted the design and construction of the higher water velocity rigs. However the results so obtained also failed to support the supersaturation model and it was necessary to seek an alternative explanation for the observations.

Having demonstrated the importance of the influent mass of phosphate and eliminated the role of the circulating water phosphate concentration as a *cause* of scaling (when other parameters were kept constant), the new problem was to identify which other parameters were important in controlling precipitation and deposition on the heated tubes.

THE BULK PRECIPITATION AND DEPOSITION MODEL

Examination of the engineering design of a station cooling system shows that the water that is purged, to prevent the concentration of salts from rising too high, is an aliquot of the circulating water. FIGURE 1 illustrated this. The purge water composition is therefore the same as the circulating water. Whatever scaling potential the circulating water has is still there when the water is purged.

If the mass of soluble phosphate entering the system is compared with that leaving the system (corrected for concentrating effects due to evaporation), it provides an estimate of the phosphate that is precipitated from solution in the process of heating and cooling. It is postulated that a similar rationale may be applied to the cations calcium and magnesium thus providing a method of quantifying the precipitation of calcium/magnesium carbonates.

Thus in terms of concentration,

$$[PO_4]_{lost\ from\ solution} = [PO_4]_{influent} - ([PO_4]_{effluent}/CF_{Na}) \quad \ldots(3)$$

$$[PO_4]_{lost\ from\ solution} = [PO_4]_{influent} - ([PO_4]_{cw}/CF_{Na}) \quad \ldots(4)$$

and to convert concentrations to mass per unit time, both sides are multiplied by the mass of water entering the system in that time.

The calculation of the rate of deposition anywhere in the cooling system depends upon the knowledge of the rate of precipitation (loss of soluble phosphate from solution, as defined in equation 4).

THE PRECIPITATION PROCESS

The precipitation rate may be measured, as in (4) above, or predicted.

As mentioned in the Introduction, precipitation is not seen below a CW concentration of about 2mgPO4/kg, such a concentration although not in true equilibrium appears to be "robust" within the time of exposure to the conditions of heating, cooling, recirculation and storage in the rigs and on the plant that they mimic. This "robust" concentration is some two orders of magnitude greater than the published values for the most soluble of the calcium phosphates under the pHs considered, [3].

The observed precipitation rate (loss from solution), as defined in equation 4, was found to correlate with:-

A) the influent mass of phosphate in excess of that portion which would form the robust solution when concentrated,

B) the circulating water pH, its calcium concentration, and a function of the highest temperature that it experienced.

These are quantified in the following equation:-

$$
\begin{aligned}
\text{Log } (PO_4)_{cppt} = {} & 1.0 * \log \{M_u * ([PO_4]_{in} - (2.0/CF_{Na}))\} \\
& + 0.3333 * pH_{cw} + 1.0 * \log[Ca]_{cw} \\
& - 2000 * [1/(T_{out} + 273) - 1/273] - 6.0 \quad\quad .....(5)
\end{aligned}
$$

The multiple correlation coefficient is 0.95 and of the graph is shown in FIGURE 4.

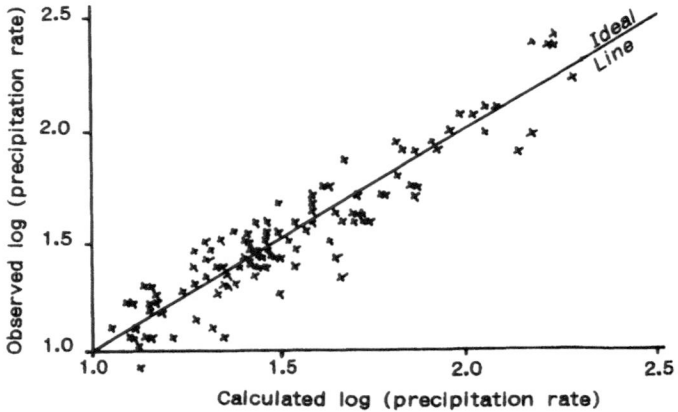

FIGURE 4.   Observed log (precipitation rate) versus calculated.

It is postulated that the mechanism by which the temperature, calcium concentration and pH control precipitation is by their effects on the solubility attained during the residence time of water in the cooling system and which may be called temporal (non-equilibrium) solubility [10].

THE DEPOSITION PROCESS

It was found that not all of the phosphate lost from solution, "precipitated phosphate", adhered to the rigs' heated tubes as "deposited phosphate". Some was found in the tower ponds as a sludge; some was removed from the system in suspension in the purged water and some was found on the cooling tower packs. This is a particular problem with the newer plastic film packings and has been observed in operating power stations [11].

Water velocity was found to have a large influence on deposition in the heated condenser tubes, as can be seen from TABLE 3.

TABLE 3
The Effect of Water Velocity on Phosphate Deposition from
Broadly Similar Water Conditions.

Code *	Vel m/s	PO₄dep mg/h	Transit Time s	Tout °C	[Ca]cw	[PO4]cw	(PO4)in mg/hr
307032	0.24	21.7	8.6	36.6	163	5.6	138
105020	1.37	2.6	5.4	37.4	156	5.8	146
406030	2.12	1.4	7.6	36.6	160	6.7	119

Following the convention established by Cleaver and Yates [12], the effects of water velocity on particle deposition were ascribed to:

A) a term describing the transport of particles to the heated surface, log Reynolds Number, and

B) a term describing the re-entrainment of deposited particles, log turbulent shear stress at the tube wall.

The quantity that deposited on the heated tubes of the high, medium and low flow rig in the 143 experiments is described by equation 6.

$$\log(PO_4)_{dep} = 0.9 * \{\log(PO_4)_{oppt} + 0.4(pH)_{cw} + 0.2\log t_{1/2} + 0.025(T_{out}-T_{in}) + 1.5\log N_{Re} - 1.0\log \Gamma - 11.0\} \quad .....(6)$$

748

The multiple correlation coefficient is 0.99 and the graph is FIGURE 5.

The above equation can be interpreted as indicating that the proportion of precipitated material that sticks to the heated brass surface depends on the pH of the water the material is suspended in, the time available for sticking and the temperature rise that it experiences; promoted by transport to the tube surface yet restricted by turbulent shear stress.

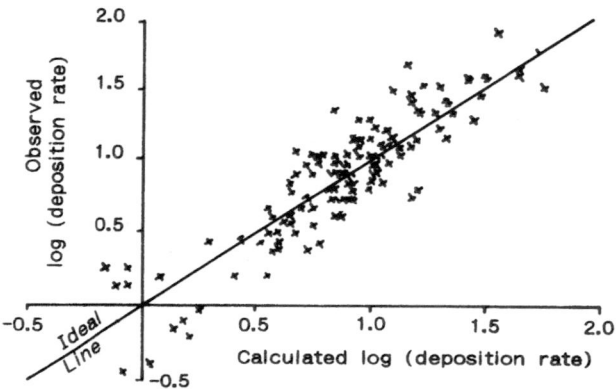

FIGURE 5.   Observed log (deposition rate) versus calculated.

DISCUSSION

The equations that describe the two models for scaling, the "supersaturation index" model and the "bulk precipitation and transport" model, differ in their foundations and their derivation. Attempts to validate the "supersaturation scaling index" model have failed. The "precipitation and deposition" model has been validated by more than a year's operating power plant monitoring as well as by the rig results reported here. However, the precipitation reaction may be capable of further refinement by interpreting precipitation in terms of temporal (non-equilibrium) solubility.

It is suggested that any investigation of carbonate scaling should not ignore the possibility of relating it to the influent mass of bicarbonates and to the mechanisms which control the loss of carbon dioxide from the system., such as cooling tower packing design, temperature and pressure, both hydraulic and atmospheric.

As with any production process, the production of calcium phosphate scale is directly influenced by the mass entering. The role of the circulating mother-liquor is influential, only as it contains the stable concentration of the material which stays in solution and is destined to be discharged, not precipitated.

CONCLUSIONS

The Saturation Index approach to phosphate scaling was based on the rationale that the rate of scale deposition must be controlled by the degree of supersaturation above the true equilibrium concentration and that this was subject to determination by chemical analysis of the concentration of the phosphate in solution. Therefore, the rate was postulated to be proportional to the phosphate in solution in the circulating water.

The bulk precipitation and transport approach is diametrically opposite, it postulates that the scale deposition rate is proportional to *what does not stay in solution* in the circulating water; it is proportional to the mass that precipitates. The goodness of fit of the equations and their confirmation on plant lead to the conclusion that this model is valid for phosphate scaling and it is postulated that the model can be extended to other foulants in cooling water, notably calcium carbonate.

This paper supports Langelier's original statement that "...the Saturation Index is an indication of directional tendency and driving force but it is in no way a measure of capacity" to deposit.

ACKNOWLEDGEMENTS

The author wishes to acknowledge the assistance of many CEGB colleagues in the initial investigation, the CEGB for releasing the rig data bank and the recent support of *min*-DEP Research, a non-profit making organisation, for supporting research at City University (London) and Nottingham Polytechnic.

REFERENCES

[1]   Parry, D.J., Hawthorn, D. and Rantell, A.,
        Fouling of power station condensers within the Midlands
        Region of the CEGB. Conference on Fouling of Heat
        Transfer Equipment, Rensselaer Polytechnic Institute,
        Troy, New York, 1979. Eds. E.F.C. Somerscales &
        J.G. Knudsen, Hemisphere, Washington.

[2]   Rantell, A., Improving the confidence in plant
        trials of cooling water treatments. Progress in the
        Prevention of Fouling. Nottingham University, 1981.

[3]   Hawthorn, D., Investigation of the parameters which
        govern the deposition of calcium phosphate from
        recirculated cooling water. Thesis for Master of
        Philosophy. The City University (London), 1991.

[4]   Conference Discussion. Eurotherm Seminar No. 23, Fouling
        Mechanisms; Theoretical and Practical Aspects.
        Grenoble April 1992.

[5]   Kingerley, D.G., Rantell, A. and Willett, M.J.,
        A rate equation for the scalng of condensers by calcium
        phosphate: first estimates from plant data. Power
        Industry Research 1, 17, 1981.

[6]   de Boice, J.N. and Thomas, J.F., Chemical
        treatment for phosphate control. Journal of Water
        Pollution Control Federation, 47, 2246, 1975.

[7]   Reddy, M.M. and Nancollas G.H., The
        crystalisation of calcium carbonate. Journal of
        Colloid Interface Science, 37, 824, 1971.

[8]   Ryznar, J.S., A new index for determining the
        amount of calcium carbonate formed by a water. Journal
        of the American Water Works Association, 36, 472, 1944.

[9]   Langelier, W.F., Chemical equilibbria in water.
        Journal of American Water Works Association, 38, 169, 1946.

[10]   Hawthorn, D., The scaling index is dead: long live the
        scaling equation. Eurotherm Seminar No. 23. Fouling
        Mechanisms; Theoretical and Practical Aspects.
        Grenoble April 1992. Editions Européennes Thermique &
        Industrie, Paris.

[11]   E. Hobson and G.A.Fitchett., The effects of cooling
        water quality on cooling tower plastic pack fouling in
        National Power electricity power generating plants.
        Paper IWC-92-9. 53rd Annual International Water
        Conference, Engineers Society of Western Pennsylvania.
        October 1992.

[12]   Cleaver (J.W.) and Yates (B.) Effect of re-entrainment
        on particle deposition. Chemical Engineering Science,
        31, 147-151, 1976.